RNA ribonucleic acid
RV residual volume
SD systolic discharge
SDA specific dynamic action
STH somatotropic hormone
SV stroke volume
SWS slow-wave sleep
TH thyrotropic hormone
TMR total metabolic rate
TSH thyroid-stimulating hormone
TV tidal volume
VC vital capacity
WBC, wbc white blood cell

Prefixes

a- without
ab- away from
ad- to, toward
adeno- glandular
amphi- on both sides
an- without
ante- before, forward
anti- against
bi- two, double, twice
circum- around, about
contra- opposite, against
de- away from, from
dys- difficult
ecto- outside
endo- in, within
ento- inside, within
epi- on
eu- well
ex- from out of, from
extra- outside, beyond, in addition

hemi- half
hyper- over, excessive, above
hypo- under, deficient
infra- underneath, below
inter- between, among
intra- within, on the side
iso- equal, like
para- beside, to side of
peri- round about, beyond
post- after, behind
pre- before, in front of
pro- before, in front of
retro- backward, back
semi- half
sub- under, beneath
super- above, over
supra- above, on upper side
syn- with, together
trans- across, beyond

Suffixes

-algia pain, painful
-blast young cell
-cele swelling
-cide killer
-cyte cell
-ectomy cutting out
-emia blood
-itis inflammation
-kinin motion (action)
-ology knowledge of
-oma tumor

ANATOMY
AND
PHYSIOLOGY

ANATOMY
AND
PHYSIOLOGY

GARY A. THIBODEAU, Ph.D.

Chancellor and Professor of Biology
University of Wisconsin–River Falls
River Falls, Wisconsin

With 811 illustrations, including 510 in color

Illustration program by

Ernest W. Beck

Joan M. Beck

James M. Krom

Times Mirror/Mosby College Publishing

St. Louis • Toronto • Santa Clara 1987

Senior Editor Donald G. Mason
Editorial Assistants June Heath and Monica Lauer
Production Editors Helen C. Hudlin, Linda Kocher, and Jeanne Genz
Art Director Kay Michael Kramer
Designer William A. Seabright
Illustrators Ernest W. Beck, Joan M. Beck, and James M. Krom
Cover and Half-Title The J. Paul Getty Museum, Bronze Statue of a
Victorious Athlete, 4th c. BC, attributed to Lysippos.
Frontispiece and Unit Openers British Museum, Greek Vase Paintings.

Library of Congress Cataloging-in-Publication Data

Thibodeau, Gary A., 1938-
     Anatomy and physiology

     Bibliography: p.
     Includes index.
     1. Human physiology.   2.   Anatomy, Human.
I.   Title.
QP34.5.T49   1987   611   86-28511
ISBN   0-8016-4925-0

GW/CA/VH/VH   9 8 7 6 5 4 3 2 1   03/A/319

PREFACE

Anatomy and Physiology is a new text that is greatly indebted to a highly successful classic in the field—*Textbook of Anatomy and Physiology* by Catherine Parker Anthony. Based on my experience as coauthor of several editions of *Textbook* and with over 20 years of introducing students to human structure and function for the first time, I have used the successful content framework provided by *Textbook* to write this modern new text, designed to meet the needs of today's courses and students. *Anatomy and Physiology* incorporates the features of *Textbook* that have proven successful in over 40 years of classroom use; yet as a new text it offers a broader, more contemporary content in a highly student-oriented format.

The selection, sequencing, and method of presentation of the material in this text are the result of literally thousands of comments and suggestions from college instructors in the field and their—as well as my own—long experience in helping students face the challenge of learning about human anatomy and physiology for the first time. Instructors looking for an up-to-date and contemporary presentation of anatomy and physiology that covers essentials without wordiness, places emphasis on concepts rather than descriptions, and correlates structure with function will welcome the approach used in this new text.

Textbooks, however, cannot replace teachers. A textbook that attempts to mix non-essential facts with basic information or to include highly specialized subject matter having little to do with introductory level anatomy and physiology can impede rather than assist the classroom instructor. Texts that attempt to provide information about anything and everything associated with anatomy and physiology often discourage learning. *Anatomy and Physiology*, in contrast, is designed to help the teacher teach and the student learn. The text is one that students will read—and learn from. It is an accurate and up-to-date presentation. Concise and yet comprehensive, the content is enhanced by numerous pedagogical aids, outstanding artwork, and many full-color micrographs and illustrations that should reinforce learning and enable the student to understand concepts better.

Organization

The thirty chapters of *Anatomy and Physiology* are divided into seven major units and follow the sequence most commonly used in teaching the subject at the undergraduate level. This sequence may be modified by the instructor, depending upon the needs of the course and the students, since each unit, following Unit One, is designed and written to stand alone.

Unit One: The Body as a Whole (Chapters 1 to 5) presents important basic information relevant to all of the following units. Chapter 1 (Organization of the Body) provides an introduction to the basic terminology and key concepts in the field as well as an overview of all the body systems. For students who have had courses in introductory chemistry and biology, Chapters 2 (The Chemical Basis of Life) and 3 (Cells) will serve as brief reviews. For those students who have not had such courses, these chapters provide the minimum background necessary to follow material in later chapters. Chapters 4 (Tissues) and 5 (The Skin and its Appendages) contain comprehensive information on the major body tissue types and the external surface of the body.

Unit Two: Support and Movement (Chapters 6 to 9) discusses the key information related to the anatomy and physiology of skeletal tissue, the skeletal system, its articulation, and the important muscles.

Unit Three: Communication, Control, and Integration (Chapters 10 to 14) covers the anatomy and physiology of nerve tissue, the elements of the central and autonomic nervous systems, the sense

v

organs, and the closely associated endocrine system.

Unit Four: Transportation (Chapters 15 to 18) provides detailed discussions of the blood, the cardiovascular system, and the lymphatic system, presenting the essential anatomy and physiology of these key aspects of the body.

Unit Five: Respiration, Nutrition, and Excretion (Chapters 19 to 26) details the anatomy and physiology of three important areas: the respiratory, digestive, and urinary systems, with special attention paid to metabolism, fluid and electrolyte balance, and acid-base balance.

Unit Six: Reproduction (Chapters 27 and 28) discusses the male and female reproductive systems.

Unit Seven: Defense and Adaptation (Chapters 29 and 30) covers the body's defense through the immune system and its response to stress.

Special Features

Content: As noted earlier, *Anatomy and Physiology* does not attempt to cover every detail of the subject, rather it provides coverage of the basics of human anatomy and physiology. Appropriate physiological content balances the detail of the anatomy. For example, the discussion of physiology in Chapters 4 to 6 on general tissue types, the skin, and skeletal tissues carefully complements and supports the descriptive anatomy, so that the student not only sees the detail of the structure but also understands the mechanisms of the function involved. Additional support material beyond the basics has been provided, but every effort has been made to identify the less critical material through the use of boxed inserts and boxed essays, and at every step the relevance and relation of the material to the important information as a whole has been stressed. At the same time, instructors will recognize and students will benefit from the careful coupling of structural information at every level of organization with the important functional concepts. The result for the student is a more integrated understanding of both human structure and function. In every chapter examples that stress the "complementarity of structure and function" were consciously selected to reinforce and emphasize the importance of such unifying concepts as homeostasis, physiological feedback, and whole body communication and control mechanisms. Supportive information has been integrated by use of carefully selected clinical examples and other aids so that topics related to each characteristic of life can be applied to every organ system in the body. Style of presentation, readability, accuracy, and appropriate level of coverage have been carefully and consciously developed to meet the needs of today's students and provide them with information on human anatomy and physiology that they will need for future courses in their chosen fields of interest.

Pedagogy: *Anatomy and Physiology* is a student-oriented text. Written in a very readable style, the text was designed with numerous pedagogical aids to maintain interest and motivation. Every chapter contains the following elements to facilitate learning and retention of information in the most effective and efficient manner:

Chapter Outline: An overview outline begins each chapter and enables the student to preview the content and direction of the chapter at the major concept level prior to beginning the detailed reading.

Chapter Objectives: Each chapter opening page contains a reasonable number of clearly identified and measurable objectives for the student. These objectives are not overwhelming in number and clearly identify for the student at the beginning what the key goals should be and what information should be mastered in the chapter.

Key Terms: The most important key terms to be introduced are first presented at the beginning of the chapter, along with a guide to their correct pronunciation. These key terms, along with additional new terms, are then introduced and defined in the text body and are identified in **boldface** to highlight their importance.

Boxed Inserts: Numerous brief boxed inserts appear in every chapter. These inserts include information ranging from brief sidelights on recent research to clinical applications or additional information on important clinical conditions and definitions. They appear close to the relevant text discussion of the normal condition. The extensive use of clinical applications should help students better understand the relationship between normal structure and function.

Boxed Essays: Throughout the text there are longer boxed essays on topics of special interest to students. These include such items as impact of diet, current research in fields such as AIDS and cancer, applications of clinical procedures, and material on sports injuries and their treatment.

Medical Terminology: Medical terminology is introduced and defined as appropriate and necessary throughout the text. However, at the end of particularly relevant chapters (4, 5, and 25), special sections are included that review medical terminology in more detail.

Chapter Summaries: Extensive and detailed end-of-chapter summaries in outline format provide excellent guides for students as they review the text materials for examination preparation. Many students may also find such detailed guides useful as a chapter preview in conjunction with the chapter outline.

Review Questions: Numerous end-of-chapter questions support student review and enable the student to test his or her knowledge after completing the chapter. Complete answers are available from instructors, who may wish to hand these out as self-study exercises rather than use them exclusively in classroom discussion.

Additional learning and study aids at the end of the text include a comprehensive *Glossary* of terms that have been introduced; an up-to-date, chapter-by-chapter list of suggested **Supplementary Readings** for students who wish to delve deeper into areas of special interest; and an **Index** that is extremely detailed and is intended to serve as a ready reference for locating information in the text. In addition, important **Abbreviations, Prefixes, and Suffixes** appear on the inside front cover.

Illustrations: A major strength of *Anatomy and Physiology* is its illustration program. Extensive use has been made of full-color line drawings, micrographs, and dissection photographs throughout the text. Illustrations proven pedagogically effective in *Textbook of Anatomy and Physiology* have been complemented by many additional figures to provide both accurate information and visual appeal. Careful attention has been paid to placement and sizing of the illustrations for maximum usefulness and clarity. A unified style has been maintained to eliminate the subtle but distracting differences that often detract from effective use of visuals having differing styles and methods of presentation. Instructors often hear students talk of vague and ill-defined "problems" with illustrations. In most cases the problem is simply the multiple styles and perspectives in the artwork that the student must interpret. In developing the art program for *Anatomy and Physiology*, an effort has been made to avoid anything that might distract from the very important reinforcement provided by the visual learning aspect of the text. For example, the labeled line drawings in Chapter 4 (Tissues) were specifically developed to match the color micrographs and reinforce for the student the important structures in each tissue emphasized in the text body. The illustrations are all carefully referenced in the text body and are designed to support the text discussion. They are an integral and important part of the learning process and should be carefully studied by the student.

Supplements

The supplements available for student and instructor that accompany *Anatomy and Physiology* include:

Laboratory Manual by Gary A. Thibodeau: The laboratory manual contains a balanced mix of hands-on and self-assessment exercises that can be used in classroom activities and discussions or required as outside assignments.

A separate *Instructor's Manual* for the *Laboratory Manual:* Available to adopters of the laboratory manual. It contains answers to exercises where appropriate, suggestions and details for laboratory preparations, and offers tips and guidelines for using each exercise effectively.

A combined *Instructor's Manual and Testbank* for the text including:

The *Instructor's Manual* portion, developed by Professor Jay Templin of Widener University, which provides extensive support for instructors in using *Anatomy and Physiology* most effectively in the classroom. Based on a sound educational model, the manual contains the following planning and instructional features for each chapter:

Detailed "Chapter Overview" and "Chapter Outline" sections to provide support for lecture preparation and rapid review of the chapter organization.

"Chapter Objectives," based on six levels of knowledge (recall, comprehension, application, analysis, synthesis, and evaluation) to help identify key facts and concepts students should master.

"Key Terms" and "Terminology Comprehension" with pronunciation guides to identify new terms, prefixes, and suffixes students should master and comprehend for use throughout the course.

"Suggested Demonstrations and Review Activities" to provide suggestions for additional in-class activities instructors might want to use to supplement text and lecture assignments.

"Ideas for Further Consideration" that suggests topics for possible independent student research and further study.

"Additional Resources" that gives suggestions for supplementary audiovisual materials available from outside sources.

The "Chapter Summary" that helps identify key points in the chapter that need emphasis and an "Answers to Review Questions" that helps conclude the chapter discussion effectively.

The *Testbank* portion, developed by Professors Sydney and Mary Walston of Central Michigan University, averages over 50 test items per chapter. The more than 1600 questions include a diversity of test item types: multiple choice, true/false, matching, and short answer. A convenient answer key provides the correct answer as well as a page reference in the text where the answer may be found. Carefully reviewed by outside instructors for accuracy and appropriateness, the *Testbank* was designed to test general knowledge, vocabulary mastery, and key concept understanding.

Overhead Transparency Acetates: Reproduced and enlarged from the text, the 75 four-color transparency acetates include the most important illustrations and drawings of key anatomical and physiological concepts and relationships and enable the instructor to illustrate and clarify difficult concepts and points during a lecture.

MicroTest Computerized Testbank: A microcomputer version of the printed *Testbank, MicroTest II* lets you edit, add, delete, or scramble questions according to your own testing format and needs. *MicroTest II* is available for the Apple II and II/e, IBM PC and XT, and compatible microcomputers.

A Word of Thanks

To Catherine Parker Anthony I extend my very special thanks and gratitude. Her contributions to students and to the teaching and learning process are legend. This book reflects her ongoing commitment to education and, hopefully, her charm and great sense of style.

Many people have contributed to the development of *Anatomy and Physiology*. The many instructors and students who have used *Textbook of Anatomy and Physiology* and provided comments over the years have been indirect but valuable sources of information and guidance for this new text. A great debt of gratitude is owed them, which I hope will be repaid by the advantage future students and instructors will have in using the resulting product.

A specific "Thank You" goes to the following instructors who critiqued in detail the most recent edition of *Textbook*. Their invaluable comments were very instrumental in the development of this text.

Minyon L. Bond, Phoenix College
Joseph R. Powell, Florida Junior College—
    North Campus
Carl Summerlin, Pensacola Junior College—
    Warrington Campus
Leslie Wiemerslage, Belleville Area College
Roberta M. O'Dell-Smith, University of New Orleans—
    Lakefront Campus
William Bednar, Mott Community College
Gus Demas, Macomb Community College
Robert Friar, Ferris State College
Clifton Lewis, Wayne County Community College—
    Downtown Campus
Sydney Walston, Central Michigan University
Judith Cole, Gaston College
Jim Hall, Central Piedmont Community College
Jeff Gerst, North Dakota State University
Rose Morgan, Minot State College
John Humphrey, Community College of Allegheny
    County—South Campus
Wayne Carley, Lamar University

For their contributions to the illustration program my thanks go to Ernest W. Beck, Joan M. Beck, and James M. Krom. Professor Robert Calentine, University of Wisconsin—River Falls, produced the outstanding color photomicrographs. A special acknowledgment is also due Dr. Branislav Vidić, M.D., and his late colleague, Dr. Faustino R. Suarez, M.D., for the use of illustrations from their excellent text, *Photographic Atlas of the Human Body*. The support provided the project by Professor Jay Templin of Widener University and Professors Sydney and Mary Walston of Central Michigan University, through their very capable development of the *Instructor's Manual and Testbank*, has been most helpful and is deeply appreciated.

At Times Mirror/Mosby College Publishing thanks are due all who have worked with me in bringing this text to completion. I wish especially to acknowledge the support and effort of my editor, Donald Mason; project editor, Carlotta Seely; and production editors, Helen Hudlin, Linda Kocher, and Jeanne Genz, all of whom were instrumental in bringing this text to its present form.

Gary A. Thibodeau
River Falls, Wisconsin

CONTENTS IN BRIEF

CONTENTS

UNIT TWO

SUPPORT AND MOVEMENT

UNIT ONE

THE BODY AS A WHOLE

1 ORGANIZATION OF THE BODY

OBJECTIVES

After you have completed this chapter, you should be able to:

1 Define the terms *anatomy* and *physiology.*
2 List and discuss in order of increasing complexity the levels of organization of the body.
3 Identify the classic "characteristics of life."
4 List and briefly discuss the major body organ systems and the function of each.
5 Discuss the concept of body type (somatotype).
6 Identify and contrast the three principal somatotype components.
7 Define the anatomical position.
8 Discuss and contrast the axial and appendicular subdivisions of the body by identifying the specific anatomical regions in each area.
9 List the nine abdominal regions and the four abdominal quadrants.
10 Define bilateral symmetry.
11 List and define the principal directional terms and body sections (planes) employed in describing the body and the relationship of its parts.
12 List the cavities of the body and identify the major organs found in each.
13 List and discuss five generalizations about body function.
14 Define homeostasis and by giving examples explain its role in maintaining normal body function.
15 Discuss the concept of complementarity of structure and function.

What a piece of work is man! how noble in reason!
how infinite in faculty! in form and moving how
express and admirable! in action how like an angel!
in apprehension how like a god! the beauty of the
world! the paragon of animals!

Hamlet, in *Hamlet, Prince of Denmark,* Act II, Scene 2

KEY TERMS

Articulation (ar-TIC-u-LA-shun)

Atrophy (AT-ro-fe)

Appendicular (ap-en-DIK-u-lar)

Axial (AK-see-al)

Bilateral symmetry (bi-LAT-er-al   SIM-e-tree)

Contralateral (kon-trah-LAT-er-al)

Homeostasis (ho-me-o-STA-sis)

Ipsilateral (ip-si-LAT-er-al)

Mediastinum (ME-de-as-TI-num)

Parietal (pah-RI-e-tal)

Peritoneum (per-i-to-NE-um)

Protoplasm (PRO-toe-plazm)

Somatotype (so-MAT-o-type)

Visceral (VIS-er-al)

This is a book about that most wondrous of all structures—the human body. The study of human anatomy and physiology is, in reality, a study of the complex interdependence of structure and function in the living organism. The science of the structure of organisms is called **anatomy** (from the Greek *ana*, "up," and *temnein*, "to cut"). **Physiology** (from the Greek *physis*, "nature," and *logos*, "discourse") is the science of the functions of organisms.

This unit begins with an overview of the body as a whole. In subsequent chapters, the body will be dissected, both structurally and functionally, into "levels of organization" so that its component parts can be more easily studied and then "fit together" into a living and integrated whole.

LEVELS OF ORGANIZATION

Before you begin the study of the structure and function of the human body and its many parts, it is important to think about how those parts are organized and how they might logically fit together and function effectively. Examine Figure 1-1. It illustrates the differing levels of organization that influence body structure and function.

CHEMICAL LEVEL—BASIS FOR LIFE

Note in Figure 1-1 that organization of the body begins at the chemical level. It is the *organization* of chemical constituents that separates living from nonliving or **inanimate** material. Even the term *organism,* used to denote a living thing, implies organization.

The unique and complex relationships that exist between atoms and molecules in living material form **protoplasm**—the essential material of all cells. Maintaining the type of chemical organization in protoplasm required for life requires the expenditure of energy, and, unless proper relationships between chemical elements are maintained, a living person soon becomes a lifeless corpse.

Figure 1-1 Structural levels of organization in the body.

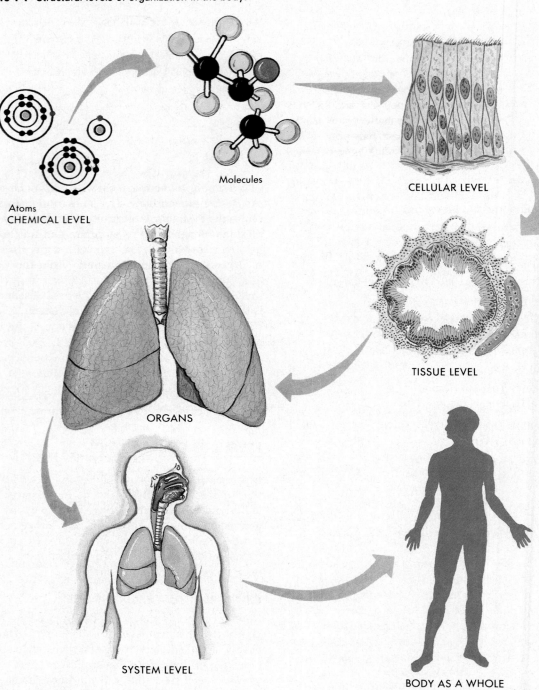

Certain characteristics of protoplasm, in addition to its specific organization, characterize it as a living substance. These so called **characteristics of life** include **metabolism** (a generalized term used to describe such physical and chemical processes as nutrition, digestion, secretion, absorption, respiration and excretion), **irritability, conductivity, contractility, growth,** and **reproduction.** Of all these properties, reproduction is considered the most fundamental characteristic of a living thing. The mechanisms of cellular reproduction will be covered in Chapter 3. Each characteristic of life—its functional manifestation in the body, its integration with other body functions and structures, and its mechanism of control—will be the subject of study in subsequent chapters of the text.

CELLULAR LEVEL

The characteristics of life ultimately result from a hierarchy of structure and function, which begins with the organization of atoms and molecules. However, when viewed from the perspective of the anatomist, the most important function of the chemical level of organization is to furnish the basic building blocks for the next higher level of body structure—the **cellular level.** Cells are the smallest and most numerous structural units that possess and exhibit the basic characteristics of living matter. How many cells are there in the body? One estimate places the number of cells in a 150-pound adult human body at 100,000,000,000,000. In case you cannot translate this number—1 with 14 zeroes after it—it is 100 trillions! or 100,000 billions! or 100 million millions!

Each cell is surrounded by a membrane and is characterized by a single **nucleus** surrounded by protoplasm **(cytoplasm)** that contains numerous structures required for specialized activity. Although all cells have certain features in common, they specialize or *differentiate* in order to perform unique functions. Fat cells, for example, are structurally modified to permit the storage of lipid material. The specialized cells shown in Figure 1-1 line the tubes of the respiratory tract. Muscle, bone, nerve, and blood cells are other examples of structurally and functionally specialized cells.

TISSUE LEVEL

As you can see in Figure 1-1, the next higher level of organization beyond the cell is the **tissue level.** Tissues represent another step in the progressive or hierarchical organization of living matter. By definition, a *tissue* is an organization of a great many similar cells that are specialized to perform a certain function. Tissue cells are surrounded by varying amounts and kinds of nonliving, intercellular substances or **matrix.**

There are four major or principal tissue types: **epithelial, connective, muscle,** and **nervous.** Considering the complex nature of the human body, this is a surprisingly short list of major tissues. Each of the four major tissues, however, can be subdivided into a number of specialized subtypes. Together, the body tissues are able to meet all of the structural and functional needs of the body.

The tissue used as an example in Figure 1-1 is a specialized type of **epithelium** that lines the tubes of the respiratory tract. There are several types of epithelial tissues. Some are specialized to form sheets that cover the body surface, while other types line body cavities or protect the passageways of the respiratory or digestive tracts. The details of tissue structure and function will be covered in Chapter 4.

Growing success in the area of tissue and organ transplantation during the past 5 years has resulted in large part from innovative surgical techniques, new antirejection drugs, and advances in immunology. Skin, blood, and cornea transplants have been in common medical practice for years. More recently, organ transplants involving the heart, lungs, kidneys, pancreas, and liver have been increasing dramatically.

The term **allograft** is used to describe a transplant between individuals of the same species. Until 1985, **xenografts,** or transplants between animals of different species, were restricted to experimental animals. Currently, a number of experimental-type xenografts between baboons and other animals and humans are under study. Xenograft procedures are topics of heated scientific and medical controversy.

ORGAN LEVEL

Organs are more complex units than tissues. An organ is an organization of several different kinds of tissues so arranged that together they can perform a special function. For example, each lung is an example of organization at the **organ level.** Muscle and specialized connective tissues form the many tubes that convey air, epithelial tissues line the microscopic air sacs, and nervous tissues permit control of air flow and muscular contraction.

Tissues seldom exist in isolation. Instead, joined together, they form organs that represent discrete but functionally complex operational units. Each organ has a unique shape, size, appearance, and placement in the body, and each can be identified by the pattern of tissues that form it. The lungs, heart, brain, kidneys, liver, and spleen are all examples of organs.

SYSTEM LEVEL

Systems are the most complex of the component units of the body. The **system level** of organization involves varying numbers and kinds of organs so arranged that together they can perform complex functions for the body. Ten major systems compose the human body: **integumentary, skeletal, muscular, nervous, endocrine, circulatory** (cardiovascular and lymphatic), **respiratory, digestive, urinary,** and **reproductive** (Figure 1-2).

Figure 1-2 Organ systems of the human body.

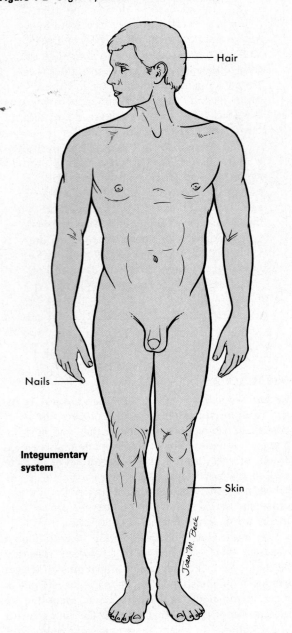

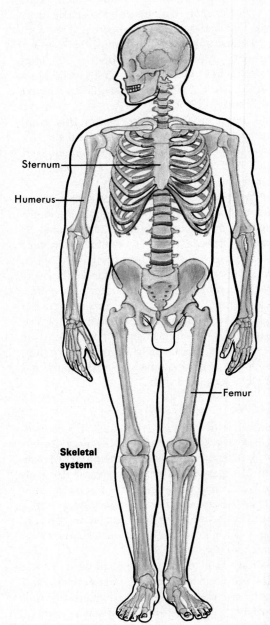

Integumentary System

The skin or integumentary system is crucial to survival itself. Its primary function is protection. It protects underlying tissue against invasion by harmful microorganisms, bars entry of most chemicals, and minimizes mechanical injury of underlying structures. In addition, the skin serves to regulate body temperature, synthesize important chemicals and hormones, and function as a sophisticated sense organ. The integumentary system includes both the layers of the skin proper and its appendages: hair, nails, and specialized glands.

Skeletal System

The skeletal system consists of bones and related tissues such as cartilage and ligaments that together provide the body with a rigid framework for support and protection. In addition, the skeletal system, through joints or **articulations,** makes movement possible. Bones also serve as reservoirs for mineral storage and function in **hemopoiesis** or blood cell formation.

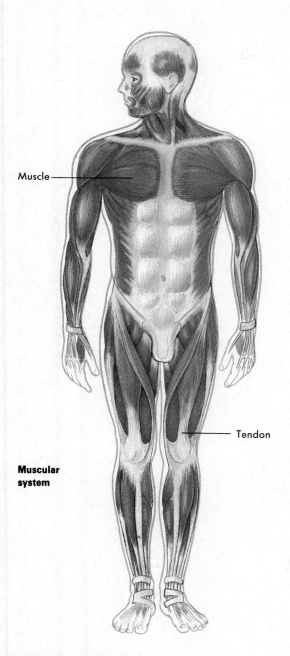

Muscle

Tendon

Muscular system

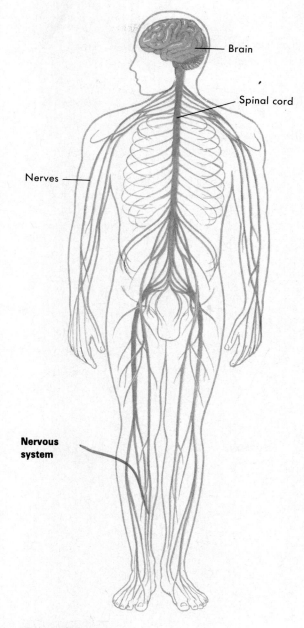

Brain

Spinal cord

Nerves

Nervous system

Muscular System

Individual muscles are the organs of the muscular system. In addition to **voluntary** or **skeletal** muscles, which have the ability to contract when stimulated and are under conscious control, the muscular system also contains **smooth** or **involuntary** muscles and the **cardiac muscle** of the heart. Muscles not only produce movement (or maintain body posture) but are also responsible for generating heat required for maintenance of a constant core temperature.

Nervous System

The brain, spinal cord, and nerves are the organs of the nervous system. The primary functions of this complex system include **communication, integration,** and **control** of body functions. These functions are accomplished by the generation, transmission, integration, and recognition of specialized **nerve impulses.** It is the nerve impulse that permits the rapid and precise control of diverse body functions. In addition, elements of the nervous system serve to recognize certain **stimuli**—such as heat, light, pressure, or temperature—which affect the body. Nervous impulses may then be generated to

Figure 1-2, cont'd Organ systems of the human body.

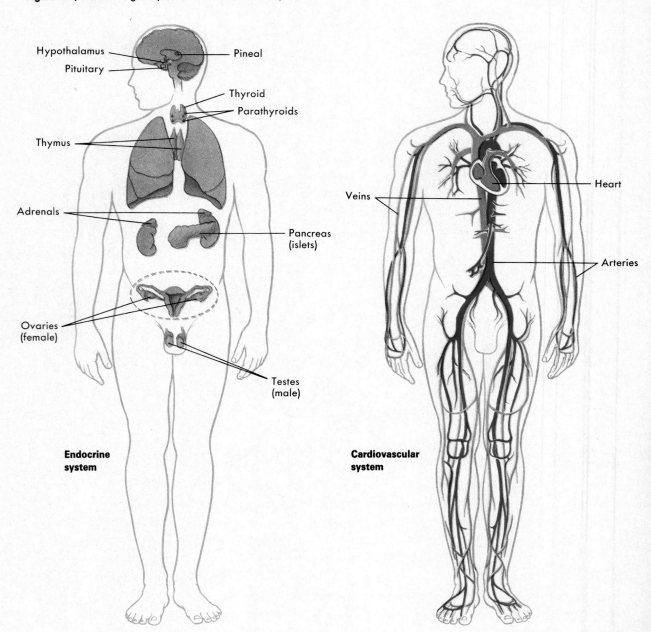

Hypothalamus
Pituitary
Pineal
Thyroid
Parathyroids
Thymus
Adrenals
Pancreas
(islets)
Ovaries
(female)
Testes
(male)

**Endocrine
system**

Veins
Heart
Arteries

**Cardiovascular
system**

convey this information to the brain where it can be analyzed and where appropriate action can be initiated. Other types of nerve impulses cause muscles to contract and glands to secrete.

Endocrine System

The endocrine system is composed of specialized glands that secrete chemicals known as **hormones** directly into the blood. Sometimes called *ductless glands*, the organs of the endocrine system perform the same general functions as the nervous system— namely, communication, integration, and control. The nervous system provides rapid, brief control by

fast-traveling nerve impulses. The endocrine system provides slower but longer-lasting control by secretion of hormones. The organs that are acted on and respond in some way to a particular hormone are referred to descriptively as **target organs.**

Hormones are the main regulators of metabolism, growth and development, reproduction, and many other body activities. They play roles of the utmost importance in such areas as fluid and electrolyte balance, acid-base balance, and energy metabolism.

The pituitary gland, pineal gland, hypothalamus, thyroid, parathyroids, adrenals, pancreas, ovaries,

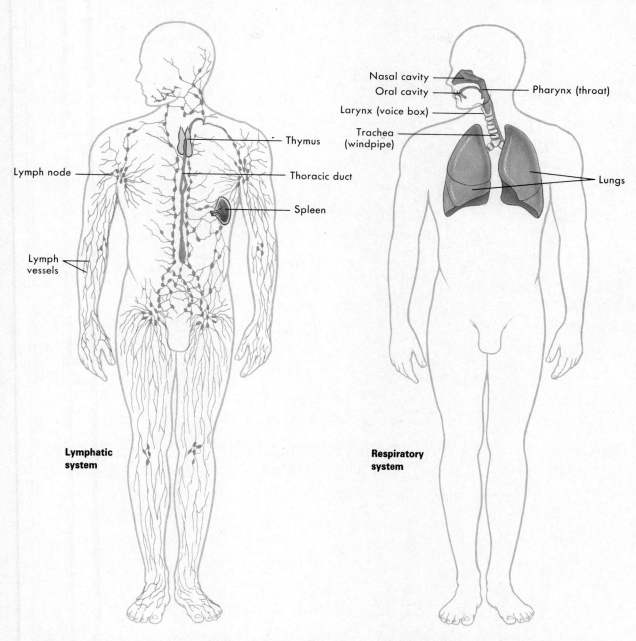

Lymph node

Lymph
vessels

Thymus

Thoracic duct

Spleen

**Lymphatic
system**

Nasal cavity

Oral cavity

Pharynx (throat)

Larynx (voice box)

Trachea
(windpipe)

Lungs

**Respiratory
system**

testes, thymus, and placenta all function as endocrine glands.

Circulatory System

The **cardiovascular** and **lymphatic systems** are both components of the circulatory system.

The cardiovascular system consists of the **heart** and a closed system of vessels called **arteries, veins,** and **capillaries.** As the name implies, blood contained in the circulatory system is pumped by the heart around a closed circle or circuit of vessels as it passes through the body.

The primary function of the cardiovascular system is *transportation*. The need for an efficient transportation system in the body is obvious. Critical transportation needs include movement of oxygen and carbon dioxide, nutrients, hormones, and other important substances on a continuing basis.

The lymphatic portion of the circulatory system is composed of **lymph, lymphatic vessels, lymph nodes,** and specialized lymphatic organs such as the **thymus** and **spleen.**

The functions of the lymphatic system include movement of fluids and certain large molecules from the tissue spaces surrounding the cells and movement of fat-related nutrients from the digestive tract

Figure 1-2, cont'd Organ systems of the human body.

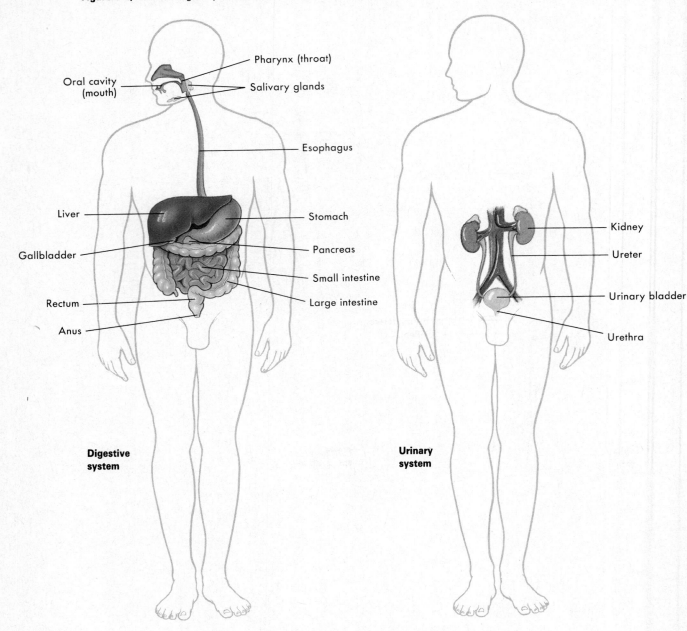

**Digestive
system**

**Urinary
system**

back to the blood. The lymphatic system is also involved in the overall functioning of the immune system, which plays a critical role in the defense mechanism of the body against disease.

Respiratory System

The organs of the respiratory system include the nose, pharynx, larynx, trachea, bronchi, and lungs. Together these organs permit the movement of air into the tiny, thin-walled sacs of the lungs called **alveoli.** It is in the alveoli that oxygen from the air is exchanged for the waste product carbon dioxide, which is carried to the lungs by the blood so it can be eliminated from the body.

Digestive System

The main organs of the digestive system include the mouth, pharynx (throat), esophagus, stomach, small intestine, large intestine, rectum, and anal canal. Accessory organs include the teeth, tongue, salivary glands, liver, gallbladder, and pancreas. The digestive system's main organs form a tube, open at both ends, called the **gastrointestinal** or **GI tract.** Food that enters the tract is digested, nutrients are absorbed, and the undigested residue is eliminated from the body as waste material called **feces.**

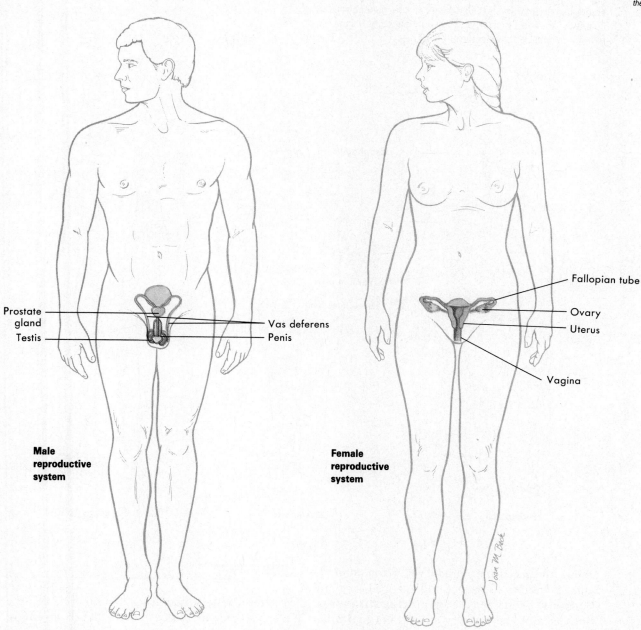

Prostate gland

Testis

Vas deferens

Penis

Male reproductive system

Fallopian tube

Ovary

Uterus

Vagina

Female reproductive system

Urinary System

The organs of the urinary system include the two kidneys, the ureters, and the bladder and urethra.

The kidneys function to "clear" or clean the blood of the many waste products that are continually produced by metabolism of foodstuffs in the body cells. The kidneys also play an important role in maintaining the electrolyte, water, and acid-base balance in the body.

The waste product produced by the kidneys is called **urine.** Once produced it flows out of the kidneys through the two ureters into the urinary bladder where it is stored. Urine passes from the bladder to the outside of the body through the urethra.

Reproductive System

The importance of normal reproductive system function is notably different from the end result of "normal function" as measured in any other organ system of the body. The proper functioning of the reproductive system ensures survival, not of the individual but of the species. In addition, production of hormones that permits development of the sexual characteristics occurs as a result of normal reproductive system activity.

In the male, the reproductive system organs include the **gonads** (testes), which produce the sex cells or sperm; the **genital ducts,** which include the epididymis, vas deferens, ejaculatory ducts, and the

Figure 1-3 Examples of extreme somatotypes. Note the smooth, soft contours of the endomorph; the defined muscularity of the mesomorph; and the dominance of height over fat or muscle in the ectomorph.

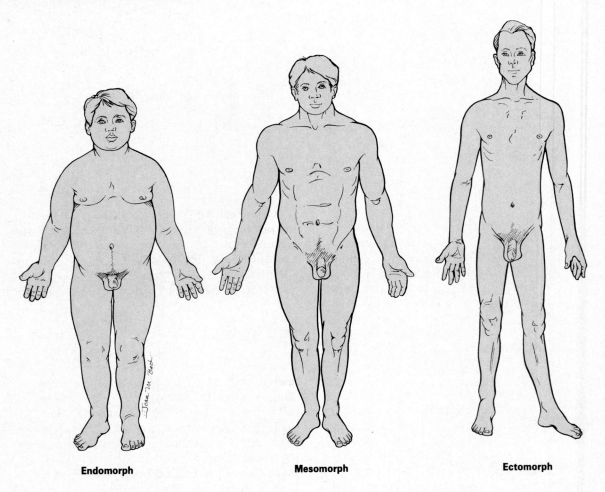

Endomorph   Mesomorph   Ectomorph

urethra; the **accessory glands,** which consist of the seminal vesicles, prostate, and the bulbourethral (Cowper's) glands; and the **supporting structures,** which include the scrotum and penis **(genitalia)** and the spermatic cords. Functioning together, these structures produce, transfer, and ultimately introduce sperm into the female reproductive tract where fertilization can occur.

The female **gonads** are the **ovaries.** The accessory organs include the **uterus, uterine (fallopian) tubes, vagina,** and **vulva** (genitalia). The breasts or **mammary glands** are also classified as external accessory sex organs in the female. The reproductive organs in the female are intended to produce the sex cells or ova, receive the male sex cells, permit fertilization and transfer of the sex cells to the uterus, and allow for development, birth and nourishment of the offspring.

BODY TYPE

The term **somatotype** is used to describe a particular category or type of body build or **physique.** Although the human body comes in many sizes and shapes, every individual can be classified as belonging to one of three basic body types, or somatotypes. The names used to describe these body types are **endomorph, mesomorph,** and **ectomorph.** In Figure 1-3, extreme examples of the three somatotype categories are shown. By carefully studying the body build of large numbers of individuals, scientists have found that the basic components that determine the different categories of physique occur in varying degrees in every person—both male and female. Only in very rare instances does an individual show almost total dominance by a single somatotype component.

Certain characteristics of body build or physique often predispose an individual to injury or disease. In endomorphs, for example, accumulation of body fat high on the torso appears to be associated with the early development of diabetes—especially in women. Endomorphic individuals of the same height and weight but with a lower body fat distribution pattern (over the hips and thighs) develop the disease in larger numbers than mesomorphs and ectomorphs but less frequently than endomorphs with a high body fat distribution pattern. This information explains why screening procedures for early detection of diabetes should include both somatotype identification procedures and analysis of body fat distribution patterns. Endomorphic individuals "at high risk" can be advised to watch closely for signs of diabetes.

ENDOMORPHY

Endomorphic persons tend to be fat. Typically, they have a heavy torso with a protruding abdomen and slightly smaller chest. Smoothness of contours caused by accumulation of fat under the skin all but eliminates muscular definition. In the extreme endomorph the neck is short and the head is almost always large and spherical. The limbs are relatively short in most endomorphs, with tapering and rounding of the thighs and upper arms.

MESOMORPHY

Mesomorphs are heavily muscled and have large, prominent bones. In most individuals with this body type the distal segments of the arms and legs are prominent and massive. The shoulders of a mesomorphic person are usually well defined and tend to project outward from the torso. In addition, the chest segment of the trunk predominates over the abdominal segment, and the waist is low.

ECTOMORPHY

In ectomorphs there is a relative predominance of linearity (height) over fat or muscle. These persons tend to be tall and thin. Typically, they have a relatively short trunk, long limbs, and poorly developed musculature. A shoulder drop is common in ectomorphs. The distal segments of the arms and legs tend to be relatively long and thin. In addition, the neck tends to be long and slender, and the face is small, with sharp, fragile features.

ANATOMICAL POSITION

Discussions about the body, how it moves, its posture, or the relationship of one area to another, assume that the body as a whole is in a specific position called the **anatomical position.** In this reference position (Figure 1-4) the body is in an erect or standing posture with the arms at the sides and palms turned forward. The head and feet are also pointing forward. The anatomical position is a reference position that gives meaning to the directional terms used to describe the body parts and regions.

BODY REGIONS

Identification of an object begins with overall or generalized recognition of its structure and form. Initially, it is in this way that the human form can be distinguished from other creatures or objects. Recognition occurs as soon as you can identify overall shape and basic outline. In order for more specific identification to occur, details of size, shape, and appearance of individual body areas must be described. Individuals differ in overall appearance because specific body areas such as the face or torso have unique identifying characteristics. Detailed descriptions of the human form require that specific regions be identified and appropriate terms be employed to describe them (Table 1-1, p. 16).

The body as a whole can be subdivided into two major portions or components: (1) **axial** and (2) **appendicular.** The axial portion of the body consists of the head, neck, and torso or trunk; the appendicular portion consists of the upper and lower extremities. Each major area is subdivided as shown in Figure 1-5. Note, for example, that the torso is composed of thoracic, abdominal, and pelvic areas and the upper extremity is divided into arm, forearm, wrist, and hand components. Although most terms used to describe gross body regions are familiar, misuse is common. The term *leg* is a good example. It refers to the area of the lower extremity between the knee and ankle and *not* to the entire lower extremity.

Figure 1-4 Anatomical position: the body is in an erect or standing posture with the arms at the sides and palms forward. The head and feet are also pointing forward.

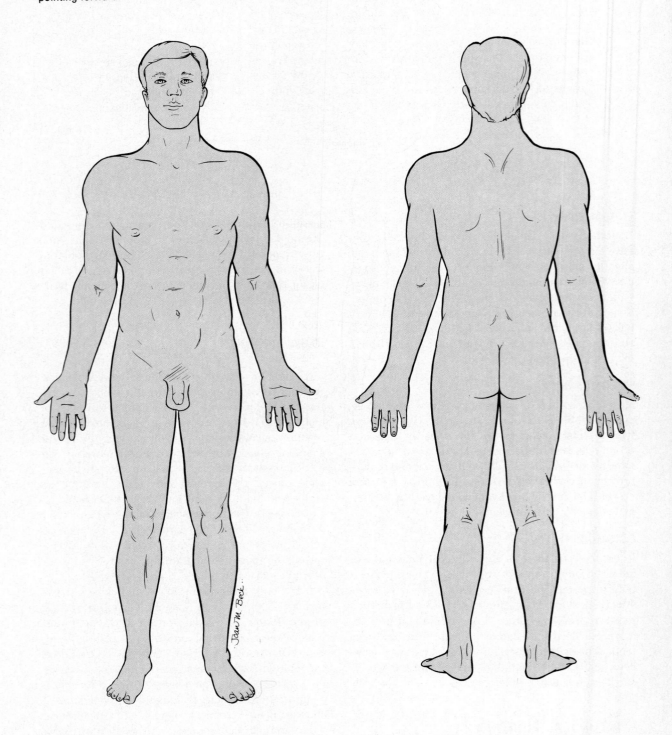

Figure 1-5 Specific body regions.

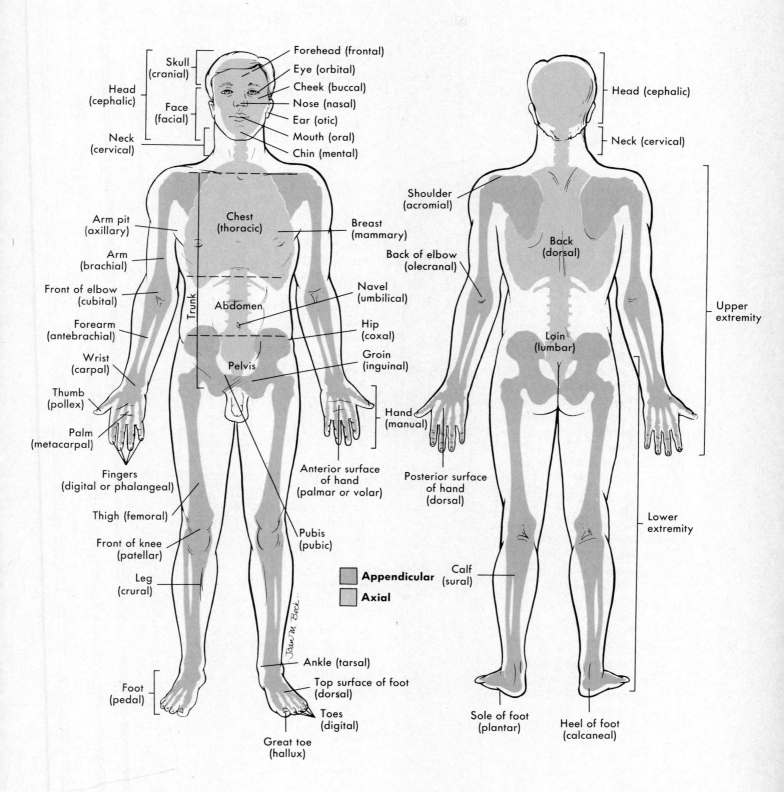

Skull (cranial)

Head (cephalic)

Face (facial)

Neck (cervical)

Forehead (frontal)

Eye (orbital)

Cheek (buccal)

Nose (nasal)

Ear (otic)

Mouth (oral)

Chin (mental)

Arm pit (axillary)

Arm (brachial)

Front of elbow (cubital)

Forearm (antebrachial)

Wrist (carpal)

Thumb (pollex)

Palm (metacarpal)

Fingers (digital or phalangeal)

Thigh (femoral)

Front of knee (patellar)

Leg (crural)

Foot (pedal)

Chest (thoracic)

Trunk

Abdomen

Pelvis

Breast (mammary)

Navel (umbilical)

Hip (coxal)

Groin (inguinal)

Hand (manual)

Anterior surface of hand (palmar or volar)

Pubis (pubic)

Ankle (tarsal)

Top surface of foot (dorsal)

Toes (digital)

Great toe (hallux)

Appendicular

Axial

Head (cephalic)

Neck (cervical)

Shoulder (acromial)

Back of elbow (olecranal)

Back (dorsal)

Loin (lumbar)

Posterior surface of hand (dorsal)

Upper extremity

Lower extremity

Calf (sural)

Sole of foot (plantar)

Heel of foot (calcaneal)

Table 1-1 Body regions

Body region	Area or example
Abdominal (ab-DOM-in-al)	Anterior torso below diaphragm
Antebrachium (an-te-BRA-ke-um)	Forearm
Antecubital (an-te-KU-bital)	Depressed area just in front of elbow
Arm	Upper extremity between shoulder and elbow
Axillary (AK-si-ler-e)	Armpit
Brachial (BRA-ke-al)	Arm
Buccal (BUK-al)	Mouth
Carpal (CAR-pal)	Wrist
Celiac (SEE-le-ak)	Area of abdomen
Cephalic (se-FAL-ik)	Head
Cervical (SER-vi-kal)	Neck
Costal (KOS-tal)	Ribs
Cranial (CRA-ne-al)	Skull
Crural (KROOR-al)	Leg
Cubital (KU-bi-tal)	Elbow
Cutaneous (ku-TANE-e-us)	Skin (or body surface)
Digital (DIJ-i-tal)	Fingers or toes
Dorsum (DOR-sum)	Back
Epigastric (ip-i-GAS-trik)	Upper middle area of abdomen
Facial (FA-shal)	Face
Frontal (FRON-tal)	Forehead
Oral (OR-al)	Mouth
Orbital or **ophthalmic** (OR-bi-tal or op-THAL-mik)	Eyes
Nasal (NA-sal)	Nose
Vestibular (ves-TIB-u-lar)	Open area at base of nose
Zygomatic (zi-go-MAT-ik)	Cheek
Femoral (FEM-or-al)	Thigh
Forearm	Upper extremity between elbow and wrist
Gluteal (GLOO-te-al)	Buttock
Groin	Root of thigh between lower extremity and abdomen
Inguinal (ING-gwi-nal)	Groin
Leg	Lower extremity between knee and ankle
Lumbar (LUM-bar)	Lower back between ribs and pelvis
Mammary (MAM-er-e)	Breast
Mastoid (MAST-oid)	Area of skull just below and behind the ear
Occipital (ok-SIP-i-tal)	Back of lower skull
Palmar (PAHL-mar)	Palm of hand
Pectoral (PEK-tor-al)	Chest
Pedal (PED-al)	Foot
Pelvic	Lower portion of torso
Perineal (per-i-NE-al)	Area (perineum) between anus and genitals
Precordial (pre-COR-di-al)	Chest area over heart
Plantar (PLAN-tar)	Sole of foot
Popliteal (pop-li-TEA-al)	Behind knee
Supraclavicular (supra-cla-VIC-u-lar)	Above clavicle
Tarsal (TAR-sal)	Ankle
Temporal	Side of skull
Thoracic (tho-RAS-ik)	Chest
Umbilical (um-BIL-i-cal)	Area around umbilicus
Volar (VO-lar)	Palm or sole

Figure 1-6 The nine regions of the abdominopelvic cavity showing the most superficial organs.

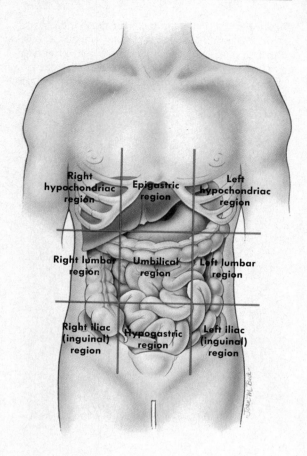

ABDOMINAL REGIONS

For convenience in locating abdominal organs, anatomists divide the abdomen into nine imaginary regions. The following is a listing of the nine regions (Figure 1-6) identified from right to left and from above downward:

1 Right hypochondriac region
2 Epigastric region
3 Left hypochondriac region
4 Right lumbar region
5 Umbilical region
6 Left lumbar region
7 Right iliac region
8 Hypogastric region
9 Left iliac region

The most superficial organs located in each of the nine abdominal regions are shown in Figure 1-6. In the **right hypochondriac** region the right lobe of the liver and the gallbladder are visible. In the **epigastric** area parts of the right and left lobes of the liver and a large portion of the stomach can be seen. Viewed superficially, only a small portion of the stomach and the left colic (splenic) flexure of the large intestine are visible in the **left hypochondriac** area. The **right lumbar** region in Figure 1-6 shows the ascending colon, the right colic (hepatic) flexure of the large intestine, and part of the small intestine. The most superficial organs seen in the **umbilical** region include a portion of the transverse colon and loops of the small intestine. Additional loops of the small intestine and the descending colon can be seen in the **left lumbar** region. The **right iliac** region contains the appendix, cecum, and parts of the small intestine. Only loops of the small intestine and the urinary bladder are seen in the **hypogastric** region. The **left iliac** region in Figure 1-6 shows portions of the descending and sigmoid colon and a portion of the small intestine.

The umbilicus is perhaps the "landmark" most frequently used when describing the surface anatomy of the abdomen. In most individuals the umbilicus is located at the level of the fourth lumbar vertebra. Although it varies greatly in position, it is always lower rather than higher when abnormally located.

Figure 1-7 Division of the abdomen into four quadrants. **A,** Photo showing surface outlines of *1,* right upper quadrant (RUQ); *2,* left upper quadrant (LUQ); *3,* right lower quadrant (RLQ); *4,* left lower quadrant (LLQ). **B,** Diagram showing relationship of internal organs to the four abdominopelvic quadrants.

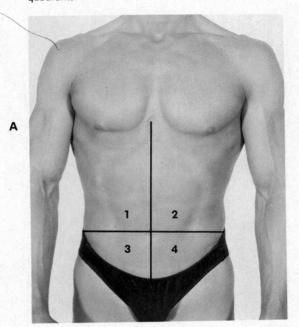

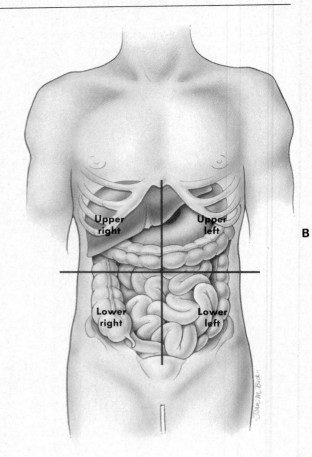

ABDOMINOPELVIC QUADRANTS

Physicians and other health professionals frequently divide the abdomen into **four quadrants** (Figure 1-7) in order to describe the site of abdominopelvic pain or locate some type of internal pathology such as a tumor or abscess. As you can see in Figure 1-7, a horizontal and vertical line passing through the umbilicus divides the abdomen into **right** and **left upper quadrants** and **right** and **left lower quadrants.**

BILATERAL SYMMETRY

Bilateral symmetry is one of the most obvious of the external organizational features in humans. To say that humans are **bilaterally symmetrical** simply means that the right and left sides of the body are mirror images of each other (Figure 1-8). One of the most important features of bilateral symmetry is balanced proportions. There is a remarkable correspondence in size and shape when comparing similar anatomical parts or external areas on opposite sides of the body.

The terms **ipsilateral** and **contralateral** are often used to identify placement of one body part with respect to another on the same or opposite sides of the body. These terms are used most frequently in describing injury to an extremity. Ipsilateral simply means "on the same side" and contralateral means "on the opposite side" of the body. Injuries to an arm or leg require careful comparison of the injured with the noninjured side. Minimal swelling or deformity on one side of the body is often apparent only to the trained observer who compares a suspected area of injury with its corresponding part on the opposite side of the body. If the right knee were injured, for example, the left knee would be designated as the contralateral knee.

Figure 1-8 Bilateral symmetry. As a result of this organizational feature, the right and left sides of the body are mirror images of each other.

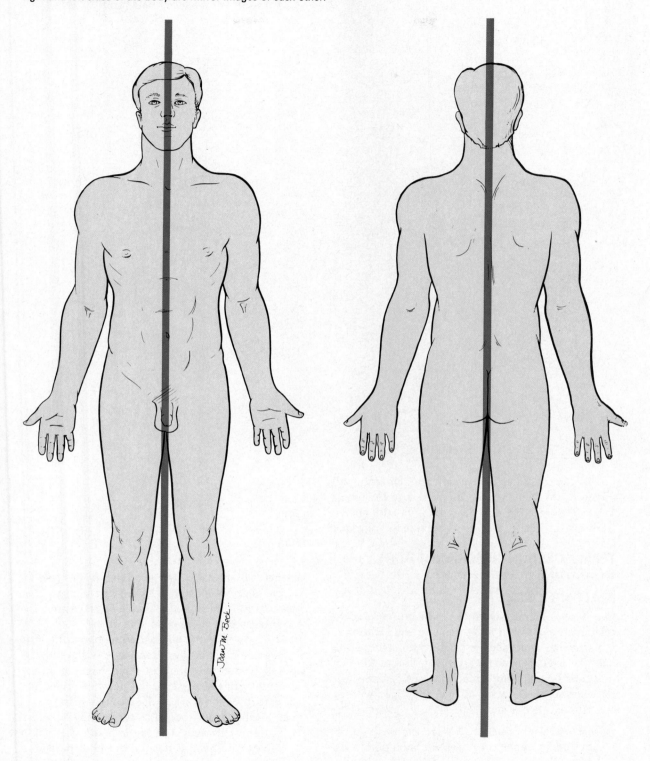

Figure 1-9 Directions and planes of the body.

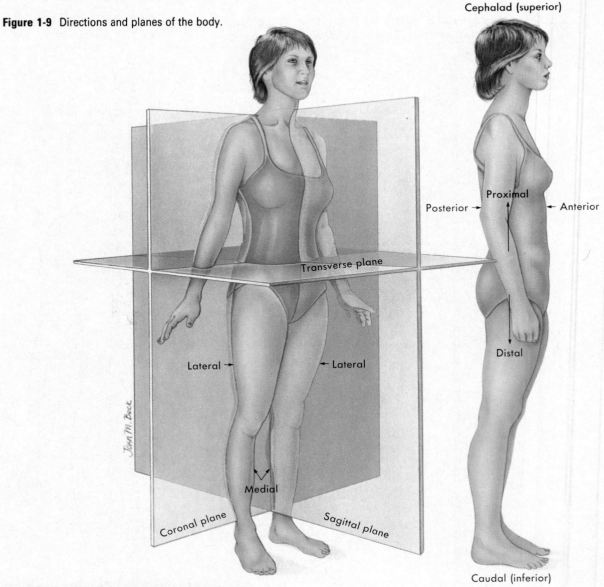

TERMS USED IN DESCRIBING BODY STRUCTURE

DIRECTIONAL TERMS

In order to minimize confusion when discussing the relationship between body areas or the location of a particular anatomical structure, specific terms must be used. When the body is in the anatomical position, the following directional terms can be used to describe the location of one body part with respect to another (Figure 1-9).

superior (cephalad or **craniad)** Toward the head end of the body; upper (the eyes are superior to the mouth)

inferior (caudad) Away from the head; lower (the bladder is inferior to the stomach)

anterior (ventral) Front (the eyes are located on the anterior surface of the head)

posterior (dorsal) Back (the shoulder blades are located on the posterior side of the body)

medial Toward the midline of the body (the great toe is located at the medial side of the foot)

lateral Away from the midline of the body (the little toe is located at the lateral side of the foot)

proximal Nearer the point of attachment of an extremity to the trunk or the point of origin of a part (the elbow is proximal to the wrist)

distal Farther from the point of attachment of an extremity to the trunk or the point of origin of a part (the hand is located at the distal end of the forearm)

superficial Nearer the surface (the skin of the arm is superficial to the muscles below it)

deep Farther from the body surface (humerus of the arm is deep to the muscles that surround it)

Figure 1-10 **A**, Transverse or horizontal plane or cut through the abdomen. **B**, Line drawing of the photograph.

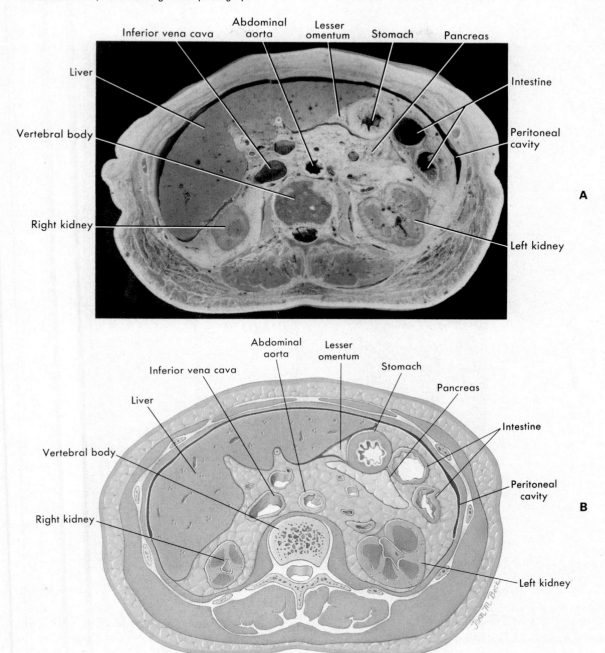

Section viewed from below

The thin, filmy-like membranes that line body cavities or cover the surface of organs within body cavities also have specialized names. The term **parietal** refers to the actual wall of a body cavity or the lining membrane that covers its surface. **Parietal peritoneum** is the membrane that lines the inside of the abdominal cavity. **Visceral** refers not to the wall or lining of a body cavity but to the thin membrane that covers the organs or **viscera** within

a cavity. **Visceral peritoneum** refers to the thin membrane that covers the organs contained in the abdominal cavity. Note in Figure 1-10 that a space or opening called the **peritoneal cavity** exists between the two membranes in the abdomen. The importance of parietal and visceral membranes will be stressed when the anatomy of the various body cavities and the viscera or organs contained within those cavities are studied in subsequent chapters.

Figure 1-11 The human brain. **A,** The brain as a whole—superior aspect. **B,** Lateral view of right half of brain. **C,** Medial view of right half of brain.

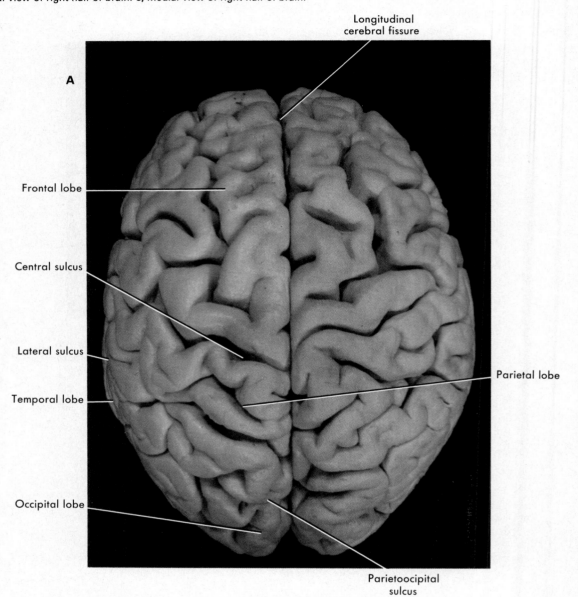

Longitudinal cerebral fissure

Frontal lobe

Central sulcus

Lateral sulcus

Temporal lobe

Occipital lobe

Parietal lobe

Parietoocipital sulcus

PLANES OR BODY SECTIONS

Read the following definitions and identify each term in Figure 1-9.

sagittal A lengthwise plane running from front to back; divides the body or any of its parts into right and left sides

median Sagittal plane through midline; divides the body or any of its parts into right and left halves

coronal or **frontal** A lengthwise plane running from side to side; divides the body or any of its parts into anterior and posterior portions

transverse or **horizontal** A crosswise plane; divides the body or any of its parts into upper and lower parts

Figure 1-10 shows the organs of the abdominal cavity as they would appear in the transverse or horizontal plane or "cut" through the abdomen represented in Figure 1-9. In addition to the actual photograph a simplified line diagram helps in identifying the primary organs. Note that organs near the bottom of the photo or line drawing are in a posterior position. The cut vertebra of the spine, for example, can be identified in its position behind or *posterior* to the stomach. The kidneys are located on either side of the vertebrae—they are *lateral* and the vertebra is *medial*.

The brain is an internal organ that shows bilateral symmetry. If cut into two halves using a mid-

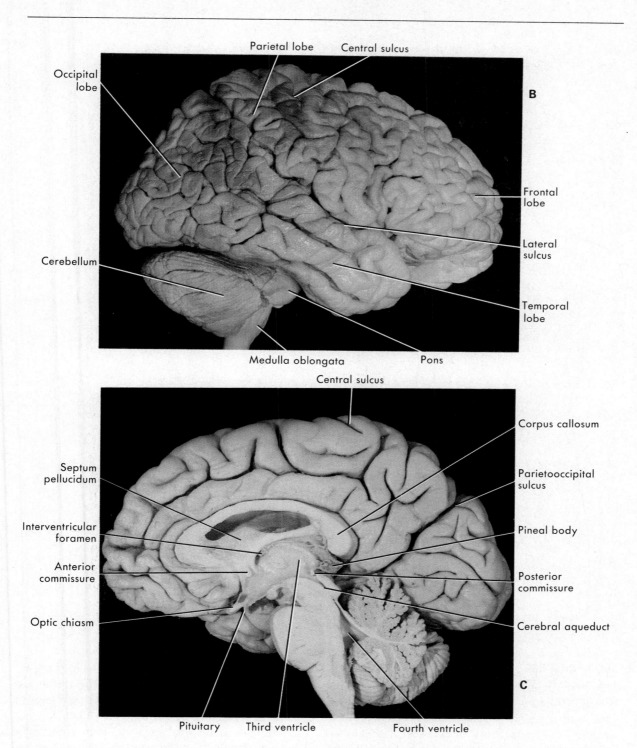

sagittal section, the right and left sides will be almost identical (mirror images) in their appearance. In Figure 1-11, a series of photographs begins with a view of the brain as seen from above (Figure 1-11, *A*). You are looking down at the top or most *superior* surface of the organ after its removal from the skull. Once removed, it can be divided into halves by a midsagittal cut or section.

Figure 1-11, *B* shows the right half of the brain after sectioning, as viewed from the side or *laterally*. The *lateral surface* is the external surface. In Figure 1-11, *C* the same right half is viewed showing its *medial surface*. The medial surface, of course, can only be seen if the organ is cut in half. As you can see, an understanding of directional terms and body planes or sections is required in order to fully appreciate and interpret the anatomical illustrations.

Figure 1-12 Location and subdivisions
of the major body cavities.

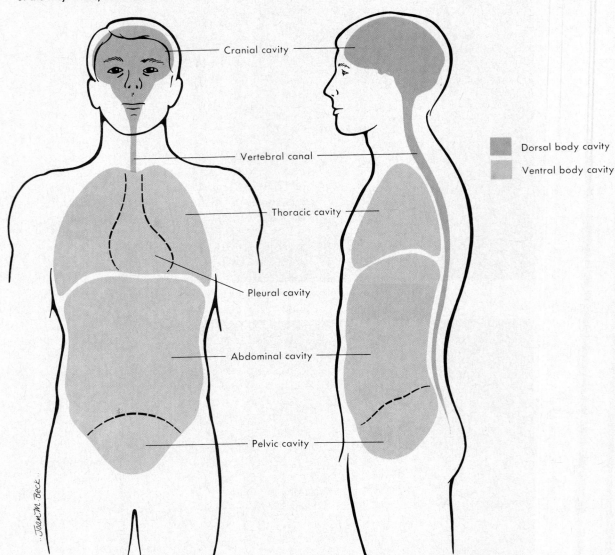

Cranial cavity

Vertebral canal

Thoracic cavity

Pleural cavity

Abdominal cavity

Pelvic cavity

Dorsal body cavity

Ventral body cavity

BODY CAVITIES

The body, contrary to its external appearance, is not a solid structure. It contains two major cavities that are in turn subdivided and contain compact, well-ordered arrangements of internal organs. The two major cavities are called the **ventral** and **dorsal** body cavities. The location and outlines of the body cavities are illustrated in Figure 1-12.

The **ventral cavity** consists of the **thoracic** or **chest cavity** and the **abdominopelvic cavity.** The *thoracic cavity* consists of a **right** and a **left pleural cavity** and a midportion called the **mediastinum.** Fibrous tissue forms a wall around the mediastinum, completely separating it from the right pleural sac, in which the right lung lies, and from the left pleural sac, in which the left lung lies. Thus the only organs in the thoracic cavity that are not located in the mediastinum are the lungs. Organs that are located in the mediastinum are the following: the heart (enclosed in its pericardial sac), the trachea and right and left bronchi, the esophagus, the thymus, various blood vessels (e.g., thoracic aorta, superior vena cava), the thoracic duct and other lymphatic vessels, various lymph nodes, and nerves (such as the phrenic and vagus). The *abdominopelvic cavity* consists of an upper portion, the **abdominal cavity,** and a lower portion, the **pelvic cavity.** The abdominal cavity contains the liver, gallbladder, stomach, pancreas, intestines, spleen, kidneys, and ureters. The bladder, certain reproductive organs (uterus, uterine tubes,

Table 1-2 Organs in ventral body cavities

Area	Organs
THORACIC CAVITY	
Right pleural cavity	Right lung (in pleural sac)
Mediastinum	Heart (in pericardial sac)
	Trachea
	Right and left bronchi
	Esophagus
	Thymus gland
	Aortic arch and thoracic aorta
	Venae cavae
	Various lymph nodes and nerves
	Thoracic duct
Left pleural cavity	Left lung (in pleural sac)
ABDOMINOPELVIC CAVITY	
Abdominal cavity	Liver
	Gallbladder
	Stomach
	Pancreas
	Intestines
	Spleen
	Kidneys
	Ureters
Pelvic cavity	Urinary bladder
	Female reproductive organs
	Uterus
	Uterine tubes
	Ovaries
	Male reproductive organs
	Prostate gland
	Seminal vesicles
	Parts of vas deferens
	Part of large intestine, namely, sigmoid colon and rectum

and ovaries in the female; prostate gland, seminal vesicles, and part of the vas deferens in the male), and part of the large intestine (namely, the sigmoid colon and rectum) lie in the pelvic cavity (Table 1-2).

The **dorsal cavity** consists of the **cranial** and **spinal cavities.** The *canial cavity* lies in the skull and houses the brain. The *spinal cavity* lies in the spinal column and houses the spinal cord.

An important generalization about body structure is that every organ, regardless of location or function, undergoes change over the years. Organs develop and grow during the years before maturity (young adulthood). After maturity, they age and, in general, **atrophy** or waste away as function de-

creases. We shall mention many specific age changes in the chapters that follow.

HOMEOSTASIS

More than a century ago a great French physiologist, Claude Bernard (1813-1878), made a remarkable observation. He noted that body cells survived in a healthy condition only when the temperature, pressure, and chemical composition of their fluid environment remained relatively constant. He called the environment of cells the *milieu interne* (internal environment). We call it **extracellular fluid.** Extracellular fluid is really two fluids, **interstitial fluid** and **blood plasma.** Interstitial fluid fills in the microscopic spaces between cells and so is also called *intercellular fluid.*

Many years after Claude Bernard's observation a famous American physiologist, Walter B. Cannon (1871-1945), suggested the name **homeostasis** for the relatively constant states maintained by the body. *Homeostasis* is a key word in modern physiology. It comes from two Greek words—*homoios*, meaning "the same," and *stasis*, meaning "standing." "Standing or staying the same" then is the literal meaning of homeostasis. However, as Cannon emphasized, homeostasis does not mean something set and immobile that stays exactly the same all the time. In his words, homeostasis "means a condition that may vary, but which is relatively consant." For example, homeostasis of blood temperature means that it remains relatively consant at about 37° C but varies slightly above and below this point. Homeostasis of blood glucose concentration means a relative constancy between 80 to 100 mg of glucose per 100 ml of blood. Most briefly, then, *homeostasis* may be defined as relative constancy of the extracellular fluid.

HOMEOSTATIC MECHANISMS

Devices for maintaining or restoring homeostasis are known as **homeostatic mechanisms.** They involve the functioning of virtually all of the body's organs and systems. Hence homeostatic mechanisms constitute the major theme of physiology and of this book. As an example, here is a brief description of a homeostatic mechanism that acts to restore homeostasis of blood carbon dioxide concentration. This particular mechanism is activated, or turned on, by a slight increase in blood carbon dioxide concentration above its homeostatic level. The increase acts as a stimulant to certain nerve cells in the brain causing them to send out more nerve impulses to certain breathing muscles. Faster breathing follows, and more carbon dioxide leaves the blood for the expired air. Blood carbon dioxide concentration,

Figure 1-13 Homeostasis of blood carbon dioxide level.

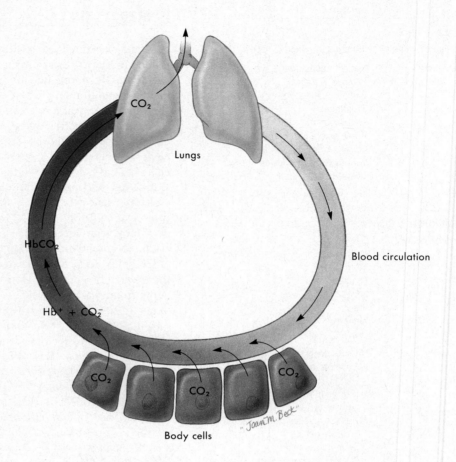

CO₂

Lungs

HbCO₂

Blood circulation

Hb⁺ + CO₂⁻

CO₂   CO₂   CO₂

Body cells

"Joan M. Beck"

therefore, necessarily decreases back toward its homeostatic level. Note the following general facts
about homeostatic mechanisms illustrated by this
example in Figure 1-13:

1 Homeostatic mechanisms are activated, or
turned on, by changes in the extracellular
fluid away from homeostasis.

2 Changes away from homeostasis are de-
tected by so-called sensor cells. Sensor cells
act by way of nerve impulses or hormones
to initiate responses that make up a homeostatic mechanism.

3 A homeostatic mechanism consists of responses that reverse the initial change that
turned on the mechanism. By reversing the
initial change, a homeostatic mechanism
tends to maintain or to restore homeostasis.

4 Homeostatic mechanisms operate on the
negative feedback principle. Essentially
this means that a change in one direction

feeds back in one way or another to bring
about an opposite or negative change. Thus
an increase in blood carbon dioxide concentration feeds back to cause a decrease
in blood carbon dioxide concentration back
towards its homeostatic level.

5 The responses that make up a homeostatic
mechanism are called appropriately **adaptive responses.** They enable the body to
adapt to changes in its environment in ways
that tend to maintain homeostasis and to
promote healthy survival. Adaptive responses make it possible for the body to
thrive as well as survive. Collectively,
adaptive responses bring about adaptation.
Adaptation is the successful coping of an
organism with its environment. Successful
adaptation means healthy survival.
Unsuccessful adaptation means disease
or death.

GENERALIZATIONS ABOUT BODY FUNCTION

1 Survival is the body's most important business—survival of itself and survival of the human species.

2 Survival depends on the body's maintaining or restoring **homeostasis,** a state of relative constancy of its internal environment. (Homeostasis is discussed in detail on pp. 25-26.)

3 Homeostasis depends on the body's ceaselessly carrying on many activities. Its major activities or functions include responding to changes in the body's environment, exchanging materials between the environment and cells, metabolizing foods, and integrating all of the body's diverse activities (see figure).

4 The body's functions are ultimately its cells' functions.

5 The body's ability to perform many of its functions changes gradually over the years. In general, the body performs its functions least well at both ends of life—in infancy and in old age. During childhood, body functions gradually become more and more efficient and effective. During late maturity and old age the opposite is true. They gradually become less and less efficient and effective. During young adulthood, they normally operate with maximum efficiency and effectiveness. The changes in functions that occur during the early years are called **devel-**

opmental processes. Those that occur during the late years are called **aging processes.** Developmental processes improve the functional capacities of the body. Aging processes, in contrast, diminish many functions and have no effect on others. Functional age changes will be noted in almost every chapter of this book.

Schematic illustrating that survival, the body's ultimate function, depends on homeostasis, which, in turn, depends on the functions listed in the lower box.

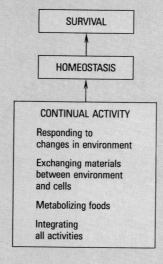

INTERACTION OF STRUCTURE AND FUNCTION

One of the most unifying and important concepts in the study of anatomy and physiology is the principle of **complementarity of structure and function.** In the chapters that follow, you will note again and again that anatomical structures seem "designed" to perform specific functions. They have a particular size, shape, form, or placement in the body because they are intended to perform a unique and specialized activity.

Indeed, structure and function are so intimately related that it is often inappropriate to totally separate the two disciplines for study purposes.

The relationship between the levels of structural organization shown in Figure 1-1 will take on added meaning as you study the respiratory system in Chapter 21. The specialized chemicals and the unique hairlike projections that cover cells in the lungs

Continued.

INTERACTION OF STRUCTURE AND FUNCTION—cont'd

have specific functions. The chemicals help prevent collapse of the lungs as air moves out during exhalation and the hairs trap and help eliminate contaminants such as dust. You will learn that the structure of the respiratory tubes and of the lungs not only assist in the efficient and rapid movement of air but also make possible the exchange of critical respiratory gases such as oxygen and carbon dioxide between the air in the lungs and the blood. Working together as the respiratory system—specialized chemicals, cells, tissues, and organs supply every cell of our body with necessary oxygen and constantly remove carbon dioxide.

The specialized nature of the respiratory system in humans is different from that found in other living organisms. Biologists now understand more fully what scientists for hundreds of years failed to grasp in trying to find an answer as to why that difference exists. The difference exists because as functional needs of the human organism developed during the course of evolution, specialized anatomical structures (such as the lungs) came into being and were continually modified to permit the body to function more effectively in a changing and often hostile environment. However, the anatomy of the respiratory system and, in particular, its relationship to

the body as a whole also changed as the body's functional needs changed. Those changes in the structure and function of the system most needed for survival evolved and became "inherited," and were passed on from one generation to the next, thus contributing to our understanding of the relationship between structure and function. This illustrates that not only does structure determine function but that function itself influences the actual anatomy of an organism over time. Current research in the study of human biology is now focused in large part on the integration, interaction, development, modification, and control of functioning body structures. The basic disciplines of anatomy and physiology undergird that research.

By applying the principle of complementarity of structure and function as you study the structural and functional levels of the body's organization in each of the remaining chapters of the text, you will be able to integrate otherwise isolated factual information into a cohesive and understandable whole. A memorized set of individual and isolated facts is soon forgotten—the component parts of an understandable anatomical structure that can be related to functional activity are not.

Outline Summary

INTRODUCTION

  A Anatomy—science of the structure of an organism and the relationship of its parts
  B Physiology—science of the functions of organisms

LEVELS OF ORGANIZATION (Figure 1-1)

  A Chemical level
    1 Organization of chemical constituents separates living from nonliving material
    2 Organization of chemical constituents results in living matter or protoplasm
    3 Characteristics of life
      a Metabolism
      b Irritability

      c Conductivity
      d Contractility
      e Growth
      f Reproduction
  B Cellular level
    1 Cells—smallest and most numerous units that possess and exhibit characteristics of living matter
    2 Cell—nucleus, surrounded by cytoplasm within a limiting membrane
    3 Cells differentiate in order to perform unique functions
  C Tissue level
    1 Tissue—an organization of similar cells specialized to perform a certain function

Outline Summary—cont'd

C Bilateral symmetry confers balanced proportions
D Remarkable correspondence of size and shape between body parts on opposite sides of the body

TERMS USED IN DESCRIBING BODY STRUCTURE

A Directional terms (Figure 1-9)
  1 Superior (cephalad or craniad)
  2 Inferior (caudad)
  3 Anterior (ventral)
  4 Posterior (dorsal)
  5 Medial
  6 Lateral
  7 Proximal
  8 Distal
  9 Superficial
  10 Deep
B Other important terms
  1 Ipsilateral
  2 Contralateral
  3 Visceral
  4 Parietal
C Planes or body sections (Figures 1-9 to 1-11)
  1 Sagittal
  2 Midsagittal (median)
  3 Coronal (frontal)
  4 Transverse (horizontal)
D Body cavities (Figure 1-12)
  1 Dorsal body cavity
    a Cranial cavity
    b Spinal cavity

  2 Ventral body cavity (Table 1-2)
    a Thoracic cavity
      (1) Right and left pleural cavities
      (2) Mediastinum
        (a) Pericardial cavity
    b Abdominopelvic cavity
      (1) Abdominal cavity
      (2) Pelvic cavity

GENERALIZATIONS ABOUT BODY FUNCTION

A Survival is most critical function
B Survival depends on homeostasis
C Homeostasis depends on complex and multiple body functions
D Body's functions are ultimately cellular functions
E Relationship of developmental to aging functions changes with time

HOMEOSTASIS (Figure 1-13)

A Internal environment surrounding cells (extracellular fluid) remains constant
B Extracellular fluid
  1 Interstitial fluid
  2 Blood plasma
C Intracellular fluid—fluid contained in cells
D Examples of homeostasis
  1 Temperature regulation
  2 Regulation of blood carbon dioxide levels
  3 Regulation of blood glucose levels

Review Questions

1 Define the terms *anatomy* and *physiology.*
2 Discuss the concept of organization in living things.
3 List and briefly describe the levels of organization that relate the structure of an organism to its function. Give examples characteristic of each level.
4 Identify the "characteristics of life."
5 Distinguish between the terms *allograft* and *xenograft.*
6 Give examples of each system level of organization in the body and briefly discuss the function of each.
7 What does the term *somatotype* mean?
8 List the three major somatotype categories and briefly describe the generalized characteristics of each.
9 Do characteristics of body build or physique predispose an individual to injury or disease? If so, give an example.
10 What is meant by the term *anatomical position?* How do the specific anatomical terms of position or direction relate this body orientation?

11 Identify the two major subdivisions of the body as a whole and list the primary anatomical areas or components of each.
12 List by name the nine abdominal regions, the four abdominal quadrants, and identify the major organs located in each.
13 What is bilateral symmetry? What terms are used to identify placement of one body part with respect to another on the same or opposite sides of the body?
14 Define briefly each of the following terms: *anterior, distal, frontal plane, medial, dorsal, coronal plane, organ, parietal peritoneum, superior, tissue.*
15 Name the two major body cavities and the subdivisions of each.
16 Locate the mediastinum and list the organs found there.
17 What does the term *homeostasis* mean? Illustrate some generalizations about body function using homeostatic mechanisms as examples.
18 Discuss in general terms the principle of complementarity of structure and function.

2 THE CHEMICAL BASIS OF LIFE

CHAPTER OUTLINE

Basic chemistry
    Structure of atoms
    Chemical reactions
    Radioactivity
Selected facts from biochemistry
    Definition of biochemistry
    Biomolecules
    Bioenergy
    Biosynthesis

OBJECTIVES

After you have completed this chapter, you should be able to:

1 Define the term *atom* and list the primary subatomic particles.
2 Explain the relationship between an element's atomic number and the number of protons in its nucleus.
3 List the major elements and major mineral elements found in protoplasm.
4 Compare and contrast electrovalent or ionic bonds with covalent bonds.
5 List and describe the four basic types of chemical reactions that occur in living material.
6 Define the term *radioactivity* and identify three kinds of radiations.
7 Discuss the relationship of biochemistry to the structural levels of organization in the body.
8 Define the term *electrolyte* and explain the electrical characteristics of anions and cations in electrolyte solutions.
9 Compare the dissociation products of a typical acid and base compound.
10 Discuss the chemical composition of a typical protein and identify the major functions of human protein compounds.
11 Identify the three major types of carbohydrate compounds and give examples of each.
12 Classify the major categories of lipids found in the body.
13 Define the term *bioenergy* and identify the most important of the bioenergy molecules.
14 Compare and contrast anabolism and catabolism.

Life itself depends on proper levels and proportions of chemical substances in the protoplasm of cells. The various structural levels of organization described in Chapter 1 are based, ultimately, on the existence and interrelationships of atoms and molecules. In order to understand life processes it is necessary to understand the basic types of interactions that may occur between the chemical substances of living matter. Modern **biochemistry** is a specialized area that deals directly with the chemical composition of living matter and the processes that underlie such life activities as growth and maintenance.

Modern biochemistry is in reality many disciplines. It is closely related to the other life sciences and to modern medicine. Biochemists use many different chemical, physical, biological, nutritional and immunological techniques to probe life processes at every level of organization. Understanding a number of basic definitions related to chemistry in general is necessary in order to understand the chemical components of and the chemical reactions in living matter.

BASIC CHEMISTRY

STRUCTURE OF ATOMS

Early in the nineteenth century an English chemist, John Dalton, proposed the concept that matter is composed of **atoms** (from the Greek *atomos*—"indivisible"). He conceived of atoms as solid, indivisible particles, and for about 100 years this was believed to be true. Then experiments done in the laboratory of another English scientist, Ernest T. Rutherford, revealed quite opposite facts. Atoms are not solid particles but are small spheres made up mostly of open space. Atoms are not indivisible but have smaller particles present within them. Startling discoveries indeed, considering the infinitesimal size of atoms. One hundred million of them lined up would measure barely an inch. Is your imagination powerful enough to visualize a particle less than a 100 millionth of an inch across, consisting mostly of space? And having still smaller particles present in it?

Atoms contain several different kinds of smaller particles. Collectively these are called *subatomic particles.* The basic ones are protons, neutrons, and electrons. **Protons** are positively charged particles. **Neutrons** are uncharged or electrically neutral particles. **Electrons** are negatively charged particles. Protons and neutrons together form the atomic **nucleus,** a tiny central core located deep inside the atom. Because protons are positively charged and

KEY TERMS

Atom (AT-um)

Biochemistry (bi-o-KEM-is-tree)

Bioenergy (bi-o-EN-er-ge)

Biomolecule (bi-o-MOL-e-kul)

Biosynthesis (bi-o-SIN-the-sis)

Carbohydrate (kar-boh-HIGH-drate)

Compound (kom-pownd)

Electrolyte (e-LEK-tro-lite)

Electron (e-LEK-tron)

Element (EL-e-ment)

Enzyme (EN-zime)

Isotope (EYE-so-tope)

Lipid (LIP-id)

Matter (MAT-er)

Molecule (MOL-e-kul)

Neutron (NU-tron)

Protein (PRO-teen)

Proton (PRO-ton)

Radioactivity (ra-de-o-ak-TIV-i-tee)

Synthesis (SIN-the-sis)

Valence (VA-lens)

DEFINITIONS

matter Anything that occupies space, that has mass, whether it is large or small, living or nonliving, is matter.

elements An element is a simple form of matter, a substance that cannot be broken down into two or more different substances. There are approximately 100 known chemical elements, 24 of which are present in the human body. Of these, by far the most abundant are hydrogen, oxygen, carbon, and nitrogen.

atoms An atom is the smallest unit of an element.

compounds A compound is a substance composed of two or more elements.

molecules A molecule is the smallest unit of a compound; a molecule consists of two or more atoms joined to each other by chemical bonds.

chemical symbols Chemical symbols are abbreviations for the names of elements. (For symbols of the elements in the body, see Table 2-1.)

Table 2-1 Elements in the human body

Name	Symbol
MAJOR ELEMENTS*	
Oxygen	O
Carbon	C
Hydrogen	H
Nitrogen	N
MAJOR MINERAL ELEMENTS	
Calcium	Ca
Phosphorus	P
Potassium	K (Latin, *kalium*)
Sulfur	S
Sodium	Na (Latin, *natrium*)
Chlorine	Cl
Magnesium	Mg
TRACE ELEMENTS	
Iron	Fe (Latin, *ferrum*)
Iodine	I
Copper	Cu (Latin, *cuprum*)
Zinc	Zn
Manganese	Mn
Cobalt	Co
Chromium	Cr
Selenium	Se
Molybdenum	Mo
Fluorine	F
Tin	Sn (Latin, *stannum*)
Silicon	Si
Vanadium	V

*These four elements compose more than 96% of body weight.

neutrons are neutral, the nucleus of an atom bears a positive electrical charge equal to the number of protons present in it. Electrons spin around the atom's nucleus in shells of space or energy levels. The number of negatively charged electrons moving around an atom's nucleus always equals the number of positively charged protons in the nucleus. The opposite charges, therefore, cancel or neutralize each other, and atoms are electrically neutral particles.

The number of protons in an atom's nucleus identifies the kind of element it is and is known as an element's **atomic number.** Hydrogen, for example, has an atomic number of 1; this means that all hydrogen atoms—and only hydrogen atoms—have 1 proton in their nuclei. All carbon atoms, and only carbon atoms, contain 6 protons and have an atomic number of 6. All oxygen atoms, and only oxygen atoms, have 8 protons and an atomic number of 8. In short, each element is identified by its own unique number of protons, that is, by its own unique atomic number. If two atoms contain a different number of protons, they necessarily have different atomic numbers and are different elements.

Protons, neutrons, and electrons all have extremely small mass or weight. (*Mass equals weight*

under ordinary conditions.) A proton weighs almost exactly the same amount as a neutron, and lightweight as protons are, each weighs 1,836 times as much as an electron! For all practical purposes, therefore, an atom's weight is considered equal to the number of protons plus the number of neutrons in its nucleus, as shown in the following equations:

$$\text{Atomic weight} = (p^+ + n)$$
$$(\text{Atomic weight} - p^+) = n$$

All atoms of the same element contain the same number of protons but do not necessarily contain the same number of neutrons. This brings us to a new term—*isotope*. An **isotope** is an element whose atoms contain a different number of neutrons from most atoms of that element. For example, atoms of a hydrogen isotope, called deuterium, contain 1 neutron in their nuclei whereas most hydrogen atoms contain no neutrons (Figure 2-1).

The total number of electrons in an atom equals the number of protons in its nucleus (Figures 2-1

Figure 2-1 Structure of hydrogen atoms. *Left,* ordinary hydrogen atom: one proton and no neutrons present in its nucleus. *Middle,* deuterium atom, an isotope of hydrogen. How does it differ from the ordinary hydrogen atom? *Right,* tritium, another isotope of hydrogen. How does it differ from deuterium? From ordinary hydrogen?

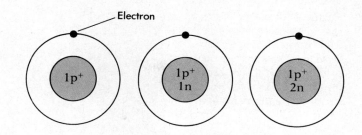

Figure 2-2 Atomic structure of carbon and one of its isotopes. *Left,* ordinary carbon (atomic number, 6; atomic weight, 12; symbols, C^{12} or C^{12}). *Right,* carbon isotope (atomic number, 6; atomic weight, 14; symbols, $_6C^{14}$ or C^{14}).

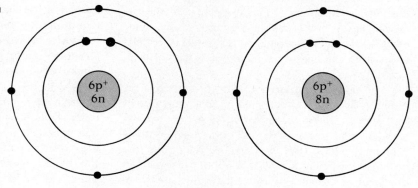

Figure 2-3 Diagram of electron transfer type of chemical reaction. Unpaired electron *(red)* in sodium atom's outer shell transfers to chlorine's outer shell. This forms an electrovalent or ionic bond between the two atoms, which changes them into a single molecule of sodium chloride.

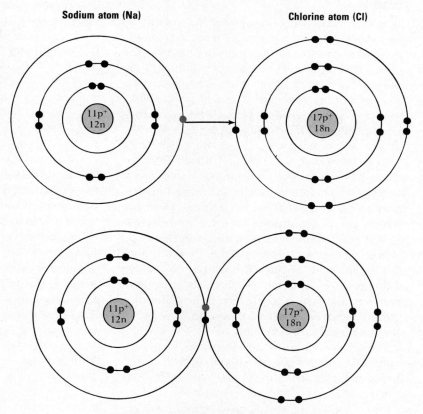

Sodium chloride molecule (Na^+Cl^- or NaCl)

Figure 2-4 Diagram of electron sharing type of chemical reaction. Two atoms of hydrogen share their two single electrons to form a covalent bond between the two atoms and convert them into one molecule of hydrogen.

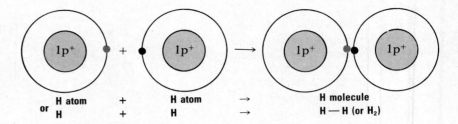

and 2-2). Each shell of space around the nucleus can accommodate a certain maximum number of orbiting electrons. The innermost shell can accommodate the fewest electrons—1 single electron in the hydrogen atom, but 1 pair of electrons in all other elements. Both the second and the outermost shells of an atom can hold a maximum of 4 pairs of electrons.

The number and arrangement of electrons orbiting in an atom's outer shell of space have great importance. They determine whether or not the atom is chemically active. When the outer shell contains 1 or more single (that is, unpaired) electrons, the atom is chemically active. It is able to take part in chemical reactions. But when the outer shell contains no single electrons but only pairs of electrons, the atom is chemically inactive or stable.

CHEMICAL REACTIONS

Chemical reactions involve unpaired electrons in the outer shells of atoms. A **chemical reaction** consists of either the transfer of unpaired electrons from the outer orbit of one atom to the outer orbital of another or the sharing of one atom's unpaired electrons with those of another atom. Figure 2-3 illustrates an electron transfer type of chemical reaction. In examining this figure, pay attention first to the number and arrangement of electrons in the outer shells of the sodium and chlorine atoms. Note that sodium has only 1 unpaired electron in its outer shell in contrast to chlorine, which has 1 unpaired electron plus 3 paired electrons, or a total of 7 electrons. Sodium transfers or donates its 1 unpaired electron to chlorine. Chlorine accepts it and pairs it with its 1 unpaired electron, thereby filling its outer shell with the maximum of 4 electron pairs. The formation of this new electron pair in chlorine's outer orbit creates an attractive force that binds the sodium and chlorine atoms to each other. In short, it creates a **chemical bond** between them.

The electron transfer constitutes a chemical reaction that changes the two atoms of the elements sodium and chlorine into one molecule of the compound, sodium chloride (ordinary table salt). A **compound** is a substance made up of two or more elements joined by chemical bonds to form units called

molecules. A chemical bond formed by the transfer of electrons from one atom to another is called an **electrovalent** or **ionic bond**. A chemical bond formed by the sharing of a pair of electrons between two atoms is called a **covalent bond** (Figure 2-4).

Valence is the number of unpaired electrons in an atom's outer shell, and therefore valence indicates the number of chemical bonds an atom can form with other atoms. For example, sodium and chlorine atoms each have 1 unpaired electron in their outer shells, so each has a valence of 1 and can form one chemical bond with another atom. Hydrogen also has 1 unpaired electron in its only shell, so it, too, has a valence of 1 and can form one chemical bond with another atom. Look again at Figure 2-2. How many unpaired electrons are in the carbon atom? What is carbon's valence? How many chemical bonds can it form with other atoms? Answers: The carbon atom contains 4 unpaired electrons in its outer shell. It has a valence of 4 and can form chemical bonds with four other atoms. For example, one carbon atom can combine with four hydrogen atoms to form one *molecule* of methane (CH_4).

Four basic types of chemical reactions that you will learn to recognize as you study physiology are:

1 Synthesis
2 Decomposition
3 Exchange
4 Reversible reactions

Synthesis (from the Greek *syn*, "together," and *thesis*, "putting") reactions put together or combine two or more substances to form a more complex substance. In so doing, synthesis reactions form new chemical bonds. Many such reactions occur in the body. Every one of its cells, for example, combines amino acid molecules to form complex protein compounds. The term **decomposition** suggests what such reactions do. They decompose or break down a substance into two or more simpler substances. In so doing, decomposition reactions break chemical bonds. Decomposition and synthesis, in a word, are opposites. Synthesis builds up; decomposition breaks down. Synthesis forms chemical bonds; decomposition breaks chemical bonds. One of the many examples of decomposition reactions in the body is the digestion or breakdown of large fat mol-

Before we leave the subject of chemical reactions, we must state the following important principle about them: Every chemical change is coupled with an energy change. The formation of chemical bonds is coupled with energy use. The breaking of chemical bonds is coupled with energy release. Synthesis reactions, therefore, use energy, and decomposition reactions release energy. Example in living cells of synthesis with energy use: reactions that combine simple food compounds to form complex compounds. Example in living cells of decomposition with release of energy: reactions that break food compounds down into simpler compounds.

Figure 2-5 Diagram shows that emission of an alpha particle transforms an element into another element.

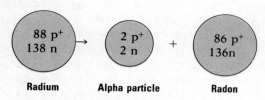

Figure 2-6 Diagram shows that emission of a beta particle from an atom of iodine transforms it into an atom of xenon.

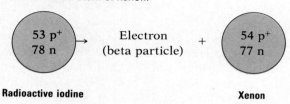

ecules into two simpler substances, fatty acids and glycerol. The nature of **exchange reactions** is also suggested by their name. They break down or decompose two compounds and in exchange synthesize two new compounds. Certain exchange reactions take place in the blood. One example is the reaction between lactic acid and sodium bicarbonate. The decomposition of both substances is exchanged for the synthesis of sodium lactate and carbonic acid. These changes can be seen more easily in an equation.

$$H \cdot lactate + NaHCO_3 \rightarrow Na \cdot lactate + H \cdot HCO_3$$

The formula "H · lactate" represents lactic acid; "NaHCO$_3$" is the formula for sodium bicarbonate; "Na · lactate" represents sodium lactate; and "H · HCO$_3$" represents carbonic acid.

Reversible reactions, as the name suggests, proceed in both directions. A great many chemical reactions are reversible, and we shall cite a number of them in later chapters of the book.

RADIOACTIVITY

Radioactivity is the emission of radiations from an atom's nucleus. Alpha particles, beta particles, and gamma rays are the three kinds of radiations. **Alpha particles** are relatively heavy particles consisting of two protons plus two neutrons. They shoot out of a radioactive atom's nucleus at a reported speed of 18,000 miles per second. **Beta particles** are electrons formed in a radioactive atom's nucleus by one of its neutrons breaking down into a proton and an electron. The proton remains behind in the nucleus, and the electron is ejected from it as a beta particle. Beta particles, since they are electrons, are much smaller than alpha particles, which consist of two

protons and two neutrons. Also, beta particles travel at a much greater speed than alpha particles. **Gamma rays** are electromagnetic radiations, a form of light energy.

How does radioactivity change an atom? To find out how the emission of an alpha particle changes the nucleus of a radium atom, examine Figure 2-5. Note that after the ejection of an alpha particle, the atom's nucleus contains fewer protons and neutrons. Basic principles about atoms, you will recall, are these: All atoms of the same element contain the same number of protons; atoms that contain different numbers of protons are, therefore, different elements. After an alpha particle is ejected from the nucleus of a radium atom, the atom contains 86 instead of 88 protons and has been changed into an atom of radon. Figure 2-6 shows how the emission of a beta particle changes an atom's nucleus. We started this paragraph with a question and shall end it with an answer. Radioactivity changes the chemical identity of an atom. It transforms an atom of one element into an atom of a different element by changing the number of protons in the atom's nucleus.

Our bodies are continually exposed to low levels of radiation present in the environment. When alpha or beta particles or gamma rays score direct hits on atoms present in living cells, they ionize the atoms by knocking electrons out of their outer shells of space. This ionization, in turn, injures or kills the cells. Knowledge of this fact underlies the use of radiation therapy to kill cancer cells. But radiation can also have an opposite effect. It can lead to the development of cancer cells. Abundant evidence indicates that leukemia and other types of cancer may result from exposure to very high levels of radiation or to lower levels of radiation for prolonged periods of time. For example, many more survivors of the atomic bombings of Hiroshima and Nagasaki developed leukemia than did individuals not exposed to the bombs.

SELECTED FACTS FROM BIOCHEMISTRY

DEFINITION OF BIOCHEMISTRY

Biochemistry is the chemistry of living organisms. It is a young but extensive science—so extensive that some biochemistry textbooks contain more than a thousand pages. Out of this abundance, we have tried to select for inclusion in this chapter the facts and principles you will need most to know to understand the rest of the book. We have organized this information around three topics: biomolecules, bioenergy, and biosynthesis.

BIOMOLECULES

A textbook by the distinguished biochemist Albert L. Lehninger opens with the sentence, "Living things are composed of lifeless molecules." Biomolecules are these lifeless molecules. By definition then, **biomolecules** are compounds that compose living organisms. Most abundant of them all is water. The thousands of other biomolecules fall into seven classes of compounds: acids, bases, salts, proteins, carbohydrates, lipids, and nucleic acids. To learn some basic facts about each of these kinds of biomolecules, study the following paragraphs.

Water is the body's most abundant compound. It makes up some 65% of body weight and serves a host of vital functions. Many of these are related to the fact that water dissolves so many substances. By dissolving oxygen and food substances, for instance, water enables these essential materials to enter and leave the blood capillaries and to enter cells. Also, by dissolving substances, water makes possible the thousands of chemical reactions that keep us alive. Dry substances are virtually nonreactive. Another important function of water, that of enabling the body to maintain a relatively constant temperature, stems from the fact that water both absorbs and gives up heat slowly.

Acids, bases, and salts belong to a large group of compounds called electrolytes. **Electrolytes,** by definition, are compounds whose molecules consist of positive ions (cations) and negative ions (anions) and that dissociate when dissolved in water into

Table 2-2 Major functions of human protein compounds

Functions	Examples
Catalyze chemical reactions	Lactase (enzyme in intestinal digestive juice) catalyzes chemical reaction that changes lactose to glucose and galactose
Transport substances in blood	Proteins classified as albumins combine with fatty acids to transport them in form of lipoproteins
Communicate information to cells	Insulin, a protein hormone, serves as chemical message from islet cells of pancreas to cells all over body
	Binding sites of certain proteins on surfaces of cell membranes serve as receptors for insulin and various other hormones
Defend body against any harmful agents	Proteins called antibodies or immunoglobulins combine with various harmful agents to render them harmless

cations and anions. In short, electrolytes are substances that ionize in solution. Formation of an electrolyte takes place by electron transfer, as shown in Figure 2-3. Here you see a sodium atom donating the single electron in its valence shell to the valence shell of a chlorine atom. Note again what this accomplishes. The transferred electron fills chlorine's outer shell by forming a fourth electron pair in it; this pair constitutes an electrovalent or ionic bond between the sodium and the chlorine. It also does something else. It converts both the sodium and chlorine atoms into ions or electrically charged particles. Since the sodium atom has donated one of its electrons, that is, one of its negative charges, it now contains one more positive charge than negative (11 p^+ and 10 e^-). It has become a *positive ion,* a **cation,** designated by the symbol Na^+. Chlorine, after accepting the electron, bears one more negative than positive charge. It has become a *negative ion,* an **anion.** The electrovalent bond between the two ions creates out of them one molecule of the ionic compound sodium chloride. The formula Na^+Cl^- represents sodium chloride (ordinary table salt). Because chemically bound ions compose its molecules, it is an **ionic compound.** Because solutions of ionic compounds conduct an electric current, another name for them is *electrolytes.* Chapters 23 and 24 give more information about these essential biomolecules.

Figure 2-7 Basic structural formula for an amino acid. All amino acids contain the part of the molecule enclosed in the rectangle; *R* represents the part unique to each of the 20 different amino acids. Examples: In glycine (the simplest amino acid) *R* is a hydrogen atom; in alanine, it is the group CH_3.

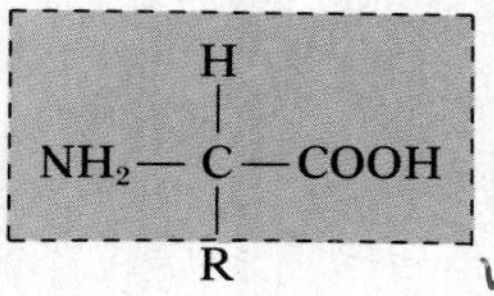

Figure 2-8 Structure of simple amino acids. The unique part of each molecule (*R* in Figure 2-7) is printed in red.

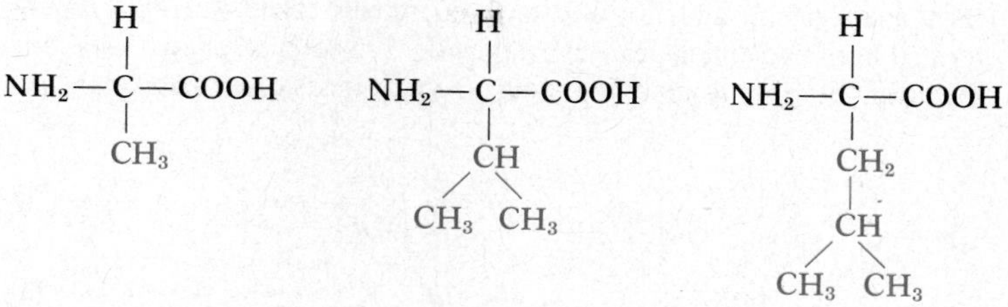

$$NH_2 - \overset{\overset{\displaystyle H}{|}}{\underset{\underset{\displaystyle CH_3}{|}}{C}} - COOH$$

Alanine

$$NH_2 - \overset{\overset{\displaystyle H}{|}}{\underset{\underset{\underset{\displaystyle CH_3 \quad CH_3}{|}}{CH}}{C}} - COOH$$

Valine

$$NH_2 - \overset{\overset{\displaystyle H}{|}}{\underset{\underset{\underset{\underset{\displaystyle CH_3 \quad CH_3}{|}}{CH}}{CH_2}}{C}} - COOH$$

Leucine

$$NH_2 - \overset{\overset{\displaystyle H}{|}}{\underset{\underset{\displaystyle R}{|}}{C}} - COOH$$

Acids, like all electrolytes, ionize when dissolved in water to form cations and anions. By definition, an *acid* is a compound that yields hydrogen ions (H^+) in solution. A *base* is a compound that yields hydroxyl or OH ions (OH^-) in solution. A **salt** is a compound that in water yields the positive ions of a base and the negative ions of an acid. A salt is formed when an acid reacts with a base. Here is an example:

HCl	+ NaOH	= NaCl +	Water
Hydrochloric acid	**Sodium hydroxide, a base**	**Salt**	

Proteins (from the Greek *proteios*, "of the first rank") are the most abundant of the carbon-containing or organic compounds in the body, and as their name implies, their functions are of first-rank importance (Table 2-2). Protein molecules are giant sized; that is, they are macromolecules. Smaller units or building blocks, namely, amino acids, make up each protein molecule. An **amino acid,** as you can see in Figure 2-7, consists of a carbon atom (called the alpha carbon) to which are bonded an amino group (NH_2), a carboxyl group (COOH), a hydrogen atom, and a side chain (R). The side chain constitutes the unique, identifying part of an amino acid. Twenty-one different amino acids enter into the composition of human proteins. In all of these except the simplest (glycine), the side chain consists of groups of atoms. Glycine's side chain is a single atom, hydrogen. Figure 2-8 shows three amino acids. What group of atoms compose alanine's side chain? Valine's side chain? Leucine's side chain? All amino acids contain at least four elements, namely, carbon, hydrogen, oxygen, and nitrogen. In addition, many amino acids contain one or more of the following: sulfur, phosphorus, and iron.

Amino acids frequently become joined by peptide bonds. A **peptide bond** is one that binds the carboxyl group of one amino acid to the amino group of another amino acid. Figure 2-9 shows how a peptide bond forms. OH from the carboxyl group of one amino acid and H from the amino group of another amino acid split off to form water plus a new compound called a peptide. A peptide made up of only two amino acids linked by a peptide bond is a **dipeptide. A tripeptide** consists of three amino acids linked by two bonds. A long sequence or chain of amino acids—usually 100 or more—linked by peptide bonds constitutes a **polypeptide.** All protein molecules are polypeptides. Some proteins consist of a single polypeptide chain and others consist of two or more chains. The protein hormone commonly known as ACTH, for example, consists of only one polypeptide chain of 39 amino acids. Insulin, perhaps the best known hormone, consists of two polypeptide chains, one of them 21 amino acids long and the other 30 amino acids long. Another of the body's many proteins is hemoglobin. It consists of four polypeptide chains with 141 amino acids in two of them and 146 amino acids in the other two.

Protein molecules come in various sizes and shapes. All of them, however, are large molecules (macromolecules) with molecular weights ranging from about 5,000 to a million or more. The sequence of amino acids in polypeptide chains is specific and unique for each different kind of protein molecule. This sequence plus coiling and folding of the polypeptide chains gives a characteristic three-dimensional shape or conformation to each kind of protein molecule.

One or more small segments of a protein molecule's surface have specific and unique contours produced by the specific and unique sequence of amino acids that constitute them. These surface re-

Figure 2-9 Peptide bond formation. A peptide bond is one that joins the carboxyl group of one amino acid to the amino group of another. Note that the removal of water from these two groups produces the peptide bond between them and yields, in addition to water, a compound called a peptide. A peptide formed from two amino acids is a dipeptide. The reaction shown here forms the dipeptide glycylalanine from the two amino acids glycine and alanine.

gions are called **binding sites** or **receptors,** for the good reason that their unique shapes enable them to bind closely to, or receive, certain other molecules or ions (called ligands). A **ligand** is a molecule or ion whose shape is complementary to the shape of a binding site on a particular protein molecule. Hence a specific ligand fits into a binding site of a specific protein—much as a specific piece of a jigsaw puzzle fits into another specific piece. A protein molecule's binding sites enable it to bind to or receive only those ligands that fit them. And this in turn makes it possible for protein molecules to perform a number of diverse and vital functions.

Various proteins transport substances in the blood. Hemoglobin transports oxygen by binding to it. Certain proteins classified as albumins bind to fatty acids to transport them in the form of lipoproteins. Many proteins communicate information to body cells. Some hormones, for example, insulin and growth hormone, are proteins. They function as chemical messengers from the endocrine glands to body cells. Many of the proteins located on the sur-

Enzymes, the largest group of proteins in the body, catalyze chemical reactions going on in the body. Each enzyme, by means of its uniquely shaped binding sites, binds to a specific ligand (called in this case a substrate). And this binding of enzyme to substrate accelerates a chemical reaction that changes the substrate into a different compound. Nearly 2,000 enzymes are known! We shall name and describe the actions of a fair number of them in ensuing chapters of this book.

Enzyme action and specificity. Enzymes are believed to have surface configurations that "fit" specific substrates. Here molecules *B* and *C* fit into enzyme surface, but *A* does not. Reactions involving *B* and *C* are speeded up by the molecules' coming in contact briefly with the enzyme. When reaction is complete, the enzyme, still unchanged, can dissociate from the substrate and is free to aid in further reactions. Molecule *A* and others not specific to this enzyme are unaffected by it.

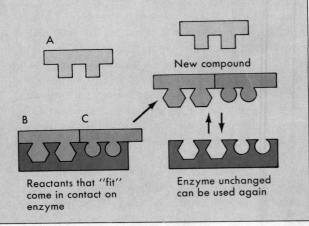

Enzyme

Reactants that "fit" come in contact on enzyme

New compound

Enzyme unchanged can be used again

Figure 2-10 Diagram shows arrangement of atoms in a glucose molecule ($C_6H_{12}O_6$). Carbon atoms that form the ring lie in one plane with the hydroxyl groups and hydrogen atoms above or below it.

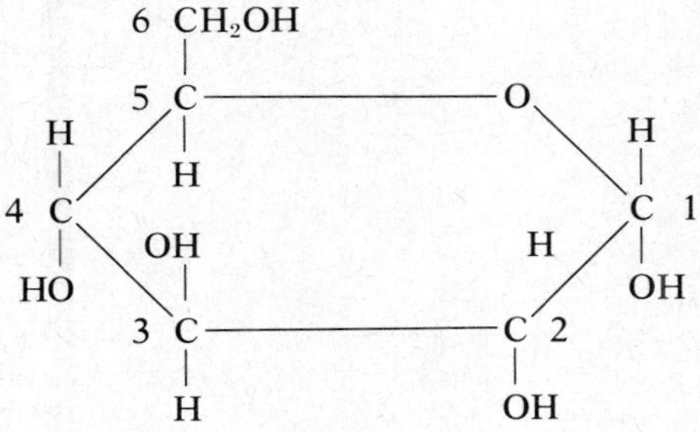

faces of cell membranes also function as communication agents. The binding sites of some of them, for example, serve as receptors for specific hormones. And this binding of protein and hormone initiates changes in the cell's activities.

Carbohydrates are the substances commonly called sugars and starches. The three types of carbohydrate compounds have long names—monosaccharides, disaccharides, and polysaccharides. In Figure 2-10, you can see the structural formula of the body's main monosaccharide, that is, glucose. Fructose and galactose are other monosaccharides. All three of these sugars consist of the same number of atoms of the same elements, but the arrangement of them within their molecules differs slightly. Look now at Figure 2-10 to note which three elements compose the glucose molecule. They also make up all other carbohydrate compounds. How many carbon atoms do you find in glucose? A monosaccharide with six carbon atoms in its molecule is called a **hexose.** The monosaccharides mentioned so far—glucose, fructose, and galactose—are all hexoses designated by the general formula, $C_6H_{12}O_6$. Not all monosaccharides, however, are hexoses. Some are **pentoses,** so-named because they contain five carbon atoms. Ribose and deoxyribose are pentose monosaccharides of great importance in the body—more about them when we discuss nucleic acids. Like all monosaccharides, ribose and deoxyribose are sugars—but strange sugars in that they are not sweet.

Disaccharides and polysaccharides are synthesized from monosaccharides. The following equation represents the synthesis of a disaccharide:

$$C_6H_{12}O_6 + C_6H_{12}O_6 \xrightarrow{\text{(synthesis)}} C_{12}H_{22}O_{11} + H_2O$$

Note that a disaccharide molecule consists of two monosaccharides minus a molecule of water. Sucrose (cane sugar), maltose, and lactose are disaccharides. Polysaccharides consist of many monosaccharides chemically joined by the removal of water to form straight or branched chains. Therefore, as you might guess, polysaccharides are large molecules. Glycogen, the main polysaccharide in the body, has an estimated molecular weight of several million—truly a macromolecule. Starch and cellulose are the predominant plant polysaccharides. Glycogen is sometimes referred to as animal starch. Any large molecule made up of many identical small molecules is called a *polymer.* Proteins are polymers of amino acids; polysaccharides are polymers of monosaccharides.

What functions do carbohydrates serve in the body? Energy release and energy storage, to answer in the fewest words. Before we try to answer more fully, we shall relate a few facts about two other kinds of molecules, namely, lipids and nucleic acids.

Lipids, according to one definition, are water-insoluble organic biomolecules. They include a large assortment of compounds that have been classified in several different ways. For our purposes, classification of lipids under these headings—triglycerides, phosphoglycerides, steroids, and prostaglandins—seems sufficient.

Triglycerides (triacylglycerols, in international nomenclature; formerly called *neutral fats*) are the most abundant lipids, and they function as the body's most concentrated source of energy. They consist of three molecules of a fatty acid chemically joined to one molecule of glycerol (glycerin). Figure 2-11 shows the structure of palmitic acid and four ways of writing its formula. Like all fatty acids, it consists of a carboxyl group attached to a tail-like chain of

$$HO{-}\overset{\overset{\displaystyle O}{\|}}{C}{-}C_{15}H_{31} \quad \text{or} \quad CH_3(CH_2)_{14}COOH$$

$$\text{or} \quad C_{15}H_{31}COOH$$

Figure 2-11 Structural formula of palmitic acid, a common saturated fatty acid. Like all fatty acids, it consists of a carboxyl group (the group that characterizes all organic acids) and a tail-like chain of hydrocarbons that terminates in a methyl group (CH_3).

$$HO{-}\overset{\overset{\displaystyle O}{\|}}{C}{-}CH_2{-}CH_2{-}CH_2{-}CH_2{-}CH_2{-}CH_2{-}CH_2{-}CH_2{-}CH_2{-}CH_2{-}CH_2{-}CH_2{-}CH_3{-}CH_2{-}CH_3$$

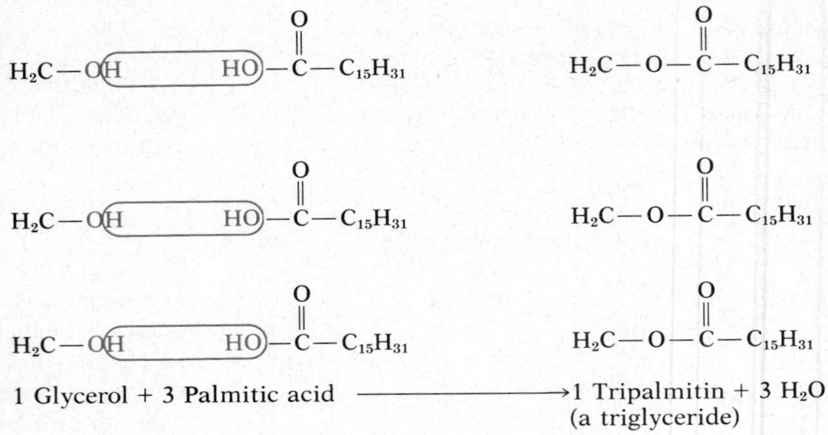

Figure 2-12 Synthesis of a triglyceride, tripalmitin, from glycerol and palmitic acid. The three water molecules resulting are not shown in the figure.

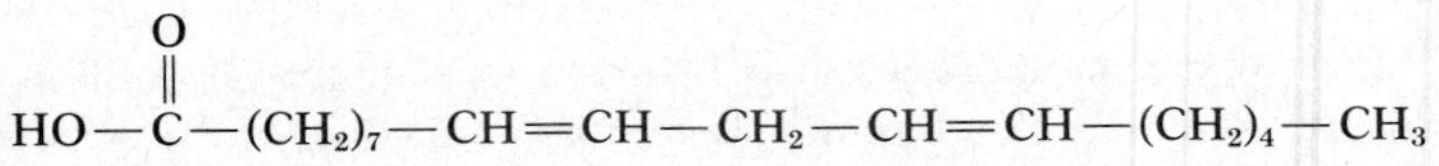

$$H_2C-OH \quad HO-\overset{\overset{\displaystyle O}{\|}}{C}-C_{15}H_{31} \qquad H_2C-O-\overset{\overset{\displaystyle O}{\|}}{C}-C_{15}H_{31}$$

$$H_2C-OH \quad HO-\overset{\overset{\displaystyle O}{\|}}{C}-C_{15}H_{31} \qquad H_2C-O-\overset{\overset{\displaystyle O}{\|}}{C}-C_{15}H_{31}$$

$$H_2C-OH \quad HO-\overset{\overset{\displaystyle O}{\|}}{C}-C_{15}H_{31} \qquad H_2C-O-\overset{\overset{\displaystyle O}{\|}}{C}-C_{15}H_{31}$$

1 Glycerol + 3 Palmitic acid $\longrightarrow$ 1 Tripalmitin + 3 H$_2$O
(a triglyceride)

Figure 2-13 Structure of linoleic acid, the most abundant polyunsaturated fatty acid in mammals.

$$HO-\overset{\overset{\displaystyle O}{\|}}{C}-(CH_2)_7-CH=CH-CH_2-CH=CH-(CH_2)_4-CH_3$$

hydrocarbons that terminates in a methyl group (CH$_3$). Because no double bonds exist between any of the hydrocarbons in its tail-like chain, palmitic acid is classified as a **saturated fatty acid.** Palmitic acid and stearic acid (C$_{17}$H$_{35}$COOH) are the two commonest saturated fatty acids.

Figure 2-12 shows the formation of a triglyceride. Its name, tripalmitin, suggests that it contains three molecules of palmitic acid. As the figure shows, tripalmitin is formed by the combining of one molecule of glycerol with three molecules of palmitic acid.

By definition a **saturated fatty acid** is one in which all available bonds of its hydrocarbon chain are filled, that is, saturated, with hydrogen atoms; the chain contains, therefore, no double bonds. In contrast, an **unsaturated fatty acid** has one or more double bonds in its hydrocarbon chain because not all of the chain carbon atoms are saturated with hydrogen atoms. Everyone today knows the word "polyunsaturated." Polyunsaturateds are triglycerides (fats) that contain fatty acids with two or more double bonds between chain carbons unsaturated with hydrogen. The most abundant polyunsaturated acid is oleic acid (C$_{17}$H$_{33}$COOH). Newspapers, magazines, and advertisements tell us over and over again to eat diets rich in polyunsaturated fats if we wish to prevent high blood pressure and heart disease. Research studies support the claim that diets high in polyunsaturated and low in saturated fats lead to a lower content of cholesterol in the blood. But less well substantiated is the claim that such diets prevent heart disease.

Two polyunsaturated fatty acids that are classified as essential fatty acids are linoleic acid (C$_{17}$H$_{31}$COOH) and linolenic acid (C$_{17}$H$_{29}$COOH). Figure 2-13 shows the structure of linoleic acid. An **essential fatty acid** is one that the body cannot synthesize and that is essential for survival. One's diet, therefore, must include essential fatty acids.

Phosphoglycerides are complex lipids that contain a phosphate group. Figure 2-14 shows the composition of one of the most abundant phosphoglycerides—an important one because it is a major structural component of cell membranes. It is perhaps best known by its now outdated name, *lecithin. Phosphatidylcholine* and *choline phosphoglyceride* are its current names. Observe in Figure 2-14 that the phosphatidylcholine molecule contains glycerol (blue letters). Joined to the glycerol are two fatty acids, palmitic and oleic (red letters), and a phosphate group (black letters). Joined to the phosphate group is an amino alcohol named choline (red letters).

Steroids are an important and large group of compounds whose molecules have as their principal component the steroid nucleus (Figure 2-15). Steroids and other simple lipids contain no fatty acids in their molecules, whereas complex lipids such as triglycerides and phosphoglycerides do contain these. Steroid compounds in the body number a famous name among them—*cholesterol.* They also include both male and female sex hormones and the adrenocortical hormones.

Prostaglandins, according to one definition, are lipids composed of a 20-carbon fatty acid that contains a 5-carbon ring (Figure 2-16). Many different kinds of prostaglandins exist in the body—14 or

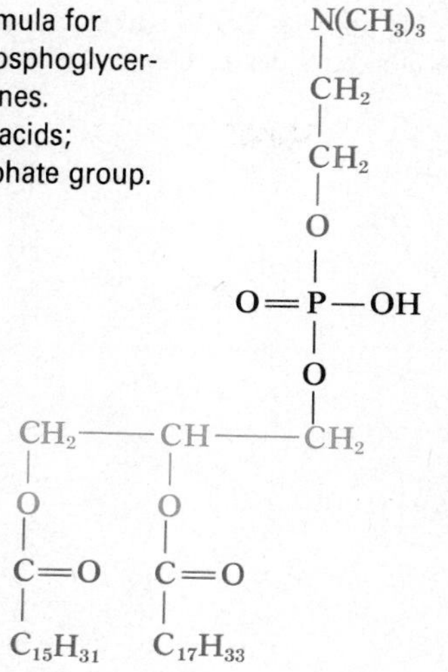

Figure 2-14 Structural formula for phosphatiolylcholine, a phosphoglyceride present in cell membranes. *Red,* choline and two fatty acids; *blue,* glycerol; *black,* phosphate group.

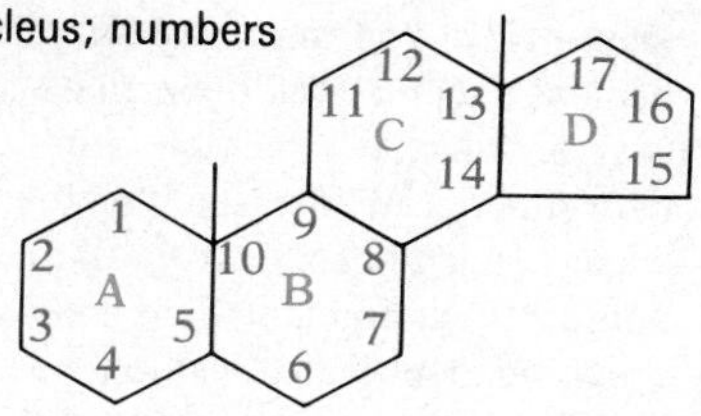

Figure 2-15 Steroid nucleus; numbers represent carbon atoms.

Figure 2-16 Structure of prostaglandin PGE$_2$, a molecule synthesized from the essential polyunsaturated fatty acid named linoleic acid (Figure 2-13).

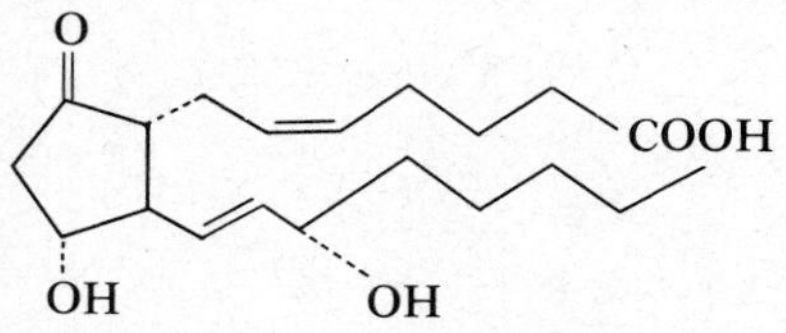

Prostaglandin PGE$_2$

more different prostaglandins in semen plus many other kinds in a number of tissues. Prostaglandins play a crucial role in various activities, including the functioning of some hormones (Chapter 14).

Survival of man as a species—and also survival of every other species—depends largely on two kinds of **nucleic acid compounds**. Their abbreviated names, DNA and RNA, almost everyone has heard or seen, but their full names are much less familiar. They are deoxyribonucleic and ribonucleic acids. Nucleic acid molecules are polymers of thousands and thousands of smaller molecules called nucleotides—deoxyribonucleotides in DNA molecules and ribonucleotides in RNA molecules. A *deoxyribonucleotide* consists of the pentose sugar named deoxyribose, a nitrogenous base (either adenine or cytosine or guanine or thymine), and a phosphate group. (*Ribonucleotides* are similar but contain ribose instead of deoxyribose and uracil instead of thymine.) The structural formula for deoxyribose appears in the upper diagram of Figure 2-17. Notice the numbering of its carbon atoms. Then observe the structural formula of a deoxyribonucleotide in the lower diagram. As in all nucleotides, the base attaches to C1 and the phosphate group to C5. Two of the bases in a deoxyribonucleotide, specifically adenine and guanine, are called **purine bases** because they derive from purine. Cytosine and thymine derive from pyrimidine, so they are known as **pyrimidine bases**. Now look at Figure 2-20. Observe that the purine bases (adenine and guanine) consist of two rings of carbon, hydrogen, and nitrogen atoms. The pyrimidine bases (cytosine and thymine) consist of only one such ring.

Nucleotides become linked together by the number 3′ carbon atom of one sugar bonding to the phosphate group on the number 5′ carbon of the

LIPOPROTEINS

There are two major classes of linked molecules of protein and fat called **lipoproteins**. So-called **high-density lipoproteins** or **HDLs** are important in the elimination and metabolism of cholesterol in the body. **Low-density lipoproteins** or **LDLs** serve to deliver cholesterol to cells and, under abnormal conditions, may contribute to deposits of the fatty material in blood vessels, thus contributing to atherosclerosis and heart disease. Because HDLs help in the transport of cholesterol to the liver for elimination, they are often described as "garbage collectors" in the body's cholesterol transport system. They are the "good" lipoproteins, which serve to maintain blood cholesterol levels within normal ranges. By increasing HDL levels and decreasing total cholesterol and LDL levels, a person's risk of heart disease can be reduced.

The 1985 Nobel Prize in Physiology or Medicine was awarded to Drs. Michael Brown and Joseph Goldstein for their research on specialized receptor sites on LDL molecules, which are elevated in the blood of individuals with certain types of heart disease. Lipid metabolism will be discussed in detail in Chapter 23.

Figure 2-17 *Top,* structural formula of the pentose sugar named deoxyribose. *Below,* it is an abbreviated structural formula for a deoxyribonucleotide. Note that it consists of a sugar (deoxyribose) with a base bonded to its number 1′ carbon atom and a phosphate group bonded to its 5′ carbon atom. Deoxyribonucleotides are the structural units of DNA.

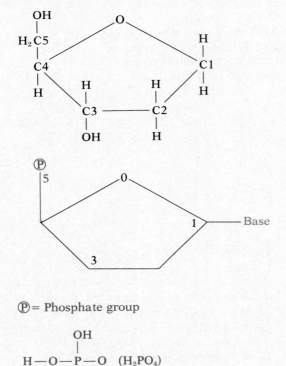

(P) = Phosphate group

$$H-O-P-O \quad (H_2PO_4)$$

Base = Purine   or   Pyrimidine
          Guanine   Cytosine
          Adenine   Thymine

Figure 2-18 Scheme to show linkage of nucleotides by phosphate groups.

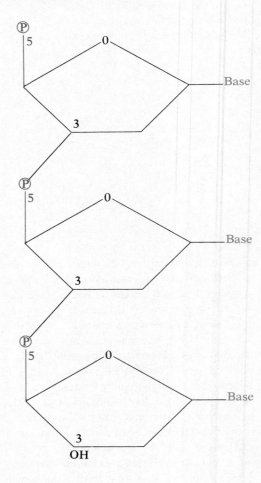

sugar below it. You can see three nucleotides linked together in Figure 2-18. As you look at this figure, try to visualize another nucleotide binding to the third one in the chain, and then another, and another, and another, to form a long chain of nucleotides—a **polynucleotide**, that is. One end of a polynucleotide chain terminates in the 5′ phosphate group and the other terminates in the 3′ hydroxyl group. Note that phosphate and sugar groups alternate to form the so-called backbone of a polynucleotide chain.

DNA molecules, the largest molecules in the body, are polymers of enormous numbers of nucleotides. Two unbelievably long polynucleotide chains compose a single DNA molecule. The chains coil around each other to form a double helix. A **helix** is a spiral shape similar to the shape of a wire in a spring. Figure 2-19 is a diagram of a double helix.

Each helical chain in a DNA molecule has its phosphate-sugar backbone toward the outside and its bases pointing inward toward the bases of the other chain. More than that, each base in one chain is joined to a base in the other chain by means of a hydrogen bond to form what is known as a **base pair**. The two polynucleotide chains of a DNA molecule are thus held together by hydrogen bonds between the two members of each base pair. One important principle to remember is that only two kinds of base pairs are present in DNA—those shown in Figure 2-20. What are they? Symbols used to represent them are A-T and G-C. Although a DNA molecule contains only these two kinds of base pairs, it contains millions of them—over 100 million pairs estimated in one human DNA molecule! Two other impressive facts are these: the millions of base pairs occur in the same sequence in all the millions of DNA molecules in one individual's body but in a different sequence in the DNA of all other individuals. In short, the base pair sequence in DNA is

Figure 2-19 Watson-Crick double-stranded helix configuration of DNA.

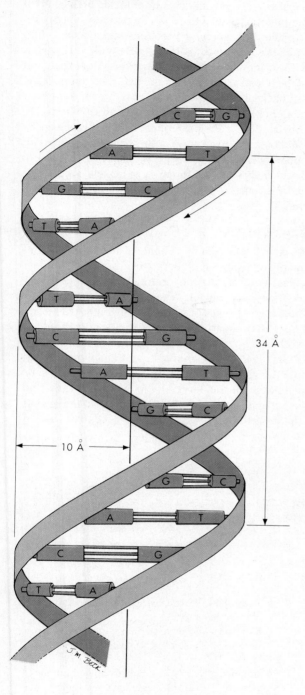

34 Å

10 Å

Figure 2-20 Base pairs. Hydrogen bonds bind a pyrimidine base (cytosine or thymine, *blue*) in one strand of a DNA molecule to a purine base (adenine or guanine, *red*) in its other strand.

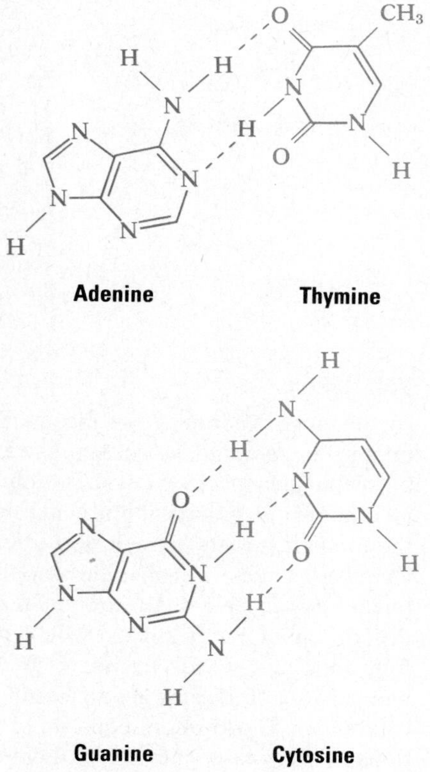

Adenine   **Thymine**

Guanine   **Cytosine**

Table 2-3 Comparison of DNA and RNA structure

	DNA	RNA
Number of polynucleotide strands in molecule	2	1
Sugar	Deoxyribose	Ribose
Base pairs	Adenine-thymine Guanine-cytosine	Adenine-uracil Guanine-cytosine

unique to each individual. This fact has momentous significance, but what it is we shall save to reveal in Chapter 3. For now, we shall merely state that DNA functions as the heredity molecule. It carries out a weighty responsibility, that of passing the traits of one generation on to the next. How it accomplishes this is a long story to be told in other portions of this text.

RNA and DNA structures differ in certain respects. To discover what these differences are, examine Table 2-3 and answer the following questions. How many polynucleotide strands compose the DNA molecule? The RNA molecule? What sugar is present in DNA? In RNA? What base is present in RNA but not in DNA? What base is present in DNA but not in RNA?

Figure 2-21 Adenosine triphosphate (ATP).

essentially decomposition reactions; they break chemical bonds. This not only breaks relatively complex compounds down into simpler ones but also releases energy from them. Catabolism breaks down the building blocks of food compounds—monosaccharides, fatty acids, glycerol, and amino acids—into simpler compounds, namely, carbon dioxide, water, and certain nitrogenous compounds. More than half of the released energy is immediately recaptured and put back into storage in one of the most important of all biomolecules—adenosine triphosphate (ATP) (Figure 2-21). The rest of the energy released by catabolism is heat energy, the heat that keeps our bodies warm. The energy stored in ATP is used to do the body's work—the work of muscle contraction and movement, of active transport, and of biosynthesis.

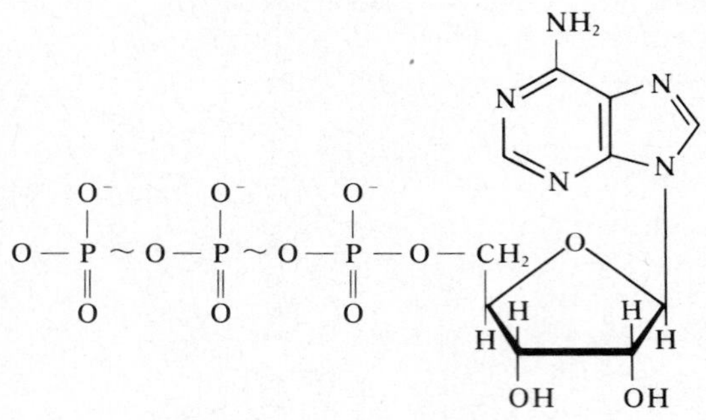

BIOENERGY

By the term, **bioenergy,** we mean energy that living organisms generate, store, and use to do the work of keeping themselves alive. The human body, like all other forms of animal life, generates energy from the foods it ingests. How it does this is one of biochemistry's major, best understood, and most extensive topics. We shall attempt only the briefest description of it here since it will be discussed more fully in Chapter 23. The name of the process by which living organisms generate and store energy is catabolism. **Catabolism** consists of chemical reactions—a very large number of them that take place in all cells in a highly coordinated fashion. They are

BIOSYNTHESIS

Biosynthesis means a chemical reaction that puts simple molecules together to form complex biomolecules, notably, carbohydrates, lipids, proteins, nucleotides, and nucleic acids. Literally thousands of biosynthesis reactions take place continually in the body. They constitute the vast, highly integrated network of reactions that make up the process of **anabolism.** Together, catabolism and anabolism constitute the process of **metabolism.** Here are two short definitions of this term. Metabolism is all the chemical changes absorbed foods undergo in living cells. Metabolism is, most simply, the body's use of foods.

Outline Summary

BASIC CHEMISTRY

A Definitions
  1 Matter—anything that occupies space and has mass
  2 Element—simple form of matter; cannot be broken down into two or more different substances
  3 Atom—smallest unit of an element
  4 Compound—substance composed of two or more elements
  5 Molecule—smallest unit of a compound; consists of two or more atoms joined by chemical bonds
  6 Chemical symbols—abbreviations for the names of elements
B Structure of atoms
  1 Atoms contain several different kinds of subatomic particles, there being three basic types
    a Protons—positively charged particles in nucleus of atom

  b Neutrons—electrically neutral particles in atom's nucleus
  c Electrons—negative particles; move around nucleus in shells of space (energy levels)
  2 Protons
    a Number of protons in nucleus unique for each element; no two elements have the same number of protons; number of protons in nucleus known as an element's atomic number; a different atomic number identifies each element
    b Nucleus of atom bears a positive charge equal to number of protons present in it
    c Weight of 1 proton equals the weight of 1 neutron, which is equal to 1,836 times the weight of 1 electron
    d Atomic weight = $(p^+ + n)$

Outline Summary—cont'd

3 Neutrons

  a Number of neutrons not necessarily the same in all atoms of an element; atoms containing a different number of neutrons from most atoms of an element are classified as *isotopes* of that element

  b Number of neutrons in atomic nucleus of an element equals the element's atomic weight minus the number of protons in its nucleus; or stated as an equation, n = (Atomic weight − p)

4 Electrons

  a Total number of electrons in an atom's shell of space equals the number of protons in its nucleus; therefore, negative charges of electrons cancel positive charges of protons, and atoms are electrically neutral structures

  b Each shell of space around an atom's nucleus can accommodate a certain maximum number of orbiting electrons; maximum number for innermost shell is 2 electrons, for outermost shell is 4 pairs of electrons

  c Presence of 1 or more single, that is, unpaired, electrons in an atom's outer shell enables the atom to take part in chemical reactions

C Chemical reactions

  1 Consist of either transfer of unpaired electrons from outer shell of one atom to outer shell of another or sharing of one atom's unpaired electrons with those of another atom

  2 Electrovalent or ionic bond—formed by transfer of electrons

  3 Covalent bond—formed by sharing of electron pairs of atoms

  4 Compound—consists of one or more elements joined by chemical bonds to form molecules, the structural units of compounds

  5 Valence—number of unpaired electrons in an atom's outer shell

  6 Types of chemical reactions

    a Synthesis—combining of two or more substances to form more complex substance; formation of new chemical bonds

    b Decomposition—breaking down of substance into two or more simpler substances; breaking of chemical bonds

    c Exchange reactions—decomposition of two substances and, in exchange, synthesis of two new compounds from them

    d Reversible reactions—proceed in both directions, although not necessarily at the same rate in both directions

  7 Chemical reactions and energy changes

    a Always coupled; formation of chemical bonds coupled with use of energy

    b Breaking of bonds coupled with release of free energy

D Radioactivity

  1 Radioactivity is emission from atom's nucleus of alpha particles, beta particles, or gamma rays

  2 Alpha particles—two protons plus two neutrons emitted from an atom's nucleus

  3 Beta particles—an electron emitted from an atom's nucleus; this electron is formed from a neutron that breaks down into a proton and an electron; proton remains in nucleus, and electron is ejected as a beta particle

  4 Gamma rays—electromagnetic radiations emitted from atom's nucleus

  5 Radioactivity changes number of protons in atom's nucleus, thereby changing it into a different element

  6 Radiations—that is, alpha particles, beta particles, or gamma rays—ionize atoms when they hit them head on; ionization of atoms in living cells injures or destroys the cells—hence, use of radiation therapy to treat cancer; in contrast, exposure to very high levels of radiation or to lower levels for prolonged periods of time may lead to development of cancer

SELECTED FACTS FROM BIOCHEMISTRY

A Biochemistry—chemistry of living organisms

B Biomolecules—compose living things; some are *inorganic compounds*, those which with few exceptions contain no carbon; most biomolecules are *organic* or *carbon-containing compounds* belonging to following classifications: proteins, carbohydrates, lipids, and nucleic acids

  1 Water—body's most abundant compound; serves many vital functions; that is, dissolves substances, thereby enabling them to enter into vital chemical reactions

  2 Electrolytes—compounds whose molecules consist of positive cations and negative anions held together by electrovalent chemical bond

  3 Proteins

    a Proteins are polypeptides

    b Sequence of amino acids in polypeptide chains composing a protein determine the protein's conformation or shape

    c Conformation of a protein molecule determines its functions; small regions (called *binding sites* or *receptors*) on the surface of a protein molecule enable it to bind closely to another molecule or ion (called a *ligand*) whose shape is complementary to shape of binding site; binding of protein to ligand enables proteins to perform various functions (Table 2-2); enzymes, the largest group of proteins, function as catalysts for vital chemical reactions

  4 Carbohydrates—compounds commonly called sugars and starches; the three types of carbohydrates are monosaccharides,

Outline Summary—cont'd

disaccharides, and polysaccharides; monosaccharides with 6 carbon atoms are called *hexoses* (e.g., glucose), those with 5 carbon atoms are called *pentoses* (e.g., ribose and deoxyribose); disaccharides are composed of two monosaccharides minus molecule of water (e.g., sucrose); polysaccharides (e.g., glycogen) are polymers of many monosaccharides chemically joined by removal of water; carbohydrate functions are energy release and energy storage

5 Lipids—water-insoluble organic biomolecules of many types, chiefly, triglycerides, phosphoglycerides, steroids, and prostaglandins

  a Triglycerides—composed of three molecules of a fatty acid chemically joined to one molecule of glycerol (Figures 2-11 and 2-12); most common saturated fatty acids are palmitic and stearic; polyunsaturated essential fatty acids are linoleic and linolenic

  b Phosphoglycerides—composed of glycerol with two fatty acids and a phosphate group joined to the glycerol and an amino alcohol joined to the phosphate; phosphatidyl choline (formerly called lecithin) is one of most abundant phosphoglycerides; phosphoglycerides function as major structural components of cell membranes

  c Steroids—main component of steroid compounds is steroid nucleus (Figure 2-15); includes cholesterol, male and female sex hormones, and adrenocortical hormones

  d Prostaglandins—lipids composed of a 20-carbon fatty acid that contains 5-carbon ring; many different kinds of prostaglandins in body; reported function is to help regulate hormone activity and metabolism

6 DNA (deoxyribonucleic acid)—composed of *deoxyribonucleotides,* that is, structural units composed of the pentose sugar (deoxyribose), phosphate group, and nitrogenous base—either cytosine or thymine or guanine or adenine; DNA molecule consists of two long chains of deoxyribonucleotides coiled into double-helix shape; alternating deoxyribose and phosphate units form backbone of the chains; base pairs, that is, either adenine-thymine or guanine-cytosine bonded together by hydrogen bonds, hold two chains of DNA molecule together; specific sequence of over 100 million base pairs constitute one human DNA molecule; sequence of base pairs is identical in all DNA of one individual and different in DNA of all other individuals; DNA functions as heredity molecule

C Bioenergy—energy that living organisms generate, store, and use to do work of muscle contraction, active transport, and biosynthesis; catabolism is process consisting of many highly coordinated chemical reactions that break down food compounds into carbon dioxide, water, and certain nitrogenous compounds, thereby releasing energy from them; immediately more than half is put back into storage in ATP; rest of energy is released as heat energy

D Biosynthesis—chemical reaction that puts simple molecules together to form complex biomolecules, notably carbohydrates, lipids, proteins, nucleotides, and nucleic acids; anabolism is process consisting of thousands of highly integrated biosynthesis reactions; metabolism is catabolism plus anabolism; it is all chemical changes absorbed foods undergo in living cells; most simply, metabolism is body's use of foods

Review Questions

1 Define the following terms: *element, compound, atom, molecule.*

2 Compare early nineteenth century and present-day concepts of atomic structure.

3 Name and define three kinds of subatomic particles.

4 Are atoms electrically charged particles? Give reason for answer.

5 What four elements make up approximately 96% of the body's weight?

6 Define and contrast meanings of the terms *atomic number* and *atomic weight.*

7 Define and give example of an isotope.

8 Explain what the term *chemical reaction* means.

9 Explain what the term *radioactivity* means.

10 How does radioactivity differ from chemical activity?

11 Explain the difference between an ionic bond and a covalent bond. Give another name for ionic bonds.

12 Define the terms: *alpha particles, beta particles,* and *gamma rays.*

13 Explain how radioactive atoms become transformed into atoms of a different element.

14 Explain the rationale for using radiation therapy to treat cancer.

15 What are electrolytes and how are they formed?

16 What is a cation? An ion?

17 What are the structural units or building blocks of proteins? Of carbohydrtates? Of triglycerides? Of DNA?

18 Explain what a protein molecule's binding site is. What function does it serve?

19 What does the term *ligand* mean?

Review Questions—cont'd

20 Describe some of the functions protein compounds perform.

21 Proteins, carbohydrates, lipids—which of these are insoluble in water? Contain nitrogen? Include prostaglandins? Include phosphoglycerides?

22 Differentiate between saturated and unsaturated fatty acids.

23 What groups compose a nucleotide?

24 What pentose sugar is present in a deoxyribonucleotide?

25 Describe the size, shape, and chemical structure of the DNA molecule.

26 What base is thymine always paired with in the DNA molecule? What other two bases are always paired together in DNA?

27 In one word, what is the function of DNA?

28 What is catabolism? What function does it serve?

29 Compare catabolism, anabolism, and metabolism.

30 ATP is the abbreviation for what important biomolecule? Why is it important?

3

CELLS

OBJECTIVES

After you have completed this chapter, you should be able to:

1 Discuss the size, shape, and generalized structure of a "typical" cell.
2 Describe the molecular structure and function of both plasma membranes and internal cell membranes.
3 Identify by name the membranous and nonmembranous organelles of the cell.
4 Discuss the structure and function of the following cell units: nucleus, nucleoli, endoplasmic reticulum, ribosomes, Golgi apparatus, mitochondria, lysosomes, and centrioles.
5 Discuss the organization and generalized function of chromatin material in the nucleus.
6 Compare and contrast the processes of diffusion, dialysis, facilitated diffusion, osmosis, and filtration.
7 Discuss and compare the factors that determine potential osmotic pressure of electrolyte and nonelectrolyte solutions.
8 Discuss the "active" cell transport mechanisms responsible for movement of materials through cell membranes.
9 Discuss the molecular structure of DNA, or deoxyribonucleic acid.
10 Discuss how genes control protein synthesis and determine hereditary characteristics.
11 Discuss and compare the five phases of mitosis in somatic cell division.
12 Explain how recombinant DNA or gene-splicing techniques can transfer complex biological traits or capabilities from one life form to another.
13 Define the term *karyotype* and explain its importance in clinical medicine.
14 Compare and contrast mitosis and meiosis.
15 Discuss the processes of spermatogenesis and oogenesis.

Cells—the smallest structures capable of maintaining life and reproducing—compose all living things from single-celled plants to multibillion-celled animals. Some 75 trillion of them make up the human body. For more than 100 years now, the study of cells has captivated the interest and occupied the time of countless scientists. But as yet these small structures have not yielded all their secrets, not even to the probing tools of present day researchers. The world of cells is indeed a fascinating one.

To introduce you to it, this chapter begins by describing cell structures and then goes on to discuss some cell functions and the highly specialized process of cell reproduction.

CELL SIZE, SHAPE, AND STRUCTURE

Almost all human cells are microscopic in size. Their diameters range from 7.5 micrometers (μm) (red blood cells) to 300 μm (female sex cell). The period at the end of this sentence measures about 100 μm— roughly 13 times as large as our smallest cells and one third the size of the largest ovum (Figure 3-1). To further appreciate the miniature world of the cell, learn the names and abbreviations for the **metric units** in Table 3-1. By international agreement, the size of cells and cell structures is now designated in micrometers and the size of atoms in nanometers. Micron (μ) is an older name for micrometer (Table 3-1).

Ideas about cell structure have changed considerably over the years. Early biologists saw cells as simple membranous sacs containing fluid and a few floating particles. Today's biologists know that cells are infinitely more complex than this. They consist of a surface membrane (namely, the plasma membrane), a relatively large centrally located structure (the nucleus), and between the two, a semifluid substance (the cytoplasm) (Table 3-2). Within the cytoplasm lie intricate arrangements of fine fibers and hundreds or even thousands of miniscule but distinct structures called organelles. Briefly then, the main cell structures (Figure 3-2) are these:

1 **Plasma membrane**
2 **Nucleus**
3 **Cytoplasm**
4 **Organelles**
5 **Fibers**

Protoplasm, a term now largely outdated, denotes the entire substance of a cell.

KEY TERMS

Centriole (SEN-tree-ole)

Chromosome (KRO-muh-sowm)

Cytoplasm (SIGH-toe-plasm)

Deoxyribonucleic acid
    (dee-ox-see-rye-bo-new-KLEE-ik ASS-id)

Dialysis (dye-AL-i-sis)

Diffusion (di-FU-zhun)

Endoplasmic reticulum
    (en-doh-PLAS-mik re-TIC-u-lum)

Filtration (fil-TRA-shun)

Hypertonic (high-per-TON-ik)

Hypotonic (high-poh-TON-ik)

Isotonic (eye-so-TON-ik)

Lysosome (LI-so-som)

Meiosis (my-O-sis)

Mitochondrion (my-toe-CHON-dree-on)

Mitosis (my-TOE-sis)

Nucleolus (nu-KLE-o-lus)

Nucleus (NU-kle-us)

Oogenesis (oh-o-JEN-e-sis)

Osmosis (oz-MO-sis)

Ribosome (RYE-bo-som)

Spermatogenesis (spur-mat-toe-JEN-e-sis)

Figure 3-1 Types of cells.

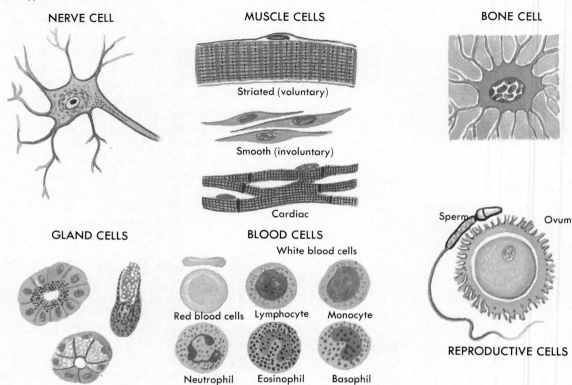

NERVE CELL

MUSCLE CELLS

BONE CELL

Striated (voluntary)

Smooth (involuntary)

Cardiac

GLAND CELLS

BLOOD CELLS

White blood cells

Red blood cells Lymphocyte Monocyte

Neutrophil Eosinophil Basophil

Sperm Ovum

REPRODUCTIVE CELLS

Table 3-1 Metric units

Abbreviation	Name	Fractional part of 1 meter		Length equivalent to 1 inch
m	Meter*			0.025 m
cm	Centimeter	1/100 m	(0.01 m)	2.5 cm
mm	Millimeter	1/1000 m	(0.001 m)	25 mm
μm	Micrometer (micron)	1 millionth m	(0.000001 m)	25,000 μm
nm	Nanometer	1 billionth m	(0.000000001 m)	25 million nm
Å	Angstrom	1/10 billionth m	(0.0000000001 m)	250 million Å

*1 meter = 39.37 inches

CELL MEMBRANES

The **plasma membrane,** because it encloses the cytoplasm, is also called the *cytoplasmic membrane.* **Internal cell membranes** form the walls of the nucleus and certain organelles in the cell. Essentially the same structure and the same thickness (75 Å or about 3/10 millionth of an inch) characterize both internal membranes and the plasma membrane.

Lipid and protein molecules compose both membranes. Of the lipid molecules in cell membranes, phosphoglycerides are the most abundant but cholesterol molecules are also present. Just how the molecules are arranged in the membranes is a subject that has long intrigued and challenged researchers. According to the concept most accepted today, they are arranged so as to form a "fluid mosaic." Lipid molecules, arranged in two layers, constitute

Figure 3-2 An artist's interpretation of cell structure as seen under an electron microscope. Note the many mitochondria, popularly known as the "power plants of the cell." Note, too, the innumerable dots bordering the endoplasmic reticulum. These are ribosomes, the cell's "protein factories."

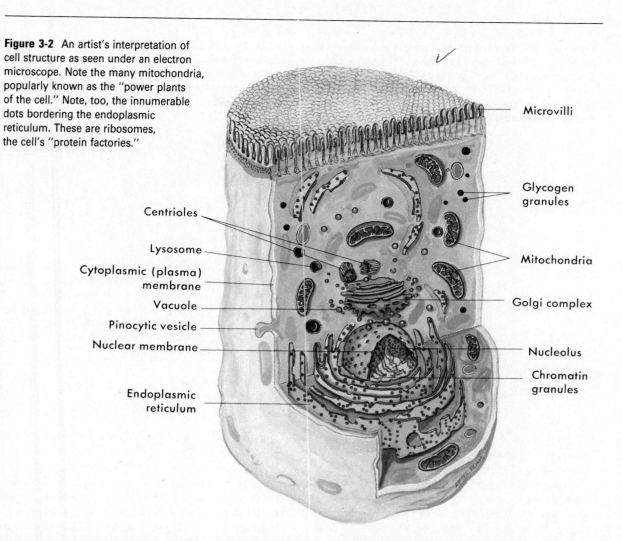

Microvilli

Glycogen granules

Centrioles

Lysosome

Cytoplasmic (plasma) membrane

Vacuole

Pinocytic vesicle

Nuclear membrane

Endoplasmic reticulum

Mitochondria

Golgi complex

Nucleolus

Chromatin granules

Table 3-2 Some major cell structures and their functions

Cell structures	Functions
Plasma membrane	Serves as the boundary of the cell, maintaining its integrity; protein molecules on outer surface of plasma membrane perform various functions; for example, they serve as markers that identify cells of each individual, receptor molecules for certain hormones, neurotransmitters that provide means of communication between cells, and receptor molecules for foreign proteins that function to produce immunity
Endoplasmic reticulum (ER)	Ribosomes attached to rough ER synthesize proteins that leave cells via the Golgi complex; smooth ER synthesizes lipids incorporated in cell membranes, steroid hormones, and certain carbohydrates used to form glycoproteins
Golgi apparatus	Synthesizes carbohydrate, combines it with protein, and packages the product as globules of glycoprotein
Mitochondria	Catabolism; ATP synthesis; a cell's "powerhouses"
Lysosomes	A cell's "digestive system"
Ribosomes	Synthesize proteins; a cell's "protein factories"
Nucleus	Dictates protein synthesis, thereby playing essential role in other cell activities, namely, active transport, metabolism, growth, and heredity
Nucleoli	Play an essential role in the formation of ribosomes

Figure 3-3 Postulated structure of plasma and internal cell membranes. Note the following features of membrane's molecular structure: the bilayer of phospholipid molecules arranged with their "tails" pointing toward each other; the location of protein molecules in all possible positions with relation to the phospholipid bilayer—at the outer surface, at the inner surface, partially penetrating the bilayer, and extending all the way through the membrane.

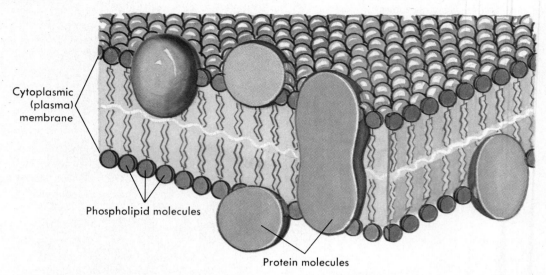

Cytoplasmic (plasma) membrane

Phospholipid molecules

Protein molecules

a fluid portion of cell membranes; protein molecules, arranged in a mosaic pattern, are embedded on and in this lipid bilayer.

Observe in Figure 3-3 the peculiar shape of phospholipid molecules. They have round "heads" and long, flexible "tails." The heads contain phosphate, and the tails consist of two fatty acids. The heads are **hydrophilic** (water-loving) and the tails are **hydrophobic** (water-fearing). This dual nature of phospholipid molecules causes them to position themselves in two rows with their tails pointing at each other. This novel arrangement allows all the water-loving heads and none of the water-fearing tails to contact water. Heads of the outer layer of molecules face into fluid outside the membrane and heads of the inner layer of molecules face into fluid inside the membrane. But the water-fearing tails of neither the outer or inner layer of molecules face into fluid. They face each other.

Now let us consider further the arrangement of protein molecules in cell membranes. They are positioned asymmetrically. Observe this in Figure 3-3. Some protein molecules attach to the surface layer of phospholipid molecules, some attach to the inner layer, and others extend all the way through the lipid bilayer. Many different kinds of protein molecules are present in the outer and inner surfaces of cell membranes. For example, glycoproteins (a carbohydrate molecule joined to a protein) occur in the outer but not the inner surfaces of cell membranes.

The cell's plasma membrane performs a variety of functions essential for healthy cell survival. Briefly, it maintains the cell's integrity, it identifies the cell as belonging to a particular individual, it receives communications from other cells, it transports some substances into or out of the cell but bars others, and it plays an important role in producing immunity. The plasma membrane is sturdy enough to maintain the cell's integrity and to preserve the arrangement of its internal structures necessary for the carrying on of activities that sustain the cell's life. This sounds impossible for a membrane only slightly more than a 10 millionth of an inch thick, yet it is true. If a cell's plasma membrane becomes torn, the cell loses its integrity, its organized wholeness. The arrangement of its internal structures becomes disorganized as the cell's contents leak out. Result? The cell dies.

The plasma membrane identifies a cell as coming from one particular individual. Many of its surface proteins serve as identification tags, since they occur only in the cells of that individual. A practical application of this fact is made in tissue typing—a procedure carried out, for example, before an organ from one individual is transplanted into another.

Certain proteins embedded in the plasma membrane's outer surface enable it to receive chemical communications from other cells. Endocrine gland cells, for example, communicate with their target cells by means of hormones that bind to the receptors (binding sites) of specific proteins on the surface of the target cells' plasma membranes. Nerve cells communicate chemically with other nerve cells. One nerve cell releases a chemical (called a **neurotransmitter**) that then binds to specific receptors on the surface of another nerve cell's plasma membrane.

Transportation, on a selective basis, is a function of both plasma membranes and internal cell membranes. They allow some substances, but not all, to move through them. Later in this chapter we shall discuss in some detail membrane transportation mechanisms.

Plasma membranes play a crucial role in defending the body against microbes and other harmful agents. Very briefly, some of the plasma membrane proteins on certain white blood cells (lymphocytes) function as antibodies. The **binding sites** of antibodies bind to potentially harmful foreign proteins called *antigens* and thereby initiate a series of events that render the antigens harmless (discussed in Chapter 29).

Internal cell membranes serve as the boundaries of various internal structures and also perform other functions. Many of the proteins in them, for example, are enzymes that accelerate (catalyze) an enormous variety of chemical reactions. Enzymes make possible the chemical reactions that keep our cells and our bodies alive.

CYTOPLASM, ORGANELLES, AND FIBERS

Cytoplasm is the part of a cell between its surface membrane and its nucleus. Far from being the homogeneous substance once thought, the cytoplasm contains hundreds or even thousands of "little organs" or **organelles** and varying numbers of **inclusions** or bits of nonliving substances plus many fibrils and microtubules. Some organelles have *membranous walls*, namely, the endoplasmic reticulum, Golgi apparatus, mitochondria, and lysosomes. Ribosomes and centrioles are *nonmembranous organelles.*

Endoplasmic Reticulum

Endoplasm means the cytoplasm located toward the center of a cell. Reticulum means network. Therefore the name **endoplasmic reticulum (ER)** means literally a network located deep inside the cytoplasm. And when first seen, it appeared to be just that. Later on, however, more highly magnified views under the electron microscope showed the endoplasmic reticulum distributed throughout the cytoplasm. It consists of membranous-walled canals and flat, curving sacs arranged in parallel rows.

There are two types of endoplasmic reticulum:

rough and *smooth.* Innumerable small granules—ribosomes by name—dot the outer surfaces of the membranous walls of the rough type and give it its "rough" appearance. Ribosomes are themselves organelles. The canals of the endoplasmic reticulum wind tortuously through the cytoplasm, extending all the way from the plasma membrane to the nucleus. The membrane forming the walls of the endoplasmic reticulum has essentially the same molecular structure as the plasma membrane.

The structural fact that the endoplasmic reticulum is an interconnected system of canals suggests that it might function as a miniature circulatory system for the cell. And in fact, proteins do move through the canals. The ribosomes attached to the rough endoplasmic reticulum synthesize proteins. These proteins enter the canals, move through them to the Golgi apparatus, and eventually leave the cell. Thus the rough endoplasmic reticulum functions in both protein synthesis and intracellular transportation.

No ribosomes border the membranous wall of the smooth endoplasmic reticulum—hence its smooth appearance and its name. Its functions are less well established and probably more varied than those of the rough type. The smooth endoplasmic reticulum is now believed to synthesize certain lipids and carbohydrates. Included among these are the lipids incorporated into cell membranes, the steroid hormones, and some of the carbohydrates used to form glycoproteins.

Ribosomes

Every cell contains thousands of **ribosomes.** Many of them are attached to the rough endoplasmic reticulum and many of them lie free, scattered through the cytoplasm. Find them in both locations in Figure 3-2. Because ribosomes are too small to be seen with a light microscope, no one knew they existed until the electron microscope revealed them. Research has now yielded information about their molecular structure. A ribosome consists of two subunits, a larger and a smaller one, both composed of ribonucleic acid bonded to protein. The symbol for the ribonucleic acid in ribosomes is rRNA. Two other types of ribonucleic acid are messenger RNA (mRNA) and transfer RNA (tRNA) (see pp. 70-71).

The function of ribosomes is protein synthesis. Ribosomes are the molecular machines that make proteins. Or to use a popular term, they are the cell's "protein factories." The ribosomes attached to the endoplasmic reticulum, as already mentioned, synthesize proteins for "export." The ribosomes free in the cytoplasm make proteins for the cell's own domestic use. They make both its structural and its functional proteins (enzymes). Working ribosomes,

Figure 3-4 **A**, Artist's interpretation of a mucus-se-creting cell, showing a typical Golgi apparatus, basal nucleus, and mitochondria. **B**, Enlargement of Golgi apparatus to show small bubble-like open areas called *vacuoles* and folds in the inner membrane called *cisternae.*

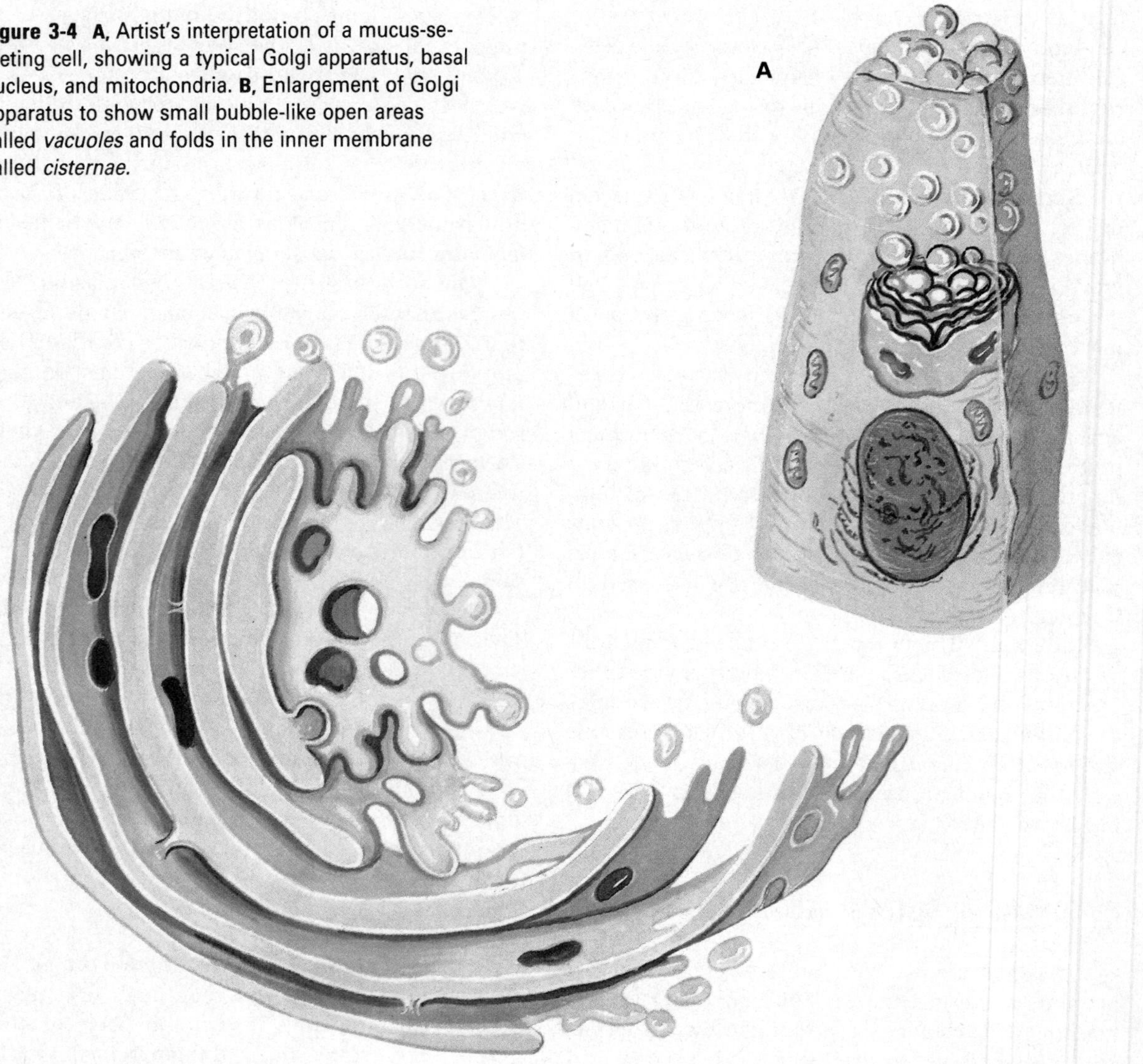

One might think that only in the secreting cells of glands would the Golgi apparatus function as just described. However, various other kinds of cells also make products for "export," products that move out of the cells that make them. In other words, various nonglandular cells secrete substances. Some examples: liver cells secrete blood proteins, plasma cells secrete antibodies, and connective tissue cells of certain types secrete substances used in bone and cartilage formation. The Golgi apparatus in all of these cells is believed to synthesize carbohydrate, to combine it with protein, and to package the product in globules for secretion.

those that are actually making proteins, appear to function in groups called *polyribosomes.* Under the electron microscope, polyribosomes look like short strings of beads.

Golgi Apparatus

The **Golgi apparatus** consists of tiny sacs stacked one on the other and located near the nucleus. Note in Figure 3-4, *A,* that the sacs look more and more distended in successive layers of the pile—as if some material were filling them up to the bulging point. And this actually is the case. Several research teams have presented evidence that the Golgi sacs synthesize large carbohydrate molecules and then combine them with proteins (brought to them through the canals of the endoplasmic reticulum) to form compounds called *glycoproteins.* As the amount of glycoproteins in the sacs increases, their shape changes from flat to "fat." They turn into perfect

Figure 3-5 Mitochondrion. Cutaway diagram showing outer and inner membranes that form wall of mitochondrion. Note folded extensions *(cristae)* of inner membrane and knobs attached to them *(inset).*

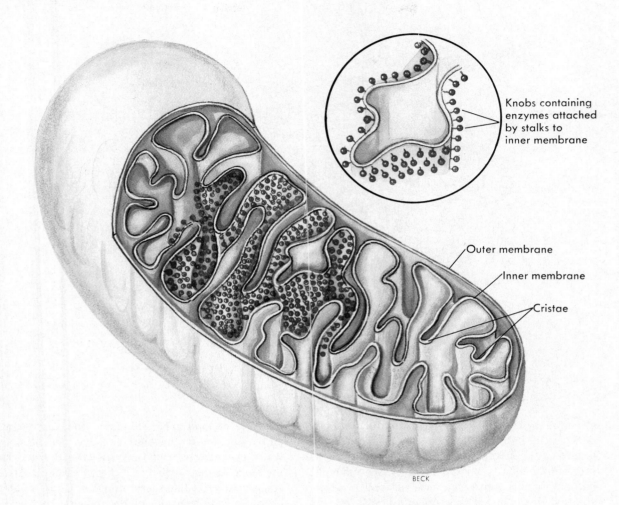

Knobs containing
enzymes attached
by stalks to
inner membrane

Outer membrane

Inner membrane

Cristae

BECK

little spheres or globules. Then one by one, they pinch away from the top of the stack. The Golgi apparatus, in other words, not only synthesizes carbohydrate and combines it with protein, but it even packages the product! Neat little globules of glycoprotein migrate outward away from the Golgi apparatus to and through the cell membrane. Once outside the cell the globules break open, releasing their contents. The cell has secreted its product.

Mitochondria

Note the **mitochondria** shown in Figures 3-2, 3-5, and 3-6. Magnified thousands of times, as they are there, they look like small, partitioned sausages—if you can imagine sausages only 1.5 μm long and one half as wide. (In case you can visualize inches better than micrometers, 1.5 μm equals about 3/50,000 of an inch.) Yet, like all organelles, and tiny as they are, mitochondria have a highly organized molecular

structure. Their membranous walls consist not of one but of two delicate membranes. They form a sac within a sac. The inner membrane folds into a number of extensions called **cristae.** Note in Figure 3-5 how they jut into the interior of the mitochondrion like so many little partitions. With the very powerful magnification of an electron microscope, one can see small round knobs attached to the cristae by short stalks and projecting inward from them. A single mitochondrion may contain thousands of these tiny knobs. Each knob, in turn, contains enzymes essential for the making of one of the most important chemicals in the world. Without this compound, life cannot exist. Its long name, *adenosine triphosphate,* and its abbreviation, *ATP,* were mentioned in Chapter 2. Chapter 21 presents more detailed information about this vital substance.

Both inner and outer membranes of the mitochondrion have essentially the same molecular

Figure 3-6 Diagram of microtrabecular lattice and microtubules that form a cell's supporting framework.

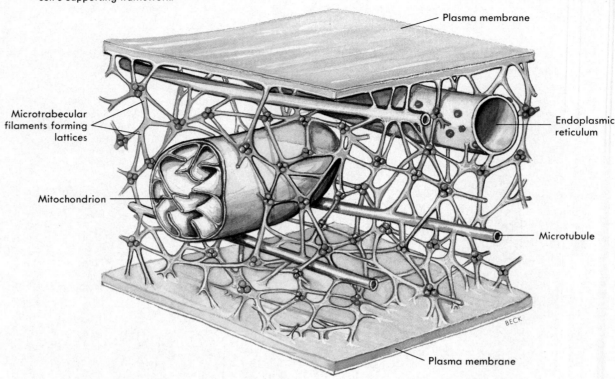

Plasma membrane

Microtrabecular filaments forming lattices

Endoplasmic reticulum

Mitochondrion

Microtubule

BECK

Plasma membrane

structure as the cell's plasma membrane. All evidence so far indicates that the proteins in the membranes of the cristae are arranged precisely in the order of their functioning. This is another example, but surely an impressive one, of the principle stressed in Chapter 1 that organization is a foundation stone and a vital characteristic of life.

Enzymes in the mitochondrial inner membrane catalyze oxidation reactions. These are the chemical reactions that provide cells with most of the energy that does all of the many kinds of work that keeps them and the body alive. Thus do mitochondria earn their now familiar title, the "power plants" of cells.

The fact that mitochondria generate most of the power for cellular work suggests that the number of mitochondria in a cell might be directly related to its amount of activity. This principle does seem to hold true. In general, the more work a cell does, the more mitochondria its cytoplasm contains. Liver cells, for example, do more work and have more mitochondria than sperm cells. A single liver cell contains 1,000 or more mitochondria whereas only about 25 mitochondria are present in a single sperm cell.

Lysosomes

Like the endoplasmic reticulum, Golgi apparatus, and mitochondria, **lysosomes** have membranous walls. The size and shape of lysosomes change with the stage of their activity. In their earliest, inactive stage, they look like mere granules. Later, as they become active, they take on the appearance of small vesicles or sacs (see Figure 3-2) and often contain tiny particles such as fragments of membranes or pigment granules. The interior of the lysosome contains various kinds of enzymes capable of breaking down all the main components of cells. These enzymes can, and under some circumstances actually do, destroy cells by digesting them. The graphic nickname "suicide bags," therefore, seems appropriate for lysosomes. Little wonder that these powerful and dangerous substances are usually kept sealed up in lysosomes. However lysosomal enzymes more often protect than destroy cells. Large molecules and large particles (for example, bacteria) that find their way into cells enter lysosomes, and their enzymes dispose of them by digesting them. Therefore "digestive bags" and even "cellular garbage disposals" are other nicknames for lysosomes. White blood cells serve as scavenger cells for the body, engulfing bacteria and destroying them in their lysosomes.

Centrioles

Under the light microscope, **centrioles** appear as two dots located near the nucleus. The electron microscope, however, reveals them not as mere dots but as tiny cylinders (see Figure 3-2). The walls of the cylinders consist of nine bundles of microtubules, with three tubules in each bundle. A curious fact about these two tubular-walled cylinders is their position at right angles to each other. One wonders why—a question as yet unanswered. Precisely how centrioles function is another unsolved riddle. They act in some way to form the spindle that appears during mitosis, or cell division (see Figures 3-15 and 3-16).

Cell Fibers

No one knew much about **cell fibers** until the development of two new research methods; one uses fluorescent molecules and the other uses stereomicroscopy, that is, three-dimensional pictures of whole, unsliced cells made with high-voltage electron microscopes. Using these techniques, investigators discovered intricate arrangements of fibers of varying widths. The smallest fibers seen have a width of about 3 to 6 nm—less than one millionth of an inch! Particularly striking about these fine fibers is their arrangement. They form a three-dimensional, irregularly shaped lattice, a kind of scaffolding in the cell. Because the appearance of the lattice reminded investigators of the trabeculae seen in spongy bone, they named it the *microtrabecular lattice.* Its fibers appear to support parts of the cell formerly thought to float free in the cytoplasm, namely, the endoplasmic reticulum, mitochondria, and so-called free ribosomes (Figure 3-6). Microtrabecular fibers also serve as cellular "muscles." (Actin and myosin, major contractile proteins of muscle cells, have been identified in microtrabecular fibers.) By contracting and expanding, microtrabecular fibers control cell shape and produce internal cell movements. Slightly larger fibers are suspended in the lattice. They are called *microtubules* and *microfilaments.* They and the microtrabecular lattice together form a supporting framework for the cell. Microtubules are postulated to become the mitotic spindle during cell division.

NUCLEUS

The **nucleus,** the largest cell structure (see Figure 3-2), occupies the central portion of the cell. Both the shape of the nucleus and the number of nuclei present in a cell vary. One spherical nucleus per cell, however, is common. Electron micrographs show that two membranes perforated by openings or pores enclose the nucleoplasm (nuclear fluid).

These nuclear membranes have essentially the same structure as other cell membranes and appear to be extensions of the membranous walls of the endoplasmic reticulum. Probably the most important fact to remember about the nucleus is this: it contains DNA molecules, the famous heredity molecules. In nondividing cells they appear as granules or threads, which are named *chromatin* (from the Greek *chroma,* "color") because they readily take the color of dyes. When the process of cell division begins, DNA molecules become tightly coiled. They then look like short, rodlike structures and are called **chromosomes.** All normal human cells (except mature sex cells) contain 46 chromosomes, and each chromosome consists of one DNA molecule plus some protein molecules.

The functions of the nucleus are actually functions of DNA molecules. In general, DNA molecules dictate both the structure and function of cells and they determine heredity. The most prominent structures visible in the nucleus are small bodies that stain densely and are called **nucleoli.** Like chromosomes, they consist chiefly of a nucleic acid, but the nucleic acid is *not* DNA. It is RNA, ribonucleic acid (see pp. 70-71).

Nucleoli function to synthesize ribosomal RNA and combine it with protein to form ribosomes, the protein synthesizers of cells. You might guess, therefore, and correctly so, that the more protein a cell makes, the larger its nucleoli will appear. Cells of the pancreas, to cite just one example, make large amounts of protein and have large nucleoli.

SPECIAL CELL STRUCTURES

Microvilli, cilia, and flagella are special cell structures present only in certain types of cells. **Microvilli,** for example, are special structures of epithelial cells that line the intestines. Microvilli consist of extensions of cytoplasm and plasma membrane. Like tiny fingers crowded close against each other, they project from the surface of the cell (see Figure 3-2). A single microvillus measures about 0.5 μm long and only 0.1 μm or less across. Since one cell has hundreds of these projections, the surface area of the cell is increased manyfold—a structural fact that enables the cell to perform its function of absorption at a faster rate.

Cilia, someone said, look like eyelashes attached to one surface of a cell. Among the cells possessing cilia are the epithelial cells forming the surface of the mucous membrane that lines some parts of the upper respiratory tract. The electron microscope has revealed the fine structure of a cilium—a fascinating example of the organization characteristic of living things. Each cilium is essentially a projection of a cell's cytoplasm and plasma membrane; it consists of a very tiny cylinder (0.2 μm in diameter) made up of nine double microtubules arranged around two single microtubules in the center. One cell may have a hundred or more cilia. They move together in such a way that they propel a fluid in one direction over the surface of the cell. Ciliated cells in the respiratory mucous membrane, for example, propel mucus upward in the respiratory tract.

A **flagellum** is a single, hairlike projection from the surface of a cell. Each male sex cell (a spermatozoon) has a flagellum, commonly called a tail. A flagellum moves in such a way that it propels the spermatozoon forward in its fluid environment.

CELL PHYSIOLOGY—MOVEMENT OF SUBSTANCES THROUGH CELL MEMBRANES

Every cell carries on a number of functions that maintain its own life—transportation and metabolism, for example. If a cell is to survive, it must continually move substances through its membranes and must metabolize foods (use them for energy and for building complex compounds.) Also, from time to time, all but a few kinds of cells perform another function, that of reproducing themselves. In addition to self-serving activities, every cell in the body also performs some special function that serves the body as a whole. Muscle cells provide the function of movement, nerve cells contribute communication services, red blood cells transport oxygen, and so on. In return the body performs vital functions for all of its cells. It brings food and oxygen to them and removes waste from them, to mention only two examples. In short, a relationship of mutual interdependence exists between the body as a whole and its various parts. Optimum health of the body depends on optimum health of each of its parts, down even to the smallest cell. Conversely, optimum health of each individual part depends on optimum health of the body as a whole. In equation form, this physiological principle of mutual interdependence might be expressed as follows:

cellular health ⇆ tissue health ⇆ organ health ⇆
system health ⇆ body health

Some parts of the body, of course, are more important for healthy survival than others. Obviously the heart is far more important than the appendix. And the nerve cells that control respiration are infinitely more important for survival than muscle cells that move the little finger.

Cell physiology deals with all the different kinds of functions cells perform. It is, as you might guess, an enormous subject. Therefore we have chosen to discuss only two aspects of it at this time—a cell's transportation activities and its reproduction. Later chapters will deal with the following cell functions: contractility, conductivity, and metabolism. The rest of this chapter contains some known and some postulated answers to the question, "How do substances move through cell membranes?"

TYPES OF TRANSPORTATION PROCESSES

Heavy traffic moves continuously in both directions through cell membranes. Streaming in and out of all cells in endless procession go molecules of water, foods, gases, wastes, and many kinds of ions. Several processes carry on this mass transportation. They are classified under two general headings as **physical** (or passive) and **physiological** (or active) processes.

The main distinction between these two kinds of processes lies in the source of the energy that does the work of moving a substance through a membrane. If the energy comes from chemical reactions taking place in a living cell, the transport mechanism is classified as an **active or physiological process.** If the energy for moving a substance stems from some other source and not from a living cell's chemical reactions, the transport mechanism is classified as a **passive or physical process.** The two preceding statements, you may have noticed, implied another distinction between active and passive transport mechanisms. Active mechanisms can move substances only through living cell membranes. Passive mechanisms, on the other hand, can move materials through either living or dead cell membranes and even through artificial membranes. Major passive or physical transport processes are diffusion, osmosis, and filtration. There are several kinds of active transport processes: those called "pumps," for example, and the mechanisms of endocytosis and exocytosis. First we shall consider these *passive processes*—diffusion (including dialysis and facilitated diffusion), osmosis, and filtration—and then we shall turn our attention to the *active processes*, namely, the so-called "pumps" and endocytosis and exocytosis.

Passive Transport Processes
Diffusion

The term **diffusion** means scattering or spreading. In liquids, gases, and solids, molecules and ions move continuously, rapidly, and in all directions. Therefore they scatter or spread themselves throughout a given substance. In short, they diffuse through it. Molecules and ions also diffuse through nonliving and living membranes that are permeable to them.

Figure 3-7 Diffusion. Because the membrane separating the 10% NaCl from the 20% NaCl is freely permeable to both NaCl and H₂O, both substances diffuse rapidly through the membrane in both directions. But, as the red arrows indicate, more sodium and chloride ions move out of the 20% solution, where there are more of them, into the 10% solution, where there are fewer of them, than in the opposite direction. Simultaneously, more water molecules move from the 10% solution, where there are more of them, into the 20% solution, where there are fewer of them. Result: equilibration of the concentrations of the two solutions after an elapse of time. From then on, equal numbers of Na ions and Cl ions diffuse in both directions, as do equal numbers of H₂O molecules.

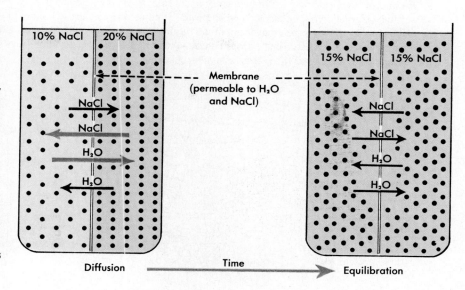

Perhaps the easiest way for you to learn the essential facts and principles about diffusion is to discover them for yourself in the following example, illustrated in Figure 3-7. Suppose a 10% sodium chloride (NaCl) solution is separated from a 20% NaCl solution by a membrane. Suppose further that the membrane is permeable to both NaCl and to water. This means that both these substances can and do pass through the membrane in both directions. NaCl particles and water molecules racing in all directions through each solution collide with each other and with the membrane. Some inevitably hit the membrane pores from the 20% side and some from the 10% side. Just as inevitably, some bound through the pores in both directions. For a while, more NaCl particles enter the pores from the 20% side simply because they are more numerous there than on the 10% side. More of these particles, therefore, move through the membrane from the 20% solution into the 10% than diffuse through it in the opposite direction. Using different words for the same thought, *net diffusion* of NaCl takes places from the solution where its concentration is greater into the one where its concentration is lesser. Thus net diffusion of NaCl occurs "down" the *NaCl concentration gradient*, that is, from the higher concentration down to the lower concentration. Net diffusion of NaCl, therefore, if given enough time, will equalize NaCl concentration on both sides of the membrane.

During the time that net diffusion of NaCl is taking place between the 20% and 10% solutions, net diffusion of water is also going on. The direction of net diffusion of any substance is always down that substance's concentration gradient. Applying this principle, net NaCl diffusion occurs down the NaCl concentration gradient, and net water diffusion occurs down the water concentration gradient. Since the greater concentration of water molecules lies in the more dilute 10% solution (where fewer water molecules have been displaced by NaCl molecules), more water molecules diffuse out of the 10% solution into the 20% solution than diffuse in the opposite direction. Thus the net diffusion of water removes water from the more dilute solution and adds it to the more concentrated one. How then does the net diffusion of water affect the concentrations of the two solutions? Does it tend to make them more equal or more different? How does the net diffusion of NaCl affect their concentrations? The answers are quite obvious. Both the net diffusion of water and the net diffusion of NaCl tend to equalize (equilibrate) the concentrations of the two solutions. Note that whereas net diffusion of NaCl and of water both go on at the same time, they go on in opposite directions.

Diffusion of NaCl and water continues even after equilibration has been achieved. But from that moment on, it is equal diffusion in both directions through the membrane and not net diffusion of either substance in either direction.

DIALYSIS. The process of **dialysis** is diffusion under certain conditions. It takes place when a solution that contains both crystalloids and colloids is sep-

Figure 3-8 Dialysis, the separation of crystalloids from colloids by means of a membrane permeable to crystalloids and impermeable to colloids. The parchment bag in **A**, which contains a solution of glucose and raw egg white, is permeable to glucose but not to albumin. Therefore glucose moves out of the bag into the surrounding water while albumin stays inside the bag. Time elapses between **A** and **B. B** shows the result of dialysis—separation of the crystalloid, glucose, from the colloid, albumin.

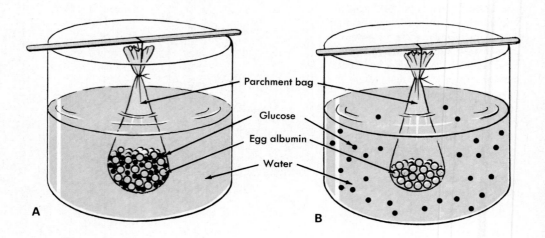

Parchment bag
Glucose
Egg albumin
Water

A   B

Crystalloids or true solutes are solute particles with diameters less than 1 nm, for example, ions, glucose, or oxygen. *Colloids* are solute particles whose diameters range from about 1 to 10 nm. Enzymes and all other proteins are colloids.

---

arated from plain water by a membrane that is permeable to crystalloids but impermeable to colloids (Figure 3-8).

When the membrane that separates a solution of crystalloids from water is permeable to crystalloids and impermeable to colloids, the crystalloids, as you would expect, diffuse through the membrane but the colloids do not. Net diffusion of crystalloids occurs down their concentration gradient, that is, out of the solution into the water. The colloids remain behind. Hence dialysis may be described as diffusion that separates crystalloids from colloids.

• • •

The following paragraphs summarize some facts and principles worth remembering about diffusion.

1 *Diffusion* is the movement of solute and solvent particles in all directions through a solution or in both directions through a membrane.

2 *Net diffusion* is the movement of more particles of a substance in one direction than in the opposite direction.

3 Net diffusion of any substance occurs down its own *concentration gradient*, which means from the higher to the lower concentration of that substance. Following are two applications of this princple: (a) net diffusion of solute particles occurs from the more concentrated to the less concentrated solution and (b) net diffusion of water molecules, in contrast, occurs from the more dilute to the less dilute solution.

4 Net diffusion of the solute in one direction through a membrane and of water in the opposite direction eventually makes the concentrations of the two solutions equal. In short, it *equilibrates* them. We can also state this principle in another way. Net diffusion of both solute and water causes the solute and water concentration gradients to gradually decrease. Eventually—at the point of equilibration—they disappear entirely. There is no difference in concentration (which is what concentration gradient means) between equilibrated solutions. Equilibrated solutions have the same, not different, concentrations.

5 Equal diffusion means that the number of solute and water particles diffusing in one direction equals the number diffusing in the opposite direction. Net diffusion means

Figure 3-9 Osmosis. Osmosis is the diffusion of water through a selectively permeable membrane. The membrane shown in this diagram is permeable to water but not to NaCl. Because there are relatively more water molecules in 10% NaCl than in 20% NaCl, more water molecules osmose from the more dilute into the more concentrated solution (as indicated by the red arrow in the left-hand diagram) than osmose in the opposite direction. The *net* direction of osmosis, in other words, is toward the more concentrated solution. Net osmosis produces the following changes in these solutions: their concentrations equilibrate, both the volume and the pressure of the originally more concentrated solution increase, and the volume and the pressure of the other solution decrease proportionately. Note the lowered level of the 15% NaCl solution on the left of the membranes in the right-hand illustration.

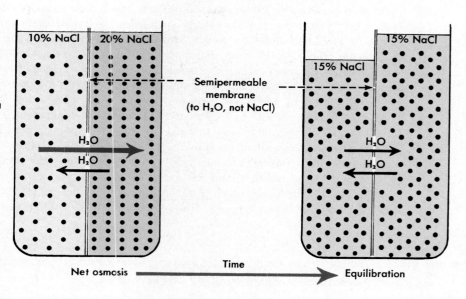

more particles diffusing in one direction than in the other.

6 Diffusion is a passive transport mechanism because cells are passive, not active and working in this process. Cellular chemical reactions do not supply the energy that moves diffusing particles. The continual random movements characteristic of all molecules and ions (molecular kinetic theory) furnish the energy for diffusion.

Think about these diffusion principles. Make sure that you understand them, for they have many applications in physiology. Our very lives, in fact, depend on diffusion. Evidence? Oxygen, the "breath of life," enters cells by diffusion through their membranes.

FACILITATED DIFFUSION. The process of **facilitated diffusion** resembles both ordinary diffusion and active transport. Like ordinary diffusion, facilitated diffusion is a passive process that moves a substance down its own concentration gradient. Like active transport, facilitated diffusion is a "carrier-mediated" process. By this we mean that a so-called carrier molecule, that is, a specific protein molecule on the outer surface of a cell membrane, binds to the substance to be carried through the membrane. The bound carrier molecule, according to one postulate, rotates in the membrane and thereby carries the substance rapidly from one side of the membrane to the other. There it dissociates from the substance, releasing it into the cytoplasm of the cell. It has

fulfilled its function of facilitating (accelerating) the substance's diffusion through the membrane.

Osmosis

The diffusion of water through a selectively permeable membrane is known as **osmosis** (Figure 3-9). As its name suggests, a selectively permeable membrane is one that is not equally permeable to all solute particles present. It permits some solutes to diffuse through it freely but hinders or prevents entirely the diffusion of others. Those solutes allowed to diffuse freely through the membrane obey the law of diffusion. Hence they eventually equilibrate across it. Their concentrations on both sides of the membrane become equal. But can particles not permitted to diffuse freely through a membrane also obey the law of diffusion? Can they, too, equilibrate across the membrane? The answer to both questions, as you can readily deduce, is "No." When a membrane hinders or prevents a substance from moving through it, the concentration of that substance necessarily remains higher on one side of the membrane than on the other. That solute cannot equilibrate across the membrane. In short, the selectively permeable membrane maintains a concentration gradient of the not freely diffusible solute.

Summarizing the preceding paragraph, a *selectively permeable membrane* may be defined either as one that does not permit free, unhampered diffusion of all the solutes present or as one that maintains at least one solute concentration gradient across itself. *Osmosis* is the diffusion of water through a selectively permeable membrane. Or, osmosis is the diffusion of water through a membrane that maintains at least one concentration gradient across itself.

Normal living cell membranes are selectively permeable membranes. Two results follow from this fact: these membranes maintain various solute concentration gradients, and water moves through them by osmosis. Here is an example that may help you understand this passive transport process. Imagine that you have a 20% NaCl solution separated from a 10% NaCl solution by a membrane. Assume that the membrane is impermeable to sodium and chloride ions but that it is freely permeable to water particles (Figure 3-9). Obviously then, sodium and chloride ions will not diffuse through this membrane. It maintains an NaCl concentration gradient across itself, in other words. Therefore water diffusion through this membrane is called osmosis. Water osmoses through the membrane in both directions but not in equal amounts.

To deduce the direction in which the greater volume of water will osmose (the direction, that is, of net osmosis), we must apply one principle and remember one fact. The principle: *net diffusion* of any substance occurs down the concentration gradient of that substance. The fact: osmosis is diffusion of water through a semipermeable or selectively permeable membrane. *Net osmosis*, therefore, ocurs down a water concentration gradient. This means that it occurs from a solution with a higher water concentration into one with a lower water concentration. In our example, 10% NaCl has a higher water concentration than 20% NaCl. Why? Because a lower solute concentration necessarily means a higher water concentration. Net osmosis, therefore, will occur from the 10% into the 20% solution. This increases the volume and decreases the solute concentration

---

FORMULAS FOR DETERMINING OSMOTIC PRESSURE

Potential osmotic pressure of nonelectrolyte (in mm Hg)	=	Molar concentration of solution	× 19,300*

Potential osmotic pressure of electrolylte (in mm Hg)	=	Molar concentration of solution	×	Number ions per molecule	× 19,300

*Experimentation has shown that a solution with a 1.0 molar concentration of any non-electrolyte has a potential osmotic pressure of 19,300 mm Hg pressure (at body temperature, 37° C).

Molar concentration	=	Grams solute in 1 liter solution divided by Molecular weight of solute

Example: Two solutions commonly used in hospitals are 0.85% NaCl and 5% glucose. What is the potential osmotic pressure of 0.85% NaCl at body temperature ? (0.85 NaCl = 8.5 gm NaCl in 1 liter solution) (Molecular weight of NaCl = 58 [NaCl yields 2 ions per molecule in solution]).

Using the formula given for computing potential osmotic pressure of an electrolyte:

Potential osmotic pressure of 0.85% NaCl	$= \dfrac{8.5}{58} \times 2 \times 19{,}300 = 5{,}658.6$ *mm Hg pressure*

Problem: What is potential osmotic pressure of 5% glucose solution at body temperature? Molecular weight glucose = 180. Glucose does not ionize. It is a nonelectrolyte.†

†5% glucose solution has potential osmotic pressure of 5,359.6 mm Hg pressure.

Figure 3-10 Factors that determine potential osmotic pressure of nonelectrolytes. Diagram shows that the number of solute particles, in this case molecules, present in a liter of a nonelectrolyte solution determines its potential osmotic pressure. The molar concentration of a solution indicates the number of molecules in it.

Figure 3-11 Factors that determine potential osmotic pressure of electrolyte solutions. Note that when the solute is an electrolyte, the number of its particles in solution is determined by both the molar concentration of the solution and the number of ions formed from one molecule of the electrolyte.

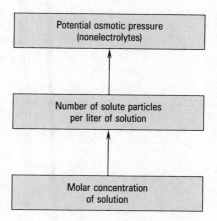

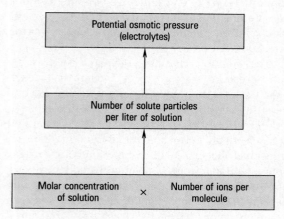

of the 20% solution. While this is happening, the volume of the 10% solution is decreasing and its solute concentration is increasing. Net osmosis, therefore, tends to equilibrate the two solutions, that is, make their concentrations equal. The principle to remember is this: net osmosis occurs down a water concentration gradient and tends to equilibrate solutions separated by a selectively permeable membrane.

Net osmosis into the originally more concentrated of our two salt solutions increases the volume of this solution. Further, the increase in its volume causes an increase in its pressure, called *osmotic pressure*—a logical term, since it is pressure caused by net osmosis. By definition, then, **osmotic pressure** is the pressure that develops in a solution as a result of net osmosis into that solution. An important principle stems from this definition: osmotic pressure develops in the solution that originally contains the higher concentration of the solute that does not diffuse freely through the membrane. For instance, in the example previously given, osmotic pressure would develop in the solution that originally had the 20% salt concentration.

Potential osmotic pressure is the maximum osmotic pressure that could develop in a solution if it were separated from distilled water by a selectively permeable membrane. (Actual osmotic pressure, on the other hand, is a pressure that already

has developed, not just one that could develop.) What determines a solution's potential osmotic pressure? Answer: the number of solute particles in a unit volume of solution directly determines its potential osmotic pressure—the more solute particles per unit volume, the greater the potential osmotic pressure.

If the solute is a nonelectrolyte, the number of solute particles in a liter of solution is determined, as Figure 3-10 indicates, solely by the **molar concentration** of the solution. (To calculate molar concentration, divide the number of grams of solute in a liter of solution by the molecular weight of the solute.) If the solute is an electrolyte, the number of solute particles per liter is determined by two factors: the molar concentration of the solution and the number of ions formed from each molecule of the electrolyte (Figure 3-11).

Since it is the number of solute particles per unit volume that directly determines a solution's potential osmotic pressure, one might at first jump to the conclusion that all solutions having the same percent concentration also have the same potential osmotic pressure. Obviously it is true that all solutions containing the same percent concentration of the same solute do also have the same potential osmotic pressures. All 5% glucose solutions, for example, have a potential osmotic pressure at body temperature of somewhat more than 5,300 mm Hg pressure. But all solutions with the same percent concentrations of different solutes do not have the same molar concentrations. And since it is molar concentration, not percent concentration that determines potential osmotic pressure, solutions with the same percent concentration of different solutes have different potential osmotic pressures. For example, 5% NaCl at body temperature has a potential osmotic pressure of approximately 33,000 mm Hg— quite different from 5% glucose's potential osmotic pressure of about 5,300 mm Hg. (If you like, you can calculate these potential osmotic pressures using the formulas given on p. 64.)

Two solutions that have the same potential osmotic pressure are said to be **isosmotic** to each other. Because they have the same potential osmotic pressure, the same amount of water will osmose in both directions between them if they are separated by a selectively permeable membrane. In short, no net osmosis occurs in either direction between isosmotic solutions. Hence no actual osmotic pressure develops in either solution. Their pressures remain the same. And because they do, isosmotic solutions are called *isotonic* (from the Greek *isos*, "the same," and *tonos*, "tension" or "pressure").

By definition **isotonic solutions** are those whose volumes and pressures will stay the same if the two solutions are separated by a membrane. For example, 0.85% NaCl is referred to as "isotonic saline," meaning that it is isotonic to the fluid inside human cells. Translated more fully, it means that if 0.85% NaCl is injected into human tissues or blood, no net osmosis occurs into or out of cells. Therefore no change in intracellular volume or pressure takes place. Cells in contact with isotonic solutions neither lose nor gain water. They become neither dehydrated nor hydrated, to use more technical terms. Physicians make use of this information frequently. For intravenous and intramuscular injections, they usually give isotonic solutions.

When two solutions have unequal potential osmotic pressures, their volumes and pressures will change if they are separated by a selectively permeable membrane. Net osmosis will occur into the solution that has the higher potential osmotic pressure. Net osmosis, as we have noted, always occurs down the water concentration gradient. Since the solution having the lower potential osmotic pressure has the lower concentration of solute particles, it necessarily has the higher concentration of water molecules. Therefore net osmosis occurs out of this solution into the solution with the higher potential osmotic pressure. Thus, in effect, the higher potential osmotic pressure acts as a water-drawing power. It draws water from the solution with the initially lower osmotic pressure and higher water concentration into the solution with the higher potential

Figure 3-12 Effects produced by existence of a water concentration gradient across a selectively permeable membrane.

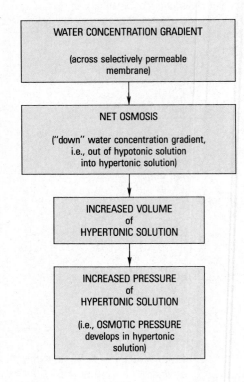

osmotic pressure and lower water concentration. This increases the volume of the solution with the initially higher potential osmotic pressure. And the increase in its volume produces an increase in its pressure. This increase in pressure is osmotic pressure—actual osmotic pressure, not just potential osmotic pressure.

When two solutions have unequal potential osmotic pressures, one of them is described as hypertonic and the other as hypotonic. Here are several definitions of the term, **hypertonic solution.**

1 A hypertonic solution is one that has a higher potential osmotic pressure than another.

2 A hypertonic solution is one into which net osmosis occurs when a selectively permeable membrane separates it from another solution.

3 A hypertonic solution is one whose volume and pressure increase when a selectively permeable membrane separates it from another solution.

4 A hypertonic solution is one in which osmotic pressure develops when a selectively permeable membrane separates it from another solution (Figure 3-12).

In summary, try to remember the following facts and principles about osmosis:

1 Osmosis is water diffusion through a semipermeable or selectively permeable membrane. (A selectively permeable membrane maintains at least one solute concentration gradient across it.)

2 Net osmosis means the osmosis of more water in one direction through a membrane than in the opposite direction. Net osmosis occurs into the solution with the higher concentration of solute particles, that is, the solution with the higher potential osmotic pressure, the hypertonic solution.

3 Net osmosis into a solution increases its volume and decreases its concentration and tends to equilibrate the two solutions separated by the membrane.

4 Net osmosis into a solution increases its pressure and produces actual osmotic pressure in the solution.

A **hypotonic solution** has characteristics opposite from those of a hypertonic solution. Example: The fluid inside human cells is hypertonic to distilled water and, conversely, distilled water is hypotonic to intracellular fluid. If, therefore, distilled water were injected into a vein, net osmosis would occur into blood cells. And eventually if intracellular volume and pressure increased beyond a certain limit, blood cell membranes would rupture and the cells would die. When this happens to red blood cells, their hemoglobin leaks out and they are said to be hemolyzed. (**Hemolysis** means the destruction of red blood cells with the escape of hemoglobin from them into the surrounding medium.)

Filtration

The physical process of **filtration** involves the passing of water and solutes through a membrane when a hydrostatic pressure gradient exists across that membrane, that is, when the hydrostatic pressure on one side of the membrane is higher than on the other. (*Hydrostatic pressure* is the force or weight of a fluid pushing against some surface.) A principle about filtration that has great physiological importance is this: filtration always occurs down a hydrostatic pressure gradient. This means that when two fluids have unequal hydrostatic pressures and are separated by a membrane, water and diffusible solutes (those to which the membrane is permeable) filter out of the solution that has the higher hydrostatic pressure into the solution that has the lower hydrostatic pressure.

An important question is: how does filtration differ from diffusion and osmosis? Both water and solutes can filter or diffuse through a membrane, whereas only water, by definition, can osmose through a membrane. Filtration occurs in only one direction through a membrane—down a hydrostatic pressure gradient. Diffusion and osmosis go on in both directions through a membrane. Net diffusion and net osmosis, however, occur only in one direction. Net diffusion occurs down the concentration gradient of the substance diffusing. Net osmosis occurs down a water concentration gradient. Filtration is a major mechanism for moving substances through the membranous walls of the blood capillaries; water and true solutes filter out of the blood into the interstitial fluid (in the microscopic spaces between cells).

Active Transport Mechanisms

An **active transport mechanism** may be defined as a device that uses energy produced by cellular chemical reactions to move molecules or ions "uphill" through a cell membrane. By "uphill" we mean that an actively transported substance moves from a region of lower concentration to one of higher concentration. In short, it moves *against* its concentration gradient. How does this compare with the direction of net diffusion of a substance? Net diffusion of a substance occurs down its concentration gradient, as if it were coasting "downhill." Active transport of a substance up its concentration gradient occurs as if it were being pushed "uphill."

Our knowledge about how active transport mechanisms operate is still incomplete. Some of them, the "pumps," are known to be carrier-mediated processes. Certain protein molecules, called *carriers*, are present in the plasma membrane of all cells. A molecule or ion to be transported through a membrane binds to a specific binding or **receptor site** of a carrier molecule on one side of a membrane. Then, according to a current postulate, the bound carrier rotates rapidly in the membrane. This moves the bound site of the carrier to the opposite surface of the membrane where the substance dissociates from the carrier. It has thus been transported through the membrane, and in the process energy has been used. Chemical reactions in the cell, mostly in the mitochondria, provide this energy (discussed in Chapter 23).

Table 3-3 Cell transport mechanisms

Transport mechanisms	Substances transported
Diffusion	Oxygen
Diffusion	Carbon dioxide
Facilitated diffusion	Glucose, through membranes of most cells
Active transport	Glucose, through membranes of epithelial cells of intestine and kidneys Amino acids Sodium, potassium and other ions
Endocytosis	
Phagocytosis	Small particles, e.g., micro-organisms
Pinocytosis	Fluid
Exocytosis	Proteins

Sodium Potassium Pump

An active transport mechanism known as the **sodium-potassium pump** operates in all of our cells. It is essential for healthy cell survival. As its name suggests, the sodium-potassium pump actively transports both sodium and potassium ions—but in opposite directions. It transports sodium ions out of cells and potassium ions into cells. By so doing, the sodium-potassium pump maintains a higher sodium concentration in the interstitial fluid outside cells than in the intracellular fluid inside them. But it maintains a higher potassium concentration in intracellular fluid than in interstitial fluid. Presumably both sodium and potassium ions bind to the same carrier, postulated to be an enzyme known as sodium-potassium ATPase. Potassium ions bind to this enzyme's potassium-binding sites at the outer surface of the plasma membrane. After the enzyme has rotated in the membrane, potassium ions are released at the inner surface of the membrane and enter the intracellular fluid. As potassium ions are being released from the enzyme's potassium-binding sites, sodium ions are being bound to its sodium-binding sites at the inner surface of the membrane. Then, following rotation of the sodium-bound enzyme, sodium ions dissociate from it at the membrane's outer surface and enter the interstitial fluid. Not only does sodium-potassium ATPase probably function as the carrier molecule in sodium and potassium transport, but it also catalyzes the splitting of ATP molecules to release energy—energy somehow used to power the sodium-potassium pump.

Table 3-3 lists the mechanisms that transport various substances through cell membranes.

Endocytosis and Exocytosis

Like the sodium-potassium pump, endocytosis and exocytosis are active transport mechanisms. They differ from the pump mechanisms in that they enable substances to enter or leave the interior of a cell without actually moving through its plasma membrane.

In **endocytosis** a segment of a cell's plasma membrane invaginates, forming a small pocket around the material to be moved into the cell, pinches off from the rest of the membrane, and migrates inward as a closed vesicle. The processes of phagocytosis and pinocytosis are two types of endocytosis. In *phagocytosis* microorganisms and other small particles are engulfed by the plasma membrane and enter the cell in vesicles that have pinched off from the membrane. Once in the cytoplasm they fuse with the membranous walls of lysosomes. Enzymes in the lysosomes then digest them. Hence the name *phagocytosis*, which means "cell eating" (from the Greek *phagein*, "to eat," and *kytos*, "a cell"). *Pinocytosis* means "cell drinking." It is through endocytosis that bits of fluid enter a cell.

Exocytosis is the process by which large molecules, notably proteins, can leave the cell even though they are too large to move out through its plasma membrane. They are first enclosed in membranous vesicles. These migrate out to the plasma membrane and by fusing with the plasma membrane, they release their contents outside the cell. Some gland cells secrete their products by exocytosis.

REPRODUCTION OF CELLS

Cell reproduction is one of the most fundamental of all living functions. On it depend the creation of all individual organisms and the continued existence of all complex, multicellular organisms. The survival, therefore, of each one of us and of our human species depends on cell reproduction. In addition, cell reproduction is one of the most fascinating stories that physiology has to tell. Advances in our understanding of this most basic but enormously complex phenomenon have been described by many as a scientific revolution with implications more far-reaching and important to the human species than the birth of the atomic age. We shall start our version of it by telling about DNA—"the most golden of all molecules" as Watson called it.*

---

*Watson, J.D.: The double helix, New York, 1969, The New American Library.

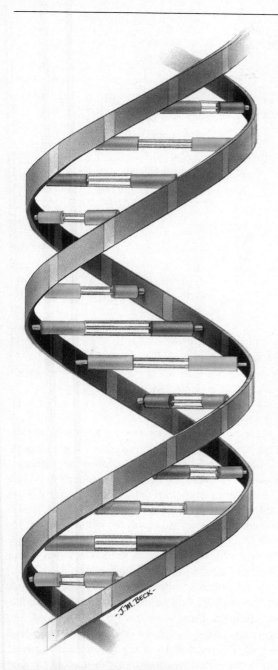

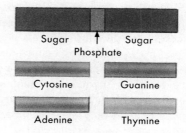

Sugar — Phosphate — Sugar

Cytosine   Guanine

Adenine   Thymine

Figure 3-13 Watson-Crick structure of DNA molecule. Note that each side of the DNA molecule consists of alternating sugar and phosphate groups. Each sugar group is united to the sugar group opposite it by a pair of nitrogenous bases, adenine-thymine or thymine-adenine and cytosine-guanine or guanine-cytosine. Differences in the sequences of base pairs establish the identity of the many different kinds of DNA.

DEOXYRIBONUCLEIC ACID (DNA)

The **deoxyribonucleic acid** molecule is a giant among molecules. Both its size and the complexity of its shape exceed those of most molecules. The importance of its function—in a word, heredity—surpasses that of any other molecule in the world. Over thirty years ago, an American, James D. Watson, and two British scientists, Francis H.C. Crick and Maurice H.F. Wilkins, solved the puzzle of DNA's molecular structure. Watson and Crick announced their discovery with an understated, one-page report that appeared in the April 25, 1953, issue of the British journal *Nature*. Nine years later the coveted Nobel Prize in Medicine was awarded all three sci-

entists for their brilliant and significant work—hailed as the greatest biological discovery of our times. Watson tells how they accomplished this in *The Double Helix*, now a classic, a book you will probably find hard to put down once you start it.

To try to visualize the shape of the DNA molecule, picture to yourself an extremely long, narrow ladder made of a pliable material (Figure 3-13). Now see it twisting round and round on its axis and taking on the shape of a steep spiral staircase millions of turns long. This is the shape of the DNA molecule—a double helix (from the Greek word for spiral).

As to its structure, the DNA molecule is a polymer. This means that it is a large molecule made up of many smaller molecules joined together in sequence. DNA is a polymer of millions of pairs of nucleotides. A nucleotide is a compound formed by combining phosphoric acid with a sugar and a nitrogenous base. In the DNA molecule there are four different kinds of nucleotides. Each nucleotide consists of a phosphate group that attaches to the sugar deoxyribose that attaches to one of four bases. Nucleotides differ, therefore, in their nitrogenous base component—containing either adenine or guanine (purine bases) or cytosine or thymine (pyrimidine bases). (Deoxyribose is a sugar that is not sweet and one whose molecules contain only five carbon atoms.) Notice what the name *deoxyribonucleic acid* tells you that this compound contains deoxyribose, that it occurs in nuclei, and that it is an acid.

Figure 3-13 reveals additional and highly significant facts about DNA's molecular structure. First, observe which compounds form the sides of the DNA spiral staircase—a long line of phosphate and deoxyribose units joined alternately one after the other. Look next at the stair steps. Notice two facts about them: that two bases join (loosely bound by a hydrogen bond) to form each step, and that only two combinations of bases occur. The same two bases invariably pair off with each other in a DNA molecule. Adenine always goes with thymine (or vice versa, thymine with adenine), and guanine always goes with cytosine (or vice versa). This fact about DNA's molecular structure is called **obligatory base-pairing.** Pay particular attention to it, for it is the key to understanding how a DNA molecule is able to duplicate itself. DNA duplication, or replication as it is usually called, is one of the most important of all biological phenomena because it is an essential and crucial part of the mechanism of heredity.

Another fact about DNA's molecular structure that has great functional importance is the sequence of its base pairs. Although the base pairs in all DNA molecules are the same, the sequence of these base pairs is not the same in all DNA molecules. For instance, the sequence of the base pairs composing the seventh, eighth, and ninth steps of one DNA molecule might be cytosine-guanine, adenine-thymine, and thymine-adenine. Such a sequence of three bases forms a code word or "triplet" called a **codon.** In another DNA molecule the coding sequence of the base pairs making up these same steps might be entirely different, perhaps thymine-adenine, guanine-cytosine, and cytosine-guanine. Perhaps these seem to be minor details. But nothing could be further from the truth, since it is the sequence of the base pairs in the nucleotides composing DNA molecules that identifies each gene. Hence it is the sequence of base pairs that determines all hereditary traits.

Figure 3-14 Brief scheme to show how genes determine hereditary characteristics.

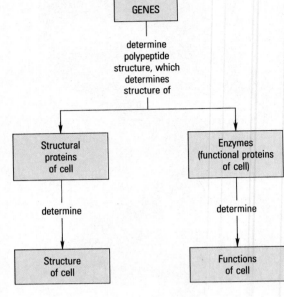

A human **gene** is a segment of a DNA molecule. One gene consists of a chain of about 1,000 pairs of nucleotides joined one after the other in a precise sequence. One gene controls the production within the cell of one polypeptide chain. Two or more polypeptides made up a cell's structural proteins and enzymes. Each enzyme catalyzes one chemical reaction. Therefore, as Figure 3-14 indicates, the millions of genes constituting a cell's DNA determine the cell's structure and its functions.

Genes control protein synthesis (anabolism). A brief synopsis of this process, as it is now understood, follows. First, a strand of RNA (ribonucleic acid) forms along a segment of one strand of a DNA molecule. RNA differs from DNA in certain respects. Its molecules are smaller than those of DNA, and RNA contains ribose instead of deoxyribose. Also, one of the four bases in RNA is uracil instead of thymine. As a strand of RNA is forming along a strand of DNA, uracil attaches to adenine and guanine attaches to cytosine. The process is known as **complementary pairing.** Thus a single-strand molecule of messenger RNA (abbreviated mRNA) is formed. The name "messenger RNA" describes its function. As soon as it is formed, it separates from the DNA strand, diffuses out of the nucleus, and carries a "message" to a ribosome in the cell's cytoplasm, directing its synthesis of a specific protein. Ribosomes contain another type of RNA called appropriately "ribosomal RNA" (abbreviated rRNA).

RECOMBINANT DNA

Each time fertilization occurs, a new and unique package of genetic material (DNA), and therefore a new individual, comes into being. Sexual reproduction involves this recombination or mingling of DNA from sperm and egg. The term **recombinant DNA,** however, means something entirely different. It refers to the joining together of hereditary material, often from different species, into new, biologically functional combinations.

Recombination or gene-splicing techniques permit the joining of DNA from different organisms—inside bacteria! Scientists split small circular rings of bacterial DNA called **plasmids** from one organism by using specialized "restriction enzymes" as chemical cutters. These same enzymes are used to cut segments of DNA from some other animal source (perhaps a mouse), and the foreign DNA fragment is then annealed into the opened plasmid ring. The recombinant plasmid is then reinserted into a bacterial "host." The result is a type of hybrid that never before existed in nature.

GENE-SPLICING CONTROVERSY

In 1971 Dr. Paul Berg of Stanford University discussed his plans to insert a DNA fragment that caused cancer in mammals into *E. coli* bacteria. He terminated the experiment at the request of a group of scientists who believed that there was at least the potential for some kind of biological catastrophe in such a study—the creation of a new mutant organism with malignant powers that might escape from the laboratory and cause widespread disease.

Further advances in recombinant DNA technology aroused both public and scientific apprehension, and in 1974 Berg and a group of other molecular biologists called for a moratorium on certain types of recombinant DNA experiments and brought to public attention the potential dangers of genetic engineering in bacteria.* Not since the congressional investigations of atomic energy and nuclear radiation dangers had science sparked such heated controversy. As a result of discussions that followed the moratorium, an elaborate set of guidelines and restrictions on recombinant DNA research was adopted by the National Institutes of Health (NIH) in 1976. Fears that recombinant DNA experiments might lead to creation of lethal toxins or disease-causing bacteria have now subsided. The guidelines that initially prohibited many experiments and allowed others to be performed only under extreme conditions of physical confinement have now been extensively revised and relaxed. At the present time hundreds of university and industrial laboratories are routinely performing gene-splicing experiments.

*Berg, P., et al.: Potential biohazards of recombinant DNA molecules, Science **185:**303, 1974.

Next, a molecule of a third type of RNA—transfer RNA, or tRNA—present in cytoplasm attaches itself to a molecule of a specific amino acid and transfers it to a ribosome. Here the amino acid fits into the position indicated by the mRNA. More tRNA molecules, one after the other in rapid sequence, bring more amino acids to the ribosome and fit them into their proper positions. Result? A chain of amino acids joined to each other in a definite sequence—a protein, in other words—is formed. In short, protein anabolism has occurred. It is one of the major kinds of cellular work. One human cell is estimated to synthesize perhaps 2,000 different enzymes! In addition to this staggering workload, it produces many different protein compounds that help form its own structures and many cells also synthesize

Table 3-4 Mitosis—somatic cell division

Interphase	Prophase	Metaphase	Anaphase	Telophase
1 Period when cell prepares for division and also grows in size	1 Chromosomes shorten and thicken (from coiling of DNA molecules that compose them); each chromosome consists of two *chromatids* attached at *centromere*	1 Chromosomes align across equator of spindle fibers; each pair of chromatids attached to spindle fiber at its centromere	1 Each centromere divides into two, thereby detaching two chromatids that compose each chromosome from each other	1 Changes occurring during telophase essentially reverse of those taking place during prophase; new chromosomes start elongating (DNA molecules start uncoiling)
2 Chromosomes elongate and become too thin to be visible as such, but chromatin granules become visible	2 Centrioles move to opposite poles of cell; spindle fibers appear and, under control of centrioles, begin to orient between opposing poles	2 Nucleoli and nuclear membrane disappear	2 Divided centromeres start moving to opposite poles, each pulling its chromatid (now called a chromosome) along with it; there are now twice as many chromosomes as there were before mitosis started	2 Nucleoli and two new nuclear membranes appear, enclosing each new set of chromosomes
3 DNA of each chromosome replicates itself, forming two chromatids attached only at centromere				3 Spindle fibers disappear
				4 *Cytokinesis,* or dividing of cytoplasm, usually occurs during telophase; starts as pinching in along equator of old cell and ends with division of old cell into two new cells
				5 Centrioles replicate

special proteins. Liver cells are a notable example; they synthesize prothrombin, fibrinogen, albumins, and globulins.

Molecular geneticists now routinely perform experiments that literally transfer complex biological traits or capabilities from one life form to another. Each time the host organism divides, so does the recombinant DNA. Such rapidly reproducing synthetic bacteria have already made human insulin, growth hormone, and interferon in the laboratory and give promise for future uses in both medicine and industry.

The ultimate goal of recombinant DNA research is to increase our understanding of mammalian gene regulation. Genes initiate and control the enormously complex mechanism that permits development of a fertilized ovum into a fully differentiated individual.

MITOSIS

Human cells, other than sex cells, reproduce by the process of **mitosis.** In this process a cell divides in order to multiply. One cell divides to form two cells.

Mitosis consists of a sequence of events plainly visible in suitably stained cells viewed with the light microscope. The events of mitosis occur in five phases: *interphase, prophase, metaphase, anaphase,* and *telophase* (Figure 3-15).

DNA replication occurs during the *interphase* of mitosis (cell reproduction). The tightly coiled DNA molecules uncoil except for small segments. (Since these remaining tight little coils are denser than the thin elongated sections, they absorb more stain and appear as chromatin granules under the microscope. The thin uncoiled sections, in contrast, are invisible because they absorb so little stain.) As the DNA molecule uncoils, its two strands come apart. Then, along each of the two separated strands of nucleotides, a complementary strand forms. Intracellular fluid contains many molecules of deoxyribose and nitrogenous bases and many phosphate ions. By the mechanism of obligatory base-pairing, these substances become attached at their right places along each DNA strand. Interpreted, this means that new thymine, that is, from the intracellular fluid, attaches to the "old" adenine in the original DNA strand. Conversely, new adenine attaches to old thymine. Also, new guanine joins old cytosine

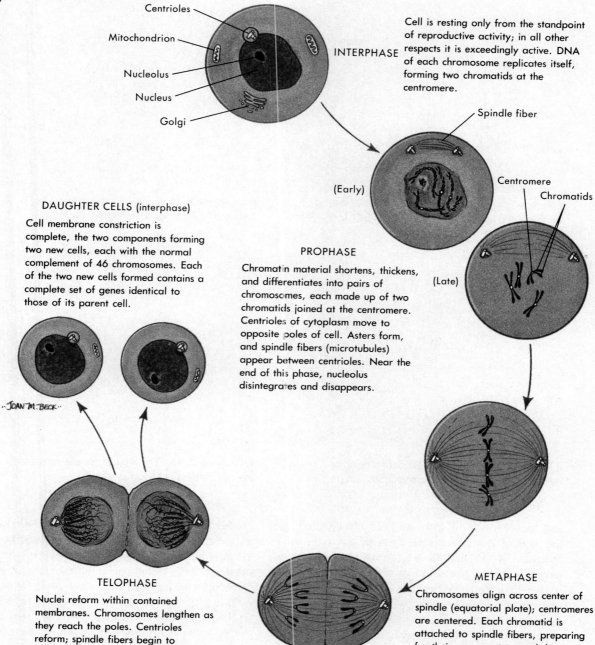

Figure 3-15 Diagram of mitosis. For simplicity, only four of the total 46 chromosomes are shown. Refer to Table 3-4 for details of each mitotic stage.

Centrioles

Mitochondrion

Nucleolus

Nucleus

Golgi

INTERPHASE

Cell is resting only from the standpoint of reproductive activity; in all other respects it is exceedingly active. DNA of each chromosome replicates itself, forming two chromatids at the centromere.

Spindle fiber

(Early)

Centromere

Chromatids

(Late)

PROPHASE

Chromatin material shortens, thickens, and differentiates into pairs of chromosomes, each made up of two chromatids joined at the centromere. Centrioles of cytoplasm move to opposite poles of cell. Asters form, and spindle fibers (microtubules) appear between centrioles. Near the end of this phase, nucleolus disintegrates and disappears.

DAUGHTER CELLS (interphase)

Cell membrane constriction is complete, the two components forming two new cells, each with the normal complement of 46 chromosomes. Each of the two new cells formed contains a complete set of genes identical to those of its parent cell.

·JOAN·M·BECK·

METAPHASE

Chromosomes align across center of spindle (equatorial plate); centromeres are centered. Each chromatid is attached to spindle fibers, preparing for their movement toward the centrioles.

TELOPHASE

Nuclei reform within contained membranes. Chromosomes lengthen as they reach the poles. Centrioles reform; spindle fibers begin to disappear. Cytoplasmic movements occur in which the mitochondria and other organelles are equally distributed between the two halves.

ANAPHASE

Each centromere divides, becoming a double structure, thereby detaching the paired chromatids that compose each chromosome. The centromeres, with the chromatids (now called chromosomes) attached, move along the spindle fibers to opposite ends of the cell. A cleavage furrow begins near the end of this phase.

Figure 3-16, A to **H**. Series of photomicrographs showing animal cells undergoing mitosis from interphase (**A**) to telophase (**H**).

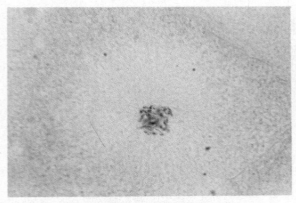

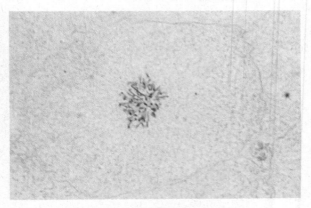

A, Early prophase. Chromosomes are beginning to form from the nuclear chromatin material. Chromosomes have already been duplicated.

B, Later prophase. Chromosomes are now fully formed and easily visible in the cell nucleus.

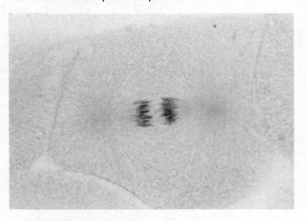

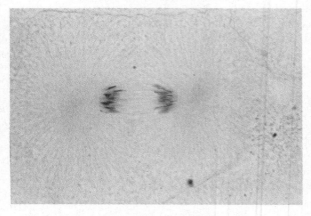

E, Later anaphase. The chromatids are moving toward the two poles of the mitotic apparatus.

F, Early telophase. The chromatids have completed their movement to the poles.

and, vice versa, new cytosine joins old guanine. By the end of the interphase, each of the two DNA strands of the original DNA molecule has a complete new complementary strand attached to it. Each half of the DNA molecule or strand, in other words, has duplicated itself to create a whole new DNA molecule. Thus two new chromosomes now replace each original chromosome. However, at this stage (end of interphase, beginning of prophase), they are called *chromatids* instead of chromosomes. The two chromatids formed from each original chromosome contain duplicate copies of DNA and, therefore, the same genes as the chromosome from which they were formed. Chromatids are present as attached pairs. *Centromere* is the name of their point of attachment.

By the time the parent cell divides to form two daughter cells, chromatids have separated and be-

come chromosomes and one set of them has become part of the nucleus of one daughter cell and the other set has become part of the nucleus of the other daughter cell. Hence each of the two new cells formed by mitosis contains a complete set of genes identical to those in its parent cell. Both daughter cells have the potentialities to become like their parent cell. Thus the function of mitosis is to enable cells to reproduce their own kind. By means of the mechanism of mitosis, a new generation of cells inherits both the structural and the functional characteristics of the preceding generation (Figure 3-14). In short, mitosis is the mechanism of heredity for cells. Table 3-4 (p. 72) summarizes the events that characterize the five phases of mitosis. As you read about the events, try to correlate them with the drawings in Figure 3-15 (p. 73) and the photomicrographs in Figure 3-16.

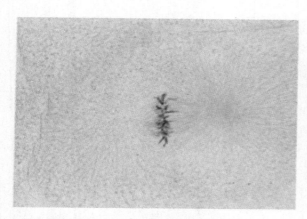

C, Metaphase. Chromosomes are lined up at the equatorial plate. Spindle fibers (microtubules) of the mitotic apparatus are distinct.

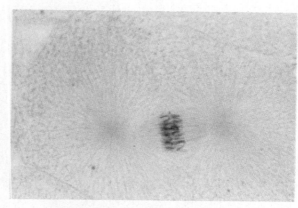

D, Early anaphase. Spindle fibers pull each of the two chromatids toward the centrioles. The centromeres have divided, separating the chromatids.

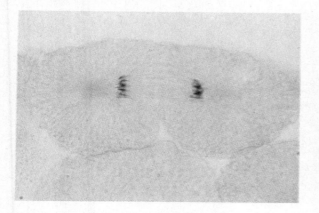

G, Telophase. The cleavage furrow that will eventually constrict the cell into two daughter cells is evident.

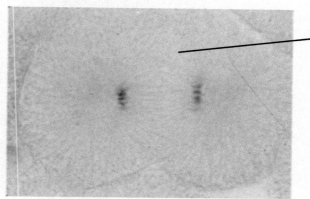

Cleavage furrow

H, Telophase. Chromosomes have completely moved to poles.

MEIOSIS

Meiosis is the type of cell division that occurs only in primitive sex cells during the process of their becoming mature sex cells. As a result of meiosis the primitive sex cells (spermatogonia in the male and oogonia in the female) become mature sex cells called **gametes.** Male gametes are named spermatozoa but are usually called sperm. Female gametes are named ova (singular, ovum). In humans, all somatic cells contain 46 chromosomes. This total of 46 chromosomes per cell is known as the **diploid** number of chromosomes. Diploid comes from the Greek *diploos,* meaning "two or pair." In somatic cells the 46 chromosomes are present in 22 homologous pairs—the remaining two being the sex chromosomes XY (male) or XX (female). During meiosis, or **reduction division,** the diploid chromosome number (46) of the primitive spermatogonium or oogonium is reduced to the haploid number of 23 found in the mature sex cells or gametes. The end result of fertilization is the fusion of two gametes, each containing the haploid number (23) of chromosomes. Fertilization results in formation of a zygote that is a diploid cell having 46 chromosomes, 23 chromosomes being contributed by each parent.

Special tissue culture techniques can be used that cause chromosomes in cells to swell and disperse, making it possible to identify each pair. Figure 3-17 is a light micrograph of the chromosome pairs seen in a single human diploid cell. In Figure 3-18 the chromosome pairs have been individually cut out of the light micrograph and arranged according to their size and the position of the centromere. When displayed in this type of array, the chromosomes are referred to as a **karyotype.**

Figure 3-17 Chromosomes of a normal human diploid cell. (× 2,000.)

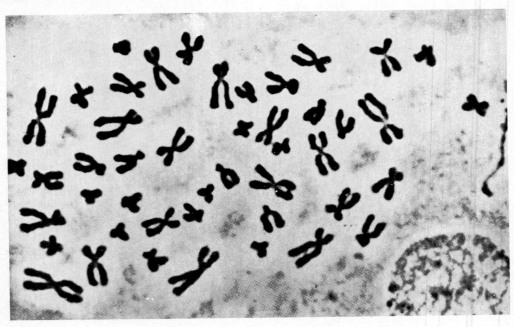

Figure 3-18 Karyotype of chromosomes seen in Figure 3-17. Chromosomes were obtained from a leukocyte or white blood cell taken from the blood of a male donor.

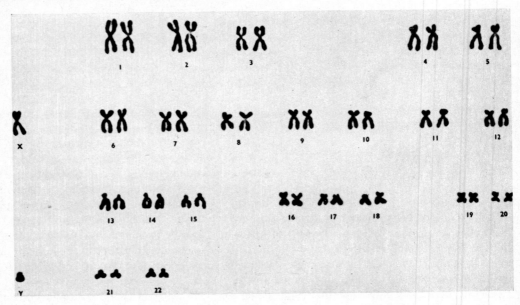

Meiosis consists of two cell divisions that take place one after the other in quick succession. They are referred to as meiotic division I and meiotic division II, and in both, an interphase, prophase, metaphase, anaphase, and telophase occur. In the interphase that precedes prophase I (of meiotic division I) the same events occur as take place in the interphase of mitosis. Specifically, the DNA of each chromosome replicates itself and thereby changes each chromosome into two chromatids, attached only at the centromere. Prophase I of meiosis consists of several stages. You can find the name of these stages in Figure 3-19. For simplicity's sake, in both Figures 3-19 and 3-20, only four of the 46 chromosomes are shown. It is in the diplotene stage of prophase I of meiosis that the phenomenon of "cross-

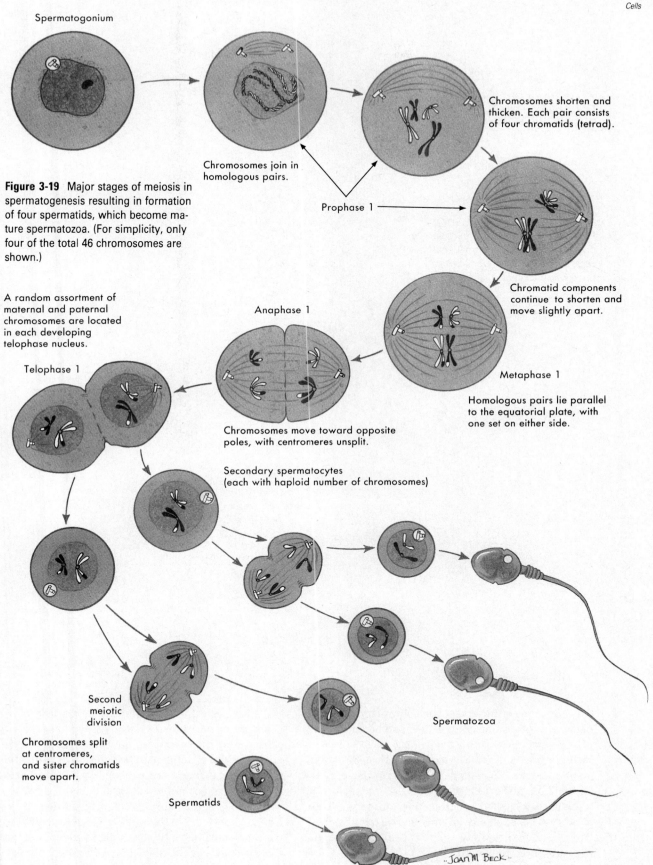

Spermatogonium

Chromosomes join in homologous pairs.

Prophase 1

Chromosomes shorten and thicken. Each pair consists of four chromatids (tetrad).

Figure 3-19 Major stages of meiosis in spermatogenesis resulting in formation of four spermatids, which become mature spermatozoa. (For simplicity, only four of the total 46 chromosomes are shown.)

Chromatid components continue to shorten and move slightly apart.

A random assortment of maternal and paternal chromosomes are located in each developing telophase nucleus.

Anaphase 1

Metaphase 1

Homologous pairs lie parallel to the equatorial plate, with one set on either side.

Telophase 1

Chromosomes move toward opposite poles, with centromeres unsplit.

Secondary spermatocytes (each with haploid number of chromosomes)

Second meiotic division

Chromosomes split at centromeres, and sister chromatids move apart.

Spermatozoa

Spermatids

JoanM Beck

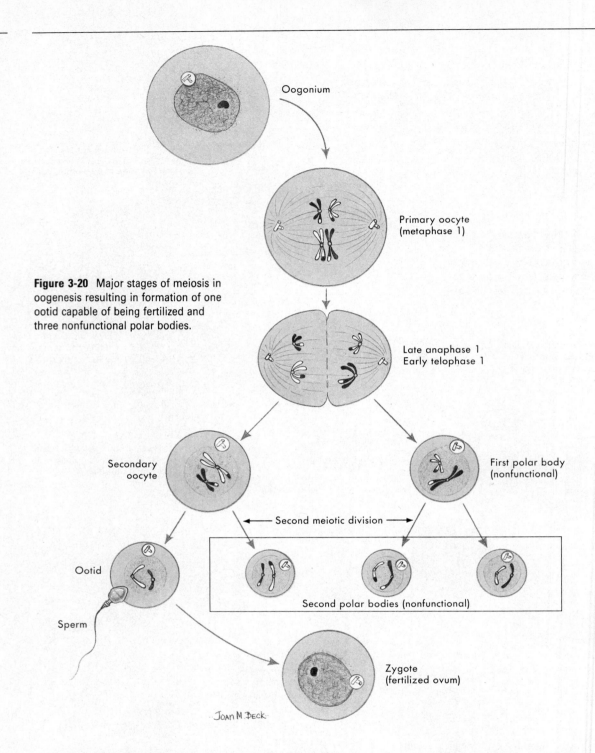

Figure 3-20 Major stages of meiosis in oogenesis resulting in formation of one ootid capable of being fertilized and three nonfunctional polar bodies.

Oogonium

Primary oocyte (metaphase 1)

Late anaphase 1
Early telophase 1

Secondary oocyte

First polar body (nonfunctional)

←——— Second meiotic division ———→

Ootid

Second polar bodies (nonfunctional)

Sperm

Zygote (fertilized ovum)

Joan M. Beck

over" occurs. During crossover a chromatid segment of each chromosome crosses over and becomes part of the adjacent chromosome in the pair. This is a highly significant event. Since each chromatid segment consits of specific genes, the crossing-over of chromatids reshuffles the genes, so to speak—it transfers some of them from one chromosome to another. This exchange of genetic material can add almost infinite variety to the ultimate genetic makeup of an individual.

Metaphase I follows the last stage of prophase I, and as in mitosis, the chromosomes align themselves along the equator of the spindle fibers as Figure 3-19 shows. But in anaphase the two chromatids that make up each chromosome do not separate from each other as they do in mitosis to form two new chromosomes out of each original one. Therefore in anaphase I of Figures 3-19 and 3-20, only two chromosomes move to each pole of the parent cell. When the parent cell divides to form two cells, each daughter cell contains two chromosomes or half as many as the parent cell had. The daughter cells formed by meiotic division I contain a **haploid** number of chromosomes, which means half as many as the diploid number present in the parent cell.

As you can see in Figures 3-19 and 3-20, meiotic division II is essentially the same as mitosis. In both spermatogenesis and oogenesis the second meiotic division reproduces each of the two cells formed by meiotic division I and so forms four cells, each with the haploid number of chromosomes.

SPERMATOGENESIS

Spermatogenesis is the process by which the primitive sex cells, or **spermatogonia,** present in the seminiferous tubules of a newborn baby boy become transformed into mature sperm. Spermatogenesis starts at about the time of puberty and usually continues throughout a man's life. Figure 3-19 diagrams some of the major steps of spermatogenesis. Meiotic division I reproduces one primary spermatocyte to form two secondary spermatocytes, each with a haploid number of chromosomes (23). One of the 23 chromosomes is either an X or a Y sex chromosome. Meiotic division II then reproduces each of the two secondary spermatocytes to form a total of four spermatids. Thus spermatogenesis forms four sperm—each with only 23 chromosomes—from one primary spermatocyte that has 23 pairs or 46 chromosomes.

OOGENESIS

Oogenesis is the process by which primitive female sex cells, or **oogonia,** become mature ova. During the fetal period, oogonia reproduce in the ovaries by mitosis to form primary oocytes—about a half million of them by the time a baby girl is born. Most of the primary oocytes develop to prophase I before birth. There they stay until puberty. Then the other steps of oogenesis shown in Figure 3-20 take place about once a month for the next 30 or 40 years, that is, until a woman reaches her menopause. Note that oogenesis forms only one mature ovum (egg) from each primitive oocyte. How many sperm are formed from each primary spermatocyte by spermatogenesis? An ovum, like a sperm, contains only 23 chromosomes, but unlike sperm, half of which have an X and half of which a Y sex chromosome, all ova have an X chromosome. This distinction between sperm and ovum makes a difference—the difference between whether the new life formed by their union will be male or female. If a sperm with an X chromosome fertilizes an ovum, the baby will be a girl. If a Y-bearing sperm fertilizes an ovum, it will be a boy.

Outline Summary

CELL SIZE, SHAPE, AND STRUCTURES

A Size—diameters of human cells vary from about 7.5 to 300 μm

B Shape—great variation (Figure 3-1)

C Structures—main ones: plasma membrane, nucleus, cytoplasm, organelles, fibers

CELL MEMBRANES

A Plasma membrane (or cytoplasmic membrane)—surface boundary of cell

B Internal cell membranes—form walls of endoplasmic reticulum, Golgi apparatus, mitochondria, lysosomes, and nucleus

C Structure (essentially the same in both plasma membrane and internal cell membranes)
   1 Composed of lipid and protein molecules
   2 Arrangement of molecules—current postulate: "fluid mosaic," that is, fluid lipid bilayer with mosaic pattern of protein molecules embedded in and on it; hydrophilic phosphate heads of lipids face into water, hydrophobic tails face away from water and toward each other

D Functions of plasma membrane
   1 Maintenance of cell's integrity
   2 Identification of cell—certain protein molecules on plasma membrane's surface are unique to cells of one individual
   3 Communication—compounds from other cells, notably hormones from endocrine gland cells and neurotransmitters from nerve cells, bind to binding sites (receptors) of specific surface proteins of plasma membrane thereby initiating changes in cell's activities
   4 Transportation—some, but not all, substances move or are moved through membranes by various mechanisms
   5 Defense—some proteins (antibodies) on surface membranes of lymphocytes bind to harmful substances (antigens), thereby initiating series of events that make antigens harmless

CYTOPLASM, ORGANELLES, AND FIBERS

A Definition—cytoplasm is protoplasm located between cell membrane and nucleus

Outline Summary—cont'd

B Cytoplasm—contains membranous organelles (endoplasmic reticulum, Golgi apparatus, mitochondria, and lysosomes), nonmembranous organelles (ribosomes and centrioles), inclusions (nonliving substances), and fibers

C Endoplasmic reticulum
   1 Structure—complicated network of canals and sacs extending through cytoplasm and opening at surface of cell; many ribosomes attached to membranes of rough endoplasmic reticulum but not to smooth
   2 Functions—ribosomes attached to rough endoplasmic reticulum synthesize proteins; canals of reticulum serve cell as its inner circulatory system, for example, proteins move through canals on way to Golgi apparatus

D Ribosomes
   1 Structure—microscopic spheres composed of RNA and protein; large numbers of them are attached to endoplasmic reticulum
   2 Function—"protein factories"; ribosomes attached to endoplasmic reticulum synthesize proteins to be secreted by cell, and those lying free in cytoplasm make proteins for cell's own use, that is, its structural proteins and enzymes; groups of ribosomes, called *polysomes,* synthesize proteins

E Golgi apparatus
   1 Structure—membranous vesicles near nucleus
   2 Function—synthesizes large carbohydrate molecules, combines them with proteins, and secretes product (glycoproteins)

F Mitochondria
   1 Structure—microscopic sacs; walls composed of inner and outer membranes separated by fluid; thousands of particles make up enzyme molecules attached to both membranes
   2 Function—"power plants" of cells; mitochondrial enzymes catalyze series of oxidation reactions that provide about 95% of cell's energy supply

G Lysosomes
   1 Structure—microscopic membranous sacs
   2 Function—cell's own digestive system; enzymes in lysosomes digest particles or large molecules that enter them; under some conditions, digest and thereby destroy cells

H Centrioles
   1 Centrioles, located near nucleus are tiny cylinders, walls of which are composed of nine groups of microtubules, three tiny tubules in each group
   2 Function—form mitotic spindle

I Cell fibers
   1 Microtrabecular fibers—form three-dimensional, irregularly shaped lattice that extends throughout cytoplasm, supports various organelles, and serves as cellular "muscle," that is, produces internal cell movements

   2 Microtubules and microfilaments—suspended from microtrabecular lattice and with it form supporting framework of cell

NUCLEUS

A Definition—spherical body in center of cell; enclosed by pore-containing membrane

B Structure
   1 Consists of nuclear membrane (two-layered, with essentially same molecular structure as plasma membrane)
   2 Contains DNA (heredity molecules), which appear as
      a Chromatin threads or granules in nondividing cells
      b Chromosomes in early stage of cell division

C Functions of nucleus are functions of DNA molecules; they determine both structure and function of cells and heredity

SPECIAL CELL STRUCTURES

A Microvilli—projections of cytoplasm and plasma membrane; increase surface area of cells whose function is absorption

B Cilia—hairlike projections of cytoplasm and plasma membrane; each cilium is a tiny cylinder made up of nine double microtubules arranged around two single microtubules; one cell may have a hundred or more cilia; they propel fluid in one direction over surface of cell, for example, upward in respiratory tract

C Flagellum—single hairlike projection from cell's surface; for example, flagellum of spermatozoon propels it forward in its fluid environment

CELL PHYSIOLOGY—MOVEMENT OF SUBSTANCES THROUGH CELL MEMBRANES

Types of transportation processes

A Physical—processes in which random, never-ceasing movements of molecules and ions supply energy for moving substances; names of physical processes: diffusion (including dialysis and facilitated diffusion), osmosis, and filtration

B Physiological or active—energy that moves substances comes from chemical reactions in living cell; names of physiological processes: active transport mechanisms, endocytosis, and exocytosis

Passive transport processes

A Diffusion
   1 Movement of solute and solvent particles in all directions through solution or in both directions through membrane
      a Net diffusion of solute particles—down solute concentration gradient, that is, from more to less concentrated solution

Outline Summary—cont'd

b Net diffusion of water—down water concentration gradient, that is, from less to more concentrated solution

2 Diffusion tends to produce equilibration of solutions on opposite sides of membrane

3 Dialysis—separation of crystalloids from colloids by diffusion of crystalloids through membrane permeable to them but impermeable to colloids

4 Facilitated diffusion—diffusion mediated by carriers (specific proteins) in cell membrane; like ordinary diffusion, facilitated diffusion moves substances down their concentration gradients

B Osmosis

1 In living systems, movement of water in both directions through membrane that maintains at least one solute concentration gradient across it

2 Net osmosis—more water osmoses in one direction through membrane than in opposite; net osmosis occurs down water concentration gradient (which is up solute concentration gradient and up potential osmotic pressure gradient); net osmosis tends to equilibrate two solutions separated by selectively permeable membrane

3 Osmotic pressure—pressure that develops in solution as result of net osmosis into it

4 Isotonic solution—one that has same potential osmotic pressure as solution to which it is isotonic; no net osmosis between isotonic solutions

5 Hypertonic solution—has greater potential osmotic pressure and higher solute concentration but lower water concentration than solution to which it is hypertonic; net osmosis into hypertonic solution from hypotonic solution

6 Hypotonic solution—has lower potential osmotic pressure and lower solute concentration but higher water concentration than solution to which it is hypotonic; net osmosis out of hypotonic solution into hypertonic solution

C Filtration

1 Definition—movement of solvent (water in body) and solutes through membrane in one direction only, that is, down hydrostatic pressure gradient

2 Comparison with diffusion

a Diffusion, like filtration, moves both water and solutes through membrane

b Diffusion occurs in both directions through membrane, filtration in only one

c Net diffusion of solute occurs down solute concentration gradient; net diffusion of water occurs down water concentration gradient; filtration occurs down hydrostatic pressure gradient

3 Comparison with osmosis

a Water is only substance that osmoses through membrane; occurs in both directions

b Net osmosis occurs down water concentration gradient; filtration occurs down hydrostatic pressure gradient

Active transport mechanisms

A Definition—devices that use energy produced by cellular chemical reactions to move molecules or ions up their concentration gradient through a cell membrane; function of sodium-potassium ATPase

B Sodium-potassium pump—actively transports both sodium and potassium ions—but in opposite directions

C Endocytosis and exocytosis

1 Endocytosis—process that moves solid particles into cell; segment of cell's plasma membrane forms pocket around particle outside cell, then pinches off from rest of membrane and migrates inward

2 Exocytosis—process by which large molecules, notably proteins, can leave the cell even though they are too large to move out through its plasma membrane

REPRODUCTION OF CELLS
Deoxyribonucleic acid (DNA)

A Structure—(see Figure 3-13)

B Relation to genes

1 One gene is a segment of a DNA molecule, consisting of a chain of about 1,000 pairs of nucleotides joined in a precise sequence

2 DNA of one human chromosome is made up of about 175,000 genes according to one cytologist's estimate

C Replication—DNA of each chromosome duplicates itself during interphase of mitosis, thereby forming two chromatids out of each chromosome

D Function—DNA replication makes possible cellular heredity

Recombinant DNA

A Definition—joining together of hereditary material, often from different species, into new, biologically functional combinations

B Techniques

1 Bacterial DNA—plasmid rings—are split

2 "Restriction enzymes" used to open plasmid rings and to cut segments of DNA from other sources

3 Foreign DNA then annealed into open plasmid ring

4 Recombinant plasmid then reinserted into a bacterial "host"

5 Each time synthetic host organism divides, so does recombinant DNA

Outline Summary—cont'd

MITOSIS (Figures 3-15 and 3-16)

A Definition—process by which somatic cells reproduce

B Phases—interphase, prophase, metaphase, anaphase, and telophase; see Table 3-4 for summary of events during each phase

C Function—transmits to daughter cells the same (diploid) number of chromosomes composed of same genes as present in parent cell and thereby makes possible inheritance of parent cell's traits

MEIOSIS

A Definition—process by which primitive sex cells reproduce in the process of becoming mature sex cells (gametes) with haploid number of chromosomes

B Stages

1 Meiotic division I—consists of interphase, prophase I, metaphase I, anaphase I, and telophase I; primary sex cell reproduces to form two secondary sex cells containing haploid number of chromosomes, that is, 23

2 Meiotic division II—consists of same phases as meiotic division I—4 sperm or 1 ovum result from this division

C Function—formation of gametes with haploid number of chromosomes

SPERMATOGENESIS

Process by which spermatogonia present at birth in seminiferous tubules undergo meiosis to become mature sperm; spermatogenesis begins at puberty and usually continues throughout a man's life; see Figure 3-19

OOGENESIS

Process by which oogonia become mature ova; oogonia in ovaries of fetus reproduce by mitosis to form primary oocytes at prophase I stage; beginning at puberty, and recurring about once a month for the next 30 or 40 years, remaining steps of oogenesis take place as shown in Figure 3-20; ovum contains 23 chromosomes, one of which is an X chromosome; sperm also contain 23 chromosomes but one of them is either an X or a Y sex chromosome

Review Questions

1 What is the term for one thousandth of a meter? One millionth of a meter?

2 What terms do the following abbreviations stand for? cm, mm, μm, nm, Å?

3 One inch equals approximately how many cm? Å? nm? μm?

4 What metric unit is used to measure cell size?

5 What is the range in human cell diameters?

6 Describe the location, molecular structure, and width of the plasma membrane.

7 Explain the communication function of the plasma membrane, its transportion function, and its identification function.

8 Describe briefly the structure and function of these organelles: endoplasmic reticulum, ribosomes, Golgi apparatus, mitochondria, centrioles, lysosomes.

9 What is the microtrabecular lattice? What functions does it serve?

10 Describe briefly the functions of the nucleus and the nucleoli.

11 Contrast passive and active transport mechanisms.

12 Define the terms *diffusion, net diffusion, facilitated diffusion,* and *dialysis.*

13 State the principle about the direction of net diffusion.

14 State the principle about the direction of active transport.

15 State the principle about the direction in which net osmosis occurs.

16 Differentiate between actual osmotic pressure and potential osmotic pressure.

17 What factor directly determines the potential osmotic pressure of a solution?

18 Explain the terms *isotonic, hypotonic,* and *hypertonic.*

19 State the principle about the solution that develops an osmotic pressure, given appropriate conditions.

20 Explain the processes of endocytosis and exocytosis.

21 Differentiate between the processes of phagocytosis and pinocytosis.

22 Describe the size and shape of a DNA molecule.

Review Questions—cont'd

23 Where are DNA molecules located?
24 Explain the steps of DNA replication.
25 When does DNA replication occur?
26 Watson called DNA "the most golden of all molecules," primarily because of its function. What is it?
27 Define the term *recombinant DNA*.
28 What is a "plasmid"?
29 What are "restriction enzymes" and how are they used in recombinant DNA research?
30 Discuss the possible applications of recombinant DNA techniques in human medicine.

31 Define mitosis and name its phases.
32 Define meiosis and name its stages.
33 When does the reduction of chromosomes from the diploid to haploid number take place? Meiosis is a part of what larger male process? Of what larger female process?
34 Define spermatogenesis. When and where does it occur?
35 Define oogenesis. When and where does it occur?

4 TISSUES

OBJECTIVES

After you have completed this chapter, you should be able to:

1 Define the term *tissue* and discuss the embryonic development of tissues from the primary germ layers.

2 List the four major categories of tissues and discuss the basic function of each type.

3 List and discuss some important structural and functional generalizations that apply to epithelium as a principal tissue type.

4 Classify membranous epithelium using both cell shape and cell layers as criteria and list the specific epithelial types that result.

5 Discuss each type of simple and stratified membranous epithelium in terms of its structure, function, and location in the body.

6 Discuss glandular epithelium and compare endocrine and exocrine glands in terms of generalized function.

7 Discuss the structural classification of exocrine glands.

8 Explain how merocrine, apocrine, and holocrine glands differ in their method of secretion.

9 Discuss and compare the two major types of epithelial membranes in terms of structure, function, and location in the body.

10 List the major types of connective tissues and contrast important structural and functional differences.

11 Discuss the major types of connective tissue fibers, cells, and matrix in terms of structure and function.

12 Compare bone and cartilage in terms of generalized function, cell types, organizational structure, and blood supply.

13 Discuss blood as a tissue.

14 Compare the three major types of muscle tissue in terms of structure and function.

15 Compare characteristics of neurons and neuroglia in terms of nervous system function.

16 Discuss the four cardinal signs of inflammation.

Tissues are organizations or communities of similar cells that are surrounded and often embedded in a nonliving intercellular material called **matrix.** Each tissue specializes in performing at least one unique function that is essential for healthy survival of the body. Tissues differ as to the size, shape, and arrangement of their cells and as to the amount and kind of intercellular matrix between the cells. It is the unique and specialized nature of the matrix in bone and cartilage, for example, that contributes to the strength and resiliency of the body. Some tissues contain almost no intercellular material. Others consist predominantly of it. Some intercellular substance may be rigid or contain fibers, some is unformed gel, and some is fluid.

PRINCIPAL TYPES OF TISSUE

Although a number of subtypes are present in the body, all tissues can be classified by their structure and function into four principal types:

1 **Epithelial tissue** covers and protects the body surface, lines body cavities, specializes in moving substances into and out of the blood (secretion and absorption), and forms many glands.

2 **Connective tissue** is specialized to support the body and its parts, to connect and hold them together, to transport substances through the body, and to protect it from foreign invaders. The cells in connective tissue are often relatively far apart and separated by large quantities of nonliving matrix.

3 **Muscle tissue** produces movement; it moves the body and its parts. Muscle cells are specialized for contractility and produce movement by the shortening of complex contractile proteins found in the cytoplasm.

4 **Nervous tissue** is the most complex tissue in the body. It specializes in communication between the various parts of the body and in integration of their activities. This tissue has as its major function the generation of complex messages for the coordination of body functions.

KEY TERMS

Adipose (AD-i-pose)

Chondrocyte (KON-dro-site)

Collagen (KOL-ah-jen)

Endocrine (EN-do-krin)

Epithelium (ep-i-THEE-le-um)

Exocrine (EK-so-krin)

Fibroblast (FIGH-bro-blast)

Hemopoiesis (HE-mo-poi-E-sis)

Histology (his-TOL-o-je)

Matrix (MAE-triks)

Neuron (NU-ron)

Osteocyte (OS-te-o-site)

Reticular (re-TIK-u-lar)

Squamous (SKWA-mus)

Vascular (VAS-ku-lar)

Histology or microscopic anatomy deals with the study of tissues. As noted in Chapter 1, the study of anatomy is often subdivided into several parts or disciplines based on methods by which the body parts are revealed or by the size of the structures studied. In gross anatomy, dissection is used to expose and separate body parts that can be studied with the naked eye. The study of histology requires the use of a microscope and the proper preparation of tissue specimens for examination.

EMBRYONIC DEVELOPMENT OF TISSUES

The four major tissues of the body appear early in the embryonic period of development (first 2 months after conception). After fertilization has occurred, repeated cell divisions soon convert the single-celled zygote into a hollow ball of cells called a **blastocyst.** The blastocyst implants in the uterus, and within 2 weeks the cells move and regroup in an orderly way into three **primary germ layers** called **endoderm, mesoderm,** and **ectoderm.** The process by which blastocyst cells move and then differentiate into the three primary germ layers is called **gastrulation.** During this process the cells in each germ layer become increasingly differentiated to form specific tissues and eventually give rise to the structures listed in Table 4-1. In summary, some epithelial tissues develop from each of the primary germ layers, whereas connective and muscle tissues arise from mesoderm, and nerve tissue develops from ectoderm. The study of how the primary germ layers develop into the different kinds of tissues is called **histogenesis.** Chapter 28 provides additional details of human development, including a discussion of differentiation of organs and body systems.

EPITHELIAL TISSUE

TYPES AND LOCATIONS

Epithelial tissue, or **epithelium,** is often subdivided into two types: (1) **membranous** (covering or lining) epithelium and (2) **glandular** epithelium. Membranous epithelium covers the body and some of its parts and lines the serous cavities (pleural, pericardial, and peritoneal), the blood and lymphatic vessels, and the respiratory, digestive, and genitourinary tracts. Glandular epithelium is grouped in solid cords or specialized follicles that form the secretory units of endocrine and exocrine glands.

Table 4-1 Structures derived from primary germ layers

Germ layers	Structures
Ectoderm	Epithelium (epidermis) of skin Lining of mouth, anus, nostrils Sweat and sebaceous glands Epidermal derivatives (hair, enamel of teeth) Nervous system (brain and spinal cord) Epithelial (sensory) parts of eye, nose, ear
Endoderm	Epithelium (lining) of digestive and respiratory systems Secretory parts of liver and pancreas Urinary bladder Epithelial lining of urethra Thyroid, parathyroid, thymus
Mesoderm	Muscles Skeleton (bones and cartilage) Blood Epithelial lining of blood vessels Dermis of skin and dentin of teeth Organs (except lining) of excretory and reproductive systems Connective tissue

FUNCTIONS

Epithelial tissues have a widespread distribution throughout the body and serve several important functions:

1 **Protection.** Generalized protection is the most important function of membranous epithelium. It is the relatively tough and impermeable epithelial covering of the skin that serves to protect the body not only from mechanical and chemical injury but also from invading bacteria and other disease-causing microorganisms.

2 **Sensory functions.** Epithelial derivatives specialized for sensory functions are found in the skin, nose, eye, and ear.

3 **Secretion.** Glandular epithelium is specialized for secretory activity. Secretory products include hormones, mucus, digestive juices, and sweat.

4 **Absorption.** The lining epithelium of the gut and respiratory tracts allows for the absorption of nutrients from the gut and the exchange of respiratory gases between air in the lungs and the blood.

5 **Excretion.** It is the specialized epithelial lining of kidney tubules that makes the excretion and concentration of excretory products in the urine possible.

GENERALIZATIONS ABOUT EPITHELIAL TISSUE

1 Most epithelial tissues are characterized by extremely limited amounts of intercellular or matrix material. This explains their characteristic appearance, when viewed under a light microscope, of a continuous sheet of cells packed tightly together. With the electron microscope, however, narrow spaces—reportedly not quite one millionth of an inch (20 nm) wide—can be seen around the cells. These spaces, like other intercellular spaces, contain interstitial fluid.

2 Sheets of epithelial cells compose the surface layer of skin and of mucous and serous membranes. The epithelial tissue attaches to an underlying layer of connective tissue by means of a thin noncellular layer of adhesive, permeable material called the **basement membrane.** Both epithelial and connective tissue cells synthesize the basement membrane, which is made up of glycoprotein material secreted by the epithelial components and a fine mesh of fibers produced by the connective tissue cells. Histologists refer to the glycoprotein material secreted by the epithelial cells as the **basal lamina** and to the connective tissue fibers as the **reticular lamina.** The union of basal and reticular lamina forms the basement membrane.

3 Epithelial tissues contain no blood vessels. As a result, epithelium is said to be **avascular** (*a,* "without"; *vascular,* "blood"). Hence oxygen and nutrients must diffuse from capillaries in the underlying connective tissue through the permeable basement membrane to reach epithelial cells.

4 At intervals between adjacent epithelial cells, their plasma membranes are modified so as to hold the cells together. These specialized intercellular structures, such as **desmosomes** and **tight junctions,** are described in Chapter 3.

5 Epithelial cells can reproduce themselves. They frequently go through the process of cell division (mitosis). Since epithelial cells in many locations meet with considerable wear and tear, this fact has great practical importance. It means, for example, that new cells can replace old or destroyed epithelial cells in the skin or in the lining of the gut or respiratory tract.

CLASSIFICATION OF EPITHELIAL TISSUE

MEMBRANOUS (COVERING OR LINING) EPITHELIUM
Classification Based on Cell Shape

The shape of membranous epithelial cells may be used for classification purposes. Four cell shapes called **squamous, columnar, pseudostratified columnar,** and **cuboidal** are used in this classification

Table 4-2 Classification of membranous epithelium

Cell layers	Shape
Simple	Simple squamous
	Simple cuboidal
	Simple columnar
	Pseudostratified columnar
Stratified	Stratified squamous
	Keratinized
	Nonkeratinized
	Stratified cuboidal
	Stratified columnar
	Transitional

scheme. Squamous (Latin, "scaly") cells are flat and platelike. Cuboidal cells, as the name implies, are cube-shaped and have more cytoplasm than the scalelike squamous cells. Columnar epithelial cells are higher than they are wide and appear narrow and cylindrical. Pseudostratified columnar epithelium has only one layer of oddly shaped columnar cells. Although each cell touches the basement membrane, the tops of some pseudostratified cells do not fully extend to the surface of the membrane. The result is a false (pseudo) appearance of layering or stratification when only a single layer of cells is present.

Classification Based on Layers of Cells

In most cases the location and function of membranous epithelium determines whether or not its cells will be stacked and layered or arranged in a sheet one-cell–layer thick. Arrangement of epithelial cells in a single layer is called **simple epithelium.** If epithelial cells are layered one on another, the tissue is called **stratified epithelium. Transitional epithelium** (described in the following) is a unique arrangement of differing cell shapes in a stratified or layered epithelial sheet.

If membranous or covering epithelium is classified both by the shape and layering of its cells, the categories or subdivisions listed in Table 4-2 are possible. Each type is described in the paragraphs that follow, and selected examples are illustrated in Figures 4-1 to 4-7.

Simple Epithelium

SIMPLE SQUAMOUS EPITHELIUM. Simple squamous epithelium consists of only one layer of flat, scalelike cells (Figure 4-1). Consequently, substances can readily diffuse or filter through this type of tissue. The microscopic air sacs (alveoli) of the lungs, for

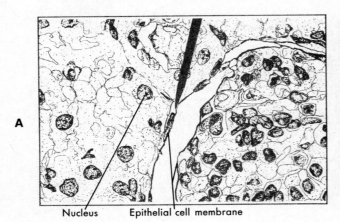

A

Nucleus   Epithelial cell membrane

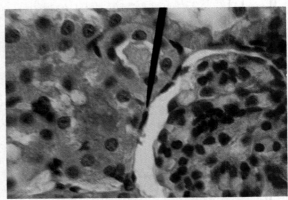

B

Figure 4-1 Simple squamous epithelial cell in parietal wall of Bowman's capsule in kidney. **A,** Sketch of photomicrograph. Arrow is touching a typical scale-like and flattened squamous epithelial cell. **B,** Lining function of these cells is seen in the surface of an open area in the tissue. (×140.)

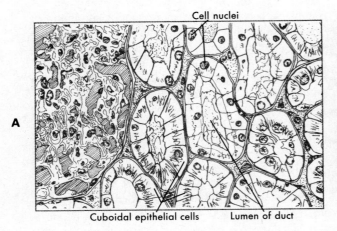

Cell nuclei

A

Cuboidal epithelial cells   Lumen of duct

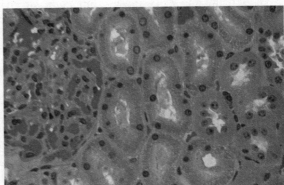

B

Figure 4-2 Simple cuboidal epithelium of the distal convoluted tubule of the kidney. **A,** Sketch of photomicrograph. The single layer of cuboidal-shaped cells touches a basement membrane. **B,** Note the cuboidal cells that enclose the duct openings. (×140.)

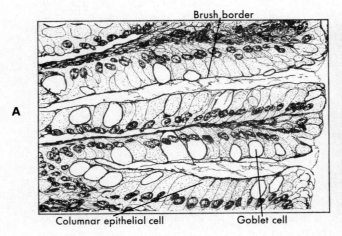

Brush border

A

Columnar epithelial cell   Goblet cell

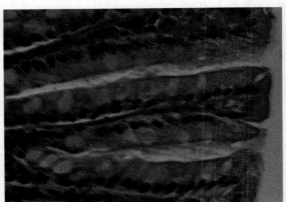

B

Figure 4-3 Simple columnar epithelium of the colon. **A,** Sketch of photomicrograph. Villi show a single layer of columnar cells covering supportive connective tissue. **B,** Note the mucus-producing goblet cells in this section of intestinal mucosa. (×140)

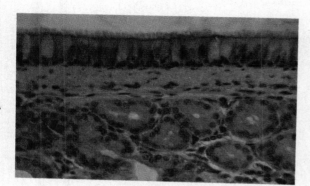

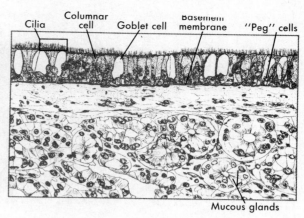

Figure 4-4 Pseudostratified ciliated epithelium of the trachea. **A,** Note that each irregularly shaped columnar cell touches the underlying basement membrane. (×140.) **B,** Sketch of photomicrograph. Placement of cell nuclei at irregular levels in the cells gives a false (pseudo) impression of stratification.

example, are composed of this kind of tissue, as are the linings of blood and lymphatic vessels and the surfaces of the pleura, pericardium, and peritoneum. (Blood and lymphatic vessel linings are called **endothelium,** and the surfaces of the pleura, pericardium, and peritoneum are called **mesothelium.** Some histologists classify these as connective tissue.)

SIMPLE CUBOIDAL EPITHELIUM. Simple cuboidal epithelium is composed of one layer of cuboidal-shaped cells resting on a basement membrane (Figure 4-2). This type of epithelium is seen in many types of glands and their ducts. Sections of thyroid gland are often used to demonstrate this type of epithelium.

SIMPLE COLUMNAR EPITHELIUM. Simple columnar epithelium composes the surface of the mucous membrane that lines the stomach, intestine, uterus, uterine tubes, and parts of the respiratory tract (Figure 4-3). It consists of a single layer of cells, many of which have a modified structure. Three common modifications are goblet cells, cilia, and microvilli. In the intestine the plasma membranes of many columnar cells extend out in hundreds and hundreds of microscopic fingerlike projections called **microvilli.** By greatly increasing the surface area of the intestinal mucosa, microvilli make it especially well suited for absorbing nutrients and fluids from the intestine.

PSEUDOSTRATIFIED COLUMNAR EPITHELIUM. Pseudostratified columnar epithelium is found lining the air passages of the respiratory system and certain segments of the male reproductive system such as the urethra

(Figure 4-4). Although appearing to be stratified, only a single layer of irregularly shaped columnar cells touches the basement membrane. The cells are of differing heights, and many are not tall enough to reach the upper surface of the epithelial sheet. This fact, coupled with placement of cell nuclei at odd and irregular levels in the cells, gives a false (pseudo) impression of stratification. Mucus-secreting goblet cells are numerous and cilia are present. In the respiratory system air passages, uniform motion of the cilia causes a thin layer of tacky mucus to move over the free surface of the epithelium. As a result, dust particles in the air are trapped and moved toward the mouth and away from the delicate lung tissues.

Stratified Epithelium

STRATIFIED SQUAMOUS (KERATINIZED) EPITHELIUM. Stratified squamous epithelium is characterized by multiple layers of cells with typical flattened squamous cells at the free or outer surface of the sheet (Figure 4-5). The presence of keratin in these cells contributes to the protective qualities of skin covering the body surface. Details of the histology of this type of epithelium are presented in Chapter 5.

STRATIFIED SQUAMOUS (NONKERATINIZED) EPITHELIUM. Nonkeratinized stratified squamous epithelium is found lining the vagina, mouth, and esophagus (Figure 4-6). Its free surface is moist, and the outer epithelial cells, unlike those found in the skin, do not contain keratin. This type of epithelium serves a protective function.

STRATIFIED CUBOIDAL EPITHELIUM. The cuboidal variety of stratified epithelium also serves a protective func-

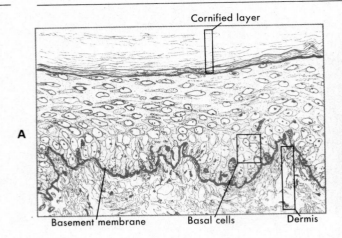

Cornified layer

Basement membrane — Basal cells — Dermis

A

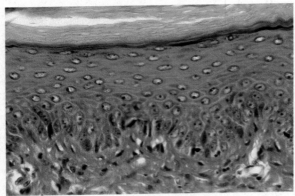

B

Figure 4-5 Cornified (keratinized) stratified squamous epithelium of the skin. **A,** Sketch of photomicrograph. Cells become progressively flattened and scale-like as they approach the surface and are lost. **B,** The outer surface of this epithelial sheet shows many flattened cells, which have lost their nuclei. (× 140.)

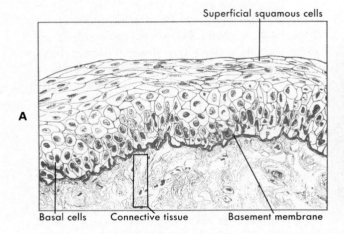

Superficial squamous cells

A

Basal cells — Connective tissue — Basement membrane

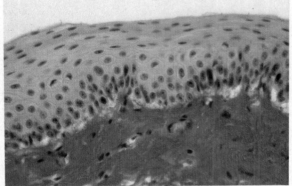

B

Figure 4-6 Stratified squamous (nonkeratinized) epithelium of the vagina. **A,** Sketch of photomicrograph. Each cell in the layer is flattened near the surface and attached to the sheet. No flaking of dead cells from the surface occurs. **B,** Cells all have nuclei. (× 140.)

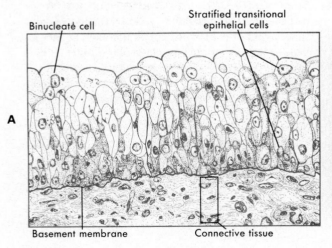

Binucleate cell

Stratified transitional epithelial cells

A

Basement membrane — Connective tissue

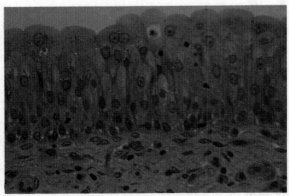

B

Figure 4-7 Stratified transitional epithelium of the urinary bladder. **A,** Sketch of photomicrograph. Cell shape is variable from cuboidal to squamous. **B,** Several layers of cells are present. Intermediate and surface cells do not touch basement membrane. (× 140.)

tion. Typically, two or more rows of low cuboidal-shaped cells are arranged randomly over a basement membrane. Stratified cuboidal epithelium can be located in the sweat gland ducts, pharynx, and over parts of the epiglottis.

STRATIFIED COLUMNAR EPITHELIUM. Although this protective epithelium has multiple layers of columnar cells, only the most superficial cells are truly columnar in appearance. Epithelium of this type is rare. It is located in segments of the male urethra and in the mucus layer near the anus.

STRATIFIED TRANSITIONAL EPITHELIUM. Stratified transitional epithelium is typically found in body areas, such as the wall of the urinary bladder, that are subjected to stress and tension changes (Figure 4-7). In many instances 10 or more layers of cuboidal-like cells of varying shape are present in the absence of stretching or tension. As pressure increases, the epithelial sheet is expanded, the number of observable cell layers will decrease, and cell shape will change from cuboidal to squamous in appearance. This ability of transitional epithelium to stretch protects the bladder wall and other distensible structures that it lines from tearing under stretch-type pressures.

GLANDULAR EPITHELIUM

Epithelium of the glandular type is specialized for secretory activity. Regardless of the secretory product produced, glandular activity is dependent on complex and highly regulated cellular activities requiring the expenditure of stored energy.

Unlike the single or layered cells of membranous epithelium that are typically found in protective coverings or linings, glandular epithelial cells may function singly as **unicellular glands** or in clusters, solid cords, or specialized follicles as **multicellular glands.** Glandular secretions may be discharged into ducts, directly into the blood, into the lumen of hollow visceral structures, or onto the body surface.

All glands in the body can be classified as either **exocrine** or **endocrine glands.** Exocrine glands, by definition, discharge their secretion products into ducts. The salivary glands are typical exocrine glands. The secretion product (saliva) is produced in the gland and then discharged into a duct that transports it to the mouth. Endocrine glands are often called **ductless glands** because they discharge their secretion products (hormones) directly into the blood or interstitial fluid. The pituitary, thyroid, and adrenal glands are typical endocrine glands.

Structural Classification of Exocrine Glands

Multicellular exocrine glands are most often classified using the complexity (branching) of their duct

Table 4-3 Structural classification of glandular epithelium

Types of exocrine glands	Structural classification
Unicellular	
Multicellular	Simple
	Simple tubular
	Simple coiled tubular
	Simple branched tubular
	Simple alveolar
	Simple branched alveolar
	Compound
	Compound tubular
	Compound alveolar
	Compound tubuloalveolar

systems and the size and form of their secretory units as distinguishing characteristics. Table 4-3 lists exocrine glands generally found when a structural classification is made. Each multicellular variety is illustrated in Figures 4-8 to 4-10.

1 **Simple tubular glands**—One duct; tube-shaped secretory portion (e.g., intestinal glands)
2 **Simple coiled tubular glands**—One duct; coiled, tube-shaped secretory portion (example, sweat glands)
3 **Simple branched tubular glands**—One duct branches into two tube-shaped secretory portions (e.g., gastric glands)
4 **Simple alveolar glands**—One duct; sac-shaped secretory portion (e.g., sebaceous glands)
5 **Simple branched alveolar glands**—One duct; branching, sac-shaped secretory portion (e.g., sebaceous glands)
6 **Compound tubular glands**—More than one duct; tube-shaped secretory portion (e.g., male sex glands, or testes)
7 **Compound alveolar glands**—More than one duct; sac-shaped secretory portion (e.g., mammary glands)
8 **Compound tubuloalveolar glands**—More than one duct; tube-shaped and sac-shaped secretory portions (e.g., salivary glands)

Functional Classification of Exocrine Glands

In addition to structural differences, exocrine glands also differ in the method by which they discharge their secretion products from the cell. Using these functional criteria, three types of exocrine glands can be identified (Figure 4-11):

1 **Apocrine**
2 **Holocrine**
3 **Merocrine**

Figure 4-8 Tubular exocrine glands. **A, B**, and **C** represent simple or single duct tubular glands. **D**, A compound gland, has more than one duct.

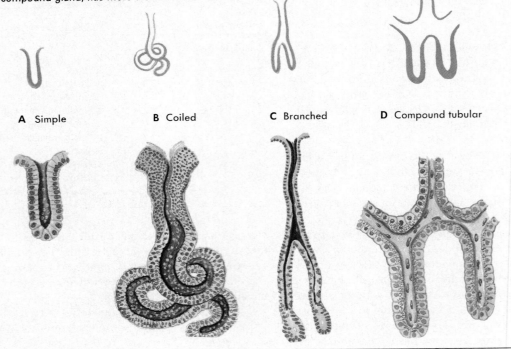

A Simple   **B** Coiled   **C** Branched   **D** Compound tubular

Figure 4-9 Alveolar exocrine glands—simple, that is, with one duct, and compound, with several ducts.

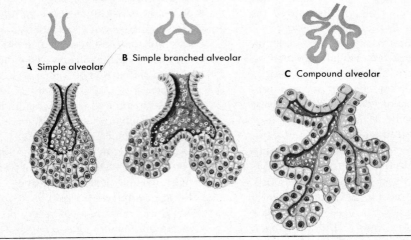

A Simple alveolar   **B** Simple branched alveolar   **C** Compound alveolar

Figure 4-10 Compound tubuloalveolar gland—more than one duct with both tube-shaped and sac-shaped secretory portions.

Figure 4-11 Classification of exocrine glands by method of secretion.

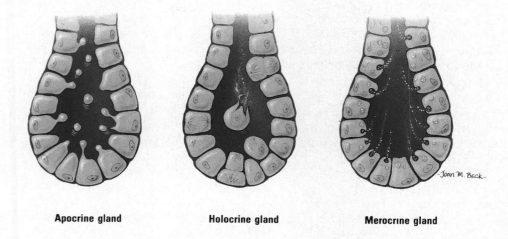

Apocrine gland   Holocrine gland   Merocrine gland

Apocrine glands collect their secretory products near the apex or tip of the cell and then release it into a duct by pinching off the distended end. This process results in some loss of cytoplasm and damage to the cell. Recovery and repair of cells is rapid, however, and continued secretion occurs. The milk-producing mammary glands are examples of apocrine-type secretion.

Holocrine glands—such as the sebaceous glands that produce oil to lubricate the skin—collect their secretory product inside the cell and then rupture completely to release it. These cells literally self-destruct to complete their function.

Merocrine glands discharge their secretion product directly through the cell or plasma membrane. This discharge process is completed without injury to the cell wall and without loss of cytoplasm. Only the secretion product passes from the glandular cell into the duct. Most secretory cells are of this type. The salivary glands are good examples of merocrine-type exocrine glands.

Pleurisy is a pathological and very painful condition characterized by inflammation of the serous membranes (pleura) that line the chest cavity and cover the lungs. Pain is caused by irritation and friction as the lungs rub against the walls of the chest cavity. In severe cases, the inflamed surfaces of the pleura fuse, and permanent adhesions may develop.

MEMBRANES

A membrane can be defined as a thin layer of tissue that covers a surface, lines a cavity, or divides a space or organ. The most common types of membranes found in the body are called **epithelial membranes** because they are formed by the union of epithelial and connective tissue components. The **cutaneous membrane** or skin constitutes the body's largest organ and is described in detail in Chapter 5. In addition to the skin, two other major types of epithelial membranes are found in the body—**serous membranes** and **mucous membranes.**

Serous membranes line closed body cavities, such as the abdomen and chest, and cover the visceral structures found in those cavities. The previously discussed pleura, peritoneum, and pericardial membranes—which are formed by sheets of simple squamous epithelium and extremely thin underlying connective tissue elements—are good examples of serous membranes. Serous-type membranes secrete a clear, waterlike lubricating fluid that serves a very important protective function. It permits visceral structures to move in body cavities with very little friction.

Mucous membranes line and protect organs that open to the exterior of the body. The ducts and passageways of the digestive and respiratory systems, for example, have a lining of mucosa or mucous membrane. Mucus produced by goblet cells helps keep the passageways moist and lubricated.

EXFOLIATIVE CYTOLOGY

Exfoliative cytology refers to the study of cells that have been shed from the body. Cancer cells are very different from normal cells in their appearance. For example, they tend to be large and have large lobulated nuclei with multiple nucleoli. Cancer cells also show a loss of distinctive tissue characteristics—a condition called **anaplasia.** As a result of these differences in appearance, pathologists can recognize the signs of cancer even in individual cells.

In addition to distinctive differences in appearance, cancer cells cling together less firmly than do normal cells. They have fewer desmosomes and tight junctions between them and are therefore less cohesive. As a result, cancer cells easily detach; they **exfoliate.**

In almost all types of cancer there is a rapid increase in the number of cells being produced. This uncontrolled proliferation of cells is called **hyperplasia.** The fact that most cancers produce large numbers of easily identifiable cells that tend to exfoliate has led to the development of exfoliative cytology as a method for cancer diagnosis. The **Pap smear** is the most common example of this technique. It is named after George Papanicolaou—the American scientist who developed it. Before the development of exfoliative cytology techniques, pathologists were required to surgically remove a section of tissue in order to examine cells for signs of cancer. Such a procedure is called a **biopsy.**

Epithelial tissue covers all the external body surfaces, lines the hollow organs, and forms the inner lining of the body cavities. As a result, epithelial membranes always have a free surface that is exposed—either to the outside or to an open space or body cavity internally. Therefore cells exfoliated from epithelial surfaces can be collected quite easily and examined. Although the Pap test is most often a smear of cells taken from the outer cervix or vagina, cells from other body secretions or from the body surface can also be examined by a Pap smear. Pap smears are routinely used to examine cells found in secretions taken from body orifices, the bronchial tubes, reproductive ducts, and the digestive tract.

Unlike biopsy techniques, which require the surgical removal of tissue, Pap smears and other exfoliative cytology methods rely instead on examination of cells that have been shed from the body. They are said to be **noninvasive** because surgical procedures are not required.

The **tympanic membrane** or eardrum is a most unusual epithelial membrane. The outer surface consists of a layer of stratified squamous epithelium that is continuous with the epithelium lining the external ear canal. The internal surface, which faces the middle ear chamber, is a layer of simple squamous epithelium. Between the two epidermal layers is a sheet of fibrous connective tissue elements.

CONNECTIVE TISSUE

Connective tissue is the most abundant tissue in the body. It exists in more varied forms than the other three basic tissues. Delicate tissue-paper webs, strong tough cords, rigid bones, and a fluid, namely, blood—all are forms of connective tissue.

TYPES, FUNCTIONS, AND CHARACTERISTICS
One scheme of classification lists the following main types of connective tissue (Table 4-4):

1 **Loose, ordinary (areolar)**
2 **Adipose**
3 **Reticular**
4 **Dense fibrous**
5 **Bone**
6 **Cartilage**
7 **Blood**

Connective tissue connects, supports, transports, and defends. It connects tissues to each other, for example. It also connects muscles to muscles, muscles to bones, and bones to bones. It forms a supporting framework for the body as a whole and for its organs individually. One kind of connective tissue—blood—transports a large array of sub-

Table 4-4 Tissues

Tissue	Location	Function
EPITHELIAL		
Simple squamous	Alveoli of lungs	Absorption by diffusion of respiratory gases between alveolar air and blood
	Lining of blood and lymphatic vessels (called *endothelium;* classed as connective tissue by some histologists)	Absorption by diffusion, filtration, osmosis
	Surface layer of pleura, pericardium, peritoneum (called *mesothelium;* classed as connective tissue by some histologists)	Absorption by diffusion and osmosis; also, secretion
Stratified squamous	Surface of mucous membrane lining mouth, esophagus, and vagina	Protection
	Surface of skin (epidermis)	Protection
Transitional	Surface of mucous membrane lining urinary bladder and ureters	Permits stretching
Simple columnar	Surface layer of mucous lining of stomach, intestines, and part of respiratory tract	Protection; secretion; absorption; moving of mucus (by ciliated columnar epithelium)
Pseudostratified	Surface of mucous membrane lining trachea, large bronchi, nasal mucosa, and parts of male reproductive tract (epididymis and vas deferens); lines large ducts of some glands (e.g., parotid)	Protection
Glandular	Glands	Secretion
CONNECTIVE (most widely distributed of all tissues)		
Loose, ordinary (areolar)	Between other tissues and organs	Connection
	Superficial fascia	Connection
Adipose (fat)	Under skin	Protection
	Padding at various points	Insulation Support Reserve food
Dense fibrous	Tendons Ligaments Aponeuroses Deep fascia Dermis Scars Capsule of kidney, etc.	Flexible but strong connection
Bone	Skeleton	Support Protection

Continued.

Table 4-4 Tissues—cont'd

Tissue	Location	Function
CONNECTIVE—cont'd		
Cartilage		
Hyaline	Part of nasal septum Covering articular surfaces of bones Larynx Rings in trachea and bronchi	Firm but flexible support
Fibrous	Disks between vertebrae Symphysis pubis	
Elastic	External ear Eustachian tube	
Hemopoietic		
Myeloid (red bone marrow)	Marrow spaces of bones	Formation of red blood cells, granular leukocytes, platelets; also reticuloendothelial cells and some other connective tissue cells
Lymphatic	Lymph nodes Spleen Tonsils and adenoids Thymus gland	Formation of lymphocytes and monocytes; also plasma cells and some other connective tissue cells
Blood	In blood vessels	Transportation Protection
MUSCLE		
Skeletal (striated voluntary)	Muscles that attach to bones Extrinsic eyeball muscles Upper third of esophagus	Movement of bones Eye movements First part of swallowing
Visceral (nonstriated, involuntary, or smooth)	In walls of tubular viscera of digestive, respiratory, and genitourinary tracts In walls of blood vessels and large lymphatic vessels In ducts of glands Intrinsic eye muscles (iris and ciliary body) Arrector muscles of hairs	Movement of substances along respective tracts Change diameter of blood vessels, thereby aiding in regulation of blood pressure Movement of substances along ducts Change diameter of pupils and shape of lens Erection of hairs (gooseflesh)
Cardiac (striated involuntary)	Wall of heart	Contraction of heart
NERVOUS	Brain Spinal cord Nerves	Irritability; conduction

stances between parts of the body. And finally, several kinds of connective tissue cells defend us against microbes and other invaders.

Connective tissue consists predominantly of intercellular material called **matrix** or **ground substance.** Embedded in the matrix are relatively few cells and varying numbers and kinds of fibers. The qualities of the matrix and fibers largely determine the qualities of each type of connective tissue. The matrix of blood, for example, is a fluid (plasma). It contains numerous blood cells but no fibers, except when it coagulates. Some connective tissues have the consistency of a soft gel, some are firm but flexible, some are hard and rigid, some are tough, others are delicate—and in each case it is their matrix and fibers that make them so.

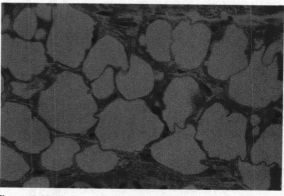

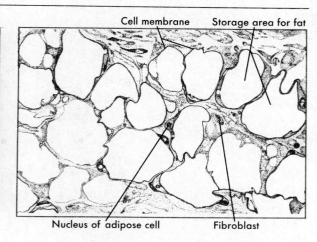

Cell membrane   Storage area for fat

Nucleus of adipose cell   Fibroblast

Figure 4-12 Adipose (fat) tissue. **A**, Human adipose tissue. The fats (lipids) have been dissolved during processing. **B**, Sketch of the photomicrograph.

A connective tissue's matrix contains one or more of the following kinds of fibers: collagenous (or white), reticular, or elastic. Fibroblasts and some other cells form these fibers. Collagenous fibers are tough and strong, reticular fibers are delicate, and elastic fibers are extensible and elastic. Collagenous or white fibers often occur in bundles—an arrangement that provides great tensile strength. Reticular fibers, in contrast, occur in networks and, although delicate, support small structures such as capillaries and nerve fibers. Collagen in its hydrated form you know as gelatin. Of all the hundreds of different protein compounds in the body, collagen is the most abundant. Biologists now estimate that it constitutes somewhat over one fourth of all the protein in the body. And interestingly, one of the most basic factors in the aging process, according to some researchers, is the change in the molecular structure of collagen that occurs gradually with the passage of the years.

Loose, Ordinary Connective Tissue (Areolar)

First, a few words of explanation about the name "loose, ordinary connective tissue." It is loose because it is stretchable and ordinary because it is one of the most widely distributed of all tissues. It is common and ordinary, not special like some kinds of connective tissue (e.g., bone and cartilage) that help form comparatively few structures. "Areolar" was the early name for the loose, ordinary connective tissue that connects many adjacent structures of the body. It acts like a glue spread between them—but an elastic glue that permits movement. The word **areolar** means "like a small space" and refers to the bubbles that appear as areolar tissue is pulled apart during dissection.

The matrix of areolar tissue is a soft, viscous gel mainly because it contains hyaluronic acid. An enzyme, hyaluronidase, can change the matrix from its viscous gel state to a watery consistency. Physicians have made use of this knowledge for a number of years. They frequently inject a commercial preparation of hyaluronidase with drugs or fluids. By decreasing the viscosity of intercellular material, the enzyme not only hastens diffusion and absorption of the injected material, but it also lessens tissue tension and pain. Some bacteria, notably pneumococci and streptococci, spread through connective tissues by secreting hyaluronidase.

The matrix of areolar tissue contains numerous fibers and cells. Typically, there are many interwoven collagenous and elastic fibers and a half dozen or so kinds of cells. **Fibroblasts** are usually present in the greatest numbers in areolar tissue, and **macrophages** are second. Fibroblasts synthesize both the gel-like ground substance and the fibers present in it. Macrophages (also known by several other names, for example, "histiocytes" and "resting wandering cells") carry on phagocytosis and so are classified as phagocytes. Phagocytosis is part of the body's vital complex of defense mechanisms. Other kinds of cells found in loose, ordinary connective tissue are mast cells, some wandering white blood cells (leukocytes), an occasional fat cell, and some plasma cells.

Adipose Tissue

Adipose tissue differs from loose, ordinary connective tissue mainly in that it contains predominantly fat cells and many fewer fibroblasts, macrophages, and mast cells (Figure 4-12, *A* and *B*). Adipose tissue forms protective pads around the kidneys and var-

Figure 4-13 Fat storage areas. Note the different distribution in the male and female bodies.

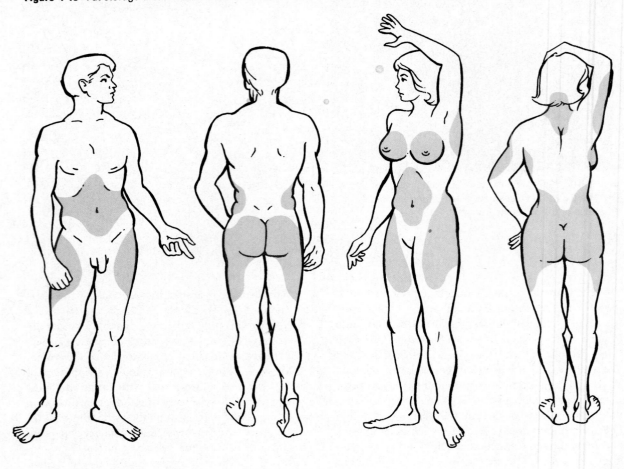

Obesity refers to an excess of body fat. Obese individuals usually have a body weight over 20% above ideal or desirable standards. Fat reserves in these individuals are doubled. Unfortunately, such a condition exists in millions of Americans. Excess fat storage of this magnitude greatly increases the risk for such diseases as diabetes, nephritis, and coronary heart disease. The relationship between heart disease and obesity is understandable when one learns that for every pound of added fat the heart must pump blood through an additional two-thirds mile of blood vessels. Although fat storage can only occur if food (caloric) intake exceeds expenditure, we now recognize obesity as a condition of multiple origins that requires multifaceted treatment.

ious other structures. It also serves two other functions—it constitutes a storage depot for excess food, and it acts as an insulating material to conserve body heat. See Figure 4-13 for the location of the main fat storage areas.

Reticular Tissue

A three-dimensional web, that is, a reticular network, identifies reticular type tissue. Slender, branching reticular fibers with reticular cells overlying them compose the reticular meshwork. Branches of the cytoplasm of reticular cells follow the branching reticular fibers.

Reticular tissue forms the framework of the spleen, lymph nodes, and bone marrow. It functions as part of the body's complex mechanism for defending itself against microbes and injurious substances. The reticular meshwork filters injurious substances out of the blood and lymph, and the reticular cells phagocytose (engulf and destroy) them. Another function of reticular cells is to make reticular fibers.

Figure 4-14 A, Tendons and ligaments of the foot are examples of dense fibrous connective tissue. **B,** Note the cordlike tendons passing to the toes and the flat, shiny tarsal ligaments on the dorsum (top) of the foot.

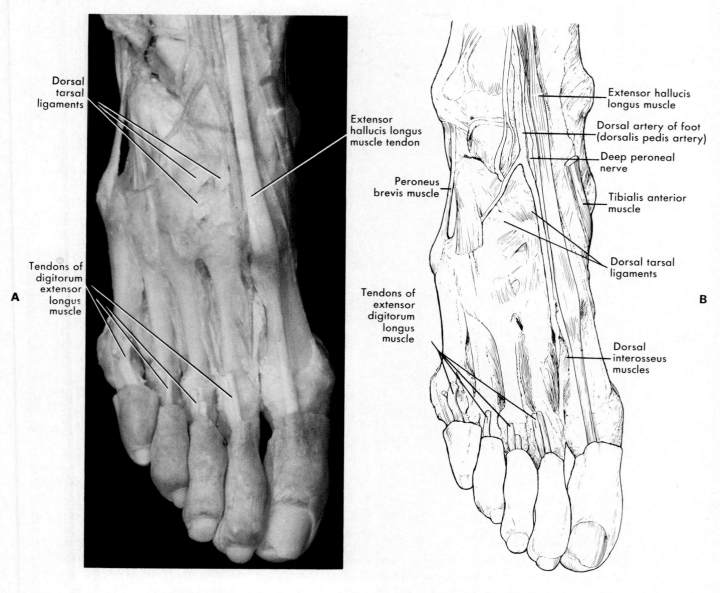

Dense Fibrous Tissue

Dense fibrous tissue consists mainly of bundles of collagenous fibers arranged in parallel rows in a fluid matrix. It contains relatively few fibroblast cells and composes tendons and ligaments (Figure 4-14). Bundles of collagenous fibers endow tendons with great tensile strength and nonstretchability—desirable characteristics for these structures that anchor our muscles to bones (Figure 4-15). In ligaments, on the other hand, bundles of elastic fibers predominate. Hence ligaments exhibit some degree of elasticity.

The terms **sprain** and **strain** should not be used interchangeably. A sprain is a ligament injury, and damage is confined to connective tissue elements at a joint. A strain, however, is an injury that may involve any component of the "musculotendinous unit." Strain injuries frequently involve a pull, stretch, or tear in a muscle, tendon, or the junction between the two, as well as their attachments to bone.

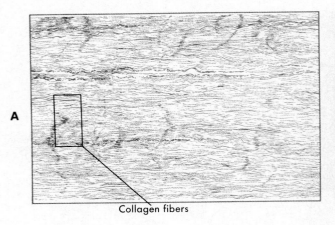

Collagen fibers

Figure 4-15 **A**, Sketch of photomicrograph. Note the multiple layers of flattened collagenous fibers arranged in parallel rows. **B**, Dense fibrous connective tissue of tendon, longitudinal section. (×35.)

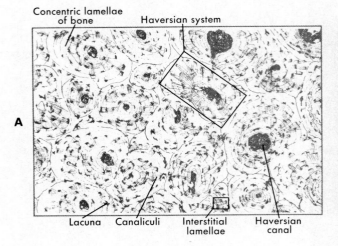

Concentric lamellae
of bone   Haversian system

Lacuna   Canaliculi   Interstitial   Haversian
                         lamellae   canal

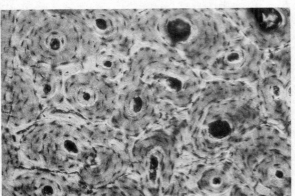

Figure 4-16 **A**, Sketch of photomicrograph. Many wheel-like structural units of bone known as haversian systems are apparent in this section. **B**, Dried, ground bone. (×35.)

Bone (Osseous) Tissue

Bone is one of the most highly specialized forms of connective tissue. The cells of bone, **osteocytes,** are imbedded in a unique matrix material containing both organic collagen material and mineral salts. The inorganic (bone salt) portion makes up about 65% of the total matrix material and is responsible for the hardness of bone.

Bones are the organs of the skeletal system. They provide support and protection for the body and serve as points of attachment for muscles. In addition, the calcified matrix of bones serves as a mineral reservoir for the body, and the red marrow functions in **hemopoiesis** or blood cell formation.

The basic organizational or structural unit of bone is the microscopic **haversian system** (Figure 4-16). Osteocytes or bone cells are located in small spaces, or **lacunae,** which are arranged in concen-

tric layers of bone matrix called **lamellae.** Small canals called **canaliculi** connect each lacuna and osteocyte with nutrient blood vessels found in the central haversian canal.

In addition to mature osteocytes, there are two additional types of bone cells: **osteoblasts** or bone-forming cells and **osteoclasts** or bone-destroying cells. Mature bone can grow and be reshaped by the simultaneous activity of osteoblasts laying down new bone as the osteoclasts break down and remove existing bone tissue.

Certain bones called **membrane bones** (e.g., flat bones of the skull) are formed within membranous tissue while others (e.g., long bones such as the humerus) are formed indirectly through replacement of cartilage in a process called **endochondral bone formation.** The details of bone formation will be presented in Chapter 6.

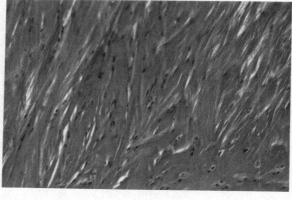

Figure 4-17 **A**, Fibrocartilage of pubic symphysis. (×90.) **B**, Sketch of photomicrograph. The strong, dense fibers that fill the matrix convey shock-absorbing qualities.

Matrix   Collagen fiber   Cartilage cell in lacuna

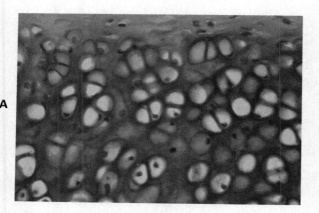
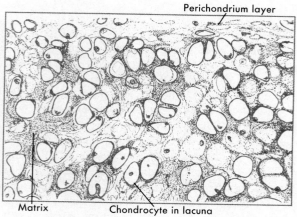

Figure 4-18 **A**, Hyaline cartilage of trachea. (×140.) **B**, Sketch of photomicrograph. Note the many spaces or lacunae in the gell-like matrix.

Perichondrium layer

Matrix   Chondrocyte in lacuna

Cartilage

Cartilage differs from other connective tissues in that only one cell type, the **chondrocyte,** is present. It is the chondrocytes that produce both the fibers and the tough, gristlike ground substance of cartilage. Chondrocytes, like bone cells, are found in small openings called lacunae. Cartilage is avascular and nutrients must reach the cells by diffusion. Movement is through the matrix from blood vessels located in a specialized connective tissue membrane, called the **perichondrium,** which surrounds the cartilage mass. Injuries to cartilage heal slowly, if at all, because of this inefficient method of nutrient delivery.

Fibrocartilage is the strongest and most durable type of cartilage (Figure 4-17). The matrix is rigid and is filled with strong white fibers. Fibrocartilage disks serve as shock absorbers between ad-

jacent vertebrae (intervertebral disks) and in the knee joint. Damage to the fibrocartilage pads or joint menisci in the knee occurs frequently as a result of sport-related injuries.

Elastic cartilage contains large numbers of very fine elastic-like fibers that give the matrix material additional strength and flexibility. This type of cartilage is found in the external ear and in the voice box or larynx.

Hyaline cartilage takes its name from the Greek word *hyalos* or "glass." The name is appropriate, since the appearance of hyaline cartilage is shiny and translucent. This is the most prevalent type of cartilage and is found in the support rings of the respiratory tubes and covering the ends of bones that articulate at joints (Figure 4-18).

Erythrocyte

A

Eosinophil   Neutrophil

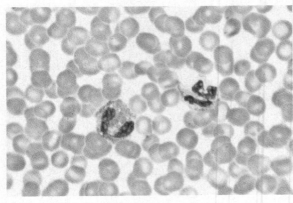

B

Figure 4-19 A, Sketch of photomicrograph. This smear shows two white blood cells or leukocytes surrounded by hundreds of smaller red blood cells. **B,** Blood smear, human. (× 350.)

Blood

Blood is perhaps the most unusual connective tissue since it exists in a liquid state and contains neither ground substance nor fibers (Figure 4-19).

Whole blood is often divided into a **liquid fraction** called **plasma** and **formed elements** or cells that can be divided into three classes: red cells, or **erythrocytes;** white cells, or **leukocytes;** and platelets. The liquid fraction makes up about 55% of whole blood, and the formed elements compose about 45%.

Blood performs many body transport functions, including movement of respiratory gases (oxygen and carbon dioxide), nutrients, and waste products. In addition, blood plays a critical role in maintaining a constant body temperature and in regulating the pH of body fluids. The white blood cells are often called the "first line of defense against disease" because of their function in destroying harmful microorganisms. Blood is described in detail in Chapter 17.

> The term **ischemia** describes a deficiency of blood supply to a body tissue or organ. Tissues deprived of an adequate blood supply quickly die. This death of tissue is called **necrosis.** If areas of cardiac muscle in the heart become ischemic because of an occluded blood vessel, localized tissue necrosis called an **infarction** will occur. The symptoms that result from an inadequate blood supply and resulting tissue necrosis in an area of the heart wall are often referred to collectively as a **heart attack.**

MUSCLE TISSUE

Three types of muscle tissue are present in the body—skeletal muscle, visceral muscle, and cardiac muscle. Their names suggest their locations. **Skeletal muscle tissue** (Figure 4-20) composes muscles attached to bones; these are the organs that we think of as our muscles. **Visceral muscle tissue** (Figure 4-21) is found in the walls of the viscera (hollow internal organs—e.g., the stomach, intestines, and blood vessels). **Cardiac muscle tissue** makes up the wall of the heart (Figure 4-22). Another name for skeletal muscle is *striated voluntary* muscle. The term *striated* refers to cross striations (stripes) visible in microscopic slides of the tissue. The term *voluntary* indicates that voluntary or willed control of skeletal muscle contractions is possible. Another name for visceral muscle is *nonstriated* (smooth) *involuntary*. It has no cross striations and cannot ordinarily be controlled by the will. Another name for cardiac muscle is *striated involuntary* muscle. Like skeletal muscle, cardiac muscle has cross striations, and, like visceral muscle, its contractions cannot ordinarily be controlled by the will.*

Look now at Figure 4-20 and observe the following structural characteristics of skeletal muscle cells: many cross striations, many nuclei per cell, and long, narrow, threadlike shape of the cells. Skeletal muscle cells may have a length of more than 3.75 cm, but they have diameters of only 10 to 100 µm. Because this gives them a threadlike appearance, muscle cells are often called muscle fibers.

---

*In recent years many individuals have learned some voluntary control over visceral and cardiac muscle contractions by using biofeedback devices.

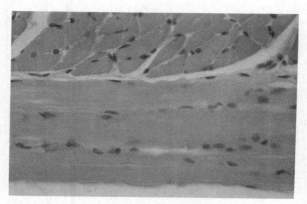

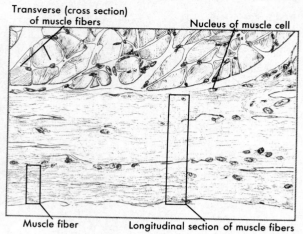

Transverse (cross section) of muscle fibers

Nucleus of muscle cell

B

Muscle fiber

Longitudinal section of muscle fibers

Figure 4-20 **A**, Skeletal muscle. (× 140.) **B**, Sketch of photomicrograph. This section of skeletal muscle shows bundles of cell fibers cut in both cross section transversely (top) and longitudinally (bottom).

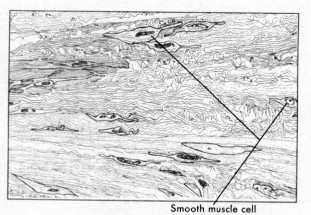

B

Smooth muscle cell

Figure 4-21 **A**, Smooth muscle, longitudinal aspect. (× 140.) **B**, Sketch of photomicrograph. Note the central placement of nuclei in the spindle-shaped (fusiform) smooth muscle fibers.

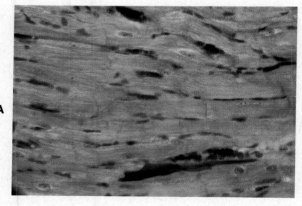

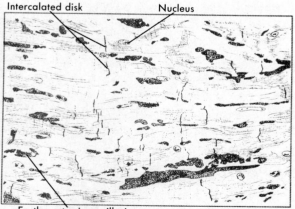

Intercalated disk

Nucleus

B

Erythrocytes in capillaries

Figure 4-22 **A**, Cardiac muscle. (× 140.) **B**, Sketch of photomicrograph. The dark bands or intercalated disks characteristic of cardiac muscle are easily identified in this tissue section.

INFLAMMATION

The terms **inflammation** or **inflammatory reaction** are used to describe the complex way in which cells and tissues react to injury. Many of the events, which are now identified as steps in the inflammatory reaction, are so dramatic that for centuries they were often thought to be a primary disease. It was in the first century AD that the Roman physician Celsus first established inflammation as an entity by describing its four cardinal signs: **rubor** (redness), **calor** (heat), **tumor** (swelling) and **dolor** (pain). His accurate and detailed description of the visible signs that signal the body's response to injury is considered a classic in the annals of medicine.

The inflammatory reaction can best be described as a series of sequenced events that occur as a result of an inflammatory stimulus or "insult." Heat, physical pressure, caustic chemicals, toxins released by harmful bacteria, or any other type of noxious stimulus will initiate an inflammatory reaction. Early studies with rabbits, using a device known as a **transparent ear chamber,** permitted scientists to view changes in living tissue after an injury for prolonged periods. Skin over an animal's external ear was subjected to very slight injury and then viewed through the chamber under a microscope.

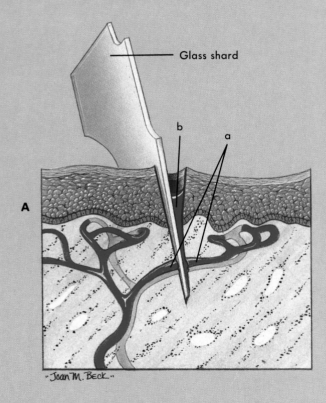

A

Glass shard

b   a

"Joan M. Beck"

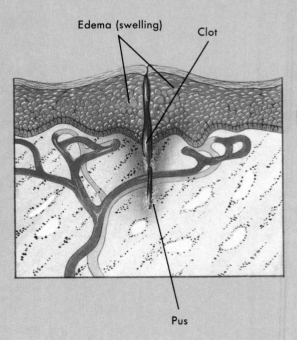

Edema (swelling)   Clot

B

Pus

Inflammation in a section of traumatized skin. **A,** Injury or insult to tissue resulting in dilation of blood vessels *(a)* and increase in blood flow *(b).* **B,** Inflammatory reaction—the body attempts to seal off an area of insult to limit bacterial invasion.

INFLAMMATION—cont'd

Immediately after an injury occurs, there is a very brief constriction of surrounding blood vessels that lasts but a moment. Then, almost immediately, blood vessels dilate, or open, and blood flow increases.

Injured tissues release a number of chemicals that affect blood vessels. These chemicals include **histamine, serotonin,** and a group of chemically related compounds called **kinins.** All of these substances result in both vasodilation and an increase in the permeability of blood vessels so that components that would normally be retained in the blood are permitted to leak out into the tissue spaces.

In the absence of injury, blood flow through a small vessel is such that the cells tend to pass in large measure within the central two thirds of the lumen with a thin layer of plasma flowing closest to the outer walls (see figure). This is called **axial flow.** After an injury, blood cells no longer pass in a central stream. Microscopic examination of vessels near an injured site shows that white cells begin to accumulate in the vessel near the point of injury and then stick, or **marginate,** to the wall. This **margination of leukocytes** continues until the endothelial surface of the vessel is covered with adherent white cells. Within minutes these cells begin to pass through the endothelial lining and out of the vessels into the interstitial spaces near the injury. One of the important functions of many white blood cells is **phagocytosis**—the process of engulfing and destroying bacteria.

Movement of white cells into the area of injury or infection is called **diapedesis.** The term **chemotaxis** describes the attraction of leukocytes, especially neutrophils, into the interstitial spaces. The attractive force is produced by the release of kinins and other chemicals by injured tissue. **Leukocytosis** means an increase in the number of leukocytes in the blood. A substance called **leukocytosis-promoting (LP) factor** is also released by injured tissue. It stimulates the release of white cells from storage areas and increases the number of circulating white blood cells.

The accumulation of dead leukocytes and tissue debris may lead to the formation of **pus** at the focal point of infection. Should this occur, an **abscess,** or cavity formed by the disintegration of tissues, may fill with pus and require surgical drainage.

Increased permeability of blood vessels, increased blood flow, and the migration and accumulation of white blood cells all contribute to the formation of **inflammatory exudate,** which accumulates in the interstitial spaces in the area of injury. The result is often swelling, or **edema,** and pain. In addition to white blood cells and tissue debris, inflammatory exudate contains the "leaked" substances normally retained in the blood but allowed to escape into the interstitial spaces because of increased capillary permeability. One such substance is a soluble protein that is soon converted into **fibrin** in the interstitial spaces. Fibrin formation results in development of a clot, which helps to seal off the infected area and decrease the spread of bacteria or other infectious material.

The cardinal signs of inflammation "make sense" when examined in the light of our understanding of the process.

1 The heat (calor) is largely the result of increased blood flow to the area of injury.
2 The redness (rubor) is also caused by increased blood flow and pooling of blood following injury.
3 Swelling (tumor) results because of edema and accumulation of inflammatory exudate and clot formation in the affected tissue spaces.
4 Pain (dolor) is caused by chemicals such as the kinins (especially **bradykinin**) and other chemical mediators that are released following tissue injury and cellular death.

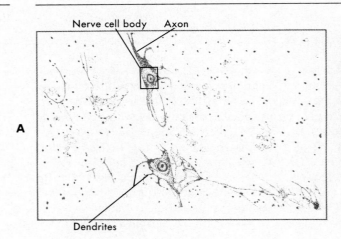

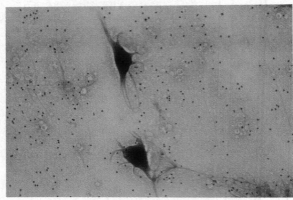

Figure 4-23 **A,** Sketch of photomicrograph. Both neurons in this slide show characteristic soma or cell bodies and multiple cell processes. **B,** Multipolar neurons in smear of spinal cord. (×35.)

Chapter 6 gives more detailed information about the structure of skeletal muscle tissue.

Smooth muscle cells are also long, narrow fibers but not nearly as long as striated fibers. One can see the full length of a smooth muscle fiber in a microscopic field but only a small part of a striated fiber. According to one estimate, the longest smooth muscle fibers measure about 500 μm and the longest striated fibers about 40,000 μm. As Figure 4-21 shows, smooth muscle fibers have only one nucleus per fiber and are nonstriated or smooth in appearance.

Under the light microscope, cardiac muscle fibers (Figure 4-22) have cross striations and unique dark bands (intercalated disks). They also seem to be incomplete cells that branch into each other to form a big continuous mass of protoplasm (a syncytium). The electron microscope, however, has revealed that the intercalated disks are actually places where the plasma membranes of two cardiac fibers abut. Cardiac fibers do branch and anastomose, but a complete plasma membrane encloses each cardiac fiber—around its end (at intercalated disks) as well as its sides.

Muscle cells are the movement specialists of the body. They have a higher degree of contractility (ability to shorten or contract) than any other tissue cells.

NERVOUS TISSUE

The basic function of the nervous system is to rapidly regulate, and thereby integrate, the activities of the different parts of the body. Functionally, rapid communication is possible because nervous tissue has much more developed excitability and conductivity characteristics than any other type of tissue.

The organs of the nervous system are the brain, the spinal cord, and the nerves. Actual nerve tissue is ectodermal in origin and consists of two basic kinds of cells: nerve cells, or **neurons,** which are the functional (conducting) units of the system, and special anatomical (connecting and supporting) cells called **neuroglia** (Figure 4-23).

All neurons are characterized by a cell body called the **soma** and at least two processes: one **axon,** which transmits nerve impulses away from the cell body, and one or more **dendrites,** which carry impulses toward the cell body. Most neurons are located within the organs of the central nervous system.

The anatomy and physiology of the nervous system is presented in Chapters 11 to 14.

Review of Medical Terms

abscess (AB-ses) A localized collection of pus in a cavity formed by the disintegration of tissues.

adhesion (ad-HE-zhun) Abnormal joining or fusion of parts to each other.

avascular (a-VAS-ku-lar) Not supplied with blood vessels.

biopsy (BI-op-se) Surgical removal and microscopic examination of tissue to establish precise diagnosis of disease.

edema (e-DE-mah) The presence of abnormally large amounts of fluid in the intercellular tissue spaces of the body.

exfoliative (EKS-fo-li-AH-tive) **cytology** (si-TOL-o-je) Microscopic study or examination of cells shed or detached from the body.

exudate (EKS-u-date) Material, such as fluid, cells, or cellular debris, that has escaped from blood vessels and has been deposited in tissues as a result of inflammation.

hemopoiesis (he-mo-poi-E-sis) The formation and development of blood cells.

hyperplasia (HI-per-PLA-ze-ah) An abnormal multiplication or increase in the number of cells in a tissue.

infarction (in-FARK-shun) An area of tissue death or necrosis caused by an inadequate blood supply.

inflammation (IN-flah-MA-shun) A localized response resulting from injury or destruction of tissues characterized by the classical signs of pain, heat, swelling, and redness.

ischemia (is-KE-me-ah) Deficiency of blood supply to a body part.

leukocytosis (LU-ko-si-TO-sis) An increase in the number of white blood cells in the circulation.

necrosis (ne-KRO-sis) Tissue death.

obesity (o-BE-si-te) An increase in body weight beyond normal standards, caused by excessive accumulation of fat.

pap smear An exfoliative cytology technique employed in the diagnosis of cancer.

phagocytosis (fag-o-si-TO-sis) Process by which white blood cells engulf and destroy bacteria.

pleurisy (PLOOR-i-se) Inflammation of the pleura around the lungs.

pus A creamy-appearing inflammation liquid of varying color made up of dead or dying white blood cells, tissue debris, and interstitial fluid products.

sprain Joint injury in which a ligament is damaged.

strain An injury to any component of a musculotendinous unit, especially overstretching or tearing of musculature, caused by excessive effort or undue exertion.

Outline Summary

INTRODUCTION

A Definition—organizations of similar cells with a specific function
B Intercellular material (matrix) is nonliving

PRINCIPAL TYPES OF TISSUE

A Epithelial tissue
B Connective tissue
C Muscle tissue
D Nervous tissue
E Histology—microscopic anatomy—deals with the study of tissues

EMBRYONIC DEVELOPMENT OF TISSUES

A Primary germ layers (Table 4-1)
  1 Endoderm
  2 Mesoderm
  3 Ectoderm
B Gastrulation—process of cell movement and differentiation, which results in development of primary germ layers
C Histogenesis—the study of specific tissue development from primary germ layers

EPITHELIAL TISSUE

A Types (principal)
  1 Membranous (covering or lining)
  2 Glandular
B Locations
  1 External body surfaces
  2 Lining hollow organs and blood vessels
  3 Forming inner lining of body cavities
  4 Forming secretory units of glands
C Functions
  1 Protection
  2 Sensory functions
  3 Secretion
  4 Absorption
  5 Excretion

Outline Summary—cont'd

D Generalizations about epithelial tissue
  1 Limited intercellular material (matrix)
  2 Membranous type attached to a basement membrane
  3 Avascular
  4 Cells are in close proximity (many desmosomes)
  5 Capable of reproduction

CLASSIFICATION OF EPITHELIAL TISSUE
Membranous (covering or lining epithelium)

A Classification based on cell shape (Table 4-2)
  1 Squamous
  2 Columnar
  3 Pseudostratified columnar
  4 Cuboidal
B Classification based on layers of cells
  1 Simple epithelium
    (a) Simple squamous epithelium (Figure 4-1)
      (1) One-cell layer of flat cells
      (2) Permeable to many substances
      (3) Examples:
        (a) Endothelium—lines blood vessels
        (b) Mesothelium—pleura
    (b) Simple cuboidal epithelium (Figure 4-2)
      (1) One-cell layer of cuboidal-shaped cells
      (2) Found in many glands
    (c) Simple columnar epithelium (Figure 4-3)
      (1) Single layer of tall (columnar-shaped) cells
      (2) Cells often modified for specialized function
        (a) Goblet cells
        (b) Cilia
        (c) Microvilli
      (3) Often lines hollow visceral structures
    (d) Pseudostratified columnar epithelium (Figure 4-4)
      (1) Columnar cells of differing heights
      (2) All cells rest on basement membrane but may not reach the free surface above
      (3) Cell nuclei at odd and irregular levels
      (4) Found lining air passages and segments of male reproductive system
      (5) Motile cilia and mucus important modifications
  2 Stratified epithelium
    a Stratified squamous (keratinized) epithelium
      (1) Multiple layers of flat squamous cells (Figure 4-5)
      (2) Cells filled with keratin
      (3) Covering outer skin on body surface
    b Stratified squamous (nonkeratinized) epithelium (Figure 4-6)
      (1) Lining vagina, mouth, and esophagus
      (2) Free surface is moist
      (3) Primary function is protection

  c Stratified cuboidal epithelium
    (1) Two or more rows of cells is typical
    (2) Basement membrane is indistinct
    (3) Located in sweat gland ducts and pharynx
  d Stratified columnar epithelium
    (1) Multiple layers of columnar cells
    (2) Only most superficial cells are typical in shape
    (3) Rare
    (4) Located in segments of male urethra and near anus
    (5) Function is protective
  e Stratified transitional epithelium (Figure 4-7)
    (1) Located lining hollow viscera subjected to stress—e.g., urinary bladder
    (2) Often 10 or more layers thick
    (3) Protects organ walls from tearing

Glandular epithelium

A Specialized for secretory activity
B Structural classification of exocrine glands (Table 4-3)
  1 Unicellular
  2 Multicellular
    a Endocrine
    b Exocrine (structural) (Figures 4-8 to 4-10)
      (1) Simple
      (2) Compound
C Functional classification of exocrine glands (Figure 4-11)
  1 Merocrine glands
    a Secrete directly through cell membrane
    b Secretion proceeds with no damage to cell wall and no loss of cytoplasm
    c Most numerous gland type
  2 Apocrine glands
    a Secretory products collect near apex of cell and are secreted by pinching off the distended end
    b Secretion process results in some damage to cell wall and some loss of cytoplasm
    c Mammary glands are good examples of this secretory type
  3 Holocrine glands
    a Secretion products when released cause rupture and death of the cell
    b Sebaceous glands are holocrine

MEMBRANES

A Thin tissue layers that cover surfaces, line cavities, and divide spaces or organs
B Epithelial membranes are most common type
  1 Serous variety
  2 Mucous variety
  3 Cutaneous membrane (skin)

Outline Summary—cont'd

C Serous membranes
  1 Line closed body cavities
  2 Cover visceral organs
  3 Pleura, peritoneum, and pericardium are examples
  4 Secrete lubricating fluids
  5 Reduce friction and serve a protective function
D Mucous membranes
  1 Line and protect organs that open to the exterior of the body
  2 Found lining ducts and passageways of respiratory and digestive tracts
  3 Goblet cells produce mucus
E Tympanic membrane (eardrum)
  1 An epithelial membrane
  2 Outer surface is stratified squamous epithelium
  3 Inner surface is simple squamous epithelium

EXFOLIATIVE CYTOLOGY

See boxed essay.

CONNECTIVE TISSUE

A General function—connects, supports, transports, and defends
B Main types
  1 Loose, ordinary connective (areolar)
  2 Adipose
  3 Reticular
  4 Dense fibrous
  5 Bone
  6 Cartilage
  7 Blood
C General characteristics—intercellular material (matrix) predominates in most connective tissues and determines their physical characteristics; consists of fluid, gel, or solid matrix, with or without fibers (collagenous, reticular, and elastic)
D Loose, ordinary connective tissue (areolar)
  1 One of the most widely distributed of all tissues; intercellular substance is prominent and consists of collagenous and elastic fibers loosely interwoven and embedded in soft viscous ground substance; several kinds of cells present, notably fibroblasts and macrophages, also mast cells, plasma cells, fat cells, and some white blood cells
  2 Function—connection
E Adipose tissue
  1 Similar to loose, ordinary connective tissue but contains mainly fat cells
  2 Functions—protection, insulation, support, and reserve food
F Reticular tissue
  1 Forms framework of spleen, lymph nodes, and bone marrow; consists of network of branching reticular fibers with reticular cells overlying them
  2 Functions—defense against microbes and other injurious substances; reticular meshwork filters out injurious particles and reticular cells phagocytose them
G Dense fibrous tissue
  1 Matrix consists mainly of bundles of collagenous fibers and relatively few fibroblast cells
  2 Locations—composes structures that need great tensile strength, such as tendons and ligaments
  3 Function—furnishes flexible but strong connection
H Bone (osseous) tissue
  1 Highly specialized connective tissue type
    a Cells—osteocytes—embedded in a calcified matrix
    b Inorganic component of matrix accounts for 65% of total bone tissue
  2 Functions
    a Protection
    b Support
    c Point of attachment for muscles
    d Reservoir for minerals
    e Hemopoiesis
  3 Haversian system (Figure 4-16)
    a Structural unit of bone
    b Spaces for osteocytes called *lacunae*
    c Matrix present in concentric rings called *lamellae*
    d Canaliculi care canals that join lacunae with the central haversian canal
  4 Cell types
    a Osteocyte—mature bone cell
    b Osteoblast—bone-forming cell
    c Osteoclast—bone-destroying cell
  5 Formation
    a In membranes—e.g., flat bones of skull
    b From cartilage (endochondral)—e.g., long bones, such as the humerus
I Cartilage
  1 Chondrocyte is only cell type present
  2 Lacunae house cells as in bone
  3 Avascular
  4 Nutrition of cells depends on diffusion of nutrients through matrix
  5 Heals slowly after injury because of slow nutrient transfer to the cells
  6 Perichondrium is membrane that surrounds cartilage
  7 Types
    a Elastic
      (1) Contains many fine elastic fibers
      (2) Provides strength and flexibility
      (3) Located in external ear and larynx
    b Fibrocartilage (Figure 4-7)
      (1) Is the strongest and most durable type of cartilage
      (2) Matrix is semirigid and filled with strong white fibers

(3) Found in intervertebral disks
(4) Serves as shock-absorbing material between bones at the knee (menisci)
  c Hyaline (Figure 4-18)
    (1) Appearance is shiny and translucent
    (2) Most prevalent type of cartilage
    (3) Located on the ends of articulating bones
J Blood
  1 A liquid tissue (Figure 4-19)
  2 Contains neither ground substance nor fibers
  3 Composition of whole blood
    a Liquid fraction or plasma is 55% of total blood
    b Formed elements contribute 45% of total blood
      (1) Red blood cells
      (2) White blood cells
      (3) Platelets
  4 Functions
    a Transportation
    b Regulation of body temperature
    c Regulation of body pH
    d White blood cells destroy bacteria

MUSCLE TISSUE

A Types
  1 Skeletal, or striated voluntary
  2 Visceral, or nonstriated involuntary
  3 Cardiac, or striated involuntary

B Microscopic characteristics
  1 Skeletal muscle—threadlike cells with many cross striations and many nuclei per cell
  2 Visceral muscle—elongated narrow cells, no cross striations, one nucleus per cell
  3 Cardiac muscle—branching cells with intercalated disks (formed by abutment of plasma membranes of two cells)

INFLAMMATION

See boxed essay.

NERVOUS TISSUE

A Functions—rapid regulation and integration of body activities
B Specialized characteristics
  1 Excitability
  2 Conductivity
C Organs
  1 Brain
  2 Spinal cord
  3 Nerves
D Cell types
  1 Neuron—functional (conducting) unit
    a Cell body or soma
    b Processes
      (1) Axon (single process)—transmits nerve impulse away from the cell body
      (2) Dendrites (two or more)—transmits nerve impulse toward the cell body

Review Questions

1 Define the term *tissue* and identify the four principal tissue types.

2 Briefly discuss the embryonic development of tissues from fertilization to 2 months of gestation.

3 What are the primary germ layers? What is the process by which blastocyst cells move and then evolve into the primary germ layers?

4 List at least three structures that are derived from each of the primary germ layers.

5 Identify the two primary subdivisions of epithelial tissue and briefly discuss the location and generalized function of each.

6 What are the five most important functions of epithelial tissue?

7 In general, how much intercellular or matrix material is found in epithelial tissue?

8 What forms the basement membrane of epithelial tissue?

9 Which of the following adjectives best describes the number of blood vessels in epithelial tissue: none, very few, very numerous?

10 How are epithelial cells held together?

11 Explain how the shape of epithelial cells is used for classification purposes. Identify the four types of epithelium described in this classification process.

12 Classify epithelium according to the layers of cells present.

13 List the types of simple and stratified epithelium and give examples of each.

14 What is glandular epithelium? Give examples.

15 Discuss the structural classification of exocrine glands. Give examples of each type.

16 Compare merocrine, apocrine, and holocrine glands by identifying the method by which they discharge their secretion products from the cell. Give an example of each type.

17 What is meant by the term *exfoliative cytology*?

18 What is a Pap smear? How does it differ from a biopsy?

19 List the types of connective tissue and briefly discuss the function and location of each variety.

20 How is adipose tissue different from most other types of connective tissue? Is fat distributed similarly or differently in male and female bodies?

21 What is the difference between a sprain and a strain?

22 Discuss and compare the microscopic anatomy of bone and cartilage tissue.

23 Compare the structure of the three major types of cartilage tissue. Locate and give an example of each type.

24 List the three major types of muscle tissue.

25 List the components of whole blood and discuss the basic function of each fraction or cell type.

26 What do the terms *ischemia, necrosis,* and *infarction* mean?

27 Identify the two basic types of cells in nervous tissue.

28 What are the four cardinal signs of inflammation? Discuss the process by which tissues respond to injury.

5

THE SKIN AND ITS APPENDAGES

OBJECTIVES

After you have completed this chapter, you should be able to:

1 Define the terms *integument* and *integumentary system.*
2 Discuss the generalized functions of the skin as an organ system.
3 Describe the three cell types and five cell layers of the epidermis in thick skin and give the function of each.
4 Discuss epidermal growth and repair.
5 Describe the layers, structural components, and functions of the dermis.
6 Discuss the three principal types of skin cancer and differentiate between them.
7 Discuss factors that influence skin color.
8 Describe the formation, structure, and growth of hair and nails.
9 Discuss and compare the structure and function of sweat (sudoriferous), sebaceous, and ceruminous glands.
10 Discuss the composition and function of skin surface film.
11 Explain how the skin functions in homeostasis of body temperature.
12 Explain the classification of burns into first, second, and third degree.
13 Discuss age changes in the skin.
14 Briefly describe or comment on the following skin conditions: blisters, albinism, vitiligo, striae, fungal infections of the scalp and feet, acne, and decubitus ulcers.

KEY TERMS

Ceruminus (see-ROO-mi-nus) gland

Corium (KO-re-um)

Cutaneous (ku-TA-ne-us) membrane

Dermis (DER-mis)

Desquamation (de-SKWA-ma-shun)

Epidermal proliferating unit (EPU)

Epidermis (ep-i-DER-mis)

Follicle (FOL-ic-kl)

Integument (in-TEG-u-ment)

Keratinocytes (ke-RAT-in-o-sites)

Keratinization (KER-rat-tin-i-ZA-shun)

Melanin (MEL-ah-nin)

Sebaceous (see-BA-shus) gland

Sudoriferous (SOO-dor-IF-er-us) gland

Vital, diverse, complex, extensive—these adjectives describe the body's largest, thinnest, and one of its most important organs, the skin. It forms a self-repairing and protective boundary between the body and an often hostile external world. The skin surface is as large as the body itself, an area in average-sized adults of roughly 1.6 to 1.9 m² (17 to 20 square feet). Its thickness varies from slightly less than 0.05 cm (1/50 inch) to slightly more than 0.3 cm (1/8 inch).

As you know, the body is characterized by a "nested" or hierarchical type of organization. Complexity progresses from cells to tissues and then to organs and organ systems. This chapter discusses the skin and its appendages—hair, nails, and skin glands—as an organ system. Ideally, by studying the skin and its appendages before you proceed to the more traditional organ systems in the chapters that follow, you will improve your understanding of how structure is related to function. **Integument** is another name for the skin. **Integumentary system** is a term used to denote the skin and its appendages.

SKIN FUNCTIONS

Skin functions are crucial to maintenance of homeostasis and to survival itself. They are also diverse. They include such different processes as protection, temperature regulation, synthesis of important chemicals and hormones (such as vitamin D), and excretion of water and salts. Also, certain substances, it is now known, can be absorbed through the skin. These include the fat-soluble vitamins (A, D, E, and K), estrogens and other sex hormones, corticoid hormones, and certain chemicals such as methyl salicylate (oil of wintergreen) and dimethyl sulfoxide (DMSO). In addition, sensory receptors in the skin enable it to function as a sophisticated sense organ. They serve as antennas that detect stimuli leading to sensations of heat, cold, pressure, touch, and pain.

The skin also produces both melanin—the pigment that serves as an extremely effective screen to potentially harmful ultraviolet light—and keratin—one of nature's most flexible yet enduring protective proteins. The keratinized stratified squamous epithelial tissue that composes the epidermis makes it a formidable barrier. It protects underlying tissues against invasion by unconquerable hordes of microorganisms, bars entry of most harmful chemicals, and minimizes mechanical injury of underlying structures.

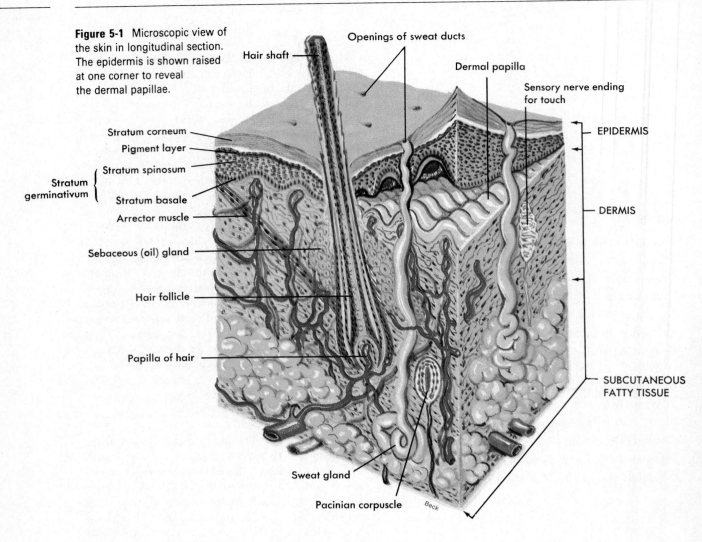

Figure 5-1 Microscopic view of the skin in longitudinal section. The epidermis is shown raised at one corner to reveal the dermal papillae.

Hair shaft

Openings of sweat ducts

Dermal papilla

Sensory nerve ending for touch

EPIDERMIS

Stratum corneum

Pigment layer

Stratum spinosum

Stratum germinativum

Stratum basale

Arrector muscle

Sebaceous (oil) gland

Hair follicle

Papilla of hair

DERMIS

SUBCUTANEOUS FATTY TISSUE

Sweat gland

Pacinian corpuscle

Beck

The skin plays a very important role in maintaining homeostasis of body temperature. (We shall discuss this in some detail later in the chapter.) Briefly, blood vessels in the dermis dilate and sweat secretion increases if body temperature rises above normal. More heat is therefore lost by radiation from the larger volume of blood in the skin and by evaporation of sweat on the skin's surface. Together these changes tend to decrease blood temperature back to its normal level. If blood temperature decreases below normal, skin blood vessels constrict and sweat secretion decreases.

STRUCTURE OF THE SKIN

The skin is a thin, relatively flat organ classified as a membrane—the **cutaneous membrane.** Two main layers compose it: an outer, thinner layer called the **epidermis** and an inner, thicker layer named the **dermis** (Figure 5-1). The cellular epidermis is an epithelial layer derived from the ectodermal germ layer of the embryo. By the seventeenth week of gestation, the epidermis of the developing baby has all the essential characteristics of the adult. The deeper dermis is a relatively dense and vascular connective tissue layer that may average just over 4 mm in thickness in some body areas. The specialized area where the cells of the epidermis meet the connective tissue cells of the dermis is called the **dermal-epidermal junction.** Beneath the dermis lies a loose **subcutaneous layer** rich in fat and areolar tissue. It is sometimes called the **hypodermis** or **superficial fascia.** The fat content of the hypodermis varies with the state of nutrition and in obese individuals may exceed 10 cm in thickness in certain areas. The density and arrangement of fat cells and collagen fibers in this area determine the relative mobility of the skin. When skin is removed from an animal by blunt dissection, separation occurs in the "cleavage plane" that exists between the superficial fascia and the underlying tissues.

EPIDERMIS

The epidermis of the skin is composed of stratified squamous epithelium. In "thin skin," which covers most of the body surface and has a total depth of 1 to 3 mm, the outer epidermis is much thinner than most of us would probably guess—less than 0.17 mm thick (1/200 inch) in most areas. Exceptions are body surfaces chronically exposed to pressure or friction, such as the soles of the feet and the palms of the hands. Here, in the "thick skin," which has a total thickness of 4 to 5 mm, the epidermis is appreciably thicker (1 to 1.3 mm) than in the thin skin that covers a majority of the body.

Cell Types

The epidermis is composed of three primary types of cells called (1) **keratinocytes,** (2) **melanocytes,** and (3) **Langerhans' cells.** Of the three cell types, keratinocytes, arranged in distinct strata or layers, are by far the most important. They comprise over 90% of the epidermal cells and form the principal structural element of the outer skin. Melanocytes contribute color to the skin and serve to filter ultraviolet light (Figure 5-2). Although they may comprise over 5% of the epidermal cells, melanocytes may be completely absent from the skin in certain nonlethal conditions. Langerhans' cells are thought to play a limited role in immunological reactions that affect the skin and may serve as a defense mechanism for the body.

Cell Layers

1 **Stratum corneum** (horny layer). The stratum corneum is the most superficial layer of the epidermis. It is composed of very thin squamous (flat) cells that at the skin surface are dead and are continually being shed and replaced. The cytoplasm in these cells has been replaced by a water-repellent protein called **keratin.** In addi-

Although the subcutaneous layer is not part of the skin itself, it carries the major blood vessels and nerves to the skin above. The rich blood supply and loose spongy texture of this area make it an ideal site for the rapid and relatively pain-free absorption of injected material. Liquid medicines, such as insulin, and pelleted implant materials are often administered by **subcutaneous injection** into this spongy and porous layer beneath the skin.

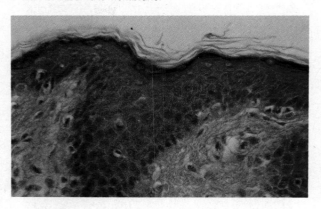

Figure 5-2 Photomicrograph of skin. Note the many-layered epidermis separated from the dermis below by a distinct basement membrane.

tion, the cell membranes become thick and chemically resistant. Specialized junctions (desmosomes) that hold adjacent keratinocytes together strengthen this layer even more and permit it to withstand considerable wear and tear. The process by which cells in this layer are formed from cells in deeper layers of the epidermis and are then filled with keratin and moved to the surface is called **keratinization.** The stratum corneum is sometimes called the **barrier area** of the skin because it functions as a barrier to water loss and to many environmental threats ranging from microorganisms and harmful chemicals to physical trauma. Once this barrier layer is damaged, the effectiveness of the skin as a protective covering is greatly reduced, and most contaminants can easily pass through the lower layers of the cellular epidermis. Certain diseases of the skin cause the stratum corneum layer of the epidermis to thicken far beyond normal limits—a condition called **hyperkeratosis.** The result is a thick, dry, scaly skin that is inelastic and subject to painful fissures (Figure 5-3).

2 **Stratum lucidum** (Latin *lucidus,* "clear"). The keratinocytes in the stratum lucidum are closely packed and clear. Typically, the nuclei are absent, and the cell outlines are indistinct. These cells are filled with a soft gel-like substance called **eleidin,** which will eventually be transformed to keratin. Eleidin is rich in protein-bound lipids and serves to block water penetration or loss. This layer is absent in thin skin but is quite apparent in sections of thick skin from the soles of the feet or palms of the hands.

3 **Stratum granulosum** (granular cell layer). The process of keratinization begins in the stratum granulosum of the epidermis. Cells are arranged in a sheet two to four layers deep and are filled

Figure 5-3 Hyperkeratosis. Note the thick, scaly-like patches of dry skin.

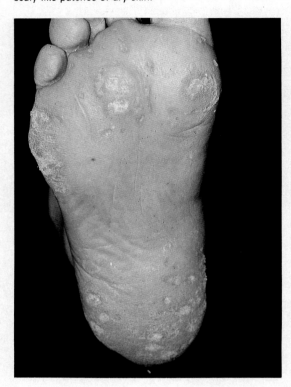

with intensely staining granules called **kerato-hyalin,** which is required for keratin formation. Cells in the stratum granulosum have started to degenerate. As a result, high levels of lysosomal enzymes are present in the cytoplasm, and the nuclei are missing or degenerate. Like the stratum lucidum, this layer of the epidermis may also be missing in some layers of thin skin.

4 **Stratum spinosum** (prickle cell layer). The stratum spinosum layer of the epidermis is formed from eight to ten layers of irregularly shaped cells with very prominent intercellular bridges or desmosomes. When viewed under a microscope, the desmosomes joining adjacent cells give the layer a spiny or prickly appearance (Latin *spinosus,* "spinelike"). Cells in this epidermal layer are rich in ribonucleic acid (RNA) and are therefore well equipped to initiate protein synthesis required for production of keratin. The term **stratum germinativum** is sometimes used to describe the stratum spinosum and the innermost layer of the epithelium called the **stratum basale** described below.

5 **Stratum basale.** The stratum basale is a single layer of columnar cells. Only the cells in this

deepest stratum of the epithelium undergo mitosis. As a result of this regenerative activity, cells transfer or migrate from the basal layer through the other layers until they are shed from the skin surface.

EPIDERMAL GROWTH AND REPAIR

The most important function of the integument—protection—largely depends on the special structural features of the epidermis and its ability to create and repair itself following injury or disease. *Turnover* and *regeneration time* are terms used to describe the time period required for a population of cells to mature and reproduce. Obviously, as the surface cells of the stratum corneum are lost, replacement of keratinocytes by mitotic activity must occur. New cells must be formed at the same rate that old keratinized cells flake off from the stratum corneum to maintain a constant thickness of the epidermis. Cells push upward from the stratum basale into each successive layer, die, become keratinized, and eventually desquamate, as did their predecesors. Incidentally, this fact nicely illustrates a physiological principle: while life continues, the body's work is never done. Even at rest it is producing millions upon millions of new cells to replace old ones.

Current research suggests that the regeneration time required for completion of mitosis, differentiation, and movement of new keratinocytes from the stratum basale to the surface of the epidermis is about 35 days. The process can be accelerated by abrasion of the skin surface, which tends to peel off a few of the cell layers of the stratum corneum. The result is an intense stimulation of mitotic activity in the stratum basale and a shortened turnover period. If abrasion continues over a prolonged period of time, the increase in mitotic activity and shortened turnover time will result in an abnormally thick stratum corneum and the development of **calluses** at the point of friction or irritation. Although callus formation is a normal and protective response of the skin to friction, a number of skin diseases are also characterized by an abnormally high mitotic activity in the epidermis. In such conditions the thickness of the corneum is dramatically increased. As a result, scales accumulate, and development of skin lesions often occurs.

Normally about 10% to 12% of all cells in the stratum basale enter mitosis each day. Cells migrating to the surface proceed upward in vertical columns from discrete groups of 8 to 10 of these basal cells that are undergoing mitosis. Each group of active basal cells together with its vertical columns of migrating keratinocytes is called an **epidermal proliferating unit** or **EPU.** Keratinization proceeds as the cells migrate toward the stratum corneum. As

Blisters (Figure 5-4) may result from injury to cells in the epidermis or from separation of the dermal-epidermal junction. Regardless of cause, they represent a basic reaction of skin to injury. Any irritant that damages the physical or chemical bonds that hold adjacent skin cells or layers together will initiate blister formation. The specialized junctions (**desmosomes**) that serve to hold adjacent cells in the epidermis together are essential for integrity of the skin. If these intercellular bridges, sometimes described as "spot-welds" between adjacent cells, are weakened or destroyed, the skin literally falls apart and away from the body. Damage to the dermal-epidermal junction will produce similar results. Blister formation follows burns, friction injuries, exposure to primary irritants, or accumulation of toxic breakdown products following cell injury or death in the layers of the skin. Typically, chemical agents that break disulfide linkages or hydrogen bonds cause blisters. Since both types of these chemical bonds are the functional connecting links in intercellular bridges (or desmosomes), their involvement in blister formation serves as a good example of the relationship between structure and function at the chemical level of organization.

Figure 5-4 Blisters resulting from contact with poison ivy.

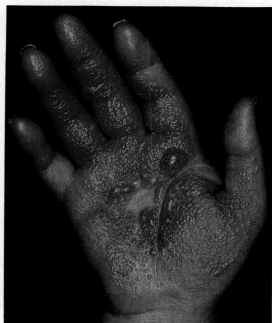

is remarkably effective in preventing separation of the two skin layers even when they are subjected to relatively high shear forces, this barrier is thought to play only a limited role in preventing passage of harmful chemicals or disease-causing organisms through the skin from the external environment.

DERMIS

The dermis, or **corium,** is sometimes called the "true skin." It is composed of a thin **papillary** and a thicker **reticular** layer. The dermis is much thicker than the epidermis and may exceed 4 mm on the soles and palms. It is thinnest on the eyelids and penis, where it seldom exceeds 0.5 mm. As a rule of thumb, the dermis on the ventral surface of the body and over the appendages is generally thinner than on the dorsal surface. The mechanical strength of the skin is in the dermis. In addition to serving a protective function against mechanical injury and compression, this layer of the skin provides a reservoir storage area for water and important electrolytes. A specialized network of nerves and nerve endings in the dermis also serves to process sensory information such as pain, pressure, touch, and temperature. At various levels of the dermis, there are muscle fibers, hair follicles, sweat and sebaceous glands, and many blood vessels. It is the rich vascular supply of the dermis that plays a critical role in regulation of body temperature—a function to be described later in the chapter.

mitosis continues and new basal cells enter the column and migrate upward, fully cornified "dead" cells are sloughed off at the skin surface.

DERMAL-EPIDERMAL JUNCTION

Recent electron microscopy and histochemical studies have now demonstrated the existence of a definite basement membrane, specialized fibrous elements, and a unique polysaccharide gel that together serve to cement the superficial epidermis to the dermis below. The area of contact between the two skin layers forms a specialized junction. It functions to "glue" the two layers together and provide a mechanical support for the epidermis, which is attached to its upper surface. In addition, it serves as a partial barrier to the passage of some cells and large molecules. Certain dyes, for example, if injected into the dermis cannot passively diffuse upward into the epidermis unless the junctional barrier is damaged by heat, enzymes, or other chemicals that change its permeability characteristics. Although the junction

The term **striae** is from the Latin word meaning "furrow" or "groove." If the elastic fibers in the skin are stretched too much—for example, by a rapid increase in the size of the abdomen during pregnancy or as a result of great obesity—these fibers will weaken and tear. The initial result is formation of pinkish or slightly bluish depressed furrows with jagged edges. These tiny linear markings (**"stretch marks"**) are really tiny tears. When they heal and lose their color, the striae that remain appear as glistening silver-white lines.

Figure 5-5 Langer's cleavage lines. **A,** If an incision "cuts across" cleavage lines, stress tends to pull the cut edges apart and may retard healing. **B,** Surgical incisions that are parallel to cleavage lines are subjected to less stress and tend to heal more rapidly.

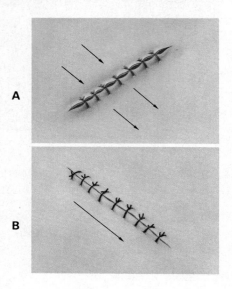

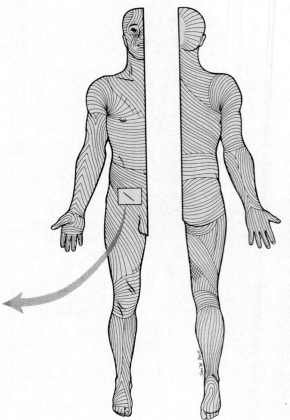

Papillary Layer

Note in Figure 5-1 that the thin superficial layer of the dermis is thrown upward into folds called **dermal papillae,** which project into the epidermis. The papillary layer takes its name from the papillae arranged in rows on its surface. Between the sculptured surface of the papillary layer and the stratum basale lies the important dermal-epidermal junction. The papillary layer and its papillae are composed essentially of loose connective tissue elements and a fine network of thin collagenous and elastic fibers.

Reticular Layer

The thick reticular layer of the dermis consists of a much more dense **reticulum,** or network of fibers, than is seen in the papillary layer above it. It is this dense layer of tough and interlacing white collagenous fibers that, when commercially processed from animal skin, results in leather. Although most of the

fibers in this layer are of the collagenous type, which give toughness to the skin, elastic fibers are also present. These make the skin stretchable and elastic (able to rebound).

The dermis contains both skeletal (voluntary) and smooth (involuntary) muscle fibers. A number of skeletal muscles are located in the skin of the face and scalp. These muscles permit a wide variety of facial expressions and are also responsible for voluntary movement of the scalp. The distribution of smooth muscle fibers in the dermis is much more extensive than the skeletal variety. Each hair follicle has a small bundle of involuntary muscles attached to it. These are the **arrector pili muscles.** Contraction of these muscles makes the hair "stand on end"—as in extreme fright, for example, or from cold. As the hair is pulled into an upright position, it raises the skin around it into what we commonly call "goose pimples." In the dermis of the skin of

Figure 5-6 Flow chart. Genes determine an individual's basic skin color by controlling the amount of melanin synthesized and deposited in the epidermis. But as the chart shows, other factors may modify the basic skin color.

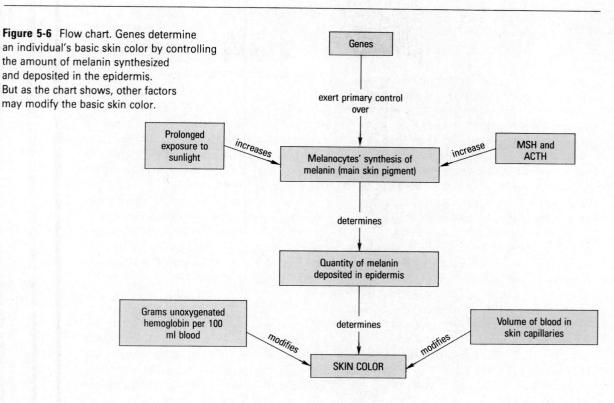

the scrotum and in the pigmented skin called the areolae surrounding the nipples, smooth muscle cells form a loose plexus. Contraction of these smooth muscle cells will wrinkle the skin and cause elevation of the testes or erection of the nipples.

Millions of specialized nerve endings called **receptors** are located in the dermis of all skin areas. They permit the skin to serve as a sense organ transmitting sensations of pain, pressure, touch, and temperature to the brain. The sensory receptors of the skin are discussed in Chapter 13.

SURFACE MARKINGS

The thin epidermal layer of the skin conforms tightly to the ridges (papillae) of the dermis below. As a result, the epidermis also has characteristic ridges on its surface. Epidermal ridges are especially well defined on the tips of the fingers and toes. In each of us they form a unique pattern—an anatomical fact made famous by the art of fingerprinting. Fingerprinting and even toe, palm, and sole printing have become widely used methods for positively identifying individuals—from innocent newborn babies and young children to hardened criminals.

The dense bundles of white collagenous fibers that characterize the reticular layer of the dermis tend to orient themselves in patterns that differ in appearance from one body area to another. The result is formation of patterns called **Langer's lines,** or **cleavage lines** (Figure 5-5). If surgical incisions are made parallel to the cleavage, or Langer's lines, the resulting wound will have less tendency to gape open and will tend to heal with a thin and less noticeable scar.

SKIN COLOR

Human skin, as everyone knows, comes in a wide assortment of colors. The basic determinant of skin color is the quantity of **melanin** deposited in the cells of the epidermis. The number of pigment-producing melanocytes that are scattered throughout the stratum basale of the epidermis in most body areas is about the same in all races. It is the amount of melanin pigment actually produced by these cells that accounts for a majority of skin color variations. Melanocytes are highly specialized and unique. Of all body cells, only melanocytes have the ability to routinely convert the amino acid **tyrosine** into the black melanin pigment. The pigment granules are then transferred to the other epidermal cells. The pigment-producing process is regulated by the enzyme **tyrosinase.** But this conversion process depends on several factors (Figure 5-6). Heredity is first of all. Geneticists tell us that four to six pairs of

SKIN CANCER

There are three principal types of skin cancer, and from a practical standpoint it is very important to differentiate among them. The most common forms are called (1) **squamous cell carcinoma,** (2) **basal cell carcinoma,** and (3) **malignant melanoma.** All three types are called epithelial cancers and apparently result from cell changes in the epidermis. Although cancers of the sweat glands, hair follicles, and sebaceous glands occur, they are relatively uncommon.

Skin cancers are the most common neoplasms, or abnormal growths, seen in humans. They account for almost one fourth of all cancer in men and nearly 14% of reported cancers in women. The higher incidence in males may be a result of more extensive and chronic occupational exposure to sunlight, the most common cause of skin cancer. Farmers and other outdoor workers are particularly susceptible. In addition to damage caused by the ultraviolet radiation in sunlight, cancer may result from x-rays or cell damage caused by such carcinogenic, or cancer-causing chemicals, as arsenic, soot, tar, and certain lubricating oils used by machinists and metal lathe workers.

Squamous cell carcinoma

Squamous cell carcinoma is the most common form of skin cancer. It often begins as a small lump with a scaly or warty appearance. Although it may appear anywhere on the body, it occurs most often in skin exposed to the sun—especially on the dorsum of the hand, ears, and the face (see figure *A*). As the tumor grows, an ulceration often develops, and underlying tissues are invaded and destroyed. If left untreated, squamous cell carcinomas will spread to other body areas, a process called **metastasis,** and death may result.

Basal cell carcinoma

This type of skin cancer arises from cells in the deep basilar layer of the epidermis. It most often appears on the skin of the lips, eyelids, chin, and nose (see figure *B*). Once established, infiltration of adjacent tissues occurs, and, although its growth is slower than squamous cell carcinoma, it is very destructive to surrounding structures. Advanced lesions can destroy even bone and cartilage. Fortunately, basal cell carcinomas do not spread by metastasis and are easily curable if treated early.

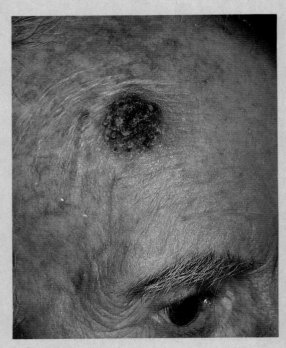

A, Squamous cell carcinoma.

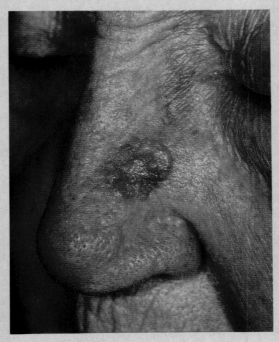

B, Basal cell carcinoma.

SKIN CANCER —cont'd

Malignant melanoma

The most common type of pigment cell skin cancer is generally preceded by a flat hairless mole that begins to darken as it undergoes malignant change (see figure C). These so-called "superficial spreading melanomas" can be cured by surgical excision if treatment occurs before deep invasion of underlying tissue occurs. Anyone who notices changes in a mole involving pigmentation or alteration of surface characteristics should seek medical evaluation immediately. A previously smooth mole that becomes rough or develops a notched edge and changes color is almost always undergoing cancerous change. If treatment is delayed, malignant melanomas metastasize and then become one of the most difficult cancers to cure.

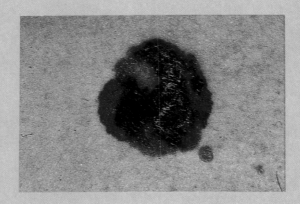

C, Malignant melanoma.

genes exert the primary control over the amount of melanin formed by melanocytes. If the enzyme tyrosinase is absent from birth because of a congenital defect, the melanocytes cannot form melanin, and a condition called **albinism** results. Albino individuals have a characteristic absence of pigment in their hair, skin, and eyes. Thus heredity determines how dark or light one's skin color will be. Other factors, however, can modify the genetic effect. Sunlight is an obvious example. Prolonged exposure to sunlight causes melanocytes to increase melanin production and darken skin color. So, too, does excess secretion of both adrenocorticotropic hormone (ACTH) or melanocyte-stimulating hormone (MSH) by the anterior pituitary gland. Increasing age may also influence melanocyte activity. In many individuals decreasing tyrosinase activity is evident in graying of the hair. In addition to melanin, the yellow pigment **carotene** also contributes to skin color, especially in Asiatic people.

An individual's basic skin color changes, as we have just observed, whenever its melanin content changes appreciably. But skin color can also change without any change in melanin. In this case the change is usually temporary and most often stems from a change in the volume of blood flowing through skin capillaries. If skin blood vessels constrict, for example, skin blood volume decreases, and the skin may turn pale. Or if skin blood vessels dilate, as they do in blushing, the skin appears to be pinker.

In general, the sparser the pigments in the epidermis, the more transparent the skin is and therefore the more vivid the change in skin color will be with a change in skin blood volume. Conversely, the richer the pigmentation, the more opaque the skin, resulting in less skin color changes with a change in skin blood volume.

In some abnormal conditions the skin color changes because of an excess amount of unoxygenated hemoglobin in the skin capillary blood. If skin contains relatively little melanin, it will appear bluish, that is, cyanotic, when 100 ml of blood contains about 5 gm of unoxygenated hemoglobin. In general, the darker the skin pigmentation, the greater the amount of unoxygenated hemoglobin that must be present before cyanosis becomes visible.

APPENDAGES OF THE SKIN

Appendages of the skin consist of the hair, nails, and skin glands.

HAIR

Only a few areas of the skin are hairless—notably the palms of the hands and the soles of the feet. Hair is also absent from lips, nipples, and some areas of the genitalia, such as the glans penis of the male and clitoris of the female. Although human body hair performs no apparent function, head hair serves a

An acquired condition called **vitiligo** results in loss of pigment in certain areas of the skin. The patches of depigmented white skin that characterize this condition contain melanocytes, but for unknown reasons they no longer produce pigment (Figure 5-7).

Figure 5-7 Vitiligo on the backs of the hands, a commonly involved site.

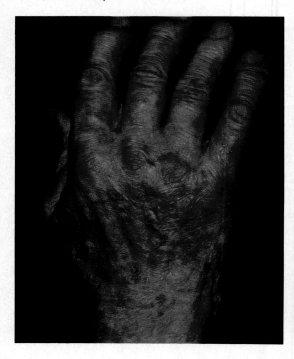

cosmetic purpose. It also offers some protection against the cold, ultraviolet rays, and mechanical injury. Eyelashes and hairs in the nose and ears keep out some dust and insects.

Many months before birth, hair follicles begin to develop in most parts of the skin. By about the sixth month of pregnancy the developing fetus is all but covered by an extremely fine and soft hair coat called **lanugo.** Most of the lanugo hair is lost before birth. Soon after birth the lanugo hair that remains is lost and then replaced by new hair that is more pigmented and strong. Replacement hair growth appears first on the scalp, eyelids, and eyebrows. The coarse pubic and axillary hair that develops at puberty is called **terminal hair.**

Hair growth begins when cells of the epidermis grow down into the dermis, forming a small tube, the **follicle.** The stratum germinativum develops into the follicle's innermost layer and forms at the bottom of the follicle a cap-shaped cluster of cells known as the **germinal matrix.** Protruding into the germinal matrix is a small mound of the dermis called the **hair papilla**—a highly important structure, since it contains the blood capillaries that nourish the germinal matrix (Figure 5-8). Cells of the germinal matrix are responsible for forming hairs. They undergo repeated mitosis, push upward in the follicle, and become keratinized to form a hair. As long as cells of the germinal matrix remain alive, hair will regenerate even though it is cut or plucked or otherwise removed. Part of the hair, namely, the root, lies hidden in the follicle. The visible part of a hair is called the **shaft,** the inner core of a hair is known as the **medulla,** and the outer portion around it is called the **cortex.** Layers of keratinized cells make up the cortex. Deposited in these cells are varying amounts of melanin, the pigment responsible for brown or black hair. Whether hair is straight or wavy depends mainly on the shape of the shaft. Straight hair has a round, cylindrical shaft. Wavy hair, in contrast, has a flat shaft that is not as strong. As a result it is more easily broken or damaged than straight hair. Two or more small sebaceous glands secrete **sebum,** an oily substance, into each hair follicle. The sebaceous gland secretions lubricate hair and keep it from becoming dry, brittle, and easily damaged.

Hair alternates between periods of growth and rest. On the average, hair on the head grows a little less than 12 mm (½ inch) per month, or about 5 inches a year. Body hair grows more slowly. Head hairs reportedly live between 2 and 6 years, then die and are shed. Normally, however, new hairs replace those lost. But baldness can develop—a fact we all know. The common type of baldness occurs only when two requirements are met: genes for baldness must be inherited and the male sex hormone, testosterone, must be present. When the right combination of causative factors exist, common baldness or **male pattern baldness** (Figure 5-9) inevitably results; no known treatment will prevent or reverse the process.

Depilatories (such as Neet) are used to remove unwanted hair. They act by dissolving the protein in hair shafts that extends above the skin surface. Since the follicle is not affected, regrowth of hair continues at a normal rate.

Figure 5-8 Hair follicle. **A,** Relationship of a hair follicle and related structures to the epidermal and dermal layers of the skin. **B,** Enlarged view of a hair root cut in longitudinal section.

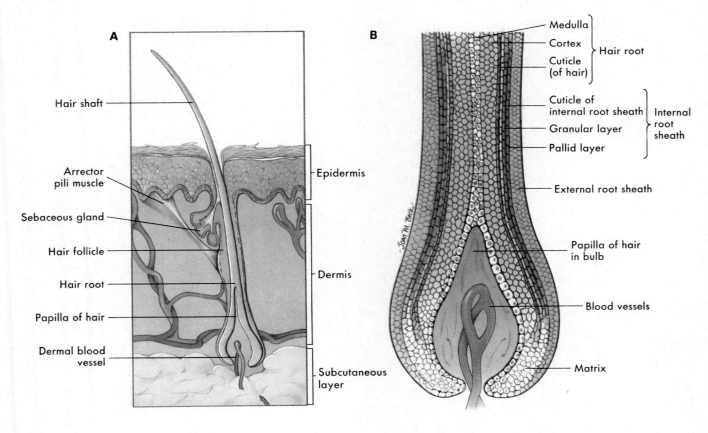

Figure 5-9 Male pattern baldness.

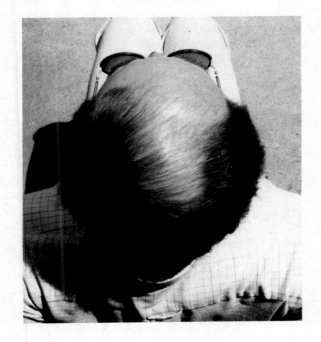

Contrary to what many people believe, hair growth is not stimulated by frequent cutting or shaving. In addition, stories about hair or beard growth continuing after death are also false. What may appear to be continuing beard growth after death is really caused by dehydration and shrinkage of the skin over the face. As a result, a "five o'clock shadow" may appear, or the beard may become more noticeable 2 or 3 days after death even if the face was carefully shaved at the time the body was initially prepared for viewing. In these cases what appears to be continued hair growth is in reality only a more visible beard in a skin surface dehydrated by environmental conditions or the embalming process.

124

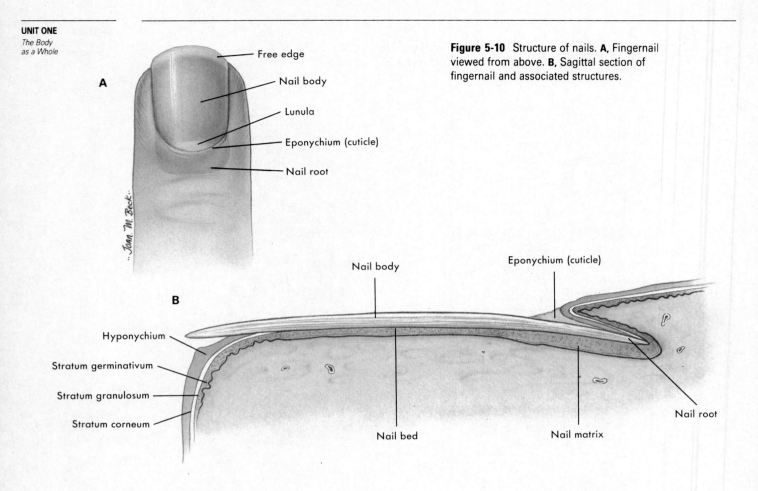

A

- Free edge
- Nail body
- Lunula
- Eponychium (cuticle)
- Nail root

Figure 5-10 Structure of nails. **A,** Fingernail viewed from above. **B,** Sagittal section of fingernail and associated structures.

B

- Nail body
- Eponychium (cuticle)
- Hyponychium
- Stratum germinativum
- Stratum granulosum
- Stratum corneum
- Nail bed
- Nail matrix
- Nail root

NAILS

Heavily keratinized epidermal cells compose fingernails and toenails. The visible part of each nail is called the **nail body.** The rest of the nail, namely, the **root,** lies in a groove hidden by a fold of skin called the **cuticle.** The nail body nearest the root has a crescent-shaped white area known as the **lunula,** or "little moon." Under the nail lies a layer of epithelium called the **nail bed** (Figure 5-10). Because it contains abundant blood vessels, it appears pink in color through the translucent nail bodies. Nails grow by mitosis of cells in the stratum germinativum beneath the lunula. On the average, nails grow about 0.5 mm a week. Fingernails, however, as you may have noticed, grow faster than toenails, and both grow faster in the summer than in the winter.

SKIN GLANDS

The skin glands include three kinds of microscopic glands, namely, sweat (sudoriferous), sebaceous, and ceruminous (Figure 5-11).

Sweat (Sudoriferous) Glands

Sweat or **sudoriferous glands** are the most numerous of the skin glands. They can be classified into two groups—**eccrine** and **apocrine**—based on type of secretion, location, and nervous system connections.

Eccrine sweat glands are by far the most numerous, important, and widespread sweat glands in the body. They are quite small, with a secretory portion less than 0.4 mm in diameter, and are distributed over the total body surface with the exception of the lips, ear canal, glans penis, and nail beds. Eccrine sweat glands are a simple, coiled, tubular type of gland. They function throughout life to produce a transparent watery liquid (**perspiration** or **sweat**) rich in salts, ammonia, uric acid, urea, and other wastes. In addition to elimination of waste, sweat plays a critical role in helping the body maintain a constant core temperature. Histologists estimate that a single square inch of skin on the palms of the hands contains about 3,000 sweat glands. Eccrine sweat glands are also very numerous on the soles of the feet, forehead, and upper torso. With a good magnifying glass you can locate the openings of these sweat gland ducts on the skin ridges of the palms and on the skin of the palmar surfaces of the fingers.

Figure 5-11 Skin glands.

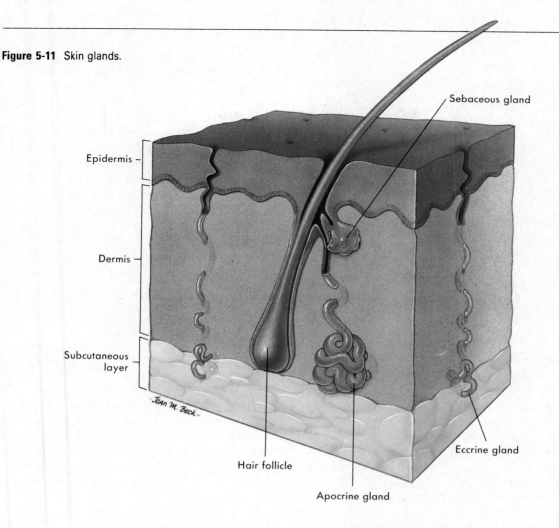

Epidermis

Dermis

Subcutaneous layer

Sebaceous gland

Hair follicle

Apocrine gland

Eccrine gland

JoAn M. Beck

Apocrine sweat glands are located deep in the subcutaneous layer of the skin in the armpit (axilla), the areola of the breast, and the pigmented skin areas around the anus. They are much larger than eccrine glands and often have secretory units that

The total number of sweat glands in humans may exceed 3 million. Although individual glands can only produce a relatively small volume of fluid each day, total body sweat production can exceed an amazing 12 liters in a 24-hour period under extreme conditions. High temperature and humidity combined with vigorous exercise and emotional stress can result in sweat production in excess of 3 liters per hour for short periods of time—a rate of loss that will almost always exceed what is replaced by drinking. If replacement of lost fluid is inadequate, serious dehydration can result.

reach 5 mm or more in diameter. They are connected with hair follicles and are classified as simple, branched tubular glands. Apocrine glands enlarge and begin to function at puberty, producing a more viscous and colored secretion than eccrine glands. In the female, apocrine gland secretions show cyclic changes that are linked to the menstrual cycle. Odor often associated with apocrine gland secretion is not caused by the secretion itself. Instead, it is caused by contamination and decomposition of the secretion by skin bacteria.

Sebaceous Glands

Sebaceous glands secrete oil for the hair and skin. Wherever hairs grow from the skin, there are sebaceous glands, at least two for each hair. The oil, or **sebum,** keeps the hair supple and the skin soft and pliant. It serves as nature's own protective skin cream by preventing excessive water loss from the epidermis. Because sebum is rich in chemicals such as triglycerides, waxes, fatty acids, and cholesterol that have an anti-fungal effect, it also contributes to the fungistatic activity of skin surface film. This property of sebum helps protect the skin from nu-

Figure 5-12 Acne.

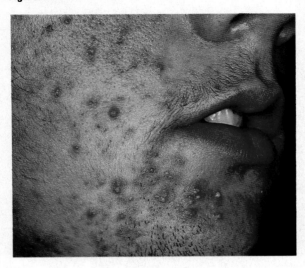

Common **acne** (Figure 5-12) or **acne vulgaris** occurs most frequently in the adolescent years as a result of overactive secretion by the sebaceous glands with blockage and inflammation of their ducts. There is an increase of more than fivefold in the rate of sebum secretion between 10 and 19 years of age. As a result sebaceous gland ducts may become plugged with sloughed skin cells and sebum contaminated with bacteria. The inflamed plug is called a **comedo** and is the most characteristic sign of acne. Pus-filled **pimples** or **pustules** result from secondary infections within or beneath the epidermis, often in a hair follicle or sweat pore.

merous types of fungal infections. Sebaceous glands are simple branched glands of varying size that are found in the dermis except in the skin of the palms and soles. Although almost always associated with hair follicles, some specialized sebaceous glands do open directly on the skin surface in such areas as the glans penis, lips, and eyelids. Sebum secretion increases during adolescence, stimulated by increased blood levels of the sex hormones. Frequently sebum accumulates in and enlarges some of the ducts of the sebaceous glands, forming white pimples. With oxidation this accumulated sebum darkens, forming a **blackhead.**

Ceruminous Glands

Ceruminous glands are a special variety or modification of apocrine sweat glands. Histologically, they appear as simple coiled tubular glands with excretory ducts that open onto the free surface of the skin in the external ear canal or with sebaceous glands into the necks of hair follicles in this area. The mixed secretions of sebaceous and ceruminous glands form a brown waxy substance called **cerumen.** Although it serves a useful purpose in protecting the skin of the ear canal from dehydration, excess cerumen can harden and cause an impaction or blockage in the ear, resulting in loss of hearing.

SURFACE FILM

The ability of the skin to act as a protective barrier against a wide array of potentially damaging assaults from the environment begins with the proper func-

tioning of a thin film of emulsified material spread over its surface. The **surface film** is produced by the mixing of residue and secretions from sweat and sebaceous glands with epithelial cells constantly being cast off from the epidermis. The shedding of epithelial elements from the skin surface is called **desquamation.** Functions of surface film include:

1 Antibacterial and antifungal activity
2 Lubrication
3 Hydration of the skin surface
4 Buffering of caustic irritants
5 Blockade of many toxic agents

The chemical composition of the surface film includes (1) amino acids, sterols, and complex phospholipids from the breakdown of sloughed epithelial cells; (2) fatty acids, triglycerides, and waxes from sebum; and (3) water, ammonia, lactic acid, urea, and uric acid from sweat. The specific chemical composition of surface film is quite variable and samples taken from skin covering one body area will often have a different "mix" of chemical components than film covering skin in another area. This difference helps explain the unique and localized distribution patterns of certain skin diseases and why the skin covering one area of the body is sometimes more susceptible to attack by certain bacteria or fungi.

HOMEOSTASIS OF BODY TEMPERATURE

Warm-blooded animals such as man maintain a remarkably constant temperature despite sizable variations in environmental temperatures.

Normally in most people, body temperature

A number of clinical skin conditions can be explained by an analysis of the quantity and chemical composition of skin surface film. In prepubertal children and elderly individuals the surface film is either deficient in certain fatty substances that originate in sebaceous gland secretions or in the total quantity of sebum produced. In the child there is not only limited production of scalp sebum but the concentration of fatty acids is much lower than in the adult. Since fatty acids, wax alcohols, and other lipid components of surface film contributed by sebaceous gland secretion inhibit fungal growth on the skin, children with their lower concentration of fatty acids are more susceptible to fungal infections, which explains the higher incidence of ringworm fungi infections *(tinea capitis)* in children. Increased postpubertal secretion of sebum results in more adequate antifungal activity in the scalp surface film of most adults and protects them from infection. In addition, the total absence of sebaceous glands on the soles of the feet also explains the high incidence and localized nature of many fungus infections in this area of the body (Figure 5-13).

The generalized decrease in sebum production in the aged makes this segment of the population particularly susceptible to skin that is dehydrated, scaly, and fissured. Decreased sebum production results in a surface film deficient in lipid components necessary for maintaining supple skin. The result is often a generalized itching called **pruritus.** The application of a lanolin- or other oil-based lotion will enable the skin to once again bind water and generally brings rapid reversal of symptoms.

Figure 5-13 Fungal infection of the feet.

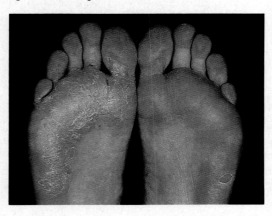

body must, of course, balance the amount of heat it produces with the amount it loses. This means that if extra heat is produced in the body, this same amount of heat must then be lost from it. Obviously if this does not occur, if increased heat loss does not follow close on increased heat production, body temperature will climb steadily upward. If body temperature increases above normal for any reason, the skin plays a critical role in heat loss by the physical phenomena of evaporation, radiation, conduction, and convection.

HEAT PRODUCTION

Heat is produced by one means—metabolism of foods. Because the muscles and glands (especially the liver) are the most active tissues, they carry on more metabolism and therefore produce more heat than any of the other tissues. So the chief determinant of how much heat the body produces is the amount of muscular work it does. During exercise and shivering, for example, metabolism and heat production increase greatly. But during sleep when very little muscular work is being done, metabolism and heat production decrease.

HEAT LOSS

Heat is lost from the body by the physical processes of evaporation, radiation, conduction, and convection. Some 80% or more of this heat transfer occurs through the skin. The rest takes place through the mucous membranes of the respiratory, digestive, and urinary tracts.

Evaporation

Heat energy must be expended to evaporate any fluid. Evaporation of water therefore constitutes one method by which heat is lost from the body, especially from the skin. Evaporation is especially important at high environmental temperatures when it

moves up and down very little in the course of a day. It hovers close to a midpoint of about 37° C, increasing perhaps to 37.6° C by late afternoon and decreasing to around 36.2° C by early morning. This homeostasis of body temperature is of the utmost importance. Why? Because healthy survival depends on biochemical reactions taking place at certain rates. And these rates in turn depend on normal enzyme functioning, which depends on body temperature staying within the narrow range of normal.

In order to maintain an even temperature, the

constitutes the only method by which heat can be lost from the skin. A humid atmosphere necessarily retards evaporation and therefore lessens the cooling effect derived from it—the explanation for the fact that the same degree of temperature seems hotter in humid climates than in dry ones. At moderate temperatures, evaporation accounts for about half as much heat loss as does radiation.

Radiation

Radiation is the transfer of heat from the surface of one object to that of another without actual contact between the two. Heat radiates from the body surface to nearby objects that are cooler than the skin and radiates to the skin from those that are warmer than the skin. This is, of course, the principle of heating and cooling systems. The amount of heat lost by radiation from the skin is made to vary as needed by dilation of surface blood vessels when more heat needs to be lost and by vasoconstriction when heat loss needs to be decreased. In cool environmental temperatures, radiation accounts for a greater percentage of heat loss from the skin than both conduction and evaporation combined. However, in hot environments no heat is lost by radiation but instead may be gained by radiation from warmer surfaces to the skin.

Conduction

Conduction means the transfer of heat to any substance actually in contact with the body—to clothing or jewelry, for example, or even to cold foods or liquids ingested. This process accounts for a relatively small amount of heat loss.

Convection

Convection is the transfer of heat away from a surface by movement of heated air or fluid particles. Usually convection causes very little heat loss from the body's surface. But it can account for considerable heat loss—as you know from experience if you have ever stepped from your bath into even slightly moving air from an open window.

THERMOSTATIC CONTROL OF HEAT PRODUCTION AND LOSS

The control mechanism that normally maintains homeostasis of body temperature consists of two parts:

1 A *heat-dissipating mechanism* that acts to increase heat loss when blood temperature increases above a certain point (Figure 5-14). This mechanism therefore prevents body temperature from rising above normal under usual circumstances.

2 A *heat-gaining mechanism* that acts to accelerate catabolism and thereby to increase heat production when blood temperature

An understanding of how heat is lost from the body by radiation is especially helpful in understanding why children are often at greater risk than adults when exposed to extremely cold weather. The body of a child has a much larger surface area to body volume ratio than an adult. Heat production is a function of body volume. The more muscle mass you have the greater the potential for heat production during exercise. Heat loss by radiation, however, is a function not of body volume but of surface area and is therefore relatively greater in the child than the adult. Children have a larger surface area when compared to body volume than adults and thus lose heat at a more rapid rate. Comparing the surface area of the head in a 5-year-old child and an adult is a good example. In the child the surface area of the head is about 13% of the total body surface area, while in the adult the percentage drops to about 9%. If left uncovered, the head serves as an effective radiator for heat loss. It is particularly important therefore for a child to wear a hat in the wintertime to protect against excess loss of body heat with resulting **hypothermia.**

decreases below a certain point. Under ordinary conditions, this mechanism prevents body temperature from falling below normal.

Heat-Dissipating Mechanism

In the anterior part of the hypothalamus (a highly specialized area of the brain to be discussed in Chapter 11) lies a group of cells referred to collectively as the "human thermostat." These neurons are thermal receptors; that is, they are stimulated by a very slight increase in the temperature of the blood above the point at which the human thermostat is set—normally about 37° C. In a sense, one might say that these cells of the hypothalamus take the temperature of the blood circulating to them. Whenever it increases by as little as 0.01° above 37° C (or some other set point), these neurons send out impulses that eventually, via sympathetic nerves, reach sweat glands and blood vessels of the skin. They stimulate the body's 2 million or more sweat glands to increase their rate of secretion. Evaporation of the larger amount of sweat causes a greater heat loss from the skin. Dilation of surface blood vessels brings a larger

Figure 5-14 Homeostasis of body temperature. Scheme to show how heat-dissipating mechanism operates to maintain normal body temperature. In principle, it cancels out any heat gain by bringing about an equal heat loss. Under usual circumstances, this mechanism succeeds in preventing body temperature from rising above the upper limit of normal.

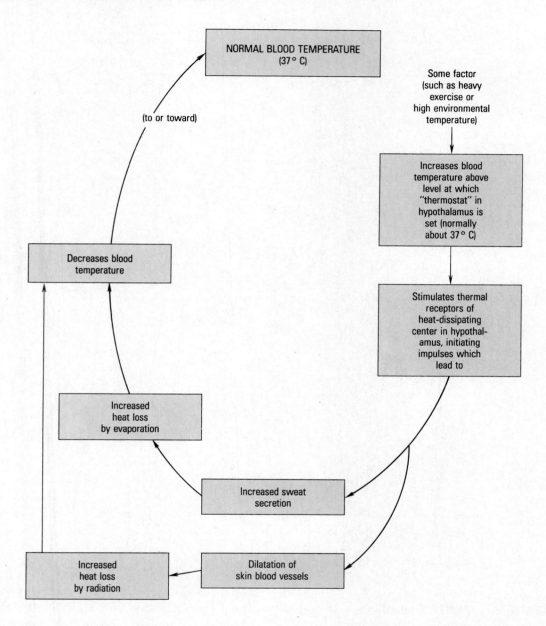

quantity of blood close to the surface for more heat loss by radiation. Blood vessel supply to the skin is profuse and far exceeds the needs of the skin. Such an abundant blood supply serves primarily for regulation of body temperature.

Heat-Gaining Mechanism

In a cold environment the mechanism that tries to maintain homeostasis of body temperature includes two kinds of responses—those that decrease heat loss and those that increase heat production. Together they almost always succeed in preventing a decrease in blood temperature below the lower limit of normal.

Decrease of heat loss results from reduced amounts of sweat secretion and from vasoconstriction of the skin blood vessels. Increased heat production occurs as a result of shivering and voluntary

Figure 5-15 "Rule of nines" is used to estimate amount of skin surface burned.

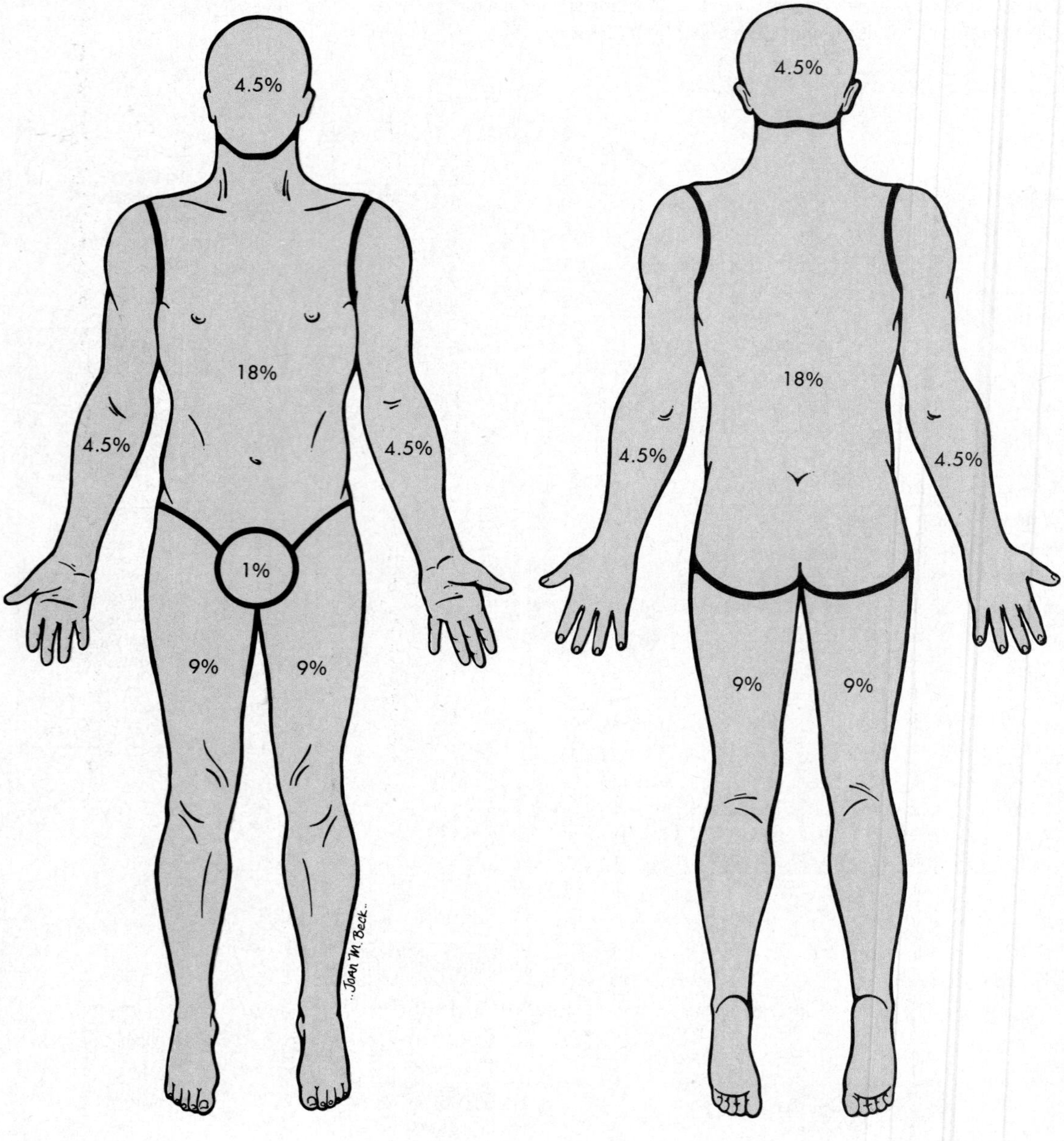

4.5%

4.5%

18%

4.5%   4.5%

1%

9%   9%

18%

4.5%   4.5%

9%   9%

John W. Beck

muscle contractions. Both kinds of muscle work accelerate catabolism and heat production.

SKIN THERMAL RECEPTORS

In addition to the thermal receptors in the hypothalamus, there are many heat and cold receptors located in the skin. Impulses initiated in skin thermal receptors travel to the cerebral cortex sensory area. Here they give rise to sensations of skin temperature and are relayed out over voluntary motor paths to produce skeletal muscle movements that affect skin temperature. For example, on a hot day you "feel hot" because of stimulation of your skin receptors.

Often you make some sort of movements to "cool yourself off." You may start fanning yourself, or perhaps you turn on an air conditioner or go swimming. As a result, your skin temperature decreases to a more comfortable level. Thus the thermal receptors in the skin will have taken part in a conscious mechanism that helps regulate skin temperature. Convincing evidence supports the view that this is their only function and that impulses from them do not travel to the heat-regulating centers in the hypothalamus, and therefore they do not take part in the automatic regulation of internal body temperature.

BURNS

Typically we think of a burn as a thermal injury or lesion caused by contact of the skin with some hot object or fire. In addition, overexposure to ultraviolet light (sunburn) or contact with an electric current or corrosive chemicals will also cause injury or death to skin cells. The injuries that result can all be classified as **burns.**

ESTIMATING BODY SURFACE AREA

When burns involve large areas of the skin, treatment and prognosis for recovery depend in large part on total area involved and severity of the burn. The severity of a burn is determined by the depth of the lesion as well as the extent (percent of body surface area burned). There are a number of ways to estimate the extent of body surface area burned. One method is called the **"rule of palms"** and is based on the assumption that the palm size of the burn victim is about 1% of the total body surface area. Therefore estimating the number of "palms" that are burned will approximate the percentage of body surface area involved.

The **"rule of nines"** (Figure 5-15) is another and more accurate method of determining the extent of a burn injury. Using this technique the body is divided up into 11 areas of 9%, with the area around the genitals called the **perineum** representing the additional 1% of body surface area. As Figure 5-15 shows, 9% of the skin covers the head and each upper extremity, including front and back surfaces. Twice as much, or 18%, of the total skin area covers the front and back of the trunk and each lower extremity, including front and back surfaces. The rule of nines works well with adults but does not reflect the differences in body surface area seen in small children. Special tables called **Lund-Browder Charts,** which take the large surface area of certain body

Older people who are confined to bed for prolonged periods are particularly susceptible to **bedsores** or **decubitus ulcers.** In this condition (Figure 5-16) skin covering bony prominences, such as the ankles, heels, and buttocks, is subjected to constant pressure caused by the body weight for long periods of time. If the patient is not turned or cannot change position in bed, the blood supply to the skin in these pressure areas will be reduced and tissue damage, infection, and ulceration will occur.

Figure 5-16 Decubitus ulcer caused by reduced blood flow.

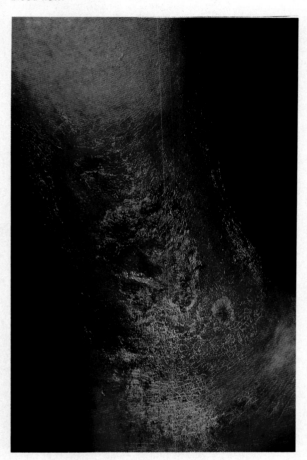

areas (such as the head) in the growing child into account, are used by physicians to estimate burn percentages in children.

The depth of a burn injury depends on the tissue layers of the skin that are involved. A **first-degree burn** (typical sunburn) will cause minor discomfort and some reddening of the skin. Although the surface layers of the burned area may peel in 1 or 2 days, no blistering occurs and actual tissue destruction is minimal. First- and second-degree burns (described below) are called **partial-thickness burns.**

Second-degree burns involve the deep epidermal layers and always cause injury to the upper layers of the dermis. In deep second-degree burns, damage to sweat glands, hair follicles, and sebaceous glands may occur but tissue death is not complete. Blisters, severe pain, generalized swelling, and edema characterize this type of burn. Scarring is common.

Third-degree or **full-thickness burns** are characterized by destruction of both the epidermis and dermis. Tissue death extends below the hair follicles and sweat glands. Burning may involve underlying muscles, fasciae, or even bone. A distinction between second- and third-degree burning is the fact that the third-degree lesion is insensitive to pain immediately after injury because of the destruction of nerve endings. Scarring is a serious problem.

AGE CHANGES IN THE SKIN

The most obvious of all age changes occur in the skin—wrinkles. Other familiar barometers of aging are dry skin, brown "age spots," gray or white hair—and less hair in both women and men. Other characteristic but less familiar changes occur in the glands of the skin and in its small blood vessels. Sebaceous glands become less active. Skin and hair that may have been oily and supple in youth become dry and brittle in old age. Sweat gland activity also decreases. This is one reason why old people adapt less well to extremely hot weather than do young people. Another age change, and one you may have noticed, is that old people seem to have more black and blue marks than do young people. The bruises stem from the increased fragility of small blood vessels in old skin.

Review of Medical Terms

albinism (AL-bi-nizm) A congenital or inherited condition characterized by a deficiency or complete absence of pigment in the eyes, hair, and skin, resulting from a defect in melanin synthesis.

blister (BLIS-ter) A thin fluid-filled vesicle that results from injury to cells in the epidermis or from separation of the dermal-epidermal junction by friction or burn injury.

callus (KAL-us) A hard, localized thickening of the horny layer of the skin caused by pressure or friction.

carcinogenic (KAR-cin-o-JEN-ik) Anything that will produce a carcinoma or cancerous growth.

carcinoma (KAR-cin-oma) A malignant or cancerous growth made up of epithelial cells that tend to spread to surrounding tissues and throughout the body.

comedo (KOM-e-do) A plug-like mass that blocks the excretory duct of a sebaceous gland in acne lesions; composed of sebum and sloughed epithelial cells and often contaminated with bacteria.

decubitus ulcer (de-KU-bi-tus) (L. "a laying down") An ulcer or sore caused by a reduced blood supply and constant pressure on the skin; often the result of prolonged bed rest in a patient who cannot move: **bedsore.**

dehydration (de-hi-DRA-shun) A condition that results from excessive loss of body water, as when fluids lost through sweating are not adequately replaced.

hyperkeratosis (HI-per-ker-ah-TO-sis) A condition characterized by a thick stratum corneum. The skin becomes thick, dry, scaly, and subject to fissures.

hypothermia (hypo-THER-me-a) Low body temperature. A disturbance in the homeostasis of body temperature caused by excessive heat loss or inadequate heat production.

melanoma (mel-AN-oma) A cancerous tumor made up of melanin-pigmented cells (malignant melanoma).

metastasis (me-TAS-tah-sis) The spread of a disease from one body region or organ to another not directly connected to it, such as the spread of cancer cells from a tumor in the lung to the brain.

pruritus (proo-RI-tus) Itching. Used to name various conditions characterized by itching, such as senile pruritus in the elderly caused by dry skin.

pustules (PUS-tuls) Small but visible elevations of the skin containing pus.

striae (STRI-e) Pinkish or purplish streaks in the skin that turn silvery-white with time, resulting from tears in the elastic fibers of the skin: **stretch marks.**

subcutaneous (SUB-ku-TA-ne-us) Literally means "beneath the skin." The area rich in fat and loose areolar tissue; also referred to as the hypodermis or superficial fascia.

vitiligo (vit-i-LI-go) A condition characterized by the absence of pigment in circumscribed areas of the skin; patches of depigmented skin.

Outline Summary—cont'd

c Empty contents into external ear canal either alone or with sebaceous glands

d Mixed secretions of sebaceous and ceruminous glands called cerumen

e Function of cerumen to protect area from dehydration but excess secretion can cause blockage of ear canal and loss of hearing

SURFACE FILM

A Emulsified protective barrier formed by mixing of residue and secretions of sweat and sebaceous glands with sloughed epithelial cells from skin surface

B Functions
1 Antibacterial, antifungal activity
2 Lubrication
3 Hydration of skin surface
4 Buffer

C Chemical composition
1 From epithelial elements
   a Amino acids
   b Sterols
   c Complex phospholipids
2 From sebum
   a Fatty acids
   b Triglycerides
   c Waxes
3 From sweat
   a Water and ammonia
   b Urea
   c Lactic acid and uric acid

HOMEOSTASIS OF BODY TEMPERATURE

A To maintain homeostasis of body temperature, heat production must equal heat loss; skin plays a critical role in this process

B Heat production
1 By metabolism of foods in skeletal muscles and liver

C Heat loss
1 By physical processes of evaporation, radiation, conduction, and convection
2 About 80% of heat loss occurs through skin; the rest takes place through mucosa of respiratory, digestive, and urinary tracts

D Thermostatic control of heat production and loss
1 Heat-dissipating mechanism—see Figure 5-14
2 Heat-gaining mechanism
   a Details not established but mechanism activated by decrease in blood temperature
   b Responses—skin blood vessel constriction, shivering, and voluntary muscle contractions

E Skin thermal receptors
1 Stimulation of skin thermal receptors gives rise to sensations of heat or cold; also initiates voluntary movements to reduce these sensations: for example, fanning to cool off or exercising to warm up

BURNS

A Defined as injury or death to skin cells caused by heat, ultraviolet light, electric current, or corrosive chemicals

B Severity of burn injury is determined by depth of lesion and percent of body surface burned

C Estimating body surface area
1 "Rule of palms"—based on assumption that palm size of burn victim is about 1% of body surface, therefore estimating the number of "palms" burned will approximate the percentage of body surface involved
2 "Rule of nines"—9% of total skin area covers head and each upper extremity, including front and back surfaces, 18% of total skin area covers each of the following: front of trunk, back of trunk, and each lower extremity, including front and back surfaces
3 Lund-Browder Charts—make allowances for the large percent of surface area in certain body regions in children (such as the head); permit more accurate estimates of burned surface area in children

D First-degree burn
1 Minor pain—no real tissue destruction
2 Some reddening of skin
3 No blistering but some peeling of surface occurs
4 No scarring

E Second-degree burn
1 Severe pain
2 Damage or destruction of epidermis and upper dermal layers
3 Blisters form with swelling and edema
4 Dermal tissue death is not complete but scarring is common

F Third-degree burn
1 Total destruction of both epidermis and dermis
2 Tissue death extends below level of hair follicles and sweat glands
3 No immediate pain; nerve endings are destroyed
4 Burning may involve deep tissues, including muscle and bone
5 Scarring is a serious problem

G Classification of burns by depth of injury
1 Partial-thickness burn—includes both first- and second-degree burns
2 Full-thickness burn—refers to third-degree burns only

AGE CHANGES IN THE SKIN

A Surface changes—wrinkles, brown "age spots," black and blue marks

B Hair color usually changes to gray or white

C Decresed secretion by sebaceous and sweat glands, resulting in dry skin and hair and poor adjustment to extremely hot weather

Review Questions

1 How do the terms *integument* and *integumentary systems* differ in meaning?
2 List and briefly discuss several of the different functions of the skin.
3 Identify the two main layers of the skin from superficial to deep. How are these layers related to the dermal-epidermal junction and the subcutaneous layer?
4 What are the three primary cell types of the epidermis?
5 List and describe the cell layers of the epidermis from superficial to deep.
6 What layer of the epidermis is sometimes called the *barrier area?*
7 Discuss the process of epidermal growth and repair.
8 What is keratin? Where is it found and how is it formed?
9 What part of the skin contains blood vessels?
10 What is the dermal-epidermal junction and how does it function?
11 Why is the process of blister formation a good example of the relationship between the skin's structure and function?
12 List the two layers of the dermis. Which layer helps make the skin stretchable and able to rebound?
13 What are arrector pili muscles?
14 What are Langer's cleavage lines? Why are these lines important to surgeons?
15 Identify and compare the three most common types of skin cancer. Discuss the possible causes of each type.

16 What is melanin? Where is it found in the skin?
17 What is the difference between albinism and vitiligo?
18 List the appendages of the skin.
19 Identify each of the following: hair papilla, germinal matrix, hair root, hair shaft, follicle.
20 List the three primary types of skin glands.
21 What is the difference between eccrine and apocrine sweat glands?
22 Discuss the importance of the surface film of the skin.
23 What is the chemical composition of skin surface film? How is the chemical composition related to its protective function?
24 Discuss the mechanism of homeostasis of body temperature. How is heat lost from the body?
25 Discuss the thermostatic control of heat production and loss.
26 How is the "rule of nines" used in determining the extent of a burn injury?
27 What are the differences between first-, second-, and third-degree burns?
28 Discuss age changes in the skin. What is a decubitus ulcer?
29 Define the following terms: *blister, callus, comedo, dehydration, hyperkeratosis, pruritus.*

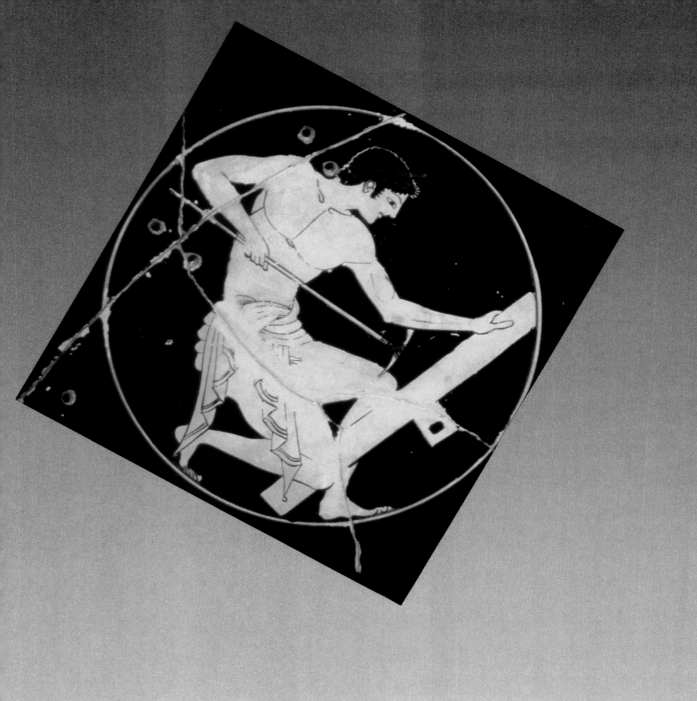

UNIT TWO

SUPPORT AND MOVEMENT

6 SKELETAL TISSUES AND PHYSIOLOGY

OBJECTIVES

After you have completed this chapter, you should be able to:

1 List and discuss the generalized functions of the skeletal system.
2 List the four types of bones and give examples of each.
3 Identify the six major structures of a typical long bone.
4 Identify each of the major constituents of bone as a tissue and discuss how structural organization contributes to function.
5 Identify by name and discuss each of the major components of a haversian system.
6 List, explain, and give examples of the most common bone markings.
7 Compare and contrast the development of intramembranous and endochondral bone.
8 Discuss bone growth, resorption, and the response of bone to stress.
9 Describe bone fractures and the steps involved in bone repair.
10 Compare the basic structural units of bone and cartilage.
11 Identify the three specialized types of cartilage, give examples of each, and summarize the structural and functional differences between them.
12 Compare the mechanism of growth in bone and cartilage.
13 Discuss osteoporosis.
14 Discuss electrically induced osteogenesis.

TISSUES OF THE SKELETAL SYSTEM

The skeletal system consists primarily of two highly specialized connective tissues—**bone** and **cartilage.** Individual bones are classified as organs and will be discussed in Chapter 7. Articulations or joints are points of contact between bones that make movement possible. Chapter 8 considers articulations.

FUNCTIONS OF BONE

Bones perform five functions for the body. Three of these you probably already know, but two of them may come as a surprise.

1 **Support.** Bones serve as supporting framework of the body much as steel girders are the supporting framework of our modern buildings.

2 **Protection.** Hard, bony "boxes" protect delicate structures enclosed by them. For example, the skull protects the brain, and the rib cage protects the lungs and heart.

3 **Movement.** Bones with their joints constitute levers. Muscles are anchored firmly to bones. As muscles contract and shorten, they pull on bones, thereby producing movement at a joint.

4 **Mineral reservoir.** Bones serve as the major storage depot for calcium, phosphorus, and certain other minerals. Homeostasis of blood calcium concentration—an essential for healthy survival—depends largely upon changes in the rates of calcium movement between the blood and bones. If, for example, blood calcium concentration increases above normal, both rates change. Calcium moves more rapidly out of the blood into bones and more slowly in the opposite direction. Result? Blood calcium concentration decreases—usually to its homeostatic level.

5 **Hemopoiesis.** Hemopoiesis, or blood cell formation, is a vital process carried on by red bone marrow (myeloid tissue). In an infant's or child's body, virtually all of the bones contain red marrow but over the years much of it becomes transformed into yellow marrow, an inactive, fatty tissue. The main bones in an adult's body that contain red marrow are those of the chest, spinal column, base of the skull, upper arm, and thigh. The very location of red marrow—hidden in bones like valuables in a safety deposit box—suggests its importance. Without its functioning, survival becomes impossible. Why? Because red marrow forms blood cells, and blood cells perform various functions essential for maintaining life (see Chapter 15).

KEY TERMS

Apatite (AP-uh-tite)

Callus (KAL-us)

Cancellous (kan-SEL-us)

Chondrocyte (KON-dro-site)

Compact (KOM-pakt)

Diaphysis (di-AF-i-sis)

Endochondral (EN-do-KON-dral)

Endosteum (en-DOS-tee-um)

Epiphysis (e-PIF-i-sis)

Fracture (FRAK-tur)

Haversian (ha-VER-shun)

Lacuna (la-KEW-nah)

Lamellae (lah-MEL-e)

Matrix (MAE-triks)

Ossification (os-i-fi-KAY-shun)

Osteoblast (OS-te-o-blast)

Osteocyte (OS-te-o-site)

Perichondrium (per-i-KON-dree-um)

Periosteum (per-e-OS-te-um)

Table 6-1 Types of bones			
Long	**Short**	**Flat**	**Irregular**
Humerus	Carpals	Frontal	Vertebrae, including sacrum and coccyx
Ulna	Tarsals	Parietal	
Femur		Temporal	Sphenoid
Tibia		Occipital	Ethmoid
Fibula		Ribs	Mandible
Metacarpals		Scapula	Face bones
Metatarsals		Patella	Ear bones
			Hyoid
			Innominate

ANATOMY OF BONES

TYPES OF BONES

There are four types of bones. Their names suggest their shapes: long bones, short bones, flat bones, and irregular bones. Table 6-1 classifies the bones of the skeleton according to these types.

MACROSCOPIC STRUCTURE
Long Bones

A long bone consists of the following structures visible to the naked eye: diaphysis, epiphyses, articular cartilage, periosteum, medullary (marrow) cavity, and endosteum. Identify each of these parts in Figure 6-1 as you read about it.

1 **Diaphysis**—main shaftlike portion. Its hollow, cylindrical shape and the thick compact bone that composes it adapt the diaphysis well to its function of providing strong support without cumbersome weight.

2 **Epiphyses**—both ends of a long bone. Epiphyses have a bulbous shape that provides generous space near joints for muscle attachments and also gives stability to joints. Look at Figure 6-2 to note the innumerable small spaces in the bone of the epiphysis. They make it look a little like a sponge—hence its name, spongy, or cancellous, bone. Yellow marrow fills the spaces of cancellous bone in most adult epiphyses but not in the proximal epiphyses of the humerus and femur. These contain red marrow.

3 **Articular cartilage**—thin layer of hyaline cartilage that covers articular or joint surfaces of epiphyses. Resiliency of this material cushions jars and blows.

4 **Periosteum**—dense white fibrous membrane that covers bone except at joint surfaces, where

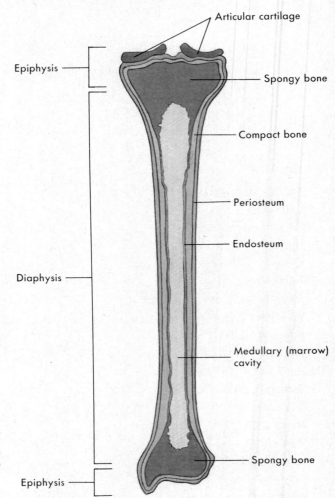

Figure 6-1 Structure of a long bone as seen in longitudinal section.

Labels (top to bottom): Articular cartilage; Epiphysis; Spongy bone; Compact bone; Periosteum; Endosteum; Diaphysis; Medullary (marrow) cavity; Spongy bone; Epiphysis

articular cartilage forms the covering. Many of the periosteum's fibers penetrate the underlying bone, welding these two structures to each other. (The penetrating fibers are called *Sharpey's fibers*.) Muscle tendon fibers interlace with periosteal fibers, thereby anchoring muscles firmly to bone. The periosteum contains many small blood vessels that send branches into the bone. Bone-forming cells called *osteoblasts* compose the inner layer of the periosteum. Because of its blood vessels and osteoblasts, the periosteum is essential for bone cell survival and for bone formation, a process that continues throughout life.

5 **Medullary (or marrow) cavity**—tubelike hollow in diaphysis of long bone. In the adult it contains yellow or fatty marrow.

6 **Endosteum**—membrane that lines medullary cavity of long bones.

Figure 6-2 A, Longitudinal section of
long bone showing both cancellous and
compact bone. **B,** Cutaway section of a
long bone.

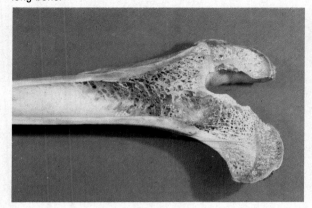

A

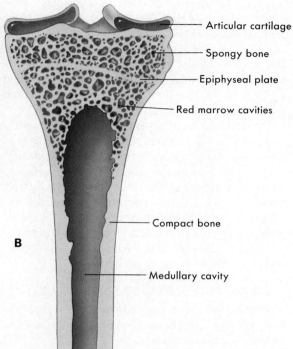

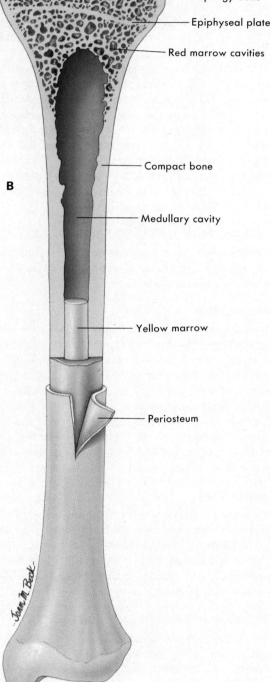

Articular cartilage

Spongy bone

Epiphyseal plate

Red marrow cavities

Compact bone

Medullary cavity

Yellow marrow

Periosteum

B

Larger quantities of red marrow can also be
aspirated for use in **marrow transplants.**
In certain types of cancer treatments or following accidental radiation poisoning, the
red marrow is destroyed and must be replaced for survival. Highly specialized techniques are required to remove large quantities of living marrow from a donor and
then to successfully reinject this delicate
tissue into a recipient without destroying
the tissue's ability to function.

Short, Flat, and Irregular Bones

Short bones, flat bones, and irregular bones all have
an inner portion of cancellous bone covered over on
the outside with compact bone. Red marrow fills the
spaces in the cancellous bone inside a few irregular
and flat bones—for example, in the vertebrae and
sternum. To help in the diagnosis of leukemia and
certain other diseases, a physician may decide to
perform a needle puncture of one of these bones. In
this type of diagnostic procedure, a needle is inserted
through the skin and compact bone into the red
marrow and a small amount of the marrow is then
aspirated and examined under the microscope for
normal or abnormal blood cells.

BONE (OSSEOUS) TISSUE

Bone is perhaps the most distinctive form of connective tissue in the body. It is typical of other connective tissues in that it consists of cells, fibers and extracellular material or **matrix.** It is unique, however, in that its extracellular components are hard and **calcified.** In bone the extracellular material or matrix predominates. It is much more abundant than the bone cells, and it contains many fibers of collagen (the body's most abundant protein). The rigidity of bone enables it to serve supportive and protective functions.

As a tissue, bone is ideally suited to its functions and the concept that structure and function are interrelated is apparent in this highly specialized tissue type. It has a tensile strength nearly equal to cast iron but at less than one third the weight. Bone is *organized* so that its great strength and minimal weight result from the interrelationships of its structural components. The relationship of structure to function is apparent in its chemical, cellular, tissue, and organ levels of organization.

CHEMICAL COMPOSITION OF BONE MATRIX

The extracellular material of bone can be subdivided into two principal chemical components: **inorganic salts** and **organic matrix.**

Inorganic Salts

The calcified nature and thus the hardness of bone results from the deposition of highly specialized chemical crystals of calcium and phosphate called **apatite.** The slender needle-like apatite crystals are about 300 Å in length by 20 Å in thickness. They are oriented in the organic components of the bone so that they can most effectively resist stress and mechanical deformation. In addition to calcium and phosphate, other mineral constituents such as magnesium and sodium are also found in bone.

Organic Matrix

The organic matrix of bone is a composite of collagenous fibers and an amorphous mixture of protein and polysaccharides called *ground substance.* Connective tissue cells secrete the gel-like and homogeneous ground substance that surrounds the fibers found in bone matrix.

MICROSCOPIC STRUCTURE OF BONE

The microscopic structure of bone is as unique as its chemical composition. Compact bone's matrix consists of structural units called **haversian systems** (in honor of Clopton Havers, a seventeenth century English anatomist who first described them). Four structures make up each haversian system: lamellae, lacunae, canaliculi, and a haversian

The disastrous nuclear accident that occurred in the Soviet Union early in 1986 released large quantities of what are described as **bone-seeking isotopes** into the environment. Nuclear reactors generate numerous radioactive elements in the fission of uranium or plutonium. Radioactive strontium is one of the most hazardous bone-seeking isotopes produced in nuclear fission reactors. Once ingested, it will quickly substitute for calcium in the apatite crystals of bone and damage the red marrow and other body tissues by radioactive emissions.

canal. As you read the following definitions, identify each structure in Figure 6-3.

lamellae Concentric, cylinder-shaped layers of calcified matrix

lacunae (Latin for "little lakes") small spaces containing tissue fluid in which bone cells (osteocytes) lie imprisoned between the hard layers of the lamellae

canaliculi Ultrasmall canals radiating in all directions from the lacunae and connecting them to each other and to a larger canal, the haversian canal

haversian canal Extends lengthwise through the center of each haversian system; contains blood vessels and lymphatic vessels; from the haversian canal, tissue fluid moves through canaliculi to the lacunae and their bone cells—a short distance of about 0.1 mm or less—bringing them nutrients and oxygen

Thousands of haversian systems make up the compact bone of a single long bone's diaphysis. Because of their cylindrical shape, haversian systems do not fit snugly together. Spaces between and around them are filled in by the **interstitial lamellae** and **circumferential lamellae.**

Cancellous (spongy) bone differs in microscopic structure from **compact bone.** In compact bone, the concentric arrangement of lamellae forms haversian systems (Figure 6-4). Cancellous bone has no haversian systems. Instead, it consists of a web-like arrangement of marrow-filled spaces separated by thin processes of bone called *trabeculae* (see Figure 6-3). An interesting fact about trabeculae is that they become arranged along the lines of stress placed on individual bones—a feature that enhances a bone's strength.

Bones are not the lifeless structures they seem to be. Within this hard, seemingly lifeless material lie the many living bone cells. They must, like all living cells, continually receive food and oxygen and excrete their wastes. So blood supply to bone is

Figure 6-3 Drawing of microscopic structure of bone. Haversian systems, several of which are shown here, compose compact bone. Note the structures that make up one haversian system: concentric lamellae, lacunae, canaliculi, and a haversian canal. Shown bordering the compact bone on the left is spongy bone, a name descriptive of the many open spaces that characterize it.

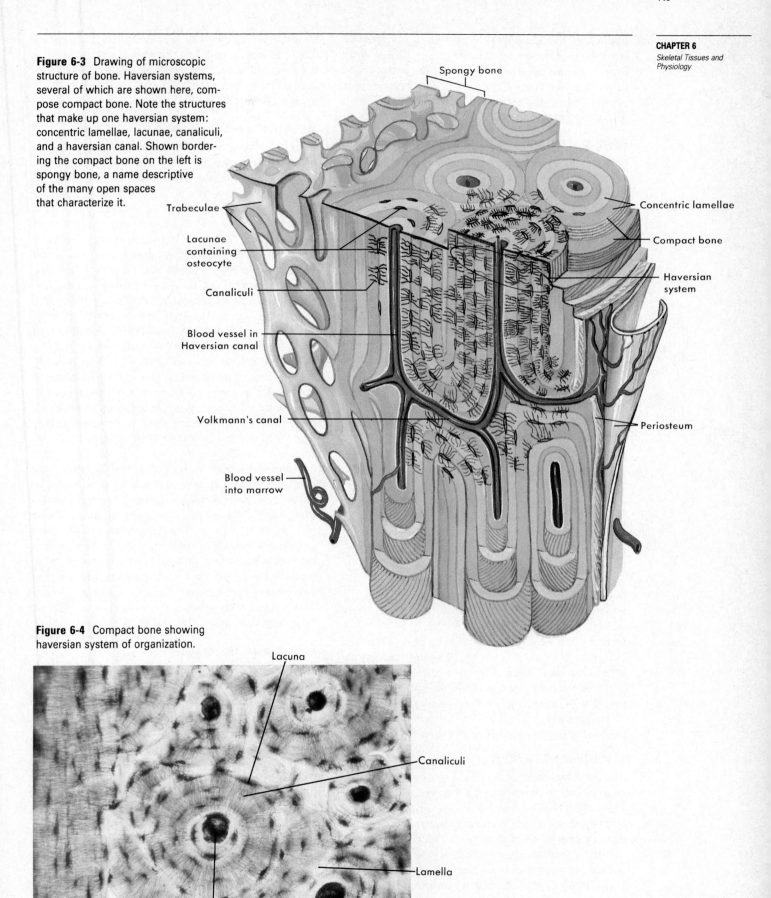

Figure 6-4 Compact bone showing haversian system of organization.

important and abundant. Numerous blood vessels from the periosteum penetrate bone by way of Volkmann's canals to connect with blood vessels of a haversian canal (see Figure 6-3). Also, one or more arteries supply the bone marrow in the internal medullary cavity of long bones.

BONE MARKINGS

Various points on bones are labeled according to the nature of their structure. This method of identifying definite parts of different bones proves helpful when locating other structures such as muscles, blood vessels, and nerves. Definitions of some common bone markings follow. Review these as you study the structure of individual bones.

Depressions and Openings

The following depressions and openings are considered bone markings:

fossa A hollow or depression (e.g., mandibular fossa of the temporal bone that serves as the socket for the lower jawbone)

sinus A cavity or spongelike space in a bone (e.g., the frontal sinus)

foramen A hole (e.g., foramen magnum of the occipital bone)

meatus A tube-shaped opening (e.g., the external auditory meatus)

Projections or Processes

The following projections or processes enter into the formation of joints:

condyle A rounded projection (e.g., condyles of the femur)

head A rounded projection beyond a narrow necklike portion (e.g., head of the femur)

Muscles attach to the following projections:

trochanter A very large projection (e.g., greater trochanter of the femur)

crest A ridge (e.g., the iliac crest); a less prominent ridge is called a *line* (e.g., ileopectineal line)

spinous process or **spine** A sharp projection (e.g., anterior superior spine of the ilium)

tuberosity A large, rounded projection (e.g., ischial tuberosity)

tubercle A small, rounded projection (rib tubercles)

DEVELOPMENT OF BONE

The term **osteogenesis** is used to describe the process of bone formation. There are two different methods by which the body can form bones. The two processes are called **intramembranous** and **endochondral ossification.**

By 3 months after fertilization of a ovum, the skeletal system has been formed, although not of bone. Most of the skeletal structures at this time consist of hyaline cartilage structures shaped like bones. Fibrous membrane rather than hyaline cartilage, however, fashions the prebone structures of the skull, including the mandible. Parts of the clavicles are also formed in membranous structures that appear during this period of development.

Intramembranous Ossification

Intramembranous ossification takes place, as its name implies, within a connective tissue membrane. The flat bones of the skull, for example, begin to take shape when groups of mesenchyme cells differentiate into **osteoblasts** or bone-forming cells. These clusters of osteoblasts are called **centers of ossification.** They secrete matrix material and collagenous fibrils. The Golgi apparatus in an osteoblast specializes in synthesizing and secreting carbohydrate compounds of the type called *mucopolysaccharides,* and its endoplasmic reticulum makes and secretes collagen, a protein. In time, relatively large amounts of the mucopolysaccharide substance or *ground substance* accumulate around each individual osteoblast. Numerous bundles of collagenous fibers become embedded in the cement substance. Together, the cement substance and collagenous fibers constitute the organic extracellular substance of bone called the **bone matrix.** Calcification of the organic bone matrix occurs when complex calcium salts are deposited in it.

As calcification of bone matrix continues, rodlike **trabeculae** appear and join in a network of interconnecting spicules to form **spongy** or **cancellous bone.** In time the core layer of spongy bone will be covered on each side by plates of compact or dense bone. Once formed, a flat bone grows in size by the addition of osseous tissue to its outer surface. The process is called **appositional growth.** Flat bones cannot grow by expansion as is the case with endochondral bone growth described in the following section.

Endochondral Ossification

Most bones of the body are formed from hyaline cartilage models. The steps of endochondral ossification are illustrated in Figure 6-5. The cartilage model of a typical long bone, such as the tibia, can be identified early in embryonic life (Figure 6-5, *A*). The cartilage model then develops a *periosteum* (Figure 6-5, *B*) that soon enlarges and produces a ring or **subperiosteal collar** of bone. Bone is deposited by osteoblasts, which differentiate from cells on the inner surface of the covering periosteum. Soon after the appearance of the ring of bone, the cartilage begins to calcify (Figure 6-5, *C*) and a **primary ossification center** forms when a blood vessel enters the rapidly changing cartilage model at the midpoint of the diaphysis (Figure 6-5, *D*). Endochondral ossification progresses from the diaph-

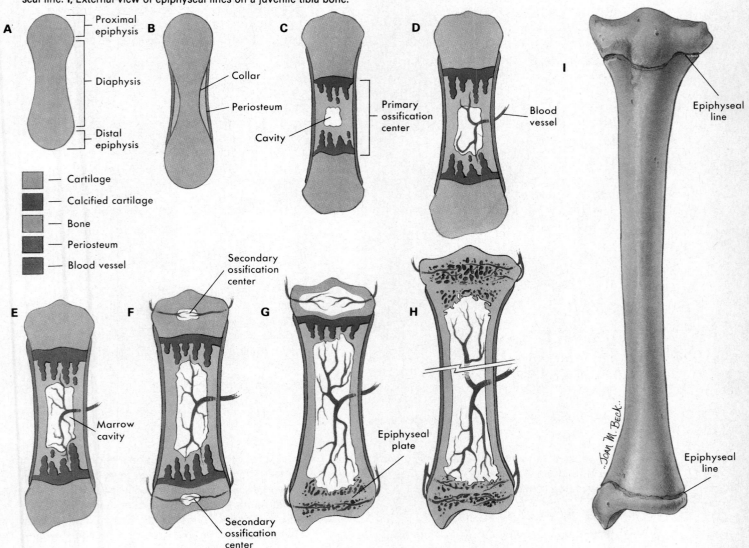

Figure 6-5 Endochondral bone formation. **A,** Cartilage model. **B,** Subperiosteal bone collar formation. **C,** Development of primary ossification center. **D,** Entrance of blood vessel. **E,** Prominent marrow cavity, with thickening and lengthening of collar. **F,** Development of secondary ossification centers in epiphyseal cartilage. **G,** With cessation of bone growth, lower, then upper, epiphyseal plates disappear. **H,** Appearance of mature bone showing continuous marrow cavity and residual epiphyseal line. **I,** External view of epiphyseal lines on a juvenile tibia bone.

ysis toward each epiphysis (Figure 6-5, *E*) and the bone grows in length. Eventually, **secondary ossification centers** appear in the epiphyses (Figure 6-5, *F*) and bone growth proceeds toward the diaphysis from each end (Figure 6-5, *H*).

Until bone growth in length is complete, a layer of cartilage known as **epiphyseal cartilage** remains between each epiphysis and the diaphysis. Proliferation of epiphyseal cartilage cells brings about a thickening of the layer of epiphyseal cartilage during periods of growth. Ossification of this additional cartilage then follows—that is, osteoblasts synthesize organic bone matrix and the matrix undergoes cal-

cification. As a result, the bone becomes longer. It is the epiphyseal plate that allows the diaphysis of a long bone to increase in length.

The progressive ossification of the distal femur and proximal tibia and fibula in the knee joint is illustrated in Figure 6-6. Note how the bones both increase in size and are "remodeled" so that maintenance of the characteristic shape and proportions of the bones can be retained during the growth period. When epiphyseal cartilage cells stop multiplying and the cartilage has become completely ossified, bone growth has ended. Radiographs can reveal any epiphyseal cartilage still present. When bones have

Figure 6-6 Drawings from radiographs showing ossification of bones forming the male knee joint.

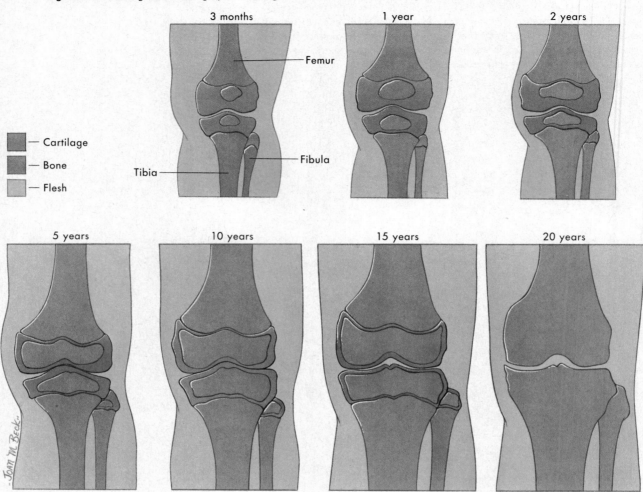

grown their full length, the epiphyseal cartilage disappears. Bone has replaced it and is then continuous between epiphyses and diaphyses.

BONE GROWTH AND RESORPTION

Bones grow in diameter by the combined action of two special kinds of cells: osteoblasts and osteoclasts. Osteoclasts enlarge the diameter of the medullary cavity by eating away the bone of its walls. At the same time, osteoblasts from the periosteum build new bone around the outside of the bone. By this dual process a bone with a larger diameter and larger medullary cavity has been produced from a smaller bone with a smaller medullary cavity.

The formation of bone tissue continues long after bones have stopped growing in size. Throughout life, bone formation (ossification) and bone destruction (resorption) go on concurrently. These opposing processes balance each other during adulthood's early to middle years. The rate of bone formation, or ossification, equals the rate of bone destruction,

or resorption. Bones, therefore, neither grow nor shrink. They stay constant in size. Not so, in the earlier years. During childhood and adolescence, ossification goes on at a faster rate than bone resorption. Bone gain outstrips bone loss, and bones grow larger. But between the ages of 35 and 40 years, the tables turn, and from that time on bone loss exceeds bone gain. Bone gain occurs slowly at the outer or periosteal surfaces of bones. Bone loss, on the other hand, occurs at the inner or endosteal surfaces and goes on at a somewhat faster pace. More bone is lost on the inside than gained on the outside, and inevitably bones become remodeled as the years go by. For instance, the thickness of the compact bone in the diaphyses of long bones decreases, and the diameter of their medullary cavities increases. Gradually they become more like hollow shells and are less able to resist compression and bending. Vertebrae, to some degree, collapse, and height decreases. Femurs frequently break, and this ominous event may even lead indirectly to death.

The point of articulation between the epiphysis and diaphysis of a growing long bone is susceptible to injury if overstressed—especially in the young child or preadolescent athlete. In these individuals, the epiphyseal plate can be separated from the diaphysis or epiphysis causing an **epiphyseal fracture** (Figure 6-7).

Osteoporosis is one of the most common and serious of all bone diseases. It is characterized by excessive loss of both calcified matrix and collagenous fibers from bone. Osteoporosis occurs most frequently in white, elderly females. Although both white and black males are also susceptible, black women are seldom affected.

Since both estrogen and testosterone serve important roles in stimulating osteoblast activity after puberty, decreasing blood levels of these hormones in the elderly reduces new bone growth and maintenance of existing bone mass. Therefore some resorption of bone and subsequent loss of bone mass is an accepted consequence of advancing years. However bone loss in osteoporosis goes far beyond the modest decrease normally seen in old age. The result is a dangerous pathological condition resulting in bone degeneration, increased susceptibility to "spontaneous fractures," and pathological curvature of the spine. Treatment may include estrogen therapy and dietary supplements of calcium and vitamin D to replace deficiencies or offset intestinal malabsorption.

RESPONSE TO STRESS

Walking, jogging, and other forms of exercise subject bones to stress. They respond by laying down more collagen fibers and mineral salts in the bone matrix. This, in turn, makes bones stronger. But inactivity and lack of exercise tend to weaken bones because of decreased collagen formation and excessive calcium withdrawal. To prevent these changes as well as many others, astronauts performed special exercises regularly in space.

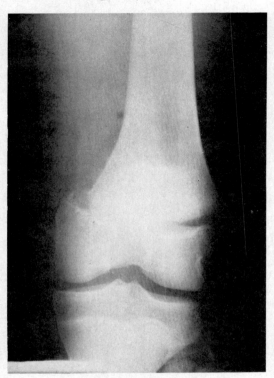

Figure 6-7 Epiphyseal fracture of the distal femur in a young athlete. Note the separation of the diaphysis and epiphysis at the level of the "growth plate."

BONE FRACTURES AND REPAIR

The term **fracture** is defined as a break in the continuity of a bone. **Fracture healing** is considered the prototype of bone repair. The complex bone repair process that follows a fracture is apparently initiated by bone death or by damage to periosteal and haversian system blood vessels.

Types of Fractures

A fracture is classified as **simple** if the skin remains unbroken and **compound** if the broken ends of the bone protrude through the skin. Compound fractures are easily infected by foreign material or skin bacteria and present special treatment problems.

In order for effective healing to take place, the broken ends of the fractured bone must be *aligned* and *immobilized*. The term **reduction** is used to signify the proper set or alignment of the fractured bone. If the fracture can be set without opening the skin, the procedure is called a *closed reduction;* if a surgical incision is required in order to view the bone fragments before aligning them, the procedure is called an *open reduction.* If circumstances permit, closed reductions are used in order to minimize the

Figure 6-8 Bone fracture and repair. **A**, Fracture of femur. **B**, Formation of fracture hematoma. **C**, Formation of internal and external callus. **D**, Bone remodeling complete.

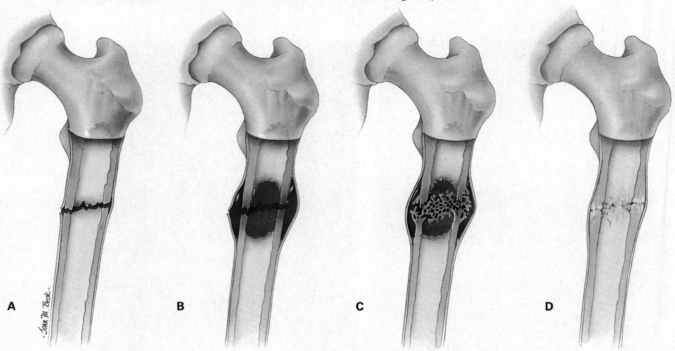

ELECTRICALLY INDUCED OSTEOGENESIS

Bone fractures are occasionally so severe that healing may be slow or impossible. These "nonunion" fractures are extremely difficult to treat. In such cases physicians have found that the passage of a weak electrical stimulus through the fracture site induces new bone formation and accelerates the process of bone repair. Small electrodes are inserted into the medullary canal of the bone near the fracture site and a weak electrical current is passed between them. Bone formation is enhanced near the cathode using pulsed DC current, DC current, and AC current. Equipment generating DC current in the range of 1 to 100 mA or an applied voltage of 1.0 to 1.5 V is most common. Evidence now suggests that changes in pH gradients near the cathode are responsible for the electrically induced osteogenesis.

risk of infection to the surrounding tissue or bone. Infection of the bone called **osteomyelitis** is a serious complication of many compound fractures.

Fractures in children are often different from fractures in adults. In children, the bones are pliable and more resilient. As a result, the fractured bone may not be completely broken into two parts. Such fractures often resemble a break in a young tree branch and are called **incomplete** (partial) or **greenstick fractures.**

Complete fractures are those that totally break or divide the bone into separate pieces. One has only to consult a medical dictionary to see how many types of fractures are identified by name.

Fracture Healing

A bone fracture invariably tears and destroys blood vessels that carry nutrients to osteocytes. It is this vascular damage that initiates the repair sequence. Eventually, dead bone is either removed by osteoclastic resorption or serves as a scaffolding or framework for the deposition of a specialized repair tissue called **callus.**

Vascular damage occurring immediately after a fracture results in hemorrhage and the pooling of blood at the point of injury. The resulting blood clot is called a **fracture hematoma** (Figure 6-8, *A*). As the hematoma is resorbed, the formation of spe-

Figure 6-9 Tomograms of a fractured right tibia. Radiographs show successive sections or slices of the fracture site from superficial **(A)** to deep **(F)**.

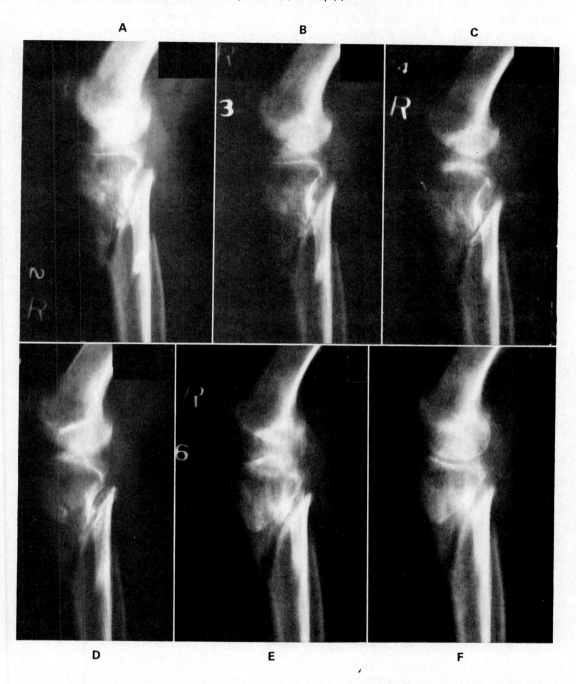

cialized callus tissue occurs. It serves to bind the broken ends of the fracture on both the outside surface and along the marrow cavity internally. The rapidly growing callus tissue effectively "collars" the broken ends and stabilizes the fracture so that healing can proceed (Figure 6-8, *B*). If the fracture is properly aligned and immobilized and if complications do not develop, callus tissue will be actively "modeled" and eventually replaced with normal bone as the injury heals completely (Figure 6-8, *C*).

The series of radiographs seen in Figure 6-9 are serial **tomograms.** Tomograms are successive "sections" or "slices" taken with a computerized x-ray machine aligned so that multiple two-dimensional views of one particular area of tissue can be visualized. In Figure 6-9 successive sections of a fractured tibia can be seen from superficial *(A)* to deep *(F)*.

CARTILAGE

TYPES

Cartilage is classified as connective tissue and consists of three specialized types called **hyaline, elastic,** and **fibrocartilage.** As a tissue, cartilage both resembles and differs from bone. Innumerable collagenous fibers reinforce the matrix of both tissues, and, like bone, cartilage consists more of extracellular substance than of cells. However in cartilage the fibers are embedded in a firm gel instead of in a calcified cement substance. Hence cartilage has the flexibility of a firm plastic material, whereas bone has the rigidity of cast iron. Another difference is that no canal system and no blood vessels penetrate the cartilage matrix. Cartilage is avascular and bone is abundantly vascular. Nevertheless, cartilage cells, like bone cells, lie in lacunae. However, because no canals and blood vessels interlace cartilage matrix, nutrients and oxygen can reach the scattered, isolated **chondrocytes** (cartilage cells) only by diffusion. They diffuse through the matrix gel from capillaries in the fibrous covering of the cartilage—the **perichondrium**—or from synovial fluid in the case of articular cartilage.

The three cartilage types differ from one another largely by the amount of matrix material that is present and also by the relative amounts of elastic and collagenous fibers that are embedded in them. *Hyaline* is the most abundant type and both elastic and fibrocartilage varieties are considered as modifications of the hyaline type. Collagenous fibers are present in all three types but are most numerous in *fibrocartilage.* Hence it has the greatest tensile strength. *Elastic* cartilage matrix contains elastic fibers as well as collagenous fibers and so has elasticity as well as firmness.

Cartilage is an excellent skeletal support tissue in the developing embryo. It forms rapidly and yet retains a significant degree of rigidity or stiffness. A majority of the bones that eventually form both the axial and appendicular skeleton described in Chapter 7 first appear as cartilage models. Skeletal maturation involves replacement of the cartilage models with bone.

After birth there is a decrease in the total amount of cartilage tissue present in the body. However it continues to play an important role in the growth of long bones until skeletal maturity and is found throughout life as the material that covers the articular surfaces of bones in joints. The three types of cartilage also serve numerous specialized functions throughout the body.

Hyaline Cartilage

Hyaline, in addition to being the most common type of cartilage, serves numerous specialized functions. It resembles milk glass in appearance (Figure 6-10).

In fact, its name is derived from the Greek word meaning "glassy." It is semitransparent and has a somewhat bluish, opalescent cast.

In the embryo, hyaline cartilage forms from differentiation of specialized mesenchymal cells that become crowded together in so-called **centers of chondrification.** As the cells enlarge, they secrete matrix material that surrounds the delicate fibrils. Eventually, the continued production of matrix separates and isolates the cells or **chondrocytes** into compartments, which, like in bone, are called *lacunae.* The two principal chemical components of matrix material (ground substance) are a mucoprotein called **chondroitin-sulfate** and a gel-like **polysaccharide.** Both substances are secreted from chondrocytes in much the same way as protein and carbohydrates are secreted from glandular cells.

In addition to covering the articular surfaces of bones, hyaline cartilage forms the costal cartilages that connect the anterior ends of the ribs with the sternum or breastbone. It also forms the cartilage rings in the trachea, bronchi of the lungs, and the tip of the nose.

Elastic Cartilage

Elastic cartilage gives form to the external ear, the epiglottis that covers the opening of the respiratory tract when swallowing, and the eustachian or auditory tubes that connect the middle ear and nasal cavity. The collagenous fibers of hyaline cartilage are also present—but in fewer numbers—in elastic cartilage. Large numbers of easily stained elastic fibers confer the elasticity and resiliency typical of this form of cartilage. In most stained sections, elastic cartilage has a yellowish color and has a greater opacity than the hyaline variety (Figure 6-10).

Fibrocartilage

Fibrocartilage is characterized by small quantities of matrix and abundant fibrous elements (Figure 6-10). It is strong, rigid and most often associated with regions of dense connective tissue in the body. It occurs in the symphysis pubis, intervertebral disks, and near the points of attachment of some large tendons to bones.

HISTOPHYSIOLOGY OF CARTILAGE

The gristle-like nature of cartilage permits it to sustain great weight when covering the articulating surfaces of bones or when serving as a shock-absorbing pad between articulating bones in the spine. In other areas, such as the external ear, nose, or respiratory passages, cartilage provides a strong yet pliable support structure that resists deformation or collapse of tubular passageways. It is cartilage that permits growth in length of long bones and is largely responsible for their adult shape and size.

Figure 6-10 Types of cartilage: **A,** Hyaline cartilage of trachea. **B,** Elastic cartilage of epiglottis. Note black elastic fibers in cartilage matrix and perichondrium layers on both surfaces. **C,** Fibrocartilage of intervertebral disk. (× 140.)

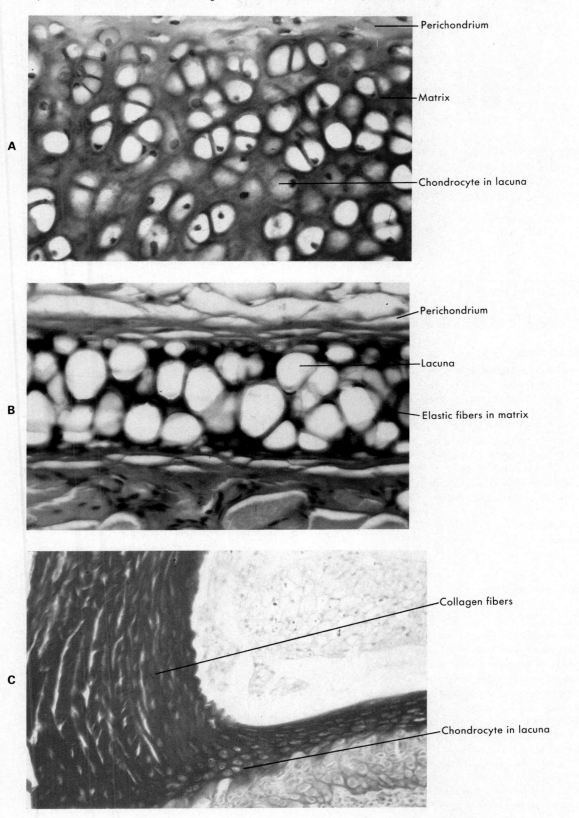

Certain nutritional deficiencies and other metabolic disturbances have an immediate and very visible impact on cartilage. It is for this reason that changes in cartilage often serve as indicators of inadequate vitamin, mineral, or protein intake. Vitamin A and protein deficiency, for example, will decrease the thickness of epiphyseal plates in the growing long bones of young children—an effect immediately apparent on x-ray examination. The opposite effect occurs in vitamin D deficiencies. As the epiphyseal cartilages increase in thickness but fail to calcify, the growing bones become deformed and bend under weight bearing. The bent long bones are a sign of **rickets.**

GROWTH OF CARTILAGE

The growth of cartilage occurs in two ways:

1 **Interstitial** or **endogenous** growth
2 **Appositional** or **exogenous** growth

During interstitial growth, cartilage cells within the substance of the tissue mass divide and begin to secrete additional matrix. Internal division of chondrocytes is possible because of the soft, pliable nature of cartilage tissue. This form of growth is most often seen during childhood and early adolescence when a majority of cartilage is still soft and capable of expansion from within.

Appositional or exogenous growth occurs when chondrocytes in the deep layer of the perichondrium begin to divide and secrete additional matrix. The new matrix is then deposited on the surface of the cartilage, causing it to increase in size. Appositional growth is unusual in early childhood but, once initiated, continues beyond adolescence and throughout life.

Outline Summary

FUNCTIONS OF BONES

A Furnish supporting framework
B Afford protection
C Movement—bones constitute levers for muscle action
D Mineral reservoir—bones serve as the major reservoir for calcium deposits and withdrawals, thereby playing an essential part in maintaining blood calcium homeostasis
E Hemopoiesis—blood cell formation by red bone marrow, that is, myeloid tissue

ANATOMY OF BONES

Types of bones

See Table 6-1

Macroscopic structure

A Long bones
  1 Diaphysis—hollow, shaftlike portion composed of thick compact or dense bone
  2 Epiphyses—both ends of long bone composed of spongy or cancellous bone; marrow fills spaces of cancellous bone (yellow marrow in most adult epiphyses but red marrow in proximal epiphyses of humerus and femur)
  3 Articular cartilage—thin layer of hyaline cartilage covering joint surfaces of epiphyses
  4 Periosteum—dense white fibrous membrane covering bone except at joint surfaces; firmly attached to underlying bone; muscle and tendons attached firmly to periosteum by interlacing fibers; periosteum contains blood vessels and bone-forming cells so essential for maintenance, growth, and repair of bones
  5 Medullary (marrow) cavity—hollow in diaphysis, filled with yellow (fatty) marrow
  6 Endosteum—membrane lining medullary cavity
B Short, flat, and irregular bones
  Cancellous bone forms inside of these bones and compact bone forms outside; red marrow in spaces of cancellous bone inside a few irregular and flat bones (e.g., vertebrae and sternum)

BONE (OSSEOUS) TISSUE

Chemical composition of bone matrix

A Consists of cells, fibers, and matrix
  1 Extracellular matrix is calcified
  2 Larger quantities of matrix than cells or fibers
  3 Tensile strength nearly equal to cast iron
  4 Highly organized structural units
B Chemical composition of bone matrix
  1 Inorganic salts
  2 Organic matrix
  3 Apatite
    a Crystals 300 Å length by 20 Å thickness
  4 Organic matrix or ground substance
    a Collagenous fibers
    b Protein
    c Polysaccharides

Microscopic structure

A Bone
  1 Mainly calcified matrix—organic matrix impregnated with calcium salts and reinforced by collagenous fibers

Outline Summary—cont'd

2 Matrix of compact bone—made up of thousands of haversian systems, i.e., structural units consisting of
  a Lamellae—concentric, cylinder-shaped layers of calcified matrix
  b Lacunae—microscopic spaces containing osteocytes (bone cells)
  c Canaliculi—microscopic canals radiating from lacunae, connecting them with haversian canals; routes by which tissue fluid reaches osteocytes
  d Haversian canal—extends lengthwise through center of each haversian system; contains blood and lymphatic vessels
3 Matrix of cancellous bone—consists of web-like arrangement of marrow-filled spaces separated by thin processes of bone called *trabeculae*

Bone markings

A Depressions and openings
  1 Fossa—hollow or depression
  2 Sinus—cavity or spongelike air spaces within bone
  3 Foramen—hole
  4 Meatus—tube-shaped opening
B Projections or processes
  1 Those that fit into joints
    a Condyle—rounded projection entering into formation of joint
    b Head—rounded projection beyond narrow neck
  2 Those to which muscles attach
    a Trochanter—very large process
    b Crest—ridge
    c Spinous process or spine—sharp projection
    d Tuberosity—large, rounded projection
    e Tubercle—small, rounded projection

Development of bone

A Osteogenesis—process of bone formation
  1 Types
    a Intramembranous—in membrane
    b Endochondral—from cartilage model
B Intramembranous ossification
  1 Occurs within connective tissue membrane
  2 Mesenchyme cells differentiate into osteoblasts
  3 Clusters of osteoblasts form centers of ossification
  4 Osteoblast secretion
    a Golgi—secrete mucopolysaccharides
    b Endoplasmic reticulum—secrete collagen
  5 Organic extracellular substance called *matrix*
  6 Calcification of matrix forms trabeculae
  7 Trabeculae form spongy (cancellous) bone
  8 Flat bones grow by "appositional growth"
  9 Examples—flat bones of skull
C Endochondral ossification
  1 Majority of bones develop from this process

2 Endochondral bones from hyaline cartilage models
3 Steps of process illustrated in Figure 6-5
  a Cartilage model
  b Development of periosteum and subperiosteal collar
  c Primary and secondary ossification centers develop
  d Epiphyseal cartilage appears
  e Progressive ossification to maturity
  f Examples—femur, humerus, tibia, and other long bones

Bone growth and resorption

A Growth in length—by continual thickening of epiphyseal cartilage followed by ossification
B Growth in diameter—medullary cavity enlarged by osteoclasts destroying bone around inner surface of medullary cavity while new bone is added around circumference by osteoblasts
C Opposing processes of bone tissue formation and destruction (resorption) go on concurrently throughout life
  1 Bone formation exceeds resorption during growth years, from infancy through adolescence
  2 Bone formation and resorption balance each other during young adulthood
  3 After young adulthood (age 35 to 40 years), more bone resorbed at endosteal surface than formed at periosteal surface; net loss of bone tissue weakens bones, causing them to fracture more easily

Response to stress

A Increased deposition of collagen fibers and calcium salts in matrix of bone
B Bone strength increased

Bone fractures and repair

A Fracture—break in continuity of a bone
B Fracture terms
  1 Simple—skin remains unbroken
  2 Compound—skin is broken
  3 Effective healing requires alignment and immobilization
  4 Reduction—proper set or alignment of fracture
    a Closed—set without opening skin
    b Open—cut skin to view fracture and align bones
  5 Osteomyelitis—bone infection
  6 Greenstick fracture—incomplete fracture in a child
  7 Complete fracture—bone divided into two separate pieces
C Fracture healing (Figure 6-8)
  1 Development of fracture hematoma

Outline Summary—cont'd

    2 Callus tissue formation
    3 Completion of healing

CARTILAGE

  A Types
    1 Hyaline
    2 Elastic
    3 Fibrocartilage
  B Generalized characteristics
    1 More extracellular matrix than cells
    2 Fibers in all types
    3 Matrix is firm but flexible
    4 Avascular
    5 Cells lie in lacunae
    6 Chondrocyte is basic cell type
    7 Perichondrium is covering membrane
    8 Nutrients to cells via diffusion
    9 Total amount in body decreases after birth
  C Hyaline cartilage
    1 Most common variety
    2 Appearance is glassy and opalescent
    3 From specialized mesenchymal cells called centers of chondrification
    4 Ground substance composed of
      a Chondroitin-sulfate (a mucoprotein)
      b Polysaccharide in a gel form

    5 Covers articular surfaces of bones
    6 Also located in larynx and respiratory tubes
  D Elastic cartilage
    1 Contains fewer collagenous fibers than hyaline type
    2 Large number of elastic fibers
    3 In external ear and auditory tubes
    4 Greater opacity than hyaline type
  E Fibrocartilage
    1 Small quantity of matrix is characteristic
    2 Is strong and rigid
    3 Has abundant fibrous elements and few chondrocytes
    4 Associated with regions of dense connective tissue
  F Histophysiology of cartilage
    1 Gristle-like consistency permits great weight bearing
    2 Structure is strong yet pliable
  G Growth of cartilage
    1 Types
      a Interstitial or endogenous
      b Appositional or exogenous

Review Questions

1 What general functions does the skeletal system perform?
2 Describe the microscopic structure of bone and cartilage.
3 Describe the structure of a long bone.
4 Explain the functions of the periosteum.
5 Compare and contrast bone formation in membranous and endochondral ossification.
6 List and discuss each of the major anatomical components that together constitute a haversian system.

7 Discuss bone fracture, list the terms used to describe the various types of fractures, and explain the sequence of steps characteristic of fracture healing.
8 Discuss the resorption of bone and the response of bone to stress.
9 Compare and contrast the basic structural elements of bone and cartilage.
10 How does the mechanism of cartilage growth differ from membranous bone growth?
11 Compare the structure and function of the three types of cartilage.

7 THE SKELETAL SYSTEM

OBJECTIVES

After you have completed this chapter, you should be able to:

1 Identify the two main subdivisions of the skeleton.
2 List the primary subdivisions of the axial skeleton.
3 Distinguish between the bones of the skull and those of the face.
4 List the sutures and fontanels of the skull.
5 Discuss the clinical significance of the cribriform plate of the ethmoid bone.
6 Name the regions of the vertebral column and give the number of vertebrae in each segment.
7 Discuss the three primary types of abnormal vertebral curvatures.
8 Discuss the bony components of the rib cage or chest.
9 List the primary subdivisions of the appendicular skeleton.
10 List the bony components of the shoulder and hip girdles.
11 Compare the structure and function of the wrist and hand with the ankle and foot.
12 Discuss the structural components and functional significance of the arches of the foot.
13 List the skeletal differences between males and females.
14 Discuss age changes in the skeleton.

KEY TERMS

Appendicular (ap-pen-DIK-u-lar)

Axial (AK-se-al)

Chondromalacia (kon-dro-mah-LA-she-ah)

Cranium (KRA-ne-um)

Fontanels (FON-ta-nels)

Kyphosis (ki-FO-sis)

Lordosis (lor-DO-sis)

Mastoiditis (mas-toi-DI-tis)

Osteoporosis (os-te-o-po-RO-sis)

Palpable (PAL-pah-ble)

Scoliosis (sco-le-O-sis)

Suture (SU-chur)

Skeletal tissues are organized to form the joined framework of living organs called *bones.* Just as skeletal tissues are organized to form bones, the bones in turn are organized or grouped to form the major subdivisions of the skeletal system described below. The rigid bones lie buried within the muscles and other soft tissues thus providing support and shape to the body as a whole. An understanding of the relationship of bones to each other and to other body structures provides a basis for understanding the function of many other organ systems. Coordinated movement, for example, is only possible because of the way bones are joined to one another in joints and the way muscles are attached to those bones. In addition, knowledge of the placement of bones within the soft tissues assists in locating and identifying other body structures.

The adult skeleton is composed of 206 separate bones. Rare variations in the total number of bones present in the body may occur as a result of certain anomalies such as extra ribs or from failure of certain small bones to fuse in the course of development.

In Chapter 6 the basic types of skeletal tissue, including cartilage, were discussed. Comparisons between the structural and functional characteristics of dense (compact) and cancellous (spongy) bone set the stage for study in this chapter of individual bones and their interrelationships in the skeleton. Chapter 8 considers articulations.

Health professionals often identify **externally palpable bony landmarks** in dealing with the sick and injured. **Palpable** bony landmarks are simply bones that can be touched and identified through the skin. They serve as reference points useful in identifying other body structures. Locating a **pressure point,** or area where a major blood vessel crosses a bone close to the body surface, is a good example. Using this knowledge, an individual can effectively control serious bleeding or **hemorrhage** in an injured person that might otherwise be fatal.

Figure 7-1 A, Skeleton, anterior view.
Axial skeleton is shown in blue. Appendicular
system is bone colored. **B,** Anterior
aspect of the right half of the thoracic,
upper limb, abdominal, and pelvic skeleton.

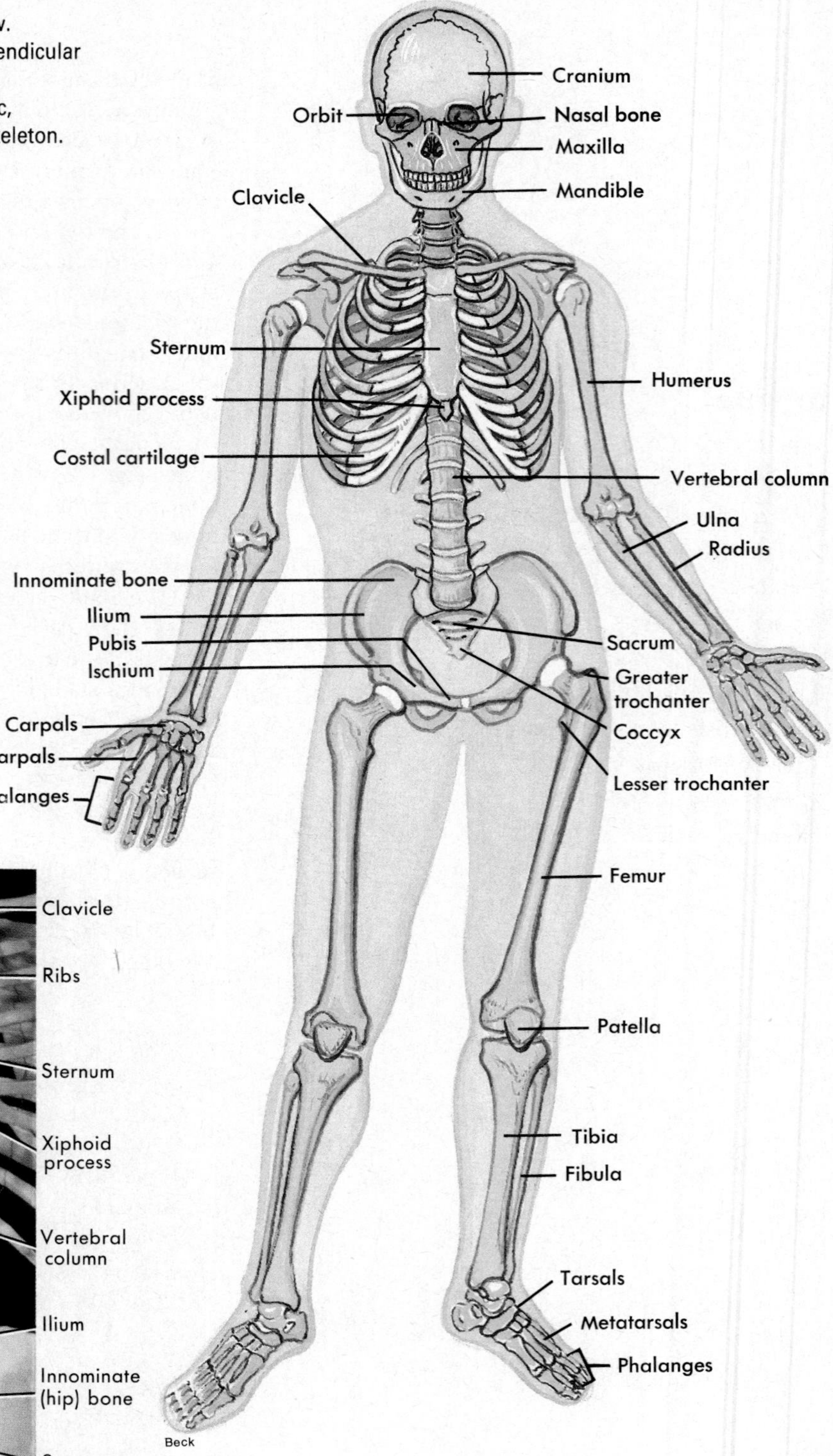

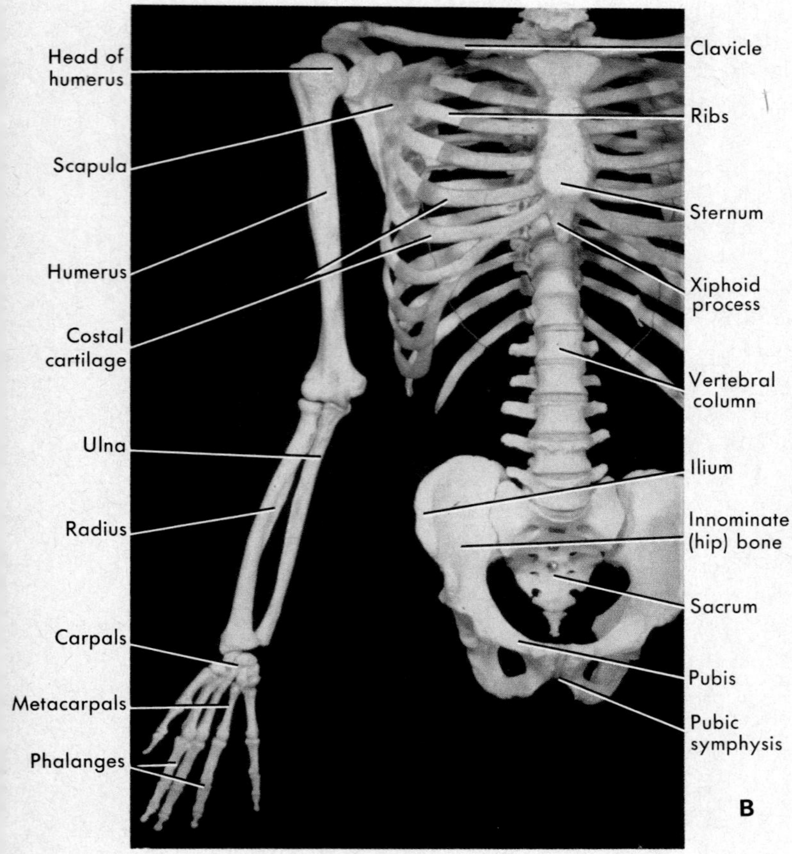

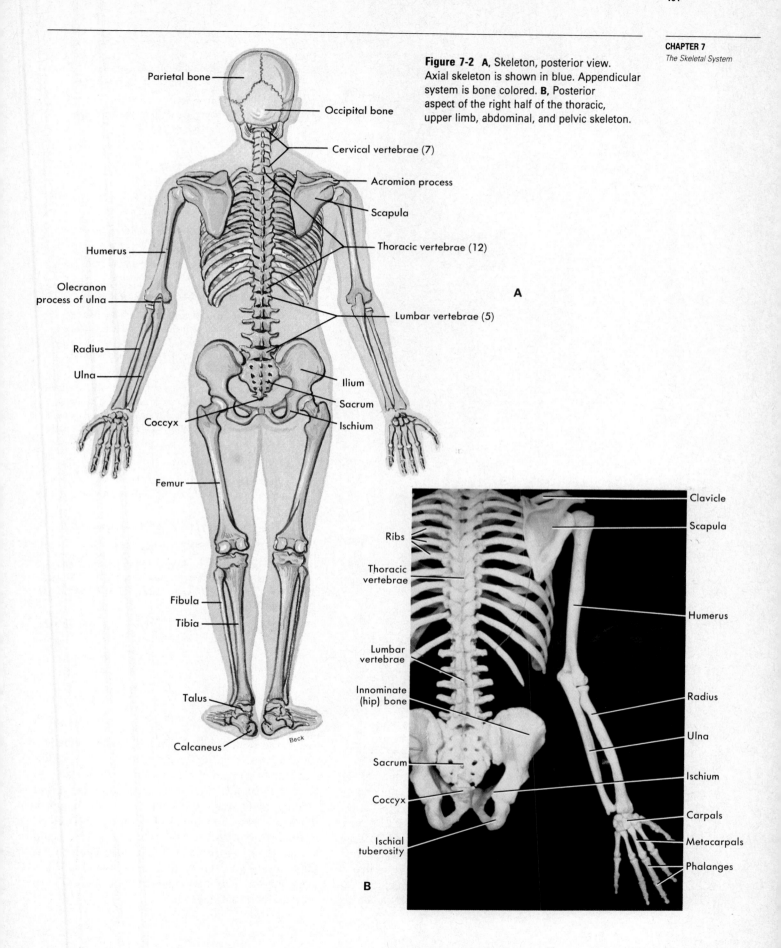

Figure 7-2 A, Skeleton, posterior view. Axial skeleton is shown in blue. Appendicular system is bone colored. **B,** Posterior aspect of the right half of the thoracic, upper limb, abdominal, and pelvic skeleton.

Parietal bone

Occipital bone

Cervical vertebrae (7)

Acromion process

Scapula

Thoracic vertebrae (12)

Humerus

Olecranon process of ulna

Lumbar vertebrae (5)

Radius

Ulna

Ilium

Sacrum

Coccyx

Ischium

Femur

Fibula

Tibia

Talus

Calcaneus

Beck

A

Ribs

Thoracic vertebrae

Lumbar vertebrae

Innominate (hip) bone

Sacrum

Coccyx

Ischial tuberosity

Clavicle

Scapula

Humerus

Radius

Ulna

Ischium

Carpals

Metacarpals

Phalanges

B

Table 7-1 Bones of skeleton

Part of body	Name of bone	Number	Identification
AXIAL SKELETON (80 bones)			Bones that form upright axis of body—skull, hyoid, vertebral column, sternum, and ribs
Skull (28 bones)			
Cranium (8 bones)			Cranium forms floor for brain to rest on and helmetlike covering over it
	Frontal	1	Forehead bone; also forms most of roof of orbits (eye sockets) and anterior part of cranial floor
	Parietal	2	Prominent, bulging bones behind frontal bone; form top sides of cranial cavity
	Temporal	2	Form lower sides of cranium and part of cranial floor; contain middle and inner ear structures
	Occipital	1	Forms posterior part of cranial floor and walls
	Sphenoid	1	Keystone of cranial floor; forms its midportion; resembles bat with wings outstretched and legs extended downward posteriorly; lies behind and slightly above nose and throat; forms part of floor and sidewalls of orbit
	Ethmoid	1	Complicated irregular bone that helps make up anterior portion of cranial floor, medial wall of orbits, upper parts of nasal septum, and sidewalls and part of nasal roof; lies anterior to sphenoid and posterior to nasal bones
Face (14 bones)	Nasal	2	Small bones forming upper part of bridge of nose
	Maxillary	2	Upper jaw bones; form part of floor of orbit, anterior part of roof of mouth, and floor of nose and part of sidewalls of nose
	Zygomatic (malar)	2	Cheekbones; form part of floor and sidewall of orbit
	Mandible	1	Lower jawbone; largest, strongest bone of face
	Lacrimal	2	Thin bones about size and shape of fingernail; posterior and lateral to nasal bones in medial wall of orbit; help form sidewall of nasal cavity, often missing in dry skull
	Palatine	2	Form posterior part of hard palate, floor, and part of sidewalls of nasal cavity and floor of orbit

DIVISIONS OF SKELETON

The human skeleton consists of two main divisions—namely, the axial skeleton and the appendicular skeleton (Figures 7-1 and 7-2). Eighty bones make up the **axial skeleton.** This includes 74 bones that form the upright axis of the body and 6 tiny middle ear bones. The **appendicular skeleton** consists of 126 bones—more than half again as many as in the axial skeleton. Bones of the appendicular skeleton form the appendages to the axial skeleton,

that is, the shoulder girdles, arms, wrists, and hands and the hip girdles, legs, ankles, and feet. One of the first things you should do in studying bones is to learn their names. Start by looking at Table 7-1 and reading the names and descriptions of the bones in the body. Then study the illustrations and text in the chapter. One picture is said to be worth a thousand words. If this is true and if you refer often to the illustrations, you should find it easy—and perhaps even fun—to learn the names of your bones and to identify their markings.

Table 7-1 Bones of skeleton—cont'd

Part of body	Name of bone	Number	Identification
	Inferior conchae (turbinates)	2	Thin scroll of bone forming kind of shelf along inner surface of sidewall of nasal cavity; lies above roof of mouth
	Vomer	1	Forms lower and posterior part of nasal septum; shaped like ploughshare
Ear bones (6 bones)	Malleus (hammer) Incus (anvil) Stapes (stirrup)	2 2 2	Tiny bones referred to as auditory ossicles in middle ear cavity in temporal bones; resemble, respectively, miniature hammer, anvil, and stirrup
Hyoid bone		1	U-shaped bone in neck between mandible and upper part of larynx; claims distinction as only bone in body not forming a joint with any other bone; suspended by ligaments from styloid processes of temporal bones
Spinal column (26 bones)			Not actually a column but a flexible segmented rod shaped like an elongated letter S; forms axis of body; head balanced above, ribs and viscera suspended in front, and lower extremities attached below; encloses spinal cord
	Cervical vertebrae	7	First or upper 7 vertebrae
	Thoracic vertebrae	12	Next 12 vertebrae; 12 pairs of ribs attached to these
	Lumbar vertebrae	5	Next 5 vertebrae
	Sacrum	1	Five separate vertebrae until about 25 years of age; then fused to form 1 wedge-shaped bone
	Coccyx	1	Four or 5 separate vertebrae in child but fused into 1 in adult
Sternum and ribs (25 bones)			Sternum, ribs, and thoracic vertebrae together form bony cage known as *thorax*; ribs attach posteriorly to vertebrae, slant downward anteriorly to attach to sternum (see description of false ribs below)
	Sternum	1	Breastbone; flat dagger-shaped bone
	True ribs	7 pairs	Upper 7 pairs; fasten to sternum by costal cartilages
	False ribs	5 pairs	False ribs do not attach to sternum directly; upper 3 pairs of false ribs attach by means of costal cartilage of seventh ribs; last 2 pairs do not attach to sternum at all; therefore called "floating"
APPENDICULAR SKELETON (126 bones)			Bones that are appended to axial skeleton; upper and lower extremities, including shoulder and hip girdles
Upper extremities (including shoulder girdle) (64 bones)	Clavicle	2	Collar bones; shoulder girdle joined to axial skeleton by articulation of clavicles with sternum; scapula does not form joint with axial skeleton
	Scapula	2	Shoulder blades; scapulae and clavicles together comprise shoulder girdle
	Humerus	2	Long bone of upper arm
	Radius	2	Bone of thumb side of forearm
	Ulna	2	Bone of little finger side of forearm; longer than radius

Continued.

Table 7-1 Bones of skeleton—cont'd

Part of body	Name of bone	Number	Identification
APPENDICULAR SKELETON—cont'd			
	Carpals (scaphoid, lunate, triquetrum, pisiform, trapezium, trapezoid, capitate, and hamate)	16	Arranged in two rows at proximal end of hand (Figures 7-30 and 7-31)
	Metacarpals	10	Long bones forming framework of palm of hand
	Phalanges	28	Miniature long bones of fingers, 3 in each finger, 2 in each thumb
Lower extremities (62 bones)	Ossa coxae or innominate bones	2	Large hip bones; with sacrum and coccyx, these 3 bones form basinlike pelvic cavity; lower extremities attached to axial skeleton by pelvic bones
	Femur	2	Thigh bone; largest strongest bone of body
	Patella	2	Kneecap; largest sesamoid bone of body*; embedded in tendon of quadriceps femoris muscle
	Tibia	2	Shin bone
	Fibula	2	Long, slender bone of lateral side of lower leg
	Tarsals (calcaneus, talus, navicular, first, second, and third cuneiforms, cuboid)	14	Bones that form heel and proximal or posterior half of foot (Figure 7-39)
	Metatarsals	10	Long bones of feet
	Phalanges	28	Miniature long bones of toes; 2 in each great toe, 3 in other toes
TOTAL		**206***	

*An inconstant number of small, flat, round bones known as **sesamoid bones** (because of their resemblance to sesame seeds) are found in various tendons in which considerable pressure develops. Because the number of these bones varies greatly between individuals, only two of them, the patellae, have been counted among the 206 bones of the body. Generally, two of them can be found in each thumb (in flexor tendon near metacarpophalangeal and interphalangeal joints) and great toe plus several others in the upper and lower extremities.
Wormian bones, the small islets of bone frequently found in some of the cranial sutures, have not been counted in this list of 206 bones either because of their variable occurrence.

STRUCTURE OF INDIVIDUAL BONES

AXIAL SKELETON

Skull

Twenty-eight irregularly shaped bones form the skull (Figures 7-3 to 7-11, pp. 172-178). It consists of two major divisions: the **cranium,** or brain case, and the **face.** The cranium is formed by eight bones, namely, frontal, two parietal, two temporal, occipital, sphenoid, and ethmoid. The 14 bones that form the face are: two maxillary, two zygomatic (malar), two nasal, mandible, two lacrimal, two palatine, two inferior nasal conchae (turbinates), and vomer. Note that all the face bones are paired except for the mandible and vomer. All cranial bones, on the other hand, are single or unpaired except for the parietal and temporal bones, which are paired. The frontal and ethmoid bones of the skull help shape the face but are not numbered among the face bones.

The **frontal bone** forms the forehead and the anterior part of the top of the cranium. It contains mucosa-lined air-filled spaces, the *frontal sinuses,* and it forms the upper part of the orbits. It unites with the two parietal bones posteriorly in an immovable joint, the *coronal suture.* Several of the more prominent frontal bone markings are described in Table 7-2. *Text continued, p. 179.*

Table 7-2 Identification of bone markings

Marking	Description	Marking	Description
FRONTAL		**TEMPORAL**	
• Supraorbital margin	Arched ridge just below eyebrow	Squamous portion	Thin, flaring upper part of bone
• Frontal sinuses	Cavities inside bone just above supraorbital margin; lined with mucosa; contain air	**Mastoid portion**	Rough-surfaced lower part of bone posterior to external auditory meatus
Frontal tuberosities	Bulge above each orbit; most prominent part of forehead	**Petrous portion**	Wedge-shaped process that forms part of center section of cranial floor between sphenoid and occipital bones; name derived from Greek word for stone because of extreme hardness of this process; houses middle and inner ear structures
Superciliary arches	Ridges caused by projection of frontal sinuses; eyebrows lie over these ridges	• Mastoid process	Protuberance just behind ear
Supraorbital notch (sometimes foramen)	Notch or foramen in supraorbital margin slightly mesial to its midpoint; transmits supraorbital nerve and blood vessels	• Mastoid air cells	Air-filled mucosa-lined spaces within mastoid process
Glabella	Smooth area between superciliary ridges and above nose	• External auditory meatus (or canal)	Opening into ear and tube extending into temporal bone
SPHENOID		• Zygomatic process	Projection that articulates with malar (or zygomatic) bone
• Body	Hollow, cubelike central portion	• Internal auditory meatus	Fairly large opening on posterior surface of petrous portion of bone; transmits eighth cranial nerve to inner ear and seventh cranial nerve on its way to facial structures
• Greater wings	Lateral projections from body; form part of outer wall of orbit	• Mandibular fossa	Oval-shaped depression anterior to external auditory meatus; forms socket for condyle of mandible
• Lesser wings	Thin, triangular projections from upper part of sphenoid body; form posterior part of roof of orbit	• Styloid process	Slender spike of bone extending downward and forward from undersurface of bone anterior to mastoid process; often broken off in dry skull; several neck muscles and ligaments attach to styloid process
• Sella turcica (or *Turk's saddle*)	Saddle-shaped depression on upper surface of sphenoid body; contains pituitary gland	Stylomastoid foramen	Opening between styloid and mastoid processes where facial nerve emerges from cranial cavity
• Sphenoid sinuses	Irregular air-filled mucosa-lined spaces within central part of sphenoid	Jugular fossa	Depression on undersurface of petrous portion; dilated beginning of internal jugular vein lodged here
Pterygoid processes	Downward projections on either side where body and greater wing unite; comparable to extended legs of bat if entire bone is likened to this animal; form part of lateral nasal wall	Jugular foramen	Opening in suture between petrous portion and occipital bone; transmits lateral sinus and ninth, tenth, and eleventh cranial nerves
• Optic foramen	Opening into orbit at root of lesser wing; transmits second cranial nerve	Carotid canal (or foramen)	Channel in petrous portion; best seen from undersurface of skull; transmits internal carotid artery
Superior orbital fissure	Slitlike opening into orbit; lateral to optic foramen; transmits third, fourth, and part of fifth cranial nerves		
Foramen rotundum	Opening in greater wing that transmits maxillary division of fifth cranial nerve		
Foramen ovale	Opening in a greater wing that transmits mandibular division of fifth cranial nerve		

• Those that seem particularly important to remember.

Continued.

Table 7-2 Identification of bone markings—cont'd

Marking	Description	Marking	Description
OCCIPITAL		**MANDIBLE**	
• Foramen magnum	Hole through which spinal cord enters cranial cavity	Body	Main part of bone; forms chin
• Condyles	Convex, oval processes on either side of foramen magnum; articulate with depressions on first cervical vertebra	Ramus	Process, one on either side, that projects upward from posterior part of body
External occipital protuberance	Prominent projection on posterior surface in midline short distance above foramen magnum; can be felt as definite bump	• Condyle (or head)	Part of each ramus that articulates with mandibular fossa of temporal bone
Superior nuchal line	Curved ridge extending laterally from external occipital protuberance	Neck	Constricted part just below condyles
		• Alveolar process	Teeth set into this arch
Inferior nuchal line	Less well-defined ridge paralleling superior nuchal line short distance below it	• Mandibular foramen	Opening on inner surface of ramus; transmits nerves and vessels to lower teeth
Internal occipital protuberance	Projection in midline on inner surface of bone; grooves for lateral sinuses extend laterally from this process and one for sagittal sinus extends upward from it	• Mental foramen	Opening on outer surface below space between two bicuspids; transmits terminal branches of nerves and vessels that enter bone through mandibular foramen; dentists inject anesthetics through these foramina
ETHMOID		Coronoid process	Projection upward from anterior part of each ramus; temporal muscle inserts here
• Horizontal (cribriform) plate	Olfactory nerves pass through numerous holes in this plate	Angle	Juncture of posterior and inferior margins of ramus
• Crista galli	Meninges attach to this process (Figures 7-9 and 7-11)		
• Perpendicular plate	Forms upper part of nasal septum (Figures 7-9 and 7-11)	**MAXILLA**	
		• Alveolar process	Arch containing teeth
• Ethmoid sinuses	Honeycombed, mucosa-lined air spaces within lateral masses of bone (Figures 7-10 and 7-11)	• Maxillary sinus or antrum of Highmore	Large air-filled mucosa-lined cavity within body of each maxilla; largest of sinuses
• Superior and middle turbinates (conchae)	Help to form lateral walls of nose (Figures 7-9 and 7-11)	• Palatine process	Horizontal inward projection from alveolar process; forms anterior and larger part of hard palate
Lateral masses	Compose sides of bone; contain many air spaces (ethmoid cells or sinuses); inner surface forms superior and middle conchae	Infraorbital foramen	Hole on external surface just below orbit; transmits vessels and nerves
		Lacrimal groove	Groove on inner surface; joined by similar groove on lacrimal bone to form canal housing nasolacrimal duct
PALATINE			
Horizontal plate	Joined to palatine processes of maxillae to complete part of hard palate		

• Those that seem particularly important to remember.

Table 7-2 Identification of bone markings—cont'd

Marking	Description	Marking	Description
SPECIAL FEATURES OF SKULL		• Nasal septum (Figure 7-7) formed by	Partition in midline of nasal cavity; separates cavity into right and left halves
• Sutures (Figure 7-4)	Immovable joints between skull bones	Perpendicular plate of ethmoid bone	Forms upper part of septum
Squamous	Line of articulation between two parietal bones	Vomer bone	Forms lower, posterior part
Coronal	Joint between parietal bones and frontal bone	Cartilage	Forms anterior part
Lambdoidal	Joint between parietal bones and occipital bone	• Wormian bones	Small islets of bones within suture
• Fontanels (Figures 7-12 and 7-13)	"Soft spots" where ossification is incomplete at birth; allow some compression of skull during birth; also important in determining position of head before delivery; 6 such areas located at angles of parietal bones	**VERTEBRAL COLUMN**	
		General features	Anterior part of vertebrae (except first two cervical) consists of body; posterior part of vertebrae consists of neural arch, which, in turn, consists of 2 pedicles, 2 laminae, and 7 processes projecting from laminae
Anterior (or frontal)	At intersection of sagittal and coronal sutures (juncture of parietal bones and frontal bone); diamond shaped; largest of fontanels; usually closed by 1½ years of age	• Thoracic vertebrae	
Posterior (or occipital)	At intersection of sagittal and lambdoidal sutures (juncture of parietal bones and occipital bone); triangular in shape; usually closed by second month	Body	Main part; flat, round mass located anteriorly; supporting or weight-bearing part of vertebra
Anterolateral (or sphenoid)	At juncture of frontal, parietal, temporal, and sphenoid bones	Pedicles	Short projections extending posteriorly from body
Posterolateral (or mastoid)	At juncture of parietal, occipital, and temporal bones; usually closed by second year	Laminae	Posterior part of vertebra to which pedicles join and from which processes project
• Sinuses		Neural arch	Formed by pedicles and laminae; protects spinal cord posteriorly; congenital absence of one or more neural arches is known as *spina bifida* (cord may protrude right through skin)
Air (or bony)	Spaces or cavities within bones; those that communicate with nose called *paranasal sinuses* (frontal, sphenoidal, ethmoidal, and maxillary); mastoid cells communicate with middle ear rather than nose, therefore not included among paranasal sinuses	Spinous process	Sharp process projecting inferiorly from laminae in midline
		Transverse processes	Right and left lateral projections from laminae
Blood	Veins within cranial cavity (Figure 16-23, p. 463)	Superior articulating processes	Project upward from laminae
• Orbits (Figure 7-8) formed by		Inferior articulating processes	Project downward from laminae; articulate with superior articulating processes of vertebrae below
Frontal	Roof of orbit	Spinal foramen	Hole in center of vertebra formed by union of body, pedicles, and laminae; spinal foramina, when vertebrae superimposed one on other, form spinal cavity that houses spinal cord
Ethmoid	Medial wall		
Lacrimal	Medial wall		
Sphenoid	Lateral wall		
Zygomatic	Lateral wall		
Maxillary	Floor		
Palatine	Floor		

• Those that seem particularly important to remember.

Continued.

Table 7-2 Identification of bone markings—cont'd

Marking	Description	Marking	Description
VERTEBRAL COLUMN—cont'd		Secondary	Concavities in *cervical* and *lumbar* regions; cervical concavity results from infant's attempts to hold head erect (3 to 4 months); lumbar concavity, from balancing efforts in learning to walk (10 to 18 months)
• Cervical vertebrae (Figures 7-15 to 7-17)			
General features	Foramen in each transverse process for transmission of vertebral artery, vein, and plexus of nerves; short bifurcated spinous processes except on seventh vertebrae, where it is extra long and may be felt as protrusion when head bent forward; bodies of these vertebrae small, whereas spinal foramina large and triangular	Abnormal	*Kyphosis,* exaggerated convexity in thoracic region (hunchback); *lordosis,* exaggerated concavity in lumbar region, a very common condition; *scoliosis,* lateral curvature in any region
		STERNUM (Figure 7-20)	
		• Body	Main central part of bone
		• Manubrium	Flaring, upper part
Atlas	First cervical vertebra; lacks body and spinous process; superior articulating processes concave ovals that act as rockerlike cradles for condyles of occipital bone; named *atlas* because supports head as Atlas was thought to have supported world (Figure 7-16)	• Xiphoid process	Projection of cartilage at lower border of bone
		RIBS (Figures 7-21 and 7-22)	
		• Head	Projection at posterior end of rib; articulates with corresponding thoracic vertebra and one above, except last three pairs, which join corresponding vertebrae only
Axis (epistropheus)	Second cervical vertebra, so named because atlas rotates about this bone in rotating movements of head; *dens,* or odontoid process, peglike projection upward from body of axis, forming pivot for rotation of atlas (Figure 7-17)	Neck	Constricted portion just below head
		Tubercle	Small knob just below neck; articulates with transverse process of corresponding thoracic vertebra; missing in lowest 3 ribs
• Lumbar vertebrae (Figures 7-18 and 7-19)	Strong, massive; superior articulating processes directed inward instead of upward; inferior articulating processes, outward instead of downward; short, blunt spinous process (Figure 7-18)	Body or shaft	Main part of rib
		• Costal cartilage	Cartilage at sternal end of true ribs; attaches ribs (except floating ribs) to sternum
• Sacral promontory	Protuberance from anterior, upper border of sacrum into pelvis; of obstetrical importance because its size limits anteroposterior diameter of pelvic inlet		
		SCAPULA (Figures 7-23 and 7-25)	
• Intervertebral foramina	Opening between vertebrae through which spinal nerves emerge	• Borders	
		Superior	Upper margin
		Vertebral	Margin toward vertebral column
• Curves	Curves have great structural importance because they increase carrying strength of vertebral column, make balance possible in upright position (if column were straight, weight of viscera would pull body forward), absorb jars from walking (straight column would transmit jars straight to head), and protect column from fracture	Axillary	Lateral margin
		• Spine	Sharp ridge running diagonally across posterior surface of shoulder blade
		• Acromion process	Slightly flaring projection at lateral end of scapular spine; may be felt as tip of shoulder; articulates with clavicle
Primary	Column curves at birth from head to sacrum with convexity posteriorly; after child stands, convexity persists only in *thoracic* and *sacral* regions, which therefore are called *primary curves*	• Coracoid process	Projection on anterior surface from upper border of bone; may be felt in groove between deltoid and pectoralis major muscles, about 1 inch below clavicle
		• Glenoid cavity	Arm socket

• Those that seem particularly important to remember.

Table 7-2 Identification of bone markings—cont'd

Marking	Description	Marking	Description
HUMERUS (Figures 7-26 and 7-27)		**ULNA** (Figures 7-28 and 7-29)	
• Head	Smooth, hemispherical enlargement at proximal end of humerus	• Olecranon process	Elbow
Anatomical neck	Oblique groove just below head	• Coronoid process	Projection on anterior surface of proximal end of ulna; trochlea of humerus fits snugly between olecranon and coronoid processes
Greater tubercle	Rounded projection lateral to head on anterior surface		
Lesser tubercle	Prominent projection on anterior surface just below anatomical neck	• Semilunar notch	Curved notch between olecranon and coronoid process into which trochlea fits
Intertubercular	Deep groove between greater and lesser tubercles; long tendon of biceps muscle lodges here	• Radial notch	Curved notch lateral and inferior to semilunar notch; head of radius fits into this concavity
Surgical neck	Region just below tubercles; so named because of its liability to fracture	Head	Rounded process at distal end; does not articulate with wrist bones but with fibrocartilaginous disk
Deltoid tuberosity	V-shaped, rough area about midway down shaft where deltoid muscle inserts		
Radial groove	Groove running obliquely downward from deltoid tuberosity; lodges radial nerve	Styloid process	Sharp protuberance at distal end; can be seen from outside on posterior surface
• Epicondyles (medial and lateral)	Rough projections at both sides of distal end	**RADIUS** (Figures 7-28 and 7-29)	
• Capitulum	Rounded knob below lateral epicondyle; articulates with radius; sometimes called *radial head* of humerus	• Head	Disk-shaped process forming proximal end of radius; articulates with capitulum of humerus and with radial notch of ulna
• Trochlea	Projection with deep depression through center similar to shape of pulley; articulates with ulna	Radial tuberosity	Roughened projection on ulnar side, short distance below head; biceps muscle inserts here
• Olecranon fossa	Depression on posterior surface just above trochlea; receives olecranon process of ulna when lower arm extends	Styloid process	Protuberance at distal end on lateral surface (with forearm supinated as in anatomical position)
• Coronoid fossa	Depression on anterior surface above trochlea; receives coronoid process of ulna in flexion of lower arm		

• Those that seem particularly important to remember.

Continued.

Table 7-2 Identification of bone markings—cont'd

Marking	Description	Marking	Description
OSSA COXAE (Figures 7-32 to 7-36)		**OSSA COXAE**—cont'd	
• Ilium	Upper, flaring portion	Superior pubic ramus	Part of pubis lying between symphysis and acetabulum; forms upper part of obturator foramen
• Ischium	Lower, posterior portion	Inferior pubic ramus	Part extending down from symphysis; unites with ischium
• Pubic bone or pubis	Medial, anterior section	• Pubic arch	Angle formed by two inferior rami
• Acetabulum	Hip socket; formed by union of ilium, ischium, and pubis	Pubic crest	Upper margin of superior ramus
• Iliac crests	Upper, curving boundary of ilium	Pubic tubercle	Rounded process at end of crest
• Iliac spines		• Obturator foramen	Large hole in anterior surface of os coxa; formed by pubis and ischium; largest foramen in body
Anterior superior	Prominent projection at anterior end of iliac crest; can be felt externally as "point" of hip	• Pelvic brim (or inlet; Figures 7-32 and 7-34)	Boundary of aperture leading into true pelvis; formed by pubic crests, iliopectineal lines, and sacral promontory; size and shape of this inlet has great obstetrical importance, since if any of its diameters too small, infant skull cannot enter true pelvis for natural birth
Anterior inferior	Less prominent projection short distance below anterior superior spine		
Posterior superior	At posterior end of iliac crest		
Posterior inferior	Just below posterior superior spine		
Greater sciatic notch	Large notch on posterior surface of ilium just below posterior inferior spine	• True (or lesser) pelvis	Space below pelvic brim; true "basin" with bone and muscle walls and muscle floor; pelvic organs located in this space
Gluteal lines	Three curved lines across outer surface of ilium—posterior, anterior, inferior, respectively	• False (or greater) pelvis	Broad, shallow space above pelvic brim, or pelvic inlet; name "false pelvis" is misleading, since this space is actually part of abdominal cavity, not pelvic cavity
Iliopectineal line	Rounded ridge extending from pubic tubercle upward and backward toward sacrum		
Iliac fossa	Large, smooth, concave inner surface of ilium above iliopectineal line	Pelvic outlet (Figures 7-33 and 7-34)	Irregular circumference marking lower limits of true pelvis; bounded by tip of coccyx and two ischial tuberosities
• Ischial tuberosity	Large, rough, quadrilateral process forming inferior part of ischium; in erect sitting position body rests on these tuberosities	Pelvic girdle (or bony pelvis)	Complete bony ring; composed of two hip bones (ossa coxae), sacrum, and coccyx; forms firm base by which trunk rests on thighs and for attachment of lower extremities to axial skeleton
Ischial spine	Pointed projection just above tuberosity		
• Symphysis pubis	Cartilaginous, amphiarthrotic joint between pubic bones		

• Those that seem particularly important to remember.

Table 7-2 Identification of bone markings—cont'd

Marking	Description	Marking	Description
FEMUR (Figure 7-37)		**TIBIA** (Figure 7-38)	
• Head	Rounded, upper end of bone; fits into acetabulum	• Condyles	Bulging prominences at proximal end of tibia; upper surfaces concave for articulation with femur
Neck	Constricted portion just below head	Intercondylar eminence	Upward projection on articular surface between condyles
• Greater trochanter	Protuberance located inferiorly and laterally to head	• Crest	Sharp ridge on anterior surface
• Lesser trochanter	Small protuberance located inferiorly and medially to greater trochanter	Tibial tuberosity	Projection in midline on anterior surface
Linea aspera	Prominent ridge extending lengthwise along concave posterior surface	Popliteal line	Ridge that spirals downward and inward on posterior surface of upper third of tibial shaft
Gluteal tubercle	Rounded projection just below greater trochanter; rudimentary third trochanter	• Medial malleolus	Rounded downward projection at distal end of tibia; forms prominence on inner surface of ankle
Supracondylar ridges	Two ridges formed by division of linea aspera at its lower end; medial supracondylar ridge extends inward to inner condyle, lateral ridge to outer condyle	**FIBULA** (Figure 7-38)	
• Condyles	Large, rounded bulges at distal end of femur; one on medial and one on lateral surface	• Lateral malleolus	Rounded prominence at distal end of fibula; forms prominence on outer surface of ankle
Adductor tubercle	Small projection just above inner condyle; marks termination of medial supracondylar ridge	**TARSALS** (Figures 7-39 and 7-42)	
Trochlea	Smooth depression between condyles on anterior surface; articulates with patella	• Calcaneus	Heel bone
Intercondyloid notch	Deep depression between condyles on posterior surface; cruciate ligaments that help bind femur to tibia lodge in this notch	• Talus	Uppermost of tarsals; articulates with tibia and fibula; boxed in by medial and lateral malleoli
		• Longitudinal arches	Tarsals and metatarsals so arranged as to form arch from front to back of foot
		Medial	Formed by calcaneus, talus, navicular, cuneiforms, and 3 medial metatarsals
		Lateral	Formed by calcaneus, cuboid, and 2 lateral metatarsals
		• Transverse (or metatarsal) arch	Metatarsals and distal row of tarsals (cuneiforms and cuboid) so articulated as to form arch across foot; bones kept in 2 arched positions by means of powerful ligaments in sole of foot and by muscles and tendons

• Those that seem particularly important to remember.

Figure 7-3 A, Anterior view of the skull. **B,** Anterior aspect of the skull and mandible.

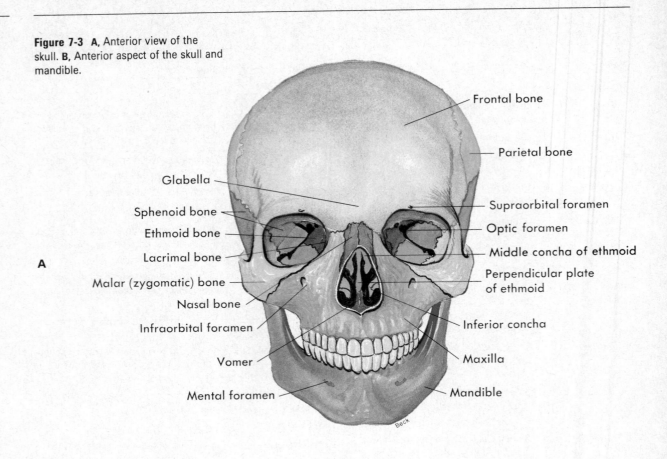

Frontal bone

Parietal bone

Glabella

Sphenoid bone

Ethmoid bone

Lacrimal bone

Malar (zygomatic) bone

Nasal bone

Infraorbital foramen

Vomer

Mental foramen

Supraorbital foramen

Optic foramen

Middle concha of ethmoid

Perpendicular plate of ethmoid

Inferior concha

Maxilla

Mandible

A

Figure 7-4 A, Skull viewed from the right side. **B,** Right lateral aspect of the skull and mandible.

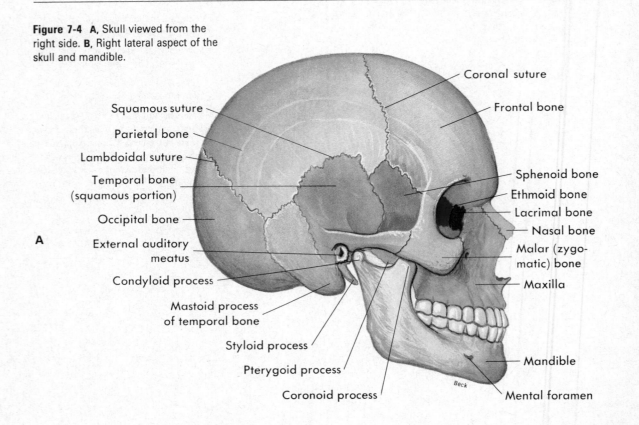

Squamous suture

Parietal bone

Lambdoidal suture

Temporal bone (squamous portion)

Occipital bone

External auditory meatus

Condyloid process

Mastoid process of temporal bone

Styloid process

Pterygoid process

Coronoid process

Coronal suture

Frontal bone

Sphenoid bone

Ethmoid bone

Lacrimal bone

Nasal bone

Malar (zygo-matic) bone

Maxilla

Mandible

Mental foramen

A

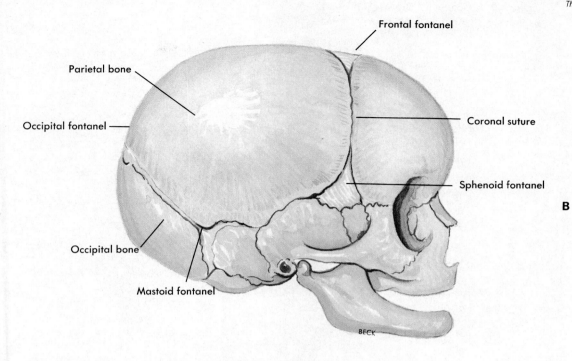

Frontal fontanel

Parietal bone

Occipital fontanel

Coronal suture

Sphenoid fontanel

Occipital bone

Mastoid fontanel

B

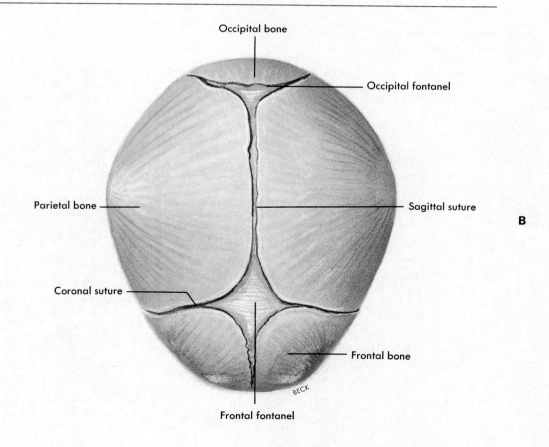

Occipital bone

Occipital fontanel

Parietal bone

Sagittal suture

Coronal suture

Frontal bone

Frontal fontanel

B

Figure 7-5 **A**, Floor of the cranial cavity.
B, Superior aspect of the base of the skull.

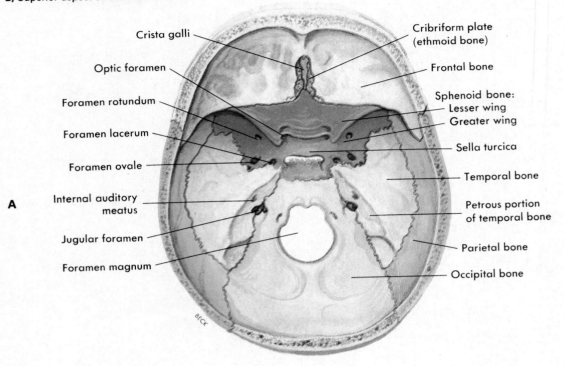

- Crista galli
- Optic foramen
- Foramen rotundum
- Foramen lacerum
- Foramen ovale
- Internal auditory meatus
- Jugular foramen
- Foramen magnum

- Cribriform plate (ethmoid bone)
- Frontal bone
- Sphenoid bone: Lesser wing
- Greater wing
- Sella turcica
- Temporal bone
- Petrous portion of temporal bone
- Parietal bone
- Occipital bone

A

Figure 7-6 **A**, Skull viewed from below.
B, Inferior aspect of the base of the skull.

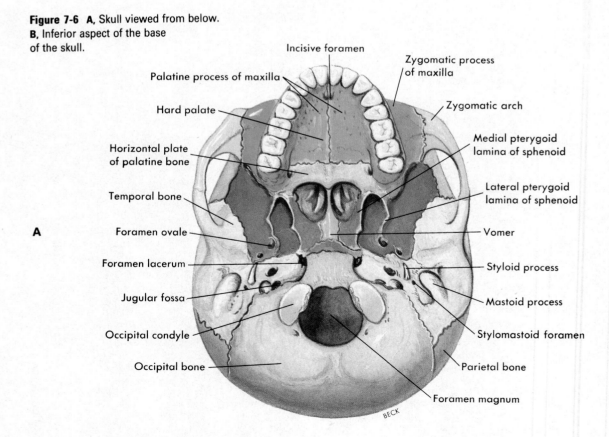

- Incisive foramen
- Palatine process of maxilla
- Hard palate
- Horizontal plate of palatine bone
- Temporal bone
- Foramen ovale
- Foramen lacerum
- Jugular fossa
- Occipital condyle
- Occipital bone

- Zygomatic process of maxilla
- Zygomatic arch
- Medial pterygoid lamina of sphenoid
- Lateral pterygoid lamina of sphenoid
- Vomer
- Styloid process
- Mastoid process
- Stylomastoid foramen
- Parietal bone
- Foramen magnum

A

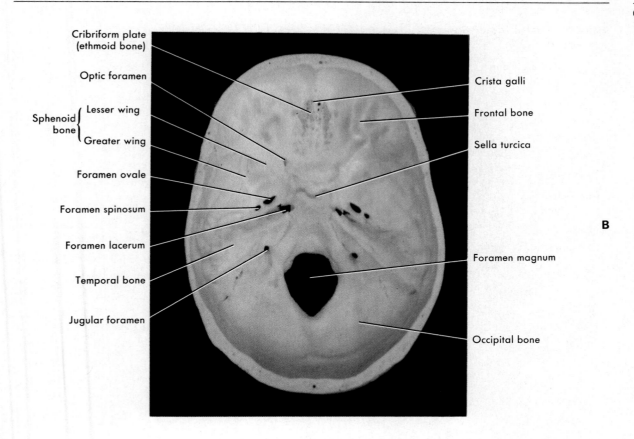

Cribriform plate (ethmoid bone)

Optic foramen

Sphenoid bone {
Lesser wing

Greater wing
}

Foramen ovale

Foramen spinosum

Foramen lacerum

Temporal bone

Jugular foramen

Crista galli

Frontal bone

Sella turcica

Foramen magnum

Occipital bone

B

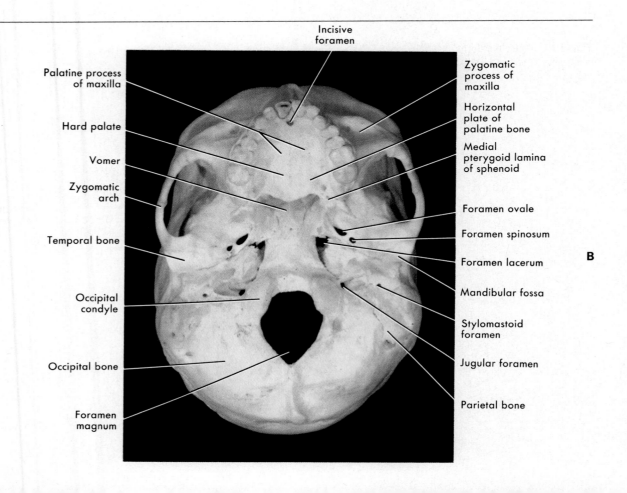

Incisive foramen

Palatine process of maxilla

Hard palate

Vomer

Zygomatic arch

Temporal bone

Occipital condyle

Occipital bone

Foramen magnum

Zygomatic process of maxilla

Horizontal plate of palatine bone

Medial pterygoid lamina of sphenoid

Foramen ovale

Foramen spinosum

Foramen lacerum

Mandibular fossa

Stylomastoid foramen

Jugular foramen

Parietal bone

B

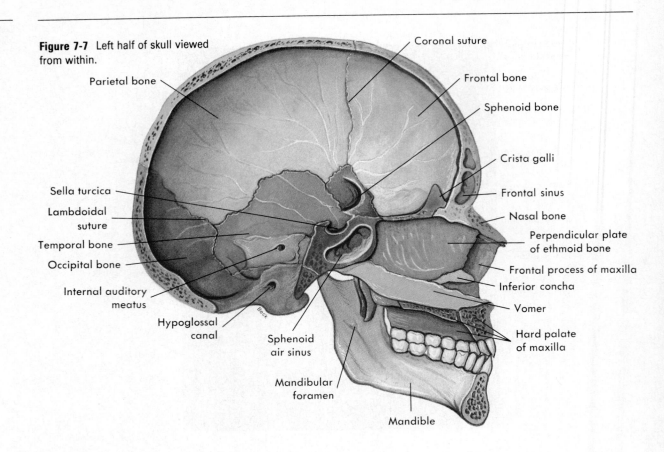

Figure 7-7 Left half of skull viewed from within.

Parietal bone
Coronal suture
Frontal bone
Sphenoid bone
Crista galli
Frontal sinus
Nasal bone
Perpendicular plate of ethmoid bone
Frontal process of maxilla
Inferior concha
Vomer
Hard palate of maxilla
Sella turcica
Lambdoidal suture
Temporal bone
Occipital bone
Internal auditory meatus
Hypoglossal canal
Sphenoid air sinus
Mandibular foramen
Mandible

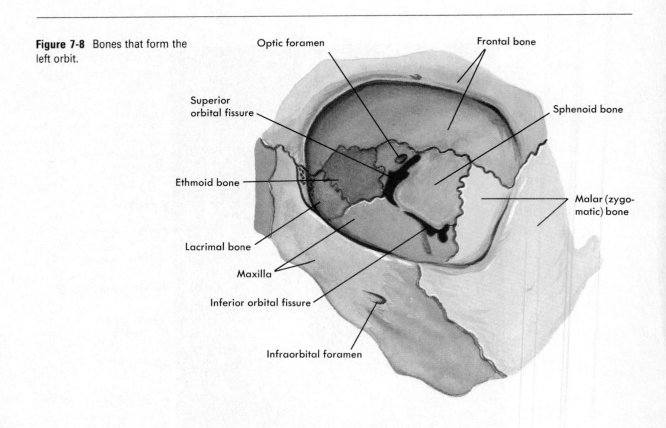

Figure 7-8 Bones that form the left orbit.

Optic foramen
Frontal bone
Sphenoid bone
Superior orbital fissure
Ethmoid bone
Malar (zygomatic) bone
Lacrimal bone
Maxilla
Inferior orbital fissure
Infraorbital foramen

Figure 7-14 Hyoid bone with muscle attachments.

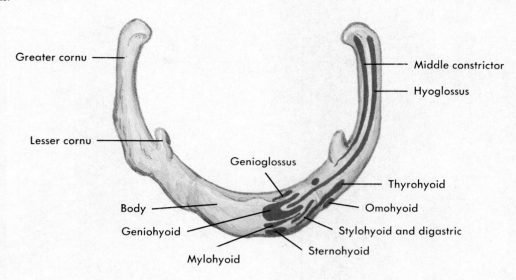

Greater cornu

Middle constrictor

Hyoglossus

Lesser cornu

Genioglossus

Thyrohyoid

Body

Omohyoid

Geniohyoid

Stylohyoid and digastric

Sternohyoid

Mylohyoid

of the temporal bone, makes the zygomatic arch. It articulates with four other facial bones: the maxillary, temporal, frontal, and sphenoid bones.

Shape is given to the nose by the two **nasal bones,** which form the upper part of the bridge of the noise, and by *cartilage,* which forms the lower part. Although small in size, the nasal bones enter into several articulations: with the perpendicular plate of the ethmoid bone, the cartilaginous part of the nasal septum, the frontal bone, the maxillae, and each other.

An almost paper-thin bone, shaped and sized about like a fingernail, lies just posterior and lateral to each nasal bone. It helps form the sidewall of the nasal cavity and the medial wall of the orbit. Because it contains a groove for the nasolacrimal (tear) duct, this bone is called the **lacrimal bone** (see Figures 7-4 and 7-8). It joins the maxilla, frontal bone, and ethmoid bone.

The two **palatine bones** join to each other in the midline like two L's facing each other. Their united horizontal portions form the posterior part of the hard palate (see Figure 7-6). The vertical portion of each palatine bone forms the lateral wall of the posterior part of each nasal cavity. The palatine bones articulate with the maxillae and the sphenoid bone.

There are two **inferior nasal conchae** (turbinates). Each one is scroll-like in shape and forms a kind of ledge projecting into the nasal cavity from its lateral wall. In each nasal cavity there are three such ledges. The superior and middle conchae (which are projections of the ethmoid bone) form the upper and middle ledges. The inferior concha (which is a separate bone) forms the lower ledge. They are mucosa covered and divide each nasal cavity into three narrow, irregular channels, the *nasal meati.* The inferior nasal conchae form immovable joints with the ethmoid, lacrimal, maxillary, and palatine bones.

Two structures that enter into the formation of the nasal septum have already been mentioned—the perpendicular plate of the ethmoid bone and the septal cartilage. One other structure, the **vomer bone,** completes the septum posteriorly (see Figure 7-7). It is usually described as being shaped like a ploughshare. It forms immovable joints with four bones: the sphenoid, ethmoid, palatine, and maxillae.

For a description of the *ear bones,* see Table 7-1, p. 163.

Special features of the skull include sutures, fontanels (Figures 7-12 and 7-13), sinuses, orbits, nasal septum, and wormian bones, all of which are described on p. 167.

Hyoid Bone

The hyoid bone is a single bone in the neck—a part of the axial skeleton. Its U shape may be felt just above the larynx and below the mandible where it is suspended from the styloid processes of the temporal bones. Several muscles attach to the hyoid bone. Among them are an extrinsic tongue muscle (hyoglossus) and certain muscles of the floor of the mouth (mylohyoid, geniohyoid) (Figure 7-14). The hyoid claims the distinction of being the only bone in the body that articulates with no other bones.

Figure 7-15 The spinal column from three views.

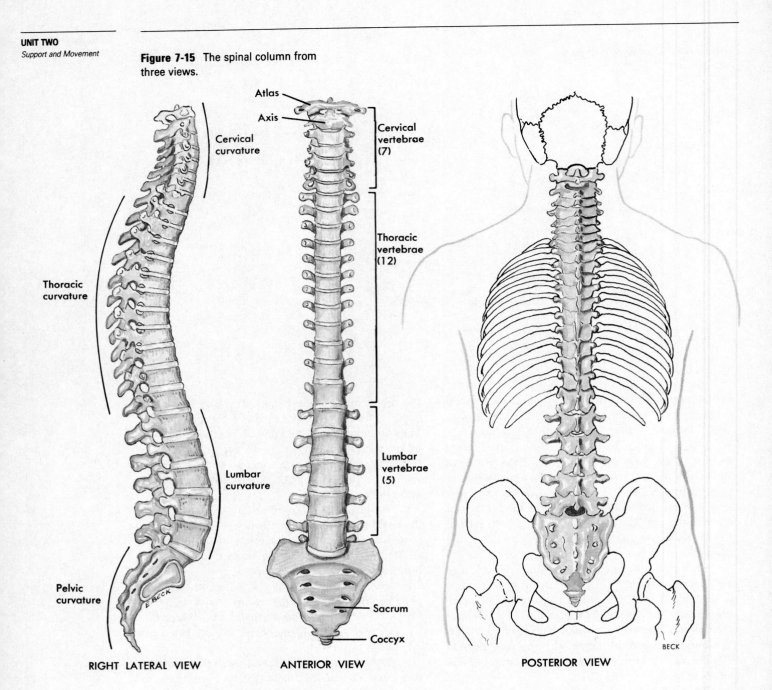

Atlas

Axis

Cervical curvature

Cervical vertebrae (7)

Thoracic curvature

Thoracic vertebrae (12)

Lumbar curvature

Lumbar vertebrae (5)

Pelvic curvature

E. BECK

Sacrum

Coccyx

RIGHT LATERAL VIEW

ANTERIOR VIEW

POSTERIOR VIEW

BECK

Spinal Column

The spinal column constitutes the longitudinal axis of the skeleton. It is a flexible rather than a rigid column because it is segmented. As Figure 7-15 shows, the spinal column consists of 24 vertebrae plus the sacrum and coccyx. Joints between the vertebrae permit forward, backward, and sideway movement of the column. Consider too these further facts about the spinal column. The head is balanced on top, the ribs and viscera are suspended in front, the lower extremities are attached below, and the spinal cord is enclosed within. It is indeed the "backbone" of the body.

The 7 **cervical vertebrae** constitute the skeletal framework of the neck (Figure 7-15). The next 12 vertebrae are called **thoracic vertebrae** because of their location in the posterior part of the chest or thoracic region. The next 5 spinal bones, the **lumbar vertebrae,** support the small of the back. Below the lumbar vertebrae lie the *sacrum* and *coccyx.* In the adult the sacrum is a single bone that has resulted from the fusion of 5 separate vertebrae, and the coccyx is a single bone that has resulted from the fusion of 4 or 5 vertebrae.

All the vertebrae resemble each other in certain features and differ in others. For example, all except

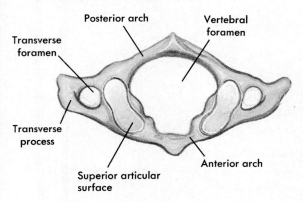

Figure 7-16 First cervical vertebra (atlas) viewed from above (superior).

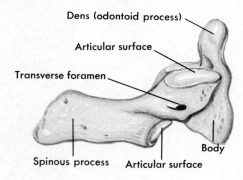

Figure 7-17 Second cervical vertebra (axis) viewed from the side (lateral).

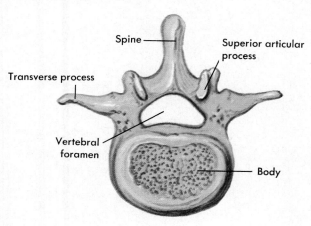

Figure 7-18 Third lumbar vertebra viewed from above (superior).

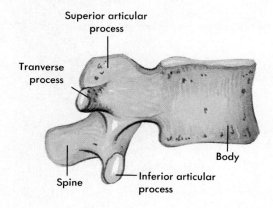

Figure 7-19 Third lumbar vertebra viewed from the side (lateral).

the first cervical vertebra have a flat, rounded body placed anteriorly and centrally, plus a sharp or blunt *spinous process* projecting inferiorly in the posterior midline and two transverse processes projecting laterally (Figures 7-15 to 7-19). All but the sacrum and coccyx have a central opening, the *vertebral foramen.* An upward projection (the *dens*) from the body of the second cervical vertebra furnishes an axis for rotating the head. A long, blunt spinous process that can be felt at the back of the base of the neck characterizes the seventh cervical vertebra. Each thoracic vertebra has articular facets for the ribs. More detailed descriptions of separate vertebrae are given in Table 7-2, p. 167. The vertebral column as a whole articulates with the head, ribs, and iliac bones. Individual vertebrae articulate with each other in joints between their bodies and between their articular processes. A description of intervertebral joints appears in Table 7-1, p. 163.

To increase the carrying strength of the vertebral column and to make balance possible in the upright position, the spinal column is curved. At birth there is a continuous posterior convexity from head to coccyx. Later, as the child learns to sit and stand, secondary posterior concavities necessary for balance develop in the cervical and lumbar regions.

Not uncommonly, spinal curves deviate from the normal. For example, the lumbar curve frequently shows an exaggerated concavity called **lordosis,** whereas any of the regions may have an abnormal lateral curvature called **scoliosis.** The so-called hunchback is an exaggerated convexity in the thoracic region called **kyphosis.**

Figure 7-20 The costal cartilages and their articulations with the body of the sternum by radiate ligaments.

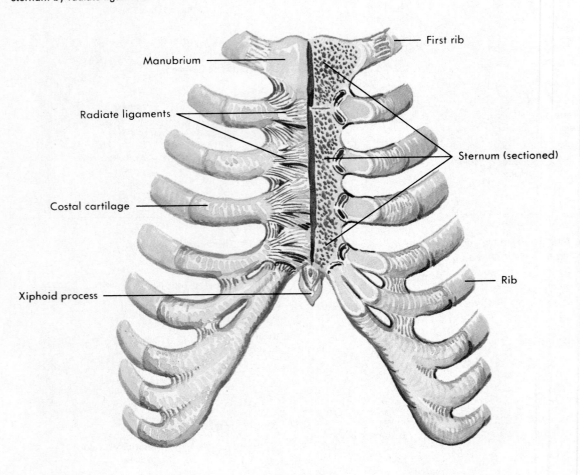

Manubrium

Radiate ligaments

Costal cartilage

Xiphoid process

First rib

Sternum (sectioned)

Rib

Sternum

The *medial part* of the anterior chest wall is supported by the **sternum,** a somewhat dagger-shaped bone consisting of three parts: the upper handle part, the **manubrium;** the middle blade part, the **body;** and a blunt cartilaginous lower tip, the **xiphoid process.** The latter ossifies during adult life. The manubrium articulates with the clavicle and first rib, whereas the next nine ribs join the body of the sternum, either directly or indirectly, by means of the *costal cartilages* (Figure 7-20).

Ribs

Twelve pairs of ribs, together with the vertebral column and sternum, form the bony cage known as the **thorax.** Each rib articulates with both the body and the transverse process of its corresponding thoracic vertebra. The head of each rib articulates with the body of the corresponding thoracic vertebra, and the tubercle of each rib articulates with the vertebra's transverse process (Figures 7-21 and 7-22). In addition, the second through the ninth ribs articulate with the body of the vertebra above. From its vertebral attachment, each rib curves outward, then forward and downward (Figures 7-2 and 7-21), a mechanical fact important for breathing. Anteriorly, each of the first seven ribs joins a costal cartilage that attaches it to the sternum. Each of the costal cartilages of the next three ribs, however, joins the cartilage of the rib above to be thus indirectly attached to the sternum. Because the eleventh and twelfth ribs do not attach even indirectly to the sternum, they are designated floating ribs.

Figure 7-21 A rib of the right side seen from behind (posterior).

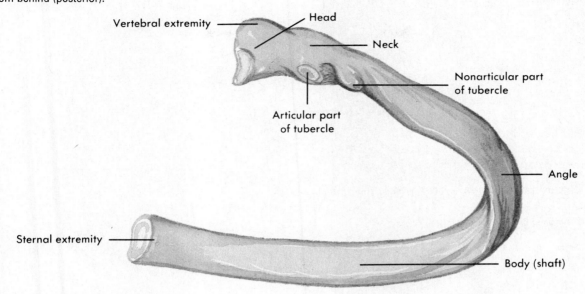

Vertebral extremity

Head

Neck

Nonarticular part
of tubercle

Articular part
of tubercle

Angle

Sternal extremity

Body (shaft)

Figure 7-22 Articulation of the ribs with thoracic vertebra.

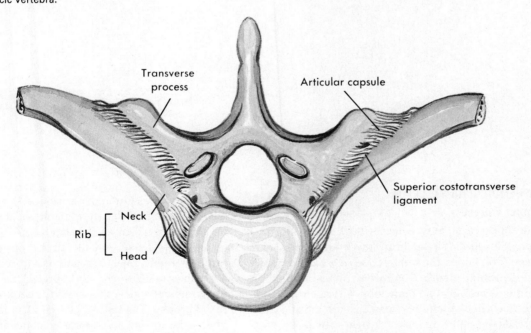

Transverse
process

Articular capsule

Superior costotransverse
ligament

Rib

Neck

Head

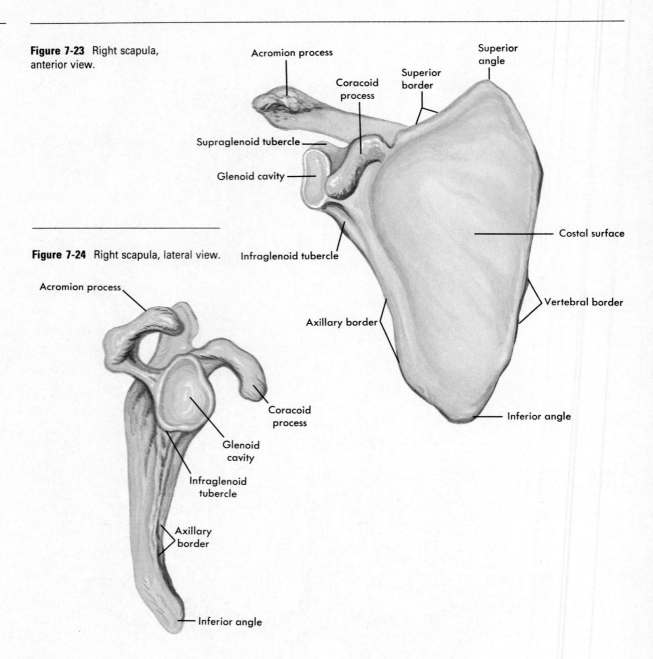

Figure 7-23 Right scapula, anterior view.

Figure 7-24 Right scapula, lateral view.

APPENDICULAR SKELETON
Upper Extremity

The upper extremity consists of the bones of the shoulder girdle, upper arm, lower arm, wrist, and hand. Two bones, the *clavicle* and *scapula*, compose the **shoulder girdle.** Contrary to appearances, this girdle forms only one bony joint with the trunk: the sternoclavicular joint between the sternum and clavicle. At its outer end, the clavicle articulates with the scapula, which attaches to the ribs by muscles and tendons, not by a joint. All shoulder movements therefore involve the sternoclavicular joint. Various markings of the scapula are described in Table 7-2, p. 168 (see also Figures 7-23 to 7-25).

The *humerus* or upper arm bone, like other long bones, consists of a shaft or diaphysis and two ends or epiphyses (Figures 7-26 and 7-27). The upper epiphysis bears several identifying structures: the head, anatomical neck, greater and lesser tubercles, intertubercular groove, and surgical neck. On the diaphysis are found the deltoid tuberosity and the radial groove. The distal epiphysis has four projections—the medial and lateral epicondyles, the capitulum, and the trochlea—and two depressions—the olecranon and coronoid fossae. For descriptions of all of these markings, see Table 7-2. The humerus articulates proximally with the scapula and distally with both the radius and ulna.

Figure 7-25 A, Right scapula, posterior view. **B,** Right scapula, posterior view showing articulation with clavicle.

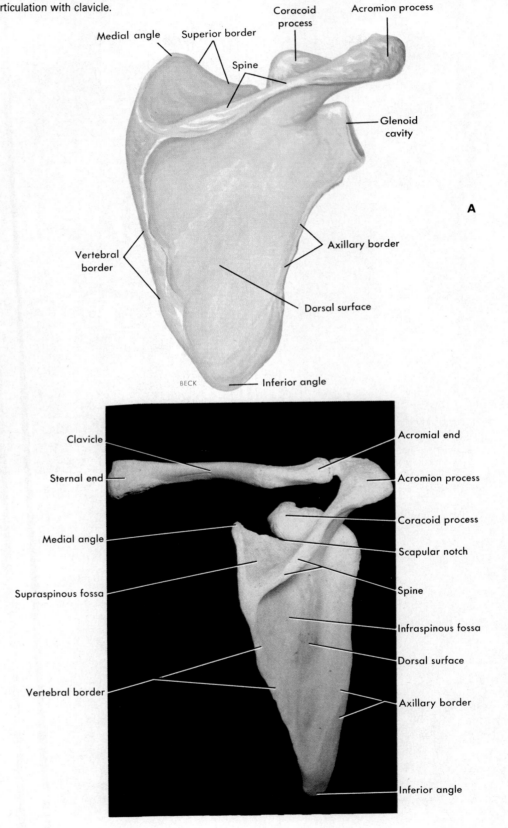

Figure 7-26 Right humerus. **A,** Anterior view. **B,** Anterior aspect of the right elbow skeleton. (See also Figures 7-1 and 7-2, **A** and **B.**)

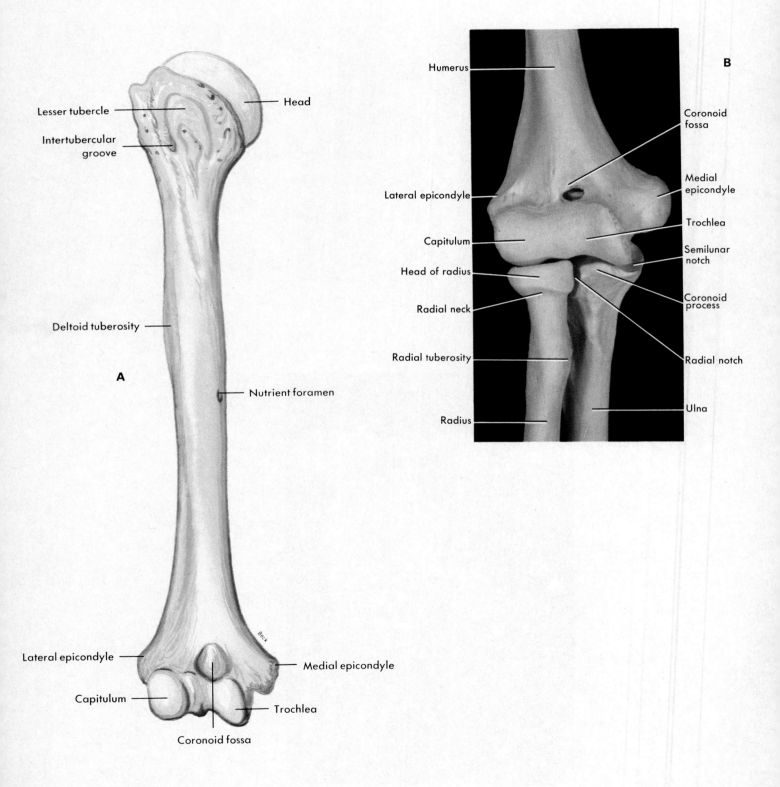

Figure 7-27 Right humerus. **A,** Posterior view. **B,** Posterior aspect of the right elbow skeleton. (See also Figures 7-1 and 7-2, **A** and **B**.)

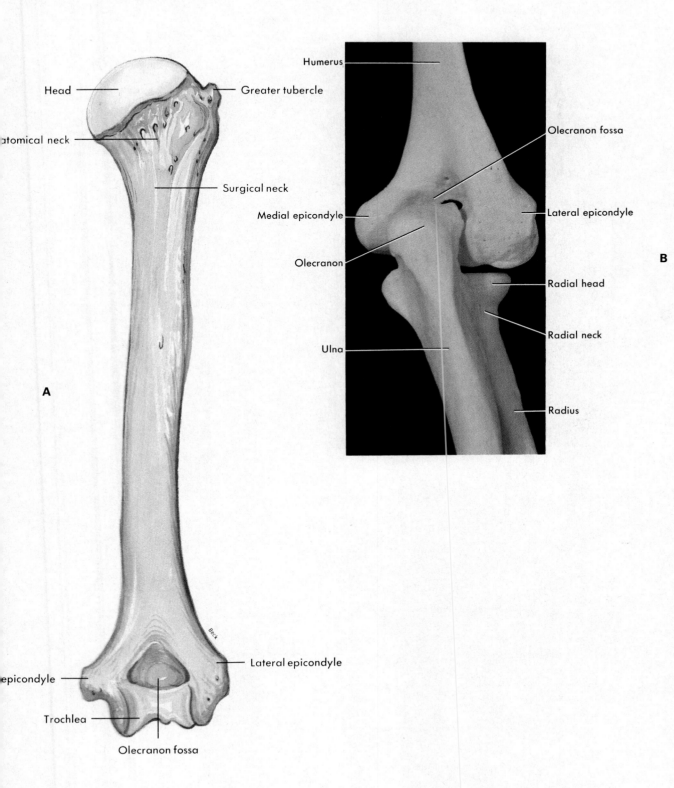

Head

Greater tubercle

atomical neck

Surgical neck

Humerus

Olecranon fossa

Medial epicondyle

Lateral epicondyle

Olecranon

Radial head

Ulna

Radial neck

Radius

A

B

epicondyle

Lateral epicondyle

Trochlea

Olecranon fossa

Figure 7-28 Right radius and ulna. Anterior surfaces.

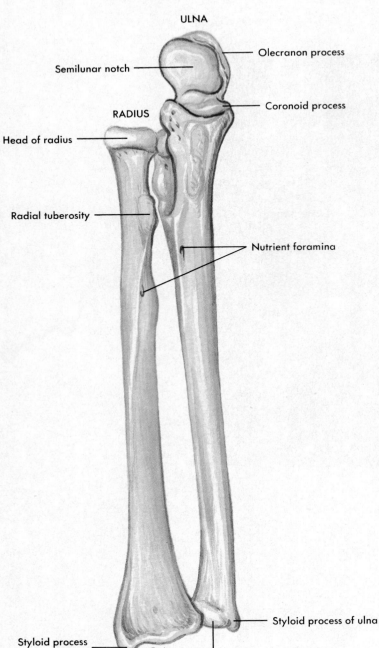

ULNA

Semilunar notch

Olecranon process

RADIUS

Coronoid process

Head of radius

Radial tuberosity

Nutrient foramina

Styloid process of radius

Beck

Head of ulna

Styloid process of ulna

Figure 7-29 Right radius and ulna. Posterior surfaces.

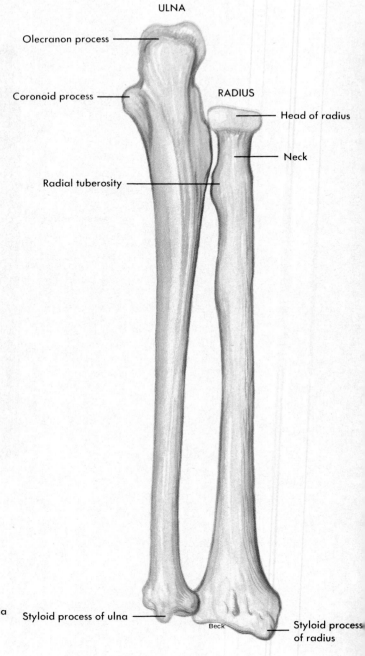

ULNA

Olecranon process

Coronoid process

RADIUS

Head of radius

Neck

Radial tuberosity

Styloid process of ulna

Beck

Styloid process of radius

Two bones form the framework for the lower arm: the **radius** on the thumb side and the **ulna** on the little finger side. At the proximal end of the ulna the olecranon process projects posteriorly and the coronoid process anteriorly. There are also two depressions: the semilunar notch on the anterior surface and the radial notch on the lateral surface. The distal end has two projections: a rounded head and a sharper styloid process. For more detailed identification of these markings, see Table 7-2 (p. 169). The ulna articulates proximally with the humerus and radius and distally with a fibrocartilaginous disk but not with any of the carpal bones.

The radius has three projections: two at its proximal end, the head and radial tuberosity, and one at its distal end, the styloid process (Figures 7-28 and 7-29). There are two proximal articulations: one with the capitulum of the humerus and the other with the radial notch of the ulna. The three distal articulations are with the scaphoid and lunate carpal bones and with the head of the ulna.

The eight **carpal bones** (Figures 7-30 and 7-31) form what most people think of as the upper part of the hand but what, anatomically speaking, is the wrist. Only one of these bones is evident from the outside, the *pisiform bone*, which projects posteriorly on the little finger side as a small rounded elevation. Ligaments bind the carpals closely and firmly together in two rows of four each: proximal row (from little finger toward thumb)—pisiform, triquetrum, lunate, and scaphoid bones; distal row—hamate, capitate, trapezoid, and trapezium bones. The joints between the carpals and radius permit wrist and hand movements.

Of the five **metacarpal bones** that form the framework of the hand, the thumb metacarpal forms the most freely movable joint with the carpals. This fact has great significance. Because of the wide range of movement possible between the thumb metacarpal and the trapezium, particularly the ability to oppose the thumb to the fingers, the human hand has much greater dexterity than the forepaw of any animal and has enabled man to manipulate his environment effectively. The heads of the metacarpals, prominent as the proximal knuckles of the hand, articulate with the phalanges.

Lower Extremity

Bones of the hip, thigh, lower leg, ankle, and foot constitute the lower extremity. Strong ligaments bind each hip bone (*os coxae* or *os innominatum*) to the sacrum posteriorly and to each other anteriorly to form the **pelvic girdle** (Figures 7-32 to 7-34), a stable, circular base that supports the trunk and attaches the lower extremities to it. In early life, each innominate bone is made up of three separate bones. Later on, they fuse into a single, massive irregular bone that is broader than any other bone in the body. The largest and uppermost of the three bones is the **ilium;** the strongest and lowermost, the **ischium;** and the anteriormost, the **pubis.** Numerous markings are present on the three bones (identified on pp. 170—see also Figures 7-35 and 7-36).

The two thigh bones, or **femurs,** have the distinction of being the longest and heaviest bones in the body. Several prominent markings characterize them. For example, three projections are conspicuous at each epiphysis: the head and greater and lesser trochanters proximally and the medial and lateral condyles and adductor tubercle distally (Figure 7-37). Both condyles and the greater trochanter may be felt externally. For a description of the various femur markings, see p. 171.

Text continued on p. 198.

Chondromalacia patellae is a degenerative process that results in a softening (degeneration) of the articular surface of the patella. The symptoms associated with chondromalacia of the patella are a common cause of knee pain in many individuals—especially young athletes. The condition is usually caused by an irritation of the patellar groove, with subsequent changes in the cartilage on the underside of the patella. The most common complaint is of pain arising from behind or beneath the kneecap, especially during activities that require flexion of the knee, such as climbing stairs, kneeling, jumping, or running.

Figure 7-30 Bones of the right hand and wrist. **A,** Dorsal surface. **B,** Dorsal aspect of the right wrist and hand skeleton.

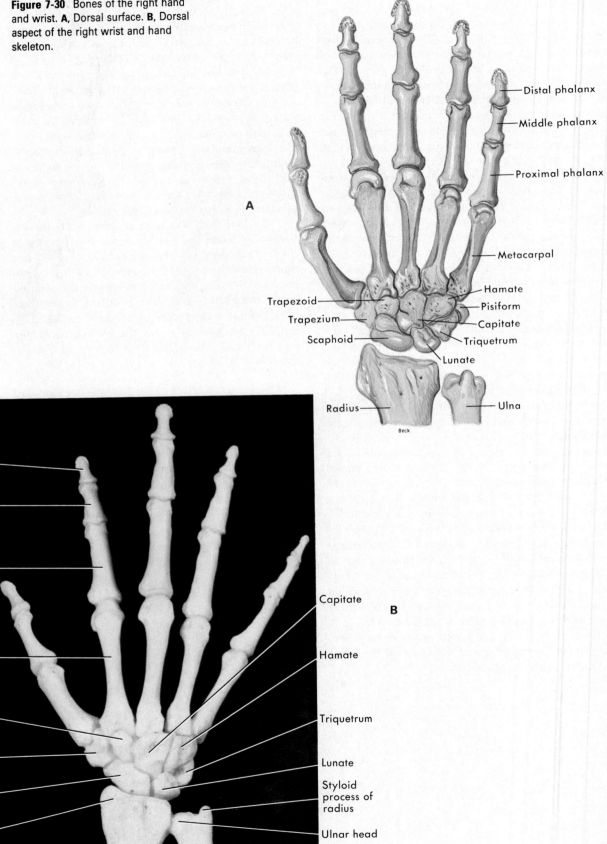

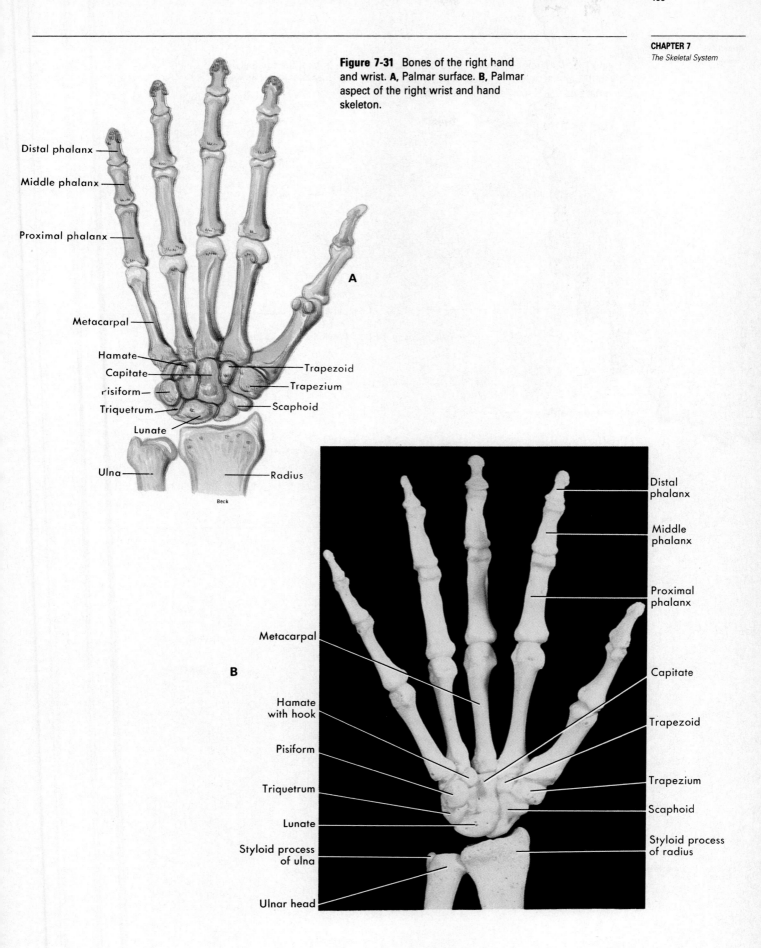

Figure 7-31 Bones of the right hand and wrist. **A**, Palmar surface. **B**, Palmar aspect of the right wrist and hand skeleton.

Figure 7-32 The female pelvis viewed from above. Note the brim of the true pelvis (dotted line) that marks the boundary between the false pelvis (pelvis major) above and the true pelvis (pelvis minor) below it.

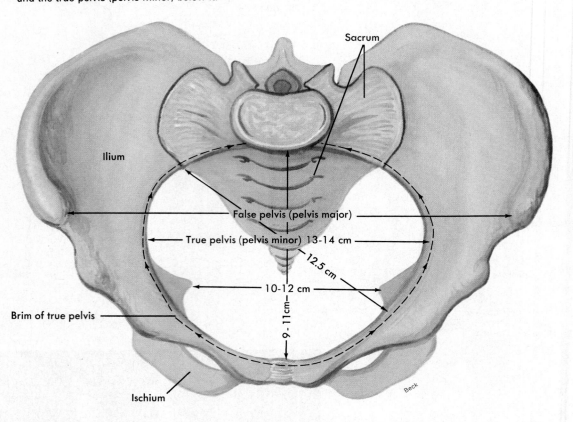

Figure 7-33 Comparison of male and female bony pelvis.

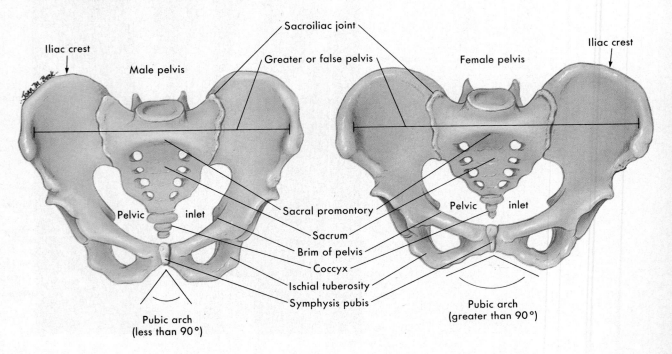

Figure 7-34 A, The female pelvic outlet viewed from below (inferior).
B, Inferior aspect of the bony pelvis.

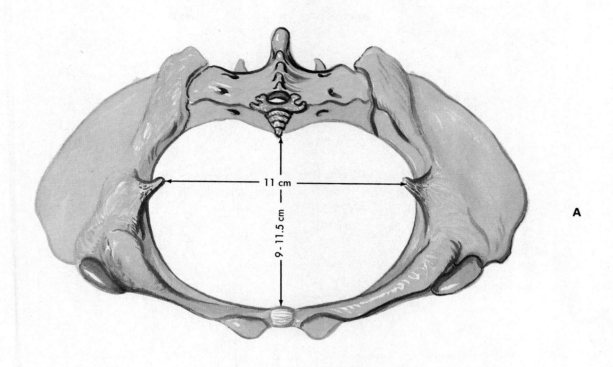

11 cm

9 - 11.5 cm

A

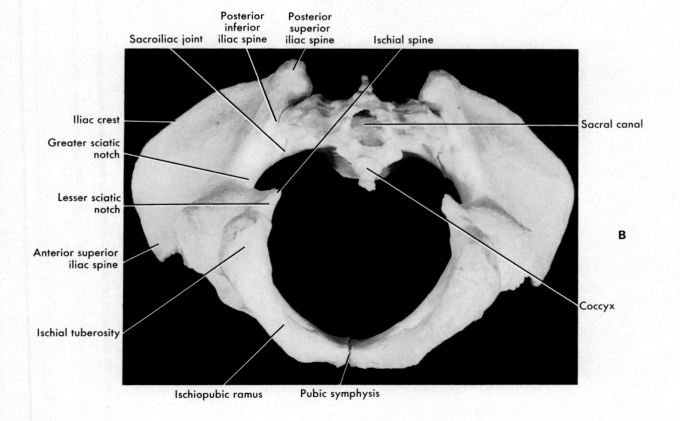

Posterior
inferior
iliac spine

Posterior
superior
iliac spine

Sacroiliac joint

Ischial spine

Iliac crest

Sacral canal

Greater sciatic
notch

Lesser sciatic
notch

Anterior superior
iliac spine

B

Ischial tuberosity

Coccyx

Ischiopubic ramus

Pubic symphysis

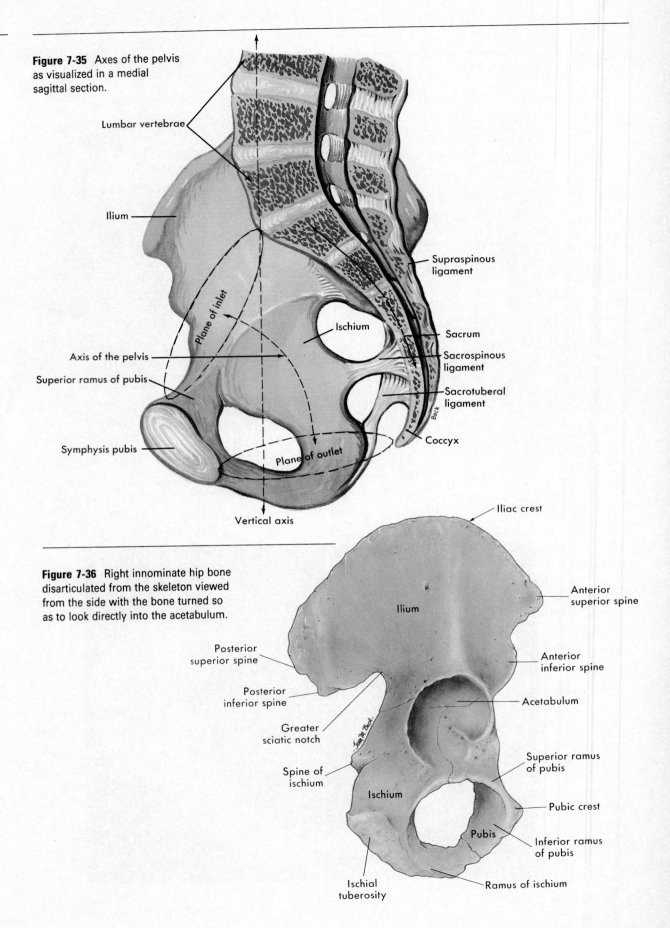

Figure 7-35 Axes of the pelvis as visualized in a medial sagittal section.

Lumbar vertebrae

Ilium

Supraspinous ligament

Plane of inlet

Ischium

Axis of the pelvis

Superior ramus of pubis

Sacrum

Sacrospinous ligament

Sacrotuberal ligament

Symphysis pubis

Coccyx

Plane of outlet

Vertical axis

Beck

Figure 7-36 Right innominate hip bone disarticulated from the skeleton viewed from the side with the bone turned so as to look directly into the acetabulum.

Iliac crest

Posterior superior spine

Ilium

Anterior superior spine

Posterior inferior spine

Greater sciatic notch

Anterior inferior spine

Acetabulum

Spine of ischium

Superior ramus of pubis

Ischium

Pubic crest

Pubis

Inferior ramus of pubis

Ischial tuberosity

Ramus of ischium

Figure 7-41 Transverse arch in the tarsal and metatarsal region of the right foot (phalanges removed).

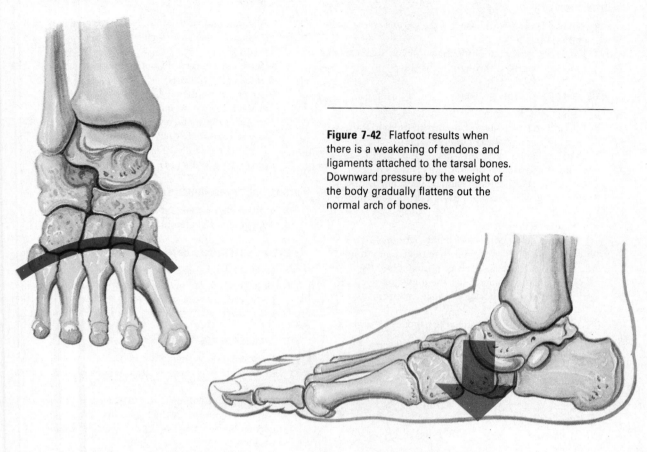

Figure 7-42 Flatfoot results when there is a weakening of tendons and ligaments attached to the tarsal bones. Downward pressure by the weight of the body gradually flattens out the normal arch of bones.

Figure 7-43 High heels throw the weight forward, causing the heads of the metatarsals to bear most of the body's weight.

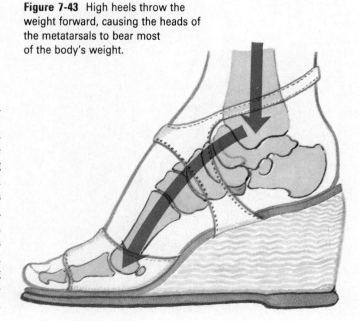

SKELETAL DIFFERENCES IN MEN AND WOMEN

Both general and specific differences exist between male and female skeletons. The general difference is one of size and weight, the male skeleton being larger and heavier. The specific differences concern the shape of the pelvic bones and cavity. Whereas the male pelvis is deep and funnel shaped with a narrow pubic arch (usually less than 90 degrees), the female pelvis, as Figures 7-33 to 7-35 show, is shallow, broad, and flaring, with a wider pubic arch (usually greater than 90 degrees). The childbearing function obviously explains the necessity for these and certain other modifications of the female pelvis.

Outline Summary

INTRODUCTION

A Skeletal tissues form bones—the organs of the skeletal system

B Externally palpable landmarks—bones that can be touched/identified through the skin

DIVISIONS OF SKELETON

A Names and definitions

  1 Axial skeleton—bones that form the upright axis of the body plus ear bones

  2 Appendicular skeleton—bones appended to axial skeleton, that is, bones of upper and lower extremities, including shoulder and hip girdles

B Names and numbers of bones in axial skeleton (80 bones)

  1 Skull (28 bones)

   a Cranium (8 bones)—frontal, parietal (2), temporal (2), occipital, sphenoid, and ethmoid

   b Face (14 bones)—nasal (2), maxillary (2), zygomatic or malar (2), mandible, lacrimal (2), palatine (2), inferior conchae or turbinates (2), and vomer

   c Ear bones (6)—malleus (2), incus (2), and stapes (2)

  2 Hyoid

  3 Spinal column (24 vertebrae plus sacrum and coccyx)

   a Cervical (7)

   b Thoracic (12)

   c Lumbar (5)

   d Sacrum

   e Coccyx

  4 Sternum and ribs (25 bones)—sternum, true ribs (7 pairs), false ribs (5 pairs, 2 pairs of which are floating)

C Names and numbers of bones in appendicular skeleton (126 bones)

  1 Upper extremities (64 bones)—clavicle (2), scapula (2), humerus (2), radius (2), ulna (2), carpals (16), metacarpals (10), and phalanges (28)

  2 Lower extremities (62 bones)—ossa coxae (2), femur (2), patella (2), tibia (2), fibula (2), tarsals (14), metatarsals (10), and phalanges (28)

STRUCTURE OF INDIVIDUAL BONES

A Axial skeleton (see Tables 7-1 and 7-2)

B Appendicular skeleton

SKELETAL DIFFERENCES IN MEN AND WOMEN

A Male skeleton larger and heavier

B Male pelvis deep and funnel shaped with narrow pubic arch; female pelvis shallow, broad, and flaring with wider pubic arch and larger iliosacral notch

AGE CHANGES IN THE SKELETON

A From infancy to young adulthood—bones increase in absolute and relative sizes; shapes of some bones change

B From young adulthood to old age—bones lose calcium, contours of their margins change, and certain bone diseases (notably, oseteoporosis) develop

Review Questions

1 Describe the skeleton as a whole and identify its two major subdivisions.
2 Identify the bones in the cranium and face.
3 Name and locate the fontanels and sutures of the skull.
4 Name the five pairs of bony sinuses in the skull.
5 Discuss the clinical (medical) importance of the cribriform plate of the ethmoid bone and the mastoid air cells in the temporal bone.
6 Identify and discuss the normal primary and secondary curves of the spine.
7 List and explain the three most important pathological curvatures of the spine.
8 Identify the bony components of the thorax.
9 Identify the bones of the shoulder and hip girdles.
10 Identify and compare the bones of the arm, forearm, wrist, and hand with those of the thigh, lower leg, ankle, and foot.
11 Discuss the arches of the foot.
12 Describe some common age changes in the skeleton.

8 ARTICULATIONS

OBJECTIVES

After you have completed this chapter, you should be able to:

1 Define the terms *articulation* and *arthrology*.
2 Compare the classification of joints according to structure and range of movement.
3 List the types of synarthroses and give an example of each.
4 Discuss the six structures that characterize diarthrotic joints.
5 Identify the types of movement at diarthroses and give examples of specific joints where each occurs.
6 Discuss the structural characteristics of uniaxial, biaxial, and multiaxial diarthroses and give examples of each.
7 Discuss the knee joint as a typical diarthrotic joint.
8 Explain the functional significance of bursae.
9 Discuss and compare osteoarthritis and gout as joint diseases.
10 Define the term *arthroscopy* and explain its significance in the diagnosis and treatment of joint diseases.

MEANING AND FUNCTIONS

Articulations in the body are joints between bones. They perform two seemingly contradictory functions. Articulations hold bones firmly bound to each other, and yet they also permit movement between them. If you, like most of us, are fortunate enough to have normal, healthy joints, you may never have been aware of these structures and their importance. To say merely that joints are important is an understatement. The existence of joints between bones makes possible movements of body parts. Movements in turn make a major contribution to the maintenance of homeostasis and therefore to survival. But they also do something more. Movements provide a large measure of our enjoyment of life. Without joints between our bones, we could make no movements; our bodies would be rigid, immobile hulks. This chapter begins by classifying joints and describing their identifying features. It goes on to discuss the structure and functions of several joints. It concludes by describing some common joint age changes and diseases.

KEY TERMS

Amphiarthroses (AM-fe-ar-THRO-sez)

Arthrology (ar-THROL-o-je)

Arthroscopy (ar-THROS-ko-pe)

Bursa (BER-sah)

Diarthroses (di-ar-THRO-sez)

Gomphoses (gom-FO-sez)

Meniscus (me-NIS-kus)

Prosthesis (pros-THEE-sis)

Sutures (SU-churs)

Symphyses (SIM-fi-sez)

Synarthroses (SIN-ar-THRO-sez)

Synchondroses (SIN-kon-DRO-sez)

Syndesmoses (SIN-des-MO-sez)

Figure 8-1 Examples of the types of synarthrotic (fibrous) joints.

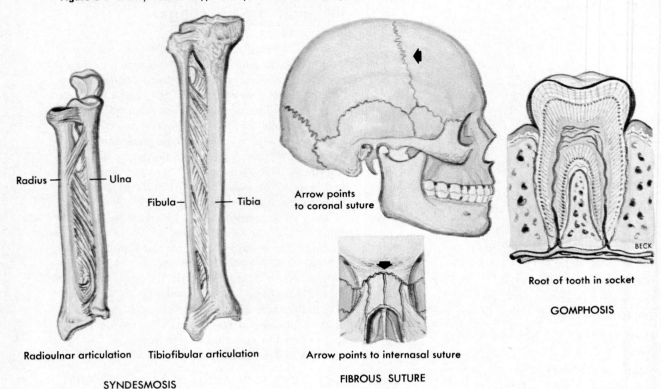

Radius — Ulna

Fibula— Tibia

Arrow points
to coronal suture

Root of tooth in socket

GOMPHOSIS

Radioulnar articulation   Tibiofibular articulation

SYNDESMOSIS

Arrow points to internasal suture

FIBROUS SUTURE

BECK

CLASSIFICATION OF JOINTS

Classified according to the degree of movement they permit, there are three main types of joints: **synarthroses,** or immovable joints; **amphiarthroses,** or slightly movable joints; and **diarthroses,** or freely movable joints.

Classified according to structure, there are also three main types of joints: fibrous joints, cartilaginous joints, and synovial joints. Fibrous tissue grows between the articulating surfaces of bones in **fibrous joints,** binding them so closely together that no movement can occur between them. Fibrous joints therefore are synarthroses. The bones of **cartilaginous joints** are connected by cartilage that permits slight movement between them. In short,

cartilaginous joints are amphiarthroses. **Synovial joints** are characterized by synovial membrane lining a joint cavity, that is, a space separating the articular surfaces of the bones. Synovial joints are diarthroses. Table 8-1 classifies joints according to both structure and range of movement. Refer often to this table and to the illustrations that follow as you read about each of the major joint types.

SYNARTHROSES (FIBROUS JOINTS)

There are three types of synarthroses, or immovable joints, namely, **sutures, syndesmoses,** and **gomphoses.** Identify each of these in Figure 8-1. *Sutures* are found only in the skull. In most sutures, teethlike projections jut out from adjacent bones and interlock with each other with only a thin layer of fibrous tissue between them. In adults, bone has replaced the fibrous tissue in some sutures. A *syndesmosis* is a joint in which fibrous bands (ligaments) connect two bones. Examples: joint between the distal ends of the radius and ulna; joint between the distal ends of the tibia and fibula. A *gomphosis* is a joint between the root of a tooth and the alveolar process of the mandible or maxilla. The fibrous tissue between the tooth's root and the alveolar process is called the *periodontal membrane.*

> **Arthrology** is a specialized area of anatomy that deals with the study and description of joints or articulations. By definition a joint or articulation exists where two or more skeletal components, whether bone or cartilage, come together or meet.

Table 8-1 Classification of joints

Types	Examples	Structural features	Movements
SYNARTHROSES (fibrous or immovable joints)			
Sutures	Joints between skull bones	Teethlike projections of articulating bones interlock with thin layer of fibrous tissue connecting them	None (synarthroses)
Syndesmoses	Joints between distal ends of radius and ulna	Fibrous bands (ligaments) connect articulating bones	None (synarthroses)
Gomphoses	Joints between roots of teeth and jaw bones	Fibrous tissue connects roots of teeth to alveolar processes	None (synarthroses)
AMPHIARTHROSES (cartilaginous or slightly movable joints)			
Symphyses	Symphysis pubis; joints between bodies of vertebrae	Fibrocartilage between articulating bones	Slight (amphiarthroses)
Synchondroses	Costal cartilage attachments of first 10 ribs to sternum	Hyaline cartilage connects articulating bones	Ordinarily none
DIARTHROSES (synovial or freely movable joints)			
Uniaxial			Around one axis; in one plane
Hinge (ginglymus)	Elbow joint	Spool-shaped process fits into concave socket	Flexion and extension only
Pivot (trochoid)	Joint between first and second cervical vertebrae	Arch-shaped process fits around peglike process	Rotation
Biaxial			Around two axes, perpendicular to each other; in two planes
Saddle	Thumb joint between first metacarpal and carpal bone	Saddle-shaped bone fits into socket that is concave-convex-concave	Flexion, extension in one plane; abduction, adduction in other plane; opposing thumb to fingers
Condyloid (ellipsoidal)	Joint between radius and carpal bones	Oval condyle fits into elliptical socket	Flexion, extension in one plane; abduction, adduction in other plane
Multiaxial			Around many axes
Ball and socket (spheroid)	Shoulder joint and hip joint	Ball-shaped process fits into concave socket	Widest range of movements; flexion, extension, abduction, adduction, rotation, circumduction
Gliding	Joints between carpal and tarsal bones	Relatively flat articulating surfaces	Gliding movements without any angular or circular movements

Figure 8-2 Examples of the types of amphiarthrotic (cartilaginous) joints.

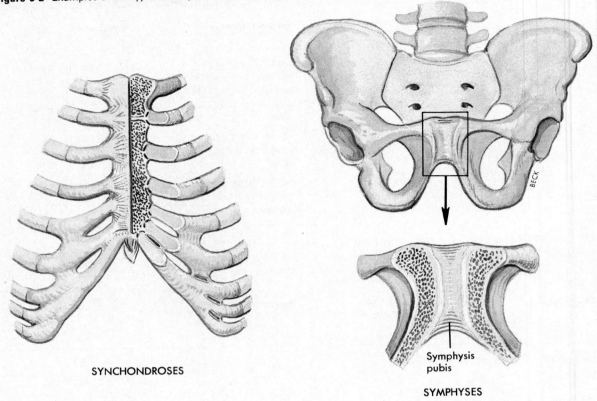

SYNCHONDROSES

Symphysis
pubis

SYMPHYSES

AMPHIARTHROSES (CARTILAGINOUS JOINTS)

Amphiarthroses are subdivided into two types: **symphyses** and **synchondroses** (Figure 8-2).

A *symphysis* is a joint in which a pad or disk of fibrocartilage connects two bones whose articular surfaces are covered with a thin layer of hyaline cartilage. The cartilages permit slight movement between the bones. Symphyses are located in the midline of the body. Examples: symphysis pubis between the two pubic bones; joints between the bodies of the vertebrae. A *synchondrosis* is a joint in which hyaline cartilage connects two bones. The joints between the first 10 pairs of ribs and the sternum are synchondroses. (Here the pieces of cartilage are called costal cartilages.) Other synchondroses are those joints present during the growth years between the epiphyses of a long bone and its diaphysis. In adults, bone has replaced the cartilage of these joints. The slight movement possible at the symphyses and synchondroses sometimes assumes great importance. During childbirth, for example, slight movement at the symphysis pubis facilitates the baby's passage through the pelvis. Slight movement at the synchondroses between the ribs and sternum allows the chest cavity to enlarge during breathing.

DIARTHROSES (SYNOVIAL JOINTS)
Definition

Diarthroses are freely movable joints. Not only are they the body's most mobile but also its most numerous and most complex joints.

Structure

The following six structures characterize diarthrotic, or freely movable, joints (Figure 8-3).

1 **Joint capsule**—Sleevelike extension of the periosteum of each of the articulating bones. The capsule forms a complete casing around the ends of the bones, thereby binding them to each other.
2 **Synovial membrane**—Moist, slippery membrane that lines the inner surface of the joint capsule. It attaches, as the red line in Figure 8-3 shows, to the margins of the articular cartilage. It also secretes synovial fluid, which lubricates and nourishes the inner joint surfaces.
3 **Articular cartilage**—Thin layer of hyaline cartilage covering and cushioning the articular surfaces of bones.
4 **Joint cavity**—Small space between the articulating surfaces of the two bones of the joint. Because of this cavity with no tissue

Figure 8-3 Structure of diarthrotic joints. **A,** Illustration showing a typical diarthrotic or synovial joint. **B,** Photograph of the ankle joint as seen from in front.

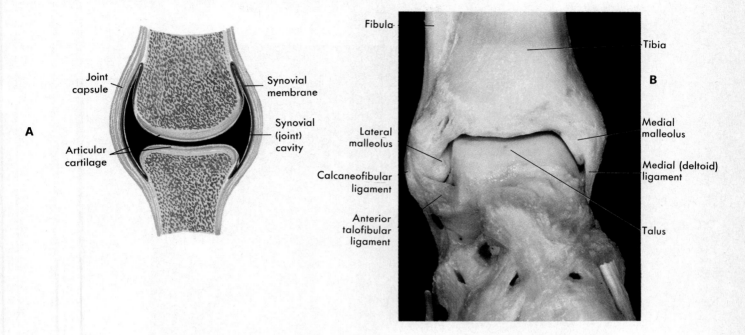

growing between the articulating surfaces of the bones, the bones are free to move against one another. Synovial joints therefore are diarthroses, or freely movable joints.

5 **Menisci (articular disks)**—Pads of fibrocartilage located between the articulating ends of bones in some diarthroses. Usually these pads divide the joint cavity into two separate cavities. The knee joint contains several menisci (see Figures 8-9 and 8-12).

6 **Ligaments**—Strong cords of dense white fibrous tissue at most synovial joints. These grow between the bones, lashing them even more firmly together than possible with the joint capsule alone.

Types of Movement at Diarthroses

The types of movement possible at diarthrotic joints depend on the shapes of the articulating surfaces of the bones and on the positions of the joint's ligaments and nearby muscles and tendons. All diarthroses, however, permit one or more of the following types of movements: angular, circular, gliding, and special movements.

Angular movements change the size of the angle between articulating bones (Figure 8-4, [a] to [m]). Flexion, extension, abduction, and adduction are some of the different types of angular movements. **Flexion** decreases the angle between bones.

The thin layer of hyaline cartilage that covers the articulating surfaces of bones in diarthrotic joints may become cracked as a result of traumatic or athletic-related injury—especially to the knee. If such an injury occurs, small fragments may become detached from the damaged layer of cartilage and enter the joint cavity. These fragments are called **loose bodies** or **"joint mice."** They move around in the joint and may cause pain and loss of motion if they become lodged between the opposing bone surfaces (see pp. 232-233).

It bends or folds one part on another. For example, if you bend your head forward on your chest, you are flexing it. Or if you bend your arm at the elbow, you are flexing your lower arm. Flexion, in short, is bending, folding, or withdrawing a part. **Extension** increases the angle between bones. It returns a part from its flexed position to its anatomical position. Extensions are straightening or stretching movements. Stretching an extended part beyond its anatomical position (as in stretching the head backward from its upright position) is called **hyperextension.** Stretching the foot down and back, as in pointing the

Text continued on p. 216.

Figure 8-4 A, Angular movements at synovial joints: (a) to (m).

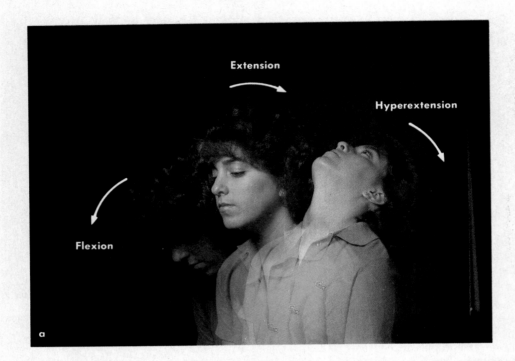

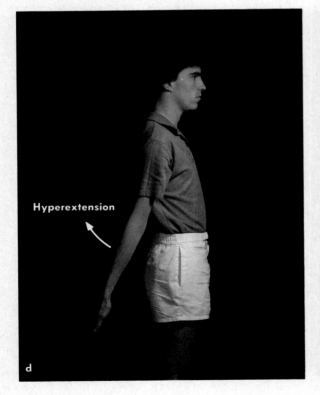

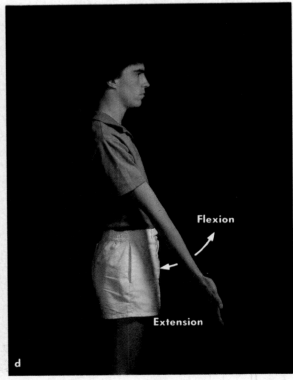

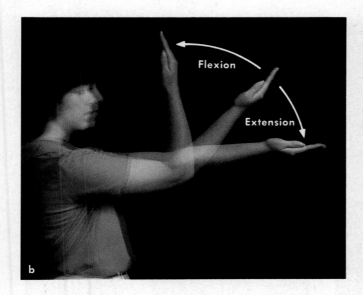

b

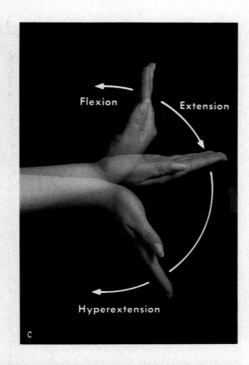

c

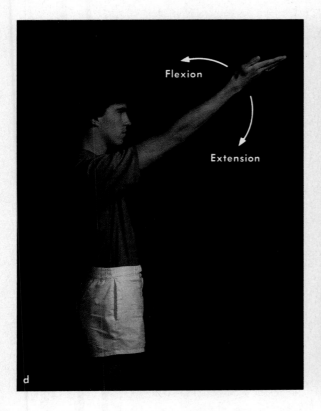

d

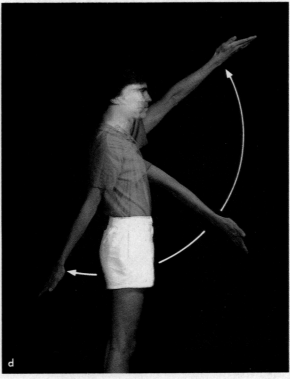

d

Continued.

Figure 8-4, cont'd **A,** Angular movements at synovial joints: (a) to (m).

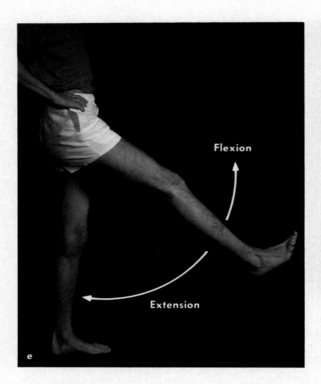

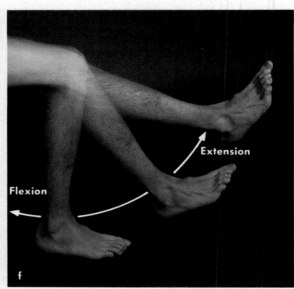

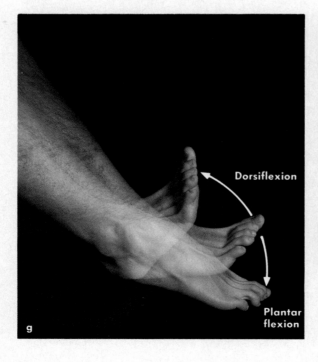

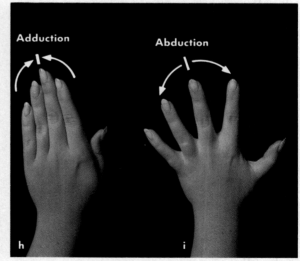

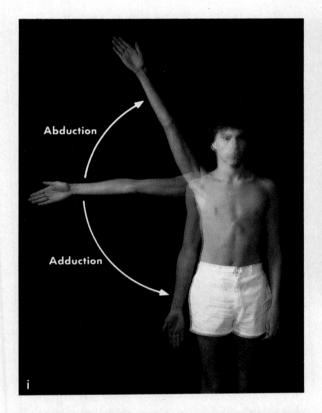

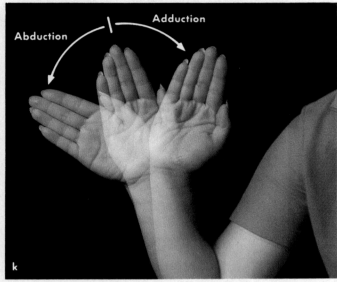

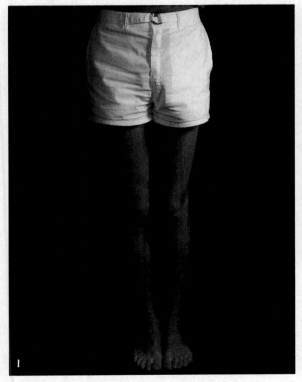

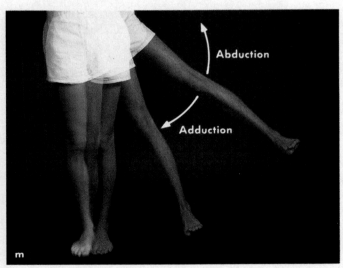

Continued.

Figure 8-4, cont'd **B,** Rotation and circumduction: (n) rotation at the atlantoaxial joint, (o) rotation of the humerus, producing supination and pronation of the hand, and (p) circumduction of the humerus at the shoulder joint.

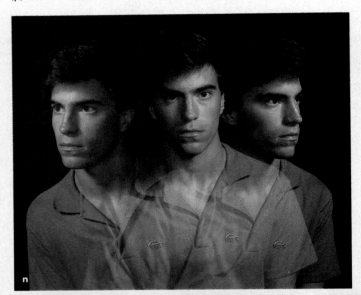

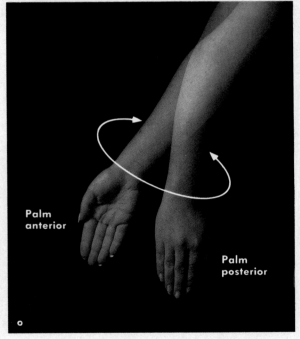

Palm
anterior

Palm
posterior

o

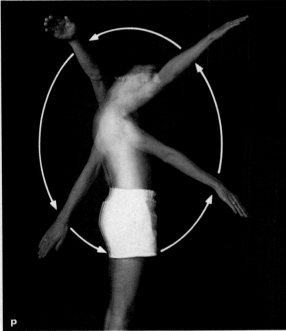

p

Figure 8-4, cont'd **C,** Special movements: (q) inversion, (r) eversion, (s) retraction, (t) protraction, (u) elevation, (v) depression.

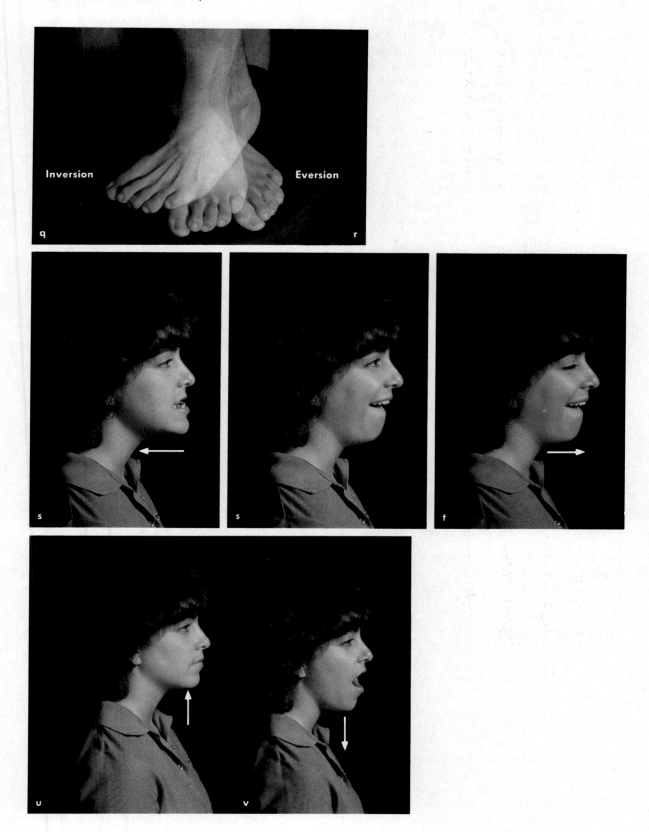

toe, increases the angle between the top of the foot and the front of the leg. It extends the foot and is usually referred to as **plantar flexion.** Flexion of the foot, that is, tilting it up, is called **dorsiflexion. Abduction** moves a part away from the median plane of the body, as in moving the arms straight out to the side or the fingers away from the midline of the hand. **Adduction** moves a part toward the median plane. Examples: bringing the arms back to the sides; moving the fingers toward the midline of the hand.

Circular movements are rotation, circumduction, supination, and pronation (Figure 8-4, [n] to [p]). **Rotation** consists of pivoting a bone on its own axis. For example, moving the head from side to side as in indicating "no." **Circumduction** moves a part so its distal end describes a circle and the rest of the part describes a cone. When a pitcher winds up to throw a ball, he circumducts his arm. If you drop your head to one shoulder, then to the chest, to the other shoulder, and toward the back, you are circumducting your head. **Supination** rotates the forearm outward, thereby turning the palm of the hand forward as in the anatomical position. **Pronation** rotates the forearm inward, thereby turning the back of the hand forward.

Gliding movements are the simplest of all movements. The articular surface of one bone barely moves over the articular surface of another without any angular or circular movement. Gliding movements can occur, for example, between the carpal bones and between the tarsals.

Special movements consist of inversion, eversion, protraction, retraction, elevation, and depression (Figure 8-4, [q] to [v]). **Inversion** turns the sole of the foot inward whereas **eversion** turns it outward. **Protraction** moves a part forward whereas **retraction** moves it back. For instance, if you stick out your jaw, you protract it, and if you pull it back, you retract it. **Elevation** moves a part up, as in shrugging the shoulders or closing the mouth. **Depression** lowers a part, moving it in the opposite direction from elevation.

Types of Diarthroses

Diarthroses are divided into three main types—uniaxial, biaxial, and multiaxial (Figure 8-5). Each of these is subdivided into two subtypes as follows:

1 Uniaxial joints are diarthroses that permit movement around only one axis and in only one plane. Hinge and pivot joints are types of uniaxial joints.

   a Hinge joints (ginglymus) are those in which the articulating ends of the bones form a hinge-shaped unit. Like a common door hinge, hinge joints permit only back and forth movements, namely, flexion and extension. If you have access to an articulated skeleton, examine the articulating end of the humerus (the trochlea) and of the ulna (the semilunar notch). Observe their interaction as you flex and extend the forearm. Do you see why you can flex and extend your forearm but cannot move it in any other way at this joint? The shapes of the trochlea and of the semilunar notch (see Figures 7-2 and 7-27) permit only the uniaxial, horizontal plane movements of flexion and extension at the elbow. Other hinge joints include the knee and interphalangeal joints.

   b Pivot joints are those in which a projection of one bone articulates with a ring or notch of another bone. Examples: a projection (dens) of the second cervical vertebra articulates with a ring-shaped portion of the first cervical vertebra; the head of the radius articulates with radial notch of the ulna.

2 Biaxial joints are diarthroses that permit movements around two perpendicular axes in two perpendicular planes. Saddle and condyloid joints are types of biaxial joints (Figure 8-5).

   a Saddle joints are those in which the articulating ends of the bones resemble reciprocally shaped miniature saddles. Only two saddle joints—one in each thumb—are present in the body. The thumb's metacarpal bone articulates in the wrist with a carpal bone (trapezium). The saddle-shaped articulating surfaces of these bones make it possible for the thumb to move over to touch the tips of the fingers—that is, to oppose the fingers. How important is this? To answer this for yourself, consider the following. Opposing the thumb to the fingers enables us to grip small objects. Were it not for this movement, we would have no manual dexterity. A surgeon could not grasp a scalpel or suture needle. None of us could hold a pen or pencil for writing.

   b Condyloid (ellipsoidal) joints are those in which a condyle fits into an elliptical socket. Examples: condyles of the occipital bone fit into elliptical depressions of the atlas; the distal end of the radius fits into depressions of the carpal bones (scaphoid, lunate, and triquetrum).

3 Multiaxial joints permit movements around three or more axes and in three or more planes.

   a Ball and socket joints (spheroid joints)

Text continued on p. 222.

Figure 8-5 Diarthroses classified according to the number of planes in which they permit uniaxial, biaxial, and multiaxial movements, with examples of each.

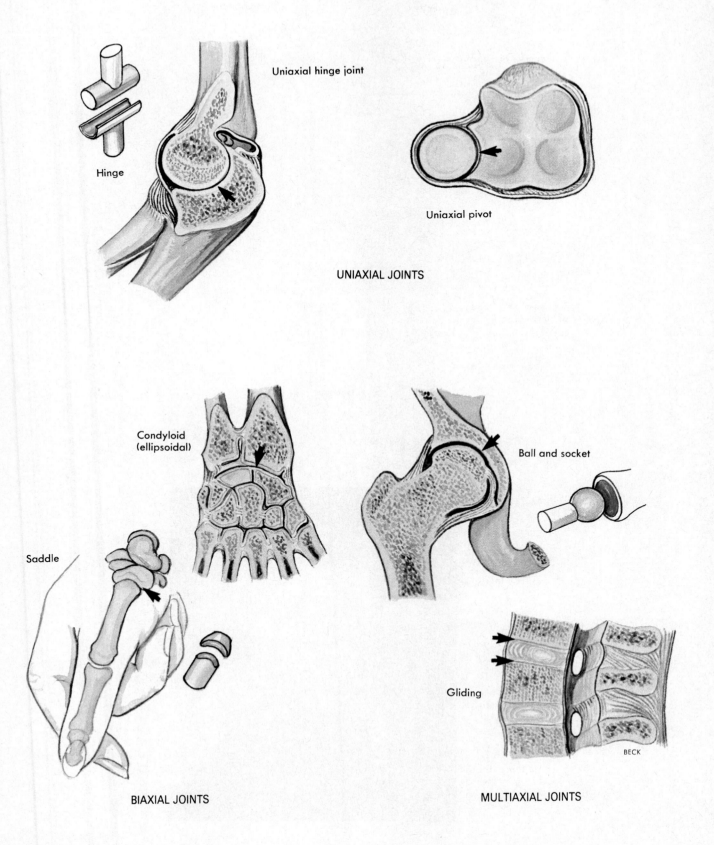

Uniaxial hinge joint

Hinge

Uniaxial pivot

UNIAXIAL JOINTS

Condyloid (ellipsoidal)

Saddle

Ball and socket

Gliding

BIAXIAL JOINTS

MULTIAXIAL JOINTS

BECK

Figure 8-6 The shoulder joint.
A and **B**, Anterior view. **C** and **D**, Viewed
from behind through shoulder joint.

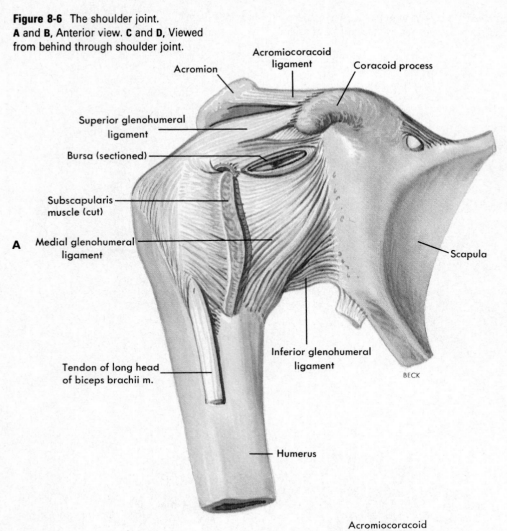

Acromion

Acromiocoracoid
ligament

Coracoid process

Superior glenohumeral
ligament

Bursa (sectioned)

Subscapularis
muscle (cut)

A Medial glenohumeral
ligament

Scapula

Inferior glenohumeral
ligament

BECK

Tendon of long head
of biceps brachii m.

Humerus

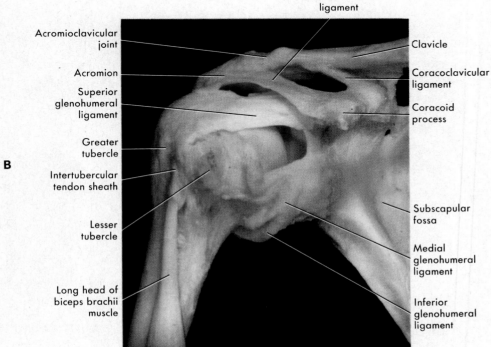

Acromiocoracoid
ligament

Acromioclavicular
joint

Clavicle

Acromion

Coracoclavicular
ligament

Superior
glenohumeral
ligament

Coracoid
process

Greater
tubercle

B

Intertubercular
tendon sheath

Subscapular
fossa

Lesser
tubercle

Medial
glenohumeral
ligament

Long head of
biceps brachii
muscle

Inferior
glenohumeral
ligament

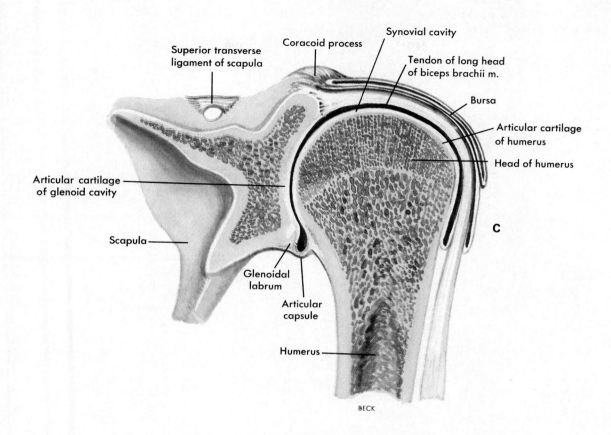

Superior transverse ligament of scapula

Coracoid process

Synovial cavity

Tendon of long head of biceps brachii m.

Bursa

Articular cartilage of humerus

Head of humerus

Articular cartilage of glenoid cavity

Scapula

Glenoidal labrum

Articular capsule

Humerus

BECK

C

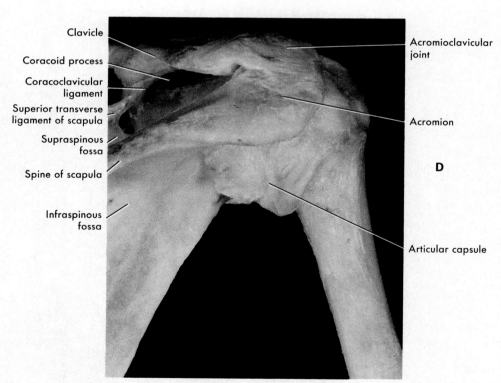

Clavicle

Coracoid process

Coracoclavicular ligament

Superior transverse ligament of scapula

Supraspinous fossa

Spine of scapula

Infraspinous fossa

Acromioclavicular joint

Acromion

Articular capsule

D

Figure 8-7 The hip joint.
A and **B**, Anterior view. **C** and **D**, Frontal
section through the hip joint.

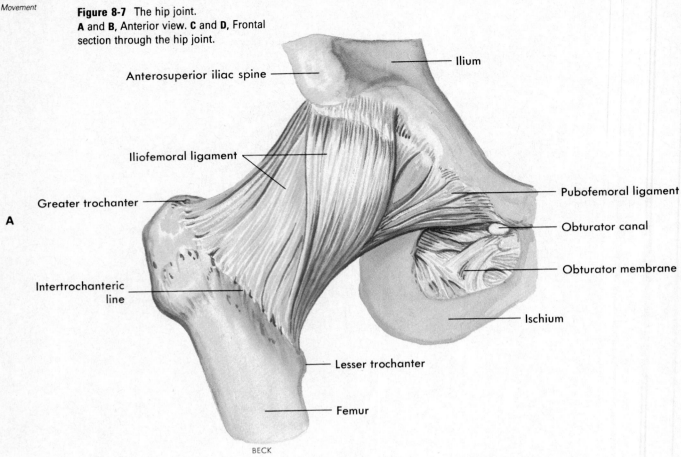

Ilium

Anterosuperior iliac spine

Iliofemoral ligament

Pubofemoral ligament

Greater trochanter

Obturator canal

Obturator membrane

Intertrochanteric
line

Ischium

A

Lesser trochanter

Femur

BECK

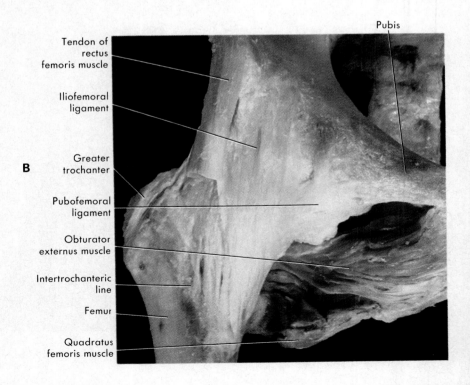

Pubis

Tendon of
rectus
femoris muscle

Iliofemoral
ligament

Greater
trochanter

B

Pubofemoral
ligament

Obturator
externus muscle

Intertrochanteric
line

Femur

Quadratus
femoris muscle

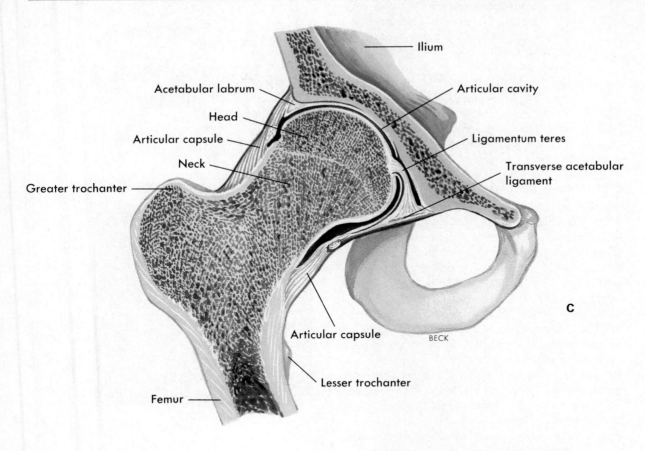

Ilium

Acetabular labrum

Head

Articular capsule

Neck

Greater trochanter

Articular cavity

Ligamentum teres

Transverse acetabular ligament

Articular capsule

BECK

Lesser trochanter

Femur

C

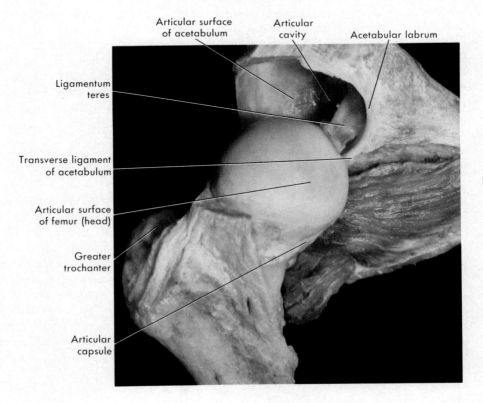

Articular surface of acetabulum

Articular cavity

Acetabular labrum

Ligamentum teres

Transverse ligament of acetabulum

Articular surface of femur (head)

Greater trochanter

Articular capsule

D

are our most movable joints. A ball-shaped head of one bone fits into a concave depression on another, thereby allowing the first bone to move in many directions. Examples: shoulder and hip joints.

b Gliding joints are characterized by relatively flat articulating surfaces that allow limited gliding movements along various axes. Examples: joints between the bodies of successive vertebrae. Gliding joints are the least movable of the diarthroses.

Representative Diarthrotic Joints

HUMEROSCAPULAR JOINT. This joint between the head of the humerus and the glenoid cavity of the scapula is the one we usually refer to as the shoulder joint (Figure 8-6). It is our most mobile joint. One anatomical fact, the shallowness of the glenoid cavity, largely accounts for this. The shallowness offers little interference to movements of the head of the humerus. Were it not for the *glenoid labrum* (Figure 8-6, *C*), a narrow rim of fibrocartilage around the glenoid cavity, it would have scarcely any depth at all. Some structures strengthen the shoulder joint and give it a degree of stability, notably several ligaments, muscles, tendons, and bursae. Note in Figure 8-6, for example, the superior, medial, and inferior glenohumeral ligaments. Each of these is a thickened portion of the joint capsule. Also in this figure, identify the subscapularis muscle, the tendon of the long head of the biceps brachii muscle, and a bursa. Shoulder muscles and tendons form a cufflike arrangement around the joint. It is called the *rotator cuff.* Baseball pitchers frequently injure the rotator cuff in the shoulder of their pitching arms. The main bursa of the shoulder joint is the *subdeltoid bursa.* It lies wedged between the inferior surface of the deltoid muscle and the superior surface of the joint capsule. Other bursae of the shoulder joint are the subscapular, subacromial, and subcoracoid bursae. All in all, the shoulder joint is more mobile than stable. Dislocations of the head of the humerus from the glenoid cavity occur rather frequently.

Ankylosis is a term used to describe the freezing or immobilizing of a joint. If, for example, the articulating bones in a joint are crushed or if inflammation results in proliferation of fibrous tissue, the potential for movement may be completely lost. In severe crushing injuries a joint may be *ankylosed* or fused surgically so that it will remain in a position of "optimum use" even if voluntary movement is not possible.

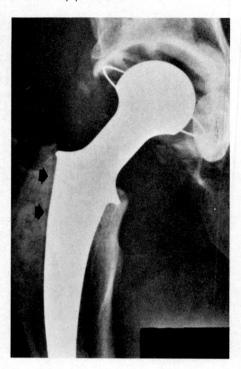

Figure 8-8 X-ray study showing placement of a total hip prosthesis.

HIP JOINT. The first characteristic to remember about the hip joint is stability; the second is mobility (Figure 8-7). The stability of the hip joint derives largely from the shapes of the head of the femur and of the acetabulum, the socket of the hip bone into which the femur head fits. Turn to Figure 7-33 and note the deep, cuplike shape of the acetabulum, and then observe the ball-like head of the femur in Figure 7-37. Compare these with the shallow, almost saucer-shaped glenoid cavity (Figure 7-24) and the head of

Total replacement of diseased knee and hip joints is now a common orthopedic procedure. Partial or total replacement of other joints is also occurring with increasing frequency. Innovative new surgical techniques coupled with the recent introduction of durable and inert materials that may be implanted in living tissues for prolonged periods of time have made such procedures highly successful. The x-ray study in Figure 8-8 shows a total hip prosthesis that has been surgically implanted following removal of a diseased joint.

Figure 8-9 Head of right tibia viewed from above, showing menisci and arrangement of ligaments.

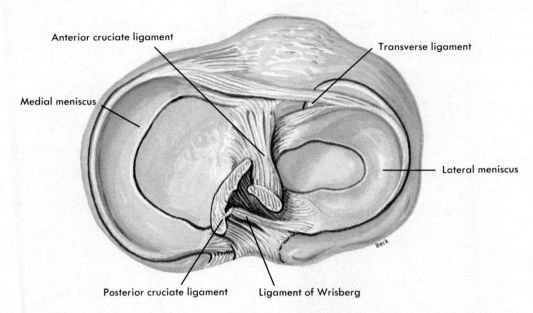

Anterior cruciate ligament

Transverse ligament

Medial meniscus

Lateral meniscus

Posterior cruciate ligament

Ligament of Wrisberg

Beck

the humerus (Figure 7-27). From these observations, do you see why the hip joint necessarily has a somewhat more limited range of movement than the shoulder joint? Both joints, however, allow multiaxial movements. Both permit flexion, extension, abduction, adduction, rotation, and circumduction.

A joint capsule and several ligaments hold the femur and hip bones together and contribute to the hip joint's stability. The iliofemoral ligament connects the ilium with the femur, and the ischiofemoral and pubofemoral ligaments join the ischium and pubic bone to the femur. The iliofemoral ligament is one of the strongest ligaments in the body.

KNEE JOINT. The knee, or tibiofemoral, joint is the largest and one of the most complex and most frequently injured joints in the body (Figures 8-9 to 8-12). The condyles of the femur articulate with the flat upper surface of the tibia. Although this is a precariously unstable arrangement, there are counteracting forces supplied by a joint capsule, cartilages, and numerous ligaments and muscle tendons. Note, for example, in Figure 8-9, the shape of the two cartilages labeled *medial meniscus* and *lateral meniscus*. They attach to the flat top of the tibia and, because of their concavity, form a kind of shallow socket for the condyles of the femur. Of the many ligaments that hold the femur bound to the tibia, four can be seen in Figure 8-9. The anterior cruciate ligament attaches to the anterior part of the tibia between its condyles, then crosses over and backward and attaches to the posterior part of the lateral condyle. The posterior cruciate ligament at-

taches posteriorly to the tibia and lateral meniscus, then crosses over and attaches to the front part of the femur's medial condyle. The ligament of Wrisberg attaches posteriorly to the lateral meniscus and extends up and over to attach to the medial condyle behind the attachment of the posterior cruciate ligament (Figures 8-9 and 8-10). The transverse ligament connects the anterior margins of the two menisci. Strong ligaments, the fibular and tibial collateral ligaments, located at the sides of the knee joint can be seen in Figure 8-10.

A baker's dozen of bursae serve as pads around the knee joint: four in front, four located laterally, and five medially. Of these, the largest is the prepatellar bursa (Figure 8-11) inserted in front of the patellar ligament, between it and the skin. The painful ailment called "housemaid's knee" is prepatellar bursitis.

Compared to the hip joint, the knee joint is relatively unprotected by surrounding muscles. Consequently the knee, more often than the hip, is injured by blows or sudden stops and turns. Athletes, for example, frequently tear a knee cartilage, that is, one of the menisci (Figure 8-12).

The structure of the knee joint permits the hingelike movements of flexion and extension. Also, with the knee flexed, some internal and external rotation can occur. In most of our day-to-day activities—even such ordinary ones as walking, going up and down stairs, and getting into and out of chairs—our knees bear the brunt of the load; they are the main weight bearers. Knee injury or disease, therefore, can be badly crippling.

Figure 8-10 The right knee joint.
A and **B,** Viewed from in front. **C** and **D,**
Viewed from behind.

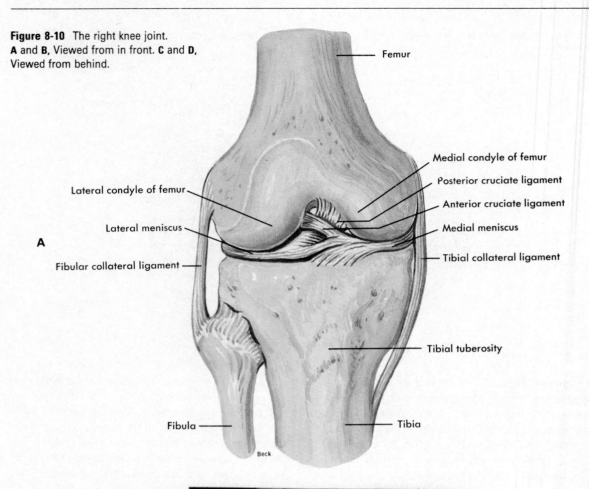

Femur

Medial condyle of femur

Posterior cruciate ligament

Anterior cruciate ligament

Lateral condyle of femur

Lateral meniscus

Medial meniscus

Fibular collateral ligament

Tibial collateral ligament

Tibial tuberosity

Fibula

Tibia

Beck

A

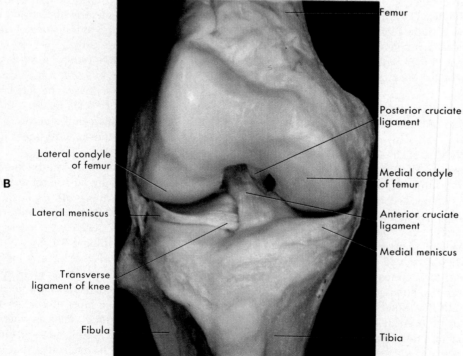

Femur

Posterior cruciate
ligament

Lateral condyle
of femur

Medial condyle
of femur

Lateral meniscus

Anterior cruciate
ligament

Medial meniscus

Transverse
ligament of knee

Fibula

Tibia

B

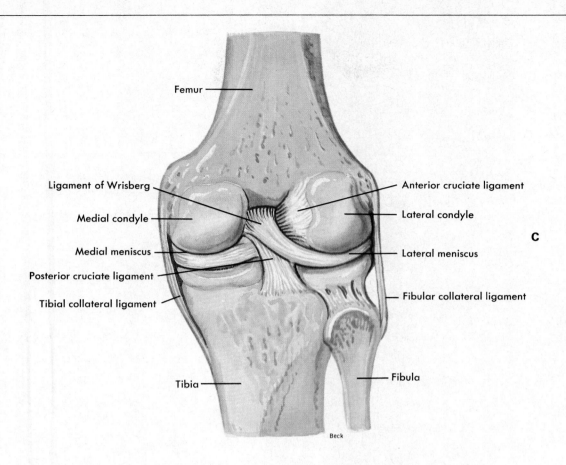

Femur

Ligament of Wrisberg

Medial condyle

Medial meniscus

Posterior cruciate ligament

Tibial collateral ligament

Anterior cruciate ligament

Lateral condyle

Lateral meniscus

Fibular collateral ligament

Tibia

Fibula

C

Beck

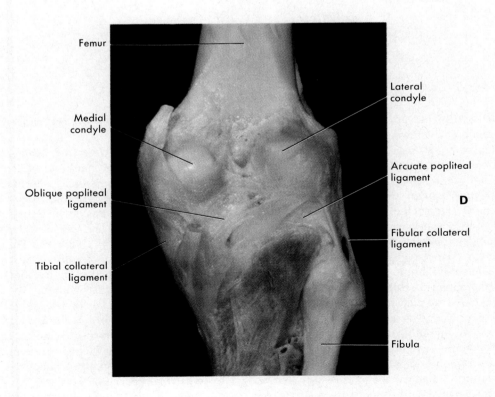

Femur

Medial condyle

Oblique popliteal ligament

Tibial collateral ligament

Lateral condyle

Arcuate popliteal ligament

Fibular collateral ligament

Fibula

D

Figure 8-11 Sagittal section through knee joint.

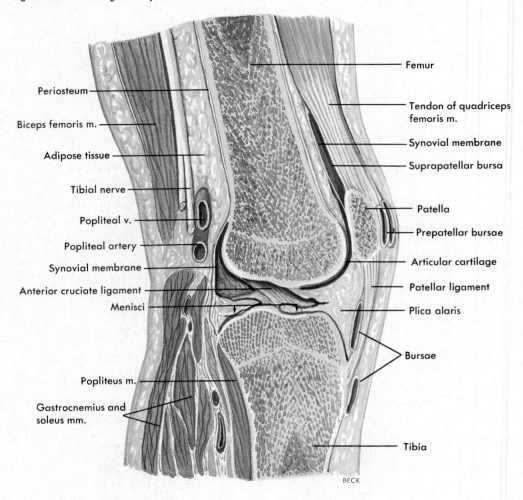

Periosteum

Biceps femoris m.

Adipose tissue

Tibial nerve

Popliteal v.

Popliteal artery

Synovial membrane

Anterior cruciate ligament

Menisci

Popliteus m.

Gastrocnemius and
soleus mm.

Femur

Tendon of quadriceps
femoris m.

Synovial membrane

Suprapatellar bursa

Patella

Prepatellar bursae

Articular cartilage

Patellar ligament

Plica alaris

Bursae

Tibia

BECK

VERTEBRAL JOINTS. One vertebra connects to another by several joints—between their bodies; laminae; and articular, transverse, and spinous processes. These joints hold the vertebrae firmly together so they are not easily dislocated, but also these joints form a flexible column. Consider how many ways you can move the trunk of your body. You can flex it forward or laterally, you can extend it, and you can circumduct or rotate it. The bodies of adjacent vertebrae are connected by intervertebral disks and strong ligaments. Fibrous tissue and fibrocartilage form a disk's outer rim (called the *annulus fibrosus*). Its central core (the *nucleus pulposus*), in contrast, consists of a pulpy, elastic substance. With age the nucleus loses some of its resiliency. It may then be suddenly compressed by exertion or trauma and pushed through the annulus, with fragments protruding into the spinal canal and pressing on spinal nerves or the spinal cord itself. Severe pain results. In medical terminology, this is called a *herniated*

disk; in popular language, it is a "slipped disk" (Figure 8-13).

In Figure 8-14 the following ligaments that bind the vertebrae together can be identified. The *anterior longitudinal ligament,* a strong band of fibrous tissue, connects the anterior surfaces of the vertebral bodies from the atlas down to the sacrum. Connecting the posterior surfaces of the bodies is the *posterior longitudinal ligament.* The *ligamenta flava* bind the laminae of adjacent vertebrae firmly together. Spinous processes are connected by *interspinous ligaments.* In addition, the tips of the spinous processes of the cervical vertebrae are connected by the *ligamentum nuchae;* its extension, the *supraspinous ligament,* connects the tips of the rest of the vertebrae down to the sacrum. And finally, *intertransverse ligaments* connect the transverse processes of adjacent vertebrae.

Table 8-2 gives brief descriptions of most of the individual joints of the body.

Figure 8-12 Knee arthrogram. **A,** Normal medial meniscus. Spot film shows normal triangular shape of meniscus *(arrows).* **B,** Linear tear of medial meniscus *(arrowheads).*

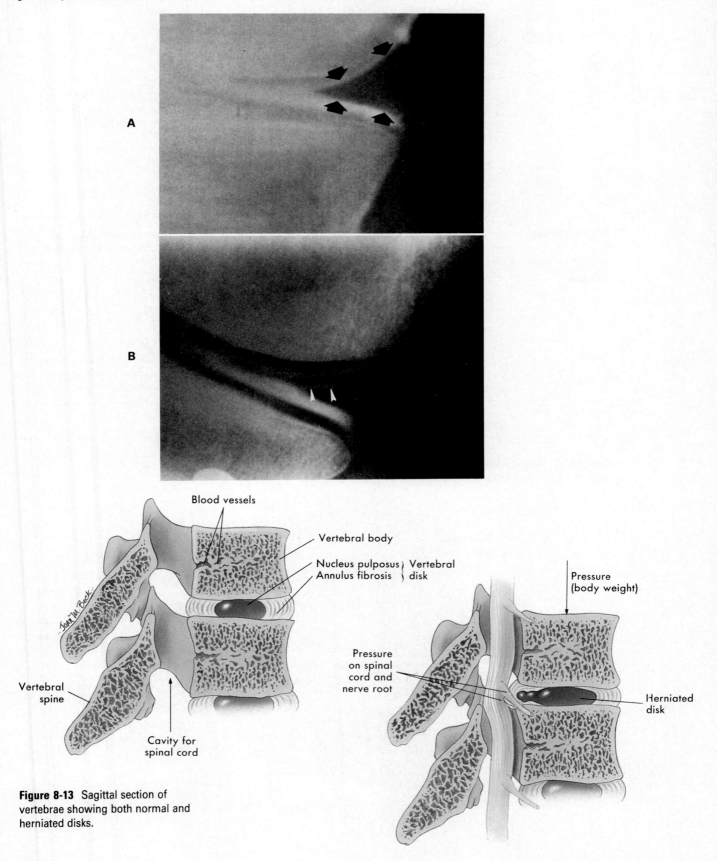

Figure 8-13 Sagittal section of vertebrae showing both normal and herniated disks.

Table 8-2 Description of individual joints

Name	Articulating bones	Type	Movements
Atlantoepistropheal	Anterior arch of atlas rotates about dens of axis (epistropheus)	Diarthrotic (pivot type)	Pivoting or partial rotation of head
Vertebral	Between bodies of vertebrae	Amphiarthrotic cartilaginous	Slight movement between any two vertebrae but considerable motility for column as whole
	Between articular processes	Diarthrotic (gliding)	
Sternoclavicular	Medial end of clavicle with manubrium of sternum	Diarthrotic (gliding)	Gliding
Acromioclavicular	Distal end of clavicle with acromion of scapula	Diarthrotic (gliding)	Gliding; elevation, depression, protraction, and retraction
Thoracic	Heads of ribs with bodies of vertebrae	Diarthrotic (gliding)	Gliding
	Tubercles of ribs with transverse processes of vertebrae	Diarthrotic (gliding)	Gliding
Shoulder	Head of humerus in glenoid cavity of scapula	Diarthrotic (ball and socket type)	Flexion, extension, abduction, adduction, rotation, and circumduction of upper arm

Figure 8-14 Sagittal section of two lumbar vertebrae and their ligaments.

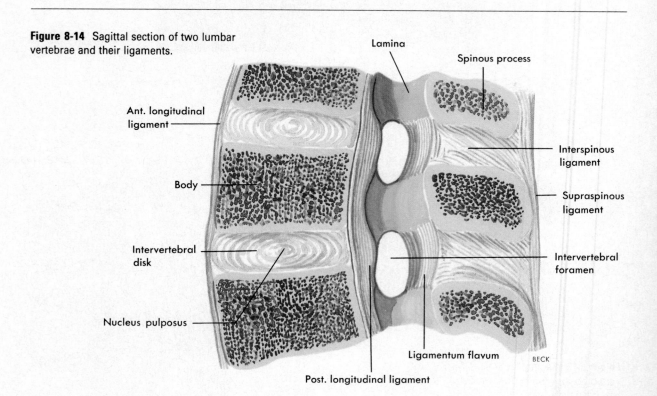

Lamina

Spinous process

Ant. longitudinal ligament

Interspinous ligament

Body

Supraspinous ligament

Intervertebral disk

Intervertebral foramen

Nucleus pulposus

Ligamentum flavum

BECK

Post. longitudinal ligament

Table 8-2 Description of individual joints—cont'd

Name	Articulating bones	Type	Movements
Elbow	Trochlea of humerus with semilunar notch of ulna; head of radius with capitulum of humerus	Diarthrotic (hinge type)	Flexion and extension
	Head of radius in radial notch of ulna	Diarthrotic (pivot type)	Supination and pronation of lower arm and hand; rotation of lower arm on upper
Wrist	Scaphoid, lunate, and triquetral bones articulate with radius and articular disk	Diarthrotic (condyloid)	Flexion, extension, abduction, and adduction of hand
Carpal	Between various carpals	Diarthrotic (gliding)	Gliding
Hand	Proximal end of first metacarpal with trapezium	Diarthrotic (saddle)	Flexion, extension, abduction, adduction, and circumduction of thumb and opposition to fingers
	Distal end of metacarpals with proximal end of phalanges	Diarthrotic (hinge)	Flexion, extension, limited abduction, and adduction of fingers
	Between phalanges	Diarthrotic (hinge)	Flexion and extension of finger sections
Sacroiliac	Between sacrum and two ilia	Diarthrotic (gliding)	None or slight
Symphysis pubis	Between two pubic bones	Synarthrotic (or amphiarthrotic), cartilaginous	Slight, particularly during pregnancy and delivery
Hip	Head of femur in acetabulum of os coxae	Diarthrotic (ball and socket)	Flexion, extension, abduction, adduction, rotation, and circumduction
Knee	Between distal end of femur and proximal end of tibia	Diarthrotic (hinge type)	Flexion and extension; slight rotation of tibia
Tibiofibular (proximal)	Head of fibula with lateral condyle of tibia	Diarthrotic (gliding type)	Gliding
Ankle	Distal ends of tibia and fibula with talus	Diarthrotic (hinge type)	Flexion (dorsiflexion) and extension (plantar flexion)
Foot	Between tarsals	Diarthrotic (gliding)	Gliding; inversion and eversion
	Between metatarsals and phalanges	Diarthrotic (hinge type)	Flexion, extension, slight abduction, and adduction
	Between phalanges	Diarthrotic (hinge type)	Flexion and extension

Figure 8-15 Bursae of the **(A)** shoulder and **(B)** elbow joints.

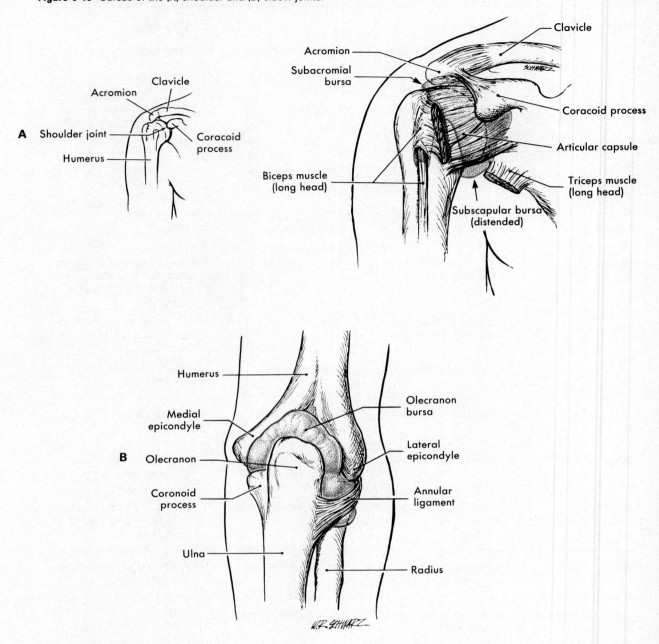

BURSAE

DEFINITION

Bursae are small connective tissue sacs lined with synovial membrane and containing synovial fluid.

LOCATIONS

Bursae are located wherever pressure is exerted over moving parts, for example, between skin and bone, between tendons and bone, or between muscles or ligaments and bone (Figure 8-15). Some bursae that fairly frequently become inflamed (bursitis) are the subacromial bursa, between the head of the humerus and the acromion process and the deltoid muscle; the olecranon bursa, between the olecranon process and the skin; and the prepatellar bursa, between the patella and the skin. Inflammation of the prepatellar bursa was previously described as "housemaid's knee," whereas olecranon bursitis is called "student's elbow."

FUNCTION

Bursae act as cushions, relieving pressure between moving parts.

Figure 8-16 Rheumatoid arthritis—hand involvement.

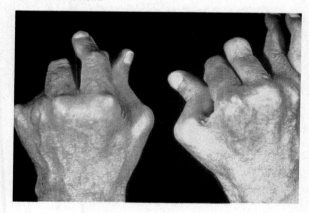

JOINT AGE CHANGES AND DISEASES

By far the most common age change in joints is a degenerative condition called **osteoarthritis.** It begins in the involved joints with a thinning of the articular cartilages. Gradually they undergo degenerative changes and new bone forms at their margins. Limitation of the joint's range of movement results from an accumulation of this new marginal bone and may develop into a major disability if knee or hip joints are involved.

Another joint disease characteristic of the middle and older years is **rheumatoid arthritis** (Figure 8-16). It is a painful, disabling condition involving inflammation of the synovial membrane (synovitis). It occurs most commonly in people between the ages of 40 and 60 years but may afflict people of any age. Granulation tissue (called *pannus*) forms on the articular cartilages of the affected joints. In time, this may erode not only the cartilages but also bone and even ligaments and tendons in the area.

Gouty arthritis, or gout, affects joints indirectly. Higher than normal levels of uric acid characterize this annoying disease, so common in older people. Sodium urate crystals form from the excess uric acid in the extracellular fluid. Most often the crystals first make their presence known by severe pain in a joint when they appear in its synovial fluid. The great toe joint between the first metatarsal and phalangeal bone is the one most frequently affected by gouty arthritis (Figure 8-17). Gout has a genetic basis and to some extent runs in families. Also, it has long been associated with the intemperate use of alcohol. An old saying declares that "in the young, wine goes to the head; in the aged, it goes to the feet."

Figure 8-17 Gout. **A,** Photograph showing involvement of metatarsophalangeal joint of great toe. **B,** X-ray study of lesion.

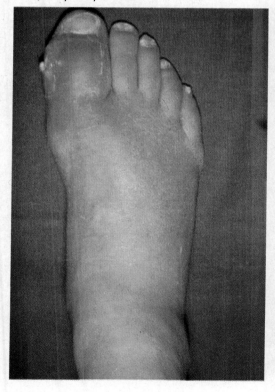

A

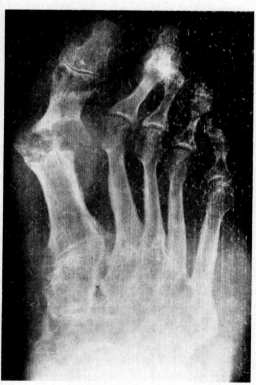

B

DIAGNOSTIC AND SURGICAL ARTHROSCOPY

Before 1970 the only way to visualize joint structures was to surgically open the joint space or inject some type of dye or contrast medium and then examine the internal structures by radiographs or x-ray studies called *arthrograms*. A technique called **arthroscopy** now permits a physician to examine the interior of a joint under local anesthesia without subjecting the patient to extensive surgery. In a typical arthroscopic procedure a small-diameter tube fitted with viewing lenses and a flexible fiberoptic light source is inserted into the joint space through a very small puncture type incision (see figure directly below). After another small needle is inserted into the joint space so that it can be distended with saline, the physician can view the interior of the joint directly or use a TV monitor to enlarge the image. If in addition to diagnostic viewing some type of surgical procedure is necessary, a third puncture-type incision is made to permit entry of specialized instruments.

In the figures (*A* to *D*) a small piece of cartilage called a *loose body* is being removed from the knee joint using arthroscopic techniques. After the procedure is completed, the small incision points heal quickly without sutures. Postoperative pain is minimal in most cases.

Inserting the tubelike arthroscope. Saline is being injected into the joint space through the upper needle.

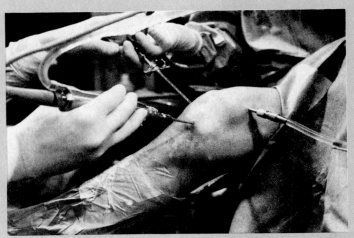

Arthroscopic surgery of the knee.

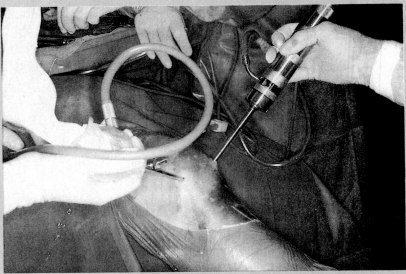

DIAGNOSTIC AND SURGICAL ARTHROSCOPY—cont'd

Arthroscopic surgery of the knee, cont'd. **A** to **C**, Photos taken through arthroscope showing forceps grasping loose body. **D**, Actual removal of loose body through incision.

A
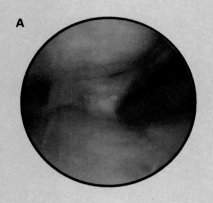

B
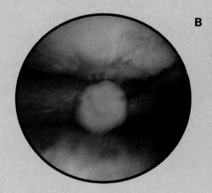

C
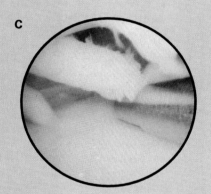

D
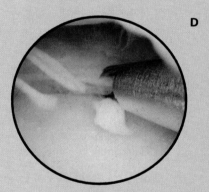

Outline Summary

MEANING AND FUNCTIONS

A Articulations are joints between bones

B Articulations hold bones together but also, in most cases, permit movement between them

CLASSIFICATION OF JOINTS

A According to degree of movement permitted

  1 Synarthroses—permit no movement

  2 Amphiarthroses—permit slight movement

  3 Diarthroses—permit free movement

B According to structure

  1 Fibrous joints—fibrous tissue connects articular surfaces of bones

  2 Cartilaginous joints—cartilage connects articular surfaces of bones

  3 Synovial joints—synovial membrane lines joint cavity (space between articular surfaces of bones)

C Synarthroses (fibrous or immovable joints)

  1 Definition—joints in which articular surfaces of bones are connected by fibrous tissue that binds them firmly together, permitting no movement at the joint

  2 Types (See Table 8-1)

D Amphiarthroses (cartilaginous or slightly movable joints)

  1 Definition—joints in which cartilage connects articular surfaces of bones; flexibility of cartilage permits slight movement at amphiarthroses

  2 Types (See Table 8-1)

E Diarthroses (synovial or freely movable joints)
  1 Definition—freely movable joints in which a space, the joint cavity, separates articulating ends of bones
  2 Structure—main features are joint capsule, synovial membrane, articular cartilage, and joint cavity between articular surfaces of bones
  3 Types of movement
    a Angular movements—change size of angle between articulating bones
     (1) Flexion—decreases angle, usually between anterior surfaces of articulating bones; exceptions: flexion of knee and toe joints decreases angle between posterior surfaces of bones
     (2) Extension—return from flexed position; increases angle between bones; extension of foot is called plantar flexion; (dorsiflexion is flexion of foot so it tilts upward)
     (3) Abduction—moves a body part away from midline (e.g., spreading fingers or toes)
     (4) Adduction—moving a body part toward the midline
    b Circular movements
     (1) Rotation—pivoting of bone on its central axis (e.g., rotating head from side to side as in saying "no")
     (2) Circumduction—moving distal end of bone in a circle causing entire bone to describe surface of a cone (e.g., moving arm, elbow and wrist extended, so hand describes circle)
     (3) Supination—rotating forearm outward to turn palm forward as in anatomical position
     (4) Pronation—rotating forearm inward to turn back of hand forward
    c Gliding movements—gliding of articular surface of one bone over articular surface of another bone
    d Special movements
     (1) Inversion—turning sole of foot inward
     (2) Eversion—turning sole outward
     (3) Protraction—moving a body part forward as in sticking lower jaw out
     (4) Retraction—reverse of protraction
     (5) Elevation—moving a part upward as in shrugging shoulders
     (6) Depression—moving a part downward; reverse of elevation
  4 Types of diarthroses (see Table 8-1)
  5 Representative diarthrotic joints—humeroscapular, hip, knee, and vertebral joints
F Descriptions of individual joints (See Table 8-2)

BURSAE

A Definition—connective tissue sacs lined with synovial membrane and containing synovial fluid
B Locations
  1 In areas where pressure is exerted over moving parts
    a Between skin and bone
    b Between tendons and bone
    c Between muscles or ligaments and bone
  2 Examples
    a Subacromial bursa
    b Olecranon bursa
    c Prepatellar bursa
     (1) Inflammation called "housemaid's knee"
C Functions—act as cushions relieving pressure between moving parts

JOINT AGE CHANGES AND DISEASES

A Osteoarthritis—most common age change in joints; a degenerative condition with loss of articular cartilage and formation at margin of articular cartilage, which enlarges and may deform joints and interfere with movements
B Rheumatoid arthritis—painful disabling inflammation of synovial membranes; pannus or granulation tissue forms on articular cartilages and may erode them and other joint structures including bones, ligaments, and tendons
C Gout—group of diseases characterized by hyperuricemia; sodium urate crystals form in extracellular fluid; their appearance in synovial fluid causes severe pain in affected joint; great toe joint between first metatarsal and phalangeal bone is most frequently affected

Review Questions

1 Classify joints according to the degree of movement permitted and according to structure.
2 What factors determine the range of movements a joint permits?
3 Define the terms *synarthroses, amphiarthroses,* and *diarthroses.*
4 Define the terms *fibrous joints, cartilaginous joints, synovial joints.*
5 Name and define three types of synarthroses. Give an example of each.
6 Name and define two types of amphiarthroses. Give an example of each.
7 Describe the characteristic structural features of diarthroses.
8 Name and define four kinds of angular movements permitted by some diarthroses.
9 Define and give an example of the following: rotation, circumduction, pronation, supination.
10 Define *hinge joint, pivot joints, saddle joints,* and *condyloid joints.* Give an example of each. Do they permit biaxial, multiaxial, or uniaxial movements? Explain.
11 What joint makes possible much of the dexterity of the human hand? Describe it and the movements it permits.
12 Describe vertebral joints.
13 Why does the range of movement vary at different joints? What factors determine this?
14 What is the most common age change in joints? Define: *osteoarthritis, rheumatoid disease, gout.*
15 Define the following terms: *arthrology, arthroscopy, ankylosis, arthrogram.*
16 How do loose bodies or cartilaginous "joint mice" differ from menisci?

9 SKELETAL MUSCLES

OBJECTIVES

After you have completed this chapter, you should be able to:

1 List and discuss the three generalized functions of skeletal muscle tissue.
2 List and describe the structural components of skeletal muscle tissue from the organ to the molecular level of organization.
3 Discuss the role of calcium in muscle contraction and relaxation.
4 Identify and explain the interaction of energy sources required for muscle contraction.
5 List five major connective tissue elements related to skeletal muscle and explain the relationship of each to the contractile elements of the muscle as a whole.
6 Discuss how a nerve impulse travels between a motoneuron and the muscle cells supplied by its axon branches.
7 Diagram and discuss a simple muscle twitch and its time components.
8 Compare and contrast the following types of muscle contractions: tonic, isotonic, isometric, and tetanic.
9 Discuss four factors that influence the strength of a muscle contraction.
10 Identify five features that may be used to name a muscle.
11 List and explain the major terms used to designate or group muscles according to function.
12 Identify major muscles, their points of attachment, and their function in the following areas:
  a Muscles that move the head
  b Muscles that move the upper and lower arm
  c Muscles that move the thigh
  d Muscles that move the lower leg and foot
  e Muscles that move the abdominal and chest wall
13 Discuss the development of a hernia or rupture.
14 Define posture and discuss its importance to the body as a whole.

KEY TERMS

Antagonist (an-TAG-oh-nist)

Aponeurosis (AP-o-nu-RO-sis)

Endomysium (EN-doh-MIZ-ee-um)

Epimysium (EP-i-MIZ-ee-um)

Insertion (in-SER-shun)

Isometric (i-so-MET-rik)

Isotonic (i-so-TON-ik)

Myofibril (mi-o-FI-bril)

Origin (OR-i-jin)

Perimysium (PER-i-MIZ-e-um)

Sarcolemma (SAR-ko-LEM-ah)

Sarcomere (SAR-ko-meer)

Sarcoplasm (SAR-ko-plazm)

Synergist (SIN-er-jist)

Tetanus (TET-ah-nus)

Tonus (TON-us)

Treppe (TREP-ee)

Humanity's survival depends in large part on the ability to adjust to the changing conditions of environment. Movements constitute the major part of this adjustment. Whereas most of the systems of the body play some role in accomplishing movement, it is the skeletal and muscular systems acting together that actually produce movements. We have investigated the architectural plan of the skeleton and have seen how its firm supports and joint structures make movement possible. However, bones and joints cannot move themselves. They must be moved by something. Muscle tissue, because of its irritability, contractility, extensibility, and elasticity, is admirably suited to this function. Our subject then for this chapter is the 40% to 50% of our body weight that is skeletal muscle—those muscle masses that attach to bones and move them about, the "red meat" of the body. (Cardiac muscle will be discussed in Chapter 17, and information about smooth muscle appears in several chapters.)

In this chapter we shall try to discover how muscles move bones, and to do this we shall try to answer many other questions—how the structure of muscles adapts them to their function, how energy is made available for their work, and how muscle activity contributes to the health and survival of the whole body—to mention only a few.

Figure 9-1 Structure of skeletal muscle. **A,** Skeletal muscle organ, composed of bundles of contractile muscle fibers and covering connective tissue components. **B,** Greater magnification of single fiber showing smaller fibers—myofibrils—in the sarcoplasm. Note sarcoplasmic reticulum and T tubules forming triad. **C,** Myofibril magnified further to show sarcomere between successive Z lines. Cross striae are visible. **D,** Molecular structure of myofibril showing thick myofilaments (myosin molecules) and thin myofilaments (actin molecules). (See also Figure 9-5.)

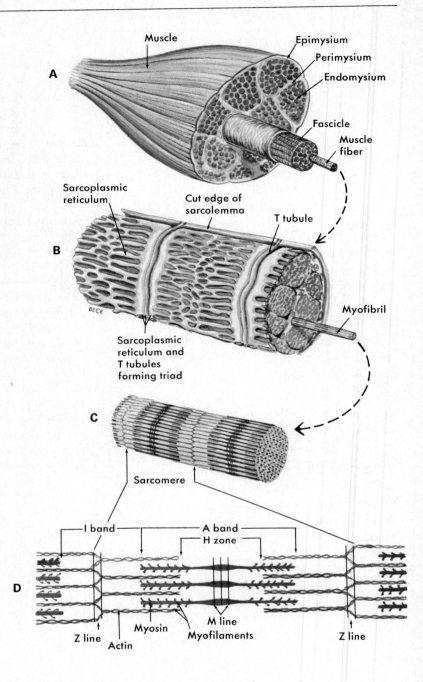

GENERAL FUNCTIONS

If you have any doubts about the importance of muscle function to normal life, you have only to observe a person with extensive paralysis—a victim of advanced muscular dystrophy, for example. Any of us possessed of normal powers of movement can little imagine life with this matchless power lost. But cardinal as it is, movement is not the only contribution muscles make to healthy survival. They also perform two other essential functions: production of a large portion of body heat and maintenance of posture.

1 **Movement.** Skeletal muscle contractions produce movements either of the body as a whole (locomotion) or of its parts.
2 **Heat production.** Muscle cells, like all cells, produce heat by the process known as catabolism (discussed in Chapter 23). But because skeletal muscle cells are both highly active and numerous, they produce a major share of total body heat. Skeletal muscle contractions therefore constitute

Figure 9-2 Electron photomicrograph of striated muscle. (× 23,800.)

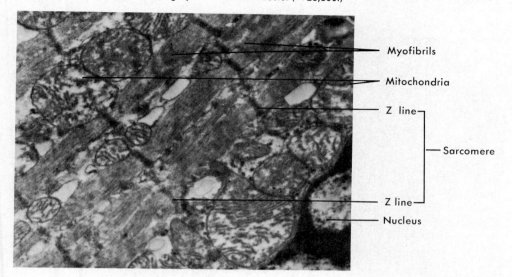

Myofibrils

Mitochondria

Z line ⌉

Sarcomere

Z line ⌋

Nucleus

one of the most important parts of the mechanism for maintaining homeostasis of temperature.

3 Posture. The continued partial contraction of many skeletal muscles makes possible standing, sitting, and other maintained positions of the body.

SKELETAL MUSCLE CELLS

MICROSCOPIC STRUCTURE

Look at Figure 9-1. As you can see there, a skeletal muscle is composed of bundles of skeletal muscle fibers that generally extend the entire length of the muscle. They are called *fibers*, instead of *cells*, because of their threadlike shape (1 to 40 mm long but with a diameter of only 10 to 100 μm). Skeletal muscle fibers have many of the same structural parts as other cells. Several of them, however, bear different names in muscle fibers. For example, **sarcolemma** is the plasma membrane of a muscle fiber. **Sarcoplasm** is its cytoplasm. Muscle cells contain a network of tubules and sacs known as the **sarcoplasmic reticulum**—a structure analogous, but not identical, to the endoplasmic reticulum of other cells. Muscle fibers contain many mitochondria, and, unlike other cells, they have several nuclei.

Certain structures not found in other cells are present in skeletal muscle fibers. For instance, bundles of very fine fibers—**myofibrils**—extend lengthwise along skeletal muscle fibers and almost fill their sarcoplasm. Myofibrils in turn are made up of still finer fibers called *thick* and *thin myofilaments* (Figure 9-1, *D*). Find the label *sarcomere* in this drawing. Note that a **sarcomere** is a segment between two

successive Z lines. Each myofibril consists of a lineup of several sarcomeres, each of which functions as a contractile unit. The A bands of the sarcomeres appear as relatively wide, dark stripes (cross striae) under the microscope, and they alternate with narrower, lighter colored stripes formed by the I bands (Figure 9-1, *D*). Because of its cross striae, skeletal muscle is also called *striated muscle.* Electron microscopy of skeletal muscle (Figure 9-2) has revolutionized our concept of both its structure and function.

Another structure unique to skeletal and cardiac muscle cells is the transverse tubular or **T system.** This name derives from the fact that this system consists of tubules that extend transversely into the sarcoplasm. T system tubules, as Figures 9-1, *B*, and 9-3 show, enter the sarcoplasm at the levels of the Z lines, that is, in the middle of the light I bands. Because invaginations of the sarcolemma form the T system tubules, they open to the exterior of the muscle fiber and may serve to carry interstitial fluid directly into each sarcomere.

The sarcoplasmic reticulum is also a system of tubules in a muscle fiber. It is separate from the T system and differs from it in that the tubules of the sarcoplasmic reticulum run parallel to muscle fibers and terminate in closed sacs at the ends of each sarcomere, that is, immediately above and below each Z line. Since T tubules constitute the Z lines, the sacs of the sarcoplasmic reticulum of one sarcomere lie just above a T tubule, whereas those of the next sarcomere lie just below it. This forms a triple-layered structure (a T tubule sandwiched between sacs of the sarcoplasmic reticulum) called a **triad** (Figures 9-1, *B*, and 9-3).

Figure 9-3 Diagram to show sarcoplasmic reticulum and T tubules of a muscle fiber.

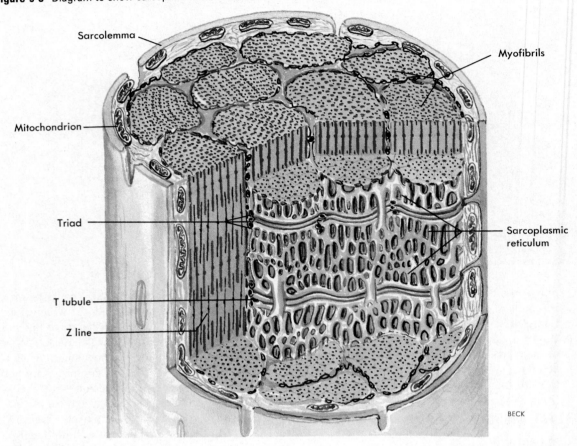

MOLECULAR STRUCTURE

In recent years researchers have made many discoveries about the molecular structure of muscle fibers. Here are a few of them. They have learned that four proteins—myosin, actin, tropomyosin, and troponin—are found only in the sarcomere units of muscle fibers and that these protein compounds interact to produce muscle contraction. Each muscle fiber may contain a thousand or more subunits, about 1 μm in diameter, called **myofibrils.** Lying side by side in each myofibril are up to 2,500 actin and myosin **myofilaments.** Researchers have established that the *thick filaments* of a myofibril consist almost entirely of myosin molecules and that there are 300 or 400 of them per filament. They know the shape of myosin molecules. Each one is a thin rod with two rounded heads at one end. They know too that these myosin molecules are arranged lengthwise in the thick filaments and that some of their heads point toward one end of the filament and some point toward the other end. Thus the heads of the myosin molecules jut out from the surface of both ends of the thick filaments in projections that are called **cross bridges.** A schematic diagram of these appears in Figure 9-1, *D*.

Thin filaments consist of a complex arrangement of three kinds of protein compounds, namely, actin, tropomyosin, and troponin. Within a myofibril the thick and thin filaments alternate, as shown in Figure 9-1, *D*. This arrangement is crucial for contraction. Another fact important for contraction is that the thin filaments attach to both Z lines of a sarcomere and that they extend in from the Z lines part way toward the center of the sarcomere. When the muscle fiber is relaxed, the thin filaments terminate at the outer edges of the H zones. In contrast, the thick myosin filaments do not attach to the Z lines and they extend only the length of the A bands of the sarcomeres.

FUNCTIONS

Skeletal muscle fibers specialize in the function of contraction. When nerve impulses arrive at a skeletal muscle fiber, they initiate impulse conduction over its sarcolemma and inward via its T tubules. This triggers the release of calcium ions from the sacs of the sarcoplasmic reticulum into the sarcoplasm. Here, calcium ions combine with the troponin molecules in the thin filaments of the myofibrils. In a resting muscle fiber, troponin prevents myosin from interacting with actin. Stated differently, troponin that is not bound to calcium prevents the cross bridges of the thick filaments from attaching to the actin molecules of the thin filaments. But calcium-bound troponin permits this action. Therefore myosin interacts with actin and pulls the thin filaments toward the center of each sarcomere. This shortens the sarcomeres and thereby shortens the myofibrils and the muscle fibers they compose. If sufficient numbers of fibers composing a skeletal muscle organ shorten, the muscle itself shortens. In a word, it contracts.

Relaxation of a muscle fiber is now believed to be brought about by a reversal of the contraction mechanism just described. The calcium-troponin combinations separate, calcium ions reenter the sacs of the sarcoplasmic reticulum, and troponin, now no longer bound to calcium, inhibits myosin-actin interaction. The sarcoplasmic reticulum is sometimes called the *relaxing factor* of muscle cells because it has a very strong affinity for calcium. Calcium ions freed by a nerve impulse to initiate contraction are available for only a few milliseconds before becoming firmly bound once again to the sarcoplasmic reticulum. This process of rapid binding of calcium ions helps to maintain muscle fibers in a relaxed state.

Muscle cells obey the all-or-none law when they contract. This means that they either contract with all the force possible under existing conditions or they do not contract at all. However, if conditions at the time of stimulation change, the force of the cell's contraction changes. Suppose, for example, that a particular muscle fiber receives an adequate oxygen supply at one time and an inadequate supply at another time. It will contract more forcefully with an adequate than with a deficient oxygen supply.

In summary, the release of calcium ions from the sacs of the sarcoplasmic reticulum turns on muscle contraction. The withdrawal of calcium ions back into the sarcoplasmic reticulum's sacs turns off muscle contraction and turns on muscle relaxation.

ENERGY SOURCES FOR MUSCLE CONTRACTION

The energy required for muscular contraction is obtained by hydrolysis of a nucleotide called **adenosine triphosphate** or ATP (Figure 9-4, *A*). Two of the three phosphate groups in ATP are attached to the molecule by *high energy bonds* (symbolized by ~P in Figure 9-4, *B*). It is the cleavage of these ~P bonds that provides the energy necessary to support contraction. As Figure 9-5 shows, nerve impulses arriving at a muscle fiber trigger both ATP breakdown and the release of calcium ions from the sacs of the sarcoplasmic reticulum into the sarcoplasm of the muscle fiber. The calcium ions combine with

Figure 9-4 **A,** Flow chart showing various energy sources for muscle contraction and how the energy is used. **B,** Structural formula of adenosine triphosphate showing high-energy phosphate bonds.

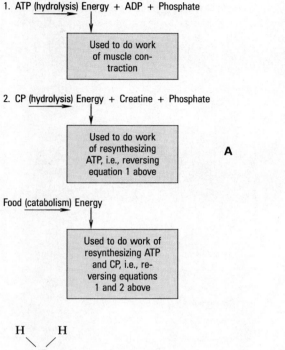

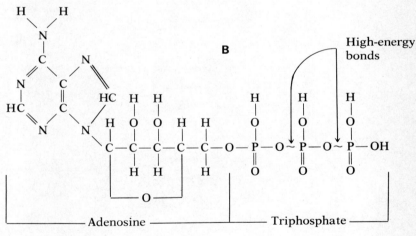

the troponin molecules of the thin filaments in myofibrils. Calcium-combined troponin acts in some way to "turn on" the thin filaments. What this means is that the actin molecules of the thin filaments become able to combine with the heads of the myosin molecules that make up the cross bridges of the thick filaments. The energy released from ATP breakdown then does the work of rotating the cross bridges to a different angle (Figure 9-6). As the cross bridges rotate, they move the thin filaments to which they are attached in toward the centers of the sarcomeres. This necessarily shortens the sarcomeres and the myofibrils. Thus energy released from ATP breakdown does the work of muscle contraction.

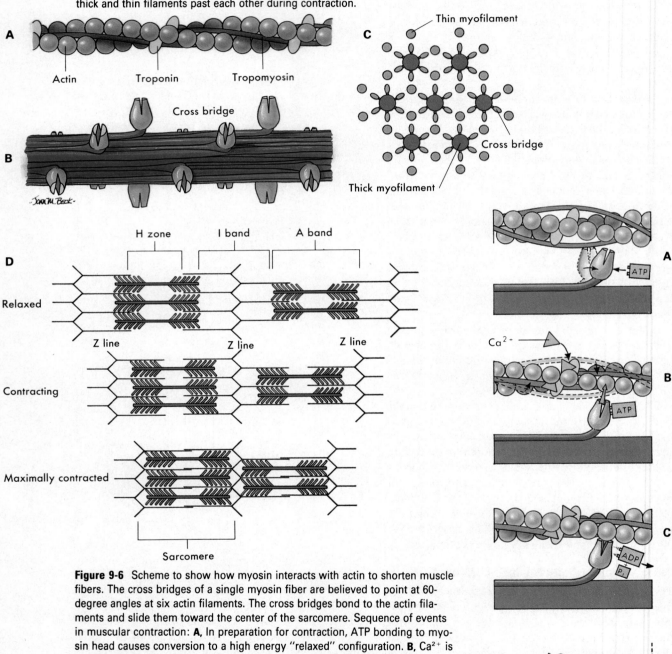

Figure 9-5 Mechanism of skeletal muscle contraction. **A,** Thin myofilament. **B,** Thick myofilament. **C,** Cross section of several thick and thin myofilaments, showing the arrangements of myofilaments and cross bridges. **D,** Demonstration of the sliding filament theory of muscle contraction, showing the changes in banding patterns from the movement of thick and thin filaments past each other during contraction.

Figure 9-6 Scheme to show how myosin interacts with actin to shorten muscle fibers. The cross bridges of a single myosin fiber are believed to point at 60-degree angles at six actin filaments. The cross bridges bond to the actin filaments and slide them toward the center of the sarcomere. Sequence of events in muscular contraction: **A,** In preparation for contraction, ATP bonding to myosin head causes conversion to a high energy "relaxed" configuration. **B,** Ca^{2+} is released from the sarcoplasmic reticulum and is bound to the troponin-tropomyosin complex, changing its conformation so that troponin no longer holds tropomyosin in a blocking position between actin and myosin. **C,** As tropomyosin is pulled out of position, the myosin head makes contact with the actin molecule. The bonding process displaces the high energy phosphate group (ADP + Pi) from the myosin head. **D,** The loss of energy causes the myosin to contract back to its original position, and, in the process, the bound actin molecule is moved along with it.

Muscle fibers must continually resynthesize ATP because they can store only small amounts of it. Immediately after ATP breaks down, energy for its resynthesis is supplied by the breakdown of another high-energy compound, creatine phosphate (CP), which is also present in small amounts in muscle fibers. But ultimately, energy for both ATP and CP synthesis comes from the catabolism of foods (see Figure 23-1 and discussion in Chapter 23).

SKELETAL MUSCLE ORGANS

STRUCTURE

Size, Shape, and Fiber Arrangement

The structures called skeletal muscles are organs. They consist mainly of skeletal muscle tissue plus important connective and nervous tissue components. Skeletal muscles vary considerably in size, shape, and arrangement of fibers. They range from extremely tiny strands, such as the stapedius muscle of the middle ear, to large masses, such as the muscles of the thigh. Some skeletal muscles are broad in shape and some narrow. Some are long and tapering and some short and blunt. Some are triangular, some quadrilateral, and some irregular. Some form flat sheets and others bulky masses.

Arrangement of fibers varies in different muscles. In some muscles the fibers are parallel to the long axis of the muscle, in some they converge to a narrow attachment, and in some they are oblique and either pennate (like the feathers in an old-fashioned plume pen) or bipennate (double-feathered, as in the rectus femoris). Fibers may even be curved, as in the sphincters of the face, for example. The direction of the fibers composing a muscle is significant because of its relationship to function. For instance, a muscle with the bipennate fiber arrangement can produce the strongest contraction.

Connective Tissue Components

A fibrous connective tissue sheath (**epimysium**) envelops each muscle and extends into it as partitions between fasicles or bundles of its fibers (**perimysium**) and between individual fibers (**endomysium**) (see Figure 9-1, *A*). Because all three of these structures are continuous with the fibrous structures that attach muscles to bones or other structures, muscles are most firmly harnessed to the structures they pull on during contraction. The epimysium, perimysium, and endomysium of a muscle, for example, may be continuous with fibrous tissue that extends from the muscle as a **tendon,** a strong tough cord continuous at its other end with the fibrous covering of bone (periosteum). Or the fibrous wrapping of a muscle may extend as a broad, flat sheet of connective tissue (**aponeurosis**) to attach it to adjacent structures, usually the fibrous wrappings of another

muscle. So tough and strong are tendons and aponeuroses that they are not often torn, even by injuries forceful enough to break bones or tear muscles. They are, however, occasionally pulled away from bones.

Tube-shaped structures of fibrous connective tissue called **tendon sheaths** enclose certain tendons, notably those of the wrist and ankle. Like the bursae, tendon sheaths have a lining of synovial membrane. Its moist smooth surface enables the tendon to move easily, almost frictionlessly, in the tendon sheath.

You may recall that a continuous sheet of loose connective tissue known as the superficial fascia lies directly under the skin. Under this lies a layer of dense fibrous connective tissue, the **deep fascia.** Extensions of the deep fascia form the epimysium, perimysium, and endomysium of muscles and their attachments to bones and other structures and also enclose viscera, glands, blood vessels, and nerves.

Nerve Supply

A nerve cell that transmits impulses to a skeletal muscle is called a **somatic motoneuron.** One such neuron plus the muscle cells in which its axon terminates constitutes a **motor unit** (Figure 9-7). The single axon fiber of a motor unit divides, on entering the skeletal muscle, into a variable number of branches. Those of some motor units terminate in only a few muscle fibers, whereas others terminate in numerous fibers. Consequently, impulse conduction by one motor unit may stimulate only a half dozen or so muscle fibers to contract at one time, whereas conduction by another motor unit may activate a hundred or more fibers simultaneously. This fact bears a relationship to the function of the muscle as a whole. As a general rule, the fewer the number of fibers supplied by a skeletal muscle's individual motor units, the more precise the movements that muscle can produce. For example, in certain

AGE CHANGES

As a person grows old, his skeletal muscles undergo a process called *fibrosis.* Gradually, some of the skeletal muscle fibers degenerate, and fibrous connective tissue replaces them. With this loss of muscle fibers and increase in connective tissue comes waning muscular strength, a common finding in elderly people. Other factors probably also contribute to decreasing muscular strength in advanced age.

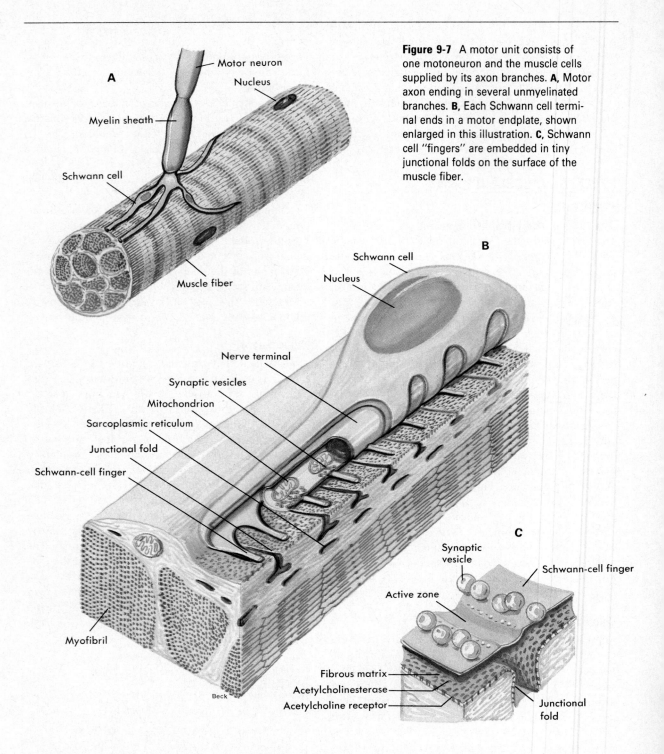

A

Motor neuron

Nucleus

Myelin sheath

Schwann cell

Muscle fiber

Figure 9-7 A motor unit consists of one motoneuron and the muscle cells supplied by its axon branches. **A,** Motor axon ending in several unmyelinated branches. **B,** Each Schwann cell terminal ends in a motor endplate, shown enlarged in this illustration. **C,** Schwann cell "fingers" are embedded in tiny junctional folds on the surface of the muscle fiber.

Schwann cell

Nucleus

B

Nerve terminal

Synaptic vesicles

Mitochondrion

Sarcoplasmic reticulum

Junctional fold

Schwann-cell finger

Myofibril

Beck

C

Synaptic vesicle

Schwann-cell finger

Active zone

Fibrous matrix

Acetylcholinesterase

Acetylcholine receptor

Junctional fold

small muscles of the hand, each motor unit includes only a few muscle fibers, and these muscles produce precise finger movements. In contrast, motor units in large abdominal muscles that do not produce precise movements are reported to include more than a hundred muscle fibers each.

The area of contact between a nerve and muscle fiber is known as the **motor endplate** or **neuromuscular junction** (Figure 9-7, *B*). When nerve im-

pulses reach the ends of the axon fibers in a skeletal muscle, small vesicles in the axon terminals release a chemical—acetylcholine—into the neuromuscular junction. Diffusing swiftly across this microscopic trough, acetylcholine contacts the sarcolemma of the adjacent muscle fibers, stimulating the fiber to contract. In addition to the many motor nerve endings, there are also many sensory nerve endings in skeletal muscles.

FUNCTION

Several methods of study have been used to amass the present-day store of knowledge about how muscles function. They vary from the traditional and relatively simple procedures, such as observing and palpating muscles in action, manipulating dissected muscles to observe movements, or deducing movements from knowledge of muscle anatomy, to the newer more complicated method of electromyography (recording action potentials from contracting muscles). As a result, there is now a somewhat overwhelming amount of knowledge about muscle functions. So perhaps we can thread our way through this maze of detail more easily if we start with general principles and then go on to the details that seem to us most useful.

Basic Principles

1 **Skeletal muscles contract only if stimulated.** They do not have the quality of automaticity inherent in cardiac and visceral muscle. Although nerve impulses are the natural stimuli for skeletal muscles, electrical and some other artificial stimuli such as heat or injury can also activate them. A skeletal muscle deprived of nerve impulses by whatever cause is a functionless mass. One therefore should think of a skeletal muscle and its motor nerve as a physiological unit, always functioning together, either useless without the other.

2 **A skeletal muscle contraction may be any one of several types.** It may be a tonic contraction, an isotonic contraction, an isometric contraction, a twitch contraction, or a tetanic contraction. And there are also other types of contractions—called *treppe, fibrillation,* and *convulsions.*

  a A **tonic contraction** (*tonus,* "tone") is a continual, partial contraction. At any one moment a small number of the total fibers in a muscle contract, producing a tautness of the muscle rather than a recognizable contraction and movement. Different groups of fibers scattered throughout the muscle contract in relays. Tonic contraction, or tone, is characteristic of the muscles of normal individuals when they are awake. It is particularly important for maintaining posture. A striking illustration of this fact is the following: when a person loses consciousness, his muscles lose their tone and he collapses in a heap, unable to maintain a sitting or standing posture. Muscles with less tone than normal are described as *flaccid muscles* and those with more than normal tone are called *spastic.* Impulses over stretch reflex arcs (see Figure 10-6) maintain tone. Specialized sensory (stretch) receptors called *muscle spindles* and *neurotendinous end-organs* are capable of detecting the degree of stretch in a muscle or at the junction of a muscle with its tendon. These receptors permit regulation of muscle tone at a reflex level. Blocking passage of impulses from these sensory end-organs by cutting the reflex arc will result in loss of muscle tonus.

  b An **isotonic contraction** (*iso,* "same"; *tonic,* "tone, pressure, or tension") is a contraction in which the tone or tension within a muscle remains the same but the length of the muscle changes. It shortens, producing movement.

  c An **isometric contraction** is a contraction in which muscle length remains the same but in which muscle tension increases. You can observe isometric contraction by pushing your arms against a wall and feeling the tension increase in your arm muscles. Isometric contractions "tighten" a muscle, but they do not produce movements or do work. Isotonic contractions, on the other hand, both produce movements and do work. Although a majority of muscles can contract either isotonically or isometrically, most body movements are a mixture of the two. Can you explain how walking or running is an example of movement resulting from both isotonic and isometric muscle contraction?

  d A **twitch contraction** is a quick, jerky contraction in response to a single stimulus. Figure 9-8 shows a record of such a contraction. It reveals that the muscle does not shorten at the instant of stimulation, but rather a fraction of a second later, and that it reaches a peak of shortening and then gradually resumes its former length. These three phases of contraction are spoken of, respectively, as the *latent period,* the *contraction phase,* and the *relaxation phase.* The entire twitch usually lasts less than $\frac{1}{10}$ of a second. Twitch contractions rarely occur in the body.

  e A **tetanic contraction** (*tetanus*) is a more sustained contraction than a twitch. It is produced by a series of stimuli bombarding the muscle in rapid succession. About 30 stimuli per second, for example, evoke a tetanic contraction in a frog's gastrocnemius muscle, but the rate varies for different muscles and different conditions. Figure 9-9 shows records of incomplete and complete tetanus. Normal movements are said to be produced by incomplete tetanic contractions.

  f **Treppe** *(staircase phenomenon)* is an event in which increasingly stronger twitch contractions occur in response to constant-strength

Figure 9-8 The simple muscle twitch and its time components.

Muscle shortening

Stimulus entry

0.01 sec. intervals

Latent period

Period of contraction

Period of relaxation

Figure 9-9 Single twitch, incomplete tetanus, and complete tetanus of muscle.

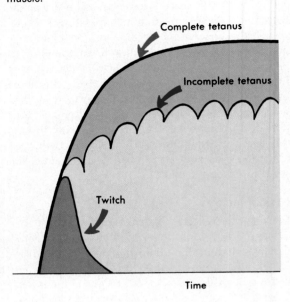

Complete tetanus

Incomplete tetanus

Twitch

Time

stimuli repeated at the rate of about once or twice a second. In other words, a muscle contracts more forcefully after it has contracted a few times than when it first contracts—a principle made practical use of by athletes when they warm up but one not yet satisfactorily explained. Presumably, it relates partly to the rise in temperature of active muscles and partly to their accumulation of metabolic products. After the first few stimuli, muscle responds to a considerable number of successive stimuli with maximal contractions (Figure 9-10). Eventually, it will respond with less and less strong contractions. The relaxation phase becomes shorter and finally disappears entirely. In other words, the muscle stays partially contracted—an abnormal state of prolonged contraction called *contracture*.

Repeated stimulation of muscle in time lessens its irritability and contractility and may result in muscle fatigue, a condition in which the muscle does not respond to the strongest stimuli. Complete muscle fatigue, however, very seldom occurs in the body but can be readily induced in an excised muscle.

g **Fibrillation** is an abnormal type of contraction in which individual fibers contract asynchronously, producing a flutter of the muscle but no effective movement. Fibrillation of the heart, for example, occurs fairly often and is a frequent cause of death.

h **Convulsions** are abnormal uncoordinated tetanic contractions of varying groups of muscles.

3 **Skeletal muscles (organs) contract according to the graded strength principle** (Figure 9-11). They do not contract according to the all-or-none principle, as do the individual muscle cells composing them. In other words, skeletal muscles contract with varying degrees of strength at different times—a fact of practical importance. (How else, for example, could we match the force of a movement to the demands of a task?)

Several generalizations may help explain the fact of graded strength contractions. The strength of the skeletal muscle contraction bears a direct relationship to the initial length of its fibers, their metabolic condition, and the number of them contracting. If a muscle is moderately stretched at the moment when contraction begins, the force of its contraction increases. This principle, established years ago, applies experimentally to heart muscle also (Starling's law of the heart, discussed in Chapter 17). Outstanding among metabolic conditions that influence contraction are oxygen and food supply.

With adequate amounts of these essentials a muscle can contract with greater force than possible with deficient amounts. The greater the number of muscle fibers contracting simultaneously, the stronger the contraction of a mus-

Figure 9-10 Record of several successive muscle contractions, showing treppe occurring in the first few.

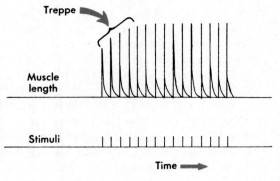

Figure 9-11 Comparison of the variation in strength of muscle contraction with strength of stimulus.

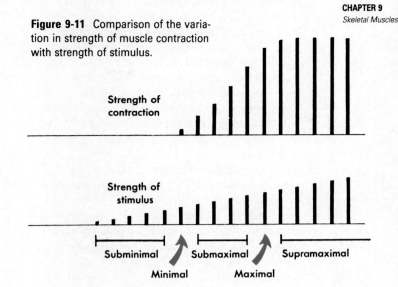

Figure 9-12 Factors that influence the strength of muscle contraction.

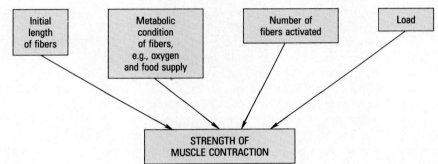

cle. How large this number is depends on how many motor units are activated, and this in turn depends on the intensity and frequency of stimulation. In general, the more intense and the more frequent a stimulus, the more motor units and therefore the more fibers are activated and the stronger the contraction. Contraction strength also relates to previous contraction, the warmup principle discussed on p. 245.

Another factor that influences the force of contraction is the size of load imposed on the muscle. Within certain limits, the heavier the load, the stronger the contraction. Lift a pencil and then a heavy book and you can feel your arm muscles contract more strongly with the book. The factors that influence muscle contraction are summarized in Figure 9-12.

4 **Skeletal muscles produce movements by pulling on bones.** Most of our muscles span at least one joint and attach to both articulating bones. When they contract, therefore, their shortening puts a pull on both bones, and this pull moves one of the bones at the joint—draws it toward the other bone, much as a pull on marionette strings moves a puppet's parts. (In

case you are wondering why both bones do not move, since both are pulled on by the contracting muscle, the reason is that one of them is normally stabilized by isometric contractions of other muscles or by certain features of its own that make it less mobile.)

5 **Bones serve as levers, and joints serve as fulcrums of these levers.** (By definition, a *lever* is any rigid bar free to turn about a fixed point called its *fulcrum.*) A contracting muscle applies a pulling force on a bone lever at the point of the muscle's attachment to the bone. This causes the bone (referred to as the *insertion bone*) to move about its joint fulcrum. We have already noted that a skeletal muscle and its motor nerve act as a functional unit. Now we can add bones and joints to this unit and describe the physiological unit for movement as a neuromusculoskeletal unit. Disease or injury of any one of these parts of the unit—of nerve or muscle or bone or joint—can, as you might surmise, cause abnormal movements or complete loss of movement. For example, poliomyelitis, multiple sclerosis, and hemiplegia all involve the neural part of the unit. In contrast, muscular

dystrophy affects the muscular part and arthritis the skeletal part.

6 **Muscles that move a part usually do not lie over that part.** In most cases the body of a muscle lies proximal to the part moved. Thus muscles that move the lower arm lie proximal to it, that is, in the upper arm. Applying the same principle, where would you expect muscles that move the hand to be located? Those that move the lower leg? Those that move the upper arm?

7 **Skeletal muscles almost always act in groups rather than singly.** Most movements are produced by the coordinated action of several muscles. Some of the muscles in the group contract while others relax. To identify each muscle's special function in the group the following classification is used:

a **Prime movers**—Muscle or muscles whose contraction actually produces the movement

b **Antagonists**—Muscles that relax while the prime mover is contracting to produce movement (exception: contraction of the antagonist at the same time as the prime mover when some part of the body needs to be held rigid, such as the knee joint when standing)

c **Synergists**—Muscles that contract at the same time as the prime mover (may help the prime mover produce its movement or may stabilize a part—hold it steady—so that the prime mover produces a more effective movement)

Hints on How to Deduce Actions

To understand muscle actions, you need first to know certain anatomical facts such as which bones muscles attach to and which joints they pull across. Then if you relate these structural facts to functional principles (for instance, those discussed in the preceding paragraphs), you may find your study of mus-

cles more interesting and less difficult than you anticipate. Some specific suggestions for deducing muscle actions follow.

1 Start by making yourself familiar with the names, shapes, and general locations of the larger muscles, using Table 9-1 as a guide.

2 Try to deduce which bones the two ends of a muscle attach to from your knowledge of the shape and general location of the muscle. For example, look carefully at the deltoid muscle as illustrated in Figures 9-13 to 9-16. To what bones does it seem to attach? Check your deductions with Table 9-4, p. 267.

3 Next, make a guess which bone moves when the muscle shortens. (The bone moved by a muscle's contraction is its **insertion** bone; the bone that remains relatively stationary is its **origin** bone.) In many cases, you can tell by trying to move one bone and then another which is the insertion bone. In some cases, either bone may function as the insertion. Although not all muscle attachments can be deduced as readily as those of the deltoid, they can all be learned more easily by using this deduction method than by relying on rote memory alone.

4 Deduce a muscle's actions by applying the principle that its insertion moves toward its origin. Check your conclusions with the text. Here, as in steps 2 and 3, the method of deduction is intended merely as a guide and is not adequate by itself for determining muscle actions.

5 To deduce which muscle produces a given action (instead of which action a given muscle produces, as in step 4), start by inferring the insertion bone (bone that moves during the action). The body and origin of the muscle will lie on one or more of the bones toward which the insertion moves—often a bone or bones proximal to the insertion bone. Couple these conclusions about origin and insertion with your knowledge of muscle names and locations to deduce the muscle that produces the action.

For example, if you wish to determine the prime mover for the action of raising the upper arms straight out to the sides, you infer that the muscle inserts on the humerus, since this is the bone that moves. It moves toward the shoulder, that is, the clavicle and scapula, so that probably the muscle has its origin on these bones. Because you know that the deltoid muscle fulfills these conditions, you conclude, and rightly so, that it is the muscle that raises the upper arms sidewise.

Antagonistic muscles have opposite actions and opposite locations. If the flexor lies anterior to the part, the extensor will be found posterior to it. For example, the pectoralis major, the flexor of the upper arm, is located on the anterior aspect of the chest, whereas the latissimus dorsi, the extensor of the upper arm, is located on the posterior aspect of the chest. The antagonist of a flexor muscle is obviously an extensor muscle and that of an abductor muscle, an adductor muscle. Some frequently used antagonists are listed in Table 9-2, p. 252.

Table 9-1 Muscles grouped according to location

Location	Muscles	Figures illustrating
Neck	Sternocleidomastoid	9-13
Back	Trapezius	9-13, 9-16
	Latissimus dorsi	9-14
Chest	Pectoralis major	9-13
	Serratus anterior	9-13
Abdominal wall	External oblique	9-13, 9-14
Shoulder	Deltoid	9-13, 9-16
Upper arm	Biceps brachii	9-15, 9-18
	Triceps brachii	9-15, 9-16, 9-19
	Brachialis	9-15, 9-21
Forearm	Brachioradialis	9-15, 9-16, 9-22
	Pronator teres	9-15, 9-20
Buttocks	Gluteus maximus	9-24, 9-27
	Gluteus minimus	9-28
	Gluteus medius	9-29
	Tensor fascia latae	9-23
Thigh		
Anterior surface	Quadriceps femoris group	
	Rectus femoris	9-23, 9-25
	Vastus lateralis	9-23, 9-25
	Vastus medialis	9-23, 9-25
	Vastus intermedius	9-25
Medial surface	Gracilis	9-23, 9-24, 9-26
	Adductor group (brevis, longus, magnus)	9-23, 9-24, 9-26
Posterior surface	Hamstring group	
	Biceps femoris	9-24, 9-30
	Semitendinosus	9-24, 9-30
	Semimembranosus	9-24, 9-30
Leg		
Anterior surface	Tibialis anterior	9-23
Posterior surface	Gastrocnemius	9-23, 9-24
	Soleus	9-24
Pelvic floor	Levator ani	9-37
	Levator coccygeus	9-37
	Rectococcygeus	9-37

6 Do not try to learn too many details about muscle origins, insertions, and actions. Remember, it is better to start by learning a few important facts thoroughly than to half learn a mass of relatively unimportant details. Remember too that trying to learn too many minute facts may well result in your not retaining even the main facts.

Figure 9-13 Superficial muscles of the anterior surface of the trunk and extremities.

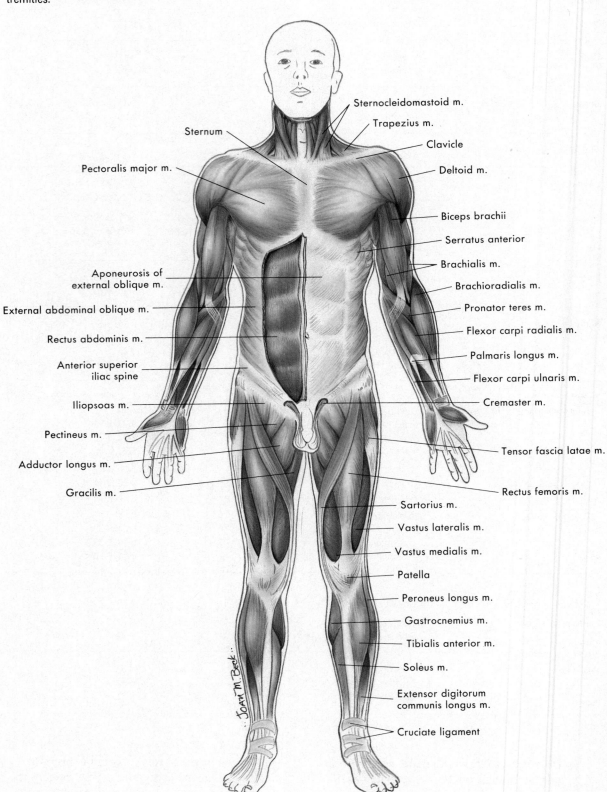

Sternocleidomastoid m.

Trapezius m.

Sternum

Clavicle

Pectoralis major m.

Deltoid m.

Biceps brachii

Serratus anterior

Brachialis m.

Aponeurosis of external oblique m.

Brachioradialis m.

External abdominal oblique m.

Pronator teres m.

Flexor carpi radialis m.

Rectus abdominis m.

Palmaris longus m.

Anterior superior iliac spine

Flexor carpi ulnaris m.

Cremaster m.

Iliopsoas m.

Pectineus m.

Tensor fascia latae m.

Adductor longus m.

Gracilis m.

Rectus femoris m.

Sartorius m.

Vastus lateralis m.

Vastus medialis m.

Patella

Peroneus longus m.

Gastrocnemius m.

Tibialis anterior m.

Soleus m.

Extensor digitorum communis longus m.

Cruciate ligament

Joan M. Beck

Figure 9-14 Superficial muscles of the posterior surface of the trunk and extremities.

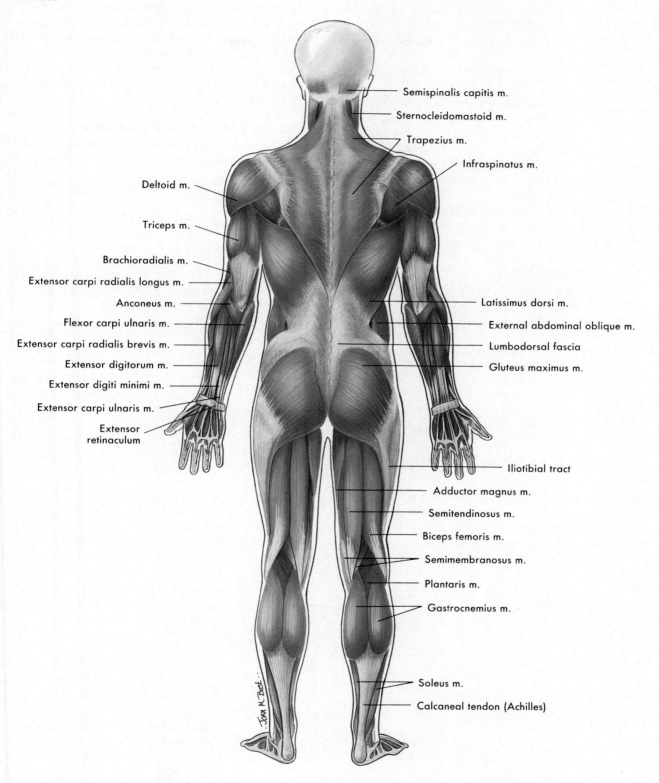

Semispinalis capitis m.

Sternocleidomastoid m.

Trapezius m.

Infraspinatus m.

Deltoid m.

Triceps m.

Brachioradialis m.

Extensor carpi radialis longus m.

Anconeus m.

Flexor carpi ulnaris m.

Extensor carpi radialis brevis m.

Extensor digitorum m.

Extensor digiti minimi m.

Extensor carpi ulnaris m.

Extensor retinaculum

Latissimus dorsi m.

External abdominal oblique m.

Lumbodorsal fascia

Gluteus maximus m.

Iliotibial tract

Adductor magnus m.

Semitendinosus m.

Biceps femoris m.

Semimembranosus m.

Plantaris m.

Gastrocnemius m.

Soleus m.

Calcaneal tendon (Achilles)

NAMES

Reasons for Names

Muscle names seem more logical and therefore easier to learn when one understands the reasons for the names. Each name describes one or more of the following features about the muscle.

1 **Its action**—as flexor, extensor, adductor, etc.
2 **Direction of its fibers**—as rectus or transversus
3 **Its location**—as tibialis or femoris
4 **Number of divisions composing a muscle**—as biceps, triceps, or quadriceps
5 **Its shape**—as deltoid (triangular) or quadratus (square)
6 **Its points of attachment**—as sternocleidomastoid

A good way to start the study of a muscle is by trying to find out what its name means.

Muscles Grouped According to Location

Just as names of people are learned by associating them with physical appearance, so the names of muscles should be learned by associating them with their appearance. As you learn each muscle name, study Figures 9-13 to 9-37 to familiarize yourself with the muscle's size, shape, and general location. To help you in this task, the names of some of the major muscles are grouped according to their location in Table 9-1, p. 249.

Muscles Grouped According to Function

The following terms are used to designate muscles according to their main actions (see Table 9-2, below, for examples).

Text continued on p. 266.

Table 9-2 Muscles grouped according to function

Part moved	Example of flexor	Example of extensor	Example of abductor	Example of adductor
Head	Sternocleidomastoid	Semispinalis capitis		
Upper arm	Pectoralis major	Trapezius Latissimus dorsi	Deltoid	Pectoralis major with latissimus dorsi
Forearm	With forearm supinated: biceps brachii With forearm pronated: brachialis With semisupination or semipronation: brachioradialis	Triceps brachii		
Hand	Flexor carpi radialis and ulnaris Palmaris longus	Extensor carpi radialis, longus, and brevis Extensor carpi ulnaris	Flexor carpi radialis	Flexor carpi ulnaris
Thigh	Iliopsoas Rectus femoris (of quadriceps femoris group)	Gluteus maximus	Gluteus medius and gluteus minimus	Adductor group
Leg	Hamstrings	Quadriceps femoris group		
Foot	Tibialis anterior	Gastrocnemius Soleus	Evertors Peroneus longus Peroneus brevis	Inverter Tibialis anterior
Trunk	Iliopsoas Rectus abdominis	Sacrospinalis		

Figure 9-15 Muscles of the flexor surface of the upper extremity.

Figure 9-16 Muscles of the extensor surface of the upper extremity.

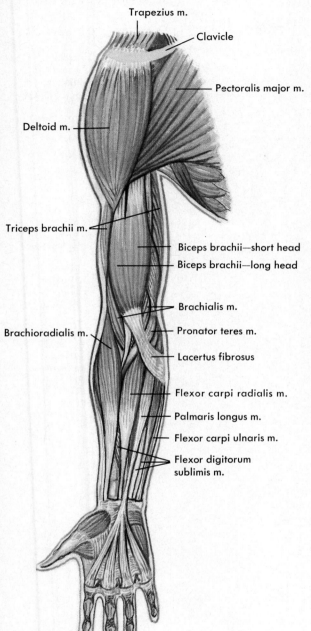

Trapezius m.
Clavicle
Pectoralis major m.
Deltoid m.
Triceps brachii m.
Biceps brachii—short head
Biceps brachii—long head
Brachialis m.
Pronator teres m.
Brachioradialis m.
Lacertus fibrosus
Flexor carpi radialis m.
Palmaris longus m.
Flexor carpi ulnaris m.
Flexor digitorum sublimis m.

Trapezius m.
Acromion process of scapula
Deltoid m.
Triceps brachii m.
Brachioradialis m.
Ulnar nerve
Anconeus
Extensor carpi radialis longus m.
Flexor carpi ulnaris m.
Extensor carpi radialis brevis m.
Extensor carpi ulnaris m.
Extensor digitorum communis m.
Extensor digiti quinti proprius m.
Extensor retinaculum

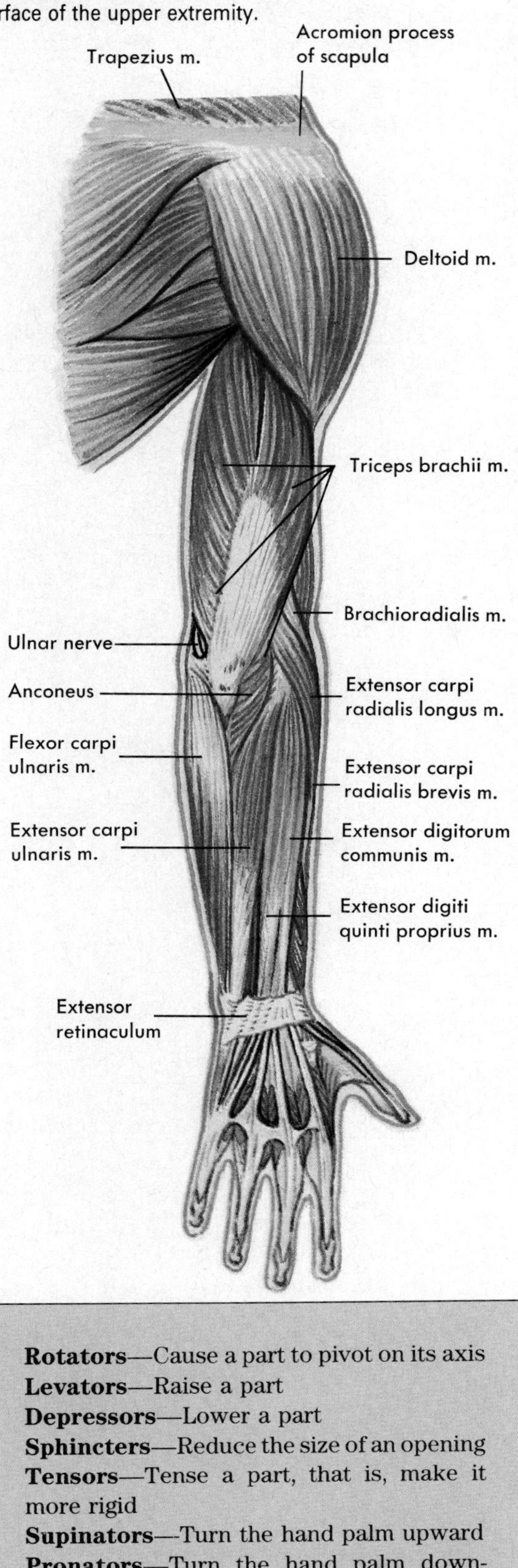

Flexors—Decrease the angle of a joint (between the anterior surfaces of the bones except in the knee and toe joints)
Extensors—Return the part from flexion to normal anatomical position; increase the angle of a joint
Abductors—Move the bone away from midline
Adductors—Move the part toward the midline

Rotators—Cause a part to pivot on its axis
Levators—Raise a part
Depressors—Lower a part
Sphincters—Reduce the size of an opening
Tensors—Tense a part, that is, make it more rigid
Supinators—Turn the hand palm upward
Pronators—Turn the hand palm downward

Figure 9-17 The anterolateral abdominal wall and inguinal area in the male. Note the placement of the superficial inguinal rings in the aponeurosis of the external oblique muscle.

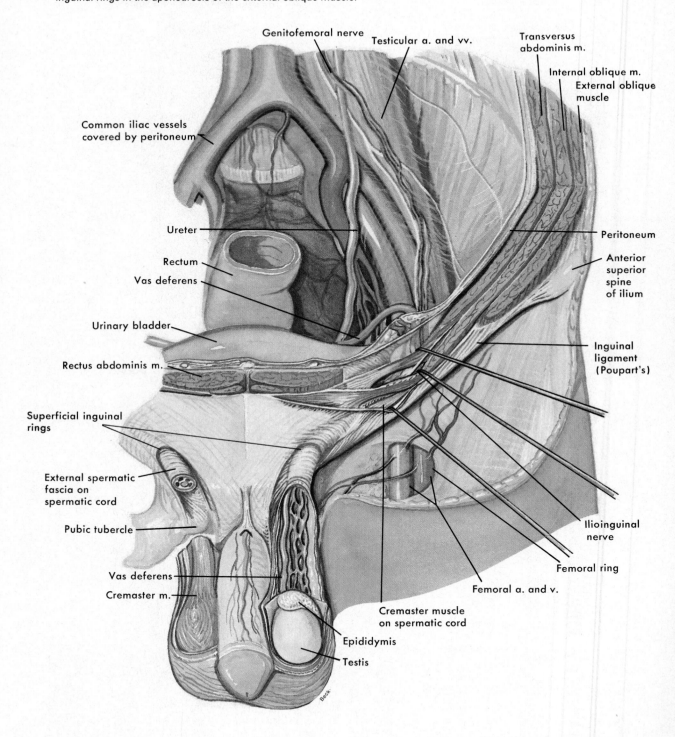

Genitofemoral nerve

Testicular a. and vv.

Transversus abdominis m.

Internal oblique m.
External oblique muscle

Common iliac vessels covered by peritoneum

Ureter

Peritoneum

Anterior superior spine of ilium

Rectum

Vas deferens

Urinary bladder

Rectus abdominis m.

Inguinal ligament (Poupart's)

Superficial inguinal rings

External spermatic fascia on spermatic cord

Pubic tubercle

Ilioinguinal nerve

Femoral ring

Vas deferens

Cremaster m.

Femoral a. and v.

Cremaster muscle on spermatic cord

Epididymis

Testis

Figure 9-18 Biceps brachii muscle.

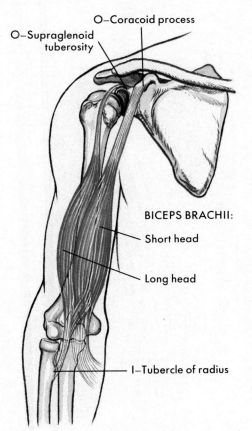

O—Supraglenoid tuberosity

O—Coracoid process

BICEPS BRACHII:

Short head

Long head

I—Tubercle of radius

Figure 9-19 Triceps brachii muscle.

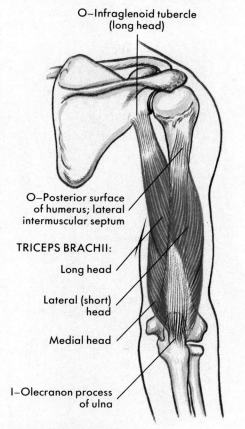

O—Infraglenoid tubercle (long head)

O—Posterior surface of humerus; lateral intermuscular septum

TRICEPS BRACHII:

Long head

Lateral (short) head

Medial head

I—Olecranon process of ulna

**O, Origin.
I, Insertion.**

Figure 9-20 Coracobrachialis and pronator teres muscles.

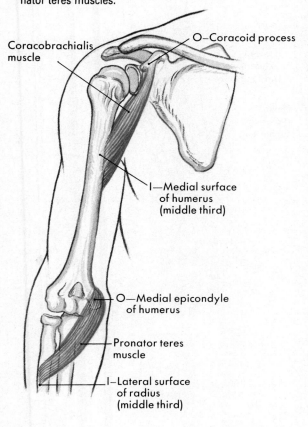

Coracobrachialis muscle

O—Coracoid process

I—Medial surface of humerus (middle third)

O—Medial epicondyle of humerus

Pronator teres muscle

I—Lateral surface of radius (middle third)

Figure 9-21 Brachialis muscle.

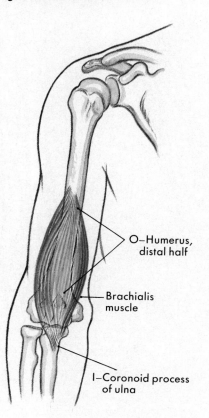

O—Humerus, distal half

Brachialis muscle

I—Coronoid process of ulna

Figure 9-22 Some muscles of the anterior (volar) aspect of the right forearm.

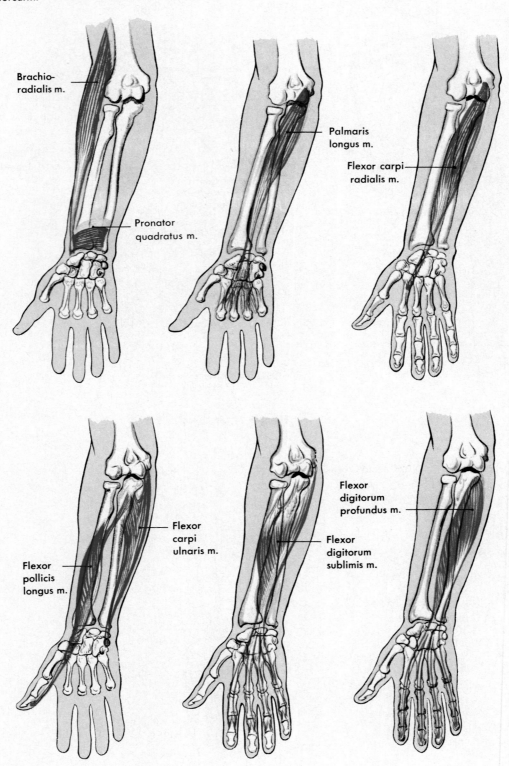

Figure 9-23 Superficial muscles of the right thigh and leg. Anterior view.

Figure 9-24 Superficial muscles of the right thigh and leg. Posterior view.

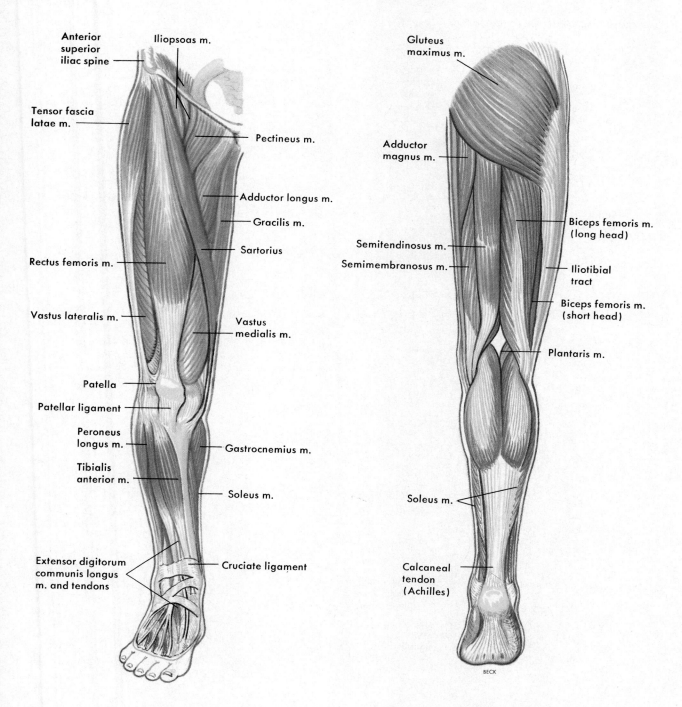

Anterior superior iliac spine

Iliopsoas m.

Tensor fascia latae m.

Pectineus m.

Adductor longus m.

Gracilis m.

Sartorius

Rectus femoris m.

Vastus lateralis m.

Vastus medialis m.

Patella

Patellar ligament

Peroneus longus m.

Gastrocnemius m.

Tibialis anterior m.

Soleus m.

Extensor digitorum communis longus m. and tendons

Cruciate ligament

Gluteus maximus m.

Adductor magnus m.

Biceps femoris m. (long head)

Semitendinosus m.

Semimembranosus m.

Iliotibial tract

Biceps femoris m. (short head)

Plantaris m.

Soleus m.

Calcaneal tendon (Achilles)

BECK

Figure 9-25 Quadriceps femoris group
of thigh muscles: rectus femoris, vastus
intermedius, vastus medialis, and
vastus lateralis. *O,* Origin. *I,* Insertion.

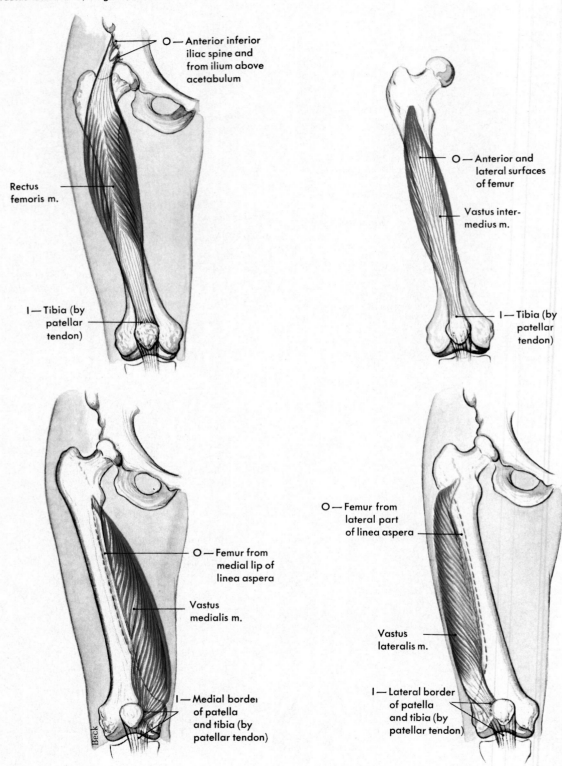

O — Anterior inferior
iliac spine and
from ilium above
acetabulum

Rectus
femoris m.

I — Tibia (by
patellar
tendon)

O — Anterior and
lateral surfaces
of femur

Vastus inter-
medius m.

I — Tibia (by
patellar
tendon)

O — Femur from
medial lip of
linea aspera

Vastus
medialis m.

I — Medial border
of patella
and tibia (by
patellar tendon)

Beck

O — Femur from
lateral part
of linea aspera

Vastus
lateralis m.

I — Lateral border
of patella
and tibia (by
patellar tendon)

Figure 9-26 Muscles that adduct the thigh. *O,* Origin. *I,* Insertion.

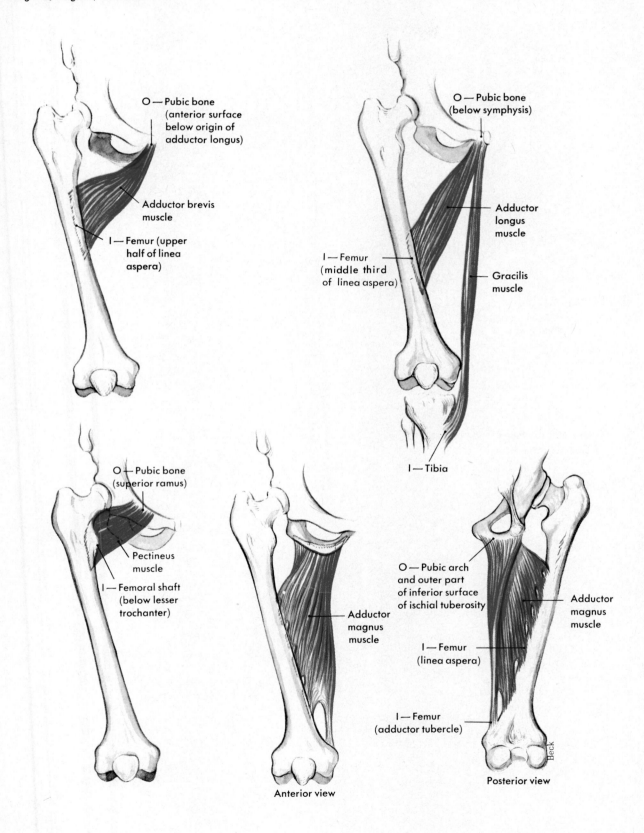

O—Pubic bone (anterior surface below origin of adductor longus)

Adductor brevis muscle

I—Femur (upper half of linea aspera)

O—Pubic bone (below symphysis)

Adductor longus muscle

I—Femur (middle third of linea aspera)

Gracilis muscle

I—Tibia

O—Pubic bone (superior ramus)

Pectineus muscle

I—Femoral shaft (below lesser trochanter)

Adductor magnus muscle

Anterior view

O—Pubic arch and outer part of inferior surface of ischial tuberosity

Adductor magnus muscle

I—Femur (linea aspera)

I—Femur (adductor tubercle)

Posterior view

Figure 9-27 Gluteus maximus muscle.
O, Origin. *I,* Insertion.

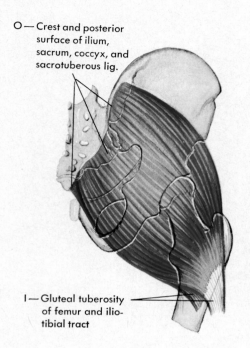

O—Crest and posterior
surface of ilium,
sacrum, coccyx, and
sacrotuberous lig.

I—Gluteal tuberosity
of femur and ilio-
tibial tract

Figure 9-28 Gluteus minimus muscle.
O, Origin. *I,* Insertion.

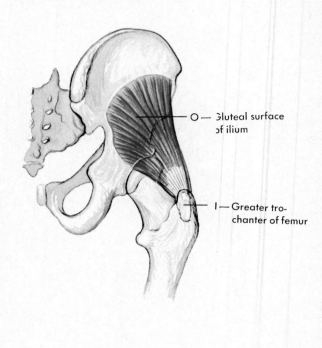

O—Gluteal surface
of ilium

I—Greater tro-
chanter of femur

Figure 9-29 Gluteus medius muscle.
O, Origin. *I,* Insertion.

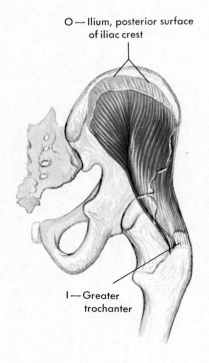

O—Ilium, posterior surface
of iliac crest

I—Greater
trochanter

Figure 9-30 Hamstring group of thigh muscles: biceps femoris, semitendinosus, and semimembranosus. *O,* Origin. *I,* Insertion.

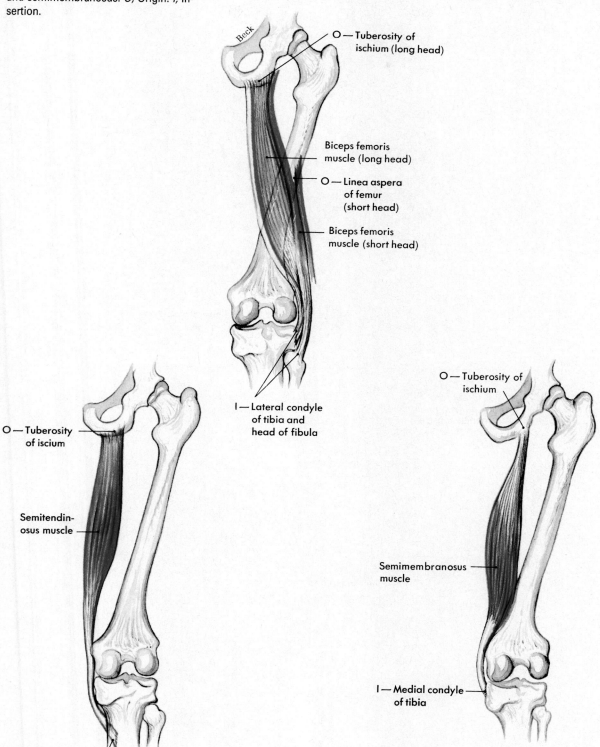

O — Tuberosity of ischium (long head)

Biceps femoris muscle (long head)

O — Linea aspera of femur (short head)

Biceps femoris muscle (short head)

I — Lateral condyle of tibia and head of fibula

O — Tuberosity of iscium

Semitendinosus muscle

I — Medial surface of tibia

O — Tuberosity of ischium

Semimembranosus muscle

I — Medial condyle of tibia

Figure 9-31 Iliopsoas muscle (iliacus, psoas major, and psoas minor muscles). *O*, Origin. *I*, Insertion.

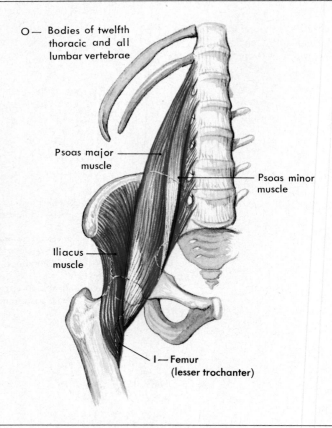

O — Bodies of twelfth thoracic and all lumbar vertebrae

Psoas major muscle

Psoas minor muscle

Iliacus muscle

I — Femur (lesser trochanter)

Figure 9-32 Quadratus lumborum muscle. *O*, Origin. *I*, Insertion.

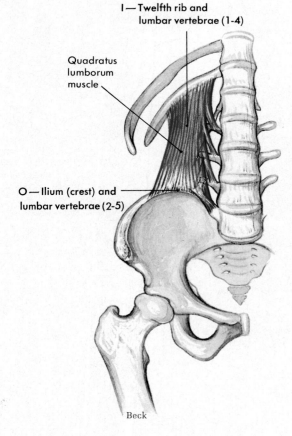

I — Twelfth rib and lumbar vertebrae (1-4)

Quadratus lumborum muscle

O — Ilium (crest) and lumbar vertebrae (2-5)

Beck

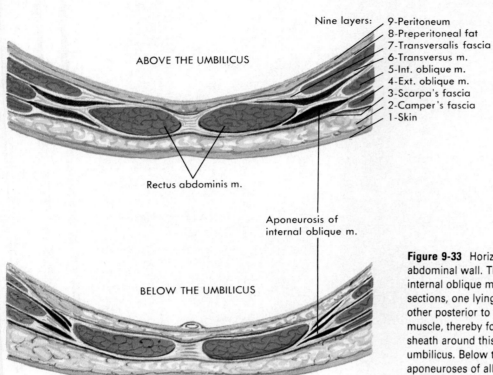

ABOVE THE UMBILICUS

Nine layers:
9-Peritoneum
8-Preperitoneal fat
7-Transversalis fascia
6-Transversus m.
5-Int. oblique m.
4-Ext. oblique m.
3-Scarpa's fascia
2-Camper's fascia
1-Skin

Rectus abdominis m.

Aponeurosis of
internal oblique m.

BELOW THE UMBILICUS

Figure 9-33 Horizontal section of the abdominal wall. The aponeurosis of the internal oblique muscle splits into two sections, one lying anterior and the other posterior to the rectus abdominis muscle, thereby forming an encasing sheath around this muscle, above the umbilicus. Below the umbilicus the aponeuroses of all muscles pass anterior to the rectus.

Figure 9-34 The diaphragm as seen from the front. Note the openings in the vertebral portion for the inferior vena cava, esophagus, and aorta.

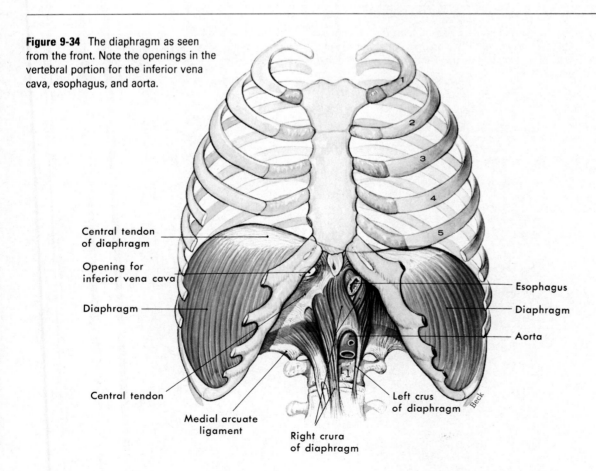

Central tendon
of diaphragm

Opening for
inferior vena cava

Diaphragm

Central tendon

Medial arcuate
ligament

Right crura
of diaphragm

Left crus
of diaphragm

Esophagus

Diaphragm

Aorta

Figure 9-35 Muscles of the head.

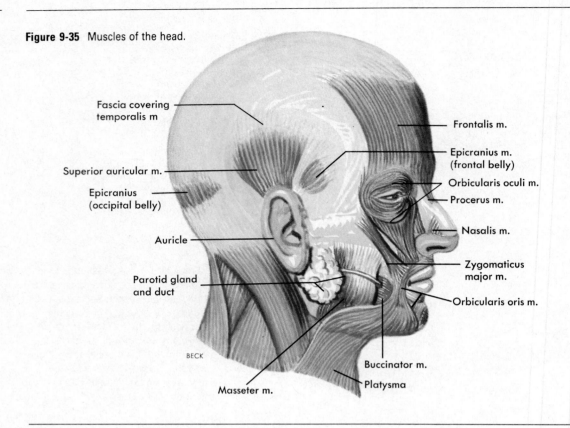

Fascia covering
temporalis m

Frontalis m.

Epicranius m.
(frontal belly)

Superior auricular m.

Orbicularis oculi m.

Procerus m.

Epicranius
(occipital belly)

Nasalis m.

Auricle

Zygomaticus
major m.

Parotid gland
and duct

Orbicularis oris m.

BECK

Buccinator m.

Platysma

Masseter m.

Figure 9-36 Muscles of facial expression. Most muscles of facial expression surround the orifices of the face: the eyes, nose, and mouth. Contraction of these muscles can produce a wide variety of facial expression and convey numerous emotions (see Table 9-15).

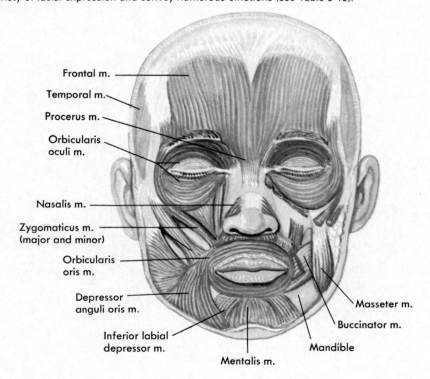

Frontal m.

Temporal m.

Procerus m.

Orbicularis
oculi m.

Nasalis m.

Zygomaticus m.
(major and minor)

Orbicularis
oris m.

Depressor
anguli oris m.

Masseter m.

Buccinator m.

Inferior labial
depressor m.

Mandible

Mentalis m.

Figure 9-37 A, Female pelvic floor viewed from above. **B,** Male pelvic floor viewed from above.

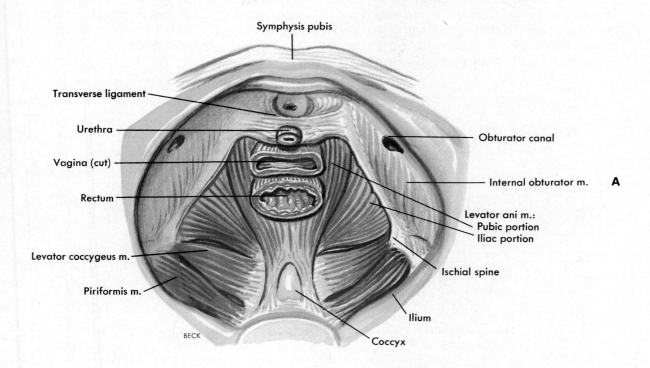

Symphysis pubis

Transverse ligament

Urethra

Vagina (cut)

Rectum

Levator coccygeus m.

Piriformis m.

Obturator canal

Internal obturator m.

Levator ani m.:
Pubic portion
Iliac portion

Ischial spine

Ilium

Coccyx

BECK

A

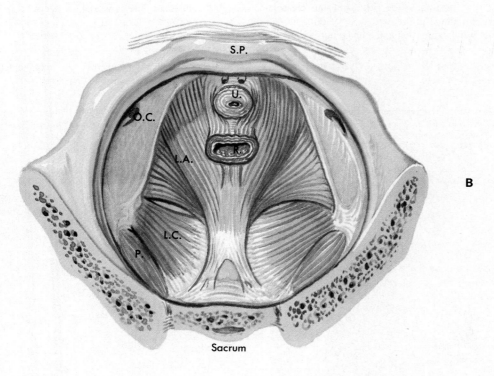

S.P.

U.

O.C.

L.A.

R.

L.C.

P.

Sacrum

B

ORIGINS, INSERTIONS, FUNCTIONS, AND INNERVATIONS OF REPRESENTATIVE MUSCLES

Basic information about many muscles is given in Tables 9-3 to 9-15. Each table has a description of a group of muscles that move one part of the body. Muscles that, in our judgment, are the most important for beginning students of anatomy to know are preceded by a bullet, and the origins and insertions so judged are set in boldface type. Remember that the actions listed for each muscle are those for which it is a prime mover. Actually, a single muscle contracting alone rarely accomplishes a given action. Instead, muscles act in groups as prime movers, synergists, and antagonists to bring about movements. As you study the muscles described in Tables 9-3 to 9-15, try to follow the hints for deducing muscle actions given on pp. 248-249.

Table 9-3 Muscles that move the shoulder*

Muscle	Origin	Insertion	Function	Innervation
• Trapezius	**Occipital bone†** (protuberance)	**Clavicle**	Raises or lowers shoulders and shrugs them	Spinal accessory, second, third, and fourth cervical nerves
	Vertebrae (cervical and thoracic)	**Scapula** (spine and acromion)	Extends head when occiput acts as insertion	
• Pectoralis minor	**Ribs** (second to fifth)	**Scapula** (coracoid)	Pulls shoulder down and forward	Medial and lateral anterior thoracic nerves
• Serratus anterior	**Ribs** (upper eight or nine)	**Scapula** (anterior surface, vertebral border)	Pulls shoulder forward; abducts and rotates it upward	Long thoracic nerve

*When trying to learn the origins and insertion of the muscles listed, refer frequently to illustrations of each muscle and to the skeleton. Also, when possible, feel each muscle on your own body.
†Origins and insertions judged to be most important for beginning students of anatomy to know are indicated by boldface type.
•Muscles judged to be most important for beginning students to know.

Table 9-4 Muscles that move the upper arm*

Muscle	Origin	Insertion	Function	Innervation
● Pectoralis major	**Clavicle** (medial half)† **Sternum** **Costal cartilages of true ribs**	**Humerus** (greater tubercle)	Flexes upper arm Adducts upper arm anteriorly; draws it across chest	Medial and lateral anterior thoracic nerves
● Latissimus dorsi	**Vertebrae** (spines of lower thoracic, lumbar, and sacral) **Ilium** (crest) Lumbodorsal fascia‡	**Humerus** (intertubercular groove)	Extends upper arm Adducts upper arm posteriorly	Thoracodorsal nerve
● Deltoid	**Clavicle** **Scapula** (spine and acromion)	**Humerus** (lateral side about halfway down—deltoid tubercle)	Abducts upper arm Assists in flexion and extension of upper arm	Axillary nerve
Coracobrachialis	Scapula (coracoid process)	Humerus (middle third, medial surface)	Adduction; assists in flexion and medial rotation of arm	Musculocutaneous nerve
Supraspinatus	Scapula (supraspinous fossa)	Humerus (greater tubercle)	Assists in abducting arm	Suprascapular nerve
Teres major	Scapula (lower part, axillary border)	Humerus (upper part, anterior surface)	Assists in extension, adduction, and medial rotation of arm	Lower subscapular nerve
Teres minor	Scapula (axillary border)	Humerus (greater tubercle)	Rotates arm outward	Axillary nerve
Infraspinatus	Scapula (infraspinatus border)	Humerus (greater tubercle)	Rotates arm outward	Suprascapular nerve

*When trying to learn the origins and insertion of the muscles listed, refer frequently to illustrations of each muscle and to the skeleton. Also, when possible, feel each muscle on your own body.

†Origins and insertions judged to be most important for beginning students of anatomy to know are indicated by boldface type.

‡Lumbodorsal fascia—extension of aponeurosis of latissimus dorsi; fills in space between last rib and iliac crest.

●Muscles judged to be most important for beginning students to know.

WEAK PLACES IN THE ABDOMINAL WALL

There are several places in the abdominal wall where rupture (hernia) with protrusion of part of the intestine may occur. At these points the wall is weakened because of the presence of an interval or space in the abdominal aponeuroses. Any undue pressure on the abdominal viscera therefore can force a portion of the parietal peritoneum, and often a part of the intestine as well, through these nonreinforced places. The weak places are (1) the inguinal canals, (2) the femoral rings, and (3) the umbilicus. Congenital defects or traumatic injury may permit abdominal viscera to project into the thorax (diaphragmatic hernia) or, on rare occasions, through the floor of the pelvis and some other areas.

Piercing the aponeuroses of the abdominal muscles are two canals, the *inguinal canals*, one on the right and the other on the left. They lie above, but

Table 9-5 Muscles that move the lower arm*

Muscle	Origin	Insertion	Function	Innervation
● Biceps brachii	**Scapula** (supraglenoid tuberosity)† **Scapula** (coracoid)	**Radius** (tubercle at proximal end)	Flexes supinated forearm Supinates forearm and hand	Musculocutaneous nerve
● Brachialis	**Humerus** (distal half, anterior surface)	**Ulna** (front of coronoid process)	Flexes pronated forearm	Musculocutaneous nerve
Brachioradialis	Humerus (above lateral epicondyle)	Radius (styloid process)	Flexes semipronated or semisupinated forearm; supinates forearm and hand	Radial nerve
● Triceps brachii	**Scapula** (infraglenoid tuberosity) **Humerus** (posterior surface—lateral head above radial groove; medial head, below)	**Ulna** (olecranon process)	Extends lower arm	Radial nerve
Pronator teres	Humerus (medial epicondyle) Ulna (coronoid process)	Radius (middle third of lateral surface)	Pronates and flexes forearm	Median nerve
Pronator quadratus	Ulna (distal fourth, anterior surface)	Radius (distal fourth, anterior surface)	Pronates forearm	Median nerve
Supinator	Humerus (lateral epicondyle) Ulna (proximal fifth)	Radius (proximal third)	Supinates forearm	Radial nerve

*When trying to learn the origins and insertion of the muscles listed, refer frequently to illustrations of each muscle and to the skeleton. Also, when possible, feel each muscle on your own body.
†Origins and insertions judged to be most important for beginning students of anatomy to know are indicated by boldface type.
●Muscles judged to be most important for beginning students to know.

parallel to, the inguinal ligaments (p. 254) and are about 5 cm long. In the male the spermatic cords extend through the canals into the scrotum (see Figure 9-17), whereas in the female the round ligaments of the uterus are in this location. The internal opening of each canal is a circular space in the aponeurosis of the transverse muscle known as the **internal (abdominal) inguinal ring.** The external openings, called the **external (superficial) inguinal rings,** are triangular spaces in the aponeuroses of the external oblique muscles. They are located inferiorly and mesially to the internal rings just above and lateral to the pubic crest. The upper surface of the inguinal ligament forms the floor of the canal as it passes obliquely downward and medially through the abdominal wall. The fact that the inguinal canal is larger in the male than in the female probably explains why inguinal hernia occurs more often in men than in women.

The **femoral (saphenous) rings** are openings in the fascia lata of the thigh below and lateral to the external inguinal rings and pubic tubercle (see Figure 9-17). The openings are covered by a thin membrane called the **cribriform fascia** that is per-

Table 9-6 Muscles that move the hand*

Muscle	Origin	Insertion	Function	Innervation
Flexor carpi radialis	Humerus (medial epicondyle)	Second metacarpal (base of)	Flexes hand Flexes forearm	Median nerve
Palmaris longus	Humerus (medial epicondyle)	Fascia of palm	Flexes hand	Median nerve
Flexor carpi ulnaris	Humerus (medial epicondyle) Ulna (proximal two thirds)	Pisiform bone Third, fourth, and fifth metacarpals	Flexes hand Adducts hand	Ulnar nerve
Extensor carpi radialis longus	Humerus (ridge above lateral epicondyle)	Second metacarpal (base of)	Extends hand Abducts hand (moves toward thumb side when hand supinated)	Radial nerve
Extensor carpi radialis brevis	Humerus (lateral epicondyle)	Second, third metacarpals (bases of)	Extends hand	Radial nerve
Extensor carpi ulnaris	Humerus (lateral epicondyle) Ulna (proximal three fourths)	Fifth metacarpal (base of)	Extends hand Adducts hand (move toward little finger side when hand supinated)	Radial nerve

*When trying to learn the origins and insertion of the muscles listed, refer frequently to illustrations of each muscle and to the skeleton. Also, when possible, feel each muscle on your own body.

forated by the great saphenous vein. They have a diameter of about 1.3 cm (½ inch) and are usually somewhat larger in females, a fact that accounts for the greater prevalence of femoral hernia in women than in men.

POSTURE

We have already discussed the major role muscles play in movement and heat production. We shall now turn our attention to a third way in which muscles serve the body as a whole—that of maintaining the posture of the body. Let us consider a few aspects of this important function.

MEANING

The term *posture* means simply position or alignment of body parts. "Good posture" means many things. It means body alignment that most favors function; it means position that requires the least muscular work to maintain, which puts the least strain on muscles, ligaments, and bones; it means keeping the body's center of gravity over its base. Good posture in the standing position, for example, means head and chest held high, chin, abdomen, and buttocks pulled in, knees bent slightly, and feet placed firmly on the ground about 6 inches apart.

HOW MAINTAINED

Since gravity pulls on the various parts of the body at all times, and since bones are too irregularly shaped to balance themselves on each other, the only way the body can be held upright is for muscles to exert a continual pull on bones in the opposite direction from gravity. Gravity tends to pull the head and trunk forward and downward; muscles (head and trunk extensors) must therefore pull backward

Table 9-7 Muscles that move the thigh*

Muscle	Origin	Insertion	Function	Innervation
• Iliopsoas (iliacus and psoas major)	**Ilium** (illiac fossa)† **Vertebrae** (bodies of twelfth thoracic to fifth lumbar)	**Femur** (Small trochanter)	Flexes thigh Flexes trunk (when femur acts as origin)	Femoral and second to fourth lumbar nerves
• Rectus femoris	**Ilium** (anterior, inferior spine)	**Tibia** (by way of patellar tendon)	Flexes thigh Extends lower leg	Femoral nerve
• Gluteal group Maximus	**Ilium** (crest and posterior surface) Sacrum and coccyx (posterior surface) Sacrotuberous ligament	**Femur** (gluteal tuberosity) **Iliotibial tract**‡	Extends thigh—rotates outward	Inferior gluteal nerve
Medius	**Ilium** (lateral surface)	**Femur** (greater trochanter)	Abducts thigh—rotates outward; stabilizes pelvis on femur	Superior gluteal nerve
Minimus	**Ilium** (lateral surface)	**Femur** (greater trochanter)	Abducts thigh; stabilizes pelvis on femur Rotates thigh medially	Superior gluteal nerve
• Tensor fasciae latae	**Ilium** (anterior part of crest)	**Tibia** (by way of **iliotibial tract**)	Abducts thigh Tightens iliotibial tract†	Superior gluteal nerve
Piriformis	Vertebrae (front of sacrum)	Femur (medial aspect of greater trochanter)	Rotates thigh outward Abducts thigh Extends thigh	First or second sacral nerves
• Adductor group Brevis	**Pubic bone**	**Femur** (linea aspera)	Adducts thigh	Obturator nerve
Longus	**Pubic bone**	**Femur** (linea aspera)	Adducts thigh	Obturator nerve
Magnus	**Pubic bone**	**Femur** (linea aspera)	Adducts thigh	Obturator nerve
Gracilis	Pubic bone (just below symphysis)	Tibia (medial surface behind sartorius)	Adducts thigh and flexes and adducts leg	Obturator nerve

*When trying to learn the origins and insertion of the muscles listed, refer frequently to illustrations of each muscle and to the skeleton. Also, when possible, feel each muscle on your own body.

†Origins and insertions judged to be most important for beginning students of anatomy to know are indicated by boldface type.

‡The iliotibial tract is part of the fascia enveloping all the thigh muscles. It consists of a wide band of dense fibrous tissue attached to the iliac crest above and the lateral condyle of the tibia below. The upper part of the tract encloses fasciae latae muscle.

•Muscles judged to be most important for beginning students to know.

Table 9-8 Muscles that move the lower leg*

Muscle	Origin	Insertion	Function	Innervation
• Quadriceps femoris group				
Rectus femoris	**Ilium** (anterior inferior spine)†	**Tibia** (by way of patellar tendon)	Flexes thigh Extends leg	Femoral nerve
Vastus lateralis	**Femur** (linea aspera)	**Tibia** (by way of patellar tendon)	Extends leg	Femoral nerve
Vastus medialis	**Femur**	**Tibia** (by way of patellar tendon)	Extends leg	Femoral nerve
Vastus intermedius	**Femur** (anterior surface)	**Tibia** (by way of patellar tendon)	Extends leg	Femoral nerve
• Sartorius	**Os innominatum** (anterior, superior iliac spines)	**Tibia** (medial surface of upper end of shaft)	Adducts and flexes leg Permits crossing of legs tailor fashion	Femoral nerve
• Hamstring group				
Biceps femoris	**Ischium** (tuberosity)	**Fibula** (head of)	Flexes leg	Hamstring nerve (branch of sciatic nerve)
	Femur (linea aspera)	**Tibia** (lateral condyle)	Extends thigh	Hamstring nerve
Semitendinosus	**Ischium** (tuberosity)	**Tibia** (proximal end, medial surface)	Extends thigh	Hamstring nerve
Semimembranosus	**Ischium** (tuberosity)	**Tibia** (medial condyle)	Extends thigh	Hamstring nerve

*When trying to learn the origins and insertion of the muscles listed, refer frequently to illustrations of each muscle and to the skeleton. Also, when possible, feel each muscle on your own body.
†Origins and insertions judged to be most important for beginning students of anatomy to know are indicated by boldface type.
• Muscles judged to be most important for beginning students to know.

and upward on them. For instance, gravity pulls the lower jaw downward; muscles must pull upward on it. Muscles exert this pull against gravity by virtue of their property of tonicity. Because tonicity is absent during sleep, muscle pull does not then counteract the pull of gravity. Hence, for example, we cannot sleep standing up.

Many structures other than muscles and bones play a part in the maintenance of posture. The nervous system is responsible for the existence of muscle tone and also regulates and coordinates the amount of pull exerted by the individual muscles. The respiratory, digestive, circulatory, excretory, and endocrine systems all contribute something toward the ability of muscles to maintain posture. This is one of many examples of the important principle that all body functions are interdependent.

IMPORTANCE TO BODY AS WHOLE

The importance of posture can perhaps be best evaluated by considering some of the effects of poor posture. Poor posture throws more work on muscles to counteract the pull of gravity and therefore leads to fatigue more quickly than good posture. Poor posture puts more strain on ligaments. It puts abnormal strains on bones and may eventually produce deformities. It interferes with various functions such as respiration, heart action, and digestion. It probably is not going too far to say that it even detracts from one's feeling of self-confidence and joy. In support of this claim, consider our use of such expressions as "shoulders squared, head erect" to denote confidence and joy and "down-in-the-mouth," "long-faced," and "bowed down" to signify dejection and anxiety. The importance of posture to the body as a whole might be summed up in a single sentence: maximal health and good posture are reciprocally related, that is, each one depends on the other.

Text continued on p. 276.

Table 9-9 Muscles that move the foot*

Muscle	Origin	Insertion	Function	Innervation
● Tibialis anterior	**Tibia** (lateral condyle of upper body)†	**Tarsal** (first cuneiform) Metatarsal (base of first)	Flexes foot Inverts foot	Common and deep peroneal nerves
● Gastrocnemius	**Femur** (condyles)	**Tarsal** (calcaneus by way of Achilles tendon)	Extends foot Flexes lower leg	Tibial nerve (branch of sciatic nerve)
● Soleus	**Tibia** (underneath gastrocnemius) **Fibula**	**Tarsal** (calcaneus by way of Achilles tendon)	Extends foot (plantar flexion)	Tibial nerve
Peroneus longus	Tibia (lateral condyle) Fibula (head and shaft)	First cuneiform Base of first metatarsal	Extends foot (plantar flexion) Everts foot	Common peroneal nerve
Peroneus brevis	Fibula (lower two thirds of lateral surface of shaft)	Fifth metatarsal (tubercle, dorsal surface)	Everts foot Flexes foot	Superficial peroneal nerve
Tibialis posterior	Tibia (posterior surface) Fibula (posterior surface)	Navicular bone Cuboid bone All three cuneiforms Second and fourth metatarsals	Extends foot (plantar flexion) Inverts foot	Tibial nerve
Peroneus tertius	Fibula (distal third)	Fourth and fifth metatarsals (bases of)	Flexes foot Everts foot	Deep peroneal nerve

*When trying to learn the origins and insertion of the muscles listed, refer frequently to illustrations of each muscle and to the skeleton. Also, when possible, feel each muscle on your own body.
†Origins and insertions judged to be most important for beginning students of anatomy to know are indicated by boldface type.
●Muscles judged to be most important for beginning students to know.

Table 9-10 Muscles that move the head*

Muscle	Origin	Insertion	Function	Innervation
• Sternoclei-domastoid	**Sternum†** **Clavicle**	**Temporal bone** (mastoid process)	Flexes head (prayer muscle) One muscle alone, rotates head toward opposite side; spasm of this muscle alone or associated with trapezius called *torticollis* or *wryneck*	Accessory nerve
Semispinalis capitis	Vertebrae (transverse processes of upper six thoracic, articular processes of lower four cervical)	Occipital bone (between superior and inferior nuchal lines)	Extends head; bends it laterally	First five cervical nerves
Splenius capitis	Ligamentum nuchae Vertebrae (spinous processes of upper three or four thoracic)	Temporal bone (mastoid process) Occipital bone	Extends head Bends and rotates head toward same side as contracting muscle	Second, third, and fourth cervical nerves
Longissimus capitis	Vertebrae (transverse processes of upper six thoracic, articular processes of lower four cervical)	Temporal bone (mastoid process)	Extends head Bends and rotates head toward contracting side	Multiple innervation

*When trying to learn the origins and insertion of the muscles listed, refer frequently to illustrations of each muscle and to the skeleton. Also, when possible, feel each muscle on your own body.
†Origins and insertions judged to be most important for beginning students of anatomy to know are indicated by boldface type.
●Muscles judged to be most important for beginning students to know.

Table 9-11 Muscles that move the abdominal wall*

Muscle	Origin	Insertion	Function	Innervation
• External oblique	**Ribs** (lower eight)	**Ossa coxae** (iliac crest and pubis by way of inguinal ligament)† **Linea alba‡** by way of an aponeurosis§‖	Compresses abdomen Important postural function of all abdominal muscles is to pull front of pelvis upward, thereby flattening lumbar curve of spine; when these muscles lose their tone, common figure faults of protruding abdomen and lordosis develop	Lower seven intercostal nerves and iliohypogastric nerves
• Internal oblique	**Ossa coxae** (iliac crest and inguinal ligament) **Lumbodorsal fascia**	**Ribs** (lower three) **Pubic bone** **Linea alba**	Same as external oblique	Last three intercostal nerves; iliohypogastric and ilioinguinal nerves
• Transversalis	**Ribs** (lower six) **Ossa coxae** (iliac crest, inguinal ligament) **Lumbodorsal fascia**	**Pubic bone** **Linea alba**	Same as external oblique	Last five intercostal nerves; iliohypogastric and ilioinguinal nerves
• Rectus abdominis	**Ossa coxae** (pubic bone and symphysis pubis)	**Ribs** (costal cartilage of fifth, sixth, and seventh ribs) Sternum (xiphoid process)	Same as external oblique; because abdominal muscles compress abdominal cavity, they aid in straining, defecation, forced expiration, childbirth, etc.; abdominal muscles are antagonists of diaphragm, relaxing as it contracts and vice versa Flexes trunk	Last six intercostal nerves

*When trying to learn the origins and insertion of the muscles listed, refer frequently to illustrations of each muscle and to the skeleton. Also, when possible, feel each muscle on your own body.

†Inguinal ligament (or Poupart's)—lower edge of aponeurosis of external oblique muscle, extending between the anterior superior iliac spine and the tubercle of the pubic bone. This edge is doubled under like a hem on material. The inguinal ligament forms the upper boundary of the femoral triangle, a large triangular area in the thigh; its other boundaries are the adductor longus muscle mesially and the sartorius muscle laterally.

‡Linea alba—literally, "a white line"; extends from xiphoid process to symphysis pubis; formed by fibers of aponeuroses of the right abdominal muscles interlacing with fibers of aponeuroses of the left abdominal muscles; comparable to a seam up the midline of the abdominal wall, anchoring its various layers. During pregnancy the linea alba becomes pigmented and is known as the linea niger.

§Aponeurosis—sheet of white fibrous tissue that attaches one muscle to another or attaches it to bone or other movable structures, for example, the right external oblique muscle attaches to the left external oblique muscle by means of an aponeurosis.

‖Origins and insertions judged to be most important for beginning students of anatomy to know are indicated by boldface type.

●Muscles judged to be most important for beginning students to know.

Table 9-12 Muscles that move the chest wall*

Muscle	Origin	Insertion	Function	Innervation
External intercostals	Rib (lower border; forward fibers)	Rib (upper border of rib below origin)	Elevate ribs	Intercostal nerves
Internal intercostals	Rib (inner surface, lower border; backward fibers)	Rib (upper border of rib below origin)	Probably depress ribs	Intercostal nerves
• Diaphragm	**Lower circumference of thorax** (of rib cage)†	**Central tendon of diaphragm**	Enlarges thorax, causing inspiration	Phrenic nerves

*When trying to learn the origins and insertion of the muscles listed, refer frequently to illustrations of each muscle and to the skeleton. Also, when possible, feel each muscle on your own body.
†Origins and insertions judged to be most important for beginning students of anatomy to know are indicated by boldface type.
•Muscles judged to be most important for beginning students to know.

Table 9-13 Muscles of the pelvic floor*

Muscle	Origin	Insertion	Function	Innervation
Levator ani	Pubis (posterior surface) Ischium (spine)	Coccyx	Together form floor of pelvic cavity; support pelvic organs; if these muscles are badly torn at childbirth or become too relaxed, uterus or bladder may prolapse, that is, drop out	Pudendal nerve
Coccygeus (posterior continuation of levator ani)	Ischium (spine)	Coccyx Sacrum	Same as levator ani	Pudendal nerve

*When trying to learn the origins and insertion of the muscles listed, refer frequently to illustrations of each muscle and to the skeleton. Also, when possible, feel each muscle on your own body.

Table 9-14 Muscles that move the trunk*

Muscle	Origin	Insertion	Function	Innervation
Sacrospinalis (erector spinae)			Extend spine; maintain erect posture of trunk Acting singly, abduct and rotate trunk	Posterior rami of first cervical to fifth lumbar spinal nerves
Lateral portion: Iliocostalis lumborum	Iliac crest, sacrum (posterior surface), and lumbar vertebrae (spinous processes)	Ribs, lower six		
Iliocostalis dorsi	Ribs, lower six	Ribs, upper six		
Iliocostalis cervicis	Ribs, upper six	Vertebrae, fourth to sixth cervical		
Medial portion: Longissimus dorsi	Same as iliocostalis lumborum	Vertebrae, thoracic ribs		
Longissimus cervicis	Vertebrae, upper six thoracic	Vertebrae, second to sixth cervical		
Longissimus capitis	Vertebrae, upper six thoracic and last four cervical	Temporal bone, mastoid process		
Quadratus lumborum (forms part of posterior abdominal wall)	Ilium (posterior part of crest) Vertebrae (lower three lumbar)	Ribs (twelfth) Vertebrae (transverse processes of first four lumbar)	Both muscles together extend spine One muscle alone abducts trunk toward side of contracting muscle	First three or four lumbar nerves
• Iliopsoas	See muscles that move thigh, p. 270		Flexes trunk	

*When trying to learn the origins and insertion of the muscles listed, refer frequently to illustrations of each muscle and to the skeleton. Also, when possible, feel each muscle on your own body.
●Muscles judged to be most important for beginning students to know.

Outline Summary

GENERAL FUNCTIONS

A Movement—sometimes locomotion, sometimes movement within given area

B Posture

C Heat production

SKELETAL MUSCLE CELLS

Microscopic structure

A Muscle cells usually called *muscle fibers*, term descriptive of their long, narrow shape

B Sarcolemma—cell membrane of muscle fiber

C Sarcoplasm—cytoplasm of muscle fiber

D Sarcoplasmic reticulum—analogous but not identical to endoplasmic reticulum of cells other than muscle fibers

E Myofibrils—numerous fine fibers packed close together in sarcoplasm

F Cross striae

  1 Dark stripes called A bands; light H zone runs across midsection of each dark A band

  2 Light stripes called I bands; dark Z line extends across center of each light I band

G Sarcomere—section of myofibril extending from one Z line to next; each myofibril consists of several sarcomeres

H T system—transverse tubules that extend into sarcoplasm at levels of Z lines; formed by invaginations of sarcolemma

I Triad—triple-layered structure consisting of T tubule sandwiched between sacs of sarcoplasmic reticulum

Table 9-15 Muscles of facial expression and of mastication*

Muscle	Origin	Insertion	Function	Innervation
Muscles of facial expression				
Epicranius (occipitofrontalis)	Occipital bone	Tissues of eyebrows	Raises eyebrows, wrinkles forehead horizontally	Cranial nerve VII
Corrugator supercilii	Frontal bone (superciliary ridge)	Skin of eyebrow	Wrinkles forehead vertically	Cranial nerve VII
Orbicularis oculi	Encircles eyelid		Closes eye	Cranial nerve VII
Procerus	Bridge of nose	Skin over epicranius	Narrows eye opening	Cranial nerve VII
Inferior labial depressor	Mandible	Skin of lower lip	Draws lower lip downwards	Cranial nerve VII
Mentalis	Incisive fossa of mandible	Skin of chin	Raises and protrudes lower lip	Cranial nerve VII
Nasalis	Maxilla	Ala of nose	Compresses nasal aperture	Cranial nerve VII
Zygomaticus major	Zygomatic bone	Angle of mouth	Laughing (elevates angle of mouth)	Cranial nerve VII
Zygomaticus minor	Zygomatic bone	Upper lip	Elevates upper lip	Cranial nerve VII
Orbicularis oris	Encircles mouth		Draws lips together	Cranial nerve VII
Platysma	Fascia of upper part of deltoid and pectoralis major	Mandible (lower border) Skin around corners of mouth	Draws corners of mouth down—pouting	Cranial nerve VII
Buccinator	Maxillae	Skin of sides of mouth	Permits smiling Blowing, as in playing a trumpet	Cranial nerve VII
Muscles of mastication				
Masseter	Zygomatic arch	Mandible (external surface)	Closes jaw	Cranial nerve V
Temporal	Temporal bone	Mandible	Closes jaw	Cranial nerve V
Pterygoids (internal and external)	Undersurface of skull	Mandible (mesial surface)	Grates teeth	Cranial nerve V

*When trying to learn the origins and insertion of the muscles listed, refer frequently to illustrations of each muscle and to the skeleton. Also, when possible, feel each muscle on your own body.

Outline Summary—cont'd

Molecular structure

A Proteins
  1 Thick filaments—composed almost entirely of myosin molecules; heads of myosin molecules are cross bridges of thick filament
  2 Thin filaments—composed of actin, tropomyosin, and troponin molecules arranged in complex fashion; thin filaments attach to Z lines, extend from them in toward center of sarcomeres; thick and thin filaments alternate in myofibrils

Functions

A Contraction—cross bridges of thick filaments attach to thin filaments and pull them toward middle of each sarcomere (Figure 9-6, p. 242)
B Muscle cells obey all-or-none law when they contract, that is, they either contract with all force possible under existing conditions or do not contract at all

Outline Summary—cont'd

C Sarcoplasmic reticulum—sometimes called "relaxing factor" of muscle cells because it has a very strong affinity for calcium

Energy sources for muscle contraction

See Figure 9-4, p. 241

SKELETAL MUSCLE ORGANS
Structure

A Size, shape, and fiber arrangement—wide variation in different muscles
B Connective tissue components
  1 Epimysium—fibrous connective tissue sheath that envelops each muscle
  2 Perimysium—extensions of epimysium, partitioning each muscle into bundles of fibers
  3 Endomysium—extensions of perimysium between individual muscle fibers
  4 Tendon—strong, tough cord continuous at one end with fibrous wrappings (e.g., epimysium) of muscle and at other end with fibrous covering of bone (periosteum)
  5 Aponeurosis—broad flat sheet of fibrous connective tissue continuous on one border with fibrous wrappings of muscle and at other border with fibrous coverings of some adjacent structure, usually another muscle
  6 Tendon sheaths—tubes of fibrous connective tissue that enclose certain tendons, notably those of wrist and ankle; synovial membrane lines tendon sheaths
  7 Deep fascia—layer of dense fibrous connective tissue underlying superficial fascia under skin; extensions of deep fascia form epimysium, etc. and also enclose viscera, glands, blood vessels, and nerves
C Nerve supply
  One motoneuron, together with skeletal muscle fibers it supplies, constitutes *motor unit;* number of muscle fibers per motor unit varies; in general, more precise movements produced by muscle in which motor units include fewer muscle fibers
D Age changes
  1 Fibrosis—with advancing years, some skeletal muscle fibers degenerate and are replaced by fibrous connective tissue
  2 Decreased muscular strength, resulting in part from fibrosis

Function

A Basic principles
  1 Skeletal muscles contract only if stimulated; natural stimulus is nerve impulses, artificial stimuli, (e.g., electrical), or injury

2 Skeletal muscle contractions of several types
  a Tonic contraction (tone, tonus)—continual, partial contractions produced by simultaneous activation of small group of motor units, followed by relaxation of their fibers and activation of another group of motor units; all healthy muscles exhibit tone when individuals are awake; specialized receptors called *muscle spindles* and neurotendinous end-organs detect degree of muscle stretch
  b Isotonic contraction—muscle shortens but its tension remains constant; isotonic contractions produce movements
  c Isometric contraction—muscle length remains unchanged but tension within muscle increases; isometric contractions "tighten" muscles but do not produce movements
  d Twitch contraction—quick, jerky contraction in response to single stimulus; consists of three phases—latent period, contraction phase, and relaxation phase; twitch contractions rare in normal body
  e Tetanic contraction (tetanus)—sustained smooth contraction produced by series of stimuli bombarding muscle in rapid succession; normal movements said to be produced by incomplete tetanic contractions
  f Treppe (staircase phenomenon)—series of increasingly stronger contractions in response to constant-strength stimuli applied at rate of one or two per second; contracture—incomplete relaxation after repeated stimulation; fatigue—failure of muscle to contract in response to strongest stimuli after repeated stimulation; true muscle fatigue seldom occurs in body
  g Fibrillation—abnormal contraction in which individual muscle fibers contract asynchronously, producing no effective movement
  h Convulsions—uncoordinated tetanic contractions of varying groups of muscles
3 Skeletal muscles contract according to graded-strength principle in contrast to individual muscle cells that compose them, which contract according to all-or-none law
4 Skeletal muscles produce movement by pulling on insertion bones across joints
5 Bones serve as levers and joints as fulcrums of these levers
6 Muscles that move part usually do not lie over that part but proximal to it
7 Skeletal muscles almost always act in groups rather than singly, that is, most movements produced by coordinated action of several muscles

Outline Summary—cont'd

B Hints of how to deduce actions
  1 Deduce bones that muscle attaches to from illustrations of muscle
  2 Make guess as to which bone moves (insertion)
  3 Deduce movement muscle produces by applying principle that its insertion moves toward its origin

Names

A Reasons for names—muscle names describe one or more of following features about muscle
  1 Its action
  2 Direction of fibers
  3 Its location
  4 Number of divisions composing it
  5 Its shape
  6 Its points of attachment
B Muscles grouped according to location—(see Table 9-1)
C Muscles grouped according to function—(see Table 9-2)
  1 Flexors—decrease angle of joint
  2 Extensors—return part from flexion to normal anatomical position
  3 Abductors—move bone away from midline of body
  4 Adductors—move bone toward midline of body
  5 Rotators—cause part to pivot on its axis
  6 Levators—raise part
  7 Depressors—lower part
  8 Sphincters—reduce size of opening

  9 Tensors—tense part or make it more rigid
  10 Supinators—turn hand palm upward
  11 Pronators—turn hand palm downward

Origins, insertions, functions, and innervations of representative muscles

See Tables 9-3 to 9-15, pp. 266-277

WEAK PLACES IN THE ABDOMINAL WALL

A Inguinal rings—right and left internal; right and left external
B Femoral rings—right and left
C Umbilicus
D Diaphragm

POSTURE

A Meaning—position or alignment of body parts
B How maintained—by continual pull of muscles on bones in opposite direction from pull of gravity, that is, posture maintained by continued partial contraction of muscles, or muscle tone; therefore indirectly dependent on many other factors, for example normal nervous, respiratory, and circulatory systems, health in general
C Importance to body as whole—essential for optima functioning of most of body, including respiration, circulation, digestion, joint action; briefly, maximal health depends on good posture, good posture depends on health

Review Questions

1 Differentiate between the three kinds of muscle tissue as to structure, location, and innervation.
2 Describe several physiological properties of muscle tissue.
3 What property is more highly developed in muscle than in any other tissue?
4 State a principle describing the usual relationship between a part moved and the location of muscles (insertion, body, and origin) moving the part.
5 Applying the principle stated in question 4, where would you expect muscles that move the head to be located? Name two or three muscles that fulfill these conditions.
6 Applying the principle stated in question 4, what part of the body do thigh muscles move? Name several muscles that fulfill these conditions.
7 What bone or bones serve as a lever in movements of the forearm? What structure constitutes the fulcrum for this lever?
8 Explain the meaning of the term *neuromusculo-skeletal unit.*

9 Name the main muscles of the back, chest, abdomen, neck, shoulder, upper arm, lower arm, thigh, buttocks, leg, and pelvic floor.
10 Name the main muscles that flex, extend, abduct, and adduct the upper arm; that raise and lower the shoulder; that flex and extend the lower arm; that flex, extend, abduct, and adduct the thigh; that flex and extend the lower leg and thigh; that flex and extend the foot; that flex, extend, abduct, and adduct the head; that move the abdominal wall; that move the chest wall.
11 Discuss the chemical reactions thought to make available energy for muscle contraction.
12 What physiological reason can you give for athletes using a warm-up period before starting a game?
13 Name several weak places in the abdominal wall where hernia may occur.
14 What and where are the inguinal canals? Of what clinical importance are they?

UNIT THREE

COMMUNICATION, CONTROL, AND INTEGRATION

10 NERVOUS SYSTEM CELLS

OBJECTIVES

After you have completed this chapter, you should be able to:

1 Describe the generalized functions of the nervous system.
2 List the primary organs of the nervous system.
3 Identify and describe the general structural and functional characteristics of the two main types of cells that compose nervous system structures.
4 List and describe the structure and function of the four types of neuroglia.
5 Identify the type of neuroglia cells found only in the nerves of the body.
6 Classify neurons according to direction of impulse conduction and type of cell processes.
7 Discuss six specialized cell structures found only in neurons and identify which one plays an essential role in regeneration of damaged nervous tissue.
8 Explain the generation and maintenance of resting membrane potentials.
9 Discuss the step-by-step mechanism of a nerve impulse represented by changes in voltage across the membrane of a neuron.
10 Compare and contrast continuous propagation of an action potential with saltatory conduction.
11 Discuss the structural and functional components of a three-neuron ipsilateral reflex arc.
12 List and describe the structural components of a synapse.
13 Explain the mechanism of conduction of an action potential across a synapse.
14 Explain the phenomena of facilitation and inhibition on threshold of stimulation.
15 Compare spatial and temporal summation.

KEY TERMS

Action potential (AK-shun po-TEN-shal)

Astrocyte (AS-tro-site)

Axon (AK-son)

Dendrite (DEN-drite)

Facilitation (fah-sel-i-TA-shun)

Inhibition (in-hi-BISH-un)

Microglia (mi-KROG-le-ah)

Myelin (MI-e-lin)

Neurilemma (nu-re-LEM-ah)

Neuroglia (nu-ROG-le-ah)

Neuron (NU-ron)

Nissl body (ni-SUL BOD-e)

Oligodendroglia (ol-i-go-den-DROG-le-ah)

Perikaryon (per-i-KAR-e-on)

Polarization (po-lar-i-ZA-shun)

Refractory (re-FRAK-to-re)

Saltatory conduction (SAL-tah-to-re kon-DUK-shun)

Schwann cell (SHWON sel)

Stimuli (STIM-u-li)

Summation (sum-MA-shun)

Synapse (SIN-aps)

The nervous system and the endocrine system together perform a vital function for the body—communication. Homeostasis and therefore survival depend on this function. Why? Because communication provides the means for controlling and integrating the many different functions performed by organs, tissues, and cells. Integrating means unifying. Unifying bodily functions means controlling them in ways that make them work together like parts of one machine to accomplish homeostasis and the one end function, survival. Communication makes possible control; control makes possible integration; integration makes possible homeostasis; homeostasis makes possible survival.

These organs—the brain, spinal cord, and nerves—make up the nervous system. Facts, theories, and questions about this, our most complex and least well understood system, are as fascinating as they are abundant. We shall approach this somewhat daunting mass of material by considering in this chapter the cells of the nervous system. Then in Chapter 11 we shall discuss the spinal cord, spinal nerves, brain, and cranial nerves. Chapter 12 discusses the autonomic nervous system and Chapter 13 deals with the special senses. The endocrine system is the subject of Chapter 14.

Figure 10-1 Neuroglia, the special connecting and supporting cells of the brain and spinal cord, are composed of astrocytes, microglia, and oligodendroglia.

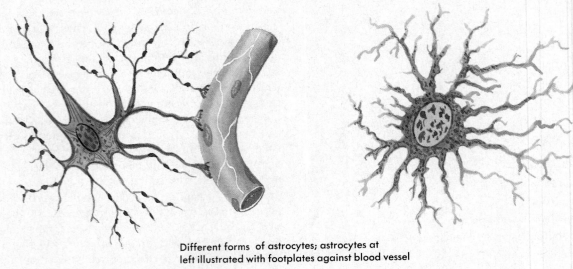

Different forms of astrocytes; astrocytes at left illustrated with footplates against blood vessel

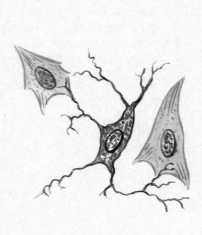

Microglia with processes extending to two nerve cell bodies

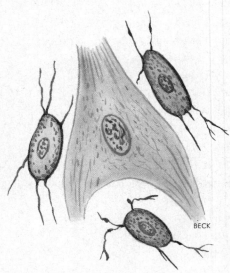

Oligodendroglia located near a nerve cell body

MAIN TYPES OF CELLS

Two main types of cells compose nervous system structures, namely, **neuroglia** and **neurons.** Neuroglia are the special connective tissue cells of the nervous system. Neurons are nerve cells. They are the specialists of the nervous system. They specialize in impulse conduction, the function that makes possible all other nervous system functions. Subtypes of neuroglia and neurons are described in the following sections.

NEUROGLIA

Number

The number of neuroglia in the human nervous system is beyond imagination. One estimate places the figure at a staggering 900 billion, or nine times the estimated number of stars in our galaxy.

Types

The four types of neuroglia (Figures 10-1 and 10-4) are:

1 **Astrocytes**
2 **Oligodendroglia**
3 **Microglia**
4 **Schwann cells**

Structure and Function

The star-shaped neuroglia, **astrocytes,** derive their name from the Greek *astron,* "star"; *kytos,* "cell." They are the largest and most numerous type of neuroglia. Webs of astrocytes form tight sheaths around the brain's blood capillaries. These sheaths and the tight junctions between the endothelial cells that form brain capillary walls together constitute the so-called *blood-brain barrier.* Small molecules (e.g., oxygen, carbon dioxide, water, alcohol) diffuse rapidly through the barrier to reach brain neurons. Larger molecules penetrate it slowly or not at all.

Microglia are small, usually stationary cells. In inflamed or degenerating brain tissue, however, microglia enlarge, move about, and carry on phagocytosis. In other words, they engulf and destroy microbes and cellular debris.

Oligodendroglia are smaller cells and have fewer processes than astrocytes. Some oligodendroglia lie clustered around nerve cell bodies; some are arranged in rows between nerve fibers in the brain and cord. They help hold nerve fibers together and also serve another and probably more important function—they produce the fatty myelin sheath that envelops nerve fibers located in the brain and cord.

Schwann cells are the type of neuroglia found only in the nerves of the body, not in the brain and spinal cord. As will be explained later, they function to form the neurilemma and the myelin sheath, the two coverings that envelop nerve fibers located outside the brain and cord in nerves.

NEURONS

Number

The human brain, according to recent estimates, contains about 100 billion (10^{11}) neurons, or about the same number as the stars in our galaxy.

Types

Neurons are classified according to two different criteria—the direction in which they conduct impulses and the number of processes they have.

Classified according to the direction in which they conduct impulses, there are three types of neurons:

1 **Afferent neurons**
2 **Efferent neurons**
3 **Interneurons**

Figure 10-7 shows one neuron of each of these types. *Afferent* (sensory) *neurons* transmit nerve impulses to the spinal cord or brain. *Efferent* (motor) *neurons* transmit nerve impulses away from the brain or spinal cord to or toward muscles or glands. *Interneurons* conduct impulses from afferent neurons toward or to motoneurons. Interneurons lie entirely within the central nervous system (brain and spinal cord).

Classified according to the number of their processes, there are also three types of neurons (Figure 10-2):

1 **Multipolar**
2 **Bipolar**
3 **Unipolar**

Multipolar neurons have only one axon but several dendrites. Most of the neurons in the brain and spinal cord are multipolar. *Bipolar neurons* have only one axon and also only one dendrite and are the least numerous kind of neuron. They are found in the retina of the eye, in the spiral ganglion of the inner ear, and in the olfactory pathway. *Unipolar neurons* originate in the embryo as bipolar neurons, but in the course of development their two processes become fused into one for a short distance beyond the cell body. Then they separate into clearly distinguishable axon and dendrite. Sensory neurons, like the one shown in Figure 10-7, are usually unipolar. How would you classify the interneuron shown in this figure? Is it unipolar, bipolar, or multipolar? Are motoneurons multipolar?

Structure

All neurons consist of a cell body (also called the *soma* or *perikaryon*) and at least two processes: one **axon** and one or more **dendrites.** Because dendrites and axons are threadlike extensions from a neuron's soma (its cell body), they are often called *nerve fibers.*

Clusters of neuron cell bodies have a slightly gray color. In the brain and spinal cord, clusters of neuron cell bodies are called *nuclei* but outside the brain and cord they are called *ganglia. Gray matter* in the brain and spinal cord consists of nuclei, that is, clusters of neuron cell bodies. In many respects the cell body, the largest part of a nerve cell, resembles other cells. It contains a nucleus, cytoplasm, and various organelles found in other cells, for example, mitochondria and a Golgi apparatus (named for Golgi, an Italian histologist [1844-1926], who first saw this apparatus in neurons). A neuron's cytoplasm extends through its cell body and its processes. A plasma membrane encloses the entire neuron.

Certain structures—dendrites, axons, neurofibrils, Nissl bodies, myelin sheath, and neurilemma—are found only in neurons. The following paragraphs describe them briefly.

Dendrites, as you can see in the multipolar neuron in Figures 10-2 and 10-3, branch extensively—like tiny trees. In fact, their name derives

Figure 10-2 Diagram of multipolar (**A**), bipolar (**B**), and unipolar neurons (**C**).

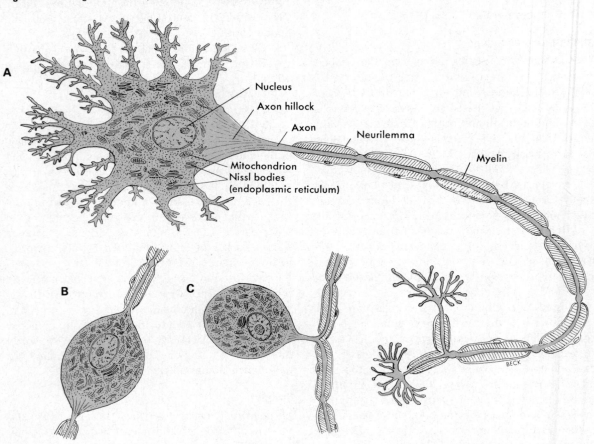

from the Greek word for tree. The distal ends of dendrites of sensory neurons are called *receptors* because they receive the stimuli that initiate conduction. Dendrites conduct impulses to the cell body of the neuron.

The **axon** of a neuron is a single process that extends out from the neuron cell body. Although a neuron has only one axon, it often has one or more side branches *(axon collaterals)*. Moreover, axons terminate in many branched filaments (telodendria). The endings of these filaments contain numerous vesicles and mitochondria.

Axons vary in both length and diameter. Some are a meter long. Some measure only a few millimeters. Axon diameters also vary considerably, from about 20 μm down to about 1 μm—a point of interest because axon diameter relates to velocity of impulse conduction. In general, the larger the diameter, the more rapid the conduction. A neuron's axon conducts impulses away from its cell body.

Neurofibrils are very fine fibers extending through dendrites, cell bodies, and axons. Electron micrographs indicate that neurofibrils consist of

bundles of still thinner fibers (i.e., microtubules and microfilaments) structures present also in the cytoplasm of other cells.

Nissl bodies consist of layers of small pieces of the endoplasmic reticulum with many ribosomes lying between them. Seen with the light microscope, Nissl bodies appear as rather large granules widely scattered through the cytoplasm of the neuron cell body but not present in the axon or the axon hillock (part of the neuron cell body from which the axon emerges). Nissl bodies specialize in protein synthesis. They provide the protein needed for maintaining and regenerating neuron processes and for renewing chemicals involved in the transmission of nerve impulses from one neuron to another.

The **myelin sheath,** a segmented wrapping around an axon, consists of a fatty substance called *myelin.* As you read the rest of this paragraph, refer frequently to Figure 10-4. One segment of a myelin sheath extends from one node of Ranvier to the next. One Schwann cell forms one segment of the myelin sheath around an axon located in a nerve. It occurs in this way. A loop of a Schwann cell's plasma mem-

Figure 10-3 Multipolar neurons.

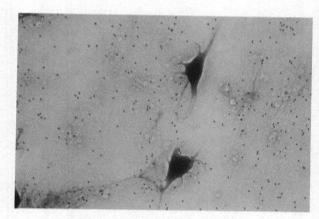

brane pushes inward toward the section of axon adjacent to it and wraps itself in jelly roll fashion around that section. This inward movement of the loop of plasma membrane squeezes the rest of the cell in the opposite direction. It pushes the Schwann cell's nucleus and a flattened layer of its cytoplasm outward to form the neurilemma. The several rolled-up layers of the Schwann cell's plasma membrane lying inside the neurilemma constitute the myelin sheath.

Oligodendroglia, instead of Schwann cells, form the myelin sheaths around central nerve fibers, that is, axons located in the brain and cord. Fibers that have a myelin sheath are called *myelinated fibers,* and those that have only a thin layer of myelin are called *unmyelinated.* Because of the high fat content of myelin, bundles of myelinated fibers have a creamy white color, and they make up the white matter of the nervous system. *White matter* in the brain and spinal cord is composed of tracts. A *tract* is a bundle of myelinated axons. White matter found outside the brain and cord consists of *nerves.* Nerves are cordlike structures composed of bundles of myelinated axons.

The **neurilemma,** or sheath of Schwann, is the delicate outer covering around axons located outside of the brain and cord in nerves (see Figure 10-4). Axons in the brain and cord have no neurilemma. The neurilemma plays an essential part in the regeneration of cut and injured axons. Therefore axons in the brain and cord do not regenerate but those in nerves do.

Figure 10-4 Diagram of a nerve fiber and its coverings. Note these features of the myelin sheath: its concentric layers surround the nerve fiber—the axon in this figure—and it is segmented. One segment is the section of myelin sheath located between two successive nodes of Ranvier. The neurilemma surrounds the myelin sheath and is continuous, not segmented.

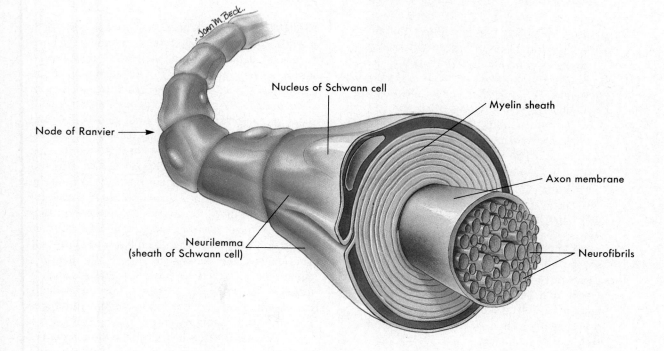

Node of Ranvier

Nucleus of Schwann cell

Myelin sheath

Axon membrane

Neurilemma (sheath of Schwann cell)

Neurofibrils

MULTIPLE SCLEROSIS (MS)

A number of diseases are associated with disorders of the oligodendroglia. Because these neuroglia cells are involved in myelin formation, the diseases as a group are called **myelin disorders.** The most common primary disease of the central nervous system is a myelin disorder called **multiple sclerosis** or **MS.** It is characterized by myelin loss and destruction accompanied by varying degrees of oligodendroglial cell injury and death. The end result is demyelination throughout the white matter of the central nervous system. Hard plaquelike lesions replace the destroyed myelin and affected areas are invaded by inflammatory cells. As the myelin surrounding axons is lost, nerve conduction is impaired and weakness, incoordination, visual impairment, and speech disturbances occur. Although the disease occurs in both sexes and all age groups, it is most common in women between 20 and 40 years.

The cause of multiple sclerosis is thought to be related to autoimmunity and to viral-type infections in some individuals. MS is characteristically relapsing and chronic in nature, but some cases of acute and unremitting disease have been reported. In most instances the disease is prolonged, with remissions and relapses occurring over a period of many years. There is no known cure.

PHYSIOLOGY OF NEURONS

DEFINITIONS

To understand the physiology of neurons, you will need to familiarize yourself with the meanings of certain terms (see boxed material below).

Potential difference—An electrical difference; an electrical gradient, to use another term. A potential difference is a difference between the amounts of electrical charge present at two points. A potential difference is a form of potential energy. It is a force that has the power to move positively charged ions down an electrical gradient, that is, from a point with a higher positive charge to a point with a lower positive charge. The magnitude of a potential difference is measured in volts or millivolts (mV).

Polarized membrane—Membrane whose outer and inner surfaces bear different amounts of electrical charges. In short, a potential difference exists across a polarized membrane.

Depolarized membrane—Membrane whose outer and inner surfaces bear equal amounts of electrical charges. A potential difference does not exist across a depolarized membrane; it is zero.

RESTING POTENTIAL

DEFINITION. When a neuron is not conducting impulses (when it is "resting"), the inner surface of its plasma membrane is slightly negative—typically about 70 mV negative—to its outer surface (Figure 10-5, *top*). This potential difference across a nonconducting neuron's plasma membrane is called the *resting potential.*

MECHANISM. The mechanism that produces the resting potential is one that creates an ionic imbalance across the neuron's plasma membrane. Specifically, it produces a minute excess of negative ions on the membrane's inner surface and of positive ions on its outer surface. The mechanism consists primarily of the sodium-potassium pump built into the plasma membrane. This pump actively transports positive sodium and potassium ions through the plasma membrane in opposite directions and at different rates. It moves three sodium ions out of a neuron for every two potassium ions it moves into it. If, for instance, the pump transports 100 potassium ions into a nerve cell from the extracellular fluid, it concurrently transports 150 sodium ions out of the cell. This makes the inner surface of the neuron's membrane slightly less positive—that is, slightly negative—to its outer surface. Thus the sodium-potassium pump creates the potential difference known as the resting potential.

Other factors that help produce the resting potential by making the inner surface of the neuron's plasma membrane slightly negative to its outer sur-

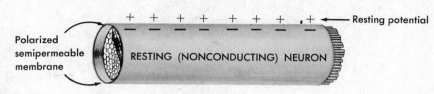

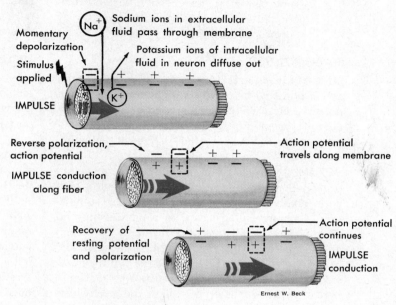

Figure 10-5 Upper diagram represents polarized state of the membrane of a nerve fiber when it is not conducting impulses. The other diagrams represent nerve impulse conduction followed by repolarization.

face are the membrane's permeability characteristics. For one thing, it is much more permeable to potassium than to sodium ions. Therefore potassium diffuses more rapidly out of the neuron's intracellular fluid than sodium ions diffuse into it from the extracellular fluid. This tends to make the inner surface of the neuron membrane less positive (i.e., somewhat negative) to the outer surface. Also, the neuron's plasma membrane is almost impermeable to the negative ions inside the neuron. Therefore fewer negative than positive ions diffuse out of the neuron's intracellular fluid. This leaves a slight excess of negative ions on the inner surface of the neuron's membrane. Briefly, then, both the high permeability of the neuron's plasma membrane to potassium ions and its relative impermeability to negative ions help create the resting potential.

ACTION POTENTIAL

DEFINITION. An action potential is, as the term itself suggests, the potential across the membrane of an active neuron, that is, one that is conducting an impulse. A synonym commonly used for action potential is nerve impulse.

MECHANISM. A step-by-step description of the mechanism that produces an action potential follows. Refer to Figure 10-5 as you read each step.

1 When an adequate stimulus is applied to a neuron, it greatly increases the permeability of its membrane to sodium ions at the point of stimulation.

2 The positive sodium ions rush inward at the stimulated point of the membrane. The excess of positive ions outside at this point, therefore, diminishes. It quickly reaches zero. In other words, the stimulated point of the membrane is no longer polarized. It has become **depolarized**—but only for an instant. As sodium ions continue streaming inward, they produce, within milliseconds, an excess of positive ions inside sufficient to produce an action potential. An **action potential** is a potential difference across a neuron's membrane of about 30 mV with the inside positive to the outside. Compared with the resting potential in which the inside of the membrane is 70 mV negative to the outside, the action potential is *reverse polarization*. Development of the action potential at the stimulated point of a neuron's membrane marks the initiation of impulse conduction by the neuron.

3 *Reverse polarization* of the stimulated region of a neuron's membrane causes a local current flow to the adjacent region of the

membrane. Here, the local current flow acts as a stimulus. It causes the adjacent membrane to depolarize and develop an action potential. Thus the action potential has moved from the point originally stimulated to the adjacent point on the membrane. As the cycle goes on repeating itself over and over again in rapid succession, the action potential travels point by point out to the neuron's axon endings.

REPOLARIZATION

DEFINITION. **Repolarization** is the replacement of the action potential by the resting potential. An action potential lasts about a millisecond at a point on a neuron's membrane before repolarization—restoration of the resting potential—occurs at that point.

MECHANISM. A change in membrane permeability brings about repolarization. The membrane's high permeability to sodium ions (induced by stimulation) lasts somewhat less than a millisecond and then changes to relative impermeability to sodium and high permeability to potassium. Because its concentration is much greater inside the cell than outside, potassium rapidly diffuses outward. Equal numbers of negative ions do not follow the positive potassium ions outward because the membrane is relatively impermeable to negative ions. An excess of positive ions therefore develops on the membrane's outer surface and an excess of negative ions on its inner surface. In a word, repolarization has occurred.

Figure 10-6 Nerve impulse represented by changes in voltage across axon membrane—the action potential.

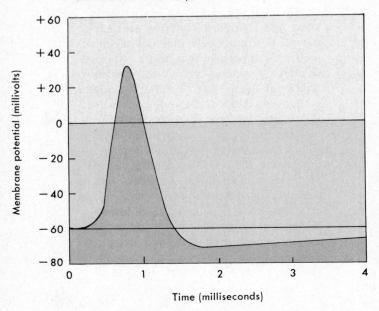

REFRACTORY PERIOD

The refractory period is a brief space of time during which a neuron is refractory, that is, resistant to stimulation. For several seconds after an action potential, a neuron will not respond to any stimulus, no matter how strong. This is called the **absolute refractory period**. The **relative refractory period** is the few milliseconds following the absolute refractory period. During the relative refractory period a neuron will respond only to very strong stimuli.

THRESHOLD OF STIMULATION

The term **threshold of stimulation** means the potential that triggers initiation of an action potential (nerve impulse). The threshold of stimulation of postsynaptic neurons under normal conditions is -59 mV. In other words, this threshold of stimulation is 11 mV less negative (or 11 mV more positive) than the usual resting potential of -70 mV. Once the threshold is reached, an action potential (about $+30$ mV) almost instantaneously develops. Figure 10-6 shows these changes in potential and the time required for them to develop. Stimuli just strong enough to produce a threshold potential and thereby initiate impulse conduction are called *threshold stimuli*. *Subthreshold stimuli* are weak stimuli that decrease a neuron's negativity slightly but not to the threshold level. Hence subthreshold stimuli do not initiate conduction.

SALTATORY CONDUCTION

Nonmyelinated fibers conduct impulses by the point-to-point progression of the axon potential described earlier. Myelinated fibers use a somewhat different method called saltatory conduction (from the Latin verb *saltare*, "to leap"). In **saltatory conduction** the action potential leaps from one node of Ranvier to the next instead of progressing more slowly from point to point along an axon. The explanation for saltatory conduction lies in the insulating character of the myelin sheath. It resists ion permeability and local current flow. Therefore action potentials do not develop beneath the myelin sheath. They occur at gaps in the myelin sheath, that is, at the nodes of Ranvier. The action potential seemingly leaps from one node to the next. Hence saltatory conduction is a faster method of impulse conduction than the point-to-point propagation of the action potential.

SPEED OF IMPULSE CONDUCTION

How fast does a neuron conduct impulses? It depends on the diameter of its axon and on its having or lacking a myelin sheath. The principles are these: the larger the diameter of an axon, the faster it conducts impulses. Myelinated fibers conduct more rapidly than unmyelinated fibers. The largest diameter

Figure 10-7 Three-neuron ipsilateral reflex arc, consisting of an afferent (sensory) neuron, an interneuron, and a motoneuron (efferent neuron). Note the presence of two synapses in this arc: (1) between sensory neuron axon terminals and interneuron dendrites and (2) between interneuron axon terminals and motoneuron dendrites and cell bodies (located in anterior gray matter). Nerve impulses traversing such arcs produce many spinal reflexes. Example: withdrawing the hand from a hot object.

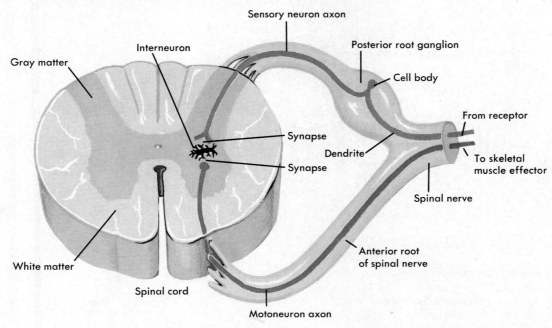

fibers—those that conduct most rapidly—are called *A fibers.* Impulses travel along them at a top speed of about 130 m/sec or close to 300 miles/hr. The smallest diameter, slowest conducting fibers are called *C fibers.* They conduct at a speed of about 0.5 m/sec, or a little faster than 1 mile/hr. *B fibers* are intermediate to A and C fibers in diameter and in conduction speed.

ROUTES OF IMPULSE CONDUCTION

The route traveled by many nerve impulses is called a *reflex arc.* Basically, a **reflex arc** is an impulse conduction route to and from the central nervous system (the brain and spinal cord). In its simplest form, a reflex arc consists of afferent neurons and efferent neurons; this is called a *two-neuron arc.* The most common form of reflex arc is the three-neuron arc. It consists of afferent neurons, interneurons, and efferent neurons. *Afferent* or *sensory neurons* conduct impulses to the central nervous system from the periphery (any part of the body outside of the central nervous system). *Efferent neurons,* or *motoneurons,* conduct impulses from the central nervous system to effectors. An *effector* is muscle tissue or glandular tissue. *Interneurons* conduct impulses from afferent neurons toward or to motoneurons. In essence, a reflex arc is an impulse

conduction route from receptors to the central nervous system and out to effectors.

Now look at Figure 10-7. Note the two labels for synapse. A **synapse** is the place where nerve impulses are transmitted from one neuron to another. Synapses are located between the axon terminals on one neuron and the dendrites or cell body of another neuron. For example, in Figure 10-7 the first synapse lies between the sensory neuron's axon terminals and the interneuron's dendrites. The second synapse lies between the interneuron's axon terminals and the motoneuron's dendrites. This reflex arc is called an *ipsilateral reflex arc* because the receptors and effectors are located on the same side of the body. Figure 10-8 shows a *contralateral reflex arc,* one whose receptors and effectors are located on opposite sides of the body.

Besides simple two-neuron and three-neuron arcs, intersegmental arcs (Figure 10-9)—even more complex multineuron, multisynaptic arcs—also exist. An important principle is this: all impulses that start in receptors do not invariably travel over a complete reflex arc and terminate in effectors. Many impulses fail to be conducted across synapses. Moreover, all impulses that terminate in effectors do not invariably start in receptors. Many of them, for example, are thought to originate in the brain.

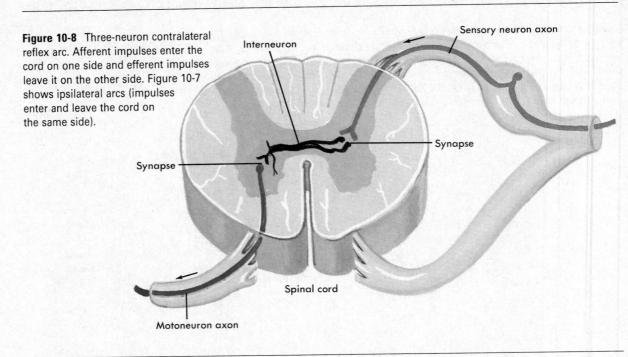

Figure 10-8 Three-neuron contralateral reflex arc. Afferent impulses enter the cord on one side and efferent impulses leave it on the other side. Figure 10-7 shows ipsilateral arcs (impulses enter and leave the cord on the same side).

Interneuron

Sensory neuron axon

Synapse

Synapse

Spinal cord

Motoneuron axon

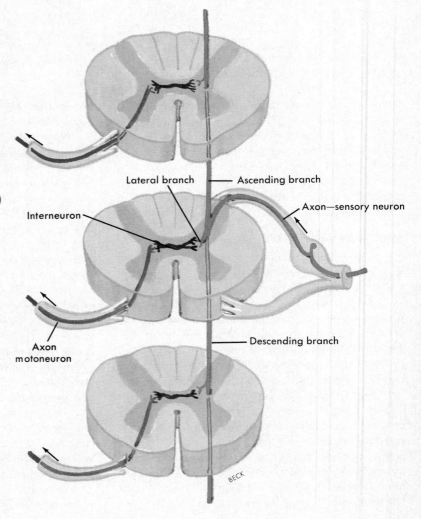

Figure 10-9 Intersegmental contralateral reflex arcs showing a sensory fiber splitting into ascending and descending branches that give rise to lateral branches that synapse with their respective interneurons. These and ipsilateral intersegmental arcs make possible the activation of more than one effector by impulses over a single sensory fiber and account for the phenomenon of *divergence*. Divergence makes it possible for a single incoming sensory impulse to "trigger" many effector cell responses. (See Figure 11-30.)

Lateral branch

Ascending branch

Axon—sensory neuron

Interneuron

Axon motoneuron

Descending branch

BECK

Figure 10-10 Synaptic knobs (end feet or end buttons) on motoneuron cell body.

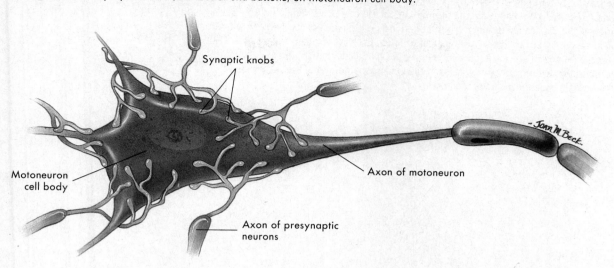

CONDUCTION ACROSS SYNAPSES

DEFINITION AND STRUCTURE OF A SYNAPSE. A **synapse** is the place where impulses are transmitted from one neuron, called the *presynaptic neuron*, to another neuron, called the *postsynaptic neuron*. Three structures make up a synapse: a synaptic knob, a synaptic cleft, and the plasma membrane of a postsynaptic neuron. A *synaptic knob* is a tiny bulge at the end of a terminal branch of a presynaptic neuron's axon (Figures 10-10 and 10-11). Each synaptic knob contains numerous small sacs or vesicles. Each vesicle contains about 10,000 molecules of a chemical compound called a **neurotransmitter.** A *synaptic cleft* is the space between a synaptic knob and the plasma membrane of a postsynaptic neuron. It is an incredibly narrow space—only 200 to 300 Å or about one millionth of an inch in width. Identify the synaptic cleft in Figure 10-11. The plasma membrane of a *postsynaptic neuron* has protein molecules embedded in it opposite each synaptic knob. These serve as receptors to which neurotransmitter molecules bind.

MECHANISM OF CONDUCTION. An action potential that has traveled the length of a neuron stops at its axon terminals. Action potentials cannot cross synaptic clefts, miniscule barriers though they are. Instead, chemical compounds called *excitatory neurotransmitters* cross the synaptic clefts and bring about conduction by postsynaptic neurons. Here is one definition of the term, **excitatory neurotransmitter:** a chemical capable of initiating impulse conduction by postsynaptic neurons. The mechanism of synaptic conduction consists of the following sequence of events:

1 When an action potential reaches a synaptic knob, calcium ions diffuse rapidly into it and its membrane permeability to calcium ions increases. As calcium concentration increases, neurotransmitter vesicles move to the knob's surface and open by fusing with its plasma membrane. Thousands of neurotransmitter molecules spurt out of the open vesicles into the synaptic cleft.

2 The released neurotransmitter molecules almost instantaneously diffuse across the narrow synaptic cleft and contact the postsynaptic neuron's plasma membrane. Here the neurotransmitters bind to their receptors. This leads to opening of sodium and potassium channels in the membrane. More sodium ions move into the postsynaptic neuron than potassium ions move out of it. Because of this net addition of positive ions to the interior of the postsynaptic neuron, its potential becomes less negative; it moves toward zero from its resting level. If a sufficient amount of excitatory neurotransmitter enters the synaptic cleft, the postsynaptic neuron's potential changes from − 70 mV (resting potential) through zero (depolarization) and then to + 30 mV (action potential or nerve impulse). By bridging the synaptic cleft, the neurotransmitter has brought about conduction across the synapse. For a summary in diagram form of this mechanism, look now at Figure 10-12, p. 295.

3 Once impulse conduction by postsynaptic neurons is initiated, neurotransmitter activity is rapidly terminated. Either one or both

Figure 10-11 Diagram of a synapse showing its components: a synaptic knob (axon terminal) of a presynaptic neuron, the plasma membrane of a postsynaptic neuron, and a synaptic cleft (space about one millionth of an inch wide) between the two neurons. On the arrival of an action potential at a synaptic knob, neurotransmitter molecules are released from vesicles in the knob into the synaptic cleft. The combining of neurotransmitter molecules with receptor molecules in the plasma membrane of the postsynaptic neuron initiates impulse conduction by it.

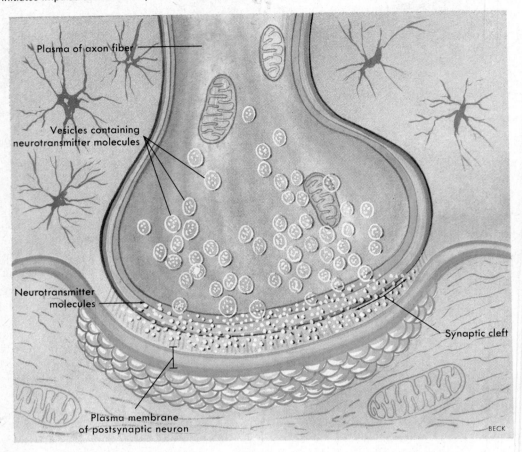

of two mechanisms bring this about. Some neurotransmitter molecules diffuse out of the synaptic cleft back into synaptic knobs. Other neurotransmitter molecules are metabolized into inactive compounds by specific enzymes.

FACILITATION

Excitatory neurotransmitters released at a synapse may either initiate impulse conduction by the postsynaptic neuron (as described previously) or they may facilitate it. A **facilitated neuron** is one that has developed an excitatory potential—usually referred to as EPSP (excitatory postsynaptic potential). The EPSP lies closer to zero than the usual resting potential but farther from zero than the threshold of stimulation. Here are some examples: resting potential −70 mV, EPSP −65 mV, threshold of stimulation −59 mV, and action potential +30 mV.

The effect of an excitatory neurotransmitter on a postsynaptic neuron—either facilitation or impulse conduction—depends largely on the amount of neurotransmitter released into a synapse. In general, the small amount of neurotransmitter released from a few synaptic knobs facilitates the postsynaptic neuron, and the larger amounts released from many knobs initiate impulse conduction.

INHIBITION

Not all neurotransmitters produce an excitatory effect, an EPSP, in postsynaptic neurons. Some are **inhibitory neurotransmitters.** They produce an *inhibitory postsynaptic potential*, an IPSP. An IPSP is more negative, that is, lies farther from zero than a resting potential. For example, an IPSP might be

Figure 10-12 A rapid-acting mechanism by which excitatory neurotransmitters may open sodium channels in a postsynaptic neuron's membrane and thereby either facilitate the neuron or initiate an action potential, that is, nerve impulse conduction by it.

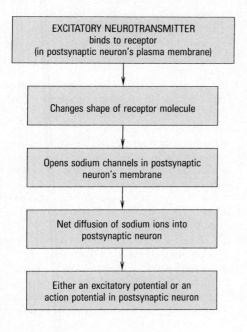

EXCITATORY NEUROTRANSMITTER
binds to receptor
(in postsynaptic neuron's plasma membrane)

Changes shape of receptor molecule

Opens sodium channels in postsynaptic neuron's membrane

Net diffusion of sodium ions into postsynaptic neuron

Either an excitatory potential or an action potential in postsynaptic neuron

-75 mV whereas a typical resting potential is -70 mV. In our next chapter we shall name some excitatory and inhibitory neurotransmitters and present some current information about them.

SUMMATION

Spatial summation is the effect produced by simultaneous stimulation of a number of synaptic knobs on the same postsynaptic neuron. The effect produced by stimulation in rapid succession of knobs on the same neuron is called **temporal summation.** At least several, usually thousands, and in some cases more than 100,000 knobs synapse with a single postsynaptic neuron. The amount of excitatory neurotransmitter released by one knob does not trigger impulse conduction. It does, however, change the postsynaptic neuron's potential in the direction of its threshold level. It produces an EPSP in the neuron. When a number of knobs are activated simultaneously or in rapid succession, the effects of the neurotransmitter released add together—they summate—and thereby may initiate impulse conduction. Usually both excitatory and inhibitory transmitters are released at the same synapse. Summation of their opposing effects occurs. If the excitatory transmitter predominates, impulse conduction may result. If the inhibitory transmitter predominates, inhibition of the postsynaptic neuron results. An inhibitory potential (IPSP) develops in it.

Outline Summary

MAIN TYPES OF CELLS

A Neuroglia—connective tissue cells in nervous system organs

B Neurons—nerve cells

Neuroglia

A Types—astrocytes, oligodendroglia, microglia, and Schwann cells

B Structure and function

1 Astrocytes—large, star shaped; form tight sheaths around brain capillaries which with tight junctions between capillary endothelial cells constitute blood-brain barrier

2 Oligodendroglia—fewer processes than other two types of neuroglia; support neurons; produce myelin sheath around nerve fibers in the brain and cord

3 Microglia—small cells but enlarge and move about in inflamed brain tissue; carry on phagocytosis

4 Schwann cells—type of neuroglia found only in nerves; form myelin sheath and neurilemma around nerve fibers outside brain and cord

Neurons

A Types

1 Classified according to direction of impulse conduction

  a Afferent or sensory—conduct impulses to cord or brain

  b Efferent or motoneurons—conduct impulses away from brain or cord to or toward muscle or glandular tissue

  c Interneurons—conduct from afferent neurons to motoneurons

2 Classified according to number of processes

  a Multipolar—one axon and several dendrites

  b Bipolar—one axon and one dendrite

  c Unipolar—one process comes off neuron cell body but divides almost immediately into one axon and one dendrite

B Structure

1 Soma, or perikaryon—cell body of neuron

2 Gray matter—clusters of neuron cell bodies

3 Nerve fibers—dendrites or axons

4 Dendrites—branching process of a neuron; conduct impulses to its cell body

5 Receptors—distal ends of dendrites of sensory neurons

Outline Summary—cont'd

6 Axon—single process of a neuron, but may have collateral branches; conducts impulses away from neuron cell body

7 Neurofibrils—fine fibers extending through dendrites, cell bodies, and axons; seem to consist of bundles of thinner fibers, that is, microtubules and microfilaments

8 Nissl bodies—layered fragments of the endoplasmic reticulum; appear as good-sized granules scattered through cytoplasm of neuron cell body; specialize in synthesizing proteins for maintaining and regenerating neuron processes and for renewing neurotransmitters

9 Myelin sheath—segmented wrapping around a nerve fiber; segments separated by nodes of Ranvier; Schwann cells form myelin sheaths around peripheral nerve fibers; oligodendroglia form myelin sheaths around central nerve fibers, that is, those located in the brain and cord

10 White matter—consists of myelinated fibers

11 Tracts—bundles of myelinated fibers located in the brain and cord

12 Nerves—bundles of myelinated fibers located outside the brain and cord

13 Neurilemma or sheath of Schwann—continuous sheath enclosing myelin sheath; neurilemma plays essential part in regenerating injured or cut nerve fibers; brain and cord fibers do not have neurilemma, so presumably do not regenerate

PHYSIOLOGY OF NEURONS

A Definitions
1 Potential difference—difference in electrical charge or voltage
2 Polarized membrane—one whose outer and inner surfaces bear different electrical charge
3 Depolarized membrane—one with zero potential difference across its membrane

B Resting potential
1 Definition—potential across nonconducting neuron's plasma membrane; usually −70 mV (inner surface of membrane negative)
2 Mechanism
  a Sodium pump transports sodium ions out of neuron's intracellular fluid into surrounding extracellular fluid, whereas potassium pump transports about one third as many potassium ions into neuron's intracellular fluid as sodium pump transports out of it
  b Because neuron cell membrane is much more permeable to potassium than to sodium, more potassium diffuses back out of the intracellular fluid than sodium diffuses back in
  c Because neuron cell membrane is relatively impermeable to intracellular negative ions, the number of negative ions diffusing out of the cell is less than the number of positive (potassium)

ions diffusing out; result—minute excess of negative ions inside neuron cell membrane and minute excess of positive ions outside it; this ionic imbalance across neuron cell membrane produces resting potential; steps a and b also contribute to minute excess of positive ions outside neuron cell membrane

C Action potential
1 Definition—potential across membrane of an active (conducting) neuron; synonym for action potential is nerve impulse
2 Mechanism
  a Stimulus increases permeability of neuron membrane to sodium ions
  b Rapid inward diffusion of positive sodium ions causes membrane's inner surface to become positive to outer surface (usually about +30 mV); this reverse polarization of membrane is the action potential
  c +30 mV at point of stimulation; this reverse polarization is the action potential
  d Reverse potential of stimulated region of membrane causes local current flow to adjacent region where it acts as a stimulus, causing adjacent region to develop an action potential; cycle repeats itself over and over in rapid succession, causing action potential to travel point to point to neuron's axon terminals

D Repolarization
1 Definition—replacement of action potential by resting potential
2 Mechanism—neuron's plasma membrane again becomes relatively impermeable to sodium but highly permeable to potassium; more positive ions (K) move out through membrane than move in, thereby restoring resting potential

E Refractory period—absolute refractory period follows action potential, lasts a few milliseconds; during this time neuron will not respond to any strength stimulus; during relative refractory period that follows the absolute refractory period, neuron will respond to strong stimuli

F Threshold of stimulation—potential that triggers impulse conduction, example, −59 mV; once threshold is reached, potential changes almost instantaneously to an action potential (example, +30 mV)

G Saltatory conduction—occurs in myelinated fibers; action potential moves from one node of Ranvier to the next instead of from point to point on neuron's membrane

H Speed of impulse conduction—directly related to diameter of axon and to myelinization; largest diameter axons (A fibers) conduct most rapidly, smallest fibers (C fibers) conduct most slowly; myelinated fibers conduct more rapidly than unmyelinated fibers

Outline Summary—cont'd

I Routes of impulse conduction

  1 Many, but by no means all, impulses are conducted over route known as reflex arc, that is, two or more neurons that conduct impulses from periphery to spinal cord or brain stem and back to periphery; impulse begins in receptors and ends in effectors; receptors—distal ends of sensory neurons; effectors—muscle or glandular cells

  2 Two-neuron or monosynaptic reflex arc—simplest arc possible; consists of at least one sensory neuron, one synapse, and one motoneuron (synapse is contact region between axon terminals of one neuron and dendrites or cell body of another neuron)

  3 Three-neuron arc—consists of at least one sensory neuron, synapse, interneuron, synapse, and motoneuron; two synapses in three-neuron arc

  4 Complex, multisynaptic neural pathways also exist; many not clearly understood

J Conduction across synapses

  1 Definition of synapse—site where impulses are transmitted from a presynaptic to a postsynaptic neuron

  2 Structure of synapse—consists of

    a Synaptic knobs—tiny distentions at presynaptic neuron's axon terminals; numerous vesicles containing neurotransmitter molecules present in each synaptic knob

    b Synaptic cleft—microscopic space between a synaptic knob and a postsynaptic neuron's dendrite or cell body; neuroeffector junctions—microscopic spaces between axon terminals of motoneurons and effector cells; neuromuscular junctions (nerve-muscle synapses) and neuroglandular junctions are types of neuroeffector junctions

    c Postsynaptic neuron's plasma membrane—protein molecules embedded in membrane function as neurotransmitter receptors and as enzymes

  3 Synaptic conduction mediated by excitatory neurotransmitters

    a When action potential reaches axon terminals, neurotransmitter molecules are released from synaptic knob vesicles into synaptic cleft

    b Neurotransmitter molecules diffuse across cleft, bind to specific receptors in postsynaptic neuron's membrane, opening channels in it, through which sodium ions diffuse into and potassium ions diffuse out of interior of neuron, thereby depolarizing it and giving rise to excitatory postsynaptic potential (EPSP) or to action potential

    c Neurotransmitter activity is rapidly terminated at synapse; some neurotransmitter molecules reenter synaptic knobs, some are metabolized into inactive compounds by specific enzymes

K Facilitation—decrease in negativity of postsynaptic neuron's potential to level below its resting potential level but above its threshold of stimulation

L Inhibition—increase in the negativity of the postsynaptic neuron's membrane potential above its usual resting potential; this increased potential called the *inhibitory postsynaptic potential* (IPSP); result of summation with excess of inhibitory transmitter in synaptic cleft

M Summation—adding together of effects of excitatory and inhibitory neurotransmitters released from various knobs that synapse with a single postsynaptic neuron; effect of summation may be either facilitation, impulse conduction, or inhibition of postsynaptic neuron

Review Questions

1 In a word or two, what general function does the nervous system perform for the body?

2 Name another system that serves the same general function.

3 Compare neurons and neuroglia as to numbers, types, structure, and function.

4 Differentiate between afferent neurons, efferent neurons, and interneurons.

5 Differentiate between myelin sheath and neurilemma.

6 Differentiate between white matter and gray matter, tracts and nerves.

7 Differentiate between polarized membrane and depolarized membrane.

8 Differentiate between resting potential, excitatory potential, inhibitory potential, and action potential.

9 The term *receptor* has one meaning when used in reference to sensory neurons and another meaning when used in reference to postsynaptic neurons. Differentiate between these two meanings.

10 What are effectors?

11 Define briefly: facilitation, inhibition, summation.

12 Explain the meanings of these terms: repolarization, refractory period, saltatory conduction.

13 Describe the series of events that mediate conduction across synapses.

14 Describe the structure of a synapse.

THE CENTRAL NERVOUS SYSTEM, SPINAL NERVES, AND CRANIAL NERVES

OBJECTIVES

After you have completed this chapter, you should be able to:

1 List the primary divisions of the nervous system.
2 Identify and locate the layers of the meninges.
3 Discuss the formation, circulation, and functions of cerebrospinal fluid.
4 Discuss the location and generalized structure of the spinal cord.
5 List and give one primary function of the major ascending and descending tracts of the spinal cord.
6 Discuss the generalized structure or branchings of a typical spinal nerve.
7 List the six major divisions of the brain.
8 Discuss the structure and functions of the brain stem.
9 Identify the cranial nerves by name and give the generalized function of each.
10 Discuss the functions of the cerebellum related to the control of skeletal muscles.
11 Identify and discuss the primary functions of the two major components of the diencephalon.
12 Describe the structure and functions of the cerebrum.
13 Discuss the synthesis, storage, release, activation, and inactivation of neurotransmitters.
14 Compare and contrast somatic sensory and somatic motor pathways.
15 List and discuss several of the somatic reflexes of clinical importance.

KEY TERMS

Afferent (AF-er-ent)

Autonomic nervous system
(aw-to-NOM-ik NER-vus SIS-tem)

Central nervous system (SEN-tral NER-vus SIS-tem)

Cerebellum (ser-e-BEL-um)

Cerebrospinal fluid (ser-e-bro-SPI-nal floo-id)

Cerebrum (SER-e-brum)

Consciousness (KON-shus-nes)

Ganglion (GANG-gle-on)

Hypothalamus (hi-po-THAL-ah-mus)

Medulla (me-DUL-ah)

Meninges (me-NIN-jez)

Neurotransmitter (nu-ro-TRANS-mit-er)

Peripheral nervous system
(PER-if-er-al NER-vus SIS-tem)

Plexus (PLEK-sus)

Reflex (RE-fleks)

Sensory (SEN-so-re)

Somatic nervous system
(so-MAT-ik NER-vus SIS-tem)

Thalamus (THAL-ah-mus)

DIVISIONS OF NERVOUS SYSTEM

From the titles of Chapters 11 and 12 you might infer that your body has more than one nervous system. It does not. It has only one nervous system, but to discuss this highly complicated set of organs, we clearly must divide it up some way. Nervous system organs may be separated by location into two divisions, namely, the **central nervous system (CNS)** and the **peripheral nervous system (PNS).** The CNS consists of the centrally located nervous system organs, that is, the brain and spinal cord. All other nervous system organs—cranial nerves, spinal nerves, autonomic nerves, and ganglia—compose the PNS.

The PNS organs consist of afferent and efferent neurons. Afferent (sensory) neurons conduct impulses to the CNS from receptors. Efferent (motor) neurons conduct impulses away from the CNS to effectors.

The nervous system may also be divided into the somatic nervous system and the autonomic nervous system, based on the type of effectors supplied. The **somatic nervous system** supplies somatic effectors (skeletal muscles). It consists of the brain, spinal cord, cranial nerves, and spinal nerves. The **autonomic nervous system** supplies autonomic effectors (smooth muscle, cardiac muscle, and glandular epithelial tissue). It consists, as the next chapter will relate, of certain nerves and ganglia.

Our plan for this chapter is to start with a description of the coverings of the brain and spinal cord. Then we shall discuss the spinal cord and its nerves and proceed upward through the lower parts of the brain to its highest part, the cerebrum. We shall also discuss brain and cord neurotransmitters and somatic sensory and motor pathways as well as some reflexes.

Figure 11-1 Meninges of the brain as seen in coronal section through the skull.

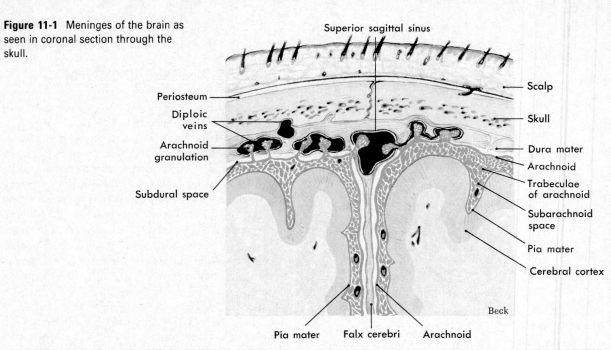

Superior sagittal sinus

Scalp

Periosteum

Skull

Diploic veins

Dura mater

Arachnoid granulation

Arachnoid

Subdural space

Trabeculae of arachnoid

Subarachnoid space

Pia mater

Cerebral cortex

Beck

Pia mater   Falx cerebri   Arachnoid

Figure 11-2 Spinal cord showing meninges, formation of the spinal nerves, and relations to a vertebra and to the sympathetic trunk and ganglia.

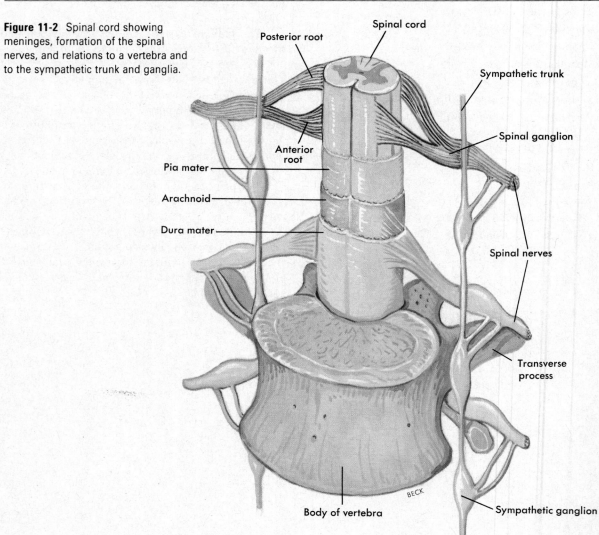

Posterior root

Spinal cord

Sympathetic trunk

Spinal ganglion

Anterior root

Pia mater

Arachnoid

Dura mater

Spinal nerves

Transverse process

Body of vertebra

BECK

Sympathetic ganglion

BRAIN AND CORD COVERINGS

Because the brain and spinal cord are both delicate and vital, nature has provided them with two protective coverings. The outer covering consists of bone: cranial bones encase the brain and vertebrae encase the cord. The inner covering consists of membranes known as **meninges.** Three distinct layers compose the meninges: the **dura mater,** the **arachnoid membrane,** and the **pia mater.** Observe their respective locations in Figures 11-1 and 11-2. The dura mater, made of strong white fibrous tissue, serves both as the outer layer of the meninges and also as the inner periosteum of the cranial bones. The arachnoid membrane, a delicate, cobwebby layer, lies between the dura mater and the pia mater or innermost layer of the meninges. The transparent pia mater adheres to the outer surface of the brain and cord and contains blood vessels.

Three extensions of the dura mater should be mentioned: the falx cerebri, falx cerebelli, and tentorium cerebelli. The **falx cerebri** projects downward into the longitudinal fissure to form a kind of partition between the two cerebral hemispheres. The **falx cerebelli** separates the two cerebellar hemispheres. The **tentorium cerebelli** separates the cerebellum from the occipital lobe of the cerebrum. It takes its name from the fact that it forms a tentlike covering over the cerebellum.

Between the dura mater and the arachnoid membrane is a small space called the subdural space, and between the arachnoid and the pia mater is the subarachnoid space; it contains cerebrospinal fluid. Inflammation of the meninges is called *meningitis.* It most often involves the arach-

noid and pia mater or the *leptomeninges,* as they are sometimes called.

The meninges of the cord continue on down inside the spinal cavity for some distance below the end of the spinal cord. The pia mater forms a slender filament known as the **filum terminale.** At the level of the third segment of the sacrum, the filum terminale blends with the dura mater to form a fibrous cord that disappears in the periosteum of the coccyx.

BRAIN AND CORD FLUID SPACES

In addition to the bony and membranous coverings, nature has further fortified the brain and spinal cord against injury by providing a cushion of fluid both around them and within them. The fluid is called **cerebrospinal fluid,** and the spaces containing it are as follows:

1 The subarachnoid space around the brain
2 The subarachnoid space around the cord
3 The ventricles and aqueduct inside the brain
4 The central canal inside the cord

The **ventricles** are cavities or spaces inside the brain. They are four in number. Two of them, the lateral (or first and second) ventricles, are located one in each cerebral hemisphere. Note in Figure 11-3 the shape of these ventricles—roughly like the hemispheres themselves. The third ventricle is little more than a lengthwise slit between the right and left thalamus. It lies beneath the midportion of the corpus callosum and longitudinal fissure. The fourth ventricle is a diamond-shaped space between the cerebellum posteriorly and the medulla and pons anteriorly. Actually, it is an expansion of the central canal of the cord after the cord enters the cranial cavity and becomes enlarged to form the medulla.

FORMATION, CIRCULATION, AND FUNCTIONS OF CEREBROSPINAL FLUID

Formation of cerebrospinal fluid occurs mainly by filtration from blood in the **choroid plexuses.** Choroid plexuses are networks of capillaries that project from the pia mater into the lateral ventricles and into the roofs of the third and fourth ventricles. From each lateral ventricle the fluid seeps through an opening, the interventricular foramen (of Monro), into the third ventricle, then through a narrow channel, the cerebral aqueduct (or aqueduct of Sylvius, Figure 11-3), into the fourth ventricle. Some of the fluid moves from the fourth ventricle directly into the central canal of the cord. Some of it moves out of the fourth ventricle through openings in its roof, two located laterally (foramina of Luschka) and one in the midline (foramen of Magendie). These openings allow cerebrospinal fluid to move into the *cisterna magna,* a space behind the medulla that is

The extension of the meninges beyond the cord is convenient for performing lumbar punctures without danger of injuring the spinal cord. A **lumbar puncture** is a withdrawal of some of the cerebrospinal fluid from the subarachnoid space in the lumbar region of the spinal cord. The physician inserts a needle just above or below the fourth lumbar vertebra, knowing that the spinal cord ends an inch or more above that level. The fourth lumbar vertebra can be easily located because it lies on a line with the iliac crest. Placing a patient on the side and arching the back by drawing the knees and chest together separates the vertebrae sufficiently to introduce the needle.

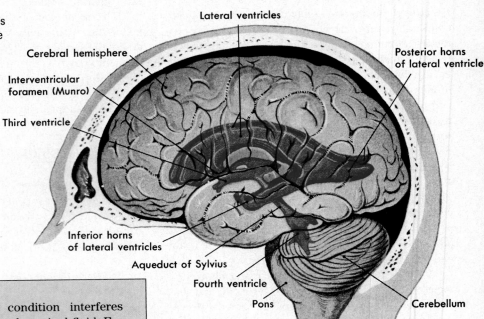

Figure 11-3 Cerebral ventricles projected on the lateral surface of the cerebrum. The smaller drawing shows the ventricles from above.

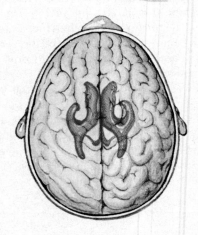

Occasionally, some condition interferes with circulation of cerebrospinal fluid. For example, a brain tumor may press against the cerebral aqueduct, shutting off the flow of fluid from the third to the fourth ventricle. In such an event the fluid accumulates within the lateral and third ventricles because it continues to form even though its drainage is blocked. This condition is known as **internal hydrocephalus** (Figure 11-4). If the fluid accumulates in the subarachnoid space around the brain, **external hydrocephalus** results. Subarachnoid hemorrhage, for example, may lead to formation of blood clots that block drainage of the cerebrospinal fluid from the subarachnoid space. With decreased drainage an increased amount of fluid remains in the space.

continuous with the subarachnoid space around the brain and cord. The fluid circulates in the subarachnoid space, then is absorbed into venous blood through the arachnoid villi (fingerlike projections of the arachnoid membrane into the brain's venous sinuses). Briefly, here is the circulation route of cerebrospinal fluid: it is formed by filtration of fluid from blood in the choroid plexuses into the ventricles of the brain, circulates through the ventricles and into the central canal and subarachnoid spaces, and is absorbed back into blood.

The amount of cerebrospinal fluid in the average adult is about 140 ml (about 23 ml in the ventricles and 117 ml in the subarachnoid space of brain and cord).

Cerebrospinal fluid serves as a protective cushion around and within the brain and cord. However, it is now known to function in other ways as well. For instance, changes in its carbon dioxide content affect neurons of the respiratory center in the medulla and thereby help control respiration.

SPINAL CORD

STRUCTURE

The spinal cord lies within the spinal cavity, extending from the foramen magnum to the lower border of the first lumbar vertebra (Figure 11-5), a distance of 17 or 18 inches in the average body. The cord does not completely fill the spinal cavity—it also contains the meninges, spinal fluid, a cushion of adipose tissue, and blood vessels.

The spinal cord is an oval-shaped cylinder that

Figure 11-4 Internal hydrocephalus.

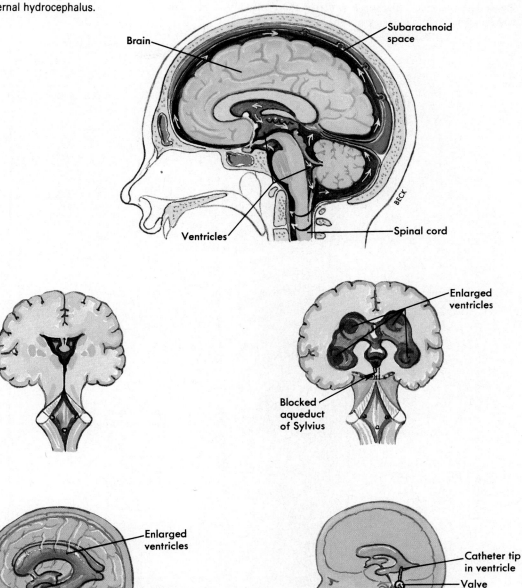

tapers slightly from above downward and has two bulges, one in the cervical region and the other in the lumbar region. Two deep grooves, the anterior median fissure and the posterior median sulcus, just miss dividing the cord into separate symmetrical halves. The anterior fissure is the deeper and the wider of the two grooves—a useful factor to remember when you examine spinal cord diagrams. It enables you to tell at a glance which part of the cord is anterior and which is posterior. **Gray matter** composes the inner core of the cord. Although it looks like a flat letter H in cross-section views of the cord,

it actually has three dimensions, since the gray matter extends the length of the cord. The limbs of the H are called anterior, posterior, and lateral horns of gray matter, or lateral gray columns. They consist predominantly of cell bodies of interneurons and motoneurons.

White matter surrounding the gray matter is subdivided in each half of the cord into three columns (or funiculi): the anterior, posterior, and lateral white columns. Each white column or funiculus consists of a large bundle of nerve fibers (axons) divided into smaller bundles called tracts, as shown in Figure

Figure 11-5 Relation of the spinal cord, part of the brain, and some of the spinal nerves to surrounding structures.

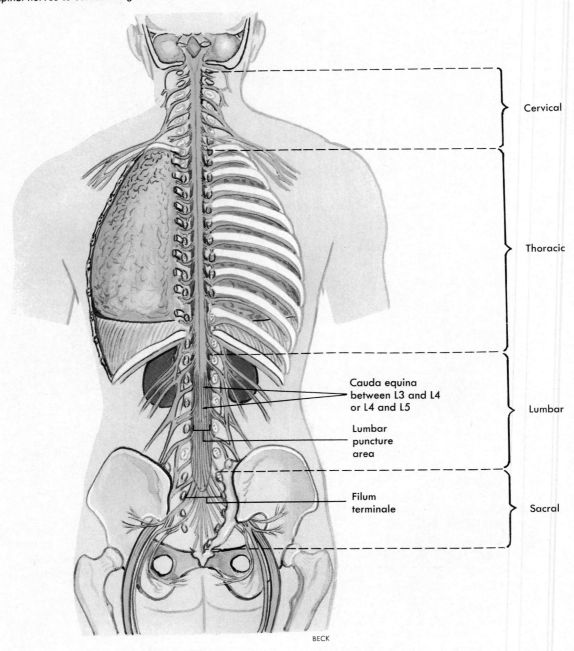

Cervical

Thoracic

Cauda equina
between L3 and L4
or L4 and L5

Lumbar
puncture
area

Lumbar

Filum
terminale

Sacral

BECK

11-6. The names of most spinal cord tracts indicate the white column in which the tract is located, the structure in which the axons that make up the tract originate, and the structure in which they terminate. Examples: the lateral corticospinal tract is located in the lateral white column of the cord, and the axons that compose it originate from neuron cell bodies located in the cortex (of the cerebrum) and terminate in the spinal cord. The ventral spinothalamic tract lies in the ventral (anterior) white column, and the axons that compose it originate from neuron cell bodies located in the spinal cord and terminate in the thalamus.

FUNCTIONS

The spinal cord performs two general functions. Briefly, it provides the two-way conduction routes to and from the brain and serves as the reflex center for all spinal reflexes.

Spinal cord tracts provide two-way conduction

Figure 11-6 Location in the spinal cord of some major projection tracts. Ascending or sensory tracts described in Table 11-1. Table 11-2 describes descending or motor tracts.

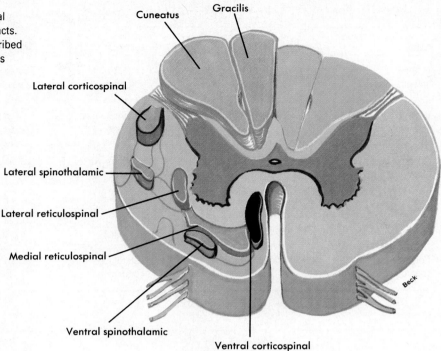

Cuneatus

Gracilis

Lateral corticospinal

Lateral spinothalamic

Lateral reticulospinal

Medial reticulospinal

Ventral spinothalamic

Ventral corticospinal

Beck

paths to and from the brain. **Ascending tracts** conduct impulses up the cord to the brain. **Descending tracts** conduct impulses down the cord from the brain. Bundles of axons compose all tracts. Tracts are both structural and functional organizations of these nerve fibers. They are structural organizations in that all of the axons of any one tract originate from neuron cell bodies located in the same structure and all of the axons terminate in the same structure. For example, all the fibers of the spinothalamic tract are axons originating from neuron cell bodies located in the spinal cord and terminating in the thalamus. Tracts are functional organizations in that all the axons that compose one tract serve one general function. For instance, fibers of the spinothalamic tracts serve a sensory function. They transmit impulses that produce our sensations of crude touch, pain, and temperature.

Because so many different tracts make up the white columns of the cord, we shall mention only a few that seem most important in humans. Locate each tract in Figure 11-6. Consult Tables 11-1 and 11-2 for a brief summary of information about tracts. Four important ascending or sensory tracts and their functions, stated very briefly, are as follows:

1 **Lateral spinothalamic tracts**—Crude touch, pain, and temperature
2 **Ventral spinothalamic tracts**—Crude touch, pain, and temperature
3 **Fasciculi gracilis and cuneatus**—Discriminating touch and conscious kinesthesia

4 **Spinocerebellar tracts**—Unconscious kinesthesia

Further discussion of the sensory neural pathways may be found on pp. 335-338.

Four important descending or motor tracts and their functions in brief are as follows:

1 **Lateral corticospinal tracts**—Voluntary movement; contraction of individual or small groups of muscles, particularly those moving hands, fingers, feet, and toes on opposite side of body
2 **Ventral corticospinal tracts**—Same as preceding except mainly muscles of same side of body
3 **Lateral reticulospinal tracts**—Transmits facilitatory impulses to anterior horn motoneurons to skeletal muscles
4 **Medial reticulospinal tracts**—Mainly inhibitory impulses to anterior horn motoneurons to skeletal muscles

The spinal cord also serves as the reflex center for all spinal reflexes. The term *reflex center* means the center of a reflex arc or the place in the arc where incoming sensory impulses become outgoing motor impulses. They are structures that switch impulses from afferent to efferent neurons. In two-neuron arcs, reflex centers are merely synapses between neurons. In all other arcs, reflex centers consist of interneurons interposed between afferent and efferent neurons. Spinal reflex centers are located in the gray matter of the cord.

Table 11-1 Major ascending tracts of spinal cord

Name	Function	Location	Origin*	Termination†
Lateral spinothalamic	Pain, temperature, and crude touch opposite side	Lateral white columns	Posterior gray column opposite side	Thalamus
Ventral spinothalamic	Crude touch, pain, and temperature	Anterior white columns	Posterior gray column opposite side	Thalamus
Fasciculi gracilis and cuneatus	Discriminating touch and pressure sensations, including vibration, stereognosis, and two-point discrimination; also conscious kinesthesia	Posterior white columns	Spinal ganglia same side	Medulla
Dorsal spinocerebellar	Unconscious kinesthesia	Lateral white columns	Posterior gray column	Cerebellum

*Location of cell bodies of neurons from which axons of tract arise.
†Structure in which axons of tract terminate.

Table 11-2 Major descending tracts of spinal cord

Name	Function	Location	Origin*	Termination†
Lateral corticospinal (or crossed pyramidal)	Voluntary movement, contraction of individual or small groups of muscles, particularly those moving hands, fingers, feet, and toes of opposite side	Lateral white columns	Motor areas of cerebral cortex (mainly areas 4 and 6, Figure 11-24) opposite side from tract location in cord	Intermediate or anterior gray columns
Ventral corticospinal (direct pyramidal)	Same as lateral corticospinal except mainly muscles of same side	Lateral white columns	Motor cortex but on same side as tract location in cord	Intermediate or anterior gray columns
Lateral reticulospinal	Mainly facilitatory influence on motoneurons to skeletal muscles	Lateral white columns	Reticular formation, midbrain, pons, and medulla	Intermediate or anterior gray columns
Medial reticulospinal	Mainly inhibitory influence on motoneurons to skeletal muscles	Anterior white columns	Reticular formation, medulla mainly	Intermediate or anterior gray columns

*Location of cell bodies of neurons from which axons of tract arise.
†Structure in which axons of tract terminate.

A **reflex** is an action that results from impulse conduction over a reflex arc. A reflex, therefore, is a response to a stimulus. Usually only unwilled or involuntary responses are called reflexes. Of all reflexes—and there are many of them—probably the most familiar is the knee jerk or patellar reflex. Two-neuron arcs mediate this spinal cord reflex. The reflex centers of these arcs lie in the cord's gray columns, in the second, third, and fourth lumbar segments.

SPINAL NERVES

STRUCTURE

Thirty-one pairs of nerves are connected to the spinal cord. They have no special names but are merely numbered according to the level of the spinal column at which they emerge from the spinal cavity. Thus there are 8 cervical, 12 thoracic, 5 lumbar, and 5 sacral pairs and 1 coccygeal pair of spinal nerves. The first cervical nerves emerge from the cord in the space above the first cervical vertebra (between it and the occipital bone). The rest of the cervical and all of the thoracic nerves pass out of the spinal cavity horizontally through the intervertebral foramina of their respective vertebrae. For example, the second cervical nerves emerge through the foramina above the second cervical vertebra. In Figure 11-7 the cervical segment of the spinal cord is viewed from its posterior aspect. The dura mater and arachnoid have been removed and the bony cervical vertebrae have been cut to better expose the emerging cervical nerves.

Lumbar, sacral, and coccygeal nerve roots, on the other hand, descend from their point of origin at the lower end of the cord (which terminates at the level of the first lumbar vertebra) before reaching the intervertebral foramina of their respective vertebrae, through which the nerves then emerge. This gives the lower end of the cord, with its attached spinal nerve roots, the appearance of a horse's tail. In fact, it bears the name **cauda equina** (Latin equivalent for horse's tail). (See Figure 11-5.)

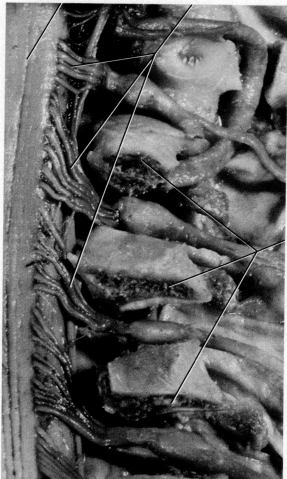

Figure 11-7 Dissection of the cervical segment of the spinal cord showing emerging cervical nerves. The cord is viewed from its posterior aspect.

Posterior median sulcus of spinal cord

Dorsal roots of C₂, C₃, and C₄ nerves

Transverse processes of vertebrae (cut)

The **brachial plexus** is located in the shoulder region from the neck to the axilla. It has clinical importance. Stretching or tearing it during birth causes paralysis and numbness of the baby's arm on that side. If untreated, it results in a withered arm. Branching from this plexus are several nerves to the skin and to voluntary muscles.

The right and left **phrenic nerves,** whose fibers come from the third and fourth or fourth and fifth cervical spinal nerves above the formation of the brachial plexus, also have considerable clinical interest, since they supply the diaphragm muscle. If the neck is broken in a way that severs or crushes the cord above this level, nerve impulses from the brain can no longer reach the phrenic nerves, and therefore the diaphragm stops contracting. Unless artificial respiration of some kind is provided, the patient dies of respiratory paralysis as a result of the broken neck. Any disease or injury that attacks the cord between the third and fifth cervical segments also paralyzes the phrenic nerve and, therefore, the diaphragm.

Each spinal nerve, instead of attaching directly to the cord, attaches indirectly by means of two short roots, anterior and posterior. The posterior roots are readily recognized by a swelling on them, named the posterior root ganglion or **spinal ganglion** (Figure 11-7). The roots and ganglia lie, respectively, within the spinal cavity in the intervertebral foramina (Figure 11-8). Figure 11-9 shows the structure of a nerve in a cross-section view.

After each spinal nerve emerges from the spinal cavity, it divides into an anterior, a posterior, and a white ramus. Anterior and posterior rami supply fibers to skeletal muscles and skin. White rami contain fibers of the autonomic nervous system. The posterior rami subdivide into lesser nerves that extend into the muscles and skin of the posterior surface of the head, neck, and trunk.

The anterior rami (except those of the thoracic nerves) subdivide and supply fibers to skeletal muscles and skin of the extremities and of anterior and lateral surfaces. Subdivisions of the anterior rami form complex networks or plexuses (Table 11-3, pp. 310-311). For example, fibers from the lower four cervical and first thoracic nerves intermix in such a way as to form a fairly definite, although apparently hopelessly confused, pattern called the **brachial plexus** (Figure 11-10). Emerging from this plexus are smaller nerves bearing names descriptive of their locations, such as *the median nerve*, the *radial nerve*, and the *ulnar nerve*. These nerves (each con-

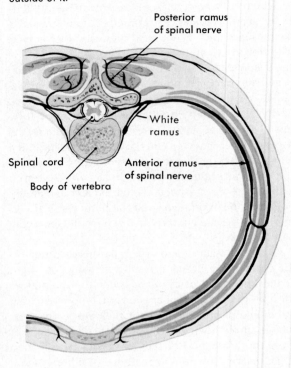

Figure 11-8 Branchings of a spinal nerve. Note the anterior and posterior roots of the nerve within the vertebral foramen and anterior and posterior rami outside of it.

Figure 11-9 Cross section of a nerve trunk. Fibers are myelinated and separated by a perineural sheath. Endoneurium surrounds individual nerve fibers; it is an extension of the perineurium.

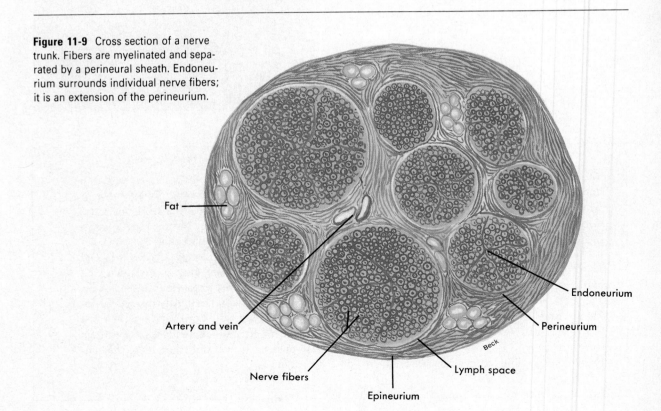

Figure 11-10 Brachial plexus. Intermixing of fibers from the lower four cervical and first thoracic nerves.

Figure 11-11 Main nerves of the lower extremity.

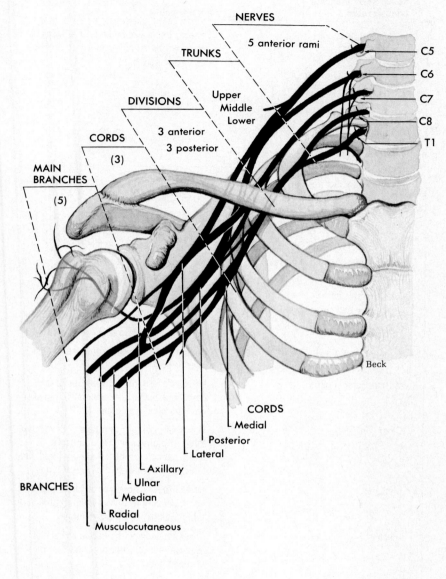

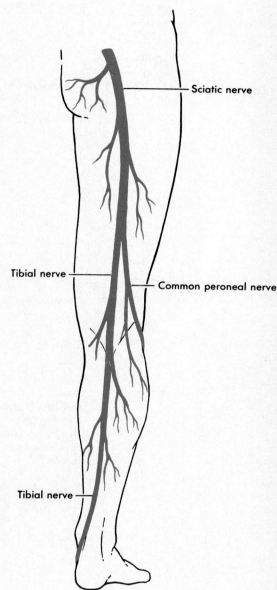

Fibers from the fourth and fifth lumbar nerves and the first, second, and third sacral nerves form the **sacral plexus.** It lies in the pelvic cavity on the anterior surface of the piriformis muscle. Among other nerves that emerge from the sacral plexus are the tibial and common peroneal nerves. In the thigh, they form the largest nerve in the body, the great sciatic nerve (Figure 11-11). It pierces the buttocks and runs down the back of the thigh. Its many branches supply nearly all the skin of the leg, the posterior thigh muscles, and the leg and foot mus-

taining fibers from more than one spinal nerve) divide further into smaller and smaller branches that completely innervate the hand and most of the arm.

Another spinal nerve plexus is the **lumbar plexus,** formed by the intermingling of fibers from the first four lumbar nerves. This network of nerves is located in the lumbar region of the back in the psoas muscle. The large femoral nerve is one of several nerves emerging from the lumbar plexus. It divides into many branches supplying the thigh and leg.

Table 11-3 Spinal nerves and peripheral branches

Spinal nerves	Plexuses formed from anterior rami	Spinal nerve branches from plexuses	Parts supplied
Cervical 1 2 3 4	Cervical plexus	Lesser occipital Great auricular Cutaneous nerve of neck Anterior supraclavicular Middle supraclavicular Posterior supraclavicular Branches to muscles	Sensory to back of head, front of neck, and upper part of shoulder; motor to numerous neck muscles
Cervical 5 6 7 8 Thoracic (or dorsal) 1 2	Brachial plexus	Phrenic (branches from cervical nerves before formation of plexus; most of its fibers from fourth cervical nerve)	Diaphragm
		Suprascapular and dorsoscapular	Superficial muscles* of scapula
		Thoracic nerves, medial and lateral branches	Pectoralis major and minor
		Long thoracic nerve	Serratus anterior
		Thoracodorsal	Latissimus dorsi
		Subscapular	Subscapular and teres major muscles
		Axillary (circumflex)	Deltoid and teres minor muscles and skin over deltoid
3 4 5 6 7 8 9 10 11 12	No plexus formed; branches run directly to intercostal muscles and skin of thorax	Musculocutaneous	Muscles of front of arm (biceps brachii, coracobrachialis, and brachialis) and skin on outer side of forearm
		Ulnar	Flexor carpi ulnaris and part of flexor digitorum profundus; some of muscles of hand; sensory to medial side of hand, little finger, and medial half of fourth finger
		Median	Rest of muscles of front of forearm and hand; sensory to skin of palmar surface of thumb, index, and middle fingers
		Radial	Triceps muscle and muscles of back of forearm; sensory to skin of back of forearm and hand
		Medial cutaneous	Sensory to inner surface of arm and forearm

*Although nerves to muscles are considered motor, they do contain some sensory fibers that transmit proprioceptive impulses.

Table 11-3 Spinal nerves and peripheral branches—cont'd

Spinal nerves	Plexuses formed from anterior rami	Spinal nerve branches from plexuses	Parts supplied
Lumbar 1 2 3 4 5 Sacral 1 2 3 4 5 Coccygeal 1	Lumbosacral plexus	Iliohypogastric Ilioinguinal — Sometimes fused	Sensory to anterior abdominal wall
			Sensory to anterior abdominal wall and external genitalia; motor to muscles of abdominal wall
		Genitofemoral	Sensory to skin of external genitalia and inguinal region
		Lateral cutaneous of thigh	Sensory to outer side of thigh
		Femoral	Motor to quadriceps, sartorius, and iliacus muscles; sensory to front of thigh and medial side of lower leg (saphenous nerve)
		Obturator	Motor to adductor muscles of thigh
		Tibial* (medial popliteal)	Motor to muscles of calf of leg; sensory to skin of calf of leg and sole of foot
		Common peroneal (lateral popliteal)	Motor to evertors and dorsiflexors of foot; sensory to lateral surface of leg and dorsal surface of foot
		Nerves to hamstring muscles	Motor to muscles of back of thigh
		Gluteal nerves, superior and inferior	Motor to buttock muscles and tensor fasciae latae
		Posterior cutaneous nerve	Sensory to skin of buttocks, posterior surface of thigh, and leg
		Pudendal nerve	Motor to perineal muscles; sensory to skin of perineum

*Sensory fibers from the tibial and peroneal nerves unite to form the *medial cutaneous* (or sural) *nerve* that supplies the calf of the leg and the lateral surface of the foot. In the thigh the tibial and common peroneal nerves are usually enclosed in a single sheath to form the *sciatic nerve*, the largest nerve in the body with a width of approximately ¾ of an inch. About two thirds of the way down the posterior part of the thigh, it divides into its component parts. Branches of the sciatic nerve extend into the hamstring muscles.

Figure 11-12 Segmental dermatome distribution of spinal nerves to the front of the body. *C*, Cervical segments; *T*, thoracic segments; *L*, lumbar segments; *S*, sacral segments.

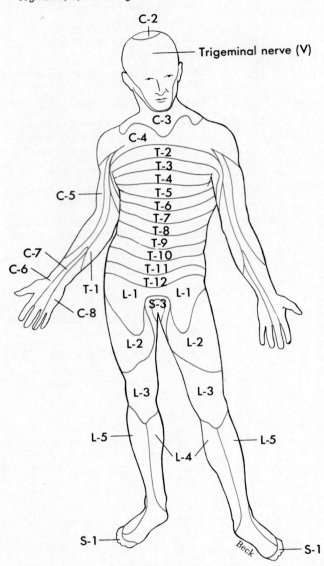

Figure 11-13 Segmental dermatome distribution of spinal nerves to the back of the body. *C*, Cervical segments; *T*, thoracic segments; *L*, lumbar segments; *S*, sacral segments.

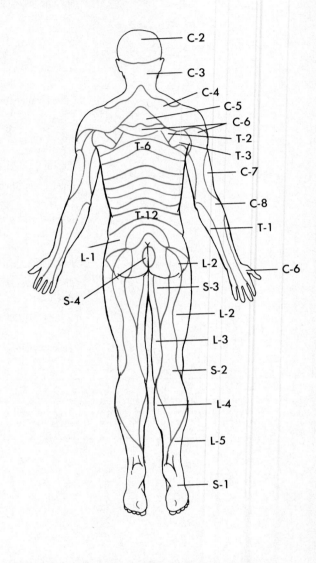

cles. Sciatica or neuralgia of the sciatic nerve is a fairly common and very painful condition.

At first glance the distribution of spinal nerves does not appear to follow an ordered arrangement. But detailed mapping of the skin surface has revealed a close relationship between the source on the cord of each spinal nerve and the level of the body it innervates (Figures 11-12 and 11-13). Knowledge of the segmental arrangement of spinal nerves has proved useful to physicians. For instance, a neurologist can identify the site of spinal cord or nerve abnormality from the area of the body insensitive to a pinprick. Skin surface areas supplied by a single spinal nerve are called **dermatomes.**

MICROSCOPIC STRUCTURE AND FUNCTIONS

Look back now at Figure 10-7, in Chapter 10. Note the microscopic structure located in the posterior root ganglion—the cell body of a sensory neuron. Note, too, that the sensory neuron's dendrite lies in the spinal nerve leading to the ganglion. Its axon lies in the posterior root of the spinal nerve. Thus all of these structures—spinal nerves, posterior root ganglia, and posterior roots of spinal nerves—conduct impulses toward the cord. In short, they serve a sensory function.

Shift your attention now to the anterior root of the spinal nerve in Figure 10-7. Notice that it contains a motoneuron axon. The cell body for this axon lies

Figure 11-14 Herpes zoster or shingles. Dermatome involvement in a 13-year-old boy.

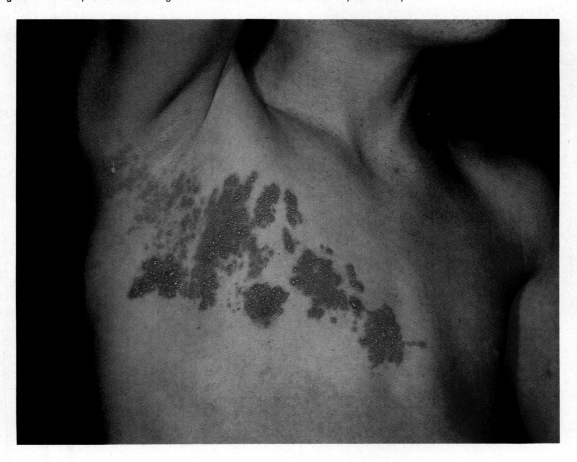

Herpes zoster or **shingles** is a unique viral infection that almost always affects the skin of a single dermatome. It is caused by the varicella zoster virus of chicken pox. About 3% of the population will suffer from shingles at some time in their lives. In most cases the disease results from reactivation of the varicella virus. The virus most likely traveled through a cutaneous nerve and remained dormant in a dorsal root ganglion for years after an episode of chicken pox. If the body's immunological protective mechanism becomes diminished in the elderly, or following stress, or in individuals undergoing radiation therapy or taking immunosuppressive drugs, the virus may reactivate. If this occurs, the virus will travel over the sensory nerve to the skin of a single dermatome. The result is a painful eruption (Figure 11-14) of red swollen plaques or vesicles that eventually rupture and crust before clearing in 2 to 3 weeks. In severe cases extensive inflammation, hemorrhagic blisters, and secondary bacterial infection may lead to permanent scarring. The herpes zoster vesicles seen in Figure 11-15 appear in the dermatome area involving the mandibular branch of the fifth cranial nerve. In most cases of shingles, the eruption of vesicles is preceded by 4 to 5 days of preeruptive pain, burning, and itching in the affected dermatome. Unfortunately, an attack of herpes zoster does not confer lasting immunity. Many individuals suffer three or four episodes in a lifetime.

Figure 11-15 Herpes zoster or shingles. Single-dermatome distribution involving the mandibular branch of the fifth cranial nerve on the right side.

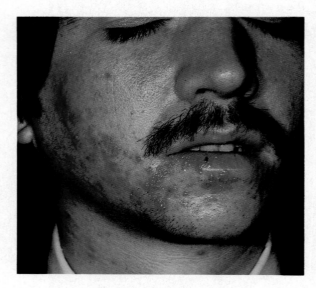

in the anterior column of gray matter. The axon terminates in a skeletal muscle, a somatic effector. Thus the anterior gray column of the cord and the anterior roots of spinal nerves conduct impulses out of the cord. In short, they serve a motor function. Neurons whose dendrites and cell bodies lie in the anterior gray columns are known by three names—*anterior horn neurons*, *somatic motoneurons*, and *lower motoneurons*. Summarizing: spinal nerves are *mixed nerves*, that is, they contain both sensory dendrites and motor axons. Posterior root ganglia contain sensory cell bodies. Posterior roots contain sensory axons. Anterior gray columns of the cord contain dendrites and cell bodies of somatic motoneurons. Anterior roots of spinal nerves contain the axons of somatic motoneurons.

DIVISIONS AND SIZE OF BRAIN

The brain is one of the largest of adult organs. It consists, in round numbers, of 100 billion neurons and 900 billion neuroglia. In most adults, it weighs about 3 pounds but generally is smaller in women than in men and in older people than in younger people. Neurons of the brain undergo mitosis only during the prenatal period and the first few months

Figure 11-16 Sagittal section through midline of the brain showing structures around the third ventricle. Only a small portion of the cerebrum is shown.

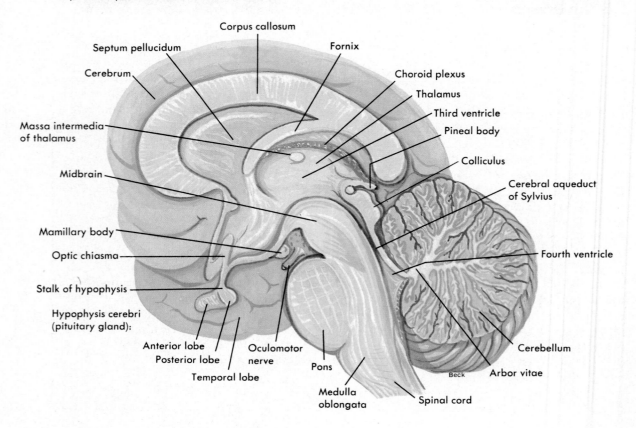

Figure 11-17 Right half of brain viewed from its medial aspect.

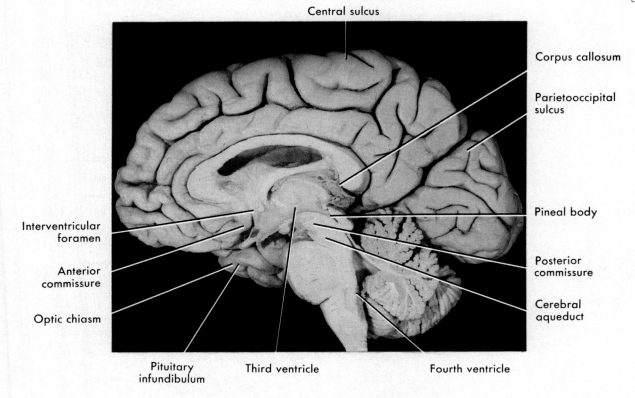

Central sulcus

Corpus callosum

Parietooccipital sulcus

Interventricular foramen

Anterior commissure

Optic chiasm

Pineal body

Posterior commissure

Cerebral aqueduct

Pituitary infundibulum

Third ventricle

Fourth ventricle

of postnatal life. Although they grow in size after that, they do not increase in number. Malnutrition during those crucial prenatal months of neuron multiplication is reported to hinder the process and result in fewer brain cells. The brain attains full size by about the eighteenth year but grows rapidly only during the first 9 years or so.

Six major divisions of the brain, named from below, upward, are as follows: *medulla oblongata, pons, midbrain, cerebellum, diencephalon,* and *cerebrum.* Very often the medulla, pons, and midbrain are referred to collectively as the brain stem. Look at these three structures in Figures 11-16 and 11-17. Do you agree that they seem to form a stem for the rest of the brain?

BRAIN STEM

STRUCTURE

Three divisions of the brain make up the brain stem. The **medulla oblongata** forms the lowest part of the brain stem, the **midbrain** forms the uppermost part, and the **pons** lies between them, that is, above the medulla and below the midbrain.

Medulla

The medulla or bulb is the part of the brain that attaches to the spinal cord. It is, in fact, an enlarged extension of the cord located just above the foramen magnum. It measures only slightly more than an inch in length and is separated from the pons above by a horizontal groove. It is composed of white matter (projection tracts) and a mixture of gray and white matter called the *reticular formation.*

The *pyramids* (Figure 11-18) are two bulges of white matter located on the ventral surface of the medulla. Fibers of the pyramidal projection tracts form the pyramids.

The *olive* (Figure 11-18) is an oval projection appearing one on each side of the ventral surface of the medulla.

Located in the medulla's reticular formation are various *nuclei,* or clusters of neuron cell bodies. Some nuclei are called centers—for example, the cardiac, respiratory, and vasomotor centers. (Often, for obvious reasons, these are spoken of as "the vital centers.") Some medullary nuclei bear individual names: *nucleus gracilis, nucleus cuneatus, olivary nuclei.*

Figure 11-18 Ventral surface of the brain showing attachment of the cranial nerves.

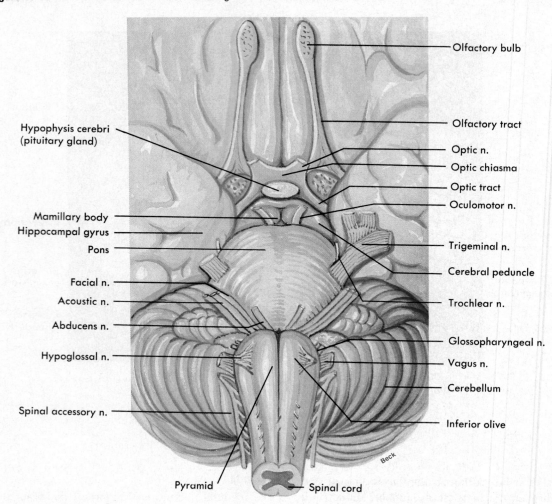

- Olfactory bulb
- Olfactory tract
- Optic n.
- Optic chiasma
- Optic tract
- Oculomotor n.
- Trigeminal n.
- Cerebral peduncle
- Trochlear n.
- Glossopharyngeal n.
- Vagus n.
- Cerebellum
- Inferior olive

- Hypophysis cerebri (pituitary gland)
- Mamillary body
- Hippocampal gyrus
- Pons
- Facial n.
- Acoustic n.
- Abducens n.
- Hypoglossal n.
- Spinal accessory n.
- Pyramid
- Spinal cord

Beck

Pons

Just above the medulla lies the pons, composed, like the medulla, of white matter and reticular formation. Fibers that run transversely across the pons and through the brachia pontis (middle cerebellar peduncles) into the cerebellum make up the external white matter of the pons and give it its bridgelike appearance.

Midbrain (Mesencephalon)

The midbrain is appropriately named. It forms the midsection of the brain, since it lies above the pons and below the cerebrum. Both white matter (tracts) and reticular formation compose the midbrain. Extending divergently through it are two ropelike masses of white matter named *cerebral peduncles* (Figure 11-18). Tracts in the peduncles conduct impulses between the midbrain and cerebrum. In addition to the cerebral peduncles, another landmark

of the midbrain is the *corpora quadrigemina* (literally, "four bodies"). The corpora quadrigemina are two *inferior colliculi* and two *superior colliculi.* Find the label *colliculus* in Figure 11-16. It points to an inferior colliculus. Notice the location of the four colliculi or, in other words, the corpora quadrigemina. They form the posterior, upper part of the midbrain, the part that lies just above the cerebellum. Certain auditory centers are located in the inferior colliculus. The superior colliculus contains visual centers. Two other midbrain structures are the *red nucleus* and the *substantia nigra.* Each of these consists of clusters of cell bodies of neurons involved in muscular control. Pigment in cells of the substantia nigra gives the area a dark color.

FUNCTIONS

The brain stem, like the spinal cord, performs sensory, motor, and reflex functions. The spinothalamic

Because the cardiac, vasomotor, and respiratory centers are essential for survival, they are called the *vital centers.* They serve as the centers for various reflexes controlling heart action, blood vessel diameter, and respiration. Because the medulla contains these centers, it is the most vital part of the entire brain—so vital, in fact, that injury or disease of the medulla often proves fatal. Blows at the base of the skull and bulbar poliomyelitis, for example, cause death if they interrupt impulse conduction in the vital respiratory centers.

tracts are important sensory tracts that pass through the brain stem on their way to the thalamus. The fasciculi cuneatus and gracilis and the spinoreticular tracts are sensory tracts whose axons terminate in the gray matter of the brain stem. Corticospinal and reticulospinal tracts are two of the major tracts present in the white matter of the brain stem.

Nuclei in the medulla contain a number of reflex centers. Of first importance are the cardiac, vasomotor, and respiratory centers. Other centers present in the medulla are those for various nonvital reflexes such as vomiting, coughing, sneezing, hiccupping, and swallowing.

The pons contains centers for reflexes mediated by the fifth, sixth, seventh, and eighth cranial nerves (see Table 11-4, pp. 320-321, for functions). In addition, the pons contains the pneumotaxic centers that help regulate respiration.

The midbrain, like the pons, contains reflex centers for certain cranial nerve reflexes, for example, pupillary reflexes and eye movements, mediated by the third and fourth cranial nerves, respectively.

CRANIAL NERVES

Twelve pairs of nerves arise from the undersurface of the brain (Figure 11-18), mostly from the brain stem. (This is our reason for discussing them here following the discussion of the brain stem.) After leaving the cranial cavity by way of small foramina in the skull, they extend to their respective destinations. Both names and numbers identify the cranial nerves. Their names suggest their distribution or function. Their numbers indicate the order in which they emerge from front to back. Like all nerves, cranial nerves are made up of bundles of axons. *Mixed cranial nerves* contain axons of sen-

sory and motor neurons. *Sensory cranial nerves* consist of sensory axons only and *motor cranial nerves* consist mainly of motor axons. Examine Table 11-4 to check the following summary:

1 **Mixed cranial nerves**—fifth, seventh, ninth, and tenth
2 **Sensory cranial nerves**—first, second, eighth
3 **Motor cranial nerves**—third, fourth, sixth, eleventh, and twelfth

Now learn the names of the cranial nerves, using the memory aid given in the footnote to Table 11-4. For more information about each cranial nerve, read the following pages on the nerve pairs.

First (Olfactory)

The olfactory nerves are composed of axons of neurons whose dendrites and cell bodies lie in the nasal mucosa, high up along the septum and superior conchae (turbinates). Axons of these neurons form about 20 small fibers that pierce each cribriform plate and terminate in the olfactory bulbs. Here they synapse with olfactory neurons II, whose axons compose the olfactory tracts. Summarizing:

Olfactory neurons I
Dendrites } In nasal mucosa
Cell body }
Axons   In small fibers that extend through cribriform plate to olfactory bulb

Olfactory neurons II
Dendrites } In olfactory bulb
Cell body }
Axons   In olfactory tracts

Second (Optic)

Axons from the third and innermost layer of neurons of the retina compose the second cranial nerves. After entering the cranial cavity through the optic foramina, the two optic nerves unite to form the *optic chiasma,* in which some of the fibers of each nerve cross to the opposite side and continue in the *optic tract* of that side (see Figure 13-21). Thus each optic nerve contains fibers only from the retina of the same side, whereas each optic tract has fibers in it from both retinas, a fact of importance in interpreting certain visual disorders. Most of the optic tract fibers terminate in the thalamus (in the portion known as the lateral geniculate body). From here a new relay of fibers runs to the visual area of the occipital lobe cortex. A few optic tract fibers terminate in the superior colliculi of the midbrain, where they synapse with motor fibers to the external eye muscles (third, fourth, and sixth cranial nerves).

Third (Oculomotor)

Fibers of the third cranial nerve originate from cells in the oculomotor nucleus in the ventral part of the

Figure 11-19 Trigeminal (fifth cranial) nerve and its three main divisions, the ophthalmic, maxillary, and mandibular nerves.

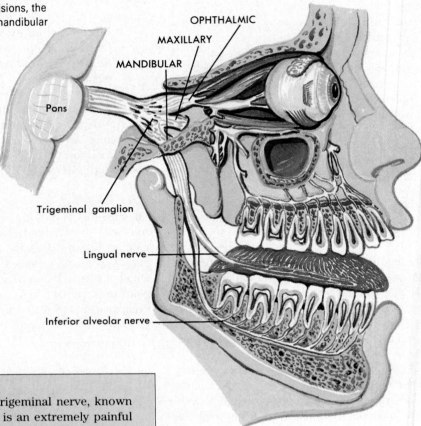

Neuralgia of the trigeminal nerve, known as *tic douloureux*, is an extremely painful condition. One method of doing away with this pain is to remove the gasserian (or trigeminal) ganglion on the posterior root of the nerve (Figure 11-19). This large ganglion contains the cell bodies of the nerve's afferent fibers. After such an operation the patient's face, scalp, teeth, and conjunctiva on the side treated show anesthesia. Special care, such as wearing protective goggles and irrigating the eye frequently, is therefore prescribed. The patient is instructed also to visit the dentist regularly, since one can no longer experience a toothache as a warning of diseased teeth.

midbrain and extend to the various external eye muscles, with the exception of the superior oblique and the lateral rectus. Autonomic fibers are also present in the oculomotor nerves; they are axons of neuron cell bodies located in nuclei of the midbrain. These autonomic fibers terminate in the ciliary ganglion. Here, they synapse with cells whose postganglionic fibers supply the intrinsic eye muscles (ciliary and iris). Still a third group of fibers is found in the third cranial nerves, namely, sensory fibers from proprioceptors in the eye muscles.

Fourth (Trochlear)

Motor fibers of the fourth cranial nerve have their origin in cells in the midbrain, from which they extend to the superior oblique muscles of the eye. Afferent fibers from proprioceptors in these muscles are also contained in the trochlear nerves.

Fifth (Trigeminal, Trifacial)

Three sensory branches (ophthalmic, maxillary, and mandibular nerves) carry afferent impulses from the skin and mucosa of the head and from the teeth to cell bodies in the trigeminal or gasserian ganglion (a swelling on the nerve, lodged in the petrous portion of the temporal bone) (Figure 11-19). Fibers extend from the ganglion to the main sensory nucleus of the fifth cranial nerve situated in the pons. A smaller motor root of the trigeminal nerve originates in the trifacial motor nucleus located in the pons just medial to the sensory nucleus. Fibers run from the motor root to the muscles of mastication by way of the mandibular nerve.

Sixth (Abducens)

The sixth cranial nerve is a motor nerve with fibers originating from a nucleus in the pons in the floor of the fourth ventricle and extending to the lateral rectus muscles of the eyes. It also contains some afferent fibers from proprioceptors in the lateral rectus muscles.

Seventh (Facial)

The motor fibers of the seventh cranial nerve arise from a nucleus in the lower part of the pons, from which they extend by way of several branches to the superficial muscles of the face and scalp and to the submaxillary and sublingual glands. Sensory fibers from the taste buds of the anterior two thirds of the tongue run in the facial nerve to cell bodies in the geniculate ganglion, a small swelling on the facial nerve, where it passes through a canal in the temporal bone. From the ganglion, fibers extend to the nucleus solitarius in the medulla.

Eighth (Vestibulocochlear)

The eighth cranial nerve has two distinct divisions: the vestibular nerve and the cochlear nerve. Both are sensory. Fibers from the semicircular canals run to the vestibular ganglion (in the internal auditory meatus). Here, their cell bodies are located and from them fibers extend to the vestibular nuclei in the pons and medulla. Together, these fibers constitute the *vestibular nerve.* Some of its fibers run to the cerebellum. The vestibular nerve transmits impulses that result in sensations of balance or imbalance. The *cochlear nerve* consists of dendrites starting in the organ of Corti in the cochlea. They have their cell bodies in the spiral ganglion in the cochlea, and their axons terminate in the cochlear nuclei located between the medulla and pons. Conduction by the cochlear nerve results in sensations of hearing.

Ninth (Glossopharyngeal)

Both sensory and motor fibers compose the ninth cranial nerve. This nerve supplies fibers not only to the tongue and pharynx, as its name implies, but also to other structures. One is the carotid sinus; it plays an important part in the control of blood pressure. Sensory fibers, with their receptors in the pharynx and posterior third of the tongue, have their cell bodies in the jugular (superior) and petrous (inferior) ganglia. These are located, respectively, in the jugular foramen and the petrous portion of the temporal bone. From the ganglia fibers extend to the nucleus solitarius in the medulla. The motor fibers of the ninth cranial nerve originate in cells in the nucleus ambiguus in the medulla and run to muscles of the pharynx. Also present in this nerve are secretory fibers whose cells of origin lie in the nucleus salivatorius (at the junction of the pons and medulla). These fibers run to the otic ganglion, from which postganglionic fibers extend to the parotid gland.

Tenth (Vagus)

The tenth cranial nerve is widely distributed and contains both sensory and motor fibers. Its sensory fibers supply the pharynx, larynx, trachea, heart, carotid body, lungs, bronchi, esophagus, stomach, small intestine, and gallbladder. Cell bodies for these sensory dendrites lie in the jugular and nodose ganglia, located, respectively, in the jugular foramen and just inferior to it on the trunk of the nerve. Centrally the sensory axons terminate in the medulla (in the nucleus solitarius) and in the pons (in the nucleus of the trigeminal nerve). Motor fibers of the vagus originate in cells in the medulla (in the dorsal motor nucleus of the vagus) and extend to various autonomic ganglia in the vagal plexus. From the plexus, postganglionic fibers run to muscles of the pharynx, larynx, and thoracic and abdominal viscera.

Eleventh (Accessory)

The eleventh cranial nerve is a motor nerve. Some of its fibers originate in cells in the medulla (in the dorsal motor nucleus of the vagus and in the nucleus ambiguus) and pass by way of vagal branches to thoracic and abdominal viscera. The rest of the fibers have their cells of origin in the anterior gray column of the first five or six segments of the cervical spinal cord and extend through the spinal root of the accessory nerve to the trapezius and sternocleidomastoid muscles.

Twelfth (Hypoglossal)

Motor fibers with cell bodies in the medulla (in the hypoglossal nucleus) compose the twelfth cranial nerve. They supply the muscles of the tongue. According to some anatomists, this nerve also contains sensory fibers from proprioceptors in the tongue.

• • •

The main facts about the distribution and function of each of the cranial nerve pairs are summarized in Table 11-4.

Severe head injuries often damage one or more of the cranial nerves, producing symptoms analogous to the functions of the nerve affected. For example, injury of the sixth cranial nerve causes the eye to turn in, because of paralysis of the abducting muscle of the eye. Injury of the eighth cranial nerve, on the other hand, produces deafness. Injury to the facial nerve results in a poker-faced expression and a drooping of the corner of the mouth from paralysis of the facial muscles.

Table 11-4 Cranial nerves

Nerve*	Sensory fibers†			Motor fibers†		Functions‡
	Receptors	Cell bodies	Termination	Cell bodies	Termination	
1 Olfactory	Nasal mucosa	Nasal mucosa	Olfactory bulbs (new relay of neurons to olfactory cortex)			Sense of smell
2 Optic	Retina	Retina	Nucleus in thalamus (lateral geniculate body); some fibers terminate in superior colliculus of midbrain			Vision
3 Oculomotor	External eye muscles except superior oblique and lateral rectus	?	?	**Midbrain (oculomotor nucleus and Edinger-Westphal nucleus)**	**External eye muscles except superior oblique and lateral rectus; fibers from Edinger-Westphal nucleus terminate in ciliary ganglion and then to ciliary and iris muscles**	**Eye movements, regulation of size of pupil, accommodation,** *proprioception (muscle sense)*
4 Trochlear	Superior oblique	?	?	**Midbrain**	**Superior oblique muscle of eye**	**Eye movements,** *proprioception*
5 Trigeminal	Skin and mucosa of head, teeth	Gasserian ganglion	Pons (sensory nucleus)	**Pons (motor nucleus)**	**Muscles of mastication**	Sensations of head and face, **chewing movements,** *muscle sense*
6 Abducens	Lateral rectus	?	?	**Pons**	**Lateral rectus muscle of eye**	**Abduction of eye,** *proprioception*
7 Facial	Taste buds of anterior two thirds of tongue	Geniculate ganglion	Medulla (nucleus solitarius)	**Pons**	**Superficial muscles of face and scalp**	**Facial expressions, secretion of saliva,** *taste*
8 Acoustic Vestibular branch	Semicircular canals and vestibule (utricle and saccule)	Vestibular ganglion	Pons and medulla (vestibular nuclei)			Balance or equilibrium sense
Cochlear or auditory branch	Organ of Corti in cochlear duct	Spiral ganglion	Pons and medulla (cochlear nuclei)			Hearing

Table 11-4 Cranial nerves (cont'd)

Nerve*	Sensory fibers†			Motor fibers†		Functions‡
	Receptors	Cell bodies	Termination	Cell bodies	Termination	
9 Glossopharyngeal	*Pharynx; taste buds and other receptors of posterior one third of tongue*	*Jugular and petrous ganglia*	*Medulla (nucleus solitarius)*	**Medulla (nucleus ambiguus)**	**Muscles of pharynx**	*Taste and other sensations of tongue,* **swallowing movements, secretion of saliva,** *aid in reflex control of blood pressure and respiration*
	Carotid sinus and carotid body	*Jugular and petrous ganglia*	*Medulla (respiratory and vasomotor centers)*	**Medulla at junction of pons (nucleus salivatorius)**	**Otic ganglion and then to parotid gland**	
10 Vagus	*Pharynx, larynx, carotid body, and thoracic and abdominal viscera*	*Jugular and nodose ganglia*	*Medulla (nucleus solitarius), pons (nucleus of fifth cranial nerve)*	**Medulla (dorsal motor nucleus)**	**Ganglia of vagal plexus and then to muscles of pharynx, larynx, and thoracic and abdominal viscera**	*Sensations and* **movements** *of organs supplied; for example,* **slows heart, increases peristalsis, and contracts muscles for voice production**
11 Spinal accessory	?	?	?	**Medulla (dorsal motor nucleus of vagus and nucleus ambiguus)**	**Muscles of thoracic and abdominal viscera and pharynx and larynx**	**Shoulder movements, turning movements of head, movements of viscera, voice production,** *proprioception(?)*
				Anterior gray column of first five or six cervical segments of spinal cord	**Trapezius and sternocleidomastoid muscle**	
12 Hypoglossal	?	?	?	**Medulla (hypoglossal nucleus)**	**Muscles of tongue**	**Tongue movements,** *proprioception(?)*

*The first letters of the words in the following sentence are the first letters of the names of the cranial nerves. Many generations of anatomy students have used this sentence as an aid to memorizing these names. It is "On Old Olympus Tiny Tops, A Finn, and German Viewed Some Hops." (There are several slightly differing versions of this mnemonic.)

†Italics indicate sensory fibers and functions. Boldface type indicates motor fibers and functions.

‡An aid for remembering the general function of each cranial nerve is the following 12-word saying: "Some say marry money but my brothers say bad business marry money." Words beginning with S indicate sensory function. Words beginning with M indicate motor function. Words beginning with B indicate both sensory and motor functions. For example, the first, second, and eighth words in the saying start with S, which indicates that the first, second, and eighth cranial nerves perform sensory functions.

CEREBELLUM

STRUCTURE

The cerebellum, the second largest part of the brain, is located just below the posterior portion of the cerebrum and is partially covered by it. A transverse fissure separates the cerebellum from the cerebrum. These two parts of the brain have several characteristics in common. For instance, gray matter makes up their outer portions and white matter predominates in their interiors. Turn to Figure 11-16 to observe the arbor vitae, that is, the internal white matter of the cerebellum. Note its distinctive pattern, similar to the veins of a leaf. Note, too, that the surfaces of both the cerebellum and the cerebrum have numerous grooves (sulci) and convolutions (gyri). The convolutions of the cerebellum, however, are much more slender and less prominent than those of the cerebrum. The cerebellum consists of two large lateral masses, the cerebellar hemispheres, and a central section called the *vermis*. The shape of the vermis resembles a worm coiled on itself. (For a detailed description of the several subdivisions of the cerebellum, consult a textbook of neuroanatomy.)

The internal white matter of the cerebellum is composed of some short and some long tracts. The short association tracts conduct impulses from neuron cell bodies located in the cerebellar cortex to neurons whose dendrites and cell bodies compose nuclei located in the interior of the cerebellum. The longer tracts conduct impulses to and from the cerebellum. Fibers of the longer tracts enter or leave the cerebellum by way of its three pairs of peduncles, as follows:

1 Inferior cerebellar peduncles (or restiform bodies)—composed chiefly of tracts into the cerebellum from the medulla and cord (notably, spinocerebellar, vestibulocerebellar, and reticulocerebellar tracts)

2 Middle cerebellar peduncles (or brachia pontis)—composed almost entirely of tracts into the cerebellum from the pons, that is, pontocerebellar tracts

3 Superior cerebellar peduncles (or brachia conjunctivum cerebelli)—composed principally of tracts from dentate nuclei through the red nucleus of the midbrain to the thalamus

An important pair of cerebellar nuclei are the dentate nuclei, one of which lies in each hemisphere. Tracts connect the nuclei with the thalamus and with motor areas of the cerebral cortex (the dentatorubrothalamic tracts to the thalamus and thalamocortical tracts to the cortex). By means of these tracts, cerebellar impulses influence the motor cortex. Impulses also travel the reverse direction. Corticopontine and pontocerebellar tracts enable the motor cortex to influence the cerebellum.

FUNCTIONS

The cerebellum performs three general functions, all of which have to do with the control of skeletal muscles. It acts with the cerebral cortex to produce skilled movements by coordinating the activities of groups of muscles. It controls skeletal muscles so as to maintain equilibrium. It helps control posture. It functions below the level of consciousness to make movements smooth instead of jerky, steady instead of trembling, and efficient and coordinated instead of ineffective, awkward, and uncoordinated (asynergic).

There have been many theories about cerebellar functions. One theory, based on comparative anatomy studies and substantiated by experimental methods, regards the cerebellum as three organs, each with a somewhat different function: synergic control of muscle action, excitation and inhibition of postural reflexes, and maintenance of equilibrium.

Synergic control of muscle action, which is ascribed to the neocerebellum (superior vermis and hemispheres), is closely associated with cerebral motor activity. Normal muscle action involves groups of muscles, the various members of which function together as a unit. In any given action, for example, the prime mover contracts and the antagonist relaxes but then contracts weakly at the proper moment to act as a brake, checking the action of the prime mover. Also, the synergists contract to assist the prime mover, and the fixation muscles of the neighboring joint contract. Through such harmonious, coordinated group action, normal movements are smooth, steady, and precise as to force, rate, and extent. Achievement of such movements results from cerebellar activity added to cerebral activity. Impulses from the cerebrum may start the action, but those from the cerebellum synergize or coordinate the contractions and relaxations of the various muscles once they have begun. Some physiologists consider this the main, if not the sole, function of the cerebellum.

One part of the cerebellum is thought to be concerned with both exciting and inhibiting postural reflexes.

Part of the cerebellum presumably discharges impulses important to the maintenance of equilibrium. Afferent impulses from the labyrinth of the ear reach the cerebellum. Here, connections are made with the proper efferent fibers for contraction of the necessary muscles for equilibrium.

Cerebellar disease (abscess, hemorrhage, tumors, trauma, etc.) produces certain characteristic symptoms. Predominant among them are ataxia (muscle incoordination), hypotonia, tremors, and disturbances of gait and equilibrium. One example of ataxia is overshooting a mark or stopping before reaching it when trying to touch a given point on the body (finger-to-nose test). Drawling, scanning, and singsong speech are also examples of ataxia. Tremors are particularly pronounced toward the end of the movements and with the exertion of effort. Disturbances of gait and equilibrium vary, depending on the muscle groups involved. The walk, for instance, is often characterized by staggering or lurching and by a clumsy manner of raising the foot too high and bringing it down with a clap. Paralysis does not result from loss of cerebellar function.

DIENCEPHALON

The diencephalon is the part of the brain located between the cerebrum and the mesencephalon (midbrain). Although the diencephalon consists of several structures located around the third ventricle, the main ones are the **thalamus** and **hypothalamus.**

STRUCTURE

The right thalamus is a rounded mass of gray matter about 2.5 cm wide and 3.75 cm long, bulging into the right lateral wall of the third ventricle. The left thalamus is a similar mass in the left lateral wall. Each thalamus consists of numerous nuclei. They are arranged in groups named the *anterior, lateral, intralaminar, midline* (or medial), and *posterior* groups of nuclei. Two important structures in the posterior group of nuclei are the medial and lateral geniculate bodies. Large numbers of axons conduct impulses into the thalamus from the cord, brain stem, cerebellum, basal ganglia, and various parts of the cerebrum. These axons terminate in thalamic nuclei, where they synapse with neurons whose axons conduct impulses out of the thalamus to virtually all areas of the cerebral cortex. Thus the thalamus serves as the major relay station for sensory impulses on their way to the cerebral cortex.

The hypothalamus consists of several structures that lie beneath the thalamus and form the third ventricle's floor and the lower part of its sidewall. Prominent among the structures composing the hypothalamus are the supraoptic nuclei, the paraventricular nuclei, the stalk of the hypophysis (pituitary gland), the neurohypophysis (the posterior lobe of the pituitary gland), and the mamillary bodies. Identify as many of these as you can in Figures 11-16 and 11-18. The supraoptic nuclei consist of gray matter located just above and on either side of the optic chiasma. The paraventricular nuclei of the hypothalamus are so named because of their location close to the wall of the third ventricle. The midportion of the hypothalamus consists of the stalk and the posterior lobe of the pituitary gland (neurohypophysis). The posterior part of the hypothalamus consists mainly of the mamillary bodies, in which are located the mamillary nuclei.

FUNCTIONS

The **thalamus** performs the following functions:

1 Plays two parts in the mechanism responsible for sensations
   a Impulses from appropriate receptors, on reaching the thalamus, produce conscious recognition of the crude, less critical sensations of pain, temperature, and touch
   b Neurons whose dendrites and cell bodies lie in certain nuclei of the thalamus relay all kinds of sensory impulses, except possibly olfactory, to the cerebrum
2 Plays a part in the mechanism responsible for emotions by associating sensory impulses with feelings of pleasantness and unpleasantness
3 Plays a part in the arousal or alerting mechanism
4 Plays a part in mechanisms that produce complex reflex movements

The **hypothalamus** is a small but functionally mighty area of the brain. It weighs little more than 7 gm, yet it performs many functions of the greatest importance both for survival and for the enjoyment of life. For instance, it functions as a link between the psyche (mind) and the soma (body). It also links the nervous system to the endocrine system. Certain areas of the hypothalamus function as pleasure centers or reward centers for the primary drives such as eating, drinking, and mating. The following paragraphs give a brief summary of hypothalamic functions.

1 The hypothalamus functions as a higher autonomic center or, rather, as several higher autonomic centers. By this we mean that axons of neurons whose dendrites and cell bodies lie in nuclei of the hypothalamus extend in tracts from the hypothalamus to both parasympathetic and sympathetic centers in the brain stem and cord. (Figure 12-2 indicates these tracts in blue.) Thus impulses from the hypothalamus can simultaneously or successively stimulate or inhibit few or many lower autonomic centers. In other words the hypothalamus serves as a regulator and coordinator of autonomic activities. It helps control and integrate the responses made by visceral effectors all over the body.
2 The hypothalamus functions as the major relay station between the cerebral cortex

and lower autonomic centers. Tracts conduct impulses from various centers in the cortex to the hypothalamus (also shown in blue in Figure 12-2). Then, via numerous synapses in the hypothalamus, these impulses are relayed to other tracts that conduct them on down to autonomic centers in the brain stem and cord and also to spinal cord somatic centers (anterior horn motoneurons). Thus the hypothalamus functions as the link between the cerebral cortex and lower centers—hence between the psyche and the soma. It provides a crucial part of the route by which emotions can express themselves in changed bodily functions. It is the all-important relay station in the neural pathways that makes possible the mind's influence over the body—sometimes, unfortunately, even to the profound degree of producing "psychosomatic disease."

3 Neurons in the supraoptic and paraventricular nuclei of the hypothalamus synthesize the hormones secreted by the posterior pituitary gland (neurohypophysis). Because one of these hormones affects the volume of urine excreted, the hypothalamus plays an indirect but essential role in maintaining water balance (see Figure 25-4, p. 656).

4 Some of the neurons in the hypothalamus function as endocrine glands. Their axons secrete chemicals, called *releasing hormones*, into blood, which circulates to the anterior pituitary gland. Releasing hormones control the release of certain anterior pituitary hormones—specifically growth hormone and hormones that control hormone secretion by the sex glands, the thyroid gland, and the adrenal cortex (discussed in Chapter 14). Thus indirectly the hypothalamus helps control the functioning of every cell in the body (see pp. 395-397).

The Nobel Prize in medicine for 1977 was shared by three individuals. Two of them, Andrew W. Schally and Roger C.L. Guillemin, received their awards for identifying, isolating, and synthesizing a protein hormone, somatostatin, secreted by the hypothalamus. Somatostatin has been shown to switch off the production of somatotropin (growth hormone) by the pituitary gland.

5 The hypothalamus plays an essential role in maintaining the waking state. Presumably it functions as part of an arousal or alerting mechanism. Clinical evidence of this is that somnolence characterizes some hypothalamic disorders.

6 The hypothalamus functions as a crucial part of the mechanism for regulating appetite and therefore the amount of food intake. Experimental and clinical findings seem to indicate the presence of a "feeding or appetite center" in the lateral part of the hypothalamus and a "satiety center" located medially. For example, an animal with an experimental lesion in the ventromedial nucleus of the hypothalamus will consume tremendous amounts of food. Similarly, a human being with a tumor in this region of the hypothalamus may eat insatiably and gain an enormous amount of weight.

7 The hypothalamus functions as a crucial part of the mechanism for maintaining normal body temperature. Hypothalamic neurons whose fibers connect with autonomic centers for vasoconstriction, dilation, and sweating and with somatic centers for shivering constitute heat-regulating centers. Marked elevation of body temperature frequently characterizes injuries or other abnormalities of the hypothalamus.

CEREBRUM

STRUCTURE

The cerebrum, the largest and uppermost division of the brain, consists of two halves, the right and left **cerebral hemispheres.** The surface of the cerebrum—called the *cerebral cortex*—is made up of gray matter only 2 to 4 mm (roughly $1/12$ to $1/6$ inch) thick. But despite its thinness, the cortex has six layers, each composed of millions of axon terminals synapsing with millions of dendrites and cell bodies of other neurons.

Provided one uses a little imagination, the surface of the cerebrum looks like a group of small sausages. Each "sausage" represents a *convolution,* or *gyrus.* Names of some of these are the *precentral gyrus, postcentral gyrus, cingulate gyrus, cuneus, hippocampal gyrus,* and *uncus.* Identify each of these in either Figure 11-20 or 11-21.

Between adjacent gyri lie either shallow grooves called *sulci* or deeper grooves called *fissures.* Fissures divide each cerebral hemisphere into five *lobes.* Four of them are named for the bones that lie over them: frontal lobe, parietal lobe, temporal lobe, and occipital lobe (Figure 11-20). A fifth lobe, the insula (island of Reil), lies hidden from view in the

Figure 11-20 Right hemisphere of cerebrum, lateral surface.

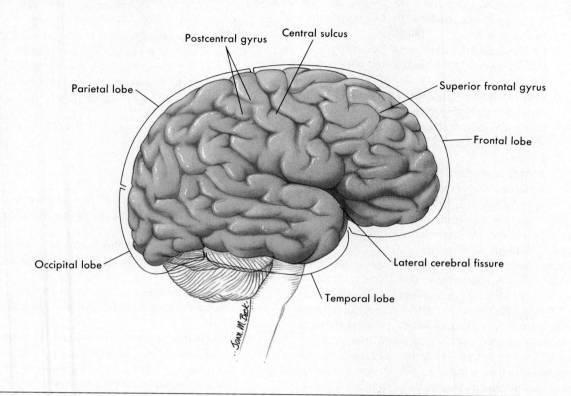

Figure 11-21 Right hemisphere of cerebrum, medial surface.

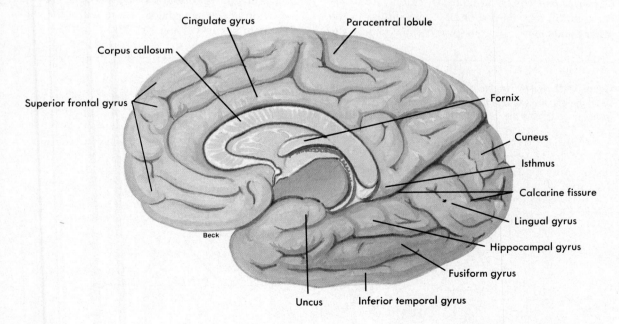

lateral fissure. Names and locations of prominent fissures are the following:

1 **Longitudinal fissure**—The deepest groove in the cerebrum; divides the cerebrum into two hemispheres

2 **Central sulcus (fissure of Rolando)**—Groove between the frontal and parietal lobes

3 **Lateral fissure (fissure of Sylvius)**—A deep groove between the temporal lobe below and the frontal and parietal lobes above; island of Reil lies deep in the lateral fissure

4 **Parietooccipital fissure**—Groove that separates the occipital lobe from the two parietal lobes

Beneath the cerebral cortex lies the large interior of the cerebrum. It consists mostly of white matter made up of numerous tracts. A few islands of gray matter, however, lie deep inside the white matter of each hemisphere. Collectively these are called *basal ganglia,* or *cerebral nuclei.*

Tracts that make up the cerebrum's internal white matter are of three types: projection tracts, association tracts, and commissural tracts. *Projection tracts* are extensions of the ascending, or sensory, spinothalamic tracts and descending or motor corticospinal tracts. *Association tracts* are the most numerous of cerebral tracts; they extend from one convolution to another in the same hemisphere. *Commissural tracts,* in contrast, extend from one convolution to a corresponding convolution in the other hemisphere. Commissural tracts compose the

When applied to the nervous system, the term *nucleus* means an area of gray matter in the brain or cord (composed mainly, as is all gray matter, of neuron cell bodies and dendrites). Such a cluster located outside the brain and cord is called a *ganglion.* So cerebral nuclei is a more accurate (but less common) name than basal ganglia.

corpus callosum (prominent white curved structure seen in Figure 11-21) and the anterior and posterior commissures.

Basal ganglia include the following masses of gray matter in the interior of each cerebral hemisphere:

1 **Caudate nucleus**—Observe in Figure 11-22 the long, curving shape of this basal ganglion

2 **Lentiform nucleus**—So named because of its lenslike shape; note in Figure 11-23 that the lentiform nucleus consists of two structures, the *putamen* and *pallidum;* the putamen lies lateral to the pallidum (also called globus pallidus)

3 **Amygdaloid nucleus**—Observe the location of the amygdaloid at the tail of the caudate nucleus (Figure 11-22)

Figure 11-22 Basal ganglia (or cerebral nuclei) within a cerebral hemisphere. Main basal ganglia are caudate nucleus, putamen, and pallidum. Together, the putamen and pallidum constitute the lenticular (lentiform) nucleus. (See Figure 11-23.)

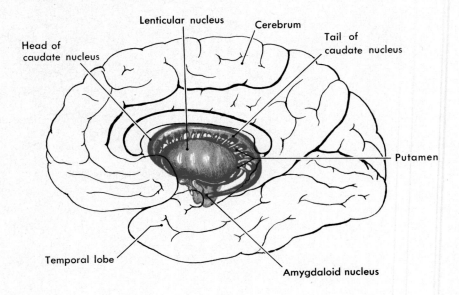

Lenticular nucleus

Cerebrum

Head of caudate nucleus

Tail of caudate nucleus

Putamen

Temporal lobe

Amygdaloid nucleus

Figure 11-23 Basal ganglia and internal capsule as seen in a transverse section through the left cerebral hemisphere. Lenticular nucleus (putamen and pallidum) plus caudate nucleus and internal capsule constitute corpus striatum.

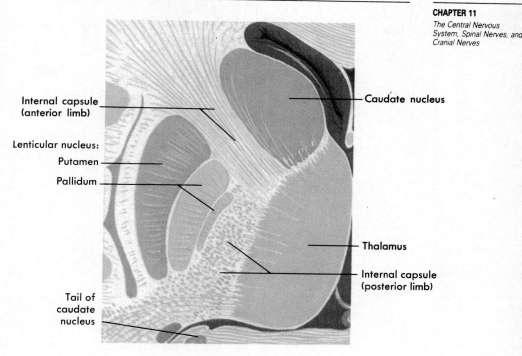

A structure associated with the basal ganglia is the *internal capsule.* It is a large mass of white matter located, as Figure 11-23 shows, between the caudate and lentiform nuclei and between the lentiform nucleus and thalamus. Together the caudate nucleus, internal capsule, and lentiform nucleus constitute the *corpus striatum.* The term means "striped body."

FUNCTIONS

During the past decade or so, research scientists in various fields—neurophysiology, neurosurgery, neuropsychiatry, and others—have added mountains of information to our knowledge about the brain. But questions come faster than answers, and clear, complete understanding of the brain's mechanisms still elude us. Perhaps it forever will. Perhaps the capacity of the human brain falls short of the ability to understand its own complexity. The following functions, however, are known to be performed by the cerebrum.

1 **Sensory functions.** Essential for normal sensations are the somatic sensory, visual, and auditory areas of the cerebral cortex (Figure 11-24).

These regions of the cortex do more than just register separate and simple sensations. They compare and evaluate them. They integrate them into perceptions of wholes. Suppose, for example, that someone blindfolded you and then put an ice cube in your hand. You would, of course, sense something cold touching your hand. But also, you would probably know that it was an ice cube because you would sense a total impression compounded of many sensations such as temperature, shape, size, weight, texture, and movement and position of your hand and arm. Discussion of somatic sensory pathways begins on p. 335.

2 **Motor functions.** Mechanisms that control voluntary movements are extremely complex and imperfectly understood. It is known, however, that for normal movements to take place, many parts of the nervous system—including certain areas of the cerebral cortex—must function. The precentral gyrus, that is, the most posterior gyrus of the frontal lobe (Figure 11-24, area 4), constitutes the primary motor area. The secondary motor area lies in the gyrus immediately anterior to the precentral gyrus. It contains motoneurons as do many other regions, including even the somatic sensory areas. Neurons in the precentral gyrus are said to control individual muscles, especially those that produce movements of distal joints (wrist, hand, finger, ankle, foot, and toe movements). Neurons in the gyrus just anterior to the precentral gyrus are thought to activate groups of muscles simultaneously. Motor pathways descending from the cerebrum through the brain stem and spinal cord are discussed on pp. 338-342.

Figure 11-24 Map of human cortex. Identity of each numbered area is determined by structural differences in neurons that compose it.

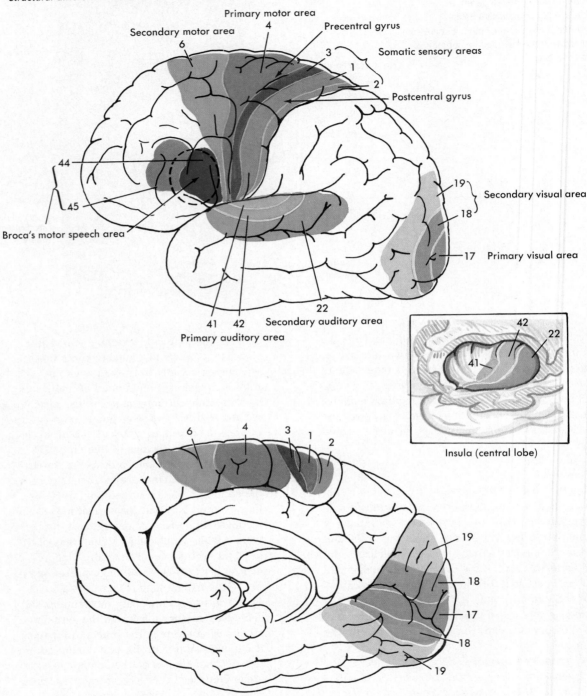

Insula (central lobe)

3 Integrative functions. "Integrative functions" is a nebulous phrase. Even more obscure, however, are the neural processes it designates. They consist of all events that take place in the cerebrum between its reception of sensory impulses and its sending out of motor impulses. Integrative functions of the cerebrum include consciousness and mental activities of all kinds. Consciousness, memory, use of language, and emotions are the integrative cerebral functions that we shall discuss briefly.

Figure 11-25 Reticular activating
system. Consists of centers in the brain
stem reticular formation plus fibers that
conduct to the centers from below and
fibers that conduct from the centers to
widespread areas of the cerebral cortex.
Functioning of the reticular activating
system is essential for consciousness.

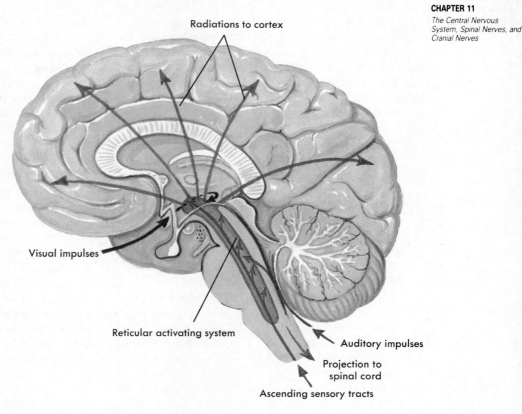

Radiations to cortex

Visual impulses

Reticular activating system

Auditory impulses

Projection to
spinal cord

Ascending sensory tracts

Consciousness may be defined as a state of awareness of one's self, one's environment, and other beings. Very little is known about the neural mechanisms that produce consciousness. One fact known, however, is that consciousness depends on excitation of cortical neurons by impulses conducted to them by a relay of neurons known as the reticular activating system. The *reticular activating system* consists of centers in the brain stem reticular formation that receive impulses from the cord and relay them to the thalamus and from the thalamus to all parts of the cerebral cortex (Figure 11-25). Both direct spinal reticular tracts and collateral fibers from the specialized sensory tracts (spinothalamic, lemniscal, auditory, and visual) relay impulses over the reticular activating system to the cortex. Without continual excitation of cortical neurons by reticular activating impulses, an individual is unconscious and cannot be aroused. Here, then, are two accepted concepts about the reticular activating system: (1) it functions as the arousal or alerting system for the cerebral cortex, and (2) its functioning is crucial for maintaining consciousness. Drugs known to depress the reticular activating system decrease alertness and induce sleep. Barbiturates, for example, act this way. Amphetamine, on the other hand, a drug known to stimulate the cerebrum and to enhance alertness and produce wakefulness, probably

acts by stimulating the reticular activating system.

Certain variations in the levels or state of consciousness are normal. All of us, for example, experience different levels of wakefulness. At times, we are highly alert and attentive. At other times, we are relaxed and nonattentive. All of us also experience different levels of sleep. Two of the best known stages are those called slow-wave sleep (SWS) and rapid eye movement (REM) sleep. SWS takes its name from the slow frequency, high-voltage delta waves that identify it. It is almost entirely a dreamless sleep. REM sleep, on the other hand, is associated with dreaming.

In addition to the various normal states of consciousness, altered states of consciousness (ASC) also occur under certain conditions. Anesthetic drugs produce an ASC, namely *anesthesia*. Disease or injury of the brain may produce the type of ASC called *coma*. Lysergic acid diethylamide (LSD), a "mind-altering" drug, induces a type of ASC known in the drug culture as a "trip."

Peoples of Eastern cultures have long been familiar with an ASC called *yoga*, or *meditation*. Yoga is a waking state but differs markedly in certain respects from the usual waking state. According to an accepted definition, yoga is a "higher" or "expanded" level of consciousness. This higher consciousness is accompanied, almost paradoxically, by a high degree of both relaxation and alertness. With training in meditation techniques and practice, an individual can enter the meditative state at will and remain in it for an extended period.

Memory is one of our major mental activities. It is an established fact that long-term memory is a function of many parts of the cerebral cortex, especially of the temporal, parietal, and occipital lobes. Findings made by Dr. Wilder Penfield, a noted Canadian neurosurgeon, first gave evidence of this in the 1920s. He electrically stimulated the temporal lobes of epileptic patients undergoing brain surgery. They responded, much to his surprise, by recalling in the most minute detail songs and events from their past. Such long-term memories are believed to consist of some kind of structural traces—called *engrams*—in the cerebral cortex. Widely accepted today is the theory that an engram consists of some kind of change in the synapses in a specific circuit of neurons. Repeated impulse conduction over a given neuronal circuit produces the synaptic change. What the change is is still a matter of speculation. Two suggestions are that it represents an increase in the number of presynaptic axon terminals or an increase in the number of receptor proteins in the postsynaptic neuron's membrane. Whatever the change is, it is such that it facilitates impulse transmission at the synapses.

A number of research findings indicate that the cerebrum's limbic system—the "emotional brain"—plays a key role in memory. To mention one role, when the hippocampus (part of the limbic system) is removed, the patient loses the ability to recall new information. Personal experience substantiates a relationship between emotion and memory.

Language functions consist of the ability to speak and write words and the ability to understand spoken and written words. Certain areas in the frontal, parietal, and temporal lobes serve as speech centers—as crucial areas, that is, for language functions. The left cerebral hemisphere contains these areas in about 90% of the population; in the remaining 10%, either the right hemisphere or both hemispheres contain them. Lesions in speech centers give rise to language defects called *aphasias*. For example, with damage to an area in the inferior gyrus of the frontal lobe (Broca's area, Figure 11-24), a person becomes unable to articulate words but can still make vocal sounds and understand words heard and read.

Emotions—both the subjective experiencing and objective expression of them—involve functioning of the cerebrum's limbic system. The name *limbic* (Latin for "border or fringe") suggests the shape of the cortical structures that make up the system. They form a curving border around the corpus callosum, the structure that connects the two cerebral hemispheres. Look now at Figure 11-21. Here on the medial surface of the cerebrum lie most of the structures of the limbic system. They are the cingulate gyrus, the isthmus, the hippocampal gyrus, the uncus, and the hippocampus (the extension of the hippocampal gyrus that protrudes into the floor of the inferior horn of the lateral ventricle). These limbic system structures have primary connections with various other parts of the brain, notably the septum, the amygdala (the tail of the caudate nucleus, one of the basal ganglia), and the hypothalamus. Some physiologists, therefore, include these connected structures as parts of the limbic system.

The limbic system (or to use its more descriptive name, the *emotional brain*) functions in some way to make us experience many kinds of emotions—anger, fear, sexual feelings, pleasure, and sorrow, for example. To bring about the normal expression of emotions, parts of the cerebral cortex other than the limbic system must also function. Considerable evidence exists that limbic activity without the modulating influence of the other cortical areas may bring on the attacks of abnormal, uncontrollable rage suffered periodically by some unfortunate individuals.

Some Generalizations About Cerebral Functions

1 Cerebral activity goes on as long as life itself. Only when life ceases (or moments before) does the cerebrum cease its functioning. Only then do all of its neurons stop conducting impulses. Proof of this has come from records of brain electrical potentials known as electroencephalograms, or EEGs, or "brain waves." These records are usually made from a number of electrodes placed on different regions of the scalp, and they consist of waves—*brain waves*, as they are called (Figure 11-26).

Four types of brain waves are recognized based on frequency and amplitude of the waves. Frequency, or the number of wave cycles per second, is usually referred to as Hertz (Hz, from Hertz, a German physicist). Amplitude means voltage. Listed in order of frequency from fastest to slowest, brain wave names are *beta, alpha, theta,* and *delta. Beta waves* have a frequency of over 13 Hz and a relatively low voltage. *Alpha waves* have a frequency of 8 to 13 Hz and a relatively high voltage. *Theta waves* have both a relatively low frequency—4 to 7 Hz—and a low voltage. *Delta waves* have the slowest frequency—less than 4 Hz—but a high voltage. Brain waves vary in different regions of the brain, in different states of awareness, and in abnormal conditions of the cerebrum.

Figure 11-26 Electroencephalography. **A,** Electrodes applied to various areas of the scalp. **B,** Resulting electroencephalogram tracings from electrodes placed on the scalp of a subject in a quiet, semidarkened room. Multiple tracings measure brain wave activity from different areas of the cerebral cortex.

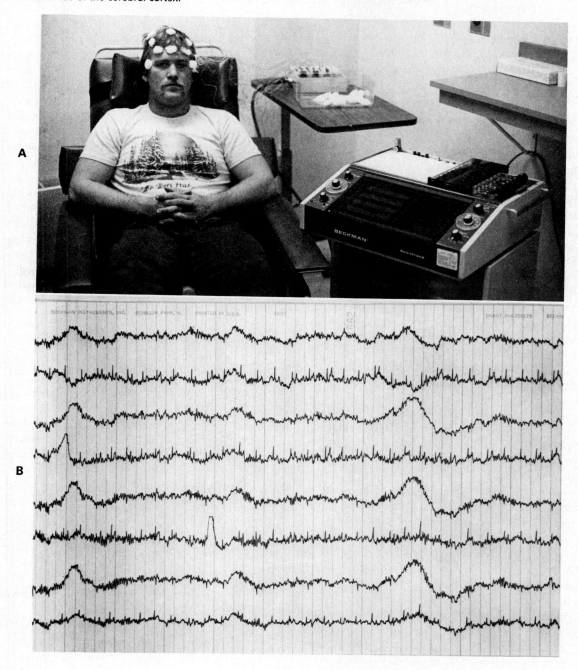

Fast, low-voltage beta waves characterize EEGs recorded from the frontal and central regions of the cerebrum when an individual is awake, alert, and attentive, with eyes open. Beta waves predominate when the cerebrum is busiest, that is, when it is engaged with sensory stimulation or mental activities. In short, beta waves are "busy waves." Alpha waves, in contrast, are "relaxed waves." They are moderately fast, relatively high-voltage waves that dominate EEGs recorded from the parietal lobe, occipital lobe, and posterior parts of the temporal lobes

Physicians use **electroencephalograms** to help localize areas of brain dysfunction, to identify altered states of consciousness, and often to establish death. Two flat EEG recordings (no brain waves) taken 24 hours apart in conjunction with no spontaneous respiration and total absence of somatic reflexes are criteria accepted as evidence of brain death.

when the cerebrum is idling, so to speak. The individual is awake but has eyes closed and is in a relaxed, nonattentive state. When drowsiness descends, moderately slow, low-voltage theta waves appear. Theta waves are "drowsy waves." "Deep sleep waves," on the other hand, are delta waves. These slowest brain waves characterize the deep sleep from which one is not easily aroused. For this reason, deep sleep is referred to as slow-wave sleep (SWS).

2 The right and left hemispheres of the cerebrum specialize in different functions. For example, as already noted, the left hemisphere specializes in language functions—it does the talking, so to speak. The left hemisphere also appears to dominate the control of certain kinds of hand movements, notably skilled and gesturing movements. Most people use their right hands for performing skilled movements, and the left side of the cerebrum controls the muscles on the right side that execute these movements. The next time you are with a group of people who are talking, observe their gestures. The chances are about 9:1 that they will gesture mostly with their right hands—indicative of left cerebral control.

Evidence that the right hemisphere of the cerebrum specializes in certain functions has rather recently been reported. As of now, it seems that one of the right hemisphere's specialties is the perception of certain kinds of auditory material. For instance, some studies have shown that the right hemisphere perceives non-speech sounds such as melodies, coughing, crying, and laughing better than does the left hemisphere. The right hemisphere may also function better at tactual perception and for perceiving and visualizing spatial relationships.

3 Certain areas of the cortex in each hemisphere of the cerebrum engage predominantly in one particular function. What that function is depends on what structures the cortical area receives impulses from or sends impulses to. For example, the postcentral gyrus (areas 3, 1, and 2 in Brodmann's map, Figure 11-24) functions mainly as a general somatic sensory area. It receives impulses from receptors activated by heat, cold, and touch stimuli. The precentral gyrus (area 4), on the other hand, functions chiefly as the somatic motor area. Impulses from neurons in this area descend over motor tracts and eventually stimulate somatic effectors, the skeletal muscles. The transverse gyrus of the temporal lobe (areas 41 and 42 in Figure 11-24) serves as the primary auditory area. The primary visual area lies in the occipital lobe (area 17). It is important to remember that no part of the brain functions alone. Many structures of the central nervous system must function in order for any one part of the brain to function.

4 For the cerebral cortex to perform its *sensory* functions, impulses must first be conducted to its sensory areas by way of relays of neurons referred to as sensory pathways (p. 335). For the cerebral cortex to perform its *motor* functions, impulses must be conducted from its motor areas to skeletal muscles by relays of neurons referred to as somatic motor pathways (p. 338).

NEUROTRANSMITTERS IN CORD AND BRAIN

Neurotransmitters are chemicals by which neurons talk to one another. At billions, or more likely, trillions, of synapses in the central nervous system, presynaptic neurons release neurotransmitters that act to facilitate, stimulate, or inhibit postsynaptic neurons. Established as neurotransmitters are at least 30 different compounds. They are not distributed diffusely or at random through the cord and brain. Instead, as we shall see, specific neurotransmitters are localized in discrete groups of neurons and released in specific pathways.

CLASSES OF NEUROTRANSMITTERS

Neurotransmitters belong mainly to the following three classes of compounds:

1 **Amino acids**—Examples: glycine (simplest of all amino acids), aspartic acid, glutamic acid, and gamma-aminobutyric acid (GABA)

2 **Monoamines** or **catecholamines** (compounds that contain one amino group and a ring of six carbon atoms called a *catechol*)—Examples: norepinephrine, dopamine, and serotonin (5-hydroxytryptamine)

3 **Neuropeptides** (short chains of 2 to 39 amino acids)—Examples: enkephalins, endorphins, substance P, vasoactive intestinal peptide, cholecystokinin, somatostatin, vasopressin, and oxytocin

Table 11-5 Neurotransmitter locations and functions in brain and cord

Neurotrans-mitter	Locations of neurons synthesizing	Locations of synapses where released	Functions
Norepinephrine	Brain stem, in area called locus cereleus ("blue place")	Hypothalamus, cerebellum, and cerebral cortex	Involved in maintaining arousal, mood regulation, pleasure or reward emotions
Dopamine	Brain stem, in substantia nigra (deeply pigmented area)	Basal ganglia, especially caudate nucleus, and limbic system	Regulation of movements; involved in emotions
Enkephalins	Brain stem, in gray matter around aqueduct of Sylvius and fourth ventricle	Substantia gelatinosa; thalamus, limbic system (amygdala), and other brain pain control areas	Inhibit conduction of pain impulses
Endorphins	Hypothalamus	Thalamus(?)	Inhibit pain conduction

GENERAL INFORMATION ABOUT NEUROTRANSMITTERS

Some general information about neurotransmitters is given in the following statements:

1 Precise locations in the central nervous system of some neurotransmitters have been identified (Table 11-5).
2 The most common excitatory neurotransmitters released into brain synapses are glutamic and aspartic amino acids.
3 The most common inhibitory neurotransmitter released into brain synapses is also an amino acid, namely, GABA.
4 An inhibitory neurotransmitter released mainly at synapses in the spinal cord is the simplest amino acid, glycine.
5 The neuropeptide known as substance P is released by axon terminals of the sensory neurons that conduct from pain receptors to the cord. It is the neurotransmitter that mediates conduction across synapses in the substantia gelatinosa to neurons that conduct pain impulses up the cord to the brain. The substantia gelatinosa is the part of the posterior gray horns of the cord identified as a pain control area.
6 Enkephalins and endorphins are neuropeptides released at various cord and brain synapses in the pain conduction pathway. These neurotransmitters inhibit conduction of pain impulses. They serve the body as its own natural opiates.
7 Some neuropeptides, namely, substance P, vasoactive intestinal peptide, and cholecystokinin, were found first in the gut. Later it was established they were neurotransmitters.
8 Some neuropeptides—for example, vasopressin (antidiuretic hormone), and oxytocin—were first identified as hormones. Later their role as neurotransmitters was established.

SYNTHESIS, STORAGE, RELEASE, ACTIVATION, AND INACTIVATION OF NEUROTRANSMITTERS

Neurotransmitter molecules are synthesized in axon terminals. Usually an amino acid is converted by enzymes to the neurotransmitter compound. Norepinephrine, for example, is synthesized from the amino acid tyrosine by three enzyme-catalyzed steps that change it first to dopa, then dopamine, and then norepinephrine. After their formation, neurotransmitter molecules are stored in vesicles in axon terminals. Arrival of a nerve impulse at the terminal induces release of the molecules into a synaptic cleft. By binding to specific receptors in the postsynaptic neuron's membrane, the neurotransmitter produces rapid changes that either excite or inhibit the postsynaptic neuron. Norepinephrine and many other transmitters operate by a somewhat different mechanism. The receptor protein for such transmitters lies coupled to the enzyme adenyl cyclase in the postsynaptic neuron's membrane. Binding of the neurotransmitter to such a receptor activates the enzyme. It then, as Figure 11-27 shows, catalyzes the conversion of ATP (adenosine triphosphate) into cyclic AMP (adenosine monophosphate or "second messenger" substance). Cyclic AMP, in turn, activates an enzyme, kinase, which catalyzes the addition of phosphate to another protein in the postsynaptic neuron's membrane. This changes the shape of the protein in such a way that sodium channels in the postsynaptic membrane become open. Net inward diffusion of sodium then occurs, producing

Figure 11-27 Mechanism by which norepinephrine and many other neurotransmitters initiate nerve impulse conduction (compare with Figure 10-12, p. 295).

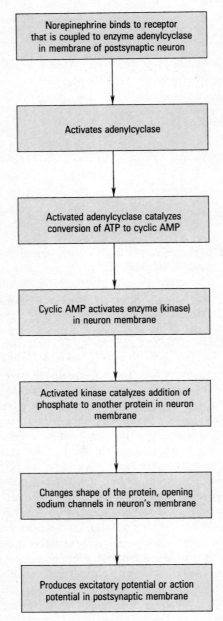

Much of our present knowledge about chemical transmitters stems from years of research by Sweden's Dr. Ulf S. von Euler, the United States' Dr. Julius Axelrod, and England's Sir Bernard Katz. For their work, these eminent scientists shared the 1970 Nobel Prize in medicine and physiology. Their most significant findings include the following:

1 Dr. von Euler identified norepinephrine (NE) as the major transmitter in some brain synapses
2 Dr. Axelrod identified the enzyme catechol-O-methyl transferase (COMT) and discovered that it inactivates norepinephrine
3 Sir Bernard Katz discovered that conducting cholinergic fibers rapidly release numerous packets of acetylcholine into synapses

This knowledge led to other discoveries and to valuable applications. For example, researchers later learned that severe psychic depression occurs when a deficit of NE exists in certain brain synapses. This finding led to the development of antidepressant drugs. Certain ones of these inhibit COMT. Because inhibited COMT does not inactivate NE, the amount of active NE in brain synapses increases, and this relieves the individual's depression. The graphic names "psychic energizers" and "mood elevators" refer to antidepressant drugs, now valuable weapons in medicine's arsenal for combatting mental disease.

either an excitatory potential or action potential in the postsynaptic neuron.

Neurotransmitter action is rapidly terminated by one or more means. For example, termination of norepinephrine action occurs by two methods. The enzyme catechol-O-transferase degrades norepinephrine in the synaptic cleft. Also, reuptake of norepinephrine into the axon terminal occurs and is followed by its degradation by the enzyme monoamine oxidase.

Clinical Applications

Abnormal amounts of dopamine at brain synapses is associated with certain diseases. Dopamine deficiency in the basal ganglia, for example, characterizes Parkinson's disease. The deficiency stems from degeneration of the neurons that synthesize dopamine, those whose cell bodies are located in the substantia nigra and whose axons terminate in the basal ganglia. Since the late 1960s, L-dopa, a precursor of dopamine, has been used to treat Parkinson's disease—often with dramatic relief of symptoms (tremors, rigidity, and other abnormalities of movements). In contrast to the dopamine deficiency of Parkinson's disease, dopamine excess, particularly in the limbic system of the brain, underlies the symptoms of schizophrenia, a major mental disease. The

Depression and mania are postulated to be associated with abnormal levels of monoamine transmitters (norepinephrine, dopamine, serotonin) at brain synapses—low levels with depression, high levels with mania. Drugs widely used to relieve depression are imipramine (Tofranil) and amitriptyline (Elavil). They increase monoamine levels at brain synapses by blocking their uptake into axon terminals. Cocaine also appears to have this action.

drug amphetamine triggers release of dopamine at brain synapses.

PAIN CONTROL AREAS IN SPINAL CORD AND BRAIN

The term *pain control area* means a place in the pain conduction pathway where impulses from pain receptors can be inhibited. The first pain control area suggested was a segment of the posterior gray horns of the spinal cord. *Substantia gelatinosa* is the name of this segment. Here, axon terminals of neurons that conduct from pain receptors to cord synapse with neurons that conduct pain impulses up the cord to the thalamus. Some 20 years ago researchers made a surprising discovery about these synapses in the substantia gelatinosa. They found they could inhibit pain conduction across them by stimulating skin touch receptors in a painful area. From this knowledge, a new theory about pain developed—the aptly named *gate-control theory of pain*. According to this theory, the substantia gelatinosa functions as a gate that can close to bar the entry of pain impulses into ascending paths to the brain. One way to close the gate is to stimulate skin touch receptors in a painful area. Today this is usually done by a device called the transcutaneous electrical stimulator. A patient uses this stimulator to apply a low level of stimulation for a long period of time. This results in closure of the spinal cord pain gate and relief of pain.

In recent years, pain control areas have been identified in the brain, notably in the gray matter around the aqueduct of Sylvius and around the third ventricle. Neurons in these areas send their axons down the spinal cord to terminate in the substantia gelatinosa. Here some evidence suggests that they release enkephalins. These act to prevent pain impulse conduction across synapses in the substantia gelatinosa. In short, enkephalins tend to close the spinal cord pain gate. Brief, intense transcutaneous stimulation at trigger or acupuncture points has been found to relieve pain in distant sites. The intense stimulation is postulated to activate brain pain control areas. They then send impulses down the cord to the substantia gelatinosa, closing the spinal cord pain gate.

SOME GENERALIZATIONS ABOUT SOMATIC SENSORY PATHWAYS

1 Most impulses that reach the sensory areas of the cerebral cortex have traveled over at least three pools of sensory neurons. We shall designate these as sensory neurons I, II, and III (Table 11-6).

   a Sensory neurons I of the relay conduct from the periphery to the central nervous system. If the receptors of these neurons lie in regions supplied by spinal nerves, their dendrites lie in a spinal nerve and their axons terminate in gray matter of the cord or brain stem. Where are their cell bodies located? (Confirm or find your answer in Figure 10-7.) If receptors of sensory neurons I lie in regions supplied by cranial nerves, their dendrites lie in a cranial nerve, their cell bodies lie in cranial nerve ganglia, and their axons terminate in gray matter of the brain stem. In either case, sensory neuron I axon terminals synapse with sensory neuron II dendrites or cell bodies (Figure 11-28).

   b Sensory neurons II conduct from the cord or brain stem up to the thalamus. Their dendrites and cell bodies are located in cord or brain stem gray matter. Their axons ascend in ascending tracts up the cord, through the brain stem, and terminate in the thalamus. Here they synapse with sensory neuron II dendrites or cell bodies (Figure 11-28).

   c Sensory neurons III conduct from the thalamus to the postcentral gyrus of the parietal lobe, the somaticosensory area. Bundles of axons of sensory neurons III form thalamocortical tracts. They extend through the portion of cerebral white matter known as the *internal capsule* to the cerebral cortex (Figure 11-28).

2 For the most part, sensory pathways to the cerebral cortex are crossed pathways. This means that each side of the brain registers sensations from the opposite side of the body. Look again at Figure 11-28. The axon that decussates (crosses from one side to the other) in this sensory pathway is part of which sensory neuron—I, II, or III? Usually it is the axon of sensory

Figure 11-28 Lateral spinothalamic tract relays sensory impulses from crude touch, pain, and temperature receptors up the cord to the thalamus. Thalamocortical tract fibers relay the impulses to the somatic sensory area of the cortex (postcentral gyrus).

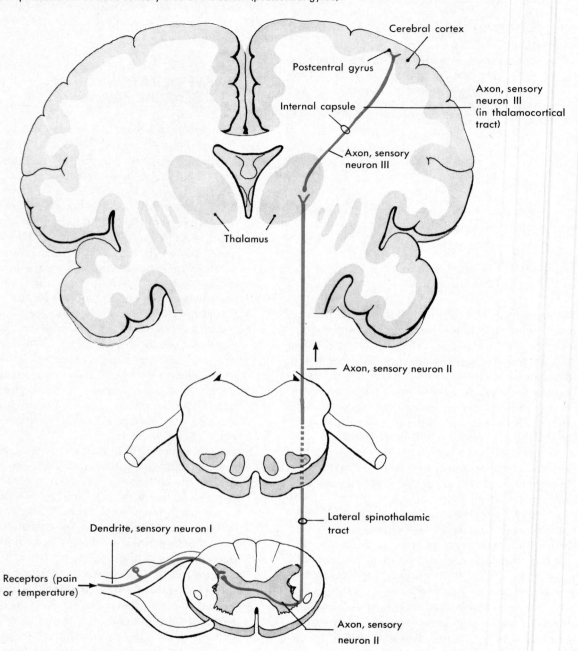

neuron II that decussates at some level in its ascent to the thalamus.

3 The sensory pathway for crude touch, pain, and temperature sensations is called the *spinothalamic pathway.*

4 Two sensory pathways conduct impulses that produce sensations of touch and pressure, namely, the medial lemniscal system and the

spinothalamic pathway. The *medial lemniscal system* consists of the tracts that make up the posterior white columns of the cord (the fasciculi cuneatus and gracilis) plus the *medial lemniscus,* a flat band of white fibers extending through the medulla, pons, and midbrain. (Derivation of the name *lemniscus* may interest you. It comes from the Greek word *lemniskos* mean-

Table 11-6 Sensory neural pathways

Neurons	Gross structures in which neuron parts are located
PAIN AND TEMPERATURE	
Sensory neuron I	
Receptors	In skin, mucosa, muscles, tendons, viscera
Dendrite	In spinal nerve and branch of spinal nerve
Cell body	In spinal ganglion, on posterior root of spinal nerve
Axon	In posterior root of spinal nerve; terminates in posterior gray column of cord
Sensory neuron II	
Dendrite	Posterior gray column
Cell body	Posterior gray column
Axon	Decussates and ascends in lateral spinothalamic tract (Figures 11-6 and 11-28); terminates in thalamus
Sensory neuron III	
Dendrite	Thalamus
Cell body	Thalamus
Axon	Thalamus via thalamocortical tract in internal capsule to general sensory area of cerebral cortex, that is, postcentral gyrus in parietal lobe
DISCRIMINATING TOUCH (TWO-POINT DISCRIMINATION, VIBRATIONS), DEEP TOUCH, AND PRESSURE AND KINESTHESIA	
Sensory neuron I	Same as sensory neuron I for pain, temperature, and crude touch stimuli, except that axon extends up cord in posterior white columns (fasciculi gracilis and cuneatus, Figure 11-6) to nucleus gracilis or cuneatus in medulla instead of terminating in posterior gray columns of cord
Sensory neuron II	
Dendrite	In nucleus gracilis or cuneatus of medulla
Cell body	In nucleus gracilis or cuneatus of medulla
Axon	Decussates and ascends in medial lemniscus (broad band of fibers extending up through medulla and midbrain) and terminates in thalamus
Sensory neuron III	Same as sensory neuron III for pain, temperature, and crude touch stimuli

ing "woolen band." Apparently, to some early anatomist, the medial lemniscus looked like a band of woolen material running through the brain stem.)

The fibers of the medial lemniscus, like those of the spinothalamic tracts, are axons of sensory neurons II. They originate from cell bodies in the medulla, decussate, and then extend upward to terminate in the thalamus on the opposite side. The function of the medial lemniscal system is to transmit impulses that produce our more discriminating touch and pressure sensations, including stereognosis (awareness of an object's size, shape, and texture), precise localization, two-point discrimination, weight discrimination, and sense of vibrations.

Crude touch and pressure sensations are functions of the spinothalamic, not the medial lemniscal, pathway. Knowing that something touches the skin is a crude touch sensation, whereas knowing its precise location, size, shape, or texture involves discriminating touch sensations.

5 The sensory pathway for *kinesthesia* (sense of movement and position of body parts) is also part of the medial lemniscal system. See Table 11-6 for a brief summary of the sensory neural pathways named in items 2, 3, and 4.

6 The first part of the cerebral cortex reached by sensory impulses conducted up spinothalamic and medial lemniscal pathways is the postcentral gyrus of the parietal lobe. This gyrus is known as the somatosensory area. In Brodmann's map of the cortex (see Figure 11-24), it is designated areas 3, 1, and 2. The somatosensory area functions to produce our general senses—not only the common ones such as pain, heat, cold, touch, and pressure, but also those

Table 11-7 Pyramidal path from cerebral cortex

Microscopic structures	Macroscopic structures in which neurons located
UPPER MOTONEURON (BETZ CELLS)	
Dendrite	Motor area of cerebral cortex
Cell body	Motor area of cerebral cortex
Axon	Motor area of cerebral cortex; descends in corticospinal (pyramidal) tract through cerebrum and brain stem; decussates in medulla and continues descent in lateral corticospinal (crossed pyramidal) tract in lateral white column; or may descend uncrossed in ventral, or direct, pyramidal tract in anterior white column and either decussate or not prior to terminating in anterior gray column
LOWER MOTONEURON	
Dendrite	Anterior gray column
Cell body	Anterior gray column
Axon	Anterior gray column to anterior root of spinal nerve to spinal nerve and branches; terminates in somatic effector, that is, skeletal muscle

less familiar senses such as stereognosis, kinesthesia, vibratory sense, and two-point and weight discrimination. General sensations of the right side of the body are predominantly experienced by the left cerebral hemisphere's somatosensory area. General sensations of the left side of the body are predominantly experienced by the right hemisphere's somatosensory area.

SOME GENERALIZATIONS ABOUT SOMATIC MOTOR PATHWAYS

Somatic motor pathways consist of motoneurons that conduct impulses from the central nervous system to somatic effectors, that is, skeletal muscles. Some motor pathways are extremely complex and not at all clearly defined. Others, notably spinal cord reflex arcs, are simple and well established. You read about these in Chapter 10. Look back now at Figure 10-7. From this diagram you can derive a cardinal principle about somatic motor pathways—the *principle of the final common path*. It is this: only one final common path, namely, the anterior horn motoneuron, conducts *impulses* to skeletal muscles. Anterior horn motoneuron axons are the only ones that terminate in skeletal muscle cells. This principle of the final common path to skeletal muscles has important practical implications. For example, it means that any condition that makes anterior horn motoneurons unable to conduct impulses also makes skeletal muscle cells supplied by these neurons unable to contract. They cannot be willed to contract nor can they contract reflexly. They are, in short, paralyzed. Most famous of the diseases that

produce paralysis by destroying anterior horn motoneurons is poliomyelitis. Numerous somatic motor paths conduct impulses from motor areas of the cerebrum down to anterior horn motoneurons at all levels of the cord.

Two methods are used to classify somatic motor pathways—one based on the location of their fibers in the medulla and the other on their influence on the lower motoneurons. The first method divides them into pyramidal (Table 11-7) and extrapyramidal tracts. The second classifies them as facilitatory and inhibitory tracts.

Pyramidal tracts are those whose fibers come together in the medulla to form the pyramids, hence their name. Because axons composing the pyramidal tracts originate from neuron cell bodies located in the cerebral cortex, they also bear another name—*corticospinal tracts*. About three fourths of their fibers decussate (cross over from one side to the other) in the medulla. After decussating, they extend down the cord in the crossed corticospinal tract located on the opposite side of the cord in the lateral white column. About one fourth of the corticospinal fibers do not decussate. Instead, they extend down the same side of the cord as the cerebral area from which they came. One pair of uncrossed tracts lies in the ventral white columns of the cord, namely, the ventral corticospinal tracts. The other uncrossed corticospinal tracts form part of the lateral corticospinal tracts (see Figure 11-6). About 60% of corticospinal fibers are axons that arise from neuron cell bodies in the precentral (frontal lobe) region of the cortex. In Figure 11-24, these are the areas numbered 4 and 6. About 40% of corticospinal fibers originate from neuron cell bodies located in postcentral

Figure 11-29 Crossed corticospinal (pyramidal) tracts. Axons that compose pyramidal tracts come from neuron cell bodies in the cerebral cortex. After they descend through the internal capsule of the cerebrum and the white matter of the brain stem, about three fourths of the fibers decussate—cross over from one side to the other—in the medulla, as shown here. Then they continue downward in the lateral corticospinal tract (Figure 11-6) on the opposite side of the cord. Each crossed corticospinal tract, therefore, conducts motor impulses from one side of the brain to interneurons or anterior horn motoneurons on the opposite side of the cord. Therefore impulses from one side of the cerebrum cause movements of the opposite side of the body.

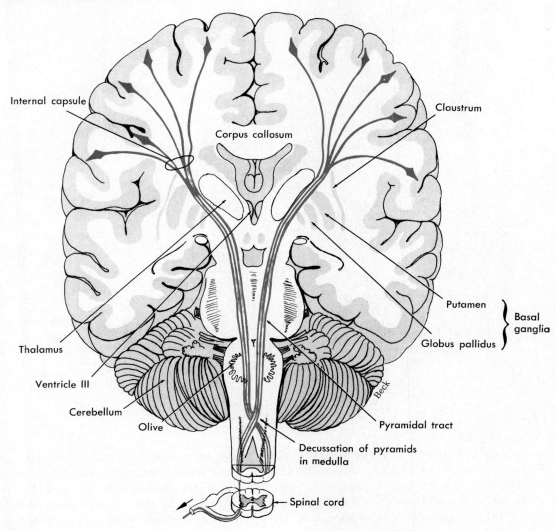

areas of the cortex, areas classified as sensory; now, more accurately, they are often called sensorimotor areas.

Relatively few corticospinal tract fibers synapse directly with anterior horn motoneurons. Most of them synapse with interneurons, which in turn synapse with anterior horn motoneurons. All corticospinal fibers conduct impulses that facilitate, that is, lower the resting negativity of anterior horn motoneurons. The effects of facilitatory impulses acting

rapidly on any one neuron add up or summate. Each impulse, in other words, decreases the neuron's resting potential a little bit more. If sufficient numbers of impulses impinge rapidly enough on a neuron, its negativity decreases to threshold level. At that moment the neuron starts conducting impulses—it is stimulated. Stimulation of anterior horn motoneurons by corticospinal tract impulses results in stimulation of individual muscle groups (mainly of the hands and feet). Precise control of their contractions

Figure 11-30 Diagram on left shows axons of three presynaptic neurons synapsing with one postsynaptic neuron (convergence). Diagram on right shows branches of one presynaptic neuron's axon synapsing with four postsynaptic neurons (divergence).

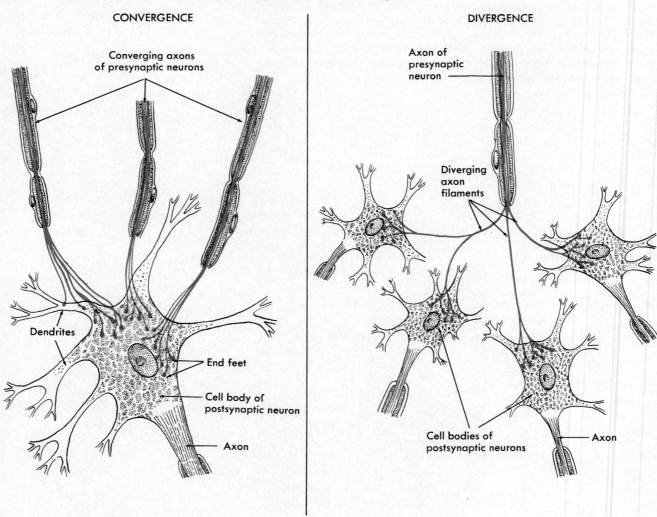

CONVERGENCE

Converging axons
of presynaptic neurons

Dendrites

End feet

Cell body of
postsynaptic neuron

Axon

DIVERGENCE

Axon of
presynaptic
neuron

Diverging
axon
filaments

Cell bodies of
postsynaptic neurons

Axon

is, in short, the function of the corticospinal tracts. Without stimulation of anterior horn motoneurons by impulses over corticospinal fibers, willed movements cannot occur. This means that paralysis results whenever pyramidal corticospinal tract conduction is interrupted. For instance, the paralysis that so often follows cerebral vascular accidents ("strokes") comes from pyramidal neuron injury—sometimes of their cell bodies in the motor areas, sometimes of their axons in the internal capsule (Figure 11-29).

Extrapyramidal tracts are much more complex than pyramidal tracts. They consist of all motor tracts from the brain to the spinal cord anterior horn motoneurons except the corticospinal (pyramidal) tracts. Within the brain, extrapyramidal tracts consist of numerous but as yet incompletely worked-out relays of motoneurons between motor areas of

the cortex, basal ganglia, thalamus, cerebellum, and brain stem. In the cord, some of the most important extrapyramidal tracts are the reticulospinal tracts.

Fibers of the *reticulospinal tracts* originate from cell bodies in the reticular formation of the brain stem and terminate in gray matter of the spinal cord, where they synapse with interneurons that synapse with lower (anterior horn) motoneurons. Some reticulospinal tracts function as facilitatory tracts, others as inhibitory tracts. Impulses over facilitatory tracts tend to decrease the lower motoneuron's resting potential. Impulses over inhibitory tracts tend to increase its negativity. Summation of these opposing influences determines the lower motoneuron's response. It initiates impulse conduction only when facilitatory impulses exceed inhibitory impulses sufficiently to decrease the lower motoneuron's negativity to its threshold level.

Conduction by extrapyramidal tracts plays a crucial part in producing our larger, more automatic movements because extrapyramidal impulses bring about contractions of groups of muscles in sequence or simultaneously. Such muscle action occurs, for example, in swimming and walking and, in fact, in all normal voluntary movements.

Conduction by extrapyramidal tracts plays an important part in our emotional expressions. For instance, most of us smile automatically at things that amuse us and frown at things that irritate us. It is extrapyramidal, not pyramidal, impulses that produce the smiles or frowns.

Axons of many different neurons converge on, that is, synapse, with each anterior horn motoneuron (Figure 11-30). Hence many impulses from diverse sources—some facilitatory and some inhibitory—continually bombard this final common path to skeletal muscles. Together the added or summated effect of these opposing influences determines lower motoneuron functioning. Facilitatory impulses reach these cells via sensory neurons (whose axons, you will recall, lie in the posterior roots of spinal nerves), pyramidal (corticospinal) tracts, and extrapyramidal facilitatory reticulospinal tracts. According to recent evidence, impulses over facilitatory reticulospinal fibers facilitate the lower motoneurons that supply extensor muscles. At the same time, they reciprocally inhibit the lower motoneurons that supply flexor muscles. Hence facilitatory reticulospinal impulses tend to increase the tone of extensor muscles and decrease the tone of flexor muscles.

Inhibitory impulses reach lower motoneurons mainly via inhibitory reticulospinal fibers that originate from cell bodies located in the *bulbar inhibitory area* in the medulla. They inhibit the lower motoneurons to extensor muscles (and reciprocally stimulate those to flexor muscles). Hence inhibitory reticulospinal impulses tend to decrease extensor muscle tone and increase flexor muscle tone—opposite effects from facilitatory reticulospinal impulses.

Brain stem inhibitory and facilitatory areas are influenced directly or indirectly by impulses from various higher motor centers. Some experimental data suggest that the basal ganglia and cerebellum send impulses to the thalamus, which in turn relays them to motor areas of the cortex (Figure 11-31). This concept, in other words, sees the basal ganglia and cerebellum functioning as the highest command centers for movements and relaying their orders through the cerebral motor cortex to skeletal muscles. If this is true, it implies that the function of the cerebral motor cortex may be to refine skeletal muscle actions rather than to will them. The traditional view is quite different. It sees the cerebral motor

Figure 11-31 Extrapyramidal motor paths, suggested by recent experimental studies.

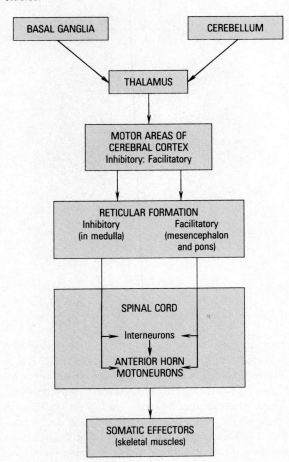

cortex functioning as the highest command center for movements and relaying some of its orders through the basal ganglia and cerebellum. If this is true, it implies that the cerebral motor cortex wills skeletal muscle actions and that the basal ganglia and cerebellum refine them. The truth is that no one yet knows the truth. Physiologists still cannot tell us positively what part of the brain initiates the commands that lead to movements, nor can they describe with assurance the neural pathways through the brain from higher to lower motor centers. Whatever these pathways may be, the ratio of facilitatory and inhibitory impulses converging on lower motoneurons (anterior horn motoneurons) is such as to maintain normal muscle tone. In other words, facilitatory impulses normally somewhat exceed inhibitory impulses. But disease sometimes alters this ratio. Parkinson's disease and "strokes," for example, may interrupt transmission by inhibitory extrapyramidal paths from basal ganglia to bulbar inhibitory centers. Facilitatory impulses then predomi-

Figure 11-32 Neural pathway involved in the patellar reflex.

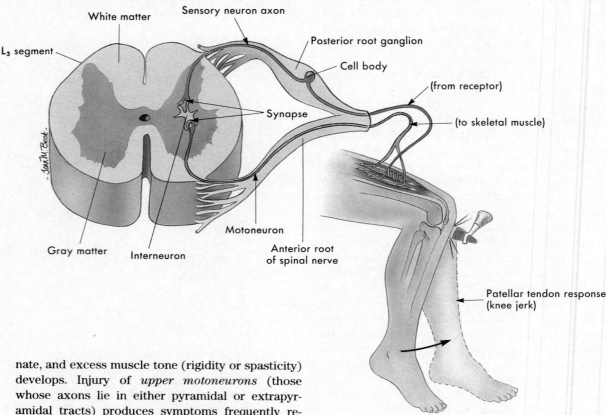

nate, and excess muscle tone (rigidity or spasticity) develops. Injury of *upper motoneurons* (those whose axons lie in either pyramidal or extrapyramidal tracts) produces symptoms frequently referred to as "pyramidal signs," notably a spastic type of paralysis, exaggerated deep reflexes, and a positive Babinski reflex (p. 343). Actually, pyramidal signs result from interruption of both pyramidal and extrapyramidal pathways. The paralysis stems from interruption of pyramidal tracts, whereas the spasticity (rigidity) and exaggerated reflexes come from interruption of inhibitory extrapyramidal pathways.

Injury to lower motoneurons produces symptoms different from those of upper motoneuron injury. Anterior horn cells or lower motoneurons, you will recall, constitute the final common path by which impulses reach skeletal muscles. This means that if they are injured, impulses can no longer reach the skeletal muscles they supply. This in turn results in the absence of all reflex and willed movements produced by contraction of the muscles involved. Unused, the muscles soon lose their normal tone and become soft and flabby (flaccid). In short, absence of reflexes and flaccid paralysis are the chief "lower motoneuron signs."

REFLEXES

DEFINITION

The action that results from a nerve impulse passing over a reflex arc is called a **reflex** (Figure 11-32 and

Table 11-8). In other words, a reflex is a response to a stimulus. It may or may not be conscious. Usually the term is used to mean only involuntary responses rather than those directly willed, that is, involving cerebral cortex activity.

A reflex consists of either muscle contraction or glandular secretion. **Somatic reflexes** are contractions of skeletal muscles. Impulse conduction over somatic reflex arcs—arcs whose motoneurons are somatic motoneurons, that is, anterior horn neurons or lower motoneurons—produces somatic reflexes. **Autonomic** (or **visceral**) **reflexes** consist either of contractons of smooth or cardiac muscle or secretion by glands; they are mediated by impulse conduction over autonomic reflex arcs, the motoneurons of which are autonomic neurons (discussed in the next chapter). The following paragraphs describe only somatic reflexes.

SOME SOMATIC REFLEXES OF CLINICAL IMPORTANCE

Clinical interest in reflexes stems from the fact that they deviate from normal in certain diseases. So the testing of reflexes is a valuable diagnostic aid. Frequently tested are the following reflexes: knee jerk,

Table 11-8 Correlation of microscopic and macroscopic structures of a three-neuron cord reflex arc*

Microscopic structures	Macroscopic structures in which neurons located
SENSORY NEURON	
Receptor	In skin or mucosa
Dendrite	In spinal nerve and branches
Cell body	In spinal ganglion on posterior root of spinal nerve
Axon	Posterior root of spinal nerve; terminates in posterior gray column of cord
INTERNEURON	
Dendrite	Posterior gray column
Cell body	Posterior gray column
Axon	Central gray matter of cord, extending into anterior gray column
MOTONEURON	
Dendrite	Anterior gray column
Cell body	Anterior gray column
Axon	Anterior gray column, extending into anterior root of spinal nerve, spinal nerve, and its branches
Effector	In skeletal muscles

*See Figures 10-7 and 10-8.

ankle jerk, Babinski reflex, corneal reflex, and abdominal reflex.

The *knee jerk* or patellar reflex is an extension of the lower leg in response to tapping of the patellar tendon. The tap stretches both the tendon and its muscles, the quadriceps femoris, and thereby stimulates muscle spindles (receptors) in the muscle and initiates conduction over the following two-neuron reflex arc:

1 *Sensory neurons*
  a Dendrites—in femoral and second, third, and fourth lumbar nerves
  b Cell bodies—second, third, and fourth lumbar ganglia
  c Axons—in posterior roots of second, third, and fourth lumbar nerves; terminate in these segments of the spinal cord; synapse directly with lower motoneurons
2 *Reflex center*—synapses in anterior gray column between axons of sensory neurons and dendrites and cell bodies of lower motoneurons
3 *Motoneurons*
  a Dendrites and cell bodies—in spinal cord anterior gray column
  b Axons—in anterior roots of second, third, and fourth lumbar spinal nerves and femoral nerves; terminate in quadriceps femoris muscle

The knee jerk can be classified in various ways as follows:

1 As a *spinal cord reflex*—because the center of the reflex arc (which transmits the impulses that activate the muscles producing the knee jerk) lies in the spinal cord gray matter
2 As a *segmental reflex*—because impulses that mediate it enter and leave the same segment of the cord
3 As an *ipsilateral reflex*—because the impulses that mediate it come from and go to the same side of the body
4 As a *stretch reflex*, or *myotatic reflex* (from the Greek *mys*, "muscle," *tasis*, "stretching")—because of the kind of stimulation used to evoke it
5 As an *extensor reflex*—because it is produced by extensors of the lower leg (muscles located on an anterior surface of the thigh, which extend the lower leg)
6 As a *tendon reflex*—because tapping of a tendon is the stimulus that elicits it
7 *Deep reflex*—because of the deep location (in tendon and muscle) of the receptors stimulated to produce this reflex (*superficial reflexes*—those elicited by stimulation of receptors located in the skin or mucosa)

When testing a patient's reflexes, a physician interprets the test results on the basis of what is known about the reflex arcs that must function to produce normal reflexes.

To illustrate, suppose that a patient has been diagnosed as having poliomyelitis. In examining him the physician finds that he cannot elicit the knee jerk when he taps the patient's patellar tendon. He knows that the poliomyelitis virus attacks anterior horn motoneurons. He also knows the information previously related about which cord segments contain the reflex centers for the knee jerk. On the basis of this knowledge, therefore, he deduces that in this patient the poliomyelitis virus has damaged the second, third, and fourth lumbar segments of the spinal cord. Do you think that this patient's leg would be paralyzed, that he would be unable to move it voluntarily? What neurons would not be able to function that must function to produce voluntary contractions?

Ankle jerk or Achilles reflex is an extension (plantar flexion) of the foot in response to tapping of the Achilles tendon. Like the knee jerk, it is a tendon reflex and a deep reflex mediated by two-neuron spinal arcs. The centers for the ankle jerk lie in the first and second sacral segments of the cord.

The *Babinski reflex* is an extension of the great toe, with or without fanning of the other toes, in response to stimulation of the outer margin of the sole of the foot. Normal infants, up until they are about 1½ years old, show this positive Babinski reflex. By about this time, corticospinal fibers have become fully myelinated and the Babinski reflex becomes suppressed. Just why this is so is not clear. But at any rate it is, and a positive Babinski reflex after this age is abnormal. From then on the normal response to stimulation of the outer edge of the sole is the *plantar reflex*. It consists of a curling under of all the toes (plantar flexion) plus a slight turning in and flexion of the anterior part of the foot. A positive Babinski reflex is one of the pyramidal signs (p. 343) and is interpreted to mean destruction of pyramidal tract (corticospinal) fibers.

The *corneal reflex* is winking in response to touching the cornea. It is mediated by reflex arcs with sensory fibers in the opthalmic branch of the fifth cranial nerve, centers in the pons, and motor fibers in the seventh cranial nerve.

The *abdominal reflex* is drawing in of the abdominal wall in response to stroking the side of the abdomen. It is mediated by arcs with sensory and motor fibers in the ninth to twelfth thoracic spinal nerves and centers in these segments of the cord. It is classified as a superficial reflex. A decrease in this reflex or its absence occurs in lesions involving pyramidal tract upper motoneurons. It can, however, be absent without any pathologic condition—in pregnancy, for example.

Outline Summary

DIVISIONS OF NERVOUS SYSTEM

A Classified by location
   1 Central nervous system (CNS)—brain and spinal cord
   2 Peripheral nervous system (PNS)—cranial nerves and ganglia, spinal nerves and ganglia, autonomic nerves and ganglia, composed of afferent and efferent neurons
B Classified by effectors innervated
   1 Somatic nervous system—innervates somatic effectors, that is, skeletal muscles; includes brain, cord, cranial nerves, and spinal nerves
   2 Autonomic nervous system—innervates visceral (autonomic) effectors, that is, smooth muscle, cardiac muscle, and glandular tissue; organs of autonomic system are autonomic nerves and ganglia

BRAIN AND CORD COVERINGS

A Bony—vertebrae around cord; cranial bones around brain

B Membranous—called meninges and consist of three layers.
   1 Dura mater—white fibrous tissue outer layer
   2 Arachnoid membrane—cobwebby middle layer
   3 Pia mater—transparent; adherent to outer surface of cord and brain; contains blood vessels, therefore, nutritive layer

BRAIN AND CORD FLUID SPACES

A Names
   1 Subarachnoid space around cord
   2 Subarachnoid space around brain
   3 Central canal inside cord
   4 Ventricles and cerebral aqueduct inside brain—four cavities within brain
      a First and second (lateral) ventricles—large cavities, one in each cerebral hemisphere
      b Third ventricle—vertical slit in cerebrum beneath corpus callosum and longitudinal fissure

Outline Summary—cont'd

c Fourth ventricle—diamond-shaped space between cerebellum and medulla and pons; expansion of central canal or cord

B Formation, circulation, and function of cerebrospinal fluid

1 Formed by plasma filtering from network of capillaries (choroid plexus) in each ventricle

2 Circulates from lateral ventricles to third ventricle, cerebral aqueduct, fourth ventricle, central canal of cord, subarachnoid space of cord and brain; venous sinuses

3 Function—protection of brain and cord

SPINAL CORD
Structure

A Size, shape, and location—about 45 cm (1½ feet) long, slightly smaller around than spinal cavity in which it is located

B Sulci—two deep grooves, anterior median fissure and posterior median sulcus, incompletely divide cord into right and left symmetrical halves; anterior median fissure deeper and wider than posterior groove

C Gray matter—shaped like three-dimensional letter **H**

D White matter—present in columns (funiculi) anterior, lateral, and posterior; composed of numerous projection tracts

Functions

A Sensory tracts conduct up to brain from peripheral nerves; motor tracts conduct down from brain to peripheral nerves; see Tables 11-1 and 11-2 for names of spinal cord tracts

B Synapses and interneurons in the cord's gray matter function as reflex centers for spinal reflexes

SPINAL NERVES
Structure

A Number of spinal nerves—31 pairs (8 cervical, 12 thoracic, 5 lumbar, 5 sacral, and 1 coccygeal)

B Origin—originate by anterior and posterior roots from cord, emerge through intervertebral foramina; spinal ganglion on each posterior root

C Distribution—branches distributed to skin, mucosa, skeletal muscles; some branches form plexuses, such as brachial plexus, from which nerves emerge to supply various parts (Table 11-3 and Figures 11-12 and 11-13)

D Microscopic structure—consist of sensory dendrites and motor axons, that is, spinal nerves are mixed nerves

Microscopic structure and functions

Conduct impulses both to cord from periphery and from cord to periphery

DIVISIONS AND SIZE OF BRAIN

A Divisions—medulla, pons, midbrain, cerebellum, diencephalon, and cerebrum

B Size—adult brain weighs about 1.4 kg (3 lb); consists of billions of neurons and an even larger number of neuroglia

BRAIN STEM
Structure

A Medulla—part formed by enlargement of cord as it enters cranial cavity through foramen magnum; consists of white matter (ascending and descending tracts) and of mixture of gray and white matter (reticular formation)

B Pons—lies just above medulla; ascending and descending tracts make up white matter of pons; contains reticular formation also

C Midbrain—lies just above pons, below diencephalon and cerebrum; contains ascending and descending tracts and few nuclei; cerebral peduncles connect pons to cerebrum (consist of tracts); corpora quadrigemina—two superior and two inferior colliculi—are rounded eminences on dorsal surface of midbrain; red nucleus lies in gray matter of midbrain; cerebral aqueduct is fluid space in midbrain

Functions

A Medulla—two-way conduction between cord and brain; cardiac, vasomotor, and respiratory centers in medulla are vital in control of heartbeat, blood pressure, and respiration; centers for reflexes of vomiting, coughing, and hiccupping also in medulla

B Pons—two-way conduction between cord and brain; centers for cranial nerves V through VIII in pons

C Midbrain—two-way conduction between cord and brain; centers for cranial nerves III and IV (pupillary reflexes and eye movements) in midbrain

CRANIAL NERVES

A Structure (see Table 11-4)

B Functions (see Table 11-4)

CEREBELLUM

A Structure—center section, vermis, lies between two hemispheres of cerebellum; surface is grooved with sulci and has slightly raised, slender convolutions; internal white matter in leaflike pattern; tracts in cerebellum are inferior, middle, and superior cerebellar peduncles; dentate nucleus in cerebellum

B Functions—synergic control of skeletal muscles; mediates postural and equilibrium reflexes

Outline Summary—cont'd

DIENCEPHALON

A Structure—thalamus and hypothalamus are two major parts of diencephalon; thalamus—large rounded mass of gray matter, one in each hemisphere of cerebrum, lateral to third ventricle; hypothalamus—includes gray matter around optic chiasma, pituitary stalk, posterior lobe of pituitary gland, mamillary bodies and adjacent regions; prominent nuclei in hypothalamus are supraoptic, paraventricular, and mamillary nuclei; tracts connect diencephalon with cerebral cortex, basal ganglia, brain stem, and cord

B Functions

  1 Thalamus—relays sensory impulses from cord to cerebral cortex; registers crude sensations of pain, temperature, and touch; emotions of pleasantness or unpleasantness associated with sensations; part of pathway for arousal or alerting and for complex reflex movements

  2 Hypothalamus—functions as higher center for both parasympathetic and sympathetic divisions of autonomic system, regulating and coordinating them and thereby integrating responses by visceral effectors; functions as link between psyche and soma by relaying impulses from cerebral cortex to autonomic centers; supraoptic and paraventricular nuclei of hypothalamus synthesize posterior pituitary hormones; many hypothalamic neurons synthesize and secrete hormones that regulate hormone secretion by anterior pituitary gland; functions as part of pathways for arousal and alerting and for regulating appetite and temperature

CEREBRUM

Structure

A Hemispheres, fissures, and lobes—longitudinal fissure divides cerebrum into two hemispheres, connected only by corpus callosum; each cerebral hemisphere divided by fissures into five lobes: frontal, parietal, temporal, occipital, and island of Reil (insula)

B Cerebral cortex—outer layer of gray matter arranged in ridges called convolutions or gyri

C Cerebral tracts—bundles of axons compose white matter in interior of cerebrum; ascending projection tracts transmit impulses toward or to brain; descending projection tracts transmit impulses down from brain to cord; commissural tracts transmit from one hemisphere to other; association tracts transmit from one convolution to another in same hemisphere

D Basal ganglia (or cerebral nuclei)—masses of gray matter embedded deep inside white matter in interior of cerebrum; caudate, putamen, and pallidum; putamen and pallidum constitute lenticular nucleus

Functions

A Sensory functions—sensory areas of cerebral cortex compare, evaluate, and integrate sensations to form total perceptions

B Motor functions—control voluntary (skeletal muscle) movements

C Integrative functions

  1 Consciousness—state of awareness of one's self, one's environment, and other beings; depends on excitation ("arousal" or "alerting") of cortical neurons by impulses conducted to them via neurons of reticular activating system; normal variations in degree of level of consciousness—waking, dream or rapid eye movement (REM) sleep, and deep, dreamless sleep; also several kinds of altered states of consciousness (ASC), for example, anesthesia, coma, and yoga or meditative state

  2 Memory—function of cerebral cortex neurons but mechanisms not definitely known; evidence that several parts of cortex (e.g., temporal, parietal, occipital lobes) store memories and that limbic system ("emotional brain") and protein synthesis play key roles in memory; protein synthesis also appears to be crucial for long-term memory

  3 Language functions—the use of language (speaking and writing) and the understanding of language (spoken and written) depend on widespread integrated cortical processes involving many parts of cortex, but certain regions in frontal, parietal, and temporal lobes called speech centers serve as focal points for integration of speech processes; speech defects (aphasias) of different kinds result from lesions of different speech centers

  4 Emotions—both subjective experience and objective expression of emotions involve functioning of cerebrum's limbic system, that is, part of the cortex located on medial surface of cerebrum that forms a border around the corpus callosum; limbic system made up of cingulate gyrus, isthmus, hippocampal gyrus, hippocampus, and uncus and tracts connecting these structures to several parts of brain, notably the septum, amygdala, and hypothalamus; for normal expression of emotions, other parts of the cortex must modulate limbic system activity

D Generalizations about cerebral functions

  1 Cerebral activity goes on as long as life itself; electroencephalograms are records of cerebrum's electrical activity, of its "brain waves"; *alpha waves*—moderately fast, moderately high-voltage waves that dominate in relaxed nonattentive state; *beta waves*—fast, low-voltage waves that dominate in attentive waking state; *delta waves*—very slow, high-voltage waves characteristic of

Outline Summary—cont'd

deep sleep; *theta waves*—moderately slow, low-voltage waves that appear as drowsiness descends; absence of brain waves ("flat EEG") generally considered best criterion of death; brain wave patterns vary in altered states of consciousness

2 Hemispheres of the cerebrum specialize in different functions; left hemisphere specializes in language functions and in the control of skilled movements and gesturing movements of right hand; right hemisphere specializes in perception of nonspeech sounds and in locating objects in space; may also function better in tactual perception and in visualizing spatial relationships

3 Certain areas of cortex in each hemisphere play dominant role in a particular function, for example, primary somatic sensory area (postcentral gyrus of parietal lobe) is crucial for experiencing general sensations (heat, cold, and touch); primary somatic motor area (precentral gyrus of frontal lobe) dominates control of somatic effectors (skeletal muscles); primary auditory area (transverse gyrus of temporal lobe) is crucial for auditory sensations; primary visual area (area 17 of occipital lobe) is crucial for vision; important to remember that any one function requires functioning of many parts of the nervous system

4 For the cerebral cortex to perform its sensory functions, impulses must first be conducted to its sensory areas by way of relays of neurons referred to as sensory pathways

5 For the cerebral cortex to perform its motor functions, impulses must be conducted from its motor areas to skeletal muscles by relays of neurons referred to as somatic motor pathways

NEUROTRANSMITTERS IN CORD AND BRAIN
Classes of neurotransmitters

A Amino acids—examples: glycine, aspartic acid, glutamic acid, and GABA

B Monoamines (catecholamines)—examples: norepinephrine, dopamine, and serotonin

C Neuropeptides—examples: enkephalins, endorphins, substance P

Some general information about neurotransmitters

A Precise locations in CNS known for only some neurotransmitters (Table 11-5)

B Glutamic and aspartic acids are most common excitatory transmitters at brain synapses

C GABA is most common inhibitory transmitter in brain

D Glycine is an inhibitory transmitter in cord

E Substance P released by axon terminals of sensory neuron I in synapses in substantia gelatinosa

F Enkephalins and endorphins are transmitters that serve as body's own opiates

G Some neuropeptides first identified in gut, namely, substance P, vasoactive intestinal peptide, and cholecystokinin

H Some neuropeptides first identified as hormones, namely, vasopressin and oxytocin

Synthesis, storage, release, action, inactivation of neurotransmitters

A Synthesis in axon terminals by chemical reactions catalyzed by enzymes; reaction usually converts an amino acid to the transmitter

B Storage in vesicles in axon terminals

C Released into synaptic cleft following arrival of nerve impulse at axon terminals

D Action: bind to receptors in postsynaptic neuron's membrane, thereby initiating events that lead to excitation or inhibition of postsynaptic neuron (see Figures 10-11 and 10-12)

E Inactivation by enzyme or reuptake into axon terminal

PAIN CONTROL AREAS IN CORD AND BRAIN

A In cord—substantia gelatinosa portion of posterior gray columns

B In brain—gray matter around aqueduct of Sylvius and third ventricle

SOME GENERALIZATIONS ABOUT SOMATIC SENSORY PATHWAYS

A Most impulses that reach sensory areas of cerebral cortex have traveled over at least three pools of sensory neurons (designated here as sensory neurons I, II, and III)

  1 Sensory neuron I conducts from periphery to cord or to brain stem

  2 Sensory neuron II conducts from cord to brain stem to thalamus

  3 Sensory neuron III conducts from thalamus to general sensory area of cerebral cortex (areas 3, 1, and 2)

B Most sensory neuron II axons decussate, so one side of brain registers mainly sensations for opposite side of body

C Pain and temperature—spinothalamic tracts to thalamus

D Touch and pressure

  1 Discriminating touch and pressure (stereognosis, precise localization, and vibratory sense)—medial lemniscal system (fasciculi cuneatus and gracilis to medulla; medial lemniscus to thalamus)

  2 Crude touch and pressure—spinothalamic tracts

E Conscious proprioception or kinesthesia—medial lemniscal system to thalamus (see also Table 11-6)

Outline Summary—cont'd

F Impulses conducted up spinothalamic and medial lemniscal pathways reach somatic sensory area of cerebral cortex (postcentral gyrus or areas 3, 1, and 2 on Brodmann's map of cortex, Figure 11-24), the primary area for general sensations

G Several secondary somatic sensory areas exist in cerebral cortex

SOME GENERALIZATIONS ABOUT SOMATIC MOTOR PATHWAYS

A Principle of final common path—motoneurons in anterior gray horns of cord constitute final common path for impulses to skeletal muscles; are only neurons transmitting impulses into skeletal muscles

B Motor pathways from cerebral cortex to anterior horn cells classified according to place where fibers enter cord

  1 Pyramidal (or corticospinal) tracts—dendrites and cells in cortex; axons descend through internal capsule, pyramids of medulla; enter cord from pyramids; descend and synapse directly with anterior horn cells or indirectly via internuncial neurons; impulses via pyramidal tracts essential for voluntary contractions of individual muscles to produce small discrete movements; also help maintain muscle tone

  2 Extrapyramidal tracts—all pathways between motor cortex and anterior horn cells except pyramidal tracts; upper extrapyramidal tracts relay impulses between cortex, basal ganglia, thalamus, and brain stem; reticulospinal tracts main lower extrapyramidal tracts; impulses via extrapyramidal tracts essential for large,

automatic movements; also for facial expressions and movements accompanying many emotions

C Motor pathways from cerebral cortex to anterior horn cells also classified according to influence on anterior horn cells

  1 Facilitatory tracts—have facilitating or stimulating effect on anterior horn cells; all pyramidal tracts and some extrapyramidal tracts facilitatory, notably facilitatory reticulospinal fibers

  2 Inhibitory tracts—have inhibiting effect on anterior horn cells; inhibitory reticulospinal fibers main ones

D Principle of convergence—axons and many neurons synapse with each anterior horn motoneuron

E Ratio of facilitatory and inhibitory impulses impinging on anterior horn cells controls activity; normally, slight predominance of facilitatory impulses maintains muscle tone

REFLEXES

A Definition—action resulting from conduction over reflex arc; reflex is response (either muscle contraction or glandular secretion) to stimulus; term *reflex* usually used to mean only involuntary responses

B Some reflexes of clinical importance (see also pp. 342 and 344)

  1 Knee jerk

  2 Ankle jerk

  3 Babinski reflex

  4 Corneal reflex

  5 Abdominal reflex

Review Questions

1 What term means the membranous coverings of the brain and cord? What three layers compose this covering?
2 Where does a physician do a lumbar puncture? Why?
3 What are the cavities inside the brain called? How many are there? What do they contain?
4 Describe the circulation of cerebrospinal fluid.
5 What function does the cerebrospinal fluid serve?
6 Explain the system for naming individual tracts.
7 Based on the explanation you gave in question 6, what is a spinothalamic tract?
8 Name several ascending or sensory tracts in the spinal cord.
9 Name two descending or motor tracts in the spinal cord.
10 Name the tracts that transmit impulses from each of the following types of receptors to the brain and state in which column of the cord each tract is located: pain and temperature receptors, crude touch receptors, proprioceptors (name two tracts for these), discriminating touch receptors (two tracts).
11 Describe the spinal cord's general functions.
12 Explain what the term *reflex center* means.
13 Describe the medulla's general functions.
14 Which cranial nerves transmit impulses that result in vision? In eye movements?
15 Which cranial nerves transmit impulses that result in hearing? In taste sensations?
16 Which cranial nerve serves as the great sensory nerve of the head?

17 Digoxin (Lanoxin), a drug that stimulates the vagus nerve, has been administered to a patient. What effect, if any, would you expect it to have on the patient's pulse rate?
18 Describe the general functions of the cerebellum.
19 Describe the general functions of the thalamus.
20 Describe the general functions of the hypothalamus.
21 Describe the general functions of the cerebrum.
22 Define the term *electroencephalogram*. What abbreviation stands for this term?
23 Identify the following kinds of brain waves according to their frequency, voltage, and the level of consciousness in which they predominate: alpha, beta, delta, theta.
24 What general functions does the cerebral cortex perform?
25 Define consciousness. Name the normal states or levels of consciousness.
26 Explain briefly what is meant by the arousal or alerting mechanism.
27 Name some altered states of consciousness.
28 Compare yoga (the meditative state) with the usual waking state.
29 Describe some of the current ideas about memory.
30 What part of the brain is called the "emotional brain"?
31 Locate the dendrite, cell body, and axon of sensory neurons I, II, and III.
32 Explain each of the following: lower motoneuron, upper motoneuron, and final common path.
33 Compare pyramidal tract and extrapyramidal tract functions.

12 THE AUTONOMIC NERVOUS SYSTEM

OBJECTIVES

After you have completed this chapter, you should be able to:

1 Identify the two major subdivisions of the autonomic nervous system.
2 Identify the three major types of visceral effector tissues in the body.
3 Compare and contrast the conduction pathway from the central nervous system to the visceral and somatic effectors.
4 Discuss the structure of the sympathetic or thoracolumbar division of the autonomic nervous system.
5 Discuss the structure of the parasympathetic or craniosacral division of the autonomic nervous system.
6 Compare the locations of the sympathetic and parasympathetic neuron cell bodies, dendrites, and axons.
7 Discuss the function of the autonomic nervous system as a whole.
8 Compare the structure and function of adrenergic and cholinergic fibers in the autonomic nervous system.
9 Explain the relationship of nicotinic and muscarinic receptors to acetylcholine activity.
10 Identify the two types of receptors that bind to norepinephrine.
11 Explain the mechanism of action that terminates the action of norepinephrine and acetylcholine.
12 Compare the specific functions of the sympathetic and parasympathetic divisions of the autonomic nervous system.

The autonomic nervous system is not, as its name suggests, a separate system. It is a division of the body's one nervous system, the division that sends efferent neurons to visceral effectors—cardiac muscle, smooth muscle, and glandular epithelial tissue (Table 12-1). The major function of the autonomic nervous system is to regulate heartbeat, smooth muscle contraction, and glandular secretion in ways that tend to maintain homeostasis.

KEY TERMS

Acetylcholine (as-e-til-KO-len)

Adrenergic (ad-ren-ER-jik)

Atropine (AT-ro-pin)

Cholinergic (ko-lin-ER-jik)

Collateral (ko-LAT-er-al)

Efferent (EF-er-ent)

Muscarinic receptor (MUS-kah-ren-ik   re-SEP-tor)

Nicotinic receptor (NIC-o-ten-ik   re-SEP-tor)

Norepinephrine (nor-ep-i-NEF-rin)

Parasympathetic (par-ah-sim-pah-THET-ik)

Postganglionic (post-gang-gle-ON-ik)

Preganglionic (pre-gang-gle-ON-ik)

Splanchnic (SPLANK-nik)

Sympathetic (sim-pah-THET-ik)

Visceral effector (VIS-er-al   ef-FEK-tor)

Table 12-1 Visceral effector tissues and organs

Cardiac muscle	Smooth muscle	Glandular epithelium
Heart	Blood vessels	Sweat glands
	Bronchial tubes	Lacrimal glands
	Stomach	Digestive glands (salivary, gastric pancreas, liver)
	Gallbladder	
	Intestines	Adrenal medulla
	Urinary bladder	
	Spleen	
	Eye (iris and ciliary muscles)	
	Hair follicles	

Figure 12-1 Diagram showing difference between the sympathetic pathway from the spinal cord to visceral effectors and the pathway from the cord to somatic effectors. A relay of two sympathetic neurons—preganglionic and postganglionic—conducts from the cord to visceral effectors. Note that, in contrast, only one somatic motoneuron (anterior horn neuron) conducts from the cord to somatic effectors with no intervening synapses. Parasympathetic impulses also travel over a relay of two neurons (parasympathetic preganglionic and postganglionic) to reach visceral effectors from the central nervous system. Note the location of the sympathetic preganglionic neuron's cell body and axon. Where in this figure do you find the sympathetic postganglionic neuron's cell body and axon located?

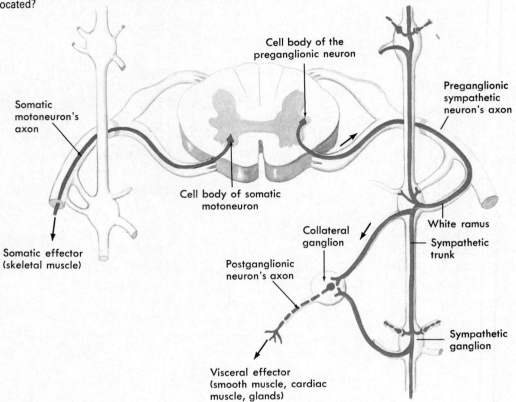

STRUCTURE

AUTONOMIC NERVOUS SYSTEM AS A WHOLE

The autonomic nervous system has two subdivisions, namely, the **sympathetic system** and the **parasympathetic system.** Each of these is made up of autonomic ganglia, nerves, and plexuses that in turn consist of autonomic neurons. All autonomic neurons are efferent neurons. They conduct impulses away from the brain stem or spinal cord to visceral effectors. Sensory neurons function in autonomic reflex arcs but few, if any, serve only autonomic arcs. Theoretically, any sensory neuron can function in either somatic or autonomic arcs. This seems to be the reason for not considering sensory neurons as autonomic neurons.

A relay of two autonomic neurons conducts from brain stem or cord to visceral effectors. The first is called a *preganglionic neuron*—an awkward

name but a descriptive one since preganglionic neurons conduct impulses before they reach an autonomic ganglion from the central nervous system. After impulses reach an autonomic ganglion, *postganglionic neurons* conduct to visceral effectors. The conduction pathway from the central nervous system to visceral effectors and to somatic effectors differs. Observe this in Figure 12-1. Conduction to somatic effectors requires only one efferent neuron, the somatic motoneuron. Conduction to visceral effectors requires a sequence of preganglionic and postganglionic autonomic neurons.

Most visceral effectors are doubly innervated by the autonomic system. Stated differently, both sympathetic and parasympathetic fibers supply most visceral effectors. Notable exceptions: only sympathetic fibers innervate sweat glands and smooth muscle of most blood vessels. All visceral effectors receive sympathetic fibers.

Table 12-2 Sympathetic and parasympathetic neurons compared as to location

Neurons	Sympathetic	Parasympathetic
PREGANGLIONIC NEURONS		
Dendrites and cell bodies	In lateral gray columns of thoracic and first four lumbar segments of spinal cord	In nuclei of brain stem and in lateral gray columns of sacral segments of cord
Axons	In anterior roots of spinal nerves to spinal nerves (thoracic and first four lumbar), to and through white rami to terminate in sympathetic ganglia at various levels or to extend through sympathetic ganglia, to and through splanchnic nerves to terminate in collateral ganglia (celiac, superior, and inferior mesenteric ganglia)	From brain stem nuclei through cranial nerve III to ciliary ganglion From nuclei in pons: through cranial nerve VII to sphenopalatine or submaxillary ganglion From nuclei in medulla through cranial nerve IX to otic ganglion or through cranial nerves X and XI to cardiac and celiac ganglia, respectively
POSTGANGLIONIC NEURONS		
Dendrites and cell bodies	In sympathetic and collateral ganglia	In parasympathetic ganglia (for example, ciliary, sphenopalatine, submaxillary, otic, cardiac, celiac) located in or near visceral effector organs
Axons	In autonomic nerves and plexuses that innervate thoracic and abdominal viscera and blood vessels in these cavities In gray rami to spinal nerves, to smooth muscle of skin blood vessels and hair follicles, and to sweat glands	In short nerves to various visceral effector organs

SYMPATHETIC (OR THORACOLUMBAR) SYSTEM

Ganglia of the sympathetic system lie on either side of the anterior surface of the spinal column. Because short fibers connect them to each other, they look a little like two chains of beads and are often referred to as the "sympathetic chain ganglia."

Each chain (usually made up of 22 ganglia) extends from the second cervical vertebra in the neck all the way down to the level of the coccyx. Sympathetic preganglionic neurons have their dendrites and cell bodies in the lateral gray horns of the thoracic and lumbar segments of the spinal cord. Observe one such neuron in Figure 12-1. Axons of sympathetic preganglionic neurons leave the cord by way of the anterior roots of the thoracic and first four lumbar spinal nerves. After extending a short distance in the spinal nerves, they enter a structure called a white ramus. Find the white ramus in Figure 12-1. Note that the sympathetic preganglionic axon extends through the white ramus to a sympathetic chain ganglion. Here it branches in several directions. It sends ascending and descending branches through the sympathetic trunk to terminate in sympathetic chain ganglia above and below the axon's point of origin. Also, it sends a branch by way of

splanchnic nerves to terminate in a collateral (prevertebral) ganglion.

Look at Table 12-2 above and review the facts just presented about sympathetic preganglionic neurons.

Sympathetic ganglia consist of three cervical ganglia (superior, middle, and inferior), eleven thoracic, four lumbar, and four sacral in each sympathetic chain. *Collateral sympathetic ganglia* are located a short distance from spinal cord. Examples include *celiac ganglia (solar plexus)*—two fairly large, flat ganglia located on each side of the celiac artery just below the diaphragm; *superior mesenteric ganglion*—a small ganglion located near the beginning of the superior mesenteric artery; and *inferior mesenteric ganglion*—a small ganglion located close to the beginning of the inferior mesenteric artery.

Figure 12-2 Autonomic system; *red,* sympathetic; *violet,* parasympathetic; *blue,* central conduction paths; *solid lines,* preganglionic fibers; *broken lines,* postganglionic fibers.

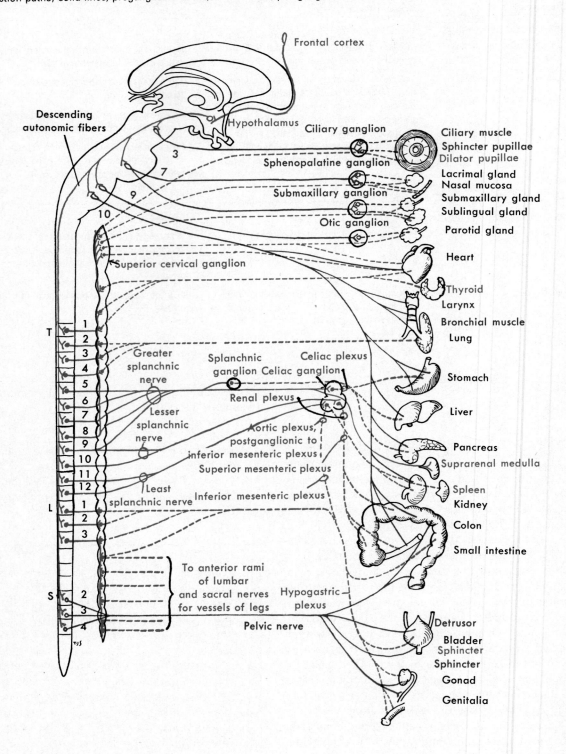

The greater, lesser, and least splanchnic nerves, shown in Figure 12-2, constitute branches from the thoracic sympathetic trunk. They form the main routes for sympathetic stimulation of abdominal viscera. Note in Figure 12-2 that preganglionic axons in the splanchnic nerves synapse in the celiac ganglion and other collateral ganglia with postganglionic sympathetic neurons to abdominal structures.

The axon of any one sympathetic preganglionic neuron synapses with many postganglionic neurons, and these frequently terminate in widely separated organs. This anatomical fact partially explains a well-known physiological principle—sympathetic responses are usually widespread, involving many organs and not just one.

Sympathetic postganglionic neurons have their dendrites and cell bodies in the sympathetic chain ganglia or in collateral ganglia. Their axons are distributed by both spinal nerves and separate autonomic nerves. They reach the spinal nerves via small filaments (gray rami) that connect the sympathetic ganglia with the spinal nerves. They then travel in the spinal nerves to blood vessels, sweat glands, and arrector hair muscles all over the body.

The course of sympathetic postganglionic axons through autonomic nerves is complex. These nerves form complicated plexuses before fibers are finally distributed to their respective destinations. For example, postganglionic fibers from the celiac and superior mesenteric ganglia pass through the celiac plexus before reaching the abdominal viscera, those from the inferior mesenteric ganglion pass through the hypogastric plexus on their way to the lower abdominal and pelvic viscera. Sympathetic postganglionic axons from the middle and inferior cervical ganglia pass through the cardiac nerves and the cardiac plexus at the base of the heart. They then leave the plexus and are distributed to the heart.

PARASYMPATHETIC (OR CRANIOSACRAL) SYSTEM

Parasympathetic preganglionic neurons have their cell bodies in nuclei in the brain stem or in the lateral gray columns of the sacral cord. Their axons are contained in cranial nerves III, VII, IX, X, and XI and in some pelvic nerves. They extend a considerable distance before synapsing with postganglionic neurons. For example, axons arising from cell bodies in the vagus nuclei (located in the medulla) travel in the vagus nerve for a distance of a foot or more before reaching their terminal ganglia in the chest and abdomen (Figure 12-2 and Table 12-2).

Parasympathetic postganglionic neurons have their dendrites and cell bodies in parasympathetic ganglia. Unlike sympathetic ganglia that lie near the spinal column, parasympathetic ganglia lie near or embedded in visceral effectors. For example, note the ciliary ganglion in Figure 12-2. This and the other ganglia shown near it are parasympathetic ganglia located in the skull. In a parasympathetic ganglion, preganglionic axons synapse with postganglionic neurons that send their short axons into the nearby visceral effector. A parasympathetic preganglionic neuron, therefore, usually synapses with postganglionic neurons to a single effector. For this reason, parasympathetic stimulation frequently involves response by only one organ. Sympathetic stimulation, on the other hand, usually evokes responses by numerous organs.

FUNCTIONS

AUTONOMIC NERVOUS SYSTEM AS A WHOLE

The autonomic nervous system as a whole functions to regulate visceral effectors in ways that tend to maintain or quickly restore homeostasis. Both sympathetic and parasympathetic divisions are *tonically active*, that is, they continually conduct impulses to visceral effectors. They exert oppostie or antagonistic influences on them—a fact that we might call the principle of autonomic antagonism. If sympathetic impulses tend to stimulate an effector, parasympathetic impulses tend to inhibit it. Doubly innervated effectors continually receive both sympathetic and parasympathetic impulses. Summation of the two opposing influences determines the dominating or controlling effect. For example, continual sympathetic impulses to the heart tend to accelerate the heart rate while continual parasympathetic impulses tend to slow it. The actual heart rate is determined by whichever influence dominates. To find other examples of autonomic antagonism, examine Table 12-3, p. 356.

The autonomic nervous system does not function autonomously as its name suggests. It is continually influenced by impulses from the so-called autonomic centers. These are clusters of neurons located at various levels in the brain whose axons conduct impulses directly or indirectly to autonomic preganglionic neurons. Autonomic centers function as a hierarchy in their control of the autonomic system. Highest ranking in the hierarchy are the autonomic centers in the cerebral cortex, for example, in the frontal lobe and limbic system (gyruses on medial surface of cerebrum that form a border around the corpus callosum). Neurons in these cen-

Table 12-3 Autonomic functions

Visceral effector	Effect of sympathetic stimulation (neurotransmitter, norepinephrine unless otherwise stated)	Effect of parasympathetic stimulation (neurotransmitter, acetylcholine)
HEART	Increased rate and strength of heartbeat (beta receptors)	Decreased rate and strength of heartbeat
SMOOTH MUSCLE OF BLOOD VESSELS		
Skin blood vessels	Constriction (alpha receptors)	No parasympathetic fibers
Skeletal muscle blood vessels	Dilation (beta receptors)	No parasympathetic fibers
Coronary blood vessels	Dilation (beta receptors)	No parasympathetic fibers
Abdominal blood vessels	Constriction (alpha receptors)	No parasympathetic fibers
Blood vessels of external genitals	Ejaculation (contraction of smooth muscle in male ducts, e.g., epididymis and vas deferens)	Dilation of blood vessels causing erection in male
SMOOTH MUSCLE OF HOLLOW ORGANS AND SPHINCTERS		
Bronchi	Dilation (beta receptors)	Constriction
Digestive tract, except sphincters	Decreased peristalsis (beta receptors)	Increased peristalsis
Sphincters of digestive tract	Contraction (alpha receptors)	Relaxation
Urinary bladder	Relaxation (beta receptors)	Contraction
Urinary sphincters	Contraction (alpha receptors)	Relaxation
Eye		
Iris	Contraction of radial muscle; dilated pupil	Contraction of circular muscle; constricted pupil
Ciliary	Relaxation; accommodates for far vision	Contraction; accommodates for near vision
Hairs (pilomotor muscles)	Contraction produces goose pimples, or piloerection (alpha receptors)	No parasympathetic fibers
GLANDS		
Sweat	Increased sweat (neurotransmitter, acetylcholine)	No parasympathetic fibers
Digestive (salivary, gastric, etc.)	Decreased secretion of saliva; not known for others	Increased secretion of saliva
Pancreas, including islets	Decreased secretion	Increased secretion of pancreatic juice and insulin
Liver	Increased glycogenolysis (beta receptors); increased blood sugar level	No parasympathetic fibers
Adrenal medulla*	Increased epinephrine secretion	No parasympathetic fibers

*Sympathetic preganglionic axons terminate in contact with secreting cells of the adrenal medulla. Thus the adrenal medulla functions, to quote someone's descriptive phrase, as a "giant sympathetic postganglionic neuron."

ters send impulses to other autonomic centers in the brain, notably in the hypothalamus. Then neurons in the hypothalamus send either stimulating or inhibiting impulses to parasympathetic and sympathetic preganglionic neurons located in the lower autonomic centers of the brain stem and cord. Turn back now to Figure 12-2. Identify the neuron whose cell body lies in the frontal cortex and whose axon de-

scends to terminate in the hypothalamus. This represents a neuron in an autonomic center of the cerebral cortex. Observe that the axon of the neuron in the hypothalamus has many branches. Some are shown synapsing in the brain stem and in the sacral segments of the cord with parasympathetic preganglionic neurons. Note that other branches synapse in the thoracic and lumbar segments of the cord with

Figure 12-3 Central nervous system hierarchy that regulates autonomic functions. (Compare with Figure 12-2.)

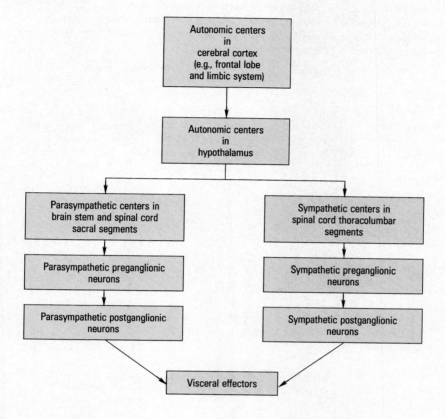

We now know that individuals can learn to control specific visceral effectors if two conditions are fulfilled. They must be informed that they are achieving the desired response and they must be rewarded for it. Various kinds of biofeedback instruments have been developed to provide these conditions. For example, a biofeedback instrument that detects slight temperature changes has been used with patients who suffer from migraine headaches. The instrument is attached to their hands and emits a high sound each time they happen to dilate their hand blood vessels. The reward for such patients is a lessening of the migraine pain—presumably because of the shunting of blood away from the head to the hands. (Migraine headaches initially involve distention of head blood vessels.)

sympathetic preganglionic neurons. Figure 12-3 shows the hierarchy of autonomic control in a diagram form.

You may be wondering why the name *autonomic system* was ever chosen in the first place if the system is really not autonomous. Originally the term seemed appropriate. The autonomic system seemed to be self-regulating and independent of the rest of the nervous system. Common observations furnished abundant evidence of its independence from cerebral control, from direct control by the will, that is. But later, even this was found to be not entirely true. Some rare and startling exceptions were discovered. A man in a brightly lighted amphitheater, for example, can change the size of his pupils from small, constricted dots (normal response to bright lights) to widely dilated circles. It is also possible to will the smooth muscle of the hairs on the arms to contract, producing gooseflesh.

Axon terminals of autonomic neurons release either of two neurotransmitters, namely, **norepinephrine** or **acetylcholine.** Axons that release norepinephrine are known as *adrenergic fibers.* Axons

Figure 12-4 Cholinergic fibers release acetylcholine (ACh). Adrenergic fibers release norepinephrine (NE). Note the three kinds of cholinergic fibers shown in the diagram: axons of both sympathetic and parasympathetic preganglionic neurons and of which postganglionic neurons? Which postganglionic neurons would you classify as adrenergic? Almost immediately after its release, ACh is inactivated by the enzyme acetylcholinesterase (AChE). The enzyme catechol-O-methyl transferase (COMT) inactivates NE.

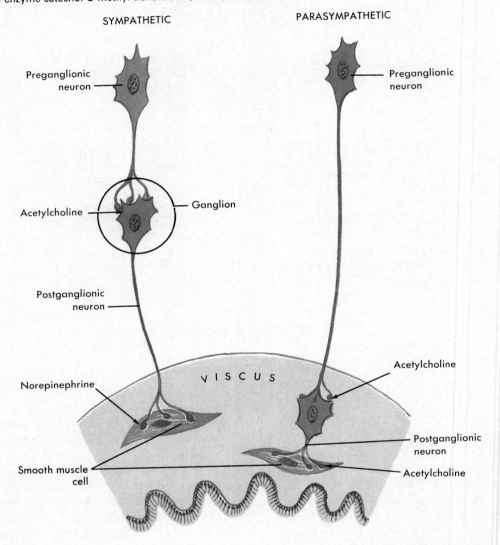

that release acetylcholine are called *cholinergic fibers*. Autonomic cholinergic fibers are the axons of preganglionic sympathetic neurons and of both preganglionic and postganglionic parasympathetic neurons. As you can see in Figure 12-4, this leaves the axons of postganglionic sympathetic neurons as the only autonomic adrenergic fibers. And not all of these are adrenergic. Sympathetic postganglionic axons to sweat glands and some blood vessels are cholinergic fibers.

Norepinephrine affects visceral effectors by first binding to receptors in their plasma membranes. The receptors are of two types, one named **alpha re-**ceptors and the other **beta receptors.** The binding of norepinephrine to alpha receptors in the smooth muscle of blood vessels has a stimulating effect on the muscle that causes the vessel to constrict. The binding of norepinephrine to beta receptors in smooth muscle produces opposite effects. It inhibits the muscle, causing the vessel to dilate. But the binding of norepinephrine to beta receptors in cardiac muscle has a stimulating effect that results in a faster and stronger heartbeat.

Drugs that bind to beta receptors block the receptors for norepinephrine-binding. Such drugs have a descriptive, informal name—beta blockers. Propranolol (Inderal) is one of these. It has long been used to treat irregular heartbeats and hypertension, and in the fall of 1981 an exciting new potential use for the drug—as a preventive for repeated heart attacks—was discovered. A nationwide, double-blind study undertaken by the National Heart, Lung, and Blood Institute revealed this potential for propranolol. The study involved 4,200 patients who had suffered at least one heart attack. Over a period of months half the group received propranolol and the other half received placebos. Startling results began appearing. So sharply lower was the number of heart attack deaths in the propranolol group than in the placebo group that the study was stopped ahead of schedule.

Norepinephrine's action is terminated in two ways—by its being broken down by the enzyme **catechol-O-methyl transferase (COMT)** and by its being transported back into axon endings, where it may be broken down by another enzyme **monoamine oxidase.**

Acetylcholine binds to two different types of receptors, namely, **nicotinic receptors** (so named because nicotine also binds to these receptors) and **muscarinic receptors.** In the autonomic ganglia, acetylcholine binds to nicotinic receptors present in the membranes of postganglionic neurons. In visceral receptors supplied by parasympathetic fibers, acetylcholine binds to muscarinic receptors. The drug atropine blocks muscarinic receptors. Acetylcholine's action is terminated by its being hydrolyzed by the enzyme, acetylcholinesterase.

SYMPATHETIC SYSTEM

Under ordinary conditions the sympathetic system functions to maintain the normal functioning of doubly innervated visceral effectors. It does this by opposing the effects of parasympathetic impulses to these structures. Here are two examples. By counteracting parasympathetic impulses that tend to slow the heart and weaken its beat, sympathetic impulses function to maintain the heartbeat's normal rate and strength. By counteracting parasympathetic impulses that tend to decrease digestive tract contractions and digestive gland secretion, sympathetic impulses function to maintain normal digestive tract contractions and secretion. The sympathetic system also serves another important function under usual conditions. Since only sympathetic fibers innervate the smooth muscle in blood vessel walls, sympathetic impulses function to maintain the normal tone of this muscle. By so doing, the sympathetic system plays a crucial role in maintaining blood pressure under usual conditions.

A unique function of the sympathetic division of the autonomic nervous system is that it serves as

an emergency system. Under stress conditions, from either physical or emotional causes, sympathetic impulses increase greatly to most visceral effectors. In fact, one of the very first steps in the body's complex defense mechanism against stress is a sudden and marked increase in sympathetic activity. This brings about a group of responses that all go on at the same time. Together they make the body ready to expend maximum energy and to engage in maximum physical exertion—as, for example, in running or fighting. Walter B. Cannon coined the descriptive and now famous phrase—the **"fight or flight" reaction**— as his name for this group of sympathetic responses. Read the middle column of Table 12-3 to find many of the "fight or flight" physiological changes. Some particularly important changes for maximum energy expenditure are faster, stronger heartbeat, dilated blood vessels in skeletal muscles, dilated bronchi, and increased blood sugar levels from stimulated glycogenolysis (conversion of glycogen to glucose). Also, sympathetic impulses to the medulla of each adrenal gland stimulate its secretion of epinephrine and some norepinephrine. These hormones reinforce and prolong effects of the norepinephrine released by sympathetic postganglionic fibers.

Sympathetic impulses usually dominate the control of most visceral effectors in times of stress, but not always. Curiously enough, parasympathetic impulses frequently become excessive to some effectors at such times. For instance, one of the first symptoms of emotional stress in many individuals is that they feel hungry and want to eat more than usual. Presumably this is partly caused by increased parasympathetic impulses to the smooth muscle of the stomach. This stimulates increased gastric contractions, which in turn may cause the feeling of hunger. The most famous disease of parasympathetic excess is peptic ulcer. In some individuals (presumably those born with certain genes), chronic stress leads to excessive parasympathetic stimulation of gastric hydrochloric acid glands, with eventual development of peptic ulcer.

PARASYMPATHETIC SYSTEM

The parasympathetic system is the dominant controller of most visceral effectors most of the time. Under quiet, nonstressful conditions, more impulses reach visceral effectors by cholinergic parasympathetic fibers than by adrenergic sympathetic fibers. Acetylcholine, the neurotransmitter of the parasympathetic system, tends to slow the heartbeat but acts to promote digestion and elimination. For example, it stimulates digestive gland secretion. It also increases peristalsis by stimulating the smooth muscle of the digestive tract. Note other parasympathetic effects in the right hand column of Table 12-3.

Outline Summary

STRUCTURE

A Autonomic nervous system as a whole
  1 Subdivisions: sympathetic system and parasympathetic system
  2 Both divisions consist of autonomic ganglia, nerves, and plexuses that are composed of autonomic neurons; all autonomic neurons are efferent
  3 Preganglionic autonomic neurons conduct from brain stem or cord to an autonomic ganglion; postganglionic neurons conduct from ganglion to visceral effector (cardiac muscle, smooth muscle, and glandular epithelial tissue); the relay of preganglionic and postganglionic neurons to conduct from central nervous system to visceral effector differs from conduction to somatic effectors, which requires only one efferent neuron, the somatic motoneuron
  4 All visceral effectors are innervated by sympathetic fibers and most are doubly innervated by sympathetic and parasympathetic fibers; exceptions: sweat glands and smooth muscle of most blood vessels, sympathetic fibers only

B Sympathetic (or thoracolumbar) system
  1 Sympathetic division consists of two chains of ganglia (one on either side of backbone) and fibers that connect ganglia with each other and with thoracic and lumbar segments of cord; other fibers extend from sympathetic ganglia out to visceral effectors
  2 Sympathetic preganglionic neurons—cell bodies in lateral gray columns of thoracic and first three or four lumbar segments of cord; axons leave cord in anterior roots of spinal nerves, leave spinal nerve by way of white ramus, and then follow one of three paths: to terminate in sympathetic ganglia; or to and through sympathetic ganglia, then up or down sympathetic trunk to terminate in a higher or lower sympathetic ganglion; or to and through sympathetic ganglia, through splanchnic nerves, to terminate in collateral ganglia
  3 Sympathetic postganglionic neurons—cell bodies in sympathetic chain or collateral ganglia (celiac, superior, and inferior mesenteric); axons in autonomic nerves and plexuses or in gray rami and spinal nerves

C Parasympathetic (or craniosacral) division
  1 Parasympathetic preganglionic neurons—cell bodies in nuclei of brain stem and of sacral segments of cord; axons from brain stem nuclei go through cranial nerve III, terminate in ciliary ganglion; from nuclei in pons through cranial nerve VII, terminate in sphenopalatine or submaxillary ganglion; from nuclei in medulla through cranial nerve IX to otic ganglion or through cranial nerves X and XI to cardiac and celiac ganglia
  2 Parasympathetic postganglionic neurons—cell bodies in ganglia on or near organs innervated; axons lie in short nerves extending into effector

FUNCTIONS

A Autonomic nervous system as a whole
  1 Autonomic system regulates visceral effectors in ways that tend to maintain or quickly restore homeostasis
  2 Both sympathetic and parasympathetic systems are tonically active, that is, they continually send impulses to visceral effectors
  3 Sympathetic and parasympathetic impulses exert antagonistic influences on visceral effectors; summation of the two determines activity of effector
  4 Autonomic system does not function autonomously but is continually influenced by impulses from the brain, from a hierarchy of autonomic centers
  5 Axons of autonomic neurons release either norepinephrine or acetylcholine as neurotransmitters, that is, are either adrenergic or cholinergic fibers; sympathetic postganglionics are only autonomic adrenergic fibers and a few of these (e.g., to sweat glands) are cholinergic
  6 Norepinephrine affects visceral effectors by first binding to alpha or beta receptors in their membranes
  7 Acetylcholine affects visceral effectors by first binding to muscarinic receptors but stimulates postganglionic parasympathetic neurons by binding to nicotinic receptors

B Sympathetic system
  1 Under normal conditions, the sympathetic system functions to maintain normal functioning of doubly innervated visceral effectors by partially counteracting parasympathetic impulses; functions to maintain normal tone of singly innervated blood vessels and thereby maintains normal blood pressure
  2 Under stress conditions, sympathetic impulses prepare body for maximum energy expenditure by producing "fight or flight" syndrome of responses

C Parasympathetic system—dominates control of most visceral effectors most of time, that is, under ordinary conditions

Review Questions

1 Compare the sympathetic and parasympathetic divisions of the autonomic nervous system as to (1) macroscopic components, (2) types of neurons, (3) locations of dendrites and cell bodies of their preganglionic and postganglionic neurons, and (4) neurotransmitters released by their preganglionic and postganglionic axons.

2 All preganglionic neurons release the same neurotransmitter. Name it.

3 Only one kind of autonomic fiber releases norepinephrine. Which kind?

4 Are all sympathetic postganglionic fibers adrenergic? If not, which ones are cholinergic?

5 Name the receptors to which acetylcholine binds in the membranes of (1) postganglionic neurons and (2) visceral effectors.

6 Name the enzyme that terminates action of acetylcholine.

7 Name two types of receptors to which norepinephrine binds in the membranes of visceral effectors.

8 Describe two mechanisms by which the action of norepinephrine is normally terminated.

9 Compare parasympathetic and sympathetic functions.

10 Explain why the name *autonomic nervous system* is misleading.

11 Classify the following structures as somatic effectors, visceral effectors, or neither: adrenal glands, biceps femoris muscle, heart, iris, skin.

12 Which of the following would indicate an increase in sympathetic impulses and which might indicate an increase in parasympathetic impulses to visceral effectors: constipation, dilated pupils, dry mouth, goose pimples, "I'm always hungry and eat too much when I am upset," rapid heartbeat?

13 Compare the "fight or flight" reaction with the physiological changes of meditation.

14 Can an individual learn to control the action of a visceral effector, such as the rate of the heartbeat? If so, explain one method for doing this.

13 SENSE ORGANS

OBJECTIVES

After you have completed this chapter, you should be able to:

1 Compare the function of general and special sense organs.
2 List the general sense organs (receptors), give their generalized location, and cite examples.
3 Classify receptors according to structure and stimuli required to activate them.
4 Identify the major anatomical structures visible in a horizontal section through the eyeball.
5 Describe the layers that compose the retina.
6 Compare the structure, function, and location of rods and cones in the retina.
7 Discuss the cavities and humors of the eye.
8 List and give the function of the extrinsic and intrinsic eye muscles.
9 Identify the accessory structures of the eye.
10 Discuss the four processes that focus light rays on the retina and describe the most common errors of refraction.
11 List and discuss the function of the major anatomical components in the external, middle, and inner ear.
12 Discuss the physiology of hearing and equilibrium.
13 Discuss the structure and function of the olfactory and gustatory sense organs.
14 Explain the rationale of radial keratotomy in the treatment of myopia (nearsightedness).

The body has millions of sense organs. They fall into two main categories, general sense organs and special sense organs. Of these, by far the most numerous are the general sense organs, or receptors. Receptors function to produce the general senses (such as touch, temperature, and pain) and to initiate various reflexes necessary for maintaining homeostasis. Special sense organs function to produce the special senses (vision, hearing, equilibrium, taste, and smell) and they too initiate reflexes important for homeostasis. In this chapter we begin with a brief description of receptors and follow with a detailed consideration of the eye and ear.

GENERAL SENSE ORGANS (RECEPTORS)

CLASSIFICATION

Receptors are commonly classified according to location, structure, and types of stimuli activating them. Classified according to location, the three types of receptors are as follows: superficial receptors, or exteroceptors; deep receptors, or proprioceptors; and internal receptors, or visceroceptors. **Superficial receptors,** or **exteroceptors,** have superficial locations; they lie near the exterior of the body. **Deep receptors,** or **proprioceptors,** lie deep under the body's surface in its muscles, tendons, and joints. **Internal receptors,** or **visceroceptors,** have internal locations; they lie in the viscera and blood vessels of the body. For the general senses resulting from stimulation of these receptors, see Table 13-1.

Classified according to structure, there are two general types of receptors, namely, free nerve endings and encapsulated nerve endings. *Free nerve endings* are free in the sense that they are not enclosed by connective tissue. Another descriptive name for them is *naked nerve endings.* See Figure 13-1 for three variations of free nerve endings. *Encapsulated nerve endings,* as the name suggests, are nerve fiber endings encased in a capsule of one sort or another. See Figures 13-2 to 13-5 and Table 13-2 for the names, locations, and functions of several types of encapsulated nerve endings.

Classified according to the kinds of stimuli that activate them, there are several types of receptors, including **mechanoreceptors,** activated by mechanical stimuli such as touch and pressure; **thermoreceptors,** activated by heat and cold; **chemoreceptors,** activated by various chemicals; and **nociceptors,** activated by intense stimuli of any kind.

Table 13-1 Receptors classified by location

Types	Locations	General senses
Superficial receptors	At or near surface of body; in skin and mucosa	Touch, pressure, heat, cold, and pain
Deep receptors (Proprioceptors)	In muscles, tendons, and joints	Proprioception (sense of position and movement) Vibration, deep pressure, deep pain
Internal receptors (Visceroceptors)	In the viscera and in blood vessel walls	Usually no sensations result from stimulation of internal receptors; exceptions: hunger, nausea, and pain from certain stimuli (notably, distention and some chemicals)

Nociceptors are commonly called pain receptors. Those located in the skin and mucosa are stimulated by any kind of intense stimuli. Those in the viscera are stimulated only by marked changes in pressure and by certain chemicals. Marked distention of the intestine, for example, causes pain, but cauterizing the uterine cervix does not. Here is another interesting fact. The pain from stimulation of nociceptors in deep structures is frequently referred to surface areas. **Referred pain** is the term for this phenomenon. Pain originating in the viscera and other deep structures is generally interpreted as coming from the skin area whose sensory fibers enter the same segment of the spinal cord as the sensory fibers from the deep structure. For example, sensory fibers from the heart enter the first to fourth thoracic segments. And so do sensory fibers from the skin areas over the heart and on the inner surface of the left arm. Pain originating in the heart is referred to those skin areas, but the reason for this is not clear.

Table 13-2 Receptors classified by structure

Types	Main locations	General senses
FREE NERVE ENDINGS (naked nerve endings)	Skin, mucosa (epithelial layers)	Pain, crude touch, possibly temperature
ENCAPSULATED NERVE ENDINGS		
Meissner's corpuscles	Skin (in papillae of dermis); numerous in fingertips and lips	Fine touch, vibration
Ruffini's corpuscles	Skin (dermal layer) and subcutaneous tissue of fingers	Touch, pressure
Pacinian corpuscles	Subcutaneous, submucous, and subserous tissues, around joints, in mammary glands and external genitals of both sexes	Pressure, vibration
Krause's end-bulbs	Skin (dermal layer), subcutaneous tissue, mucosa of lips and eyelids, external genitals	Touch
Golgi tendon receptors	Near junction of tendons and muscles	Subconscious muscle sense
Muscle spindles	Skeletal muscles	Subconscious muscle sense

Figure 13-1 Free nerve endings. **A,** In dermis of skin. **B,** Surrounding, in linear and circular fashion, the root of a hair follicle. **C,** In the cornea.

Figure 13-2 Meissner's corpuscle (tactile corpuscle) in skin papilla, an encapsulated nerve ending found in hairless portions of skin. Meissner's corpuscles mediate sensation of touch.

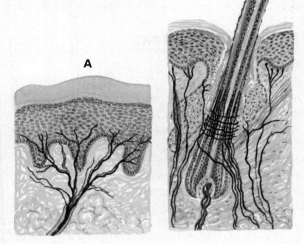

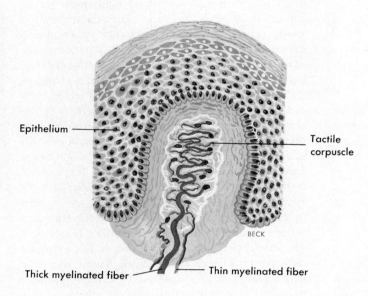

Epithelium

Tactile corpuscle

Thick myelinated fiber

Thin myelinated fiber

BECK

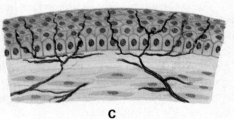

Figure 13-3 Ruffini's corpuscle, a skin receptor that probably mediates touch, rather than heat, as formerly thought.

Figure 13-4 Pacinian corpuscle, an encapsulated nerve ending widely distributed in subcutaneous tissue; mediates sensation of pressure.

Figure 13-5 Krause's end-bulb, an encapsulated nerve ending; may mediate sensation of cold, but evidence indicates that it is not the only type of receptor for cold.

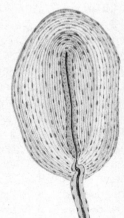

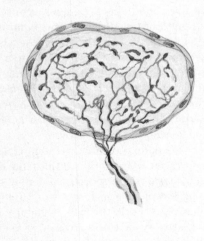

DISTRIBUTION

Receptors are widely distributed through the body in the skin, mucosa, connective tissues, muscles, tendons, joints, and viscera. Their distribution, however, is not uniform in all areas. In some, it is very dense and in others, sparse. The skin covering the fingertips, for instance, contains many more receptors to touch than does the skin on the back. A simple procedure, the two-point discrimination test, demonstrates this fact. A subject reports the number of touch points felt when an investigator touches the skin simultaneously with two points of a compass. If the skin on the fingertip is touched with the compass points barely one eighth of an inch apart, the subject senses them as two points. But if the skin on the back is touched with the compass points this close together, they will be felt as only one point. Unless they are an inch or more apart, they cannot be discriminated as two points. Why this difference? Because touch receptors are so densely distributed in the fingertips that two points very close to each other stimulate two different receptors. Hence they are sensed as two points. But the situation is quite different in the skin on the back. There, touch receptors are so widely scattered that two points have to be at least an inch apart to stimulate two receptors and be felt as two points.

GENERALIZATION ABOUT RECEPTOR FUNCTIONS

The general function of receptors is to respond to stimuli by converting them to nerve impulses. As a rule, different structural types of receptors respond to different types of stimuli. Free nerve endings are an exception to this rule; they respond to various kinds of intense stimuli. When an adequate stimulus acts on a receptor, a potential develops in the receptor's membrane. It is called a **receptor potential** or a **generator potential.** Within limits, the stronger the stimulus, the greater the magnitude of the resulting receptor potential. In other words, the receptor potential is a graded response, graded to the strength of the stimulus. This contrasts with an action potential, which, you will recall, is an all-or-none response. A nerve fiber responds to a stimulus either with an action potential of maximum magnitude (for the existing physiological conditions) or it does not respond at all. When a receptor potential reaches a certain threshold, it triggers an action potential in the sensory neuron's axon.

Receptors exhibit a functional characteristic known as adaptation. **Adaptation** means that the magnitude of the receptor potential decreases over a period of time in response to a continuous stimulus. As a result, the rate of impulse conduction by the sensory neuron's axon also decreases. So too does the intensity of the resulting sensation. Examples of adaptation abound. One familiar example is feeling the touch of your clothing when you first put it on and soon not sensing it at all. Touch receptors adapt very rapidly. In contrast, the proprioceptors in our muscles, tendons, and joints adapt very slowly. As long as stimulation of them continues, they continue sending impulses to the brain.

SPECIAL SENSE ORGANS

EYE

Coats of Eyeball

Approximately five sixths of the eyeball lies recessed in the orbit, protected by this bony socket. Only the small anterior surface of the eyeball is exposed. Three layers of tissues or coats compose the eyeball. From the outside in, they are the **sclera,** the **choroid,** and the **retina.** Both the sclera and the choroid coats consist of an anterior and a posterior portion.

Tough white fibrous tissue fashions the *sclera.* Deep within the anterior part of the sclera at its junction with the cornea lies a ring-shaped venous sinus, the *canal of Schlemm* (Figures 13-6, 13-9, and 13-10). The anterior portion of the sclera is called the *cornea* and lies over the colored part of the eye (iris). The cornea is transparent, whereas the rest of the sclera is white and opaque, a fact that explains why the visible anterior surface of the sclera is usually spoken of as the "white" of the eye. No blood vessels are found in the cornea, in the aqueous and vitreous humors, or in the lens.

The middle or *choroid coat* of the eye contains a great many blood vessels and a large amount of pigment. Its anterior portion is modified into three separate structures: the ciliary body, the suspensory ligament, and the iris (Figure 13-6).

The **ciliary body** is formed by a thickening of the choroid and fits like a collar into the area between the anterior margin of the retina and the posterior margin of the iris. The small *ciliary muscle,* composed of both radial and circular smooth muscle fibers, lies in the anterior part of the ciliary body. Attached to the ciliary body is the *suspensory lig-*

Surgical removal of opaque or deteriorating corneas and replacement with donor transplants is a common medical practice. Corneal tissue is **avascular;** that is, the cornea is free of blood vessels. Therefore corneal tissue is seldom rejected by the body's immune system. Antibodies carried in the blood have no way to reach the transplanted tissue and therefore long-term success following implant surgery is excellent.

Figure 13-6 Horizontal section through right eyeball.

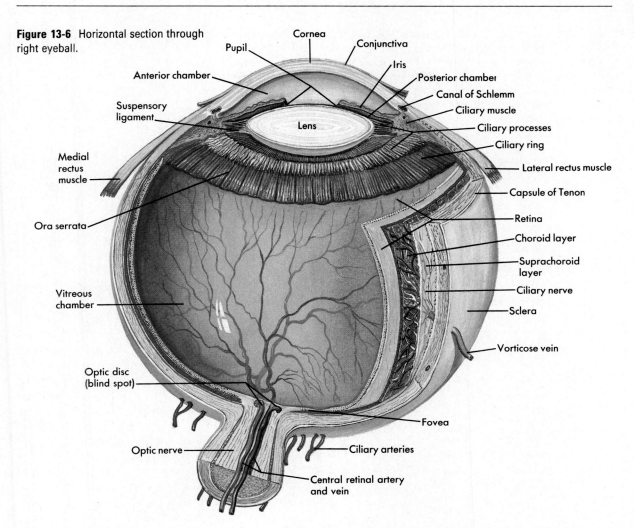

ament, which blends with the elastic capsule of the *lens* and holds it suspended in place.

The **iris,** or the colored part of the eye, consists of circular and radial smooth muscle fibers arranged so as to form a doughnut-shaped structure. (The hole in the middle is called the **pupil**). The iris attaches to the ciliary body.

The **retina** is the incomplete innermost coat of the eyeball—incomplete in that it has no anterior portion. Pigmented epithelial cells form the layer of the retina next to the choroid coat. Three layers of neurons make up the major portion of the retina. Named in the order in which they conduct impulses, they are *photoreceptor neurons, bipolar neurons,* and *ganglion neurons.* Identify each of these in Figure 13-7. The distal ends of the dendrites of the photoreceptor neurons have been given names descriptive of their shapes. Because some look like tiny rods and others like cones, they are called **rods** and **cones,** respectively. They constitute our visual receptors, structures highly specialized for stimulation by light rays (discussed on p. 377). They differ as

to numbers, distribution, and function. Cones are less numerous than rods and are most densely concentrated in the **fovea centralis,** a small depression in the center of a yellowish area, the **macula lutea,** found near the center of the retina (Figure 13-8). They become less and less dense from the fovea outward. Rods, on the other hand, are absent entirely from the fovea and macula and increase in density toward the periphery of the retina. How these anatomical facts relate to rod and cone functions is revealed on p. 377.

All the axons of ganglion neurons extend back to a small circular area in the posterior part of the eyeball known as the *optic disc* or papilla. This part of the sclera contains perforations through which the fibers emerge from the eyeball as the *optic nerves.* The optic disc is also called the *blind spot* because light rays striking this area cannot be seen. Why? Because it contains no rods or cones, only nerve fibers. For an outline summary of the coats of the eyeball see Table 13-3; for location, see Figures 13-6 and 13-8.

Figure 13-7 Layers that compose the retina. The inner limiting membrane lies nearest the inside of the eyeball and it adheres to the vitreous humor, whereas the pigment epithelium lies farthest from the inside of the eyeball and it adheres to the choroid coat. Note relay of three neurons in the retina: photoreceptor, bipolar, and ganglion neurons (named in order of impulse transmission). Light rays pass through the vitreous humor and various layers of retina to stimulate rods and cones, the receptors of the photoreceptor neurons.

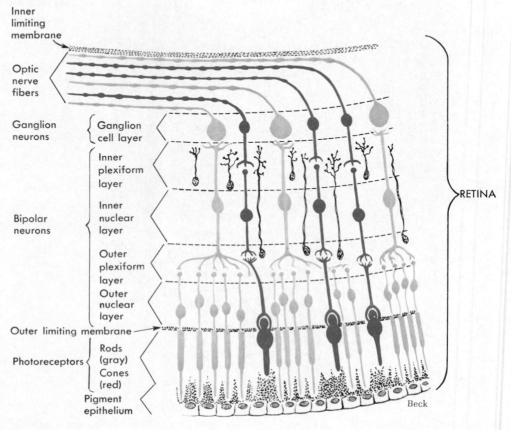

Cavities and Humors

The eyeball is not a solid sphere but contains a large interior cavity that is divided into two cavities, anterior and posterior.

The **anterior cavity** has two subdivisions known as the *anterior* and *posterior chambers.* As Figure 13-6 shows, the entire anterior cavity lies in front of the lens. The posterior chamber of the anterior cavity consists of the small space directly posterior to the iris but anterior to the lens. And the anterior chamber of the anterior cavity is the space anterior to the iris but posterior to the cornea. *Aqueous humor* fills both chambers of the anterior cavity. This substance is clear and watery and often leaks out when the eye is injured.

The **posterior cavity** of the eyeball is considerably larger than the anterior, since it occupies all the space posterior to the lens, suspensory ligament, and ciliary body (see Figure 13-6). It contains *vitreous humor*, a substance with a consistency comparable to soft gelatin. This semisolid material helps maintain sufficient intraocular pressure to prevent the eyeball from collapsing. (An obliterated artery, the *hyaloid canal*, runs through the vitreous humor between the lens and optic disc.)

An outline summary of the cavities of the eye appears in Table 13-4.

Still not established is the mechanism by which aqueous humor forms. It comes from blood in capillaries (located mainly in the ciliary body). Presumably the ciliary body actively secretes aqueous humor into the posterior chamber. But also, passive filtration from capillary blood may contribute to aqueous humor formation. From the posterior chamber, aqueous humor moves from the area between the iris and the lens through the pupil into the anterior chamber (Figure 13-9). From here, it drains into the canal of Schlemm and moves on into small veins. Normally, aqueous humor drains out of the anterior chamber at the same rate at which it enters the pos-

Figure 13-8 Right eyeground (fundus) showing vessels, optic disc, macula lutea, and layers of the eyeball.

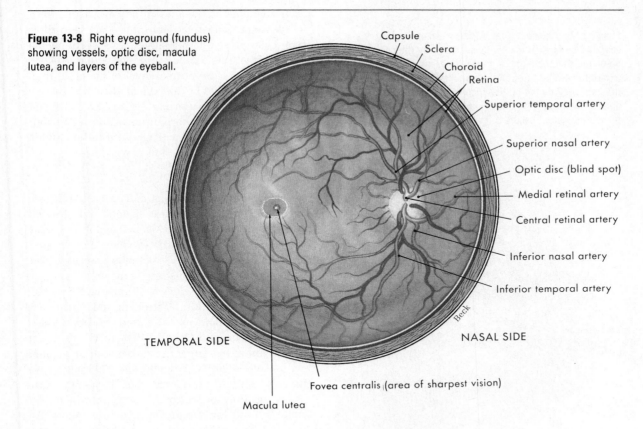

Capsule
Sclera
Choroid
Retina
Superior temporal artery
Superior nasal artery
Optic disc (blind spot)
Medial retinal artery
Central retinal artery
Inferior nasal artery
Inferior temporal artery
TEMPORAL SIDE
NASAL SIDE
Fovea centralis (area of sharpest vision)
Macula lutea

Table 13-3 Coats of the eyeball

Location	Posterior portion	Anterior portion	Characteristics
Outer coat (sclera)	Sclera proper	Cornea	Protective fibrous coat; cornea transparent; rest of coat white and opaque
Middle coat (choroid)	Choroid proper	Ciliary body; suspensory ligament; iris (pupil is hole in iris); lens suspended in suspensory ligament	Vascular, pigmented coat
Inner coat (retina)	Retina	No anterior position	Nervous tissue; rods and cones (receptors for second cranial nerve) located in retina

Table 13-4 Cavities of the eye

Cavity	Divisions	Location	Contents
Anterior	Anterior chamber Posterior chamber	Anterior to iris and posterior to cornea Posterior to iris and anterior to lens	Aqueous humor Aqueous humor
Posterior	None	Posterior to lens	Vitreous humor

Figure 13-9 Aqueous humor *(heavy arrows)* is believed to be formed mainly by secretion by the ciliary body into the posterior chamber. It passes into the anterior chamber through the pupil, from which it is drained away by the ring-shaped canal of Schlemm, and finally into the anterior ciliary veins. *Small arrows* indicate pressure of the aqueous humor.

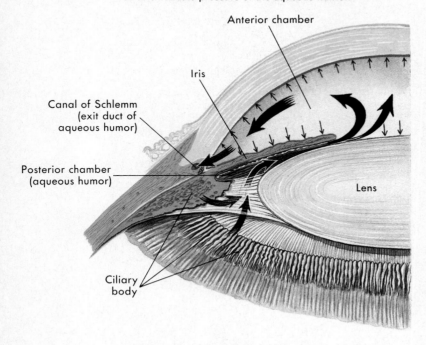

Figure 13-10 Acute glaucoma. When pressure of the aqueous humor in the anterior chamber becomes extreme, the iris is pushed forward, causing it to press on and block the canal of Schlemm from draining the fluid. In most instances, proper eyedrop medication will dilate the duct, permitting excess fluid escape. When medication fails, a new outflow duct can be created surgically.

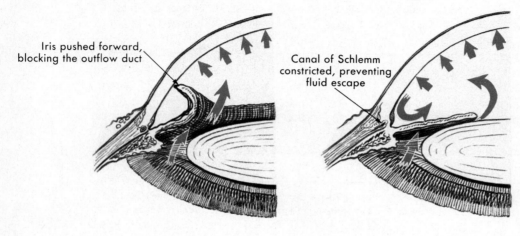

terior chamber, so the amount of aqueous humor in the eye remains relatively constant—and so too does intraocular pressure. But sometimes something happens to upset this balance, and intraocular pressure increases above the normal level of about 20 to 25 mm Hg pressure. The individual then has the eye disease known as *glaucoma* (Figure 13-10). Either excess formation or, more often, decreased absorption is seen as an immediate cause of this condition, but underlying causes are unknown.

Muscles

Eye muscles are of two types: extrinsic and intrinsic. **Extrinsic eye muscles** are skeletal muscles that attach to the outside of the eyeball and to the bones of the orbit. They move the eyeball in any desired direction and are, of course, voluntary muscles. Four of them are straight muscles, and two are oblique. Their names describe their positions on the eyeball. They are the superior, inferior, mesial, and lateral rectus muscles and superior and inferior oblique muscles (Figures 13-11 and 13-12).

Intrinsic eye muscles are smooth or involuntary muscles located within the eye. Their names are the *iris* and the *ciliary muscles.* Incidentally, the eye is the only organ in the body in which both voluntary and involuntary muscles are found. The iris regulates the size of the pupil. The ciliary muscle controls the shape of the lens. As the ciliary muscle contracts, it releases the suspensory ligament from the backward pull usually exerted on it. And this allows the elastic lens, suspended in the ligament, to bulge or become more convex—a necessary accommodation for near vision (p. 376). Some essential facts about eye muscles are summarized in Table 13-5.

Accessory Structures

Accessory structures of the eye include the eyebrows, eyelashes, eyelids, and the lacrimal apparatus.

Figure 13-11 Muscles that move the right eye as viewed from above.

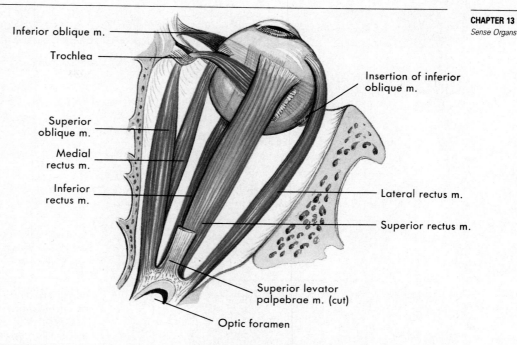

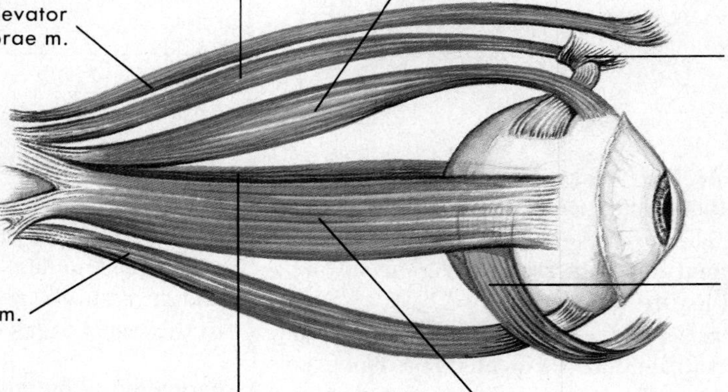

Inferior oblique m.

Trochlea

Superior oblique m.

Medial rectus m.

Inferior rectus m.

Insertion of inferior oblique m.

Lateral rectus m.

Superior rectus m.

Superior levator palpebrae m. (cut)

Optic foramen

Figure 13-12 Muscles of the right orbit as viewed from side.

Superior oblique m.

Superior rectus m.

Superior levator palpebrae m.

Trochlea

Inferior oblique m.

Inferior rectus m.

Medial rectus m.

Lateral rectus m.

EYEBROWS AND EYELASHES. The eyebrows and eyelashes serve a cosmetic purpose and give some protection against the entrance of foreign objects into the eyes. Small glands located at the base of the lashes secrete a lubricating fluid. They frequently become infected, forming a sty.

EYELIDS. The eyelids or palpebrae consist mainly of voluntary muscle and skin, with a border of thick connective tissue at the free edge of each lid known as the *tarsal plate*. One can feel the tarsal plate as a ridge when turning back the eyelid to remove a foreign object. Mucous membrane called *conjunctiva* lines each lid (Figure 13-13). It continues over the surface of the eyeball, where it is modified to give transparency. Inflammation of the conjunctiva

(conjunctivitis) is a fairly common infection. Because it produces a pinkish discoloration of the eye's surface, it is called *pinkeye*.

The opening between the eyelids bears the technical name of *palpebral fissure*. The height of the fissure determines the apparent size of the eyes. In other words, if the eyelids are habitually held widely opened, the eyes appear large, although there is very little difference in size between eyeballs of different adults. Eyes appear small if the upper eyelids droop. Plastic surgeons can correct this common aging change with an operation called *blepharoplasty*. The upper and lower eyelids join, forming an angle or corner known as a *canthus*, the inner canthus being the mesial corner of the eye and the outer canthus the lateral corner.

Table 13-5 Eye muscles

	Extrinsic muscles	Intrinsic muscles
Names	Superior rectus Inferior rectus Lateral rectus Medial rectus Superior oblique Inferior oblique	Iris Ciliary muscle
Kind of muscle	Voluntary (striated, skeletal)	Involuntary (smooth, visceral)
Location	Attached to eyeball and bones of orbit	Modified anterior portion of choroid coat of eyeball; iris, doughnut-shaped sphincter muscle; pupil, hole in center of iris
Functions	Eye movements	Iris regulates size of pupil—therefore amount of light entering eye; ciliary muscle controls shape of lens (accommodation—therefore its refractive power)
Innervation	Somatic fibers of third, fourth, and sixth cranial nerves	Autonomic fibers of third and fourth cranial nerves

LACRIMAL APPARATUS. The lacrimal apparatus consists of the structures that secrete tears and drain them from the surface of the eyeball. They are the lacrimal glands, lacrimal ducts, lacrimal sacs, and nasolacrimal ducts (Figure 13-14).

The *lacrimal glands,* comparable in size and shape to a small almond, are located in a depression of the frontal bone at the upper outer margin of each orbit. Approximately a dozen small ducts lead from each gland, draining the tears onto the conjunctiva at the upper outer corner of the eye.

The *lacrimal glands* are small channels, one above and the other below each *caruncle* (small red body at inner canthus). They empty into the lacrimal sacs. The openings into the canals are called *punctae* and can be seen as two small dots at the inner canthus of the eye. The *lacrimal sacs* are located in a groove in the lacrimal bone. The *nasolacrimal ducts* are small tubes that extend from the lacrimal sac into the inferior meatus of the nose. All the tear ducts are lined with mucous membrane, an extension of the mucosa that lines the nose. When this membrane becomes inflamed and swollen, the nasolacrimal ducts become plugged, causing the tears to overflow from the eyes instead of draining into the nose as they do normally. Hence when we have a common cold, "watering" eyes add to our discomfort.

PHYSIOLOGY OF VISION

In order for vision to occur, the following conditions must be fulfilled: an image must be formed on the retina to stimulate its receptors (rods and cones) and the resulting nerve impulses must be conducted to the visual areas of the cerebral cortex.

Formation of Retinal Image

Four processes focus light rays so that they form a clear image on the retina: *refraction* of the light rays, *accommodation* of the lens, *constriction* of the pupil, and *convergence* of the eyes.

REFRACTION OF LIGHT RAYS. Refraction means the deflection or bending of light rays. It is produced by light rays passing obliquely from one transparent medium into another of different optical density; the more convex the surface of the medium, the greater its refractive power. The refracting media of the eye are the cornea, aqueous humor, lens, and vitreous humor. Light rays are bent or refracted at the anterior surface of the cornea as they pass from the rarer air into the denser cornea, at the anterior surface of the lens as they pass from the aqueous humor into the denser lens, and at the posterior surface of the lens as they pass from the lens into the rarer vitreous humor.

Figure 13-13 Longitudinal section through eyeball and eyelids showing the conjunctiva, a lining of mucous membrane. Conjunctiva covering the cornea is called *bulbar,* and that lining the posterior surface of the eyelids is called *palpebral.*

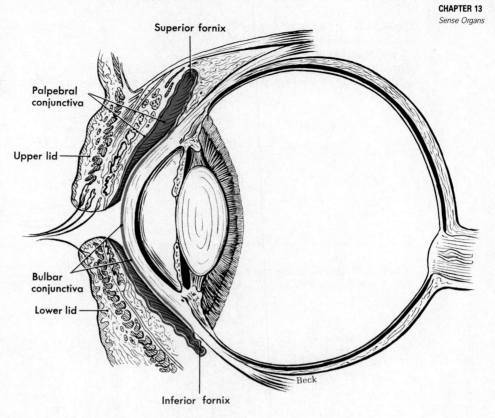

Figure 13-14 Lacrimal apparatus. Arrows indicate direction of drainage from the excretory ducts of the lacrimal glands across the eye to the naso-lacrimal duct.

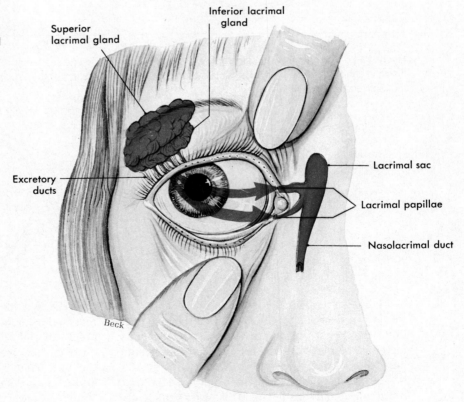

Figure 13-15 In the normal eye, light rays from an object are refracted by the cornea, aqueous humor, lens, and vitreous humor to converge on the fovea of the retina, where an inverted image is clearly formed.

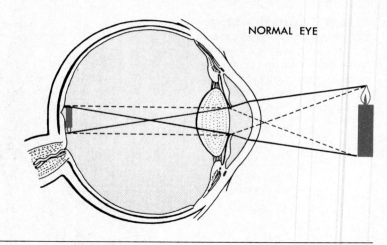

NORMAL EYE

Figure 13-16 Nearsighted or myopic eye focuses the image in front of the retina. This may occur when the eyeball is too long or the lens is too thick. Correction is by concave lens.

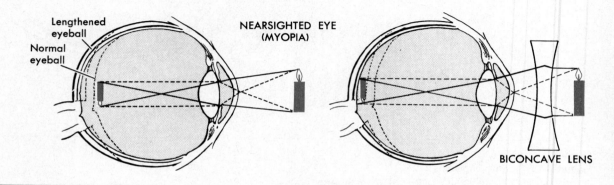

Lengthened eyeball
Normal eyeball

NEARSIGHTED EYE (MYOPIA)

BICONCAVE LENS

Figure 13-17 The farsighted or hyperopic eye can only focus the image at a hypothetical distance behind the retina. This may occur when the eyeball is too short or the lens is too thin. Correction is by convex lens.

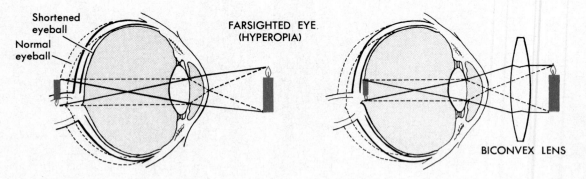

Shortened eyeball
Normal eyeball

FARSIGHTED EYE (HYPEROPIA)

BICONVEX LENS

When an individual goes to an ophthalmologist for an eye examination, the doctor does a "refraction." In other words, by various specially designed methods, he measures the refractory or light-bending power of that person's eyes.

In a relaxed normal (emmetropic) eye the four refracting media together bend light rays sufficiently to bring to a focus on the retina the parallel rays reflected from an object 20 or more feet away (Figure 13-15). Of course a normal eye can also focus on objects located much nearer than 20 feet from the eye. This is accomplished by a mechanism known as *accommodation* (discussed on p. 376). Many eyes, however, show **errors of refraction,**

RADIAL KERATOTOMY

Nearly 25% of the American population is nearsighted. **Radial keratotomy** is a relatively new surgical technique now being used to permanently improve myopia or nearsightedness in many individuals. The term *keratotomy* refers to surgical incision of the cornea. In a radial keratotomy a number of radial microscopic incisions are made in the cornea like the spokes of a wheel. The 30-minute procedure is often performed on an outpatient basis. The radial incisions (generally about eight) permit the cornea to stretch and become flatter because of normal intraocular pressure from behind. The result is better visual acuity postoperatively with apparently very little risk of infection. Many individuals see a star-shape or starburst pattern caused by light bouncing off the corneal scars immediately after surgery. By 5 to 6 months after surgery, however, the scars fade and the starburst pattern becomes insignificant. The procedure works best for people between 20-45 years old with mild to moderate myopia.

Figure 13-18 Astigmatism results most frequently from an irregular cornea. The cornea may be only slightly flattened horizontally, vertically, or diagonally to produce distortion of vision. The compensating shape of the lens largely nullifies the irregularities of the cornea.

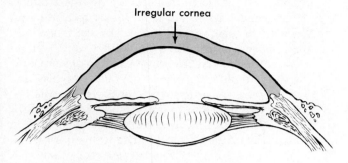

Irregular cornea

Figure 13-19 An irregular lens causes light to be bent in such a way that it does not focus the image on the sharpest area of vision on the retina. An astigmatic eye with this defect causes distorted or blurred vision.

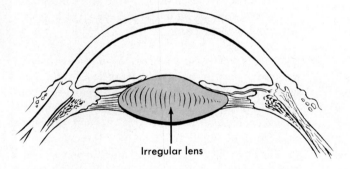

Irregular lens

that is, they are not able to focus the rays on the retina under the stated conditions. Some common errors of refraction are **nearsightedness** (myopia), **farsightedness** (hyperopia), and **astigmatism.**

The nearsighted (myopic) eye sees distant objects as blurred images because it focuses rays from the object at a point in front of the retina (Figure 13-16). According to one theory, this occurs because the eyeball is too long and the distance too great from the lens to the retina in the myopic eye. Concave glasses, by lessening refraction, can give clear distant vision to nearsighted individuals. Presumably, opposite conditions exist in the farsighted (hyperopic) eye (Figure 13-17).

Astigmatism is a more complicated condition in which the curvature of the cornea or of the lens is uneven, causing horizontal and vertical rays to be focused at two different points on the retina (Figures 13-18 and 13-19). Instead of the curvature of the cornea being a section of a sphere, it is more like that of a teaspoon, with horizontal and vertical arcs uneven. Suitable glasses correct the refraction of such an eye.

Visual acuity or the ability to distinguish form and outline clearly is indicated by a fraction that compares the distance at which an individual sees an object (usually letters of a definite size and shape) clearly with the distance at which the normal eye would see the object. Thus if a person sees clearly at 20 feet an object that the normal eye would be able to see clearly at 20 feet, visual acuity is said to be 20/20 or normal. But if an object is seen clearly at 20 feet that the normal eye sees clearly at 30 feet, visual acuity is 20/30 or two thirds normal.

As people grow older, they tend to become farsighted because lenses lose their elasticity and therefore their ability to bulge and to accommodate for near vision. This condition is called **presbyopia.**

ACCOMMODATION OF LENS. Accommodation for near vision necessitates three changes: increase in the curvature of the lens, constriction of the pupils, and convergence of the two eyes. Light rays from objects 20. or more feet away are practically parallel. The normal eye, as previously noted, refracts such rays sufficiently to focus them clearly on the retina. But light rays from nearer objects are divergent rather than parallel. So obviously they must be bent more acutely to bring them to a focus on the retina. Accommodation of the lens or, in other words, an increase in its curvature takes place to achieve this greater refraction. (It is a physical fact that the greater the convexity of a lens, the greater its refractive power.) Most observers accept Helmholtz' theory about the mechanism that produces accommodation of the lens. According to his theory, the ciliary muscle contracts, pulling the ciliary body and choroid forward toward the lens. This releases the tension on the suspensory ligament and therefore on the lens, which, being elastic, immediately bulges. For near vision, then, the ciliary muscle is contracted and the lens is bulging, whereas for far vision the ciliary muscle is relaxed and the lens is comparatively flat. Continual use of the eyes for near work produces eyestrain because of the prolonged contraction of the ciliary muscle. Some of the strain can be avoided by looking into the distance at intervals while doing close work.

CONSTRICTION OF PUPIL. The muscles of the iris play an important part in the formation of clear retinal images. Part of the accommodation mechanism consists of contraction of the circular fibers of the iris, which constricts the pupil. This prevents divergent rays from the object from entering the eye through the periphery of the cornea and lens. Such peripheral rays could not be refracted sufficiently to be brought to a focus on the retina and therefore would cause a blurred image. Constriction of the pupil for near vision is called the *near reflex* of the pupil and occurs simultaneously with accommodation of the lens in near vision. The pupil constricts also in bright light (**photopupil reflex** or **pupillary light reflex**) to protect the retina from too intense or too sudden stimulation.

CONVERGENCE OF EYES. Single binocular vision (seeing only one object instead of two when both eyes are used) occurs when light rays from an object fall on corresponding points of the two retinas. The foveas and all points lying equidistant and in the same direction from the foveas are corresponding points. Whenever the eyeballs move in unison, either with the visual axes parallel (for far objects) or converging on a common point (for near objects), light rays strike corresponding points of the two retinas. Con-

Figure 13-20 Rhodopsin cycle.

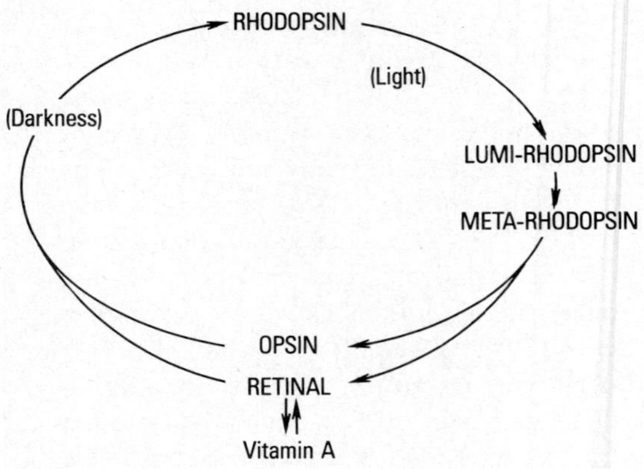

vergence is the movement of the two eyeballs inward so that their visual axes come together or converge at the object viewed. The nearer the object, the greater the degree of convergence necessary to maintain single vision. A simple procedure serves to demonstrate the fact that single binocular vision results from stimulation of corresponding points on two retinas. Gently press one eyeball out of line while viewing an object. Instead of one object, you will see two. In order to achieve unified movement of the two eyeballs, a functional balance between the antagonistic extrinsic muscles must exist. For clear distant vision, the muscles must hold the visual axes of the two eyes parallel. For clear near vision, they must converge them. These conditions cannot be met if, for example, the internal rectus muscle of one eye should contract more forcefully than its antagonist, the external rectus muscle. That eye would then be pulled in toward the nose. The movement of its visual axis would not coordinate with that of the other eye. Light rays from an object would then fall on noncorresponding points of the two retinas, and the object would be seen double (diplopia). Sometimes the individual can overcome the deviation of the visual axes by neuromuscular effort (extra conduction to and contraction of the weak muscle) and thereby achieve single vision, but only at the expense of muscular and nervous strain. The condition in which the imbalance of the eye muscles can be overcome by extra effort on the part of the weak muscle is called *heterophoria* (*esophoria* if the internal rectus muscle is stronger and pulls the eye nasalward and *exophoria* if the external rectus muscle is stronger and pulls the eye temporalward). **Strabismus** (cross-eye or squint) is an exaggerated esophoria that cannot be overcome by neuromuscular effort. An individual with strabismus usually does not have double vision, as you might expect, because he learns to suppress one of the images.

Stimulation of Retina

Both rods and cones contain photopigments, that is, light-sensitive, pigmented compounds. The photopigment present in rods is named *rhodopsin*. Rhodopsin is so highly light sensitive that even dim light causes its rapid breakdown into opsin (a protein) and retinal (a vitamin A derivative). See the right side of Figure 13-20. If the rod is then exposed to darkness for a short time, the change reverses itself and rhodopsin forms again. The breakdown of rhodopsin acts in some way to initiate impulse conduction by rods and produce vision in dim light. Objects are seen in shades of gray but not in colors.

Three types of cones are present in the retina. Each contains a different photopigment, either *erythrolabe*, *chlorolabe*, or *cyanolabe*. Each of the three primary colors (red, blue, and green) reflects light rays of a different wavelength. Each wavelength acts on one type of cone, causing its particular photopigment to break down and initiate impulse conduction by the cone. Light rays from red colors cause the breakdown of erythrolabe. Green colors stimulate chlorolabe breakdown and blue colors stimulate cyanolabe breakdown. Because these cone photopigments are less sensitive to light than rhodopsin, brighter light is necessary for their breakdown. Cones therefore function to produce vision in bright light and color vision. The fovea contains the greatest concentration of cones and is therefore the point of clearest vision in good light. For this reason, when we want to see an object clearly in the daytime, we look directly at it so as to focus the image on the fovea. But in dim light or darkness, we see an object better if we look slightly to the side of it, thereby focusing the image nearer the periphery of the retina, where rods are more plentiful.

Conduction to Visual Area

Fibers that conduct impulses from the rods and cones reach the visual cortex in the occipital lobes via the optic nerves, optic chiasma, optic tracts, and optic radiations. Look closely at Figure 13-21. Notice that each optic nerve contains fibers from only one retina but that the optic chiasma contains fibers from the nasal portions of both retinas. Each optic tract also contains fibers from both retinas. These anatomical facts explain certain peculiar visual abnormalities that sometimes occur. Suppose a person's right optic tract were injured so that it could not conduct impulses. He would be totally blind in neither eye but partially blind in both eyes. Specifically,

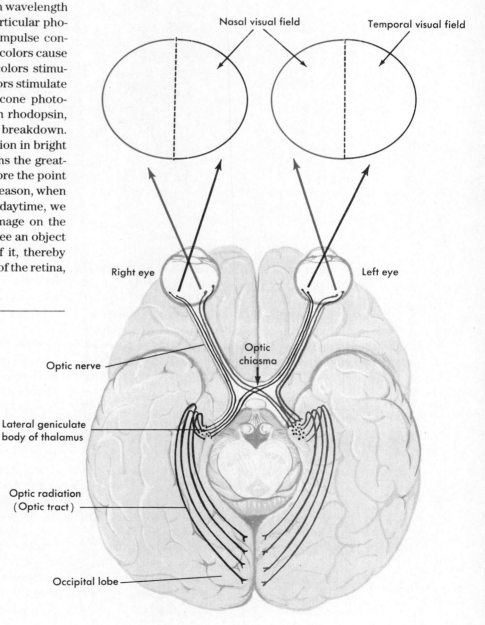

Figure 13-21 Visual pathways. Note structures that compose each pathway: optic nerve, optic chiasma, lateral geniculate body of thalamus, optic radiations, and visual cortex of occipital lobe. Fibers from nasal portion of each retina cross over to opposite side at optic chiasma, hence terminate in lateral geniculate body of opposite side. Location of a lesion in the visual pathway determines the resulting visual defect. Examples: destruction of an optic nerve produces permanent blindness in the same eye. Pressure on the optic chiasma—by a pituitary tumor, for instance—produces bitemporal hemianopia, or more simply, blindness in both temporal visual fields. Why? Because it destroys fibers from nasal sides of both retinas.

Nasal visual field

Temporal visual field

Right eye

Left eye

Optic nerve

Optic chiasma

Lateral geniculate body of thalamus

Optic radiation (Optic tract)

Occipital lobe

Figure 13-22 Components of the ear. External ear consists of auricle (pinna), external acoustic meatus (ear canal), and tympanic membrane (eardrum). Middle ear (tympanic cavity) includes malleus (hammer), incus (anvil), and stapes (stirrup). Internal ear contains semicircular canals, vestibule, and cochlea.

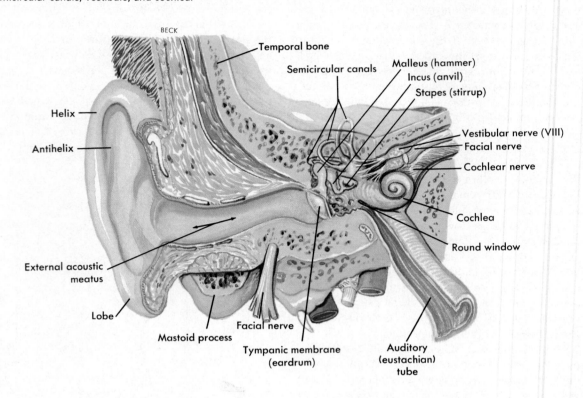

he would be blind in his right nasal and left temporal visual fields. Here are the reasons. The right optic tract contains fibers from the right retina's temporal area, the area that sees the right nasal visual field. In addition, the right optic tract contains fibers from the left retina's nasal area, the area that sees the left temporal visual field.

Earache is a common problem in young children. Pain may be because of irritation or infection of the external auditory meatus or more frequently from an infection located in the middle ear. The tympanic membrane often becomes pink and inflamed as a result of such infections. Visual inspection of the membrane using a specialized instrument called an **otoscope** enables a health professional to diagnose such conditions readily.

EAR
Anatomy

Each ear has three parts: **external, middle,** and **internal** (Figure 13-22).

External Ear

The external ear has two divisions: the flap or modified trumpet on the side of the head called the *auricle* or *pinna* and the tube leading from the auricle into the temporal bone named the *external acoustic meatus (ear canal)*. This canal is about 3.0 cm long and takes, in general, an inward, forward, and downward direction, although the first portion of the tube slants upward and then curves downward. Because of this curve in the auditory canal, the auricle should be pulled up and back to straighten the tube when medications are to be dropped into the ear. Modified sweat glands in the auditory canal secrete *cerumen* (waxlike substance), which occasionally becomes impacted and may cause pain and deafness. The **tympanic membrane** (eardrum) stretches across the inner end of the auditory canal, separating it from the middle ear.

Figure 13-23 Ossicles—malleus, incus, and stapes—of the middle ear as seen through the tympanic membrane and in relation to the internal ear's semicircular canals and cochlea. These structures are shown 4½ times their actual size.

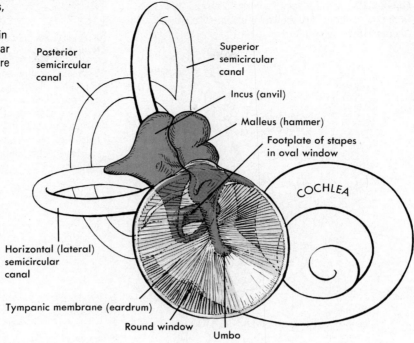

Posterior semicircular canal

Superior semicircular canal

Incus (anvil)

Malleus (hammer)

Footplate of stapes in oval window

COCHLEA

Horizontal (lateral) semicircular canal

Tympanic membrane (eardrum)

Round window

Umbo

Middle Ear

The middle ear (tympanic cavity), a tiny epithelial-lined cavity hollowed out of the temporal bone, contains the three **auditory ossicles:** the malleus, incus, and stapes (Figure 13-23). The names of these very small bones describe their shapes (hammer, anvil, and stirrup). The "handle" of the malleus is attached to the inner surface of the tympanic membrane, whereas the "head" attaches to the incus, which in turn attaches to the stapes. There are several openings into the middle ear cavity: one from the external auditory meatus, covered over with the tympanic membrane; two into the internal ear, the *fenestra ovalis* (oval window), into which the stapes fits, and the *fenestra rotunda* (round window), which is covered by a membrane; and one into the eustachian tube.

Posteriorly the middle ear cavity is continuous with a number of mastoid air spaces in the temporal bone. The clinical importance of these middle ear openings is that they provide routes for infection to travel. Head colds, for example, especially in children, may lead to middle ear or mastoid infections via the nasopharynx-eustachian tube-middle ear-mastoid path.

The **eustachian** or **auditory tube** is composed partly of bone and partly of cartilage and fibrous tissue and is lined with mucosa. It extends downward, forward, and inward from the middle ear cavity to the nasopharynx (the part of the throat behind the nose).

Internal Ear

The internal ear is also called the *labyrinth* because of its complicated shape. It consists of two main parts, a bony labyrinth and inside this a membranous labyrinth. The bony labyrinth consists of three parts:

The **eustachian tube** provides the path by which throat infections may invade the middle ear. But the eustachian tube also serves a useful function. It makes possible equalization of pressure against inner and outer surfaces of the tympanic membrane and therefore prevents membrane rupture and the discomfort that marked pressure differences produce. The way the eustachian tube equalizes tympanic membrane pressure is this: when one swallows or yawns, air spreads rapidly through the open tube. Atmospheric pressure then presses against the inner surface of the tympanic membrane. And since atmospheric pressure is continually exerted against its outer surface, the pressures are equal. You might test this mechanism sometime when you are ascending or descending in an airplane—start chewing gum to increase your swallowing and observe whether this relieves the discomfort in your ears.

Figure 13-24 Membranous labyrinth
(red) of the internal ear shown in relation
to bony labyrinth.

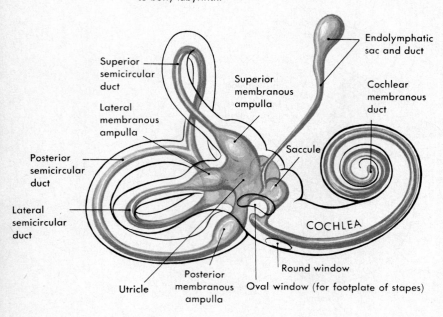

Table 13-6 Divisions of internal ear (labyrinth)

Bony labyrinth (part of temporal bone)	Membranous labyrinth (inside bony labyrinth)
Vestibule—central section of bony labyrinth; oval and round windows are openings of middle ear into vestibule; bony semicircular canals also open into vestibule	Utricle—one of the two parts of the membranous labyrinth contained in bony vestibule; utricle contains fluid endolymph, and an equilibrium sense organ, the macula; macula senses both head positions and movements and acceleration and deceleration of the body; vestibular nerve (branch of eighth cranial nerve) supplies macula
	Saccule—the other part of the membranous labyrinth within bony vestibule; contains endolymph and macula
Cochlea—spiraling bony tube that resembles a snail shell in shape	Cochlear duct—membranous tube that forms shelf across interior of bony cochlea; contains endolymph and organ of Corti, the sense organ for hearing; cochlear nerve (branch of eighth cranial nerve) supplies organ of Corti; cochlear duct separated from bony cochlea by scala vestibuli and scala tympani, spaces that contain perilymph
Bony semicircular canals—three of these semicircular-shaped canals; each lies approximately at right angles to the others	Membranous semicircular canals—separated from bony semicircular canals by perilymph; contain endolymph and the crista, an equilibrium sense organ that senses head movements; vestibular nerve supplies crista

vestibule, cochlea, and semicircular canals (Table 13-6). The membranous labyrinth, as Figure 13-24 shows, consists of the **utricle** and **saccule** inside the vestibule, the **cochlear duct** inside the cochlea, and the **membranous semicircular canals** inside the bony ones.

VESTIBULE, UTRICLE, AND SACCULE. The vestibule constitutes the central section of the bony labyrinth. Into it open both the oval and round windows from the middle ear as well as the three semicircular canals of the internal ear. The utricle and saccule have membranous walls and are suspended within the vestibule. They are separated from the bony walls of the vestibule by fluid *(perilymph)*, and both utricle and saccule contain a fluid called *endolymph*.

Located within the utricle (and also within the saccule) lies a small structure called the *macula*. It consists mainly of hair cells and a gelatinous membrane that contains *otoliths* (tiny ear "stones," that is, small particles of calcium carbonate). A few delicate hairs protrude from the hair cells and are embedded in the gelatinous membrane. Receptors for the vestibular branch of the eighth cranial nerve contact the hair cells of the macula located in the utricle. Changing the position of the head produces a change in the amount of pressure on the gelatinous membrane and causes the otoliths to pull on the hair cells. This stimulates the adjacent receptors of the vestibular nerve. Its fibers conduct impulses to the brain that produce a sense of the position of the head and also a sensation of a change in the pull of gravity, for example, a sensation of acceleration. In addition, stimulation of the macula in the utricle evokes *righting reflexes*, muscular responses to restore the body and its parts to their normal position when they have been displaced. (Impulses from proprioceptors and from the eyes also activate righting reflexes. Interruption of the vestibular or visual or

Figure 13-25 Diagram of the bony and membranous cochlea, uncoiled. Note end-organ of Corti projecting into endolymph contained in cochlear duct (membranous cochlea). Perilymph indicated above endolymph occupies the scala vestibuli. That in the lower compartment lies in the scala tympani (see also Figure 13-26).

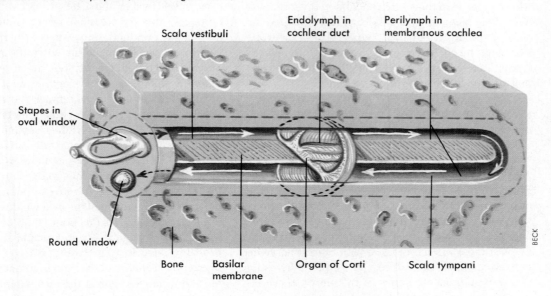

COCHLEA AND COCHLEAR DUCT. The word **cochlea,** which means snail, describes the outer appearance of this part of the bony labyrinth. When sectioned, the cochlea resembles a tube wound spirally around a cone-shaped core of bone, the *modiolus.* The modiolus houses the spiral ganglion, which consists of cell bodies of the first sensory neurons in the auditory relay. Inside the cochlea lies the membranous *cochlear duct*—the only part of the internal ear concerned with hearing. This structure is shaped like a somewhat triangular tube. It forms a shelf across the inside of the bony cochlea, dividing it into upper and lower sections all along its winding course (Figures 13-25 and 13-26). The upper section (above the cochlear duct) is called the **scala vestibuli,** whereas the lower section below the cochlear duct is the **scala tympani.** The roof of the cochlear duct is known as *Reissner's membrane* or the vestibular membrane. **Basilar membrane** is the name given the floor of the cochlear duct. It is supported by bony and fibrous projections from the wall of the cochlea. Perilymph fills the scala vestibuli and scala tympani, and endolymph fills the cochlear duct.

The hearing sense organ, the **organ of Corti,** rests on the basilar membrane throughout the whole length of the cochlear duct. The structure of the organ of Corti resembles that of the equilibrium

proprioceptive impulses that initiate these reflexes may cause disturbances of equilibrium, nausea, vomiting, and other symptoms.)

Figure 13-26 Section through one of the coils of the cochlea. Perilymph fills the scalae vestibuli and tympani. Endolymph fills the cochlear duct. Part of the spiral ganglion is shown (bulging stemlike structure at right of diagram).

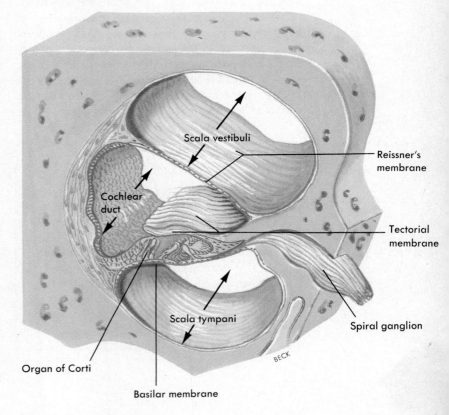

sense organ, that is, the macula in the utricle. It consists of supporting cells plus the important **hair cells** that project into the endolymph and are topped by an adherent gelatinous membrane called the **tectorial membrane.** Dendrites of the sensory neurons whose cells lie in the spiral ganglion in the modiolus have their beginnings around the bases of the hair cells of the organ of Corti. Axons of these neurons extend in the cochlear nerve (a branch of the eighth cranial nerve) to the brain. They conduct impulses that produce the sensation of hearing.

SEMICIRCULAR CANALS. Three semicircular canals, each in a plane approximately at right angles to the others, are found in each temporal bone. Within the bony semicircular canals and separated from them by perilymph are the membranous semicircular canals. Each contains endolymph and connects with the utricle, one of the membranous sacs inside the bony vestibule. Near its junction with the utricle the canal enlarges into an *ampulla.* Some of the receptors for the vestibular branch of the eighth cranial nerve lie in each ampulla. Like all receptors for both vestibular and auditory branches of this nerve, these receptors too lie in contact with hair cells in a supporting structure. Here in the ampulla the hair cells and supporting structure together are named the *crista ampullaris.* In the utricle and saccule they are called the *macula* and in the cochlear duct, the *organ of Corti.* Sitting atop the crista is a gelatinous structure called the *cupula.*

The crista with its vestibular nerve endings presumably functions as the end-organ for sensations of head movements. The macula in the utricle serves as the end-organ for sensations of head positions. Hence both the crista and the macula of the utricle function as end-organs for the sense of equilibrium.

The function of the macula in the saccule is not known. Some evidence, however, suggests that it serves as a receptor for vibratory stimuli.

> The **cochlea** is often described as the most mechanically complex body organ because of its numerous and incredibly intricate mechanical components. The number of "moving parts" in the organ exceeds 1 million. Each of the approximately 17,000 hair cells is covered with 100 or more tiny cilia, which are themselves capable of independent movement.

PHYSIOLOGY OF THE EAR

Hearing

Hearing results from stimulation of the auditory area of the cerebral cortex (temporal lobe, Figure 11-24, p. 328). Before reaching this area of the brain, however, sound waves must be projected through air, bone, and fluid to stimulate nerve endings and set up impulse conduction over nerve fibers.

Sound waves in the air enter the external auditory canal, probably without much aid from the pinna in collecting and reflecting them because of its smallness in man. At the inner end of the canal, they strike against the tympanic membrane, setting it in vibration. Vibrations of the tympanic membrane move the malleus, whose handle attaches to the membrane. The head of the malleus attaches to the incus, and the incus attaches to the stapes. So when the malleus vibrates, it moves the incus, which moves the stapes against the oval window into which it fits so precisely. At this point, fluid conduction of sound waves begins. To understand this, you will probably need to refer to Figures 13-25 and 13-26 frequently as you read the next few sentences. When the stapes moves against the oval window, pressure is exerted inward into the perilymph in the scala vestibuli of the cochlea. This starts a "ripple" in the perilymph that is transmitted through Reissner's membrane (the roof of the cochlear duct) to endolymph inside the duct and then to the organ of Corti and to the basilar membrane that supports the organ of Corti and forms the floor of the cochlear duct. From the basilar membrane the ripple is next transmitted to and through the perilymph in the scala tympani and finally expends itself against the round window—like an ocean wave expending itself as it breaks against the shore, although on a much reduced scale. Dendrites (of neurons whose cell bodies lie in the spiral ganglion and whose axons make up the cochlear nerve) terminate around the bases of the hair cells of the organ of Corti, and the tectorial membrane adheres to their upper surfaces. The movement of the hair cells against the adherent tectorial membrane somehow stimulates these dendrites and initiates impulse conduction by the cochlear nerve to the brain stem. Before reaching the auditory area of the temporal lobe, impulses pass through "relay stations" in nuclei in the medulla, pons, midbrain, and thalamus.

Equilibrium

In addition to hearing, the internal ear aids in the maintenance of equilibrium. It does this by sensing head movements and positions and also by sensing acceleration or deceleration of the body (see pp. 380 and 381).

Figure 13-27 Stimulation of olfactory cells in the nasal epithelium.

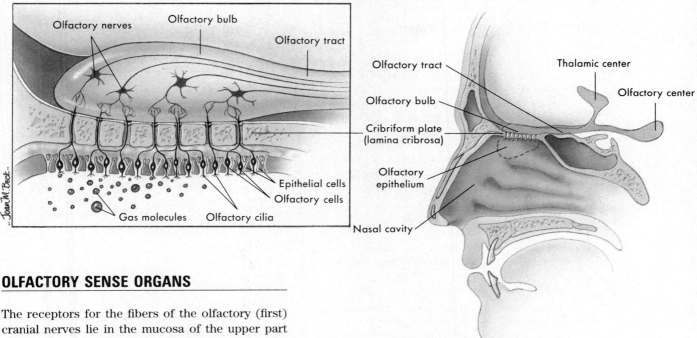

OLFACTORY SENSE ORGANS

The receptors for the fibers of the olfactory (first) cranial nerves lie in the mucosa of the upper part of the nasal cavity (Figure 13-27). Their location here explains the necessity for sniffing or drawing air forcefully up into the nose to smell delicate odors. The olfactory sense organ consists of hair cells and is relatively simple compared with the complex visual and auditory organs. Whereas the olfactory receptors are extremely sensitive, that is, stimulated by even very slight odors, they are also easily fatigued—a fact that explains why odors that are at first very noticeable are not sensed at all after a short time.

GUSTATORY SENSE ORGANS

The receptors for the taste nerve fibers (in branches of the seventh and ninth cranial nerves) are known as *taste buds* (Figure 13-28). Most of them are located on the tongue and the roof of the mouth. Taste buds are exteroceptors of the type called chemoreceptors, for the obvious reason that chemicals stimulate them. Only four kinds of taste sensations—sweet, sour, bitter, and salty—result from stimulation of taste buds. All the other flavors we sense result from a combination of taste bud and olfactory receptor stimulation. In other words, the myriads of tastes recognized are not tastes alone but tastes plus odors. For this reason a cold that interferes with the stimulation of the olfactory receptors by odors from foods in the mouth markedly dulls one's taste sensations.

Figure 13-28 Dorsal surface of tongue showing location of papillae and associated taste buds.

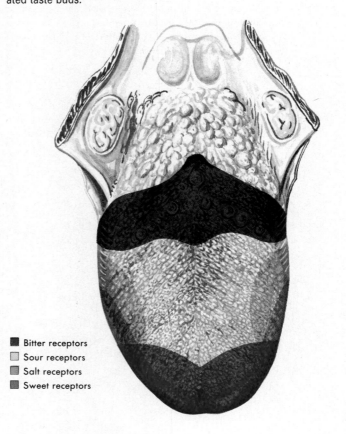

■ Bitter receptors
□ Sour receptors
▨ Salt receptors
▩ Sweet receptors

Outline Summary

GENERAL SENSE ORGANS (RECEPTORS)
Classification

A According to location—superficial, deep, and internal receptors; see Table 13-1 for summary

B According to structural characteristics—free and encapsulated nerve endings; variations of encapsulated endings summarized in Table 13-2

C According to kinds of stimuli that activate receptors—mechanoreceptors, activated by mechanical stimuli; thermoreceptors, activated by heat and cold; chemoreceptors, activated by chemicals; nociceptors (pain receptors), activated by intense stimuli of any kind

Distribution

Wide but uneven distribution in skin, mucosa, connective tissues, muscles, tendons, joints, and viscera

Generalization about receptor functions

A General function of receptors—respond to stimuli by converting them to nerve impulses; first response is development of receptor (generator) potential in receptor's membrane; receptor potential is a graded response to a stimulus—within limits, the stronger the stimulus, the greater the magnitude of the receptor potential; when receptor potential reaches a certain threshold, it triggers an action potential in the sensory neuron's axon

B Receptors exhibit adaptation, that is, in response to continuous stimulus, the magnitude of the receptor potential decreases, resulting in decreased intensity or extinction of sensation

SPECIAL SENSE ORGANS
Eye

A Coats of eyeball
1 Outer coat (sclera)
2 Middle coat (choroid)
3 Inner coat (retina)
4 For outline summary, see Table 13-3

B Cavities and humors
1 Anterior cavity
2 Posterior cavity
3 For outline summary, see Table 13-4

C Muscles
1 Extrinsic
  a Attach to outside of eyeball and to bones of orbit
  b Voluntary muscles; move eyeball in desired directions
  c Four straight (rectus) muscles—superior, inferior, lateral, and mesial; two oblique muscles—superior and inferior
2 Intrinsic
  a Within eyeball; called *iris* and *ciliary muscles*
  b Involuntary muscles

  c Iris regulates size of pupil
  d Ciliary muscle controls shape of lens, making possible accommodation for near and far objects
3 For outline summary, see Table 13-4

D Accessory structures
1 Eyebrows and eyelashes—protective and cosmetic
2 Eyelids
  a Lined with mucous membrane that continues over surface of eyeball; called *conjunctiva*
  b Opening between eyelids; called *palpebral fissure*
  c Corners where upper and lower eyelids join called *canthus, mesial* and *lateral.*
3 Lacrimal apparatus—lacrimal glands, lacrimal canals, lacrimal sacs, and nasolacrimal ducts

Physiology of vision

A Fulfillment of following conditions results in conscious experience known as vision: formation of retinal image, stimulation of retina, and conduction to visual area

B Formation of retinal image
1 Accomplished by four processes:
  a Refraction or bending of light rays as they pass through eye
  b Accommodation or bulging of lens—normally occurs if object viewed lies nearer than 20 feet from eye
  c Constriction of pupil; occurs simultaneously with accommodation for near objects and also in bright light
  d Convergence of eyes for near object so light rays from object fall on corresponding points of two retinas; necessary for single binocular vision
  e Radial keratotomy

C Stimulation of retina
1 Dim light causes rhodopsin in rods to break down; initiates impulse conduction by rods; results in vision in shades of gray
2 Bright light causes photopigments in cones to break down; initiates impulse conduction by cones; results in color vision; light reflected by red colors causes breakdown of erythrolabe, photopigment present in one type of cone; green colors cause breakdown of chlorolabe, photopigment in another type of cone; blue colors cause breakdown of cyanolabe, photopigment in another type of cone

D Conduction to visual area
Fibers that conduct impulses from rods and cones reach visual cortex in occipital lobes via optic nerves, optic chiasma, optic tracts, and optic radiations

Outline Summary—cont'd

Ear

A Anatomy
  1 External ear
    a Auricle or pinna
    b External acoustic meatus (ear canal)
  2 Middle ear
    a Separated from external ear by tympanic membrane
    b Contains auditory ossicles (malleus, incus, and stapes) and openings from external acoustic meatus, internal ear, eustachian tube, and mastoid sinuses
    c Eustachian tube, collapsible tube, lined with mucosa, extending from nasopharynx to middle ear
      (1) Equalizes pressure on both sides of eardrum
      (2) Open when yawning or swallowing
  3 Internal ear
    a Consists of bony and membranous portions, latter contained within former
    b Bony labyrinth has three divisions—vestibule, cochlea, and semicircular canals
    c Membranous cochlear duct contains receptors for cochlear branch of eighth cranial nerve (sense of hearing)
    d Utricle and membranous semicircular canals contain receptors for vestibular branch of eighth cranial nerve (sense of equilibrium)

Physiology of the ear

A Hearing—results from stimulation of auditory area of temporal lobes by impulses over cochlear nerves, which are stimulated by sound waves being projected through air, bone, and fluid before reaching auditory receptors (organ of Corti in cochlear duct)
B Equilibrium—stimulation of receptors (crista in semicircular canal and macula in utricle) leads to sense of equilibrium; also initiates righting reflexes essential for balance

OLFACTORY SENSE ORGANS

A Receptors for olfactory (first) cranial nerve located in nasal mucosa high along septum
B Receptors very sensitive but easily fatigued

GUSTATORY SENSE ORGANS

A Receptors for taste nerve fibers (in branches of seventh and ninth cranial nerves) called *taste buds* located on tongue and roof of mouth
B Four kinds of taste—sweet, sour, salty, and bitter
C All other tastes result from fusion of two or more of these tastes or from olfactory stimulation

Review Questions

1. Distinguish between superficial, deep, and internal receptors. Give another name for each of these types of receptors.
2. Distinguish between free and encapsulated nerve endings.
3. Identify several types of encapsulated nerve endings as to locations and functions.
4. Describe briefly one theory about the mechanism of referred pain.
5. Explain the phenomenon of adaptation.
6. Name the outer, middle, and inner coats of the eyeball.
7. Name two involuntary muscles in the eye. Explain their functions.
8. Define the term *refraction*. Name the refractory media of the eye.
9. Explain briefly the mechanism for accommodation for near vision.
10. Explain briefly how concave glasses correct nearsighted vision.
11. Name the receptors for vision in dim light and those for vision in bright light.
12. Name the photopigments present in rods and in cones. Explain their role in vision.
13. Explain why "night blindness" may occur in marked vitamin A deficiency. (Figure 13-21 contains a clue.)
14. Describe the main features of middle ear structure.
15. Name the parts of the bony and membranous labyrinths and describe the relationship of membranous labyrinth parts to those of the bony labyrinth.
16. In what ear structure(s) is the hearing sense organ located? The equilibrium sense organs?
17. What is the name of the hearing sense organ? Of the equilibrium sense organs?

14 THE ENDOCRINE SYSTEM

CHAPTER OUTLINE

Meaning

Prostaglandins (tissue hormones)

How hormones act

Pituitary gland (hypophysis cerebri)

    Size, location, and component glands

    Anterior pituitary gland (adenohypophysis)

        Growth hormone

        Prolactin (lactogenic or luteotropic hormone)

        Tropic hormones

        Melanocyte-stimulating hormone

        Control of secretion

    Posterior pituitary gland (neurohypophysis)

        Antidiuretic hormone

        Oxytocin

        Control of secretion

Thyroid gland

        Location and structure

        Thyroid hormone and calcitonin

        Effects of hypersecretion and hyposecretion

Parathyroid glands

        Location and structure

        Parathyroid hormone

Adrenal glands

    Location and structure

    Adrenal cortex

        Glucocorticoids

        Mineralocorticoids

        Sex hormones

        Control of secretion

    Adrenal medulla

        Epinephrine and norepinephrine

        Control of secretion

Islands of Langerhans

Ovaries

Testes

Pineal gland (pineal body or epiphysis cerebri)

Thymus

Gastric and intestinal mucosa

Placenta

OBJECTIVES

After you have completed this chapter, you should be able to:

1 Compare and contrast the general functions and mechanisms of action of the endocrine system and the nervous system.

2 Distinguish between the endocrine glands and the exocrine glands.

3 List and give the location of the primary endocrine glands in the body.

4 Discuss the chemical nature, classification, and mechanism of action of prostaglandins.

5 Briefly describe the proposed mechanism of hormone action at the cellular level.

6 Discuss the size, location, and anatomical components of the pituitary gland.

7 List and discuss anterior and posterior pituitary hormones and their specific target organs.

8 Discuss the possible clinical effects of abnormal growth hormone secretion.

9 Describe how hormone secretions are regulated by negative feedback control mechanisms.

10 Explain the relationship between the hypothalamus and the posterior lobe of the pituitary gland.

11 Discuss the structure, location, and functions of thyroid and parathyroid glands.

12 Compare the effects of hypersecretion and hyposecretion of thyroid hormone and identify the disease states associated with each.

13 Identify the hormones produced by the cell layers (zona) of the adrenal cortex and describe the pathological conditions that result from malfunction.

14 Discuss the relationship between exocrine and endocrine functions of the pancreas and briefly describe clinical problems associated with malfunction of the islands of Langerhans.

15 List the hormones associated with the ovaries, testes, pineal gland, thymus, gastric and intestinal mucosa, and the placenta.

KEY TERMS

Adrenal (ah-DRE-nal)

Angiotensin (an-je-o-TEN-sin)

Endocrine (EN-do-krin)

Exocrine (EK-so-krin)

Glucocorticoid (gloo-ko-KOR-ti-koid)

Hormone (HOR-mone)

Hypothalamus (hi-po-THAL-ah-mus)

Mineralocorticoid (min-er-al-o-KOR-ti-koid)

Pituitary (pi-TU-i-tery)

Prostaglandin (pros-tah-GLAN-din)

Target organ

Thymus (THY-mus)

MEANING

The endocrine system and the nervous system both function to achieve and maintain homeostasis or stability of our internal environment. Working alone or in concert as a single **neuroendocrine system,** they perform the same general functions for the body: communication, integration, and control. But they accomplish these general functions through different kinds of mechanisms and with somewhat different types of results. The mechanism of the nervous system consists of nerve impulses conducted by neurons from one specific structure to another. In contrast, the mechanism of the endocrine system consists of circulation of minute quantities of specific chemical messengers that, as you undoubtedly know, are called **hormones.** Hormones (from the Greek root *hormaein*, "to excite") are secreted from endocrine gland cells directly into the blood where they circulate to all parts of the body. Nerve impulses and hormones do not exert exactly the same kind of control, nor do they control precisely the same structures and functions. This is fortunate. It makes for better timing and more precision of control. For instance, nerve impulses produce rapid, short-lasting responses, whereas hormones produce slower and generally longer lasting responses. Nerve impulses control directly only two kinds of cells: muscle and gland cells. Although certain hormones such as thyroid and growth hormones have widespread physiological activity, most hormones, despite their widespread distribution by the blood to every body tissue, are highly specific in their action. Cells that are acted on and respond to a particular hormone are referred to descriptively as **target organ cells.**

Endocrine glands secrete hormones directly into the blood. Because they have no duct system, they are often called "ductless glands." Exocrine or duct glands, in contrast, release their secretions into ducts. Although the list of endocrine glands and hormone-secreting cell types is growing, the names and locations of the main endocrine glands are given in Table 14-1 and Figure 14-1.

Figure 14-1 Location of the endocrine glands in the female. Thymus gland is shown at maximum size at puberty.

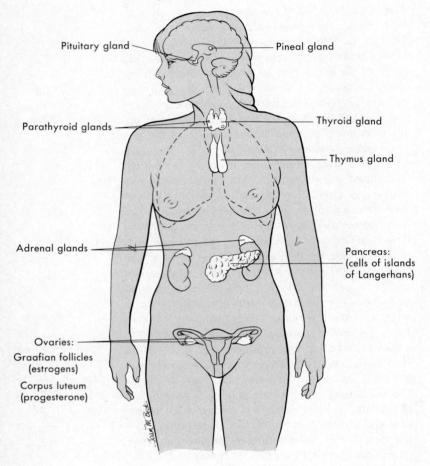

Pituitary gland

Pineal gland

Parathyroid glands

Thyroid gland

Thymus gland

Adrenal glands

Pancreas: (cells of islands of Langerhans)

Ovaries:
Graafian follicles (estrogens)

Corpus luteum (progesterone)

Table 14-1 Names and locations of endocrine glands

Name	Location
Pituitary gland (hypophysis cerebri) and hypothalamus 　Anterior lobe (adenohypophysis) 　Intermediate lobe (pars intermedia) 　Posterior lobe (neurohypophysis)	Cranial cavity
Pineal gland (epiphysis)	Cranial cavity
Thyroid gland	Neck
Parathyroid glands	Neck
Thymus	Mediastinum
Adrenal glands 　Adrenal cortex 　Adrenal medulla	Abdominal cavity (retroperitoneal)
Islands of Langerhans	Abdominal cavity (pancreas)
Gastric and intestinal mucosa	Abdominal cavity
Ovaries 　Graafian follicle 　Corpus luteum	Pelvic cavity
Testes (interstitial cells)	Scrotum
Placenta	Pregnant uterus

Endocrine gland cells synthesize hormones by the process of anabolism. Hormones are either protein compounds—compounds derived from proteins—or steroid compounds. All have relatively large molecules. For instance, human growth hormone consists of a chain of 188 amino acids and has a molecular weight of 20,500. Compared with a carbon dioxide molecule, molecular weight 44, the growth hormone molecule is indeed a giant. (Carbon dioxide is one of several "regulatory chemicals." These compounds resemble hormones in two respects—they are released into the blood from cells, and they exert a profound influence on various structures and functions. They differ from hormones, however, in that they are neither proteins nor steroids but are much smaller inorganic molecules or ions, they are present in the blood in much larger amounts than are hormones, and they are not prod-

ucts of anabolism. Instead, some regulatory chemicals, including carbon dioxide, are products of catabolism.)

Exaggerating the importance of endocrine glands is almost impossible. Hormones are the main regulators of metabolism, of growth and development, of reproduction, and of stress responses. They play roles of the utmost importance in maintaining homeostasis—fluid and electrolyte balance, for example. Excesses or deficiencies of hormones make the difference between normalcy and all sorts of abnormalities such as dwarfism, gigantism, and sterility—and even the difference between life and death in some instances.

PROSTAGLANDINS (TISSUE HORMONES)

The prostaglandins (PGs) are a unique group of biological compounds (20-carbon fatty acids that have a 5-carbon ring) that serve important and widespread integrative functions in the body but do not meet the definition of a typical hormone. Although they may be secreted directly into the bloodstream like hormones in general, they are rapidly metabolized and inactivated, so that circulating levels are

Information and understanding about basic biological and cellular processes gained as a result of prostaglandin research may soon lead to powerful new tools in medicine. The potential therapeutic use of compounds found in almost every body tissue and capable of regulating hormone activity on the cellular level has been described as the most revolutionary development in medicine since the advent of antibiotics. They may eventually play an important role in the treatment of such diverse diseases as hypertension, coronary thrombosis, asthma, and ulcers.

Figure 14-2 Mechanism of hormone action, including the postulated role of prostaglandins. Hormone acts as "first messenger," delivering its message to a membrane receptor in the target organ cell. Prostaglandins appear to regulate hormone activity by influencing adenyl cyclase and cyclic AMP activity within the cell. Cyclic AMP serves as the "second messenger."

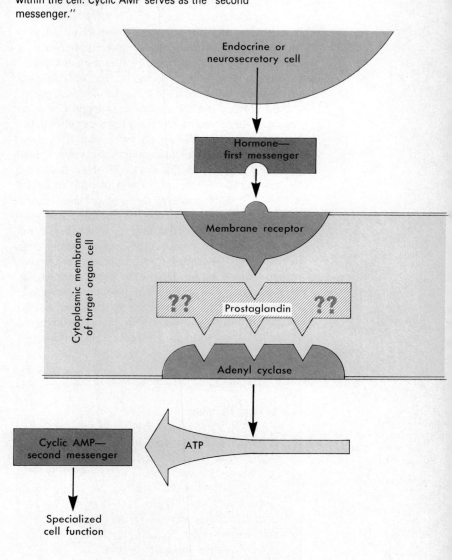

extremely low. The term **tissue hormone** is appropriate because in many instances the substance is produced in a tissue and diffuses only a short distance to act on cells within that tissue. Whereas typical hormones integrate activities of widely separated organs, typical prostaglandins integrate activities of neighboring cells.

Three classes of prostaglandins, prostaglandin A (PGA), prostaglandin E (PGE), and prostaglandin F (PGF), have been isolated and identified from a wide variety of tissues, including the seminal vesicles, kidneys, lungs, iris, brain, and thymus. As a group, the prostaglandins have diverse physiological effects and are among the most varied and potent of any naturally occurring biological compounds. They are intimately involved in overall endocrine regulation by influencing adenyl cyclase and cyclic AMP activity within the cell (Figure 14-2). Specific biological effects depend on the class of prostaglandin.

Intraarterial infusion of prostaglandins in group A (PGA) results in an immediate fall in blood pressure accompanied by an increase in regional blood flow to several areas, including the coronary and renal systems. Just how PGA infusion reduces blood pressure remains unclear. The mechanism probably involves relaxation of smooth muscle fibers in the walls of muscular arteries and arterioles.

PGE compounds play an important role in a number of systemic vascular, metabolic, and gastrointestinal functions. Vascular effects of PGE_1 and PGE_2 include regulation of both red blood cell "deformability" and platelet aggregation tendencies (see Chapter 15). PGE_1 also plays a role in certain systemic manifestations of inflammation such as fever. Evidence now suggests that aspirin and many other common antiinflammatory agents function, at least in part, by inhibition of PGE synthesis. PGE_1 is also important in the regulation of hydrochloric acid secretion in the stomach and the prevention of gastric ulcer.

Prostaglandins of the F class play an especially important role in the reproductive system. $PGF_{2\alpha}$ causes uterine muscle to contract and can be used (often with PGE_2) to induce labor and accelerate delivery. PGF compounds also affect intestinal motility and are required for normal peristalsis.

HOW HORMONES ACT

What a hormone does to its target cells to cause them to respond in particular ways has been the subject of much conjecture and research. Dr. Earl W. Sutherland isolated a chemical, adenosine 3',5'-monophosphate *(cyclic AMP)*, and suggested the part it might play in hormone action. For his brilliant work, Dr. Sutherland received the 1971 Nobel Prize in medicine and physiology. He conceived what is known as the **second messenger hypothesis** of hormone action. According to this concept a hormone acts as a "first messenger," that is, it delivers its chemical message from the cells of an endocrine gland to highly specific membrane receptor sites on cells of a target organ. On reaching a target cell, the hormone reacts with an enzyme, adenyl cyclase, present in the cell's plasma membrane, causing it to act on ATP molecules present inside the cell. Adenyl cyclase catalyzes the conversion of ATP to cyclic AMP. Cyclic AMP serves as the "second messenger," delivering information inside the cell that causes the cell to respond by performing its specialized function. For example, cyclic AMP causes thyroid cells to respond by secreting thyroid hormones. Current research involving a number of tropic hormones suggests that other intermediate messengers, especially the prostaglandins, increase the activation of adenyl cyclase and therefore serve to regulate the levels of cyclic AMP and ultimately target cell function.

Steroid Hormones

Although some protein hormones can actually enter a target organ cell intact, most are too large and must combine with receptors on the surface of the plasma membrane as shown in Figure 14-2. However steroid hormones such as estrogen or aldosterone are lipid soluble and can pass through the cell membrane with relative ease. Once inside the cell they combine with steroid-specific receptor proteins in the cytoplasm and then enter the nucleus. The combined steroid–receptor protein complex acts directly on genes in the nucleus. As a result of this activation, messenger RNA is formed and leaves the nucleus to initiate the synthesis of specific proteins. These proteins alter cell activity in specialized ways and result in the hormone's action. In this way steroid hormones influence body functions.

In summary, hormones serve as first messengers, providing communication between endocrine and target organ. Cyclic AMP then acts as the second messenger, providing communication within a hormone's target cells. Figure 14-2 summarizes the mechanism of hormone action, including the postulated role of prostaglandins.

PITUITARY GLAND (HYPOPHYSIS CEREBRI)

SIZE, LOCATION, AND COMPONENT GLANDS

The pituitary gland is truly a small but mighty structure. At its largest diameter, it measures only 1.2 to 1.5 cm (about ½ inch). By weight, it is even less impressive—only about 0.5 gm, $\frac{1}{60}$ ounce! And yet so crucial are the functions of the anterior lobe of this gland that it is called the "master gland."

The pituitary body has a protected location. It lies in the sella turcica and is covered over by an extension of the dura mater known as the *pituitary diaphragm.* The deepest part of the sella turcica (saddle-shaped depression in the sphenoid bone) is called the *pituitary fossa,* since the pituitary body lies in it (Figure 14-3). A stemlike portion of the gland, the infundibulum or pituitary stalk, juts up through a tiny perforation in the pituitary diaphragm and attaches the gland to the undersurface of the brain. More specifically, the stalk attaches the pituitary body to the hypothalamus.

Although the pituitary looks like just one gland, it actually consists of two separate glands—the **adenohypophysis,** or anterior pituitary gland, and the **neurohypophysis,** or posterior pituitary gland. The anterior portion of the gland is itself partially divided by a narrow cleft and a connective tissue remnant into the larger *pars anterior* and small *pars intermedia* (see Figure 14-7). The anterior and posterior divisions of the pituitary gland develop from different embryonic structures, have a different microscopic structure, and secrete different hormones. The anterior pituitary or adenohypophysis develops as an upward projection from the embryo's pharynx, whereas the posterior pituitary or neurohypophysis develops as a downward projection from the brain. Microscopic differences between the two glands are suggested by their names—*adeno* means "gland" and *neuro* means "nervous." The adenohypophysis has the microscopic structure of an endocrine gland, whereas the neurohypophysis has the structure of nervous tissue. Hormones secreted by the adenohypophysis serve very different functions from those released from the neurohypophysis.

ANTERIOR PITUITARY GLAND (ADENOHYPOPHYSIS)

The anterior pituitary gland (pars anterior and pars intermedia) is composed of irregular clumps or nests

Figure 14-3 Normal skull, lateral view. *a,* Coronal suture; *b,* frontal sinus; *c,* dorsum sellae; *d,* anterior clinoid process; *e,* floor of sella turcica; *f,* sphenoid sinus; *g,* shadow of earlobe; *h,* lambdoidal suture.

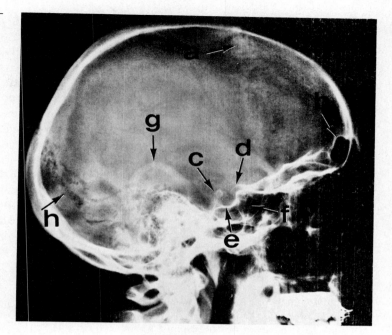

Figure 14-4 Anterior pituitary hormones and their target organs: adrenocorticotropic hormone *(ACTH);* thyroid-stimulating hormone *(TSH);* follicle-stimulating hormone *(FSH);* luteinizing hormone *(LH);* male analogue of LH *(ICSH);* melanocyte-stimulating hormone *(MSH).*

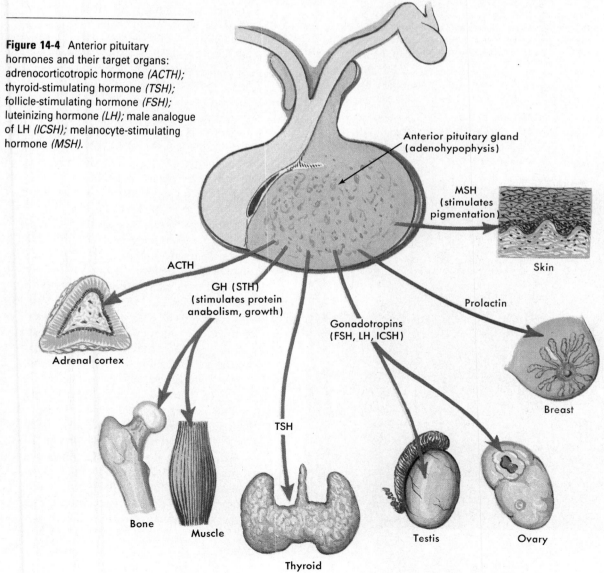

of secretory cells supported by fine connective tissue fibers and surrounded by a rich vascular network. Three types of cells can be identified according to their affinity (or lack of affinity) for certain types of stains or dyes. About half of all cells present show little affinity for any type of stain or dye and are appropriately called **chromophobes** (from the Greek *chroma*, "color," and the Latin *phobia*, "fear of"). Of the remaining cells, about 40% are called

acidophils (from the Greek *philia*, "love of") because of their affinity for acid dyes and 10% are **basophils,** which stain easily with basic dyes. Acidophils secrete growth hormone (GH) (also called somatotropin [STH]) and prolactin (also called luteotropic hormone). Basophils secrete the other five hormones of the anterior lobe: thyrotropin (TH) (also called thyroid-stimulating hormone [TSH]), adrenocorticotropin (ACTH), two gonadotropins (follicle-stimulating hormone [FSH] and luteinizing hormone [LH, or ICSH in the male]), and melanocyte-stimulating hormone (MSH). MSH is secreted by basophils in the pars intermedia of the anterior pituitary gland. Figure 14-4 summarizes the anterior pituitary hormones and their target organs.

Figure 14-5 A pituitary giant and dwarf contrasted with normal-sized men. Excessive secretion of growth hormone* by the anterior lobe of the pituitary gland during the early years of life produces giants of this type, whereas deficient secretion of this substance produces well-formed dwarfs.

*The structure of the human growth hormone molecule was identified by Dr. C.H. Li, Professor of Biochemistry at the University of California, in 1966. It consists of a chain of 188 amino acids with one loop of 93 subunits and another of 6 subunits. In January of 1971, Dr. Li announced that his laboratory had succeeded in synthesizing human growth hormone. In 1982, recombinant DNA techniques made production of human growth hormone in the laboratory possible.

GROWTH HORMONE ABNORMALITIES

If the anterior pituitary gland secretes an excess of growth hormone during the growth years, that is, before closure of the epiphyseal cartilages, bones grow more rapidly than normal and **gigantism** results (Figure 14-5). Undersecretion of the growth hormone produces **dwarfism** when it occurs during the years of skeletal growth. If oversecretion of growth hormone occurs after the individual is fully grown, the condition known as **acromegaly** develops. Characteristic of this disease are enlargement of the bones of the hands, feet, jaws, and cheeks and an increase too in their overlying soft tissues (Figures 14-6 to 14-8). The typical facial appearance in acromegaly results from the combination of bone and soft tissue overgrowth. A prominent forehead and large nose are characteristic. In addition, the skin is characterized by large, widened pores and the mandible grows in length so that separation of the lower teeth is common. The overbite or "lantern jaw" caused by continued mandibular growth and the enlarged skin pores can be seen in Figure 14-7. In most cases of acromegaly the skin on the scalp and neck also becomes thickened and develops deep folds (Figure 14-8). The x-ray film in Figure 14-9 on p. 394 shows an enlarged sella turcica in a patient with acromegaly. In this individual the enlarged pituitary fossa is filled by a tumor. Compare this x-ray film to the normal size of the sella turcica shown in Figure 14-3.

Figure 14-6 Acromegaly. Note large head, exaggerated forward projection of jaw, and protrusion of frontal bone.

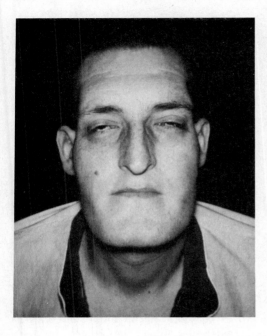

Figure 14-7 Acromegaly. Notice enlarged skin pores and separation of lower teeth.

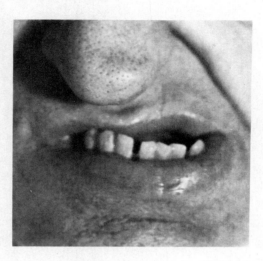

Figure 14-8 Acromegaly. Note marked thickening of skin on **A**, scalp and **B**, posterior surface of neck.

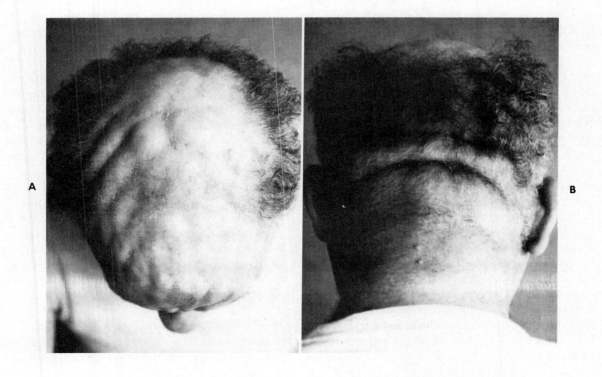

Figure 14-9 Enlargement of sella turcica in patient with acromegaly. Compare sella with that in Figure 14-3.

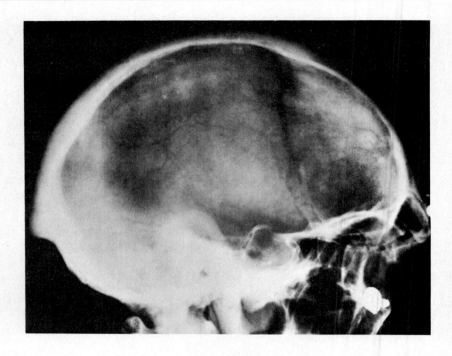

Growth Hormone

Growth hormone or **somatotropin** (from the Greek *soma*, "body," and *trope*, "turning") is thought to promote bodily growth indirectly by accelerating amino acid transport into cells—evidence: blood amino acid content decreases within hours after administration of growth hormone to a fasting animal. With the faster entrance of amino acid into cells, anabolism of amino acids to form tissue protein also accelerates. This in turn tends to promote cellular growth. Growth hormone stimulates growth of both bone and soft tissues.

In addition to its stimulating effect on protein anabolism, growth hormone also influences fat metabolism and thereby affects carbohydrate (glucose) metabolism. Briefly, growth hormone tends to accelerate both the mobilization of fats from adipose tissues and their catabolism by other tissues. In other words, it tends to cause cells to shift from glucose catabolism to fat catabolism for their energy supply. Less glucose therefore leaves the blood to enter cells and more remains in the blood. Thus growth hormone tends to increase blood's concentration of glucose. In a word, it tends to have a hyperglycemic effect. Insulin produces opposite effects. It increases carbohydrate metabolism, thereby tending to decrease the blood concentration of glucose, causing a hypoglycemic effect. Whereas adequate amounts of insulin prevent diabetes mellitus, long-continued excess amounts of growth hormone produce diabetes. In short, growth hormone and insulin function as antagonists. Or, as more commonly stated, growth hormone has an anti-insulin effect, a fact of considerable clinical importance. In short, growth hormone helps regulate metabolism in the following ways:

1 It promotes protein anabolism (synthesis of tissue proteins), so it is essential for normal growth and for tissue repair and healing.
2 It promotes fat mobilization and catabolism and decreases glucose catabolism. By decreasing glucose catabolism, an excess of growth hormone may in time produce hyperglycemia and diabetes.

Prolactin (Lactogenic or Luteotropic Hormone)

Acidophils in the anterior pituitary gland secrete two hormones: growth hormone and prolactin. Another name for **prolactin**—lactogenic hormone—suggests that it "generates," or initiates, milk secretion. It stimulates the mammary glands to start secreting soon after delivery of an infant. During pregnancy, it helps promote the breast development that makes possible milk secretion after delivery. In addition to its effect on milk secretion after childbirth, prolactin plays a supportive role (with luteinizing hormone) in maintaining the corpus luteum late in the postovulatory or premenstrual phase of each menstrual cycle. Because of this supportive role, prolactin is also called *luteotropic hormone (LTH)*.

Tropic Hormones

Basophil cells in the anterior pituitary gland secrete four tropic hormones—hormones that have a stimulating effect on other endocrine glands. Names of the tropic hormones are *thyrotropin* or *thyroid-stimulating hormone, adrenocorticotropic hormone,* and the two primary gonadotropins—*follicle-stimulating hormone* and *luteinizing hormone.*

All tropic hormones perform the same general function. Each one stimulates one other endocrine gland—stimulates it both to grow and to secrete its hormone at a faster rate. It seems to affect only this one structure as if, like a bullet, it had been aimed at a target, hence the name "target gland."

Individual tropic hormones perform the following functions:

1 **Thyrotropin** (thyroid-stimulating hormone, TSH) promotes and maintains growth and development of its target gland, the thyroid, and stimulates it to secrete thyroid hormone.

2 **Adrenocorticotropin** (ACTH) promotes and maintains normal growth and development of the adrenal cortex and stimulates it to secrete cortisol and other glucocorticoids.

3 **Follicle-stimulating hormone** (FSH) stimulates primary graafian follicles to start growing and to continue developing to maturity, that is, to the point of ovulation. FSH also stimulates follicle cells to secrete estrogens, one type of female sex hormones. In the male, FSH stimulates development of the seminiferous tubules and maintains spermatogenesis by them.

4 **Luteinizing hormone** (LH in the female, ICSH in the male), as its name suggests, stimulates formation of the corpus luteum, just one of its functions. Before this, LH acts with FSH to bring about complete maturation of the follicle; LH then produces ovulation and stimulates formation of the corpus luteum in the ruptured follicle. Finally, LH in human females stimulates the corpus luteum to secrete progesterone and estrogens. The male pituitary gland also secretes LH, but it is called **interstitial cell–stimulating hormone** (ICSH) because it stimulates interstitial cells in the testes to develop and secrete testosterone.

• • •

During childhood the anterior pituitary gland secretes insignificant amounts of gonadotropins. Then it gradually steps up their production a few years before puberty. Just before puberty begins, presumably the anterior pituitary's secretion of these hormones takes a sudden and marked spurt. As a result, the blood concentration of gonadotropins increases markedly. This first high blood con-

High estrogen and progesterone levels in pregnancy often cause darkening of the face ("mask of pregnancy") as well as the areolae, nipples, and, to a lesser extent, the genitalia. Another anterior pituitary hormone (namely, ACTH) will produce increased pigmentation of the skin. Structurally, also, the two molecules resemble each other. Several of the same amino acids occur in the same sequence in both MSH and ACTH.

centration of gonadotropins serves as the stimulus that brings on the first menstrual period and the many other changes that signal the beginning of puberty.

Melanocyte-Stimulating Hormone

Basophils in the pars intermedia of the anterior pituitary gland secrete the hormone generally believed to play the most important role in maintaining human pigmentation—**melanocyte-stimulating hormone (MSH) or intermedin.** In some species, including man, administration of MSH will cause a rapid increase in the synthesis and dispersion of melanin (pigment) granules in skin cells called *melanocytes.* The resulting hyperpigmentation resembles that which may occur when blood levels of other "darkening hormones" such as estrogen or progesterone become elevated.

Control of Secretion

Neurons in certain parts of the hypothalamus synthesize chemicals that their axons secrete into blood in a complex of small veins, the **pituitary portal system.** Via this system the neurosecretions travel from the hypothalamus to the anterior lobe of the pituitary gland. There they stimulate the gland to release various hormones; hence **releasing hormones** is the other name for neurosecretions of the hypothalamus. Here are their abbreviated names, full names, and functions.

1 GRH, or growth hormone-releasing hormone—stimulates anterior pituitary gland to release, or to secrete, growth hormone.*

2 CRF, or corticotropin-releasing factor—stimulates anterior pituitary secretion of ACTH.

---

*Current evidence indicates that the hypothalamus also secretes somatostatin, which, as the name suggests, inhibits somatotropin secretion.

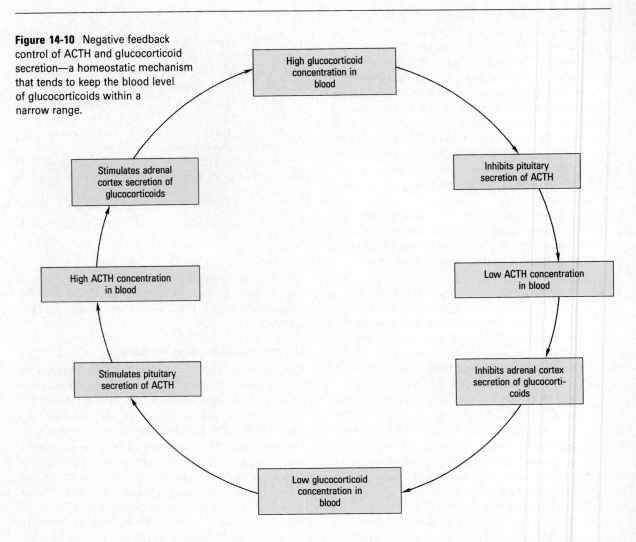

Figure 14-10 Negative feedback control of ACTH and glucocorticoid secretion—a homeostatic mechanism that tends to keep the blood level of glucocorticoids within a narrow range.

3 TRH, or thyrotropin-releasing hormone—stimulates anterior pituitary secretion of TSH (thyroid-stimulating hormone).

4 FSH-RH, or follicle-stimulating hormone-releasing hormone—stimulates anterior pituitary secretion of FSH.

5 LH-RH, or luteinizing hormone-releasing hormone—stimulates anterior pituitary secretion of LH.

6 PIF, or prolactin-inhibitory factor—hypothalamic substance that *inhibits* anterior pituitary secretion of prolactin except during pregnancy and after delivery. However, even if a woman is not pregnant or nursing, low levels of prolactin are secreted to help maintain the corpus luteum during the late stages of each menstrual cycle (luteotropic effects of prolactin). Evidence suggests that prolactin-releasing factor (PRF), also in hypothalamus extracts, acts with PIF in a dual control system to regulate prolactin blood levels.

Negative feedback mechanisms control the pituitary gland's secretion of tropic hormones. Such mechanisms operate on the following principles. A high blood level of a tropic hormone stimulates its target gland to increase its hormone secretion. This results in a high blood level of the target gland's hormone, which feeds back by way of the circulating blood to directly and indirectly inhibit secretion of the tropic hormone by the anterior pituitary gland. Figure 14-10 illustrates the negative feedback mechanism that directly inhibits pituitary secretion of ACTH. Indirect inhibition of ACTH occurs as follows. A high blood glucocorticoid concentration inhibits the hypothalamus from secreting corticotropin-releasing factor. Result? A low CRF concentration in pituitary portal vein blood inhibits pituitary secretion of ACTH. Similar mechanisms regulate the anterior lobe's secretion of thyrotropin and the thyroid gland's secretion of thyroid hormone. Also, a similar mechanism controls the anterior lobe's secretion of FSH and the ovary's secretion of estrogens.

Clinical facts furnish interesting evidence about feedback control of hormone secretion by the anterior pituitary gland and its target glands. For instance, patients who have their pituitary gland removed (hypophysectomy) surgically or by radiation must be given hormone replacement therapy for the rest of their life. If not, they will develop thyroid, adrenocortical, and gonadotropic deficiencies—deficiencies, that is, of the anterior pituitary's target gland hormones. Another well-known clinical fact is that estrogen deficiency develops in women between 40 and 50 years of age. By then the ovaries seem to have tired of producing hormones and ovulating each month. They no longer respond to FSH stimulation, so estrogen deficiency develops, brings about the menopause, and persists after the menopause. What therefore would you deduce is true of the blood concentration of FSH after the menopause? Apply the principle implied in the preceding paragraph that a low concentration of a target gland hormone stimulates tropic hormone secretion by the anterior pituitary gland.

Before leaving the subject of control of pituitary secretion, we want to call attention to another concept about the hypothalamus. It most likely functions as an important part of the body's complex machinery for coping with stress situations. For example, in severe pain or intense emotions, the cerebral cortex—especially its limbic lobe—is thought to send impulses to the hypothalamus. They stimulate it to secrete its releasing hormones into the pituitary portal veins. Circulating quickly to the anterior pituitary gland, they stimulate it to secrete more of its hormones. These in turn stimulate increased activity by the pituitary's target structures. In essence, what the hypothalamus does through its releasing hormones is to translate nerve impulses into hormone secretion by endocrine glands. Thus the hypothalamus links the nervous system to the endocrine system. It integrates the activities of these two great integrating systems—particularly, it seems, in times of stress. When healthy survival is threatened, the hypothalamus, via its releasing factors, can take over the command of the anterior pituitary gland. By so doing, it indirectly controls all of the pituitary's target glands—the thyroid, the adrenal cortex, and the gonads. Finally, by means of the hormones these glands secrete, the hypothalamus can dictate the functioning of literally every cell in our bodies.

These facts have tremendous implications. They mean that the cerebral cortex can do more than just receive impulses from all parts of the body and send out impulses to muscles and glands. They mean that the cerebral cortex—and therefore our thoughts and emotions—can, by way of the hypothalamus, influence the functioning of all our billions of cells. In

short, the brain has two-way contact with every tissue of the body. Thus the state of the body can and does influence mental processes, and, conversely, mental processes can and do influence the functioning of the body. Therefore both somatopsychic and psychosomatic relationships exist between the body and the brain.

POSTERIOR PITUITARY GLAND (NEUROHYPOPHYSIS)

The posterior lobe of the pituitary gland serves as a storage area for the release of two hormones—one known as **antidiuretic hormone** (ADH) and the other called **oxytocin.** But strangely enough, cells of the posterior lobe, called *pituicytes*, do not themselves make these hormones. Instead, neurons in two areas of the hypothalamus called the *supraoptic* and *paraventricular nuclei* synthesize them (Figure 14-11). From the cell bodies of these neurons in the hypothalamus the hormones pass down along axons (in the hypothalamohypophyseal tract) into the posterior lobe of the pituitary gland. Instead of the chemical "releasing factors" that triggered secretion of hormones from the anterior pituitary gland, release of both ADH and oxytocin into the blood is controlled by nervous stimulation.

Antidiuretic Hormone

Antidiuresis means literally "against the production of a large urine volume." And this is exactly what ADH does. It prevents the formation of a large volume of urine. The one function ADH performs for the body is to decrease water loss from it. In other words, ADH is a water-retaining hormone. It performs this function by acting on cells of the distal and collecting tubules of the kidney to make them more permeable to water. This causes faster reabsorption of water from tubular urine into blood. And this in turn automatically produces antidiuresis (smaller urine volume). In instances of dehydration or water deprivation the osmolarity of plasma will increase as levels of solutes in the blood become more and more concentrated. As plasma osmolarity increases, specialized *osmoreceptors* near the supraoptic nucleus respond by stimulating ADH release from the posterior pituitary gland. The resulting increase in water reabsorption will then expand the plasma volume and reduce its osmolarity by diluting the blood solutes. Water is conserved, and urine volume decreases.

Marked diuresis, or abnormally large urine volume, occurs if ADH secretion is inadequate, such as occurs in the disease **diabetes insipidus.** A preparation of ADH used to treat this condition is called vasopressin (Pitressin). (The term *vasopressin* is sometimes used as another name for ADH. It can be misleading. Although large [pharmacological] doses

Figure 14-11 Nerve tracts from hypothalamus to posterior lobe of pituitary gland.

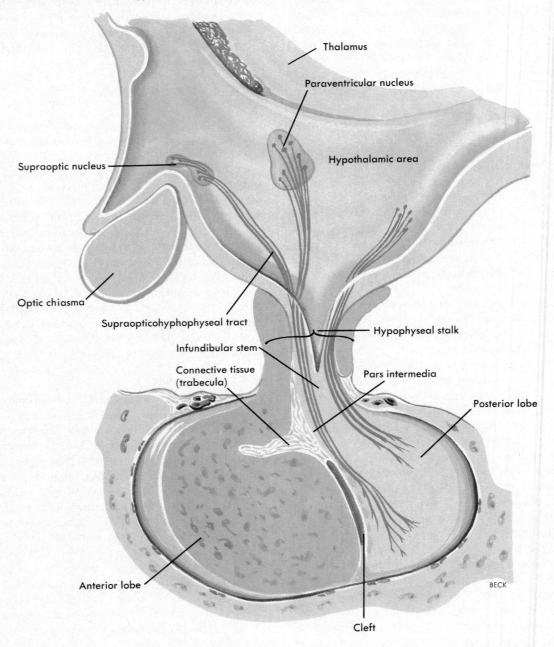

of ADH will result in vasoconstriction of blood vessels, normal [physiological] levels produce little if any vasoconstrictor effects in humans.) In recent years studies have shown that this brain peptide also functions in some unknown way to improve learning and memory. A growing number of researchers feel that ADH or one of its synthetic analogs may someday prove useful in the treatment of senility.

Oxytocin

Oxytocin has two actions: it stimulates powerful contractions by the pregnant uterus and it causes milk ejection from the lactating breast. Under the influence of this hormone from the posterior lobe of the pituitary gland, alveoli (cells that synthesize milk) release the milk into the ducts of the breast. This is a highly important function of oxytocin because milk cannot be removed by suckling unless it has first been ejected into ducts. However, it was oxytocin's other action—its stimulating effect on contractions of the pregnant uterus—that inspired its name. The term means "swift childbirth" (from the Greek *oxys*, "swift," and *tokos*, "childbirth"). Whether or not oxytocin takes part in initiating labor

Table 14-2 Production, control, and effects of pituitary hormones

Pituitary gland (hypophysis cerebri)	Hormone	Source (cell type or location)	Control mechanism	Effect
ANTERIOR PITUITARY GLAND (ADENOHYPOPHYSIS)				
Pars anterior	Growth hormone (GH, somatotropin [STH])	Acidophils	GRH (growth hormone–releasing hormone) from hypothalamus	Promotes body growth, protein anabolism, and mobilization and metabolism of fats; decreases glucose catabolism; increases blood glucose levels
	Prolactin (lactogenic or luteotropic hormone [LTH])	Acidophils	Prolactin inhibitory factor (PIF); prolactin-releasing factor (PRF) from hypothalamus and high blood levels of oxytocin	Stimulates milk secretion and development of secretory alveoli; helps maintain corpus luteum
	Thyrotropin (TH); thyroid-stimulating hormone (TSH)	Basophils	Thyrotropin-releasing hormone (TRH) from hypothalamus	Growth and maintenance of thyroid gland and stimulation of thyroid hormone secretion
	Adrenocorticotropin (ACTH)	Basophils	Corticotropin-releasing factor (CRF) from hypothalamus	Growth and maintenance of adrenal cortex and stimulation of cortisol and other glucocorticoid secretions
	Follicle-stimulating hormone (FSH)	Basophils	Follicle-stimulating hormone–releasing hormone (FSH-RH) from hypothalamus	In female—stimulates follicle growth and maturation and estrogen secretion
				In male—stimulates development of seminiferous tubules and maintains spermatogenesis
	Luteinizing hormone (LH in female, ICSH in male)	Basophils	Luteinizing hormone–releasing hormone (LH-RH) from hypothalamus	In female (LH)—induces ovulation and stimulates formation of corpus luteum and progesterone secretion
				In male (ICSH)—stimulates interstitial cell secretion of testosterone
Pars intermedia	Melanocyte-stimulating hormone (MSH); intermedin	Basophils	Unknown in humans	May cause darkening of skin by increasing melanin production
POSTERIOR PITUITARY GLAND (NEUROHYPOPHYSIS)				
	ADH (vasopressin)	Hypothalamus, mainly supraoptic nucleus	Osmoreceptors in hypothalamus stimulated by increase in blood osmotic pressure, decrease in extracellular fluid volume, and stress	Decreased urine output
	Oxytocin	Hypothalamus, paraventricular nucleus	Nervous stimulation of hypothalamus caused by stimulation of nipples (nursing)	Contraction of uterine smooth muscle and ejection of milk into lactiferous ducts

is still an unsettled question. Commercial preparations of oxytocin are given to stimulate uterine contractions after delivery of an infant in order to lessen the danger of uterine hemorrhage.

Control of Secretion

Details of the mechanism that controls the secretion of ADH by the posterior lobe of the pituitary gland have not been established. However, review of the effects of dehydration and water deprivation on ADH levels illustrates the dominant roles of two important control factors, namely, the osmotic pressure of the extracellular fluid and its total volume. Without going into a discussion of evidence or details, the general principles of control of ADH secretion are as follows.

1 An increase in the osmotic pressure of extracellular fluid stimulates ADH secretion. This leads to decreased urine output and tends to increase the volume of extracellular fluid, which in turn tends to decrease its osmotic pressure back toward normal. Opposite effects result from a decrease in the osmotic pressure of extracellular fluid.
2 A decrease in the total volume of extracellular fluid acts in some way to stimulate ADH secretion and thereby decreases urine output and increases extracellular volume back toward normal.
3 Stress, whether induced by physical or emotional factors, brings about an increased

secretion of ADH. Presumably, it acts in some way on the hypothalamus to stimulate ADH secretion.

• • •

About all that is known about the mechanism that controls the release of oxytocin from the posterior lobe of the pituitary gland is that stimulation of the nipples by the infant's nursing initiates sensory impulses that eventually reach the paraventricular nucleus of the hypothalamus, stimulating it to synthesize more oxytocin. The posterior lobe of the pituitary gland in turn increases its release of oxytocin. In addition to causing ejection of milk into the ducts of the breast, oxytocin also stimulates increased production of prolactin. Lactation is therefore controlled by two hormones: prolactin from the anterior pituitary gland, which stimulates milk production by the secretory alveoli, and oxytocin from the posterior pituitary gland, which results in ejection of milk from alveoli into the ducts. Table 14-2 summarizes the production, control, and effects of the pituitary hormones.

THYROID GLAND

Location and Structure

Two fairly large lateral lobes and a connecting portion, the isthmus, constitute the thyroid gland. Weight of the gland in the adult is somewhat variable but is usually about 30 gm. It is located in the neck

Figure 14-12 Thyroid and parathyroid glands. Note their relations to each other and to the larynx (voice box) and trachea.

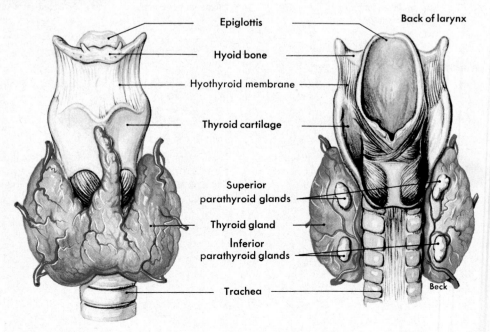

Figure 14-13 Thyroid gland. Note that each of the follicles is filled with colloid. (× 140.)

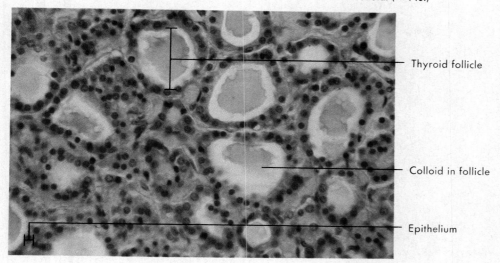

Thyroid follicle

Colloid in follicle

Epithelium

just below the larynx. The isthmus lies across the anterior surface of the upper part of the trachea. In many individuals a thin pyramidal-shaped extension of thyroid tissue can be found arising from the upper surface of the connecting isthmus (Figure 14-12). Thyroid tissue is made up of numerous structural units called **follicles.** Each follicle is a small closed sac consisting of an outer layer of simple cuboidal epithelium surrounding a central cavity. Colloid, composed largely of an iodine-containing protein known as *thyroglobulin,* fills these tiny sacs (Figure 14-13). Thyroid cells have a unique ability to actively take up and concentrate blood iodide (I⁻). Once in the follicular cells I⁻ is changed to iodine (I) and used in the synthesis of the iodine-containing thyroid hormone.

Thyroid Hormone and Calcitonin

The thyroid gland secretes thyroid hormone and calcitonin. Actually, thyroid hormone consists of two hormones. Thyroxine (tetraiodothyronine) is the name of the main one, and triiodothyronine is the name of the less abundant one. One molecule of thyroxine contains four atoms of iodine, and one molecule of triiodothyronine, as its name suggests, contains three iodine atoms. After synthesizing its hormones, the thyroid gland stores considerable amounts of them before secreting them. (Other endocrine glands do not store up their hormones.) As a preliminary to storage, thyroxine and triiodothyronine combine with a globulin in the thyroid cell to form a compound called *thyroglobulin.* Then thyroglobulin is stored in the colloid material in the follicles of the gland. Later the two hormones are

released from thyroglobulin and secreted into the blood as thyroxine and triiodothyronine. Almost immediately, however, they combine with a blood protein. They travel in the bloodstream in this protein-bound form. But in the tissue capillaries, they are released from the protein and enter tissue cells as thyroxine and triiodothyronine.

The iodine in the protein-bound thyroxine and triiodothyronine is called **protein-bound iodine** (PBI). The amount of PBI can be measured by laboratory procedure and, in fact, is widely used as a test of thyroid functioning. (With a normal thyroid gland, the PBI is about 4 to 8 μg per 100 ml of plasma.)

The main physiological actions of thyroid hormone are to help regulate the metabolic rate and the processes of growth and tissue differentiation. Thyroid hormone increases the metabolic rate. How do we know this? Because oxygen consumption increases following thyroid administration. Like pituitary somatotropin, thyroid hormone stimulates growth, but unlike somatotropin, it also influences tissue differentiation and development. For example, **cretins** (individuals with thyroid deficiency) not only are dwarfed but also may be mentally retarded because the brain fails to develop normally. Their bones and many other tissues also show an abnormal pattern of development.

Convincing evidence indicates that the thyroid gland secretes **calcitonin.** Calcitonin acts quickly to decrease blood's calcium concentration. Presumably, it produces this effect by either or both of these actions: inhibiting bone breakdown with calcium release into blood or promoting calcium deposition in

Figure 14-14 Simple goiter.

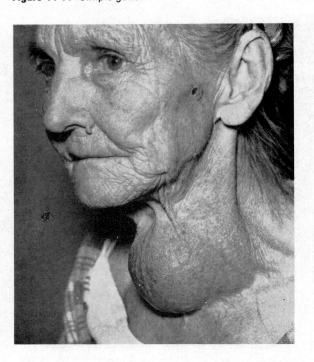

GOITER

A diffuse and painless enlargement of the thyroid gland called **simple goiter** (Figure 14-14) was once common in areas of the United States where the iodine content of the soil and water is inadequate. The highest incidence was in those states bordering the Great Lakes—Michigan, Wisconsin, and Ohio—and the Rocky Mountain area. In simple goiter the gland enlarges in an attempt to compensate for the lack of iodine in the diet necessary for synthesis of thyroid hormone. The use of iodized salt has dramatically reduced the incidence of simple goiter caused by low iodine intake.

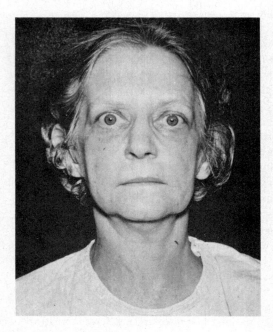

Figure 14-15 Graves' disease, caused by hypersecretion by the thyroid gland.

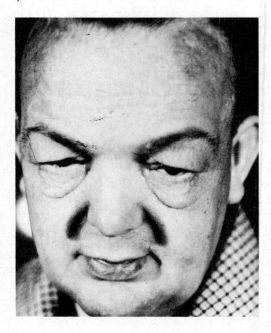

Figure 14-16 Myxedema, a condition produced by hyposecretion of the thyroid gland during the adult years. Note the edema around the eyes and facial puffiness.

bone. The function of calcitonin is to help maintain blood calcium homeostasis and prevent harmful hypercalcemia. Parathyroid hormone serves as an antagonist to calcitonin (p. 404).

Effects of Hypersecretion and Hyposecretion

Hypersecretion of thyroid hormone produces the disease **exophthalmic goiter** (known as Graves' disease, Basedow's disease, and by several other names) (Figure 14-15). It is characterized by an elevated PBI level, increased metabolism (+30 or more), increased appetite, loss of weight, increased nervous irritability, and exophthalmos. Marked edema of the fatty tissue behind the eye, attributed to the high thyrotropic hormone content in blood, produces the exophthalmos.

Hyposecretion during the formative years leads to malformed dwarfism or **cretinism,** a condition characterized by a low metabolic rate, retarded growth and sexual development, and often, too, retarded mental development. Later in life, deficient thyroid secretion produces the disease **myxedema** (Figure 14-16). The low metabolic rate that characterizes myxedema leads to lessened mental and physical vigor, a gain in weight, loss of hair, and a thickening of the skin from an accumulation of fluid in the subcutaneous tissues. Because of a high mucoprotein content, this fluid is viscous. Therefore it gives firmness to the skin, and the skin does not pit when pressed, as it does in some other types of edema.

PARATHYROID GLANDS

Location and Structure

The parathyroid glands are small round bodies attached to the posterior surfaces of the lateral lobes of the thyroid gland (Figures 14-12 and 14-17). Usually there are four or five, but sometimes there are fewer and sometimes more of these glands.

Parathyroid Hormone

The parathyroid glands secrete **parathyroid hormone.** Its chief function is to help maintain homeostasis of blood calcium concentration by promoting calcium absorption into the blood and thereby tending to prevent hypocalcemia. It acts as follows:

1 Parathyroid hormone acts on intestine, bones, and kidney tubules to accelerate calcium absorption from them into the blood. Hence parathyroid hormone tends to increase the blood concentration of calcium. Its primary action on bone is to stimulate bone breakdown or resorption. This releases calcium and phosphate, which diffuse into the blood.

2 Parathyroid hormone acts on kidney tubules to accelerate their excretion of phosphates from the blood into the urine. Note,

> Parathyroid hormone excess produces hypercalcemia or a higher than normal blood concentration of calcium. Sometimes it causes a bone disease called **osteitis fibrosa cystica.** Bone mass decreases (as a result of increased bone destruction followed by fibrous tissue replacement), decalcification occurs, and cystlike cavities appear in the bone.

Figure 14-17 Parathyroid gland within thyroid gland. (×35.)

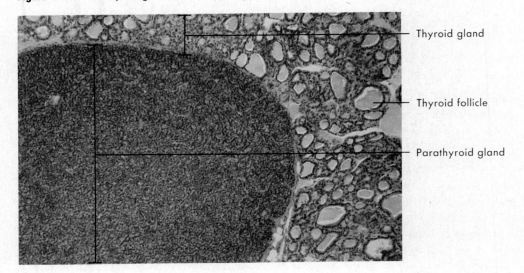

Thyroid gland

Thyroid follicle

Parathyroid gland

therefore, that parathyroid hormone has opposite effects on the kidney's handling of calcium and phosphate. It accelerates calcium reabsorption but phosphate excretion by the tubules. Consequently, parathyroid hormone tends to increase the blood concentration of calcium and to decrease the blood concentration of phosphate.

The maintenance of calcium homeostasis is highly important for healthy survival. Normal neuromuscular irritability, blood clotting, cell membrane permeability, and also normal functioning of certain enzymes all depend on the blood concentration of calcium being maintained at a normal level. Neuromuscular irritability is inversely related to blood calcium concentration. In other words, neuromuscular irritability increases when the blood concentration of calcium decreases. Suppose, for example, that a parathyroid hormone deficiency develops and causes *hypocalcemia* (lower than normal blood concentration of calcium). The hypocalcemia increases neuromuscular irritability—sometimes so much that it produces muscle spasms and convulsions, a condition called tetany.

ADRENAL GLANDS

LOCATION AND STRUCTURE

The adrenal glands are located atop the kidneys, fitting like a cap over these organs. The outer portion of the gland is called the **cortex** and the inner sub-

stance the **medulla.** Although the adrenal cortex and adrenal medulla are structural parts of one organ, they function as two separate endocrine glands.

ADRENAL CORTEX

Three different zones or layers of cells make up the adrenal cortex (Figure 14-18). Starting with the zone directly under the outer capsule of the gland, their names are **zona glomerulosa, zona fasciculata,** and **zona reticularis.** The outer zone of adrenal cells secretes hormones called mineralocorticoids. The middle zone secretes glucocorticoids, and the innermost zone secretes small amounts of both glucocorticoids and sex hormones. We shall now discuss briefly the functions of these three kinds of adrenal cortical hormones.

Glucocorticoids

The chief glucocorticoids secreted by the zona fasciculata of the adrenal cortex are **cortisol** (also called hydrocortisone) and **corticosterone.** Glucocorticoids affect literally every cell in the body. Although the precise primary actions of glucocorticoids remain unknown, their most outstanding effects are as follows:

1 Glucocorticoids tend to accelerate the breakdown of proteins to amino acids in all cells except liver cells. These "mobilized" amino acids move out of tissue cells into blood and circulate to liver cells. Here they are changed to glucose by a process that

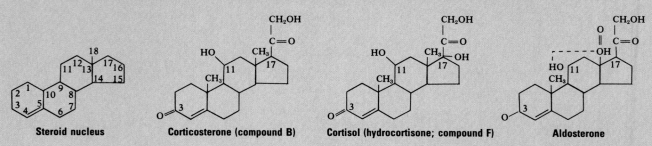

Some steroid compounds, whose common steroid nucleus is shown at the left.

Steroid nucleus **Corticosterone (compound B)** **Cortisol (hydrocortisone; compound F)** **Aldosterone**

Corticosterone (compound B) may be the parent substance of other corticoids.

Compound E or cortisone (chemical name, 17-hydroxy-11-dehydrocorticosterone, signifying that the molecule is the same as corticosterone with —OH instead of —H on C17 and H on C11).

DOC (11-desoxycorticosterone—corticosterone molecules without any oxgen at C11).

Relation of molecular structure to function:

1 Oxygen at C11 produces glucocorticoid effects described in text above.
2 OH at C17 (in addition to oxygen at C11) enhances glucocorticoid effects.
3 Aldehyde group at C18 (as in aldosterone) produces marked salt-retaining effect.

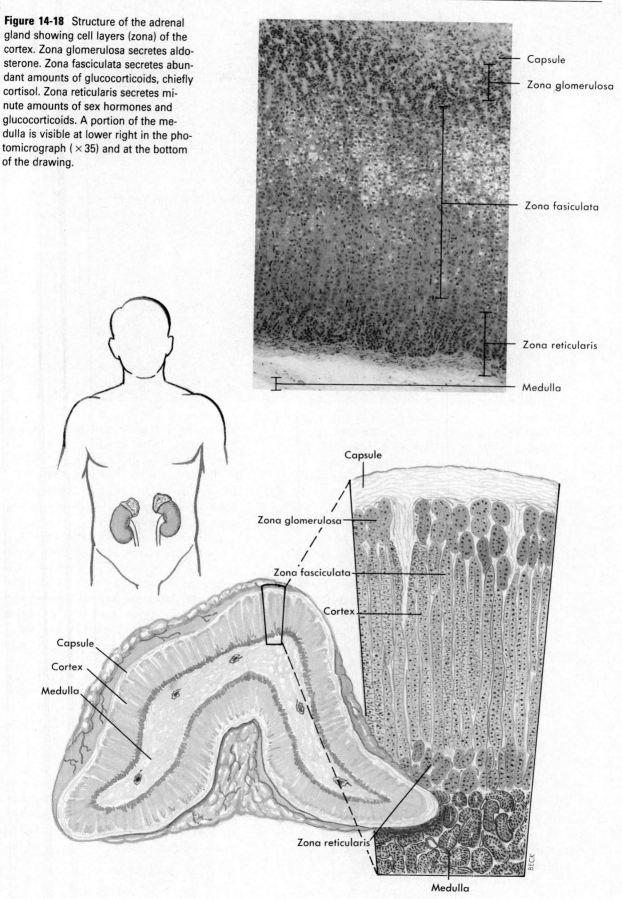

Figure 14-18 Structure of the adrenal gland showing cell layers (zona) of the cortex. Zona glomerulosa secretes aldosterone. Zona fasciculata secretes abundant amounts of glucocorticoids, chiefly cortisol. Zona reticularis secretes minute amounts of sex hormones and glucocorticoids. A portion of the medulla is visible at lower right in the photomicrograph (×35) and at the bottom of the drawing.

Capsule

Zona glomerulosa

Zona fasiculata

Zona reticularis

Medulla

Capsule

Zona glomerulosa

Zona fasciculata

Cortex

Zona reticularis

Medulla

Capsule

Cortex

Medulla

BECK

Figure 14-19 Cushing's syndrome, the result of chronic excess glucocorticoids. **A,** Preoperatively. **B,** Six months postoperatively.

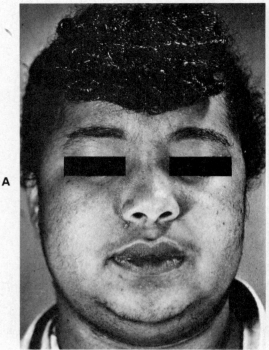

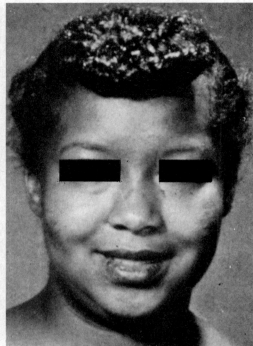

consists of a series of chemical reactions and is called *gluconeogenesis*. A prolonged, high blood concentration of glucocorticoids in the blood, therefore, results in a net loss of tissue proteins, that is, a negative nitrogen balance or "tissue wasting," and a higher than normal blood glucose concentration, that is, hyperglycemia—a common symptom in diabetes mellitus. In summary, we might describe glucocorticoids as protein-mobilizing (or protein-catabolic), gluconeogenic, and hyperglycemic (or diabetogenic) hormones.

2 Glucocorticoids tend to accelerate both the mobilization of fats from adipose cells and the catabolism of fats by almost all kinds of cells. In other words, glucocorticoids tend to cause cells to "shift" for their energy supply from their usual carbohydrate catabolism to fat catabolism. But also the fats mobilized by glucocorticoids may be used by liver cells for gluconeogenesis. Chronic excess of glucocorticoids, as in Cushing's syndrome (Figure 14-19), results in a redistribution of bony fat. It apparently accelerates fat mobilization from the arms and legs and paradoxically promotes fat deposition in the face ("moon face"), shoulders ("buffalo hump"), trunk, and abdomen.

3 Glucocorticoids are essential for maintaining a normal blood pressure. Without adequate amounts of glucocorticoids in the blood, the hormones norepinephrine and epinephrine cannot produce their vasoconstricting effect on blood vessels, and blood pressure falls precariously.

4 Glucocorticoid secretion is known to increase during stress, particularly in stress produced by anxiety or severe injury. What is not known, after more than 30 years of study and debate, is how or even whether the increased blood glucocorticoid concentration helps the body cope successfully with factors that threaten its healthy survival.

5 A high blood concentration of glucocorticoids rather quickly causes a marked decrease in the number of eosinophils in blood (eosinopenia) and marked atrophy (decrease in size) of lymphatic tissues, particularly the thymus gland and lymph nodes. This, in turn, leads to a decrease in the number of lymphocytes and plasma cells in blood. Because of the decreased number of lymphocytes and plasma cells, antibody formation also decreases. Antibody formation is an important part of both immunity and allergy.

6 Normal physiological amounts of glucocorticoids act with epinephrine, a hormone secreted by the medullary portion of the adrenal glands, to bring about a normal recovery from injury produced by many kinds of inflammatory agents. How they accomplish this antiinflammatory effect is still unsettled. Pharmacological amounts of glucocorticoids have been used for many years to relieve the symptoms of rheumatoid arthritis and some other inflammatory conditions.

Mineralocorticoids

Mineralocorticoids, as their name suggests, play an important part in regulating mineral salt (electrolyte) metabolism. In the human, aldosterone is the only physiologically important mineralocorticoid. Its primary general function seems to be to maintain homeostasis of blood sodium concentration. It does this through its action on the distal renal tubule cells. Aldosterone stimulates them to increase their reabsorption of sodium ions from tubule urine back into blood. Because the tubule cells excrete either a potassium or a hydrogen ion in exchange for each sodium ion they reabsorb, aldosterone helps maintain a normal blood potassium concentration and a normal pH.

Moreover, as each positive sodium ion is reabsorbed, a negative ion (bicarbonate or chloride) follows along, drawn by the attraction force between ions that bear opposite electrical charges. Also, the reabsorption of electrolytes causes net diffusion of water back into blood. Briefly, then, because of its primary sodium-reabsorbing effect on kidney tubules, aldosterone tends to produce sodium and water retention but potassium and hydrogen ion loss.

Sex Hormones

The normal adrenal cortex in both sexes secretes small but physiologically significant amounts of male hormones (androgens) and insignificant amounts of female hormones (estrogens). The androgens secreted by the cortex do not have strong masculinizing properties, except for testosterone, but the cortex secretes only trace amounts of this. New information suggests that the small amounts of androgens secreted by the female cortex probably support sexual behavior. Tumors of the adrenal cortex that secrete large amounts of androgens are known as **virilizing tumors**. The excessive androgens produce masculinizing effects such as those evidenced in Figure 14-20. Extreme androgen excess in a woman may cause a beard to grow.

Control of Secretion

GLUCOCORTICOIDS. Mechanisms that control secretion of glucocorticoids are discussed on pp. 395-397.

Figure 14-20 Virilizing tumor of the adrenal cortex of a young girl. The tumor secretes excess androgens, thereby producing masculinizing adrenogenital syndrome.

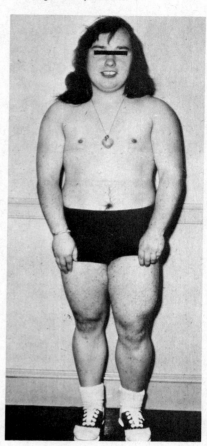

MINERALOCORTICOIDS (ALDOSTERONE MAINLY). Aldosterone secretion is largely controlled by the renin-angiotensin mechanism and by blood potassium concentration. The renin-angiotensin mechanism operates as indicated in Figure 14-21. When blood pressure in the afferent arterioles of the kidney decreases below a certain level, it acts as a stimulus to the juxtaglomerular apparatus,* causing its cells to secrete renin into the blood and interstitial fluid. Renin is an enzyme. It catalyzes reactions that change a compound called renin substrate—a normal constituent of blood—into angiotensin I. Angiotensin I is immediately converted to angiotensin II by another enzyme normally present in blood and aptly named "converting enzyme." Angiotensin II stimulates the zona glomerulosa of the adrenal cortex to increase its secretion of aldosterone.

---

*From the Latin *juxta,* "near to." Cells located in afferent arterioles near their entry into the glomeruli constitute the juxtaglomerular apparatus.

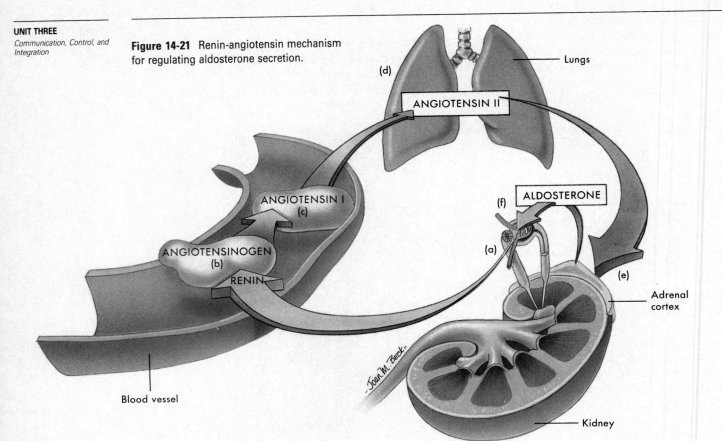

Figure 14-21 Renin-angiotensin mechanism for regulating aldosterone secretion.

Blood potassium concentration also helps regulate aldosterone secretion. Specifically, a high blood potassium concentration stimulates aldosterone secretion and a low blood potassium concentration inhibits it.

ADRENAL MEDULLA

Epinephrine and Norepinephrine

The adrenal medulla secretes two catecholamines—about 80% epinephrine and the rest norepinephrine. These hormones affect smooth muscle, cardiac muscle, and glands the way sympathetic stimulation does. They serve to increase and prolong sympathetic effects.

Control of Secretion

Increased epinephrine secretion by the adrenal medulla is one of the body's first responses to stress. Impulses from the hypothalamus stimulate sympathetic preganglionic neurons, which synapse with cells of the adrenal medulla, stimulating them to increase their output of epinephrine.

ISLANDS OF LANGERHANS

The relationship between exocrine (digestive) and endocrine (hormone) secreting glandular cells in the pancreas is discussed in Chapter 21 (see Figure 21-31). In Figure 14-22 clusters of endocrine secreting cells called **islands of Langerhans** can be identified between the acinar or exocrine secreting portions of this dual-purpose gland. In Figure 14-23 a high-power micrograph of the pancreas shows a portion of one islet of Langerhans and adjacent exocrine secreting (acinar) cells. The acinar cells that form digestive enzymes are easily identified by their secretory products, which appear as distinct granules within the cytoplasm. The endocrine secretions of the islet cells pass directly into the blood while the exocrine secretions of the acinar cells pass into ducts that eventually empty into the duodenum.

Special staining techniques permit identification of two primary endocrine secreting cell types in islet tissue—**alpha cells** and **beta cells.** The beta cells of the islands of Langerhans secrete insulin, and the alpha cells secrete glucagon. Insulin tends to accelerate the movement of glucose, amino acids, and fatty acids out of blood and through the cytoplasmic membranes of cells into their cytoplasm. Hence insulin tends to lower the blood concentrations of these food compounds and to promote their metabolism. The importance of insulin as a key regulator of cellular metabolic activity is discussed in Chapter 23.

Figure 14-22 Pancreas. Two islands of Langerhans (islets) or hormone-producing areas are evident among the serous acini that produce pancreatic enzymes. (× 140.)

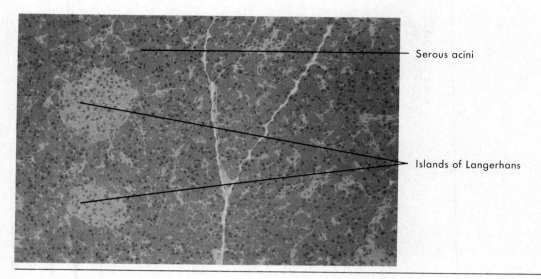

— Serous acini

— Islands of Langerhans

Figure 14-23 Pancreas, showing one island of Langerhans region *(right)* with numerous serous acini *(left)*. Note secretory granules (zymogen granules) in cytoplasm of acinar cells. (× 140.)

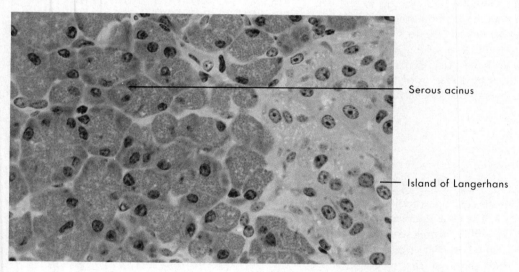

— Serous acinus

— Island of Langerhans

Glucagon, the hormone secreted by the alpha cells of the islands of Langerhans, tends to increase blood glucose concentration. Chapter 23 describes the mechanism by which glucagon produces this opposite effect from insulin. It also describes the mechanisms that regulate insulin and glucagon secretion.

There is a definite "distribution pattern" of hormone-producing cell types in the islands of Langerhans. Insulin-secreting beta cells tend to be concentrated in the central portion of each islet. Glandular cells somewhat different in appearance from the glucagon-secreting alpha cells, but found in association with them in a type of halo around the periphery of

Current research findings of some investigators, such as Dr. R.L. Hazelwood of the University of Houston, Houston, Texas, and others, have demonstrated a new polypeptide hormone secreted by the pancreas. The active component of the new hormone is a polypeptide residue containing 36 amino acids. Until the new hormone is named, scientists are referring to it simply as "pancreatic polypeptide" (PP).

Figure 14-24 Thymus. One Hassall's corpuscle (thymic corpuscle) is visible among the thymocyte (lymphoid) cells. (× 140.)

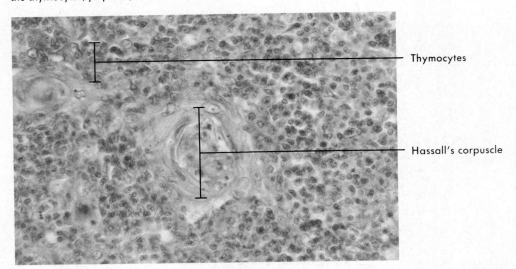

Thymocytes

Hassall's corpuscle

each clump of islet tissue, are now known to secrete pancreatic polypeptide and have been designated as PP cells.

Secretion of pancreatic polypeptide is elevated in diabetes and is influenced by blood levels of a number of compounds, including certain proteins and lipids. The hormone increases both gastric secretions and the production of glucagon. Available evidence seems to suggest that PP plays a variety of important functions in digestion of foodstuffs, distribution of nutrients to the tissues, and cellular metabolism.

OVARIES

The ovaries produce two kinds of steroid female hormones: **estrogens** (chiefly estradiol and estrone) and **progesterone.** The chief endocrine glands that secrete estrogens are microscopic structures located in the ovary and named ovarian or graafian follicles. Another endocrine gland in the ovary secretes mostly progesterone. It is the corpus luteum and has a particular claim to distinction. It is a temporary structure, replaced once a month during a woman's reproductive years, except in months when she is pregnant. In Chapter 29, we shall discuss the functions of estrogens and progesterone and the control of their secretion.

TESTES

The interstitial cells of the testes secrete **testosterone,** a steroid hormone classed as an androgen, that is, a substance that promotes "maleness." Chapter

28 presents details about testosterone's functions and the control of its secretion.

PINEAL GLAND (PINEAL BODY OR EPIPHYSIS CEREBRI)

The pineal gland has long been a mystery organ. Even now its function in the human being remains a controversial matter. A small cone-shaped structure about 1 cm long, it is located in the cranial cavity behind the midbrain and the third ventricle (see Figure 14-1).

Melatonin, the primary hormone secreted by the pineal gland, appears to inhibit luteinizing hormone secretion and ovarian function. The pineal gland may also secrete a second hormone called **adrenoglomerulotropin** because of its ability to stimulate aldosterone secretion by the adrenal's zona glomerulosa.

THYMUS

There is no longer any doubt that the thymus functions as an endocrine gland. Once relegated to a position of being a vestigial structure of little importance, the thymus is now considered the primary organ of the lymphatic system (Figures 14-1 and 14-24). As a component of the overall body immune system, the endocrine function of the thymus is not only important but essential. The anatomy of the thymus is discussed in Chapter 18.

The hormone **thymosin** has been isolated from thymus tissue and is considered responsible for its endocrine activity. Thymosin is actually a family of biologically active peptides that together play a critical role in the maturation and development of the immune system. In an individual with impaired thymic function, injection of thymosin will increase the population of specialized lymphocytes called "T cells" and activate the immune system (discussed in Chapter 29).

Suppression of the immune system sometimes occurs in certain disease states such as **AIDS** (acquired immune deficiency syndrome) and in patients who are undergoing massive chemotherapy or radiotherapy for the treatment of cancer. Such individuals are said to be "immunosuppressed" and are extremely susceptible to infections. Thymosin may prove useful as an activator of the immune system in such patients.

GASTRIC AND INTESTINAL MUCOSA

The mucosal lining of the gastrointestinal tract, like the pancreas, contains cells that produce both en-

docrine and exocrine secretions. Gastrointestinal hormones such as **gastrin, cholecystokinin-pancreozymin,** and **secretin** play important regulatory roles in controlling and coordinating the secretory and motor activities involved in the digestive process. Chapter 22 describes the hormonal control of digestion in the stomach and small intestine.

PLACENTA

The placenta functions as a temporary endocrine gland. During pregnancy, it produces **chorionic gonadotropins**—so-called because they are secreted by cells of the chorion, the outermost fetal membrane. In addition to gonadotropins, the placenta also produces estrogens and progesterone. During pregnancy the kidneys excrete large amounts of gonadotropins in the urine. This fact, discovered more than a half century ago by Aschheim and Zondek, led to the development of the now familiar pregnancy tests.

Outline Summary

MEANING

A Composed of glands that pour secretions into blood instead of into ducts

B General functions and importance of neuroendocrine system
1 Communication, integration, and control—hormones produce slower and longer lasting responses in regulation of metabolism, reproduction, and responses to stress
2 Nervous system (nerve impulses) produce rapid, short-lasting responses
3 Cells that are acted on and respond to hormones are called *target organ cells*
4 Names and locations of endocrine glands (see Table 14-1)

PROSTAGLANDINS (TISSUE HORMONES)

A Unique group of widespread and potent biological compounds (20-carbon fatty acids with 5-carbon ring)

B Regulate endocrine activity at the cellular level by influencing adenyl cyclase and cyclic AMP activity (see Figure 14-2)

C Three classes of prostaglandins
1 Prostaglandin A (PGA)—may help in regulation of blood pressure
2 Prostaglandin E (PGE)—vascular effects involving red blood cell "deformability" and platelet aggregation; effect on hydrochloric acid secretion in stomach

3 Prostaglandin F (PGF)—important in reproductive function

D May someday be used in treatment of such diverse diseases as hypertension, coronary thrombosis, asthma, and ulcers

E How proteins act
1 Second messenger hypothesis
    a Hormone acts as "first messenger"
    b Hormone reacts with adenyl cyclase
    c Adenyl cyclase converts ATP to cyclic AMP
    d Cyclic AMP is second "messenger," which initiates a specialized function in cell

F Steroid hormones
1 Steroid hormones are lipid soluble
2 Once inside cell, hormone combines with cytoplasmic receptor protein
3 Hormone-receptor protein complex enters nucleus and activates mRNA
4 mRNA enters cytoplasm and initiates synthesis of specific proteins
5 Synthesized proteins initiate hormone activity

PITUITARY GLAND (HYPOPHYSIS CEREBRI)

A Location—in sella turcica of sphenoid bone

B Consists of two endocrine glands: anterior lobe (adenohypophysis) and posterior lobe (neurohypophysis)

C Anterior portion divided into the larger pars anterior and the small pars intermedia

D Table 14-2 summarizes the production, control, and effects of the pituitary hormones

Outline Summary—cont'd

Anterior pituitary gland (adenohypophysis)

A Cell types and functions

1 Chromophobes—undifferentiated cells that do not stain well

2 Chromophils

a Acidophils—stain with acid dyes; secrete growth hormone (GH) and prolactin

b Basophils—secrete thyrotropin (TSH), adrenocorticotropin (ACTH), follicle-stimulating hormone (FSH), luteinizing hormone (LH), and melanocyte-stimulating hormone (MSH)

B Pars anterior secretes all hormones of the anterior lobe except melanocyte-stimulating hormone (MSH), which the pars intermedia secretes

C Growth hormone (somatotropin or somatotropic hormone)

1 Accelerates protein anabolism, so promotes growth

a Disorders caused by excess growth hormone—gigantism and acromegaly

b Disorders caused by deficient growth hormone—dwarfism and pituitary cachexia

2 Tends to accelerate fat mobilization from adipose cells and fat catabolism by other cells, thereby decreasing glucose catabolism and tending to increase blood glucose concentration (hyperglycemic effect); prolonged excess may produce diabetes

D Prolactin (lactogenic hormone or luteotropic hormone [LTH])

1 During pregnancy, promotes breast development

2 After delivery, initiates milk secretion

3 Helps in maintaining corpus luteum

E Four tropic hormones (hormones that have a stimulating effect on other endocrine glands) are secreted by basophil cells of anterior pituitary gland

1 Thyrotropin—thyroid-stimulating hormone (TSH)

a Promotes growth and development of thyroid gland

b Stimulates thyroid gland to secrete thyroid hormone

2 Adrenocorticotropin (ACTH)

a Promotes growth and development of adrenal cortex

b Stimulates adrenal cortex to secrete cortisol and other glucocorticoids

3 Follicle-stimulating hormone (FSH)

a Stimulates primary graafian follicle to start growing and to develop to maturity

b Stimulates follicle cells to secrete estrogens

c In male, stimulates development of seminiferous tubules and maintains spermatogenesis by them

4 Luteinizing hormone (LH)

a Acts with FSH to cause complete maturation of follicle

b Brings about ovulation

c Stimulates formation of corpus luteum (luteinizing effect)

d Stimulates corpus luteum to secrete progesterone and estrogens

e In male, LH called interstitial cell–stimulating hormone (ICSH); stimulates interstitial cells in testis to develop and secrete testosterone

f Just before puberty, presumably sudden marked increase in secretion of gonadotropins (FSH, LH) initiates first menses

F Melanocyte-stimulating hormone (MSH) or intermedin—tends to produce increased pigmentation of skin

G Control of secretion

1 Releasing hormones—neurosecretions produced in hypothalamus that reach anterior pituitary gland via blood in pituitary portal system to regulate hormone production

a Growth hormone-releasing hormone (GRH)

b Corticotropin-releasing factor (CRF)

c Thyrotropin-releasing hormone (TRH)

d Follicle-stimulating hormone-releasing hormone (FSH-RH)

e Luteinizing hormone–releasing hormone (LH-RH)

f Prolactin-inhibitory factor (PIF)

g Prolactin-releasing factor (PRF)

2 Negative feedback mechanisms operate between target glands and anterior lobe of pituitary gland

a High blood concentration of target gland hormone inhibits its pituitary secretion of tropic hormones either directly or indirectly, that is, via hypothalamic releasing hormones; conversely, low blood concentration of target gland hormones stimulates pituitary secretion of tropic hormones

b High blood concentration of tropic hormone stimulates target gland secretion of its hormone

3 Stress acts in some way to stimulate the hypothalamus to secrete releasing hormones, which in turn stimulate the anterior pituitary gland to increase its secretion of hormones

Posterior pituitary gland (neurohypophysis)

A Cell type—pituicytes do not secrete either of the hormones released from the posterior pituitary gland

B Releases antidiuretic hormone (ADH)—synthesized by neurons in supraoptic nucleus of hypothalamus—and oxytocin—synthesized by neurons in the paraventricular nucleus of hypothalamus

C Normal (physiological) levels of ADH (vasopressin or Pitressin) will stimulate water reabsorption by distal and collecting tubules and in large (pharmacological) doses will stimulate smooth muscle of blood vessels

Figure 15-1 Human blood cells. There are close to 30 trillion red blood cells in the adult. Each cubic millimeter of blood contains from 4.5 to 5.5 million red blood cells and an average total of 7,500 white blood cells.

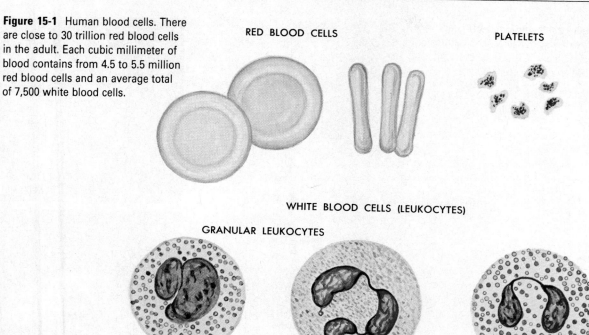

RED BLOOD CELLS

PLATELETS

WHITE BLOOD CELLS (LEUKOCYTES)

GRANULAR LEUKOCYTES

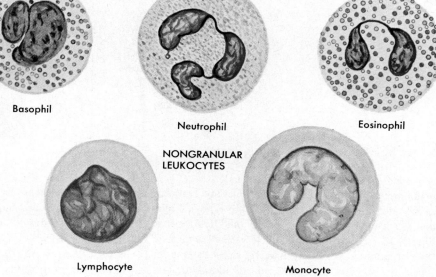

Basophil

Neutrophil

Eosinophil

NONGRANULAR LEUKOCYTES

Lymphocyte

Monocyte

Figure 15-2 Scanning electron micrograph of normal erythrocytes. (× 3,000.)

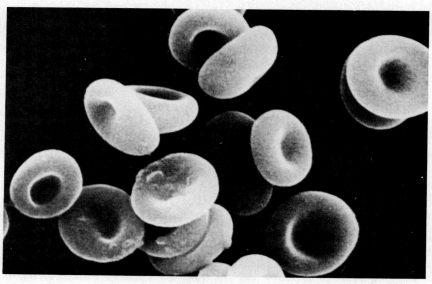

Figure 15-3 Abnormal forms of the erythrocyte. Appearance in peripheral blood smear *(left)* and with scanning electron microscope *(right).*

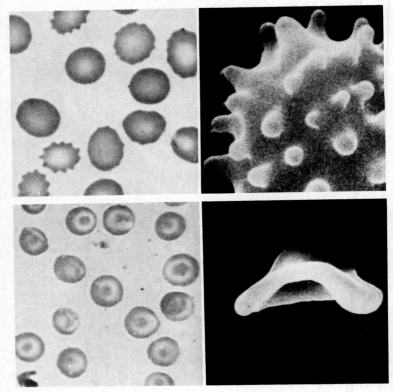

A, Crenated erythrocyte.

B, Target cell or codocyte.

prominent central dimple. Target cells are common in iron-deficiency anemia and certain types of liver disease. The abnormal red blood cells illustrated in Figure 15-3, *C*, are called **sickle cells** because of their tendency to form bizarre crescent-shaped forms if subjected to low levels of blood oxygen. Sickle cell anemia is caused by an abnormal type of hemoglobin (hemoglobin S).

The red cell fragments shown in Figure 15-3, *D*, are called **schizocytes.** These distorted and irregularly shaped red cell fragments often result from mechanical injury or trauma to red cells as they pass through the circulation. Intravascular blood clots that have exposed strands of fibrin sometimes produce schizocytes by the "clothesline effect." Passing erythrocytes are cut to pieces by the strands of fibrin as they pass the clot. Patients who have prosthetic heart valves often have schizocytes present in their blood as a result of trauma to red cells passing through the valve.

Considered together, the total surface area of all the red blood cells in an adult is enormous. It provides an area larger than a football field for the exchange of respiratory gases between hemoglobin found in circulating erythrocytes and interstitial fluid that bathes the body cells. This is an excellent example of the familiar principle that structure determines function (specifically, the transport and ex-

change of oxygen and carbon dioxide). Packed within one tiny red blood cell are an estimated 200 to 300 million molecules of the complex compound **hemoglobin,** which makes up about 95% of the dry weight of each cell. One hemoglobin molecule contains four iron atoms. This structural fact enables one hemoglobin molecule to unite with four oxygen molecules to form oxyhemoglobin (a reversible reaction). Hemoglobin can also combine with carbon dioxide to form carbaminohemoglobin (also reversible). But in this reaction the structure of the globin part of the hemoglobin molecule rather than of its heme part makes the combining possible.

A man's blood usually contains more hemoglobin than a woman's. In most normal men, 100 ml of blood contains 14 to 16 gm of hemoglobin. The normal hemoglobin content of a woman's blood is a little less—specifically, in the range of 12 to 14 gm per 100 ml. Any adult who has a hemoglobin content of less than 10 gm per 100 ml of blood is diagnosed as having **anemia** (from the Greek *a*-, "not," and *haima*, "blood"). In addition, the term may be used to describe a reduction in the number or volume of functional red blood cells in a given unit of whole blood. Anemias are classified according to the size and hemoglobin content of red blood cells. Such a morphological or structural classification of anemia would include the following types:

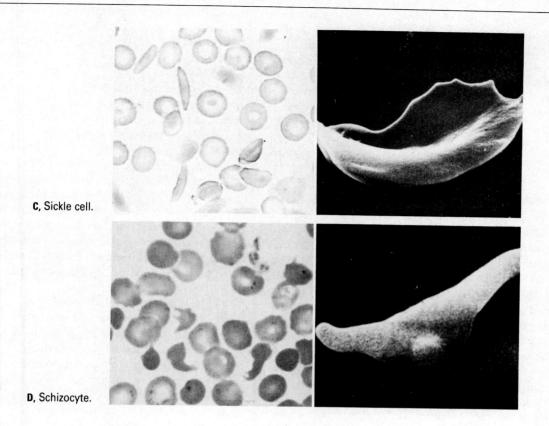

C, Sickle cell.

D, Schizocyte.

1 **Normocytic**—Red blood cells produced are reduced in number but are normal in size.
2 **Macrocytic**—Although reduced in number, those red blood cells that are produced are larger in size than normal.
3 **Microcytic**—The majority of red blood cells are smaller than normal.

The amount of hemoglobin (pigment) that a red blood cell contains will influence the intensity of its color. If the hemoglobin content of a red blood cell is high, the color will be more intense—such a cell is described as *hyperchromic* (from the Greek *hyper,* "above, over," and *chroma,* "color"). Low hemoglobin levels result in *hypochromic* red blood cells. Red blood cells that have an average (normal) color and hemoglobin content are said to be *normochromic.* Example: an injury resulting in hemorrhage with rapid loss of whole blood would cause a reduction in circulating red blood cells and hemoglobin but would not change the average size or hemoglobin content of those cells that remained. The anemia that would result from such a rapid loss of blood would be classified as *normocytic normochromic.*

The macrocytic and microcytic types of anemia are characterized by very apparent variation from the normal diameter (6.2 to 8.2 μm) of red blood cells. The term **anisocytosis** refers to abnormal vari-

BLOOD DOPING

Reports that olympic athletes have employed blood transfusions to improve performance have surfaced repeatedly during the past 20 years. The practice—called **blood doping** or **blood boosting**—is intended to increase oxygen delivery to muscles. Theoretically, infused red blood cells and elevation of hemoglobin levels following transfusion should increase oxygen consumption and muscle performance during exercise. In practice, however, questions of how effective transfusions might be in affording even world-class and professional athletes substantial competitive advantages have not been resolved. Improved performance might result but the advantage appears to be minimal.

All blood transfusions carry some risk and unnecessary or questionably indicated transfusions are medically unacceptable.

ation in red blood cell diameter *(size)* beyond these normal limits. The extent of anisocytosis is arbitrarily graded 1+ to 4+. **Poikilocytosis** is the term used to describe red blood cells that vary from the normal in *shape* (Figure 15-3, *A* to *D*). Poikilocytosis is also graded 1+ to 4+. The crenated red blood cell in Figure 15-3, *A*, would be graded 2+ anisocytosis because of its reduced size but 4+ poikilocytosis because of its extreme variation in shape from the normal, biconcave, disk-shaped erythrocyte.

Formation (Erythropoiesis)

The entire process of red blood cell formation is called *erythropoiesis* (from the Greek *poiesis,* "production"). In the adult, erythrocytes begin their maturation sequence in the red bone marrow from nucleated cells known as **hemocytoblasts** or stem cells (Figure 15-4). These stem cells divide by mitosis and go through several stages of development. Figure 15-5 lists the sequential cell types or stages that can be identified as transformation from the immature forms to the mature red blood cell occurs. Development proceeds from step to step by gradual transition. The large *rubriblast* (Figure 15-6) has a round, centrally placed and very prominent nucleus. Cytoplasm in the rubriblast is scanty and appears opaque. The more mature *rubricyte* (Figure 15-7) and *metarubricyte* (Figure 15-8) are smaller cells that have a dense nucleus and more noticeable cytoplasm. As development proceeds, the nucleus becomes smaller and evenutally disappears. Note in Figure 15-5 that overall cell size decreases as the maturation sequence progresses. Newly formed red blood cells, at the time they leave the bone marrow and enter the blood, contain hemoglobin. A small number of newly formed erythrocytes exhibit a reticulum in their cytoplasm. For this reason they are called *reticulocytes.*

Frequently a physician needs information about the rate of erythropoiesis to help in making a diagnosis or to prescribe treatment. A **reticulocyte count** gives this information. Approximately 0.5% to 1.5% of the red blood cells in normal blood are reticulocytes. A reticulocyte count of less than 0.5% of the red blood cell count usually indicates a slowdown in the process of red blood cell formation. Conversely, a reticulocyte count higher than 1.5% usually indicates an acceleration of red blood cell formation—as occurs, for example, following treatment of anemia.

Reticulocytes are almost 20% larger than mature red blood cells and are coated with a type of globulin that makes them more adhesive than adult corpuscles. This adhesive quality may help in retaining them within the bone marrow until maturity. Once in the circulation, reticulocytes pass through small blood vessels much more slowly than do mature red blood cells. After leaving the bone marrow the "life span" of a circulating reticulocyte rarely exceeds 30 hours.

Destruction

The life span of a red blood cell circulating in the bloodstream averages about 105 to 120 days. They break apart, that is, undergo fragmentation, in the capillaries. Reticuloendothelial cells in the lining of blood vessels, particularly in the liver, spleen, and bone marrow, then phagocytose the red blood cell fragments. In the process, iron is released from hemoglobin and the pigment bilirubin is formed. Both are transported to the liver, where iron is put in temporary storage and bilirubin is excreted in the bile. Eventually the bone marrow uses most of the iron over again for new red blood cell synthesis, and the liver excretes the bile pigments in the bile.

Erythrocyte Homeostatic Mechanism

Red blood cells are formed and destroyed at a breathtaking rate. Normally, every minute of every day of our adult lives, over 100 million red blood cells are formed to replace an equal number destroyed during that brief time. Since in health the number of red blood cells remains relatively constant at about 4.5 to 5.5 million per mm^3 of blood, efficient homeostatic mechanisms must operate to balance the number of cells formed against the num-

A clinical example of failure of red blood cell homeostasis seen in recent years is anemia caused by bone marrow injury by x-ray or gamma ray radiations. Other factors may also cause marrow damage. To help diagnose this condition a sample of marrow is removed, for example, from the sternum by means of a sternal puncture, and is studied microscopically for abnormalities. When damaged marrow can no longer keep red blood cell production apace with destruction, red blood cell homeostasis is not maintained. Instead, the red blood cell count falls below normal. Anemia (fewer red blood cells than normal) develops whenever the rates of red blood cell formation and destruction become unequal. Therefore either a decrease in red blood cell formation (as in pernicious anemia or bone marrow injury by radiation) or an increase in red blood cell destruction (as in infections and malignancies) can lead to anemia.

Figure 15-4 The hemocytoblast.

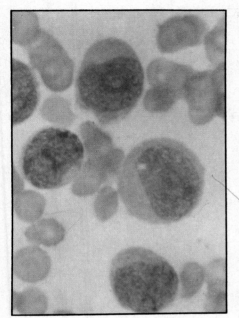

Figure 15-5 Stages in development of red blood cells.

Hemocytoblast (20 to 23 μm)
↓
Rubriblast (15 to 19 μm)
↓
Prorubricyte (10 to 15 μm)
↓
Rubricyte (9 to 12 μm)
↓
Metarubricyte (8 to 10 μm)
↓
Reticulocyte (7 to 9 μm)
↓
Erythrocyte (6 to 8 μm)

Figure 15-6 The rubriblast.

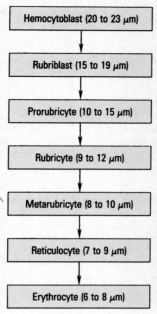

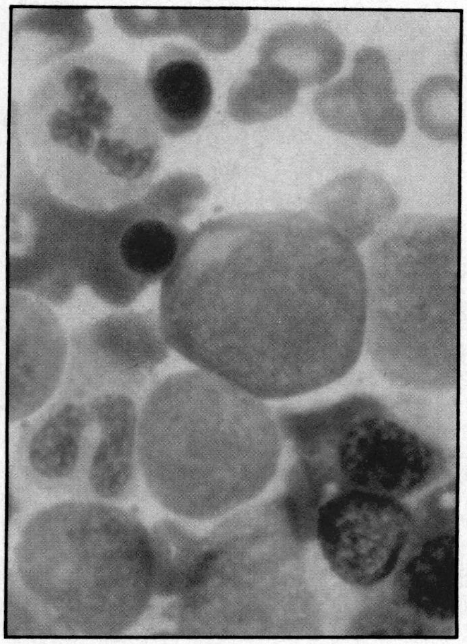

Figure 15-7 The rubricyte.

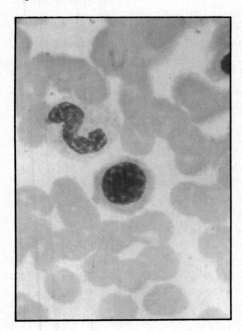

Figure 15-8 The metarubricyte.

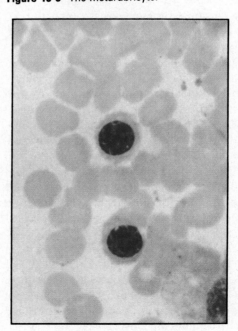

Figure 15-9 Postulated red blood cell homeostatic mechanism—a negative feedback mechanism. A decrease in the number of circulating red blood cells "feeds back" to the red blood cell–forming structure (red bone marrow) to cause an increased rate of red blood cell formation, which in turn tends to increase the number of red blood cells sufficiently to restore their normal number.

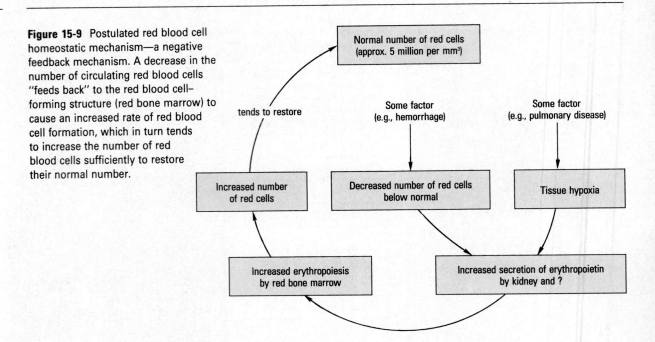

ber destroyed. The exact mechanism responsible for this constancy is not known. It is known, however, that the rate or red blood cell production soon speeds up if either the number of red blood cells decreases appreciably or tissue hypoxia (oxygen deficiency) develops. Either of these conditions acts in some way to stimulate the kidneys (and perhaps some other structures) to increase their secretion of a hormone named **erythropoietin.** The resulting increased blood concentration of erythropoietin stimulates bone marrow to accelerate its production of red blood cells (Figure 15-9). Under maximum stimulation the bone marrow can increase red blood cell production about seven times. The name *erythropoietin* makes it easy to remember its function—erythropoietin stimulates erythropoiesis (process of red blood cell formation).

For the red blood cell homeostatic mechanism to succeed in maintaining a normal number of red blood cells, the bone marrow must function adequately. To do this the blood must supply it with adequate amounts of several substances with which to form the new red blood cells—vitamin B_{12}, iron, and amino acids, for example, and also copper and cobalt to serve as catalysts. In addition, the gastric mucosa must provide some unidentified intrinsic factor necessary for absorption of vitamin B_{12} (also called **extrinsic factor** because it derives from external sources in foods and is not synthesized by the body; vitamin B_{12} is also called the **antianemic principle**).

Pernicious anemia develops when the gastric mucosa fails to produce the intrinsic factor needed for adequate vitamin B_{12} absorption. The marrow then produces fewer but larger red blood cells than normal (macrocytic anemia). Many of these cells are immature with overly fragile membranes, a fact that leads to their more rapid destruction.

The number of red blood cells in determined by the **red blood cell count** or is estimated by the hematocrit. The **hematocrit** is the volume percentage of red blood cells in whole blood. To be more specific, a hematocrit of 47 means that in every 100 ml of whole blood there is 47 ml of blood cells and 53 ml of fluid (plasma). Normally the average hematocrit for a man is about 45 (± 7, normal range) and for a woman about 42 (± 5). Healthy individuals who live and work in high altitudes often have elevated red blood cell counts and hematocrit values. The condition is called **physiological polycythemia** (from the Greek *polys,* "many," *kytos,* "cell," and *haima,* "blood"). Can you explain the stimulus for increased erythropoiesis in these individuals?

LEUKOCYTES
Appearance, Size, Shape, and Number

There are five types of leukocytes classified according to the presence or absence of granules in their cytoplasm. Granular leukocytes consist of neutrophils, eosinophils, and basophils; nongranular leukocytes are lymphocytes and monocytes.

Consult Figure 15-1. Note particularly the differences in color of the cytoplasmic granules and in the shapes of the nuclei of the granular leukocytes. *Neutrophils* (Figure 15-10) take their name from the fact that their cytoplasmic granules stain a very light purple with neutral dyes. The granules in these cells

Figure 15-10 Hematocrit tubes showing **A**, normal blood, **B**, anemia, and **C**, polycythemia.

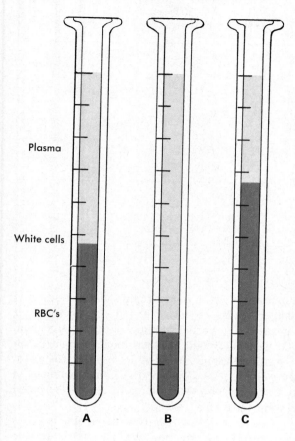

Plasma

White cells

RBC's

A   B   C

Figure 15-11 Leukocytes in human blood smears. **A**, Neutrophil. **B**, Lymphocyte. **C**, Basophil. **D**, Monocyte.

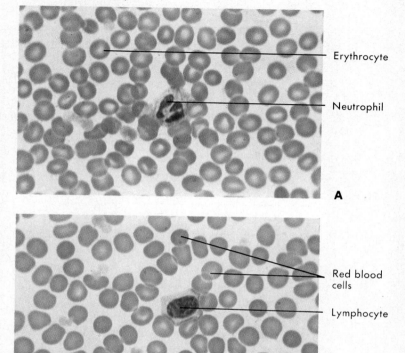

Erythrocyte

Neutrophil

A

Red blood cells

Lymphocyte

B

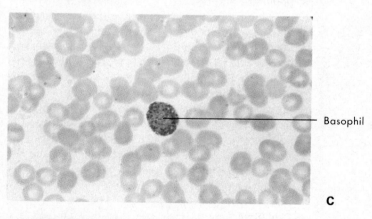

Basophil

C

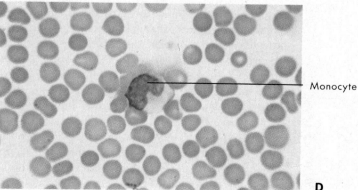

Monocyte

D

are small and numerous and tend to give the cytoplasm a "coarse" appearance. Because their nuclei have two, three, or more lobes, neutrophils are also called **polymorphonuclear leukocytes** or, to avoid that tongue twister, simply *polys*. Cytoplasmic granules in eosinophils are large and numerous and stain orange with acid dyes such as eosin. Their nuclei have two oval lobes. The relatively large but sparse cytoplasmic granules in *basophils* stain a dark purple with basic dyes. Their nuclei are roughly S shaped.

Most lymphocytes in a stained blood smear are about 8 μm in diameter—slightly larger than erythrocytes. In a typical lymphocyte a relatively large spherical nucleus with a slight indentation on one side is surrounded by a thin layer of homogeneous cytoplasm (Figure 15-11). An average monocyte—the largest of all leukocytes—measures about 15 to 20 μm in diameter. Monocytes (Figure 15-11) have a kidney-shaped nucleus surrounded by abundant quantities of pale grayish or steel blue colored cytoplasm.

Table 15-1 White blood cells (leukocytes)

Class	Differential count*	
	Normal range (%)	Typical range (%)
THOSE WITH GRANULAR CYTOPLASM AND IRREGULAR NUCLEI		
Neutrophils (neutral staining)	65 to 75	65
Eosinophils (acid staining)	2 to 5	3
Basophils (basic staining)	½ to 1	1
THOSE WITH NONGRANULAR CYTOPLASM AND REGULAR NUCLEI		
Lymphocytes (large and small)	20 to 25	25
Monocytes	3 to 8	6
TOTAL		100

*In any differential count the sum of the percentages of the different kinds of leukocytes must, of course, total 100%.

A cubic millimeter of normal blood usually contains about 5,000 to 9,000 leukocytes, with different percentages of each type. Because these numbers change in certain abnormal conditions, they have clinical significance. In acute appendicitis, for example, the percentage of neutrophils increases and so, too, does the total white blood cell count. In fact, these characteristic changes may be the deciding points for surgery.

The procedure in which the different types of leukocytes are counted and their percentage of total white blood cell count is computed is known as a **differential count.** In other words, a differential count is a percentage count of white blood cells. The different kinds of white blood cells and a normal differential count are listed in Table 15-1. A decrease in the number of white blood cells is **leukopenia.** An increase in the number of white blood cells is **leukocytosis.** (*Leukemia* is a malignant disease characterized by a marked increase in the number of white blood cells.)

Functions

White blood cells function as part of the body's defense against microorganisms (Chapter 28). All leukocytes are motile cells. This characteristic enables them to move out of capillaries by squeezing through the intercellular spaces of the capillary wall—a process called **diapedesis** (from the Greek *dia*, "through," and *pedesis*, "a moving") and to migrate by ameboid movement toward microorganisms or other injurious particles that may have invaded the tissues. Neutrophils, monocytes, lymphocytes, and basophils are highly motile, whereas eosinophils are sluggishly so. Once in the tissue spaces the most

important function of neutrophils and monocytes is **phagocytosis**—to ingest and digest microbes or other injurious particles. At any one time, large numbers of white blood cells are moving about in the tissues performing, or ready to perform, their protective functions. Red blood cells and platelets, in contrast, perform their functions within the blood vessels, not in the tissues.

You may recall that reticuloendothelial cells also perform the function of phagocytosis. In general, however, they do this work within more localized areas than do white blood cells. Reticuloendothelial cells, since they do not enter the bloodstream, cannot be transported by it to any part of the body, as white blood cells can.

Lymphocytes play a dominant and vital role in defending the body against microscopic invaders—notably bacteria, fungi, and viruses. Immunity to infectious diseases results largely from lymphocyte activities. Chapter 28 discusses the immune functions of lymphocytes.

Eosinophils are weak phagocytes and show only very limited motility. As a result, they do not play an important role in the body's defense against the usual types of infectious microorganisms. It has been suggested that eosinophils function to detoxify proteins and other harmful products that accumulate as a result of allergic reactions or cellular injury caused by infection with parasites such as hookworm.

The primary function of basophils (Figure 15-11) in the circulating blood is almost totally unknown. Heparin, a highly active anticoagulant, can be obtained from the granules of basophils and may play a role in the prevention of intravascular coagulation.

Figure 15-12 Stages in development of granular leukocytes.

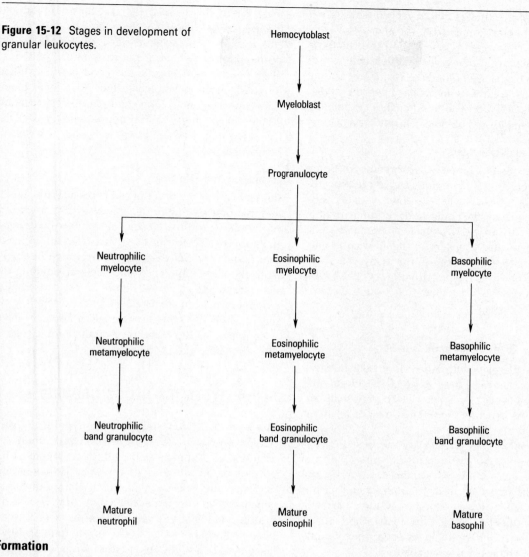

Formation

There is a substantial body of research evidence that supports the theory of a single, multipotential, blood-forming stem cell—the **hemocytoblast.** This primitive stem cell serves as the precurser of not only the erythrocytes but also the leukocytes and platelets in blood. Figure 15-12 shows the maturation sequence that results in formation of the granular leukocytes from undifferentiated hemocytoblasts. Steps in the maturation of nongranular leukocytes are listed in Figure 15-13.

Neutrophils, eosinophils, basophils, and a few lymphocytes and monocytes originate, as do erythrocytes, in red bone marrow (myeloid tissue). Most lymphocytes and monocytes derive from hemocytoblasts in lymphatic tissue. Although many lymphocytes are found in bone marrow, presumably most were formed in lymphatic tissues and carried to the bone marrow by the bloodstream.

Myeloid tissue (bone marrow) and lymphatic tissue together constitute the hemopoietic or blood cell–forming tissues of the body. Red bone marrow

Figure 15-13 Stages in development of nongranular leukocytes.

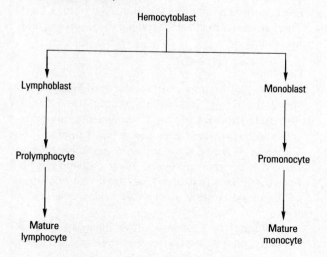

is myeloid tissue that is actually producing blood cells. Its red color comes from the red blood cells it contains. Yellow marrow, on the other hand, is yellow because it stores considerable fat. It is not active in the business of blood cell formation as long as it remains yellow. Sometimes, however, it becomes active and red in color when an extreme and prolonged need for red blood cell production occurs.

Destruction and Life Span

The life span of white blood cells is not known. Some evidence seems to indicate that granular leukocytes may live 3 days or less whereas other evidence suggests that they may live about 12 days. Some of them are probably destroyed by phagocytosis and some by microorganisms. Contrary to the assumption accepted by many hematologists for many years, we now know that a significant fraction of small lymphocytes survive a long time, 100 to 200 days or more.

PLATELETS

Appearance, Size, Shape, and Number

To compare platelets with other blood cells in terms of appearance and size, see Figure 15-1. In circulating blood, platelets are small, colorless bodies that usually appear as irregular spindles or oval disks about 2 to 4 μm in diameter.

Three important physical properties of platelets—namely, agglutination, adhesiveness, and aggregation—make attempts at classification on the basis of size or shape in dry blood smears all but impossible. As soon as blood is removed from a vessel, the platelets adhere to each other and to every surface they contact; in so doing, they assume a variety of shapes and irregular forms.

Platelet counts in adults average about 250,000 per mm³ of blood. A range of 150,000 to 350,000 per mm³ is considered normal. Newborn infants often show reduced counts, but these rise gradually to reach normal adult values at about 3 months of age. There are no differences between the sexes in platelet count.

Functions

Platelets play an important role in both **hemostasis** (from the Greek *stasis*, "a standing") and **blood clotting.** The two, although interrelated, are separate and distinct functions. Hemostasis refers to the stoppage of blood *flow* and may occur as an end result of any one of several body defense mechanisms. The role that platelets play in the operation of the blood clotting mechanism is discussed on p. 434.

Within 1 to 5 seconds after injury to a blood capillary, platelets will adhere to the damaged lining of the vessel and to each other to form a hemostatic platelet plug that helps to stop the flow of blood into the tissues. At least one of the prostaglandins (PGE_2) and certain prostaglandin-like substances called *thromboxanes* (which are both found in platelets) play roles in hemostasis and blood clotting. When released, these substances affect both local blood flow (by vasoconstriction) and platelet aggregation at the site of injury. If the injury is extensive, the blood-clotting mechanism is activated to assist in hemostasis.

Formation and Life Span

Platelets are formed in the red bone marrow, lungs, and, to some extent, in the spleen by fragmentation of very large (40 to 100 μm) cells known as **megakaryocytes.** These cells are characterized by large multilobular nuclei that are often bizarre in shape. Note the large number of platelets surrounding the megakaryocyte shown in Figure 15-14. The steps in platelet formation are listed in Figure 15-15. Platelets have a short life span, an average of about 10 days. A summary of the basic facts about blood cells is given in Table 15-2.

BLOOD TYPES (OR BLOOD GROUPS)

The term *blood type* refers to the type of antigens (called *agglutinogens*) present on red blood cell membranes. Antigens A, B, and Rh are the most important blood antigens as far as transfusions and newborn survival are concerned. Many other antigens have also been identified, but they are less important clinically and are seldom discussed. Every person's blood belongs to one of the four AB blood groups and, in addition, is either Rh positive or Rh negative. Blood types are named according to the antigens present on red blood cell membranes. Here, then, are the four AB blood types:

1 **Type A**—Antigen A on red blood cells
2 **Type B**—Antigen B on red blood cells
3 **Type AB**—Both antigen A and antigen B on red blood cells
4 **Type O**—Neither antigen A nor antigen B on red blood cells

The term *Rh-positive blood* means that Rh antigen is present on its red blood cells. *Rh-negative blood*, on the other hand, is blood whose red cells have no Rh antigen present on them.

Antigen—substance capable of stimulating formation of other substances called *antibodies* that can combine with the antigen, for example, to agglutinate or clump it.

Figure 15-14 Megakaryoctye. Note the large number of platelets surrounding the cell.

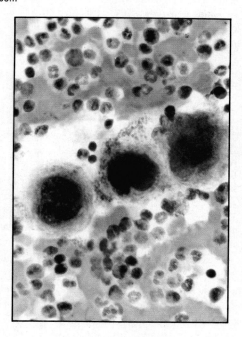

Figure 15-15 Stages in platelet formation.

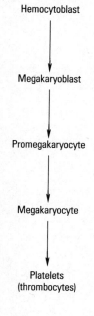

Hemocytoblast

↓

Megakaryoblast

↓

Promegakaryocyte

↓

Megakaryocyte

↓

Platelets
(thrombocytes)

Table 15-2 Blood cells

Cells	Number	Function	Formation (hemopoiesis)	Destruction
Red blood cells (erythrocytes)	4.5 to 5.5 million /mm³ (total of approximately 30 trillion in adult body)	Transport oxygen and carbon dioxide; buffers (H⁺)	Red marrow of bones (myeloid tissue)	Reticuloendothelial cells in lining of blood vessels in liver, spleen, and bone marrow phagocytose old red blood cells; live about 105 to 120 days in bloodstream
White blood cells (leukocytes)	Usually about 5,000 to 9,000/mm³	Play important part in producing immunity (phagocytosis by neutrophils and monocytes); antibody formation by lymphocytes and their descendants, plasma cells; production of heparin by basophils; detoxification by eosinophils	Granular leukocytes in red marrow; before birth and for a few months after, some lymphocytes are formed in thymus gland; later, most lymphocytes and monocytes formed in lymph nodes and other lymphatic tissues	Not known definitely; probably some destroyed by phagocytosis
Platelets (thrombocytes)	150,000 to 350,000/mm³	Initiate blood clotting and hemostasis	Red marrow, lungs, and spleen	Unknown

Blood type

Anti-A serum

Anti-B serum

O

A

B

AB

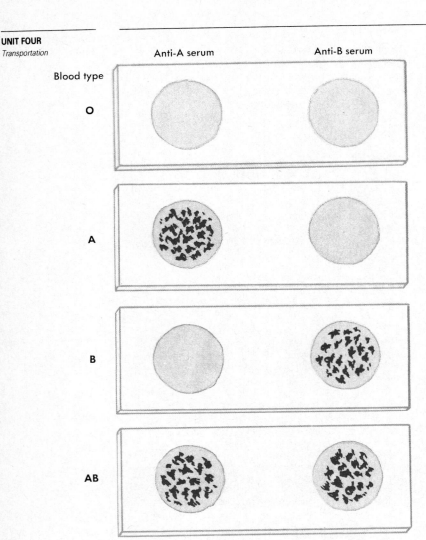

Figure 15-16 Appearance of agglutination or clumping tests used to type blood. Blood type O red cells do not clump when mixed with either anti-A or anti-B serum; blood type A red cells clump when mixed with anti-A serum; and blood type B red cells clump with anti-B serum. Red cells in type AB blood are agglutinated (clumped) by both anti-A and anti-B serums.

Blood plasma may or may not contain antibodies that can react with red blood cell antigens A, B, and Rh. An important principle about this is that plasma never contains antibodies against the antigens present on its own red blood cells—for obvious reasons. If it did, the antibody would react with the antigen and thereby destroy the red blood cells. But (and this is an equally important principle) plasma does contain antibodies against antigen A or antigen B if they are *not* present on its red blood cells. Applying these two principles: In type A blood, antigen A is present on its red blood cells; therefore its plasma contains no anti-A antibodies but does contain anti-B antibodies. In type B blood, antigen B is present on its red blood cells; therefore its plasma contains no anti-B antibodies but does contain anti-A antibodies (Figure 15-16). What antigens do you deduce are present on the red blood cells of type AB blood? What antibodies, if any, does its plasma contain?*

No blood normally contains anti-Rh antibodies. However, anti-Rh antibodies can appear in the blood of an Rh-negative person provided Rh-positive red blood cells have at some time entered the bloodstream. One way this can happen is by giving an Rh-negative person a transfusion of Rh-positive blood. In a short time, the person's body makes anti-Rh antibodies, and these remain in the blood. There is one other way in which Rh-positive red blood cells can enter the bloodstream of an Rh-negative individual—but this can happen only to a woman. If she becomes pregnant, and if her mate is Rh positive, and if the fetus, too, is Rh positive, some of the red blood cells of the fetus may find their way into her blood from the fetal blood capillaries (via the placenta). These Rh-positive red blood cells then stimulate her body to form anti-Rh antibodies. If these antibodies attack and destroy the red blood cells of the fetus, the condition is called **erythroblastosis fetalis.** Briefly, the only people who can ever have anti-Rh antibodies in their plasma are Rh-negative men or women who have been transfused with Rh-

Anti-A, anti-B, and anti-Rh antibodies are agglutinins. **Agglutinins** are chemicals that agglutinate cells, that is, make them stick together in clumps. The danger in giving a blood transfusion is that antibodies present in the plasma of the person receiving the transfusion (the recipient's plasma) may agglutinate the donor's red blood cells. If, for example, type B blood were used for transfusing a person who had type A blood, the anti-B antibodies in the recipient's blood would agglutinate the donor's type B red blood cells. These clumped cells are potentially lethal. They can plug vital small vessels and cause the recipient's death.

---

*Type AB blood—antigens A and B present on red blood cells; no anti-A nor anti-B antibodies in plasma.

ARTIFICIAL BLOOD

Transfusion of blood from one individual to another always involves some risk. For years scientists have been engaged in the search for an acceptable blood substitute that would eliminate many of the risks of transfusion and yet provide a critically ill person with one or more of the benefits of additional whole blood. One of the most important functions that transfused blood performs is additional oxygen transport to the body tissues; therefore one of the critical requirements for any artificial substitute is an ability to function in the same way. Extensive clinical trials of an artificial blood that will carry significant amounts of oxygen are currently in progress in the United States and Japan. The complex chemical **(fluosol)** is capable of delivering substantial amounts of oxygen to tissues and may eventually prove useful in treating serious anemias, carbon monoxide poisoning, and heart attacks.

whole blood is centrifuged to form plasma. This consists of a rapid whirling process that hurls the blood cells to the bottom of the centrifuge tube. A clear, straw-colored fluid—blood plasma—lies above the cells. Plasma consists of 90% water and 10% solutes. By far the largest quantity of these solutes are proteins; normally, they constitute about 6% to 8% of the plasma. Other solutes present in much smaller amounts in plasma are food substances (principally glucose, amino acids, and lipids), compounds formed by metabolism (e.g., urea, uric acid, creatinine, and lactic acid), respiratory gases (oxygen and carbon dioxide), and regulatory substances (hormones, enzymes, and certain other substances).

Some of the solutes present in blood plasma are true solutes or crystalloids. Others are colloids. **Crystalloids** are solute particles less than 1 nm in diameter (ions, glucose, and other small molecules). **Colloids** are solute particles from 1 to about 100 nm in diameter (for example, proteins of all types). Blood solutes also may be classified as **electrolytes** (molecules that ionize in solution) or **nonelectrolytes**—examples: inorganic salts and proteins are electrolytes; glucose and lipids are nonelectrolytes.

positive blood or Rh-negative women who have carried an Rh-positive fetus.

Type O blood is referred to as **universal donor** blood, a term that implies that it can safely be given to any recipient. This, however, is not true because the recipient's plasma may contain agglutinins other than anti-A and anti-B antibodies. For this reason the recipient's and the donor's blood—even if it is type O—should be cross-matched, that is, mixed and observed for agglutination of the donor's red blood cells.

Universal recipient (type AB) blood contains neither anti-A nor anti-B antibodies, so it cannot agglutinate type A or type B donor's red blood cells. This does not mean, however, that any type of donor blood may be safely given to an individual who has type AB blood without first cross matching. Other agglutinins may be present in the so-called universal recipient blood and clump unidentified antigens (agglutinogens) in the donor's blood.

BLOOD PLASMA

Plasma is the liquid part of blood—whole blood minus its cells, in other words. In the laboratory,

The proteins in blood plasma consist of three main kinds of compounds; albumins, globulins, and fibrinogen. Electrophoretic methods of measuring the amounts of these compounds indicate that 100 ml of plasma contains a total of approximately 6 to 8 gr of protein. Albumins constitute about 55% of this total, globulins about 38%, and fibrinogen about 7%.

Plasma proteins are crucially important substances. Fibrinogen, for instance, and an albumin named prothrombin play key roles in the blood-clotting mechanism. Globulins function as essential components of the immunity mechanism—circulating antibodies (immune bodies) are modified gamma globulins. All plasma proteins contribute to the maintenance of normal blood viscosity, blood osmotic pressure, and blood volume. Therefore plasma proteins play an essential part in maintaining normal circulation. Synthesis of plasma proteins takes place in liver cells. They form all kinds of plasma proteins except some of the gamma globulins—specifically, the circulating antibodies synthesized by plasma cells.

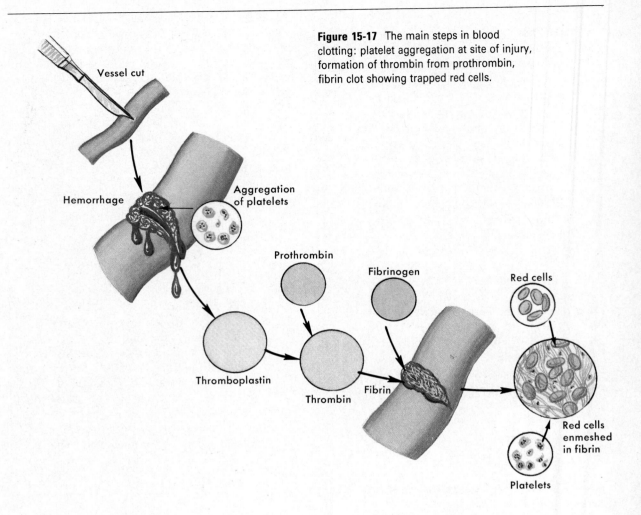

Figure 15-17 The main steps in blood clotting: platelet aggregation at site of injury, formation of thrombin from prothrombin, fibrin clot showing trapped red cells.

BLOOD COAGULATION

PURPOSE

The purpose of blood coagulation is obvious—to plug up ruptured vessels so as to stop bleeding and prevent loss of a vital body fluid.

MECHANISM

Because of the function of coagulation, the mechanism for producing it must be swift and sure when needed, such as when a vessel is cut or ruptured. Equally important, however, coagulation needs to be prevented from happening when it is not needed because clots can plug up vessels that must stay open if cells are to receive blood's life-sustaining cargo of oxygen.

What makes blood coagulate? Over a period of many years a host of investigators have searched for the answer to this question. They have tried to find out what events compose the coagulation mechanism and what sets it in operation. They have succeeded in gathering an abundance of relevant information. But still the questions about this complicated and important process outnumber the answers.

The blood-coagulation mechanism presumably consists of a series of chemical reactions that take place in a definite and rapid sequence.

The so-called "classic theory" of coagulation was advanced in 1905 and dominated research efforts in this complex area for almost half a century. It continues as the basis of our current understanding of coagulation. This theory assumed: (1) the interaction of four plasma components in the presence of calcium ions and (2) that the interaction between the components occurred in two steps. The plasma components are as follows:

1 Prothrombin
2 Thrombin
3 Fibrinogen
4 Fibrin

The following are the two basic coagulation steps (Figure 15-17):

Step I:

$$\text{Prothrombin} \xrightarrow{\frac{\text{Thromboplastin}}{\text{Ca}++}} \text{Thrombin}$$

Table 15-3 Coagulation factors—standard nomenclature and synonyms

Factor	Common synonym(s)
Factor I	Fibrinogen
Factor II	Prothrombin
Factor III	Thromboplastin Thrombokinase
Factor IV	Calcium
Factor V	Proaccelerin Labile factor
Factor VI (now obsolete)	None in use
Factor VII	Serum prothrombin conversion accelerator (SPCA)
Factor VIII	Antihemophilic globulin (AHG) Antihemophilic factor (AHF)
Factor IX	Plasma thromboplastin component (PTC), Christmas factor
Factor X	Stuart factor
Factor XI	Plasma thromboplastin antecedent (PTA)
Factor XII	Hageman factor
Factor XIII	Fibrin-stabilizing factor

Step II:

$$\text{Fibrinogen} \xrightarrow{\text{Thrombin}} \text{Fibrin}$$

It is interesting that these two basic reaction steps have been modified only by the action of a host of additional coagulation factors discovered in recent years (Table 15-3). In addition, the source of thromboplastin in step I of the current coagulation theory mechanism is now used to divide this step into *intrinsic* and *extrinsic systems*.

STEP I. In step I of the coagulation mechanism, prothrombin (a plasma protein) is converted into thrombin (a plasma enzyme) by the action of thromboplastin. In addition to thromboplastin, calcium ions are required for this reaction to occur. Thromboplastin may be released (1) from platelets or (2) from damaged tissues.

Intrinsic system. If thromboplastin is released from damaged platelets *(platelet thromboplastin)*, step I of the coagulation process proceeds via the *intrinsic system*. In this system plasma factors IV, V, VIII, IX, X, XI, and XII are also required to convert prothrombin to thrombin.

Extrinsic system. If thromboplastin is released from damaged tissues *(tissue thromboplastin)*, step I of coagulation proceeds via the *extrinsic system*. Plasma factors IV, V, VII, and X act with tissue thromboplastin to convert prothrombin to thrombin in this system.

STEP II. In step II of the coagulation mechanism, fibrinogen (a soluble plasma protein) is converted into strands of insoluble fibrin. Thrombin and plasma factors IV and XIII are required for completion of step II of the coagulation mechanism.

Unfortunately, initiation of the step I intrinsic system of coagulation often occurs when the normally very smooth endothelial lining of a blood vessel becomes "rough" because of a cut or the formation of cholesterol plaques (patchlike deposits in the vessel wall). Within a matter of seconds, clumps of platelets adhere to the injured area and disintegrate. This releases platelet thromboplastin, which then triggers the intrinsic system response in the first step of the coagulation process. As a result, prothrombin is converted to thrombin if calcium ions and the appropriate coagulation factors (IV, V, VIII, IX, X, XI, XII) are present. In step II thrombin accelerates the conversion of the soluble plasma protein fibrinogen to insoluble fibrin. Fibrin appears in blood as fine threads all tangled together. Blood cells catch in the entanglement, and because most of the cells are red blood cells, clotted blood has a red color. The pale yellowish liquid left after a clot forms is **blood serum.** How do you think serum differs from plasma? What is plasma? To check your answers, see Figure 15-18.

Liver cells synthesize both prothrombin and fibrinogen, as they do almost all other plasma proteins. For the liver to synthesize prothrombin at a normal rate, blood must contain an adequate amount of vitamin K. Vitamin K is absorbed into the blood from the intestine. Some foods contain this vitamin, but it is also synthesized in the intestine by certain bacteria (not present for a time in newborn infants). Because vitamin K is fat soluble, its absorption requires bile. If, therefore, the bile ducts become obstructed and bile cannot enter the intestine, a vitamin K deficiency develops. The liver cannot then produce prothrombin at its normal rate, and the blood's prothrombin concentration soon falls below normal. A prothrombin deficiency gives rise to a bleeding tendency. As a preoperative safeguard, therefore, patients with obstructive jaundice are generally given some kind of vitamin K preparation.

Factors that Oppose Clotting

Although blood clotting probably goes on continuously and concurrently with clot dissolution (fibrinolysis), several factors operate to oppose clot formation in intact vessels. Most important by far is the perfectly smooth surface of the normal endothelial lining of blood vessels. Platelets do not adhere to it;

Figure 15-18 Difference between blood plasma and blood serum. Plasma is whole blood minus cells. Serum is whole blood minus the clotting elements. Plasma is prepared by centrifuging blood. Serum is prepared by clotting blood.

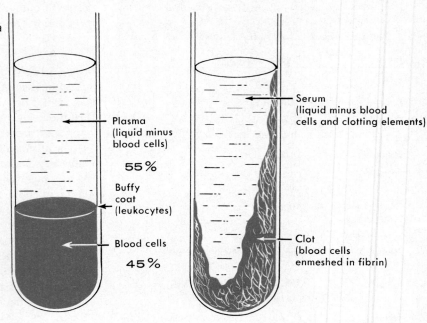

Plasma (liquid minus blood cells)

55%

Buffy coat (leukocytes)

Blood cells

45%

Serum (liquid minus blood cells and clotting elements)

Clot (blood cells enmeshed in fibrin)

consequently, they do not disintegrate and release platelet factors into the blood and, therefore, the blood-clotting mechanism does not begin in normal vessels. As an additional deterrent to clotting, blood contains certain substances called *antithrombins.* The name suggests their function—they oppose or inactivate thrombin. Thus antithrombins prevent thrombin from converting fibrinogen to fibrin. **Heparin,** a natural constituent of blood, acts as an antithrombin. It was first prepared from liver (hence its name), but other organs also contain heparin. Its normal concentration in blood is too low to have much effect in keeping blood fluid. However, injections of heparin are used to prevent clots from forming in vessels. Coumarin compounds impair the liver's utilization of vitamin K and thereby slow its synthesis of prothrombin and factors VII, IX, and X. Indirectly, therefore, coumarin compounds, such as bishydroxycoumarin, retard coagulation. Citrates keep donor blood from clotting before transfusion.

Factors that Hasten Clotting

Two conditions particularly favor thrombus formation: a rough spot in the endothelium (blood vessel lining) and abnormally slow blood flow. Atherosclerosis, for example, is associated with an increased tendency toward thrombosis because of endothelial rough spots in the form of plaques of accumulated cholesterol-lipid material. Immobility, on the other hand, may lead to thrombosis because blood flow slows down as movements decrease. Incidentally, this fact is one of the major reasons why physicians insist that bed patients must either move or be moved frequently. Presumably, sluggish blood flow

allows thromboplastin to accumulate sufficiently to reach a concentration adequate for clotting.

Once started, a clot tends to grow. Platelets enmeshed in the fibrin threads disintegrate, releasing more thromboplastin, which in turn causes more clotting, which enmeshes more platelets, and so on, in a vicious circle. Clot-retarding substances, available in recent years, have proved valuable for retarding this process.

Clot Dissolution

The physiological mechanism that dissolves clots is known as **fibrinolysis.** Newer evidence indicates that the two opposing processes of clot formation and fibrinolysis go on continuously. Dr. George Fulton of Boston University has presented one bit of dramatic evidence. He took micromovies that show tiny blood vessels rupturing under apparently normal circumstances and clots forming to plug them. Blood contains an enzyme, fibrinolysin, that catalyzes the hydrolysis of fibrin, causing it to dissolve. Many other factors, however, presumably also take part in clot dissolution, for instance, substances that activate profibrinolysin (inactive form of fibrinolysin). Streptokinase, an enzyme from certain streptococci, can act this way and so can cause clot dissolution and even hemorrhage.

Clinical Methods of Hastening Clotting

One way of treating excessive bleeding is to speed up the blood-clotting mechanism. The principle involved is apparent—to increase any of the substances essential for clotting. Application of this principle is accomplished in the following ways:

1 By applying a rough surface such as gauze, by applying heat, or by gently squeezing the tissues around a cut vessel. Each of these procedures causes more platelets to disintegrate and release more platelet factors. This, in turn, accelerates the first of the clotting reactions.

2 By applying purified thrombin (in the form of sprays or impregnated gelatin sponges that can be left in a wound). Which stage of the clotting mechanism does this accelerate?

3 By applying fibrin foam, films, and so on.

Outline Summary

BLOOD

A Primary function—transportation of various substances to and from body cells; exchange of materials between respiratory, digestive, and excretory organs and blood and between blood and cells

B Secondary functions—contributes to all bodily functions, for example
  1 Cellular metabolism
  2 Homeostasis of fluid volume
  3 Homeostasis of pH
  4 Homeostasis of temperature
  5 Defense against microorganisms

BLOOD VOLUME

A Measurement—by direct and indirect methods using radioisotopes

B Effect of body fat—blood volume per kilogram of body weight varies inversely with the amount of excess body fat

C Average volume—70 kg adult male averages 71 ml/per kg of blood—about 5,000 ml total

BLOOD CELLS

A Three main kinds—red blood cells (erythrocytes), white blood cells (leukocytes), and platelets (thrombocytes); (see Figure 15-1)

B Whole blood consists of approximately 55% fluid (blood plasma) and 45% blood cells

Erythrocytes

A Appearance, size, shape, and number—biconcave disks about 7 μm in diameter with large surface area relative to volume
  1 Biconcave shape permits "deformity" to occur without injury
    a Prostaglandin (PGE$_1$) increases deformability; PGE$_2$ decreases it
    b Numbers of RBCs—male: 5,500,000 per mm^3; female, 4,800,000 per mm^3
  2 Erythrocytes do not contain ribosomes, mitochondria, or other organelles

B Structure and functions—millions of molecules of hemoglobin inside each red blood cell makes possible red blood cell functions of oxygen and carbon dioxide transport
  1 Anemia—classification according to size and hemoglobin content

    a Size classification—normocytic, macrocytic, and microcytic
    b Classification by hemoglobin content—normochromic, hyperchromic, and hypochromic
    c Anisocytosis—abnormal variation in RBC size; arbitrarily graded 1+ to 4+
    d Poikilocytosis—abnormal variation in RBC *shape;* arbitrarily graded 1+ to 4+

C Erythrocyte abnormalities (Figure 15-3)
  a Crenated cell—caused by hypertonic solution and some anticoagulants
  b Target cell (codocyte) common in iron-deficiency anemia
  c Sickle cells—sickle cell anemia; contains abnormal (HbS) hemoglobin
  d Schizocytes—cell fragments caused by trauma

D Formation (erythropoiesis)—by myeloid tissue (red bone marrow)
  1 Stages of development of red blood cells (Figure 15-5)

E Destruction—by fragmentation in capillaries; reticuloendothelial cells phagocytose red blood cell fragments and break down hemoglobin to yield iron-containing pigment and bile pigments; bone marrow reuses most of iron for new red blood cell synthesis; liver excretes bile pigments; life span of red cells about 120 days according to studies made with radioactive isotopes

F Erythrocyte homeostatic mechanism—stimulus thought to be tissue hypoxia; (description of mechanism, see Figure 15-9)

G Functions (see Table 15-2)

Leukocytes

A Appearance and size—vary; some relatively large cells, for example, monocytes, and some small cells, for example, small lymphocytes; nuclei vary from spherical to S shaped to lobulated; cytoplasm of neutrophils, eosinophils, and basophils contains granules that take neutral, acid, and basic stains, respectively

B Functions—defense or protection, for example, by carrying on phagocytosis

C Formation—in myeloid tissue (red bone marrow) for granular leukocytes and probably monocytes; lymphocytes formed in lymphatic tissue, that is, mainly in lymph nodes and spleen
  1 Maturation of granular leucocytes (Figure 15-12)

Outline Summary—cont'd

2 Maturation of nongranular leucocytes (Figure 15-13)

D Destruction and life span—some destroyed by phagocytosis; life span unknown

E Numbers—about 5,000 to 10,000 leukocytes per mm^3 of blood; (differential count, see Table 15-1)

F Functions (see Table 15-2)

Platelets

A Appearance and size—platelets are small fragments of cells

B Functions—blood coagulation and hemostasis

C Prostaglandin (PGE_1) and thromboxane—role in hemostasis related to vasoconstriction and increase in platelet aggregation at site of injury

D Formation and life span—formed in red bone marrow, lungs, and spleen by fragmentation of large cells (megakaryocytes); life span about 10 days

E Stages in platelet formation (Figure 15-14)

F Functions (see Table 15-2)

BLOOD TYPES (OR BLOOD GROUPS)

A Names—indicate type of antigen agglutinogen on red blood cell membranes (Figure 15-16)

B Every person's blood belongs to one of the four AB blood groups (type A, type B, type AB, or type O) and, in addition, is either Rh positive or Rh negative

C Plasma does not normally contain antibodies against antigens present on its red blood cells but does normally contain antibodies against A or B antigens not present on its red blood cells

D No blood normally contains anti-Rh antibodies; anti-Rh antibodies appear in blood only of an Rh-negative person and only after Rh-positive red blood cells have entered the bloodstream, for example, by transfusion or by carrying an Rh-positive fetus

E Universal donor—type O; universal recipient—type AB, but cross matching should be done before use for safety

BLOOD PLASMA

A Liquid part of blood or whole blood minus its cells; constitutes about 55% of total blood volume

B Composition
  1 About 90% water and 10% solutes
  2 Most solutes crystalloids but some colloids
  3 Most solutes electrolytes but some nonelectrolytes
  4 Solutes include foods, wastes, gases, hormones, enzymes, vitamins, and antibodies and other proteins

C Plasma proteins
  1 Types—albumins, globulins, and fibrinogen

2 Total plasma proteins—6 to 8 gm per 100 ml of plasma

3 Albumins—about 55% of total plasma proteins

4 Globulins—about 38% of total plasma proteins

5 Fibrinogen—about 7% of total plasma proteins

6 Functions—contribute to blood viscosity, osmotic pressure, and volume; prothrombin (an albumin) and fibrinogen play key roles in blood clotting; modified gamma globulins are circulating antibodies, essential for immunity

7 All plasma proteins except circulating antibodies are synthesized in liver cells

BLOOD COAGULATION

A Purpose—to plug up ruptured vessels and thus prevent fatal hemorrhage (Figure 15-17)

B Mechanism—complicated process, still incompletely understood
  1 Triggered by appearance of rough spot in blood vessel lining
  2 Clumps of platelets adhere to rough spot; they release "platelet factors" that initiate a series of rapidly occurring chemical reactions
  3 Last two clotting reactions convert prothrombin to thrombin, an enzyme that catalyzes conversion of soluble fibrinogen to insoluble fibrin
  4 Clot consists of tangled mass of fibrin threads and blood cells; blood serum is fluid left after clot forms

C Factors that oppose clotting
  1 Smoothness of endothelium that lines blood vessels prevents platelets' adherence and consequent disintegration; some platelet disintegration continuously despite this preventive
  2 Blood normally contains certain anticoagulants, for example, antithrombins, substances that inactivate thrombin so that it cannot catalyze fibrin formation

D Factors that hasten clotting
  1 Endothelial "rough" spots, for example, cholesterol-lipid plaques in atherosclerosis
  2 Sluggish blood flow

E Pharmaceutical preparations that retard clotting—commercial heparin, bishydroxycoumarin (Dicumarol), citrates (for transfusion blood)

F Clot dissolution—process called *fibrinolysis;* presumably this and opposing process (clot formation) go on all the time

G Clinical methods of hastening clotting
  1 Apply rough surfaces to wound to stimulate platelets and tissues to liberate more thromboplastin
  2 Apply purified thrombin
  3 Apply fibrin foam, film, etc.

Review Questions

1 Compare different kinds of blood cells as to appearance and size; functions; formation, destruction, and life span; number per cubic millimeter of blood.
2 What is the effect of body fat on blood volume?
3 Explain the difference between direct and indirect methods used to measure blood volume.
4 Explain the effect of prostaglandins PGE_1 and PGE_2 on red blood cell "deformability."
5 Classify the various types of anemia according to size of red blood cell and hemoglobin content.
6 Describe the role of platelets in hemostasis and blood clotting.
7 A person suffering from a severe parasitic infestation, such as hookworms, would have an elevation in what leukocyte cell type?
8 "Graft-rejection cells" is a nickname for which kind of blood cells? Why?
9 What type of cells secrete circulating antibodies?
10 What is the function of circulating antibodies? Of cellular antibodies?
11 Since plasma cell synthesize and secrete circulating antibodies, what organelles would you expect to be prominent or abundant in plasma cells?
12 Suppose your doctor has told you that you have a lymphocyte count of 2,000 per mm^3. Would you think this was normal or abnormal?
13 Why might a surgeon give antilymphocyte serum to a patient who had had a kidney transplant?
14 Explain what the term *hematocrit* means. Is it normally less than 50? More than 55?
15 What triggers blood clotting?
16 Write two equations to show the basic chemical reactions that produce a blood clot.
17 Why and how does a vitamin K deficiency affect blood clotting?
18 Explain some principles and methods by which blood clotting may be hastened.
19 Explain what these terms mean: *type AB blood, Rh-negative blood.*
20 Suppose you have type B blood. Which two of the following kinds of blood might be used for transfusions for you: type A? type B? type AB? type O?
21 Which of the following infants would be most likely to have erythroblastosis fetalis: fourth child of Rh-positive mother and Rh-negative father? first child of Rh-negative mother and Rh-positive father? second child of Rh-negative mother and Rh-positive father? Explain your reasoning.
22 The cells of what organ synthesize most plasma proteins? What is the normal plasma protein concentration?
23 What are some of the functions served by plasma proteins?

16 ANATOMY OF THE CARDIOVASCULAR SYSTEM

OBJECTIVES

After you have completed this chapter, you should be able to:

1 List the primary organs of the cardiovascular system and relate each organ or group of organs to the movement and/or the direction of blood flow in the system.

2 Discuss the location, size, and position of the heart in the thoracic cavity.

3 Describe the structure of the pericardium, the function of each pericardial layer, the pericardial space, and the pericardial fluid.

4 List and discuss the three layers of the heart wall, the heart cavities, and the valves.

5 Trace blood through the heart and its coronary blood vessels.

6 List the anatomical components of the heart conduction system.

7 List, locate, and compare the primary coats or layers of tissue found in major arteries and veins.

8 Correlate structure of arteries, arterioles, veins, venules, and capillaries with their function.

9 List anatomical components of microcirculation and discuss the reservoir function of veins.

10 Identify the four main divisions of the aorta and list the major blood vessels that originate as branches from each aortic segment.

11 Discuss the functional significance of the portal circulation and identify the tributaries of the portal vein.

12 List and discuss the function of the six structures characteristic of the fetal circulation.

13 Discuss changes that occur in the vascular system at birth.

14 Trace a drop of blood from a superficial vein in the lower extremity to and through the heart and lungs and into the middle cerebral artery of the brain.

15 Briefly discuss the following abnormalities of the cardiovascular system: atherosclerosis, aneurysm, varicose veins, and phlebitis.

The cardiovascular system is sometimes called the *blood-vascular* or simply the *circulatory system*. It consists of the heart, which is a muscular pumping device, and a closed system of vessels called *arteries*, *veins*, and *capillaries*. As the name implies, blood contained in the circulatory system is pumped by the heart around a closed circle or circuit of vessels as it passes again and again through the various "circulations" of the body (discussed on p. 467).

As in the adult, survival of the developing embryo depends on the circulation of blood to maintain homeostasis and a favorable cellular environment. In response to this need, the cardiovascular system makes its appearance early in development and reaches a functional state long before any other major organ system. Incredible as it seems, the primitive heart begins to beat regularly early in the fourth week following fertilization!

HEART

LOCATION, SIZE, AND POSITION

The human heart is a four-chambered muscular organ, shaped and sized roughly like a man's closed fist. It lies in the mediastinum just behind the body of the sternum between the points of attachment of the second through the sixth ribs. Approximately two thirds of its mass is to the left of the midline of the body and one third to the right.

Posteriorly the heart rests on the bodies of the fifth to the eighth thoracic vertebrae. Because of its placement between the sternum in front and the bodies of the thoracic vertebrae behind, it can be compressed by application of pressure to the lower portion of the body of the sternum using the heel of the hand. Rhythmic compression of the heart in this way can maintain blood flow in cases of cardiac arrest and, if combined with effective artificial respiration, the resulting procedure, called **cardiopulmonary resuscitation (CPR),** can be lifesaving.

KEY TERMS

Anastomosis (ah-nas-tuh-MOW-sis)

Arteriole (ar-TEER-e-ole)

Artery (AR-tur-e)

Atria (A-tre-ah)

Capillary (KAP-i-lair-e)

Chordae tendineae (KOR-de TEN-di-nee)

Endocardium (en-do-KAR-de-um)

Epicardium (ep-i-KAR-de-um)

Myocardium (mi-o-KAR-de-um)

Pericardium (per-i-KAR-de-um)

Portal (POR-tal)

Pulmonary (PUL-mo-ner-e)

Systemic (sis-TEM-ik)

Vein (vain)

Ventricle (VEN-tri-kl)

Venule (VEN-ul)

Figure 16-1 Anatomical relationship of the heart to other structures in the thoracic cavity.

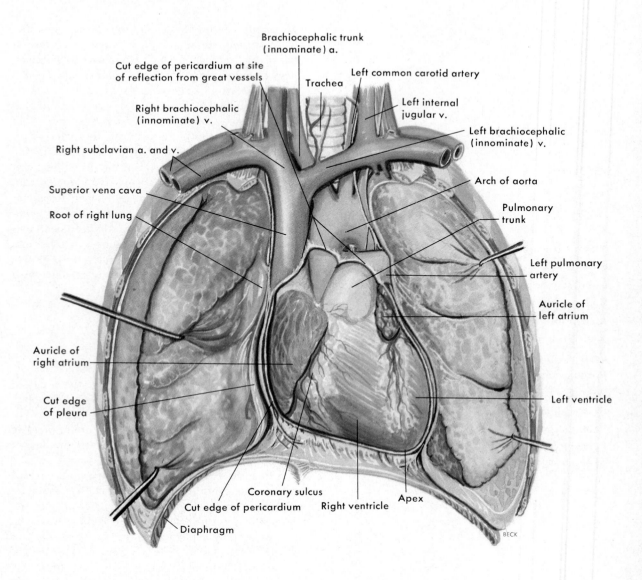

The anatomical relationship of the heart to other structures in the thoracic cavity is shown in Figure 16-1.

The lower border of the heart, which forms a blunt point known as the *apex*, lies on the diaphragm, pointing toward the left. To count the apical beat, one must place a stethoscope directly over the apex, that is, in the space between the fifth and sixth ribs (fifth intercostal space) on a line with the midpoint of the left clavicle.

The upper border of the heart, that is, its base, lies just below the second rib. The boundaries, which, of course, indicate its size, have considerable clinical importance, since a marked increase in heart size accompanies certain types of heart disease. Therefore when diagnosing heart disorders, the physician charts the boundaries of the heart. The "normal" boundaries of the heart are, however, influenced by such factors as age, body build, and state of contraction.

Figure 16-2 Posterior view of the heart and great vessels.

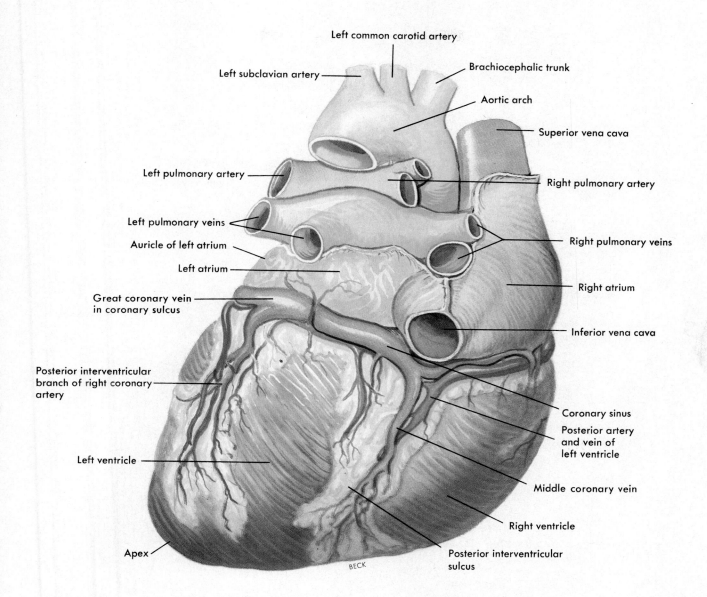

At birth the heart is said to be transverse in type and appears large in proportion to the diameter of the chest cavity. In the infant, it is $1/130$ of the total body weight compared to about $1/300$ in the adult. Between puberty and 25 years of age the heart attains its adult shape and weight—about 310 gm is average for the male and 225 gm for the female.

In the adult the shape of the heart tends to resemble that of the chest. In tall, thin individuals the heart is frequently described as elongated, whereas in short, stocky individuals it has greater width and is described as transverse. In individuals of average height and weight it is neither long nor transverse but somewhat oblique and intermediate between the two. Its approximate dimensions are length, 12 cm; width, 9 cm; and depth, 6 cm. Figure 16-2 shows the heart and great vessels in a posterior view and Figure 16-3 in an anterior view.

Figure 16-3 A, Anterior view of the heart and great vessels.

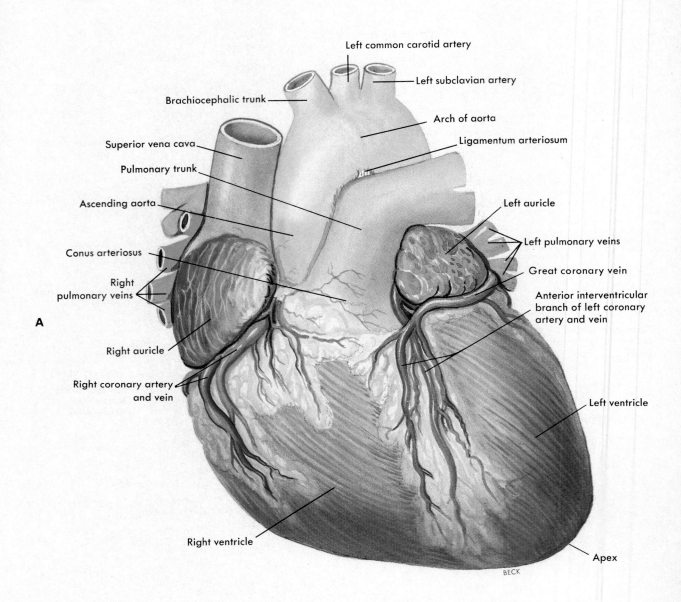

Figure 16-3—cont'd B, Anterior view of the heart and great vessels. The pericardium still covers the blood vessels at the top, which are shown in detail in **A**.

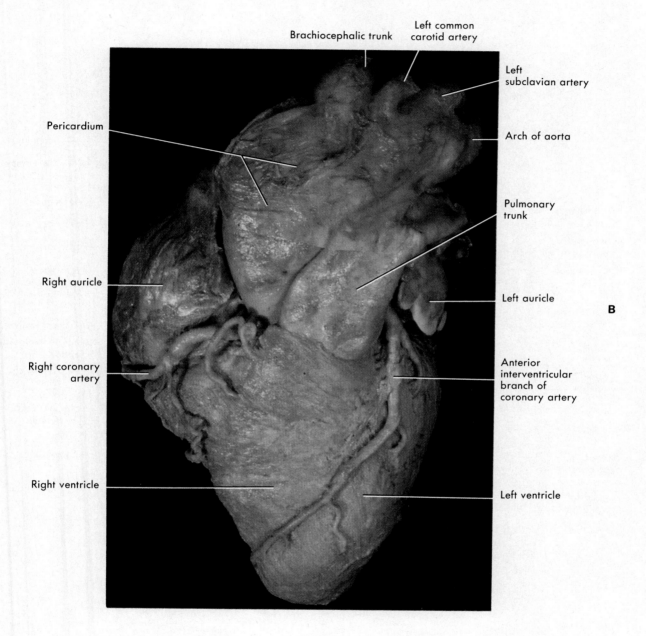

Brachiocephalic trunk

Left common carotid artery

Left subclavian artery

Arch of aorta

Pericardium

Pulmonary trunk

Right auricle

Left auricle

B

Right coronary artery

Anterior interventricular branch of coronary artery

Right ventricle

Left ventricle

Figure 16-4 Relation of heart and great vessels to anterior wall of thorax (sternum, costal cartilages, and clavicles). Valves of the heart are projected on the anterior thoracic wall.

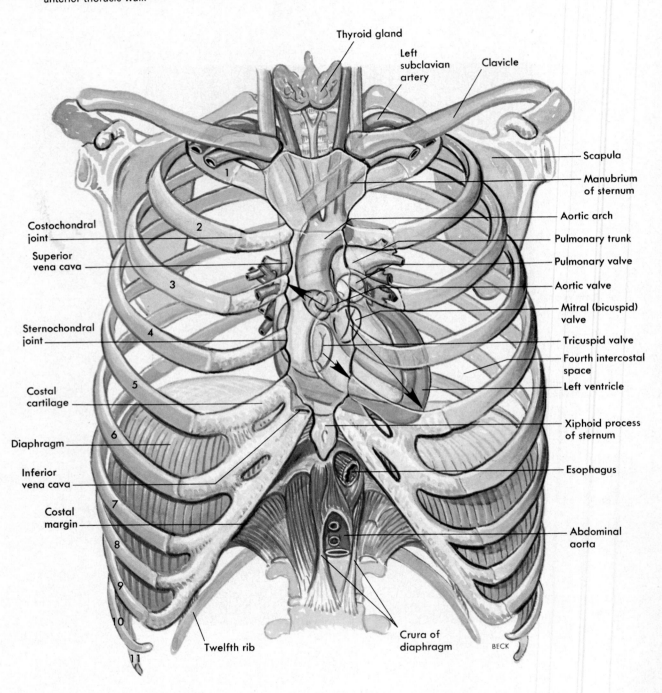

Thyroid gland

Left subclavian artery

Clavicle

Scapula

Manubrium of sternum

Aortic arch

Pulmonary trunk

Pulmonary valve

Aortic valve

Mitral (bicuspid) valve

Tricuspid valve

Fourth intercostal space

Left ventricle

Xiphoid process of sternum

Esophagus

Abdominal aorta

Costochondral joint

Superior vena cava

Sternochondral joint

Costal cartilage

Diaphragm

Inferior vena cava

Costal margin

Twelfth rib

Crura of diaphragm

BECK

Figure 16-5 Section of the heart wall showing the components of the outer pericardium (heart sac), muscle layer (myocardium), and inner lining (endocardium).

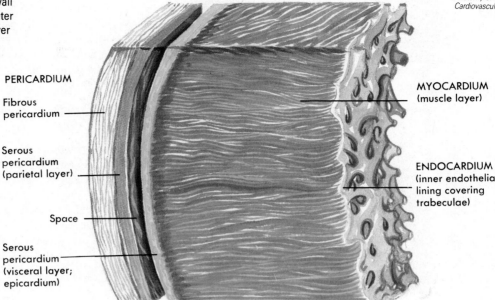

PERICARDIUM

Fibrous pericardium

Serous pericardium (parietal layer)

Space

Serous pericardium (visceral layer; epicardium)

MYOCARDIUM (muscle layer)

ENDOCARDIUM (inner endothelial lining covering trabeculae)

SURFACE PROJECTION

Figure 16-4 shows several surface relationships between the adult heart and thoracic cage. It is important to remember, however, that considerable variation within the normal range makes a precise "surface projection" outline of the boundaries or valves of the heart on the chest wall difficult.

COVERING

Structure

The heart has its own special covering, a loose-fitting inextensible sac called the **pericardium.** The pericardium consists of two parts: a fibrous portion and a serous portion (Figure 16-5). The sac itself is made of tough white fibrous tissue but is lined with smooth, moist serous membrane—the parietal layer of the serous pericardium. The same kind of membrane covers the entire outer surface of the heart. This covering layer is known as the visceral layer of the serous pericardium or as the **epicardium.** The fibrous sac attaches to the large blood vessels emerging from the top of the heart but not to the heart itself. Therefore it fits loosely around the heart, with a slight space between the visceral layer adhering to the heart and the parietal layer adhering to the inside of the fibrous sac. This space is called the **pericardial space.** It contains a few drops of lubricating fluid secreted by the serous membrane and called **pericardial fluid.**

The structure of the pericardium can be summarized in outline form as follows:

1 **Fibrous pericardium**—Tough, loose-fitting, and inelastic sac around the heart
2 **Serous pericardium**—Consisting of two layers
   a *Parietal layer*—Lining inside of the fibrous pericardium
   b *Visceral layer (epicardium)*—Adhering to the outside of the heart; between visceral and parietal layers is a potential space, the pericardial space, which contains a few drops of pericardial fluid

If the pericardium becomes inflamed (**pericarditis**) and too much pericardial fluid accumulates, or if fibrin or pus develops in the pericardial space, the visceral and parietal layers may stick together. In addition, the inability of the fibrous pericardium to stretch as fluid accumulates in the pericardial space will result in a rapid increase in pressure around the heart during acute episodes of pericarditis. If the resulting compression of the heart, called **cardiac tamponade,** seriously hampers pumping efficiency, it sometimes becomes necessary to remove the fibrous pericardium with its lining of parietal serous membrane in order for the heart to continue functioning.

Figure 16-6 Cardiac muscle, longitudinal section. High-power micrograph showing striations and branching of cardiac muscle fibers. Note the presence of numerous intercalated disks or specialized junctions between the adjacent cells, extending at right angles to the muscle fibers. (× 140.)

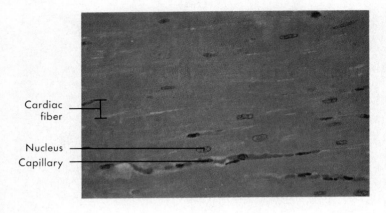

Cardiac fiber

Nucleus
Capillary

Figure 16-7 Frontal section of the heart showing the four chambers, valves, openings, and major vessels. Arrows indicate direction of blood flow.

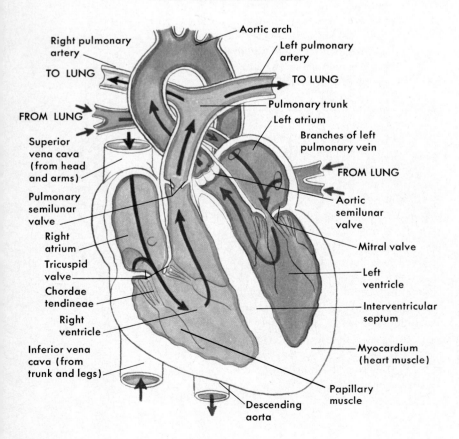

Right pulmonary artery

Aortic arch

Left pulmonary artery

TO LUNG

TO LUNG

FROM LUNG

Pulmonary trunk

Left atrium

Superior vena cava (from head and arms)

Branches of left pulmonary vein

FROM LUNG

Pulmonary semilunar valve

Aortic semilunar valve

Right atrium

Mitral valve

Tricuspid valve

Left ventricle

Chordae tendineae

Interventricular septum

Right ventricle

Myocardium (heart muscle)

Inferior vena cava (from trunk and legs)

Papillary muscle

Descending aorta

Function

The fibrous pericardial sac with its smooth, well-lubricated lining provides protection against friction. The heart moves easily in this loose-fitting jacket with no danger of irritation from friction between the two surfaces, as long as the serous pericardium remains normal.

STRUCTURE

Wall

Three distinct layers of tissue make up the heart wall (Figure 16-5) in both the atria and the ventricles.

1 The outer layer of the heart wall is called the **epicardium**—the visceral layer of the serous pericardium already described.

2 The bulk of the heart wall is the thick, contractile, middle layer of specially constructed and arranged cardiac muscle cells called the **myocardium.** The minute structure of cardiac muscle has been described in Chapter 4. Look at the electron micrograph of cardiac muscle (Figure 16-6) and note the large number of mitochondria and the arrangement of the contractile components. As you can see, many of the same microscopic structural parts found in skeletal muscle cells and described in Chapter 9 can also be identified in cardiac muscle.

   Figure 16-6 is a high-power (× 140) micrograph that shows in greater detail the minute structure of cardiac muscle described in Chapter 4. Note the appearance of cross striations and the characteristic branching of fibers typical of cardiac muscle. The specialized junctions between adjacent cells, the **intercalated disks,** are readily apparent in Figure 16-6. On its inner surface the myocardium is raised into ridgelike projections, the papillary muscles (Figures 16-7 and 16-8).

3 The lining of the interior of the myocardial wall is a delicate layer of endothelial tissue known as the **endocardium.**

Cavities

The interior of the heart is divided into four chambers, two upper and two lower. The upper cavities are named **atria*** and the lower ones **ventricles** (Figure 16-7). Of these, the ventricles are considerably larger and thicker walled than the atria because they carry a heavier pumping burden than the atria. Also, the left ventricle has thicker walls than the

---

*The atria are sometimes called *auricles*. Strictly speaking, the latter term means the earlike flaps protruding from the atria. The terms atria and auricle should not be used synonymously.

right for the same reason. It has to pump blood through all the vessels of the body, except those to and from the lungs, whereas the right ventricle sends blood only through the lungs.

Valves and Openings

The heart valves are mechanical devices that permit the flow of blood in one direction only. Four sets of valves are of importance to the normal functioning of the heart (Figures 16-7 to 16-9). Two of these, the **cuspid (atrioventricular) valves,** are located in the heart, guarding the openings between the atria and ventricles (atrioventricular orifices). The other two, the **semilunar valves,** are located inside the pulmonary artery and the great aorta just as they arise from the right and left ventricles, respectively.

The cuspid valve guarding the right atrioventricular orifice consists of three flaps of endocardium anchored to the papillary muscles of the right ventricle by several cordlike structures called **chordae tendineae.** Because this valve has three flaps, it is appropriately named the **tricuspid valve.** The valve that guards the left atrioventricular orifice is similar in structure to the tricuspid except that it has only two flaps and is, therefore, called the **bicuspid** or, more commonly, the **mitral valve.** (An easy way to remember which valve is on the right and which is on the left is this: the names whose first letters come nearest together in the alphabet go together—thus, L and M for *l*eft side, *m*itral valve, and R and T for *r*ight side, *t*ricuspid valve.)

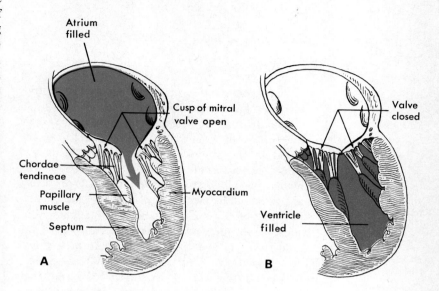

Figure 16-8 Action of the cuspid (atrioventricular) valves. **A,** When the valves are open, blood passes freely from the atria to the ventricles. **B,** Filling of the ventricles closes the valves and prevents a backflow of blood into the atria when the ventricles contract.

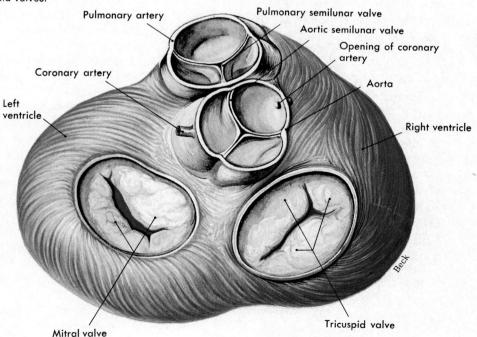

Figure 16-9 Valves of the heart viewed from above. The atria are removed to show the mitral and tricuspid valves.

Endothelial tissue resembles simple squamous epithelial tissue in that it consists of a single layer of flat cells. It differs from epithelial tissue in that it arises from the mesoderm layer of the embryo, whereas epithelial tissue arises from the ectoderm.

The construction of both cuspid valves allows blood to flow from the atria into the ventricles but prevents it from flowing back up into the atria from the ventricles. Ventricular contraction forces the blood in the ventricles hard against the cuspid valves, closing them and thereby ensuring the movement of the blood upward into the pulmonary artery and aorta as the ventricles contract (Figure 16-8).

The **semilunar valves** consist of half-moon–shaped flaps growing out from the lining of the pulmonary artery and great aorta. When these valves are closed, as in Figure 16-9, blood fills the spaces between the flaps and the vessel wall. Each flap then looks like a tiny filled bucket. Inflowing blood smooths the flaps against the blood vessel walls, collapsing the buckets and thereby opening the valves. Closure of the semilunar valves, as of the cuspid valves, simultaneously prevents backflow and ensures forward flow of blood in places where

Any one of the four valves may lose its ability to close tightly. Such a condition is known as **valvular insufficiency,** or because it permits blood to "leak" back into the part of the heart from which it came, "leakage of the heart." Leakage of blood through a diseased or malformed heart valve often creates turbulence in blood flow and abnormal sounds called **murmurs.** Identification of murmurs may contribute important information to the recognition and diagnosis of heart disease. **Mitral stenosis** is an abnormality in which the left atrioventricular orifice becomes narrowed by scar tissue that forms as a result of disease. This hinders the passage of blood from the left atrium to the left ventricle and leads to circulatory failure. But fortunately the marvel of open-heart surgery has made possible the correction of many valvular defects.

Table 16-1 Coronary arteries

Right coronary artery	Left coronary artery
Divides into two main branches: 　Posterior descending artery—sends branches to both ventricles 　Marginal artery—sends branches to right ventricle and right atrium	Divides into two main branches: 　Anterior descending artery—sends branches to ventricles 　Circumflex artery—sends branches to left ventricle and left atrium

there would otherwise be considerable backflow. Whereas the cuspid valves prevent blood from flowing back up into the atria from the ventricles, the semilunar valves prevent it from flowing back down into the ventricles from the aorta and pulmonary artery.

Blood Supply

Myocardial cells receive blood by way of two small vessels, the right and left coronary arteries. Since the openings into these vitally important vessels lie behind flaps of the aortic semilunar valve, they come off of the aorta at its very beginning and are its first branches. Both right and left coronary arteries have two main branches, as shown in Table 16-1.

More than a half million Americans die every year from coronary disease and another 3.5 million or more are estimated to suffer some degree of incapacitation from this great killer. Knowledge about the distribution of coronary artery branches, therefore, has the utmost practical importance. Here, then, are some principles about the heart's own blood supply that seem worth noting:

1 Both ventricles receive their blood supply from branches of both the right and left coronary arteries.

2 Each atrium, in constrast, receives blood only from a small branch of the corresponding coronary artery (Table 16-1).

3 The most abundant blood supply of all goes to the myocardium of the left ventricle—an appropriate amount, since the left ventricle does the most work and so needs the most oxygen and nutrients delivered to it.

4 The right coronary artery is dominant in about 50% of all hearts and the left coronary artery in about 20%, and in about 30% neither right nor left coronary artery dominates.

Another fact about the heart's own blood supply—one of life-and-death importance—is that

451

CHAPTER 16
*Anatomy of the
Cardiovascular System*

SOFT WATER AND CARDIOVASCULAR DISEASE

There is growing evidence to suggest that a causal relationship exists between water softness and risk of cardiovascular disease. The "hardness" of water is based largely on calcium and mangesium levels. As the concentrations of these two minerals increase, the water becomes harder. Naturally occurring soft water, especially water with very low levels of calcium, is associated with greater risk of heart disease. The relationship between water softness and heart disease risk apparently increases as the hardness of water consumed decreases. For reasons yet unexplained, artificial water softening using salt does not appear to contribute to risk.

only a few anastomoses exist between the larger branches of the coronary arteries. An **anastomosis** consists of one or more branches from the proximal part of an artery to a more distal part of itself or of another artery. Thus anastomoses provide detours that arterial blood can travel if the main route becomes obstructed. In short, they provide collateral circulation to a part. This explains why the scarcity of anastomoses between larger coronary arteries looms so large as a threat to life. If, for example, a blood clot plugs one of the larger coronary artery branches, as it frequently does in coronary thrombosis or embolism, too little or no blood can reach some of the heart muscle cells. They become ischemic, in other words. Deprived of oxygen, they release too little energy for their own survival. **Myocardial infarction** (death of ischemic heart muscle cells) soon results. There is another anatomical fact, however, that brightens the picture somewhat—many anastomoses do exist between very small arterial vessels in the heart, and, given time, new ones develop and provide collateral circulation to ischemic areas. In recent years, several surgical procedures have been devised to aid this process.

After blood has passed through capillary beds in the myocardium, it enters a series of coronary veins before draining into the right atrium through a common venous channel called the *coronary sinus*. Several veins that collect blood from a small area of the right ventricle do not end in the coronary sinus but instead drain directly into the right atrium. As a rule of thumb, the coronary veins (see Figures 16-2 and 16-3) follow a course that closely parallels that of the coronary arteries.

THROMBOLYSIS THERAPY

Obstruction of a coronary artery by a blood clot deprives an area of the heart wall of oxygen and other nutrients. As a result, cardiac muscle cells are damaged or die and death may result. There is evidence to suggest that if blocked arteries can be reopened and blood flow restored within 2 hours of the initial obstruction, damage can be minimized and mortality decreased.

Treatment designed to break up a clot in the coronary artery and restore blood flow is called **thrombolysis therapy.** The reopening of a blocked vessel is called **recanalization.** Treatment that involves infusion of an enzyme directly into the occluded coronary artery so it can digest a blood clot has been shown to be potentially beneficial.

Until recently thrombolytic therapy involved injection of the enzyme **streptokinase** directly into the occluded artery. This enzyme acts to dissolve the structural component of clots—fibrin. Unfortunately, it also attacks fibrinogen, which is a precursor of fibrin. Decreased fibrinogen levels greatly increases the possibility of severe bleeding and represents a serious side effect of treatment.

A new and very specific fibrinolytic, or clot-dissolving agent, has now been introduced that acts only on fibrin and not on fibrinogen. It is a complex or conjugate consisting of an antibody that binds specifically to fibrin and the enzyme **urokinase.** Since the antibody is specific for fibrin it serves to bind the enzyme directly to the blood clot, thus decreasing the likelihood of hemorrhage caused by destruction of fibrinogen. If the clot can be dissolved so that reperfusion of affected heart muscle cells occurs quickly, the severity and permanence of myocardial damage can be lessened.

Conduction System

Four structures—the sinoatrial node, atrioventricular node, atrioventricular bundle, and Purkinje fibers—compose the conduction system of the heart. Each of these structures consists of cardiac muscle modified enough in structure to differ in function from ordinary cardiac muscle. The main specialty of ordinary cardiac muscle is contraction. In this, it is like all muscle, and like all muscle, ordinary cardiac muscle can also conduct impulses. But conduction alone is the specialty of the modified cardiac muscle that composes the conduction system structures.

Sinoatrial Node

The sinoatrial node (SA node, Keith-Flack node, or pacemaker) consists of hundreds of cells located in the right atrial wall near the opening of the superior vena cava (see Figure 17-1, p. 472).

Atrioventricular Node

The atrioventricular node (AV node or node of Tawara), a small mass of special cardiac muscle tissue, lies in the right atrium along the lower part of the interatrial septum.

Atrioventricular Bundle and Purkinje Fibers

The atrioventricular bundle (AV bundle or bundle of His) is a bundle of special cardiac muscle fibers that originate in the AV node and extend by two branches down the two sides of the interventricular septum. From there, they continue as the Purkinje fibers. The latter extend out to the papillary muscles and lateral walls of the ventricles. The functioning of the conduction system of the heart will be discussed in Chapter 17.

Nerve supply

Both divisions of the autonomic nervous system send fibers to the heart. Sympathetic fibers (contained in the middle, superior, and inferior cardiac nerves) and parasympathetic fibers (in branches of the vagus nerve) combine to form **cardiac plexuses** located close to the arch of the aorta. From the cardiac plexuses, fibers accompany the right and left coronary arteries to enter the heart. Here most of the fibers terminate in the SA node, but some end in the AV node and in the atrial myocardium. Sympathetic nerves to the heart are also called *accelerator nerves*. Vagus fibers to the heart serve as inhibitory or depressor nerves.

BLOOD VESSELS

KINDS

There are three kinds of blood vessels: arteries, veins, and capillaries. By definition an **artery** is a vessel that carries blood away from the heart. After birth all arteries except the pulmonary artery and its branches carry oxygenated blood. Small arteries are called **arterioles.**

TRIGGER FOR BLOOD VESSEL GROWTH

Over 15 years ago scientists advanced a theory that certain types of solid cancerous tumors secrete a potent chemical that stimulates new growth and extension of blood vessels. It was argued that such a factor would be necessary to foster the remarkable growth of new vessels required to provide an adequate blood supply for rapidly growing tumors. The unknown factor was named **tumor angiogenesis factor (TAF).**

In 1985, after 15 years of intensive work, researchers at Harvard Medical School isolated and characterized the elusive protein that triggers and promotes the growth and extension of blood vessels. The substance has been named **angiogenin.** It is a coiled sequence of 123 amino acids and has a molecular weight of 14,400.

The isolation and purification of angiogenin provides scientists with a powerful tool to study the basic mechanisms responsible for inducing organ development. In addition, the chemical has great potential for therapeutic or clinical application. It is extremely potent in eliciting new blood vessel growth.

Angiogenin has been suggested as a treatment to increase blood supply in cases of stroke, heart disease, or extensive wound healing. On the other hand, by inhibiting its action, the blood supply to solid tumors of the colon, breast, and lung might be reduced and the disease arrested. Because it is secreted by many malignant tumors, its detection and measurement in the blood may also serve as a basis for detecting cancers or monitoring their growth.

A **vein,** on the other hand, is a vessel that carries blood toward the heart. All of the veins except the pulmonary veins contain deoxygenated blood. Small veins are called **venules.** Both arteries and veins are macroscopic structures.

Capillaries are microscopic vessels that carry blood from small arteries to small veins, that is, from arterioles to venules. They represented the "missing link" in the proof of circulation for many years—from the time William Harvey first declared that blood circulated from the heart through arteries to veins and back to the heart until the time that microscopes made it possible to find these connecting vessels between arteries and veins. Many people rejected Harvey's theory of circulation on the basis that there was no possible way for blood to get from arteries to veins. The discovery of the capillaries formed the final proof that the blood actually does circulate from the heart into arteries to arterioles, to capillaries, to venules, to veins, and back to the heart.

STRUCTURE

Consult Figures 16-10 and 16-11 and Table 16-2 for the structure of the blood vessels.

The three coats or layers of the blood vessels are considered homologous with the layers of the heart wall—the outer **tunica adventitia** with the epicardium, the **tunica media** with the myocardium, and the **tunica intima** with the endocardium.

FUNCTIONS

The capillaries, although seemingly the most insignificant of the three kinds of blood vessels because of their diminutive size, nevertheless are the most important vessels functionally. Since the prime function of blood is to transport essential materials to and from the cells and since the actual delivery and collection of these substances take place in the capillaries, the capillaries must be regarded as the most important blood vessels. Arteries serve merely as "distributors," carrying the blood to the arterioles. Arterioles, too, serve as distributors, carrying blood from arteries to capillaries. But, in addition, arterioles perform another function, one that is of great importance for maintaining normal blood pressure and circulation. They serve as resistance vessels. Note in Figure 16-11 that partial cuffs of smooth muscle called **precapillary sphincters** are located at the points of origin of two capillaries from an arteriole. The importance of precapillary sphincters in regulating resistance and controlling blood flow will be discussed in Chapter 17.

Table 16-2 Structure of blood vessels

Arteries	Veins	Capillaries
COATS		
Outer coat (tunica adventitia or externa) of white fibrous tissue; causes artery to stand open instead of collapsing when cut (see Figure 16-10)	Same three coats but thinner and fewer elastic fibers; veins collapse when cut; semilunar valves present at intervals	Only lining coat present; therefore walls only one cell thick
Muscle coat (tunica media) of smooth muscle, elastic, and some white fibrous tissues; this coat permits constriction and dilation		
Lining (tunica intima) of endothelium		
BLOOD SUPPLY		
Endothelial lining cells supplied by blood flowing through vessels; exchange of oxygen, etc., between cells of middle coat and blood by diffusion; outer coat supplied by tiny vessels known as *vasa vasorum,* or "vessels of vessels"		
NERVE SUPPLY		
Smooth muscle cells of tunica media innervated by autonomic fibers		
ABNORMALITIES		
Atherosclerosis—hardening of walls of arteries (arteriosclerosis) characterized by lipid deposits in tunica intima		
Aneurysm—saclike dilation of artery wall		
Varicose veins—stretching of walls, particularly around semilunar valves		
Phlebitis—inflammation of vein; "milk leg," phlebitis of femoral vein of women after childbirth		

Figure 16-10 **A,** Schematic drawings of an artery and vein showing comparative thick-nesses of the three coats: outer coat (tunica adventitia), muscle coat (tunica media), and lining of endothelium (tunica intima). Note that the muscle and outer coats are much thin-ner in veins than in arteries and that veins have valves. **B,** Muscular artery and vein. Note the prominent elastic membrane and tunica media in the artery *(left)* and the more exten-sive tunica adventitia in the vein *(right).* (× 140.)

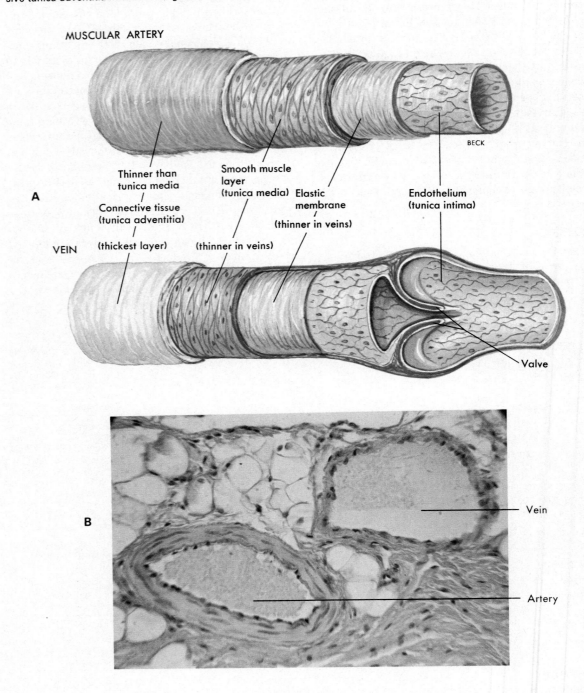

MUSCULAR ARTERY

BECK

Thinner than
tunica media

Smooth muscle
layer
(tunica media)

Elastic
membrane

Endothelium
(tunica intima)

Connective tissue
(tunica adventitia)

(thinner in veins)

A

VEIN

(thickest layer)

(thinner in veins)

Valve

B

Vein

Artery

Figure 16-11 The walls of capillaries consist of only a single layer of endothelial cells. These thin, flattened cells permit rapid movement of substances between blood and interstitial fluid. Note the thicker walls and the presence of precapillary sphincters (smooth muscle cells) on the arteriole.

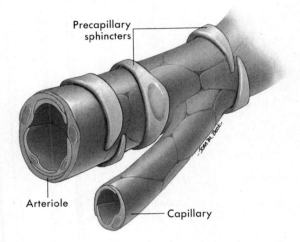

Figure 16-12 Illustration showing the reservoir function of veins. Pooled blood is moved toward the heart as valves are forced open by pressure from volume of blood from below.

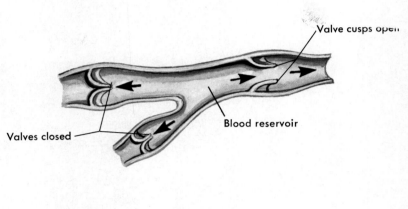

Figure 16-13 Anatomical components of microcirculation.

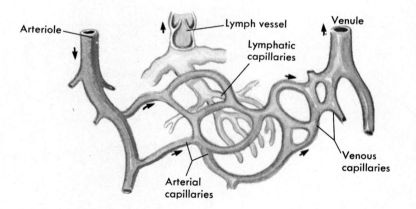

Veins function both as collectors and as reservoir vessels. Not only do they return blood from the capillaries to the heart, but they also can accommodate varying amounts of blood. This reservoir function of veins, which we shall discuss later, plays an important part in maintaining normal circulation. Figure 16-12 illustrates the potential for pooling of blood between valves in one segment of a vein. Pooled blood in each valved segment is moved toward the heart by the pressure from the moving volume of blood from below. The heart acts as a "pump," keeping the blood moving through this circuit of vessels—arteries, arterioles, capillaries, venules, and veins. In short, the entire circulatory mechanism pivots around one essential, that of keeping the capillaries supplied with an amount of blood adequate to the changing needs of the cells. All the factors governing circulation operate to this one end.

Although capillaries are very tiny (on the average, only 1 mm long or about ¹⁄₂₅ inch), their numbers are so great as to be incomprehensible. Someone has calculated that if these microscopic tubes were joined end to end, they would extend 62,000 miles, despite the fact that it takes 25 of them to reach a single inch! According to one estimate, 1 cubic inch of muscle tissue contains over 1.5 million of these

important little vessels. None of the billions of cells composing the body lies very far removed from a capillary. The reason for this lavish distribution of capillaries is, of course, apparent in view of their function of keeping the cells supplied with vital materials and rid of injurious wastes.

This vital exchange between blood and interstitial fluid that occurs at the capillary bed is referred to as the **microcirculation.** Figure 16-13 shows the anatomical components of microcirculation: arterioles, arteriole capillaries, venous capillaries, venules, lymphatic capillaries, and lymph vessels. The role of lymphatic vessels and capillaries in the process is discussed in Chapter 17.

Figure 16-14 Principal arteries of the body.

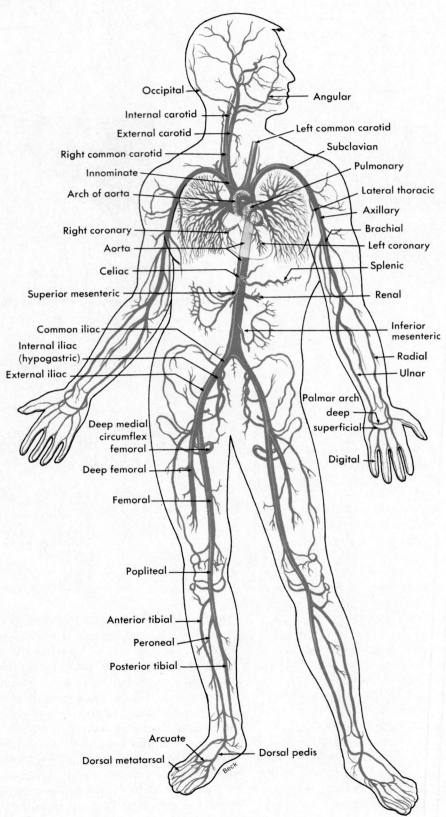

Occipital

Angular

Internal carotid

External carotid

Left common carotid

Right common carotid

Subclavian

Innominate

Pulmonary

Arch of aorta

Lateral thoracic

Axillary

Right coronary

Brachial

Aorta

Left coronary

Celiac

Splenic

Superior mesenteric

Renal

Common iliac

Inferior mesenteric

Internal iliac (hypogastric)

External iliac

Radial

Ulnar

Palmar arch
deep
superficial

Deep medial circumflex femoral

Digital

Deep femoral

Femoral

Popliteal

Anterior tibial

Peroneal

Posterior tibial

Arcuate

Dorsal metatarsal

Dorsal pedis

Beck

Figure 16-15 Main arteries of the face and head. Superficial vessels are shown in brighter color than deep vessels.

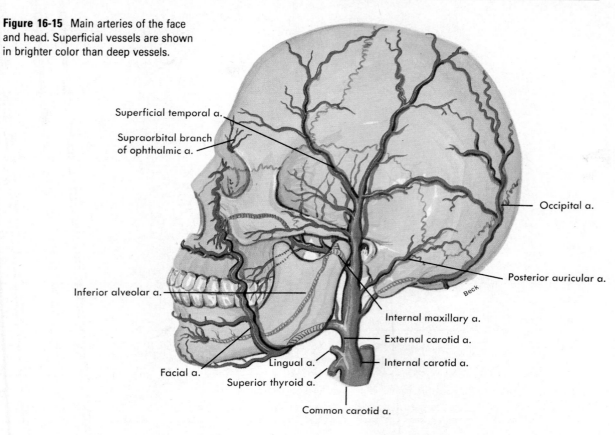

Figure 16-16 Location of the vertebral artery.

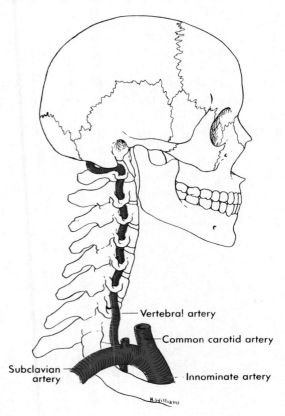

MAIN BLOOD VESSELS
Systemic Circulation
Arteries

Locate the arteries listed in Table 16-3 (see also Figures 16-14 to 16-17). You may find it easier to learn the names of blood vessels and the relation of the vessels to each other from diagrams than from descriptions.

As you learn the names of the main arteries, keep in mind that these are only the major pipelines distributing blood from the heart to the various organs and that in each organ the main artery resembles a tree trunk in that it gives off numerous branches that continue to branch and rebranch, forming ever smaller vessels (arterioles), which also branch, forming microscopic vessels, the capillaries. In other words, most arteries eventually ramify into capillaries. Arteries of this type are called **end-arteries.** Important organs or areas of the body supplied by end-arteries are subject to serious damage or death in occlusive arterial disease. As an example, permanent blindness results when the central artery of the retina, an end-artery, is occluded. Therefore occlusive arterial disease, such as atherosclerosis, is of great concern in clinical medicine when it affects important organs having an end-arterial blood supply.

Table 16-3 Main arteries

Artery	Branches (only largest ones named)
Ascending aorta	Coronary arteries (two, to myocardium)
Aortic arch	Innominate (or brachiocephalic) artery Left subclavian Left common carotid
Innominate	Right subclavian Right common carotid
Subclavian (right and left)	Vertebral* Axillary (continuation of subclavian)
Axillary	Brachial (continuation of axillary)
Brachial	Radial Ulnar
Radial and ulnar	Palmar arches (superficial and deep arterial arches in hand formed by anastomosis of branches of radial and ulnar arteries; numerous branches to hand and fingers)
Common carotid (right and left)	Internal carotid (brain, eye, forehead, and nose)* External carotid (thyroid, tongue, tonsils, ear, etc.)
Descending thoracic aorta	Visceral branches to pericardium, bronchi, esophagus, mediastinum Parietal branches to chest muscles, mammary glands, and diaphragm

Descending abdominal aorta	Visceral branches:
	  Celiac axis (or artery), branches into gastric, hepatic, and splenic arteries (stomach, liver, and spleen)
	  Right and left suprarenal arteries (suprarenal glands)
	  Superior mesenteric artery (small intestine)
	  Right and left renal arteries (kidneys)
	  Right and left spermatic (or ovarian) arteries (testes or ovaries)
	  Inferior mesenteric artery (large intestine)
	Parietal branches to lower surface of diaphragm, muscles and skin of back, spinal cord, and meninges
	Right and left common iliac arteries—abdominal aorta terminates in these vessels in an inverted Y

Right and left common iliac	Internal iliac or hypogastric (pelvic wall and viscera) External iliac (to leg)
External iliac (right and left)	Femoral (continuation of external iliac after it leaves abdominal cavity)
Femoral	Popliteal (continuation of femoral)
Popliteal	Anterior tibial Posterior tibial
Anterior and posterior tibial	Plantar arch (arterial arch in sole of foot formed by anastomosis of terminal branches of anterior and posterior tibial arteries; small arteries lead from arch to toes)

*The right and left vertebral arteries extend from their origin as branches of the subclavian arteries up the neck, through foramina in the transverse processes of the cervical vertebrae, and through the foramen magnum into the cranial cavity and unite on the undersurface of the brain stem to form the *basilar artery*, which shortly branches into the right and left *posterior cerebral arteries*. The internal carotid arteries enter the cranial cavity in the midpart of the cranial floor, where they become known as the *anterior cerebral arteries*. Small vessels, the *communicating arteries*, join the anterior and posterior cerebral arteries in such a way as to form an arterial circle (the *circle of Willis*) at the base of the brain, a good example of arterial anastomosis (Figure 16-17).

Figure 16-17 Arteries at base of the brain. Those arteries that compose the circle of
Willis are the two anterior cerebral arteries joined to each other by the anterior
communicating cerebral artery and to the posterior cerebral arteries by the posterior
communicating arteries.

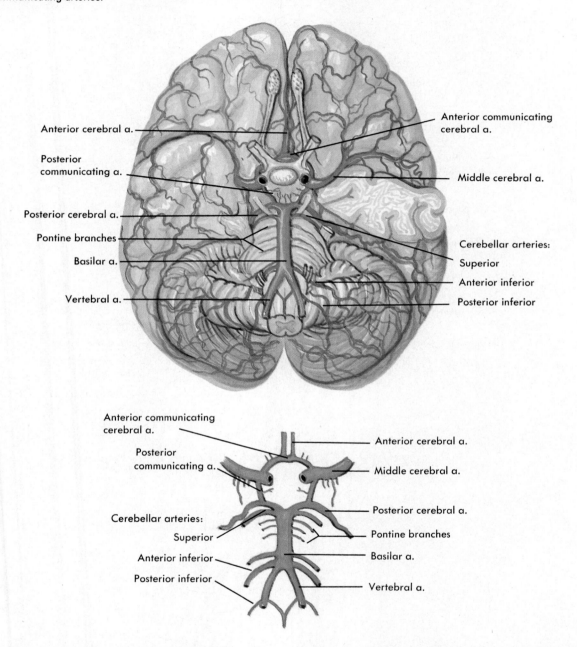

A few arteries open into other branches of the
same or other arteries. Such a communication is
termed an **arterial anastomosis.** Anastomoses, we
have already noted, fulfill an important protective
function in that they provide detour routes for blood
to travel in the event of obstruction of a main artery.
The incidence of arterial anastomoses increases as
distance from the heart increases, and smaller arterial branches tend to anastomose more often than
larger vessels. Examples of arterial anastomoses are
the palmar and plantar arches. Other examples are
found around several joints as well as in other locations.

Veins

Several facts should be borne in mind while learning
the names of veins.

   1 Veins are the ultimate extensions of capillaries, just as capillaries are the eventual
   extensions of arteries. Whereas arteries

Figure 16-18 Principal veins of the body. Only the superficial veins are shown on the forearms and hands.

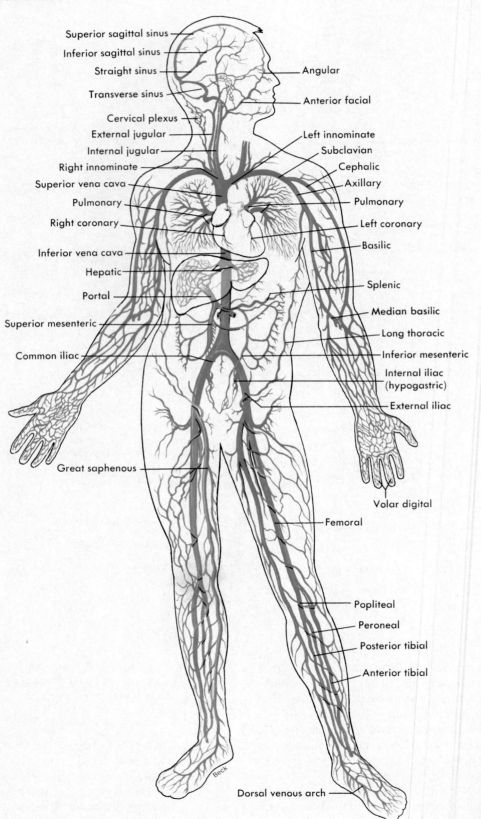

Superior sagittal sinus

Inferior sagittal sinus

Straight sinus

Transverse sinus

Cervical plexus

External jugular

Internal jugular

Right innominate

Superior vena cava

Pulmonary

Right coronary

Inferior vena cava

Hepatic

Portal

Superior mesenteric

Common iliac

Great saphenous

Angular

Anterior facial

Left innominate

Subclavian

Cephalic

Axillary

Pulmonary

Left coronary

Basilic

Splenic

Median basilic

Long thoracic

Inferior mesenteric

Internal iliac (hypogastric)

External iliac

Volar digital

Femoral

Popliteal

Peroneal

Posterior tibial

Anterior tibial

Dorsal venous arch

Beck

branch into vessels of decreasing size to form arterioles and eventually capillaries, capillaries unite into vessels of increasing size to form venules and eventually veins.

2 Many of the main arteries have corresponding veins bearing the same name and are located alongside or near the arteries. These veins, like the arteries, lie in deep, well protected areas, for the most part close along the bones—example: femoral artery and femoral vein, both located along the femur bone.

3 Veins found in the deep parts of the body are called **deep veins** in contradistinction to **superficial veins,** which lie near the surface. The latter are the veins that can be seen through the skin.

4 The large veins of the cranial cavity, formed by the dura mater, are not called veins but **sinuses.** They should not be confused with the bony sinuses of the skull.

5 Veins communicate (anastomose) with each other in the same way as arteries. Such venous anastomoses provide for collateral return blood flow in cases of venous obstruction.

6 Venous blood from the head, neck, upper extremities, and thoracic cavity, with the exception of the lungs, drains into the superior vena cava. Blood from the lower extremities and abdomen enters the inferior vena cava.

The following list identifies the major veins. Locate each one as named on Figures 16-18 to 16-23.

Veins of upper extremities (Figures 16-18 and 16-23)

a *Deep*
Palmar (volar) arch (also superficial)
Radial (partially deep, partially superficial)
Ulnar (partially deep, partially superficial)
Brachial
Axillary (continuation of brachial)
Subclavian (continuation of axillary)

b *Superficial*
Veins of hand from dorsal and volar venous arches, which, together with complicated network of superficial veins of lower arm, finally pour their blood into two large veins—cephalic (thumb side) and basilic (little finger side); these two veins empty into deep axillary vein

Veins of lower extremities (Figures 16-18, 16-20, 16-21)

a *Deep*
Plantar arch
Anterior tibial
Posterior tibial
Popliteal
Femoral
External iliac

Figure 16-19 Main superficial veins of the upper extremity, anterior view. The median basilic vein is commonly used for removing blood or giving intravenous infusions.

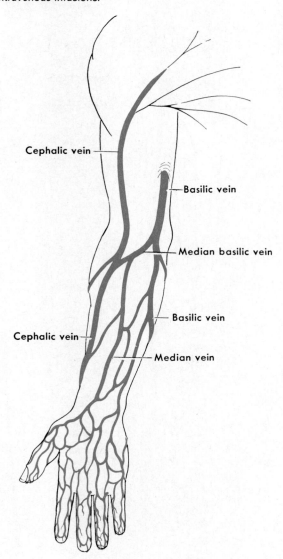

Cephalic vein

Basilic vein

Median basilic vein

Basilic vein

Cephalic vein

Median vein

b *Superficial*
Dorsal venous arch of foot
Great (or internal or long) saphenous
Small (or external or short) saphenous
(Great saphenous terminates in femoral vein in groin; small saphenous terminates in popliteal vein)

Veins of head and neck (Figures 16-22 and 16-23)

a *Deep (in cranial cavity)*
Longitudinal (or sagittal) sinus
Inferior sagittal and straight sinus
Numerous small sinuses
Right and left transverse (or lateral) sinuses
Internal jugular veins, right and left (in neck);

Figure 16-20 Main superficial veins of the lower extremity, anterior view.

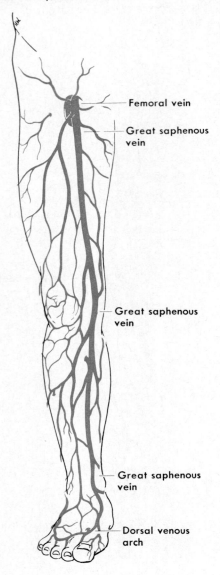
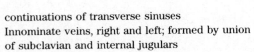

- Femoral vein
- Great saphenous vein
- Great saphenous vein
- Great saphenous vein
- Dorsal venous arch

Figure 16-21 Main superficial veins of the lower extremity, posterior view.

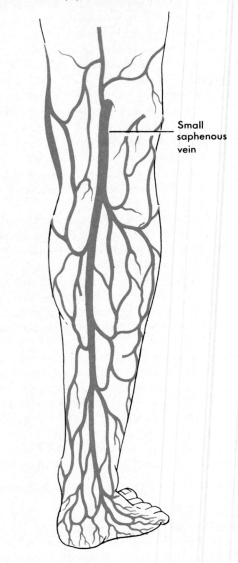

- Small saphenous vein

continuations of transverse sinuses
Innominate veins, right and left; formed by union of subclavian and internal jugulars

b *Superficial*

External jugular veins, right and left (in neck); receive blood from small superficial veins of face, scalp, and neck; terminate in subclavian veins (small emissary veins connect veins of scalp and face with blood sinuses of cranial cavity, a fact of clinical interest as a possible avenue for infections to enter cranial cavity)

Veins of abdominal organs (Figures 16-18 and 16-24)

Spermatic (or ovarian)
Renal   Drain into inferior
Hepatic   vena cava
Suprarenal

Left spermatic and left suprarenal veins usually drain into left renal vein instead of into inferior vena cava; for return of blood from abdominal digestive organs, see subsequent discussion of portal circulation

Veins of thoracic organs

Several small veins—such as bronchial, esophageal, pericardial—return blood from chest organs (except lungs) directly into superior vena cava or azygos vein; azygos vein lies to right of spinal column and extends from inferior vena cava (at level of first or second lumbar vertebra) through diaphragm to terminal part of superior vena cava; hemiazygos vein lies to left of spinal column, extending from lumbar level of inferior vena cava through diaphragm to terminate in azygos vein; accessory hemiazygos vein connects some of superior intercostal veins with azygos or hemiazygos vein

Figure 16-22 Semischematic projection of the large veins of the head. Deep veins and dural sinuses are projected on the skull. Note connections (emissary veins) between the superficial and deep veins.

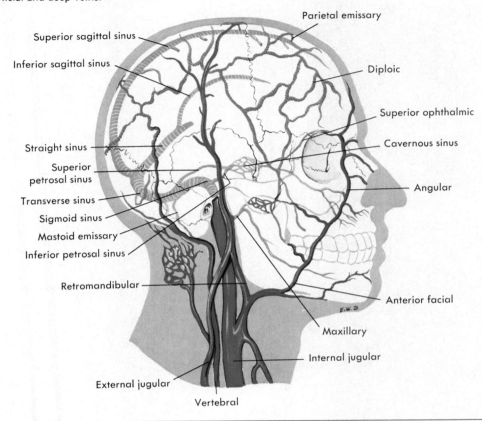

Figure 16-23 Venous sinuses shown in relation to the brain and skull.

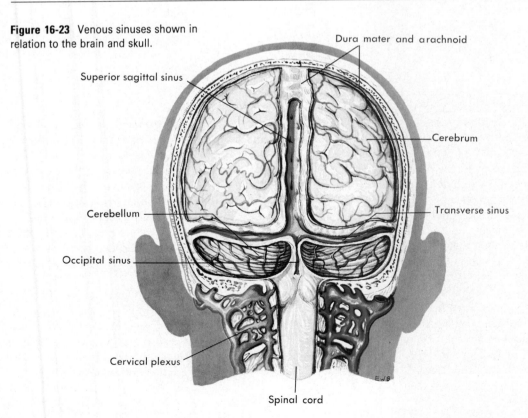

Figure 16-24 Tributaries of the portal vein (diagrammatic).

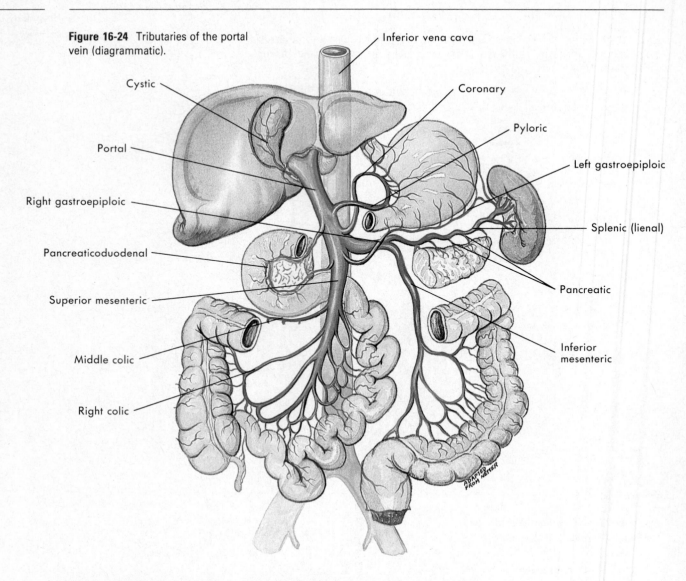

Inferior vena cava

Cystic

Coronary

Portal

Pyloric

Right gastroepiploic

Left gastroepiploic

Pancreaticoduodenal

Splenic (lienal)

Superior mesenteric

Pancreatic

Middle colic

Inferior mesenteric

Right colic

Middle ear infections sometimes cause infection of the transverse sinuses with the formation of a thrombus. In such cases the internal jugular vein may be ligated to prevent the development of a fatal cardiac or pulmonary embolism.

Intravenous injections are perhaps most often given into the median basilic vein at the bend of the elbow. Blood that is to be used for various laboratory tests is also usually removed from this vein. In an infant, however, the longitudinal sinus is more often punctured (through the anterior fontanel) because the superficial arm veins are too tiny for the insertion of a needle.

Portal Circulation

Veins from the spleen, stomach, pancreas, gallbladder, and intestines do not pour their blood directly into the inferior vena cava as do the veins from other abdominal organs. Instead, they send their blood to the liver by means of the portal vein. Here the blood mingles with the arterial blood in the capillaries and is eventually drained from the liver by the hepatic veins that join the inferior vena cava. The reason for this detouring of the blood through the liver before it returns to the heart will be discussed in the chapter on the digestive system.

Figure 16-24 shows the plan of the portal system. In most individuals the portal vein is formed by the union of the splenic and superior mesenteric veins, but blood from the gastric, pancreatic, and inferior mesenteric veins drains into the splenic vein before it merges with the superior mesenteric vein.

If either portal circulation or venous return from the liver is interfered with (as they often are in certain types of liver or heart disease), venous drainage from most of the other abdominal organs is necessarily obstructed also. The accompanying increased capillary pressure accounts at least in part for the occurrence of ascites ("dropsy" of abdominal cavity) under these conditions.

Fetal Circulation

Circulation in the body before birth necessarily differs from circulation after birth for one main reason—because fetal blood secures oxygen and food from maternal blood instead of from its own lungs and digestive organs. Obviously, then, there must be additional blood vessels in the fetus to carry the fetal blood into close approximation with the maternal blood and to return it to the fetal body. These structures are the two **umbilical arteries,** the **umbilical vein,** and the **ductus venosus.** Also, there must be some structure to function as the lungs and digestive organs do postnatally, that is, a place where an interchange of gases, foods, and wastes between the fetal and maternal blood can take place. This structure is the *placenta.* The exchange of substances occurs without any actual mixing of maternal and fetal bloods, since each flows in its own capillaries.

In addition to the placenta and umbilical vessels, three structures located within the fetus' own body play an important part in fetal circulation. One of them (ductus venosus) serves as a detour by which most of the blood returning from the placenta bypasses the fetal liver. The other two (foramen ovale and ductus arteriosus) provide detours by which blood bypasses the lungs. A brief description of each of the six structures necessary for fetal circulation follows (Figure 16-25).

1 The **two umbilical arteries** are extensions of the internal iliac (hypogastric) arteries and carry fetal blood to the placenta.

2 The **placenta** is a structure attached to the uterine wall. Exchange of oxygen and other substances between maternal and fetal blood takes place in the placenta.

3 The **umbilical vein** returns oxygenated blood from the placenta, enters the fetal body through the umbilicus, extends up to the undersurface of the liver where it gives off two or three branches to the liver, and then continues on as the ductus venosus. Two umbilical arteries and the umbilical vein together constitute the **umbilical cord** and are shed at birth along with the placenta.

4 The **ductus venosus** is a continuation of the umbilical vein along the undersurface of the liver and drains into the inferior vena cava. Most of the blood returning from the placenta bypasses the liver. Only a relatively small amount enters the liver by way of the branches from the umbilical vein into the liver.

5 The **foramen ovale** is an opening in the septum between the right and left atria. A valve at the opening of the inferior vena cava into the right atrium directs most of the blood through the foramen ovale into the left atrium so that it bypasses the fetal lungs. A small percentage of the blood leaves the right atrium for the right ventricle and pulmonary artery. But even most of this does not flow on into the lungs. Still another detour, the ductus arteriosus, diverts it.

6 The **ductus arteriosus** is a small vessel connecting the pulmonary artery with the descending thoracic aorta. It therefore enables another portion of the blood to detour into the systemic circulation without going through the lungs.

Almost all fetal blood is a mixture of oxygenated and deoxygenated blood. Examine Figure 16-25 carefully to determine why this is so. What happens to the oxygenated blood returned from the placenta via the umbilical vein? Note that it flows into the inferior vena cava.

Changes in the Vascular System at Birth

Since the six structures that serve fetal circulation are no longer needed after birth, several changes take place. As soon as the umbilical cord is cut, the two umbilical arteries, the placenta, and the umbilical vein obviously no longer function. The placenta is shed from the mother's body as the afterbirth, with part of the umbilical vessels attached. The sections of these vessels remaining in the infant's body eventually become fibrous cords that remain throughout life (the umbilical vein becomes the round ligament of the liver). The ductus venosus, no longer needed to bypass blood around the liver, eventually becomes the ligamentum venosum of the liver. The foramen ovale normally becomes functionally closed soon after a newborn takes the first breath and full circulation through the lungs becomes established. Complete structural closure, however, usually requires 9 months or more. Eventually the foramen ovale becomes a mere depression (fossa ovalis) in the wall of the right atrial septum. The ductus arteriosus contracts as soon as respiration is established. Eventually it also turns into a fibrous cord.

Figure 16-25 Scheme to show the plan of fetal circulation. Note the following essential features: (1) two umbilical arteries, extensions of the internal iliac arteries, carry blood to (2) the placenta, which is attached to the uterine wall; (3) one umbilical vein returns blood, rich in oxygen and food, from the placenta; (4) the ductus venosus, a small vessel that connects the umbilical vein with the inferior vena cava; (5) the foramen ovale, an opening in the septum between the right and left atria; and (6) the ductus arteriosus, a small vessel that connects the pulmonary artery with the thoracic aorta.

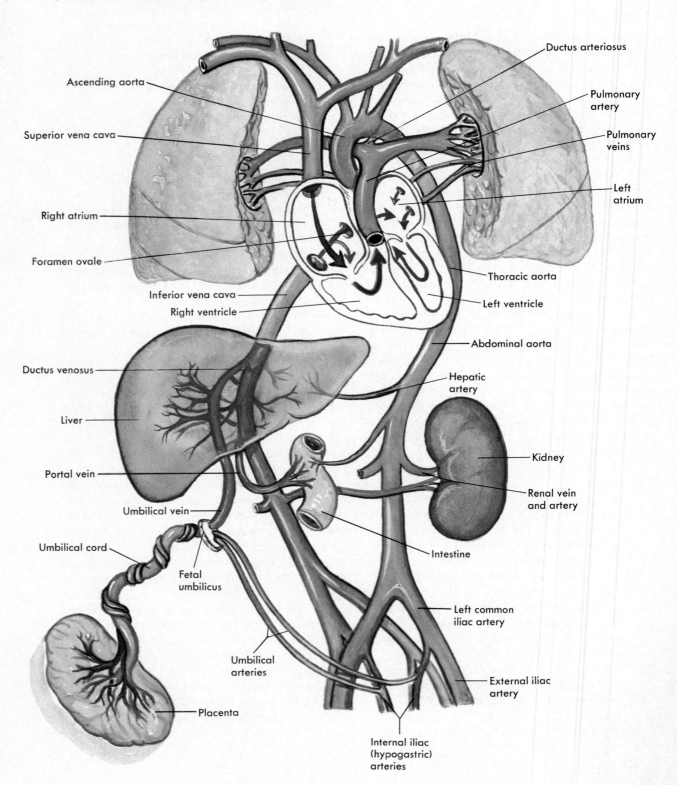

Ascending aorta

Superior vena cava

Right atrium

Foramen ovale

Inferior vena cava

Right ventricle

Ductus venosus

Liver

Portal vein

Umbilical vein

Umbilical cord

Fetal umbilicus

Umbilical arteries

Placenta

Ductus arteriosus

Pulmonary artery

Pulmonary veins

Left atrium

Thoracic aorta

Left ventricle

Abdominal aorta

Hepatic artery

Kidney

Renal vein and artery

Intestine

Left common iliac artery

External iliac artery

Internal iliac (hypogastric) arteries

CIRCULATION

DEFINITIONS

The term *circulation of blood* suggests its meaning, namely, blood flow through vessels arranged to form a circuit or circular pattern. Blood flow from the heart (left ventricle) through blood vessels to all parts of the body and back to the heart (to the right atrium) is spoken of as **systemic circulation.** The left ventricle pumps blood into the ascending aorta. From here it flows into arteries that carry it into the various tissues and organs of the body. Within each structure, blood moves, as indicated in Figure 16-26 (see also Figure 16-13), from arteries to arterioles to capillaries. Here the vital two-way exchange of substances occurs between blood and cells. Blood flows next out of each organ by way of its venules and then its veins to drain eventually into the inferior or superior vena cava. These two great veins of the body return venous blood to the heart (to the right atrium) to complete systemic circulation. But the blood has not quite come full circle back to its starting point, the left ventricle. To do this and start on its way again, it must first flow through another circuit, the **pulmonary circulation.** Observe in Figure 16-26 that venous blood moves from the right atrium to the right ventricle to the pulmonary artery to lung arterioles and capillaries. Here, exchange of gases between blood and air takes place, converting venous blood to arterial blood. This oxygenated blood then flows on through lung venules into four pulmonary veins and returns to the left atrium of the heart. From the left atrium it enters the left ventricle to be pumped again through the systemic circulation.

HOW TO TRACE

To list the vessels through which blood flows in reaching a designated part of the body or in returning to the heart from a part, one must remember the following:

1 That blood always flows in this direction— from *left ventricle* of heart to *arteries*, to *arterioles*, to *capillaries* of each body part, to *venules*, to *veins*, to *right atrium, right ventricle, pulmonary artery, lung capillaries, pulmonary veins, left atrium,* and back to *left ventricle* (Figure 16-26)
2 That when blood is in capillaries of abdominal digestive organs, it must flow through portal system (see Figure 16-24) before returning to heart
3 Names of main arteries and veins of body

For example, suppose glucose were instilled into the rectum. To reach the cells of the right little finger, the vessels through which it would pass after absorption from the intestinal mucosa into capillar-

Figure 16-26 Relationship of systemic and pulmonary circulation. As indicated by the numbers, blood circulates from the left side (ventricle) of the heart to arteries, to arterioles, to capillaries, to venules, to veins, to the right side of the heart (atrium to ventricle), to the lungs, and back to the left side of the heart, thereby completing a circuit. Refer to this diagram when tracing the circulation of blood to or from any part of the body.

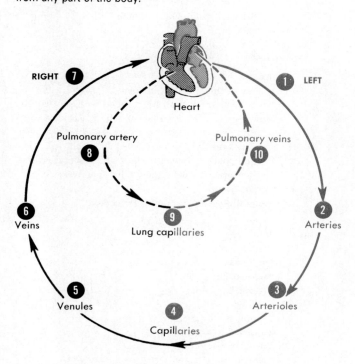

ies would be as follows: *capillaries* into venules of large intestine into inferior mesenteric *vein*, splenic vein, portal vein, capillaries of liver, hepatic veins, inferior vena cava, *right atrium* of heart, *right ventricle, pulmonary artery, pulmonary capillaries, pulmonary veins, left atrium, left ventricle, ascending aorta,* aortic arch, innominate artery, right subclavian artery, right axillary artery, right brachial artery, right ulnar artery, arteries of palmar arch, arterioles, and *capillaries* of right little finger.

NOTE: The structures italicized show the direction of blood flow as described in points **1** and **2** and illustrated in Figure 16-26. Follow this course of circulation first on Figure 16-24 and then on Figures 16-14 and 16-18. Answer review question 17 at the end of this chapter using the plan outlined.

Outline Summary

HEART

A Four-chambered muscular organ

B Lies in mediastinum with apex on diaphragm, two thirds of its bulk to left of midline of body and one third to right

C Apical beat may be counted by placing stethoscope in fifth intercostal space on line with left midclavicular point

Covering

A Structure

1 Loose-fitting, inextensible sac (fibrous pericardium) around heart, lined with serous pericardium (parietal layer), which also covers outer surface of heart (visceral layer or epicardium)

2 Small space between parietal and visceral layers of serous pericardium contains few drops of pericardial fluid

B Function—protection against friction

Structure

A Wall

1 Myocardium—muscular wall

2 Endocardium—lining

3 Pericardium—covering

B Cavities

1 Upper two—atria

2 Lower two—ventricles

C Valves and openings

1 Openings between atria and ventricles— atrioventricular orifices, guarded by cuspid valves, tricuspid on right and mitral or bicuspid on left; valves consist of three parts; flaps, chordae tendineae, and papillary muscles

2 Opening from right ventricle into pulmonary artery guarded by semilunar valves

3 Opening from left ventricle into great aorta guarded by semilunar valves

D Blood supply—from coronary arteries; branch from ascending aorta behind semilunar valves

1 Left ventricle receives blood via both major branches of left coronary artery and from one branch of right coronary artery

2 Right ventricle receives blood via both major branches of right coronary artery and from one branch of left coronary artery

3 Each atrium receives blood only from one branch of its respective coronary artery

4 Usually only few anastomoses between larger branches of coronary arteries so that occlusion of one of these produces areas of myocardial infarction; if not fatal, anastomoses between smaller vessels may grow and provide collateral circulation

E Conduction system

1 Sinoatrial node (SA node; pacemaker of heart)— small mass of modified cardiac muscle at junction of superior vena cava and right atrium; inherent rhythmicity of SA node impulses set basic rate of heartbeat; ratio of sympathetic/parasympathetic impulses to node per minute and blood concentrations of epinephrine and thyroid hormone act on node to modify its activity and alter heart rate

2 Atrioventricular node (AV node)—small mass of modified cardiac muscle in septum between two atria

3 Atrioventricular bundle (AV bundle, bundle of His)—special cardiac muscle fibers originating in AV node and extending down interventricular septum

4 Purkinje fibers—extension of bundle of His fibers out into walls of ventricles

F Nerve supply

1 Sympathetic fibers (in cardiac nerves) and parasympathetic fibers (in vagus) form cardiac plexuses

2 Fibers from plexuses terminate mainly in SA node

3 Sympathetic impulses tend to accelerate and strengthen heartbeat

4 Vagal impulses slow heartbeat

BLOOD VESSELS

Kinds

A Arteries—vessels that carry blood away from heart; all except pulmonary artery carry oxygenated blood

B Veins—vessels that carry blood toward heart; all except pulmonary veins carry deoxygenated blood

C Capillaries—microscopic vessels that carry blood from small arteries (arterioles) to small veins (venules)

Structure (Table 16-2)

Functions

A Arteries and arterioles—carry blood away from heart to capillaries

B Capillaries—deliver materials to cells (by way of tissue fluid) and collect substances from them; vital function of entire circulatory system

C Veins and venules—carry blood from capillaries back to heart

Main blood vessels

A Systemic circulation

1 Arteries (see Figure 16-14, p. 456, and Table 16-3)

2 Veins (see Figures 16-18 to 16-24, pp. 460-464)

B Portal circulation (see Figure 16-24, p. 464)

C Fetal circulation (see Figure 16-25, p. 466)

Outline Summary—cont'd

CIRCULATION
Definitions

A Circulation—blood flow through closed circuit of vessels

B Systemic circulation—blood flow from left ventricle into aorta, other arteries, arterioles, capillaries, venules, and veins of all parts of body to right atrium of heart

C Pulmonary circulation—blood flow from right ventricle to pulmonary artery to lung arterioles, capillaries, and venules, to pulmonary veins, to left atrium

How to trace (Figure 16-26)

Review Questions

1 Discuss the size, position, and location of the heart in the thoracic cavity.
2 How is compression of the heart in cardiopulmonary resuscitation effected?
3 Describe the pericardium, differentiating between the fibrous and serous portions.
4 The fibrous pericardium is tough and inelastic. Why is an understanding of this anatomical fact important in clinical medicine?
5 Exactly where is pericardial fluid found? Explain its function.
6 Describe the heart's own blood supply. Explain why occlusion of a large coronary artery branch has serious consequences.
7 Why would you expect large numbers of mitochondria to be present in cardiac muscle?
8 Identify, locate, and describe the functions of each of the following structures: SA node, AV node, bundle of His.
9 Compare arteries, veins, and capillaries as to structure and functions.
10 Differentiate between systemic, pulmonary, and portal circulations.
11 Explain why occlusion of an end-artery is more serious than occlusion of other small arteries.
12 Discuss the location, formation, and importance of the circle of Willis.
13 Name and locate the chambers and valves of the heart.

14 Trace the flow of blood through the heart.
15 Discuss the relationship between the layers of the blood vessels and the heart.
16 Explain the relationship between a heart "murmur" and valvular insufficiency.
17 Starting with the left ventricle of the heart, list the vessels through which blood flows in reaching the small intestine; the large intestine; the liver (two ways); the spleen; the stomach; the kidneys; the suprarenal glands; the ovaries or testes; the anterior part of the base of the brain; the little finger of the right hand. List the vessels through which the blood returns from these parts to the right atrium of the heart (see Figures 16-14, 16-18, and 16-26).
18 Briefly define the following terms: *aneurysm, atherosclerosis, chordae tendineae, coronary sinus, endocardium, foramen ovale, mitral stenosis, myocardium, phlebitis, Purkinje fibers.*
19 List the tributaries of the hepatic portal vein.
20 Why is the internal jugular vein sometimes ligated as a result of middle ear infections?
21 Give the general location of the following veins: longitudinal sinus, internal jugular vein, inferior vena cava, basilic vein, coronary sinus, great saphenous vein, hepatic portal vein.

17 PHYSIOLOGY OF THE CARDIOVASCULAR SYSTEM

OBJECTIVES

After you have completed this chapter, you should be able to:

1 Trace a cardiac impulse through the conduction system of the heart.

2 Discuss how the two types of artificial cardiac pacemakers can regulate the heart rate.

3 Identify and discuss normal ECG deflections and intervals and their relationship to mechanical contraction.

4 Discuss autonomic nervous system control of heart rate mediated by cardioinhibitory and cardioaccelerator centers, cardiac pressoreflexes, carotid sinus reflex, and the aortic reflex.

5 Compare the results of parasympathetic and sympathetic stimulation on the heart and explain the mechanism involved in both types of autonomic control.

6 Identify and briefly discuss several factors that influence heart rate.

7 Discuss six major events of the cardiac cycle and relate each event of the cycle to time.

8 Discuss the physical principles that govern fluid flow and circulation.

9 Identify and coordinate the areas of a typical blood pressure gradient curve with blood vessels in the circulatory system from aorta to venae cavae.

10 Discuss how arterial blood pressure is influenced by cardiac output, stroke volume, peripheral resistance, vasomotor pressoreflex, and chemoreflex control mechanisms.

11 Identify and discuss the most important factors influencing venous return to the heart.

12 Explain the main determinants of peripheral resistance.

13 Discuss the measurement of arterial blood pressure clinically.

14 Define pulse and identify the two factors most responsible for its existence.

15 Identify those body areas where the pulse can be felt and those areas where pressure may be applied to stop arterial bleeding.

KEY TERMS

Atrioventricular node (a-tree-oh-ven-TRICK-u-lar node)

Baroreceptor (bar-o-re-SEP-tor)

Chemoreceptor (ke-mo-re-SEP-tor)

Diastole (di-ah-STOL-lee)

Electrocardiogram (e-lek-tro-KAR-de-o-gram)

Hyperemia (hi-per-E-me-ah)

Pacemaker

Pulse (puls)

Purkinje system (pur-KIN-je)

Sinoatrial node (si-no-A-tre-al node)

Systole (sis-TOL-lee)

Vasoconstriction (vas-o-kon-STRIK-shun)

Vasodilation (vas-o-di-LA-shun)

FUNCTIONS AND IMPORTANCE OF CONTROL MECHANISMS

The vital role of the cardiovascular system in maintaining homeostasis depends on the continuous and controlled movement of blood through the thousands of miles of capillaries that permeate every tissue and reach every cell in the body. It is in the microscopic capillaries that blood performs its ultimate transport function. Nutrients and other essential materials pass from capillary blood into fluids surrounding the cells as waste products are removed. Blood must not only be kept moving through its closed circuit of vessels by the pumping activity of the heart, but it must be directed and delivered to those capillary beds surrounding cells that need it most. Blood flow to cells at rest is minimal. In contrast, blood is shunted to the digestive tract following a meal or to skeletal muscles during exercise. The thousands of miles of capillaries could hold far more than the body's total blood volume if it were evenly distributed. Regulation of blood pressure and flow must therefore change in response to cellular activity.

Numerous control mechanisms help to regulate and integrate the diverse functions and component parts of the cardiovascular system in order to supply blood to specific body areas according to need. These mechanisms ensure a constant *milieu intérieur*, that is, a constant internal environment surrounding each body cell regardless of differing demands for nutrients or production of waste products. This chapter presents information about several of the control mechanisms that regulate both the pumping activity of the heart and the smooth and directed flow of blood through the complex channels of the circulation.

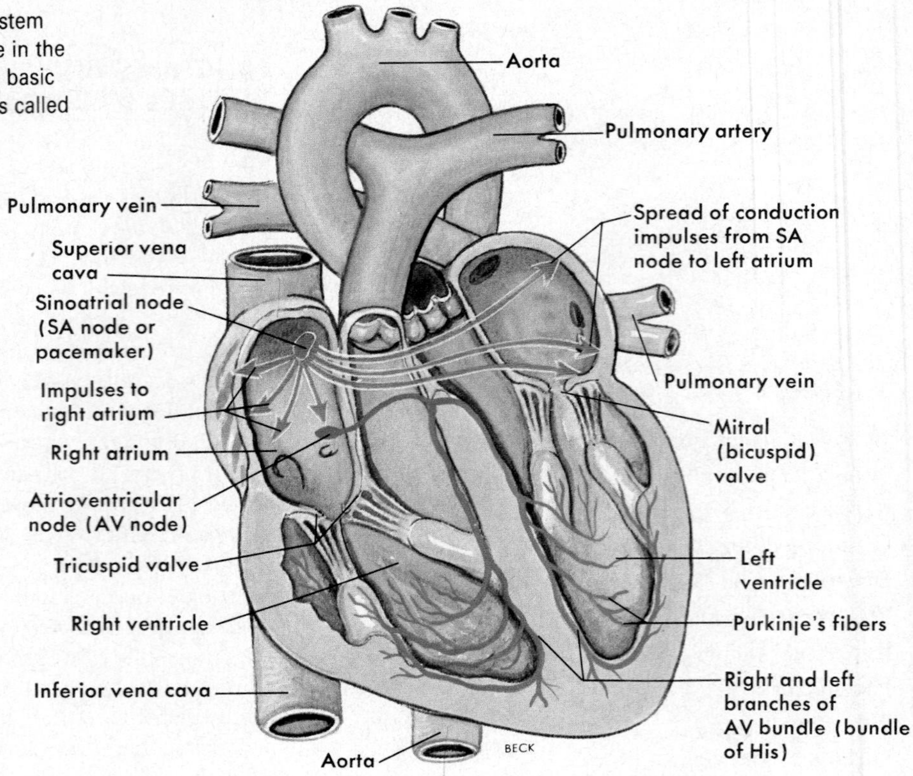

Figure 17-1 The conduction system of the heart. The sinoatrial node in the wall of the right atrium sets the basic pace of the heart's rhythm, so is called the "pacemaker."

Aorta

Pulmonary artery

Pulmonary vein

Superior vena cava

Sinoatrial node (SA node or pacemaker)

Impulses to right atrium

Right atrium

Atrioventricular node (AV node)

Tricuspid valve

Right ventricle

Inferior vena cava

Aorta

BECK

Spread of conduction impulses from SA node to left atrium

Pulmonary vein

Mitral (bicuspid) valve

Left ventricle

Purkinje's fibers

Right and left branches of AV bundle (bundle of His)

PHYSIOLOGY OF THE HEART

CONDUCTION SYSTEM

The anatomy of four structures that compose the conduction system of the heart—the sinoatrial node, atrioventricular node, AV bundle, and Purkinje system—was discussed in Chapter 16. Each of these structures consists of cardiac muscle modified enough in structure to differ in function from ordinary cardiac muscle. The main specialty of ordinary cardiac muscle is contraction. In this, it is like all muscle, and like all muscle, ordinary cardiac muscle can also conduct impulses. But the conduction system structures (Figure 17-1) are more highly specialized, both structurally and functionally, than ordinary cardiac muscle tissue. They are not contractile. Instead, they permit only generation or rapid conduction of an action potential through the heart.

The normal cardiac impulse that initiates mechanical contraction of the heart arises in the **SA node** (or pacemaker), located just below the atrial epicardium at its junction with the superior vena cava (Figure 17-1). Specialized pacemaker cells in the node possess an **intrinsic rhythm.** This means that without any stimulation by nerve impulses from the brain and cord, they themselves initiate impulses at regular intervals. Even if the heart is removed

from the body and placed in a nutrient solution, completely separated from all nervous and hormonal control, it will continue to beat!

Each impulse generated at the SA node travels swiftly throughout the muscle fibers of both atria. Thus stimulated, the atria begin to contract. As the action potential enters the AV node from the right atrium, its conduction slows markedly, thus allowing for complete contraction of both atrial chambers before the impulse reaches the ventricles. After passing slowly through the AV node, conduction velocity increases as the impulse is relayed through the AV bundle (bundle of His) into the ventricles. Here, right and left branches of the bundle fibers and the Purkinje fibers in which they terminate conduct the impulses throughout the muscle of both ventricles, stimulating them to contract almost simultaneously.

Thus the SA node initiates each heartbeat and sets its pace—it is the heart's own natural pacemaker. Normally the SA node will "discharge," or "fire," at an intrinsic rhythmical rate of 70 to 75 beats per minute. However, if for any reason the SA node loses its ability to generate an impulse, pacemaker activity will shift to another excitable component of the conduction system, such as the AV node or the Purkinje fibers. Pacemakers other than the SA node are called abnormal, or **ectopic pacemakers.** Although ectopic pacemakers fire rhythmically, their rate of discharge is generally much slower than that of the SA node. For example, a pulse of 40 to 60 beats per minute would result if the AV node were forced to assume pacemaker activity.

ARTIFICIAL CARDIAC PACEMAKERS

Today, everyone has heard about **artificial pacemakers,** devices that electrically stimulate the heart at a set rhythm (continuously discharging type) or those that fire only when the heart rate decreases below a preset minimum (demand pacemakers). They do an excellent job of maintaining a steady heart rate and of keeping many individuals with damaged hearts alive for many years. Over 140,000 people currently have permanently implanted cardiac pacemakers.

Several types of artificial pacemakers have been designed to deliver an electrical stimulus to the heart muscle. The stimulus passes through electrodes that are sewn directly to the epicardium on the outer surface of the heart or are inserted by a catheter into a heart chamber, such as the right ventricle, and placed in contact with the endocardium. Modern pacemakers generate a stimulus that lasts from 0.08 to 2 msec and produce a very low milliampere output.

One of the most common methods of inserting a permanent pacemaker is by the **transvenous approach.** In this procedure a small incision is made just above the right clavicle, and the electrode is threaded into the jugular vein and then advanced to the apex of the right ventricle. Examine Figure 17-2 to determine the position of the cardiac chambers, great vessels, and valves as seen in x-ray films of the chest taken from behind (PA, or posterior to anterior view) and from the side (lateral view). Figure 17-3, *A*, shows the battery-powered stimulus generator, which is placed in a pocket beneath the skin on the right side of the chest just below the clavicle. The proximal end of the electrical lead or catheter is then directed through the subcutaneous tissues and attached to the power pack. Figure 17-3, *B*, shows the tip of the pacemaker in the apex of the right ventricle. Although lifesaving, these devices must be judged inferior to the heart's own natural pacemaker. Why? Because they cannot speed up the heartbeat (as is necessary, for example, to make strenuous physical activity possible), nor can they slow it down again when the need has passed. The SA node, influenced as it is by autonomic impulses and hormones, can produce these changes. Discharging an average of 75 times each minute, this truly remarkable bit of specialized tissue will generate well over 2 billion action potentials in an average lifetime of some 70 years!

Figure 17-2 X-ray evaluation of the heart. **A,** Posterior to anterior (PA) view. **B,** Lateral view. *1,* Superior vena cava; *2,* right atrium; *3,* right ventricle; *4,* pulmonary outflow tract; *5,* left atrium; *6,* left ventricle; *7,* left atrial appendage; *8,* inferior vena cava; *9,* aorta; *10,* aortic valve; *11,* mitral valve.

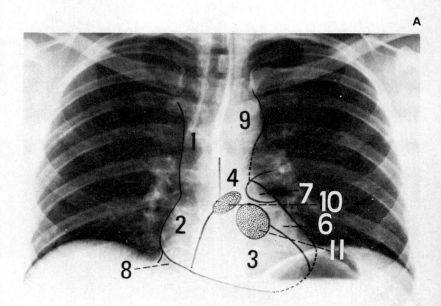

A

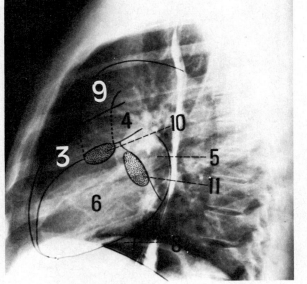

B

Figure 17-3 X-ray appearance of transvenous pacemaker. **A,** Posterior to anterior (PA) view showing stimulus generator in subcutaneous tissue of chest wall. Box contains four batteries. Pacemaker wires travel through veins to heart. **B,** Lateral view. Tip of pacemaker lead in contact with endocardium of right ventricle.

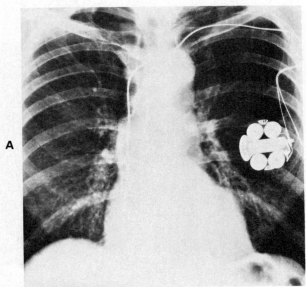

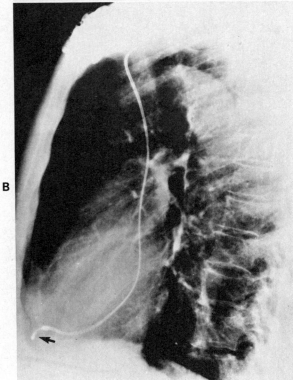

ELECTROCARDIOGRAM

Impulse conduction generates tiny electrical currents in the heart that spread through surrounding tissues to the surface of the body. This fact has great clinical importance. Why? Because from the skin, visible records of heart conduction can be made with an instrument called an electrocardiograph. Skilled interpretation of these records may sometimes make the difference between life and death.

The **electrocardiogram (ECG)** is a graphic record of the heart's action currents, its conduction of impulses. It is not a record of the heart's contractions but of the electrical events that precede them. Because electrocardiography is far too complex a science to attempt to explain here, normal ECG **deflection waves** and the **ECG intervals** between them shall be only briefly described. As shown in Figure 17-4, *A,* the normal ECG is composed of a P wave, a QRS complex, and a T wave. (The letters do not stand for any words but were chosen arbitrarily.) Briefly, the P wave represents depolarization of the atria, that is, the passage of a depolarization current from the SA node through the musculature of both atria; the QRS complex represents depolarization of the ventricles; and the T wave reflects repolarization (relaxation) of the ventricles. The principal ECG **intervals** between P, QRS, and T waves are shown in Figure 17-4, *B.* Measurement of these intervals can provide valuable information concerning the rate of conduction of an action potential through the heart. Can you predict from Figure 17-4, *C,* which interval would be lengthened if the passage of an action potential were slowed between atria and ventricles?

CONTROL OF HEART RATE
Autonomic Nervous System Control

Although the SA node normally initiates each heartbeat, the rate it sets is not an unalterable one. Various factors can and do change the rate of the heartbeat. One major modifier of SA node activity—and therefore of the heart rate—is the ratio of sympathetic and parasympathetic impulses conducted to the

Autonomic control of heart rate is the result of opposing influences between parasympathetic (chiefly vagus) and sympathetic (accelerator nerve) stimulation. The results of parasympathetic stimulation on the heart are inhibitory and are mediated by vagal release of acetylcholine, whereas sympathetic (stimulatory) effects result from the release of norepinephrine.

Figure 17-4 **A,** Normal ECG deflections. Depolarization and repolarization. **B,** Principal ECG intervals between P, QRS, and T waves. **C,** Schematic representation of ECG and its relationship to cardiac electrical activity.

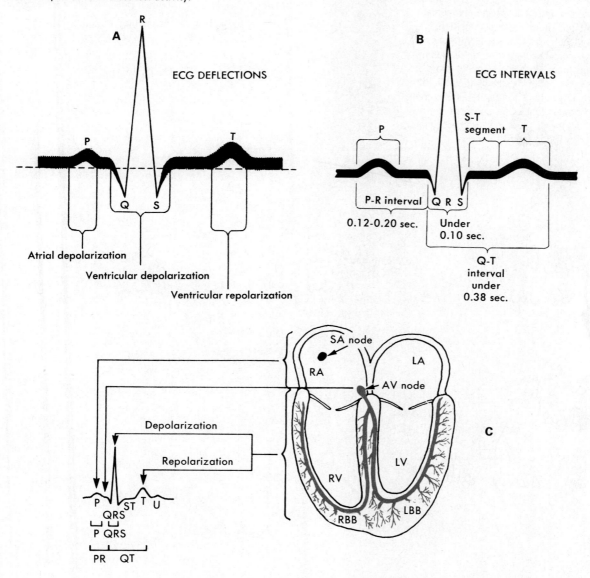

node per minute. Branches of the two divisions of the autonomic nervous system that reach the heart were discussed in Chapter 16.

CARDIOINHIBITORY CENTER. Parasympathetic impulses from a specialized area in the medulla called the *cardioinhibitory center* reach the SA node through the vagus. Normally the heart is under the restraint of vagal inhibition, as released acetylcholine decreases the rate of SA node firing. The vagus is said to act as a "brake" on the heart.

CARDIOACCELERATOR CENTER. Sympathetic impulses originate in the cardioaccelerator center of the me-

dulla and reach the heart via sympathetic fibers (contained in the middle, superior, and inferior cardiac nerves). Norepinephrine released as a result of sympathetic stimulation increases both heart rate and strength of cardiac muscle contraction. Figure 17-5 illustrates the control of cardioinhibitory and cardioaccelerator center activity, or outflow from the medulla.

CARDIAC PRESSOREFLEXES. Receptors sensitive to changes in pressure (baroreceptors) are located in two cardiovascular regions; they send afferent nerve fibers to the medullary control centers. Specifically, these stretch receptors are located in the aorta and

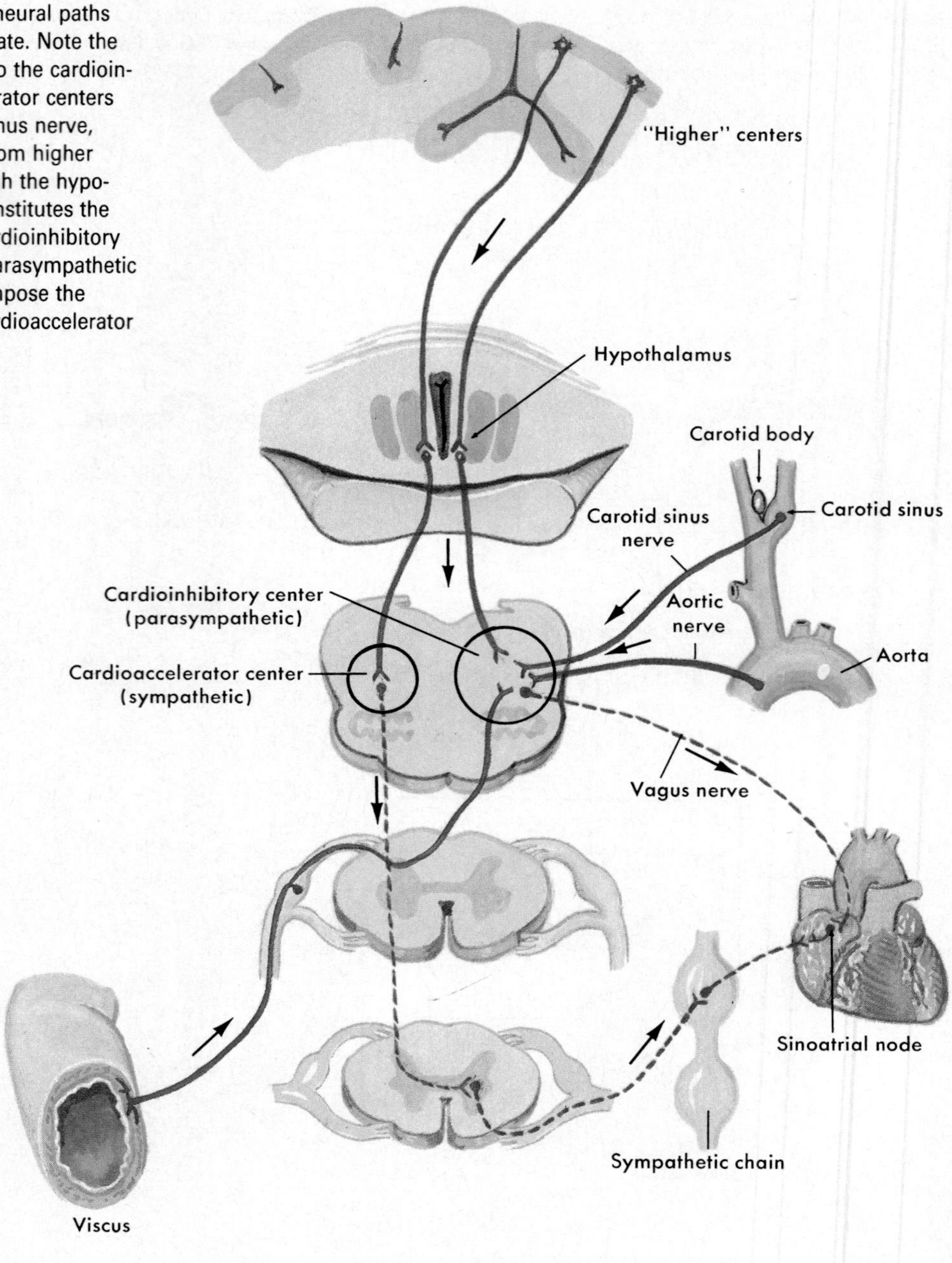

Figure 17-5 Diagram of neural paths for controlling the heart rate. Note the following afferent paths to the cardioinhibitory and cardioaccelerator centers in the medulla: carotid sinus nerve, aortic nerve, and tracts from higher autonomic centers through the hypothalamus. What nerve constitutes the efferent path from the cardioinhibitory center to the heart? Do parasympathetic or sympathetic fibers compose the efferent path from the cardioaccelerator center to the heart?

"Higher" centers

Hypothalamus

Carotid body

Carotid sinus nerve

Carotid sinus

Cardioinhibitory center (parasympathetic)

Aortic nerve

Cardioaccelerator center (sympathetic)

Aorta

Vagus nerve

Sinoatrial node

Sympathetic chain

Viscus

carotid sinus and constitute one of the most important heart rate control mechanisms because of their effect on the cardioinhibitory and accelerator centers—and therefore on parasympathetic and sympathetic outflow.

CAROTID SINUS REFLEX. The carotid sinus is a small dilation at the beginning of the internal carotid artery just above the bifurcation of the common carotid artery to form the internal and external carotid arteries. The sinus lies just under the sternocleidomastoid muscle at the level of the upper margin of the thyroid cartilage. Sensory (afferent) fibers from

carotid sinus baroreceptors run through the carotid sinus nerve (of Hering) and on through the glossopharyngeal (or ninth cranial) nerve to the cardioinhibitory center. Stimulation of these stretch receptors results in a reflex slowing of the heart.

AORTIC REFLEX. Sensory (afferent) fibers also extend from baroreceptors located in the wall of the arch of the aorta through the aortic nerve and then through the vagus (tenth cranial) nerve to terminate in the cardioinhibitory center of the medulla.

• • •

Presumably, emotions produce changes in the heart rate through the influence of impulses from the "higher centers" in the cerebrum via the hypothalamus. Such impulses can influence activity of either the cardioinhibitory or the cardioaccelerator center.

If blood pressure within the aorta or carotid sinus increases suddenly, it stimulates the aortic or carotid baroreceptors, as shown in Figure 17-3. The result of such stimulation is a reflex slowing of the heart. On the other hand, a decrease in aortic or carotid blood pressure usually initiates reflex acceleration of the heart. The lower blood presure decreases the baroreceptors' stimulation. Hence the cardioinhibitory center receives fewer stimulating impulses and the cardioaccelerator center fewer inhibitory impulses. Net result? The heart beats faster. Details of pressoreflex activity will also be included later in the chapter as part of a mechanism that tends to maintain or restore homeostasis of arterial blood pressure.

Miscellaneous Factors that Influence Heart Rate

Included in the miscellaneous category are such important factors as emotions, exercise, hormones, blood temperature, pain, and stimulation of various exteroceptors. Anxiety, fear, and anger often make the heart beat faster. Grief, in contrast, tends to slow it.

In exercise the heart normally accelerates. The mechanism is not definitely known, but it is thought to include impulses from the cerebrum through the hypothalamus to cardiac centers. Epinephrine is the hormone most noted as a cardiac accelerator.

Increased blood temperature or stimulation of skin heat receptors tends to increase the heart rate, and decreased blood temperature or stimulation of skin cold receptors tends to slow it. Sudden intense stimulation of pain receptors in such visceral structure as the gallbladder, ureters, or intestines can result in such slowing of the heart that fainting may result. Figure 17-6 shows the major factors that influence activity of the cardioinhibitory and cardioaccelerator centers.

CARDIAC CYCLE

The term **cardiac cycle** means a complete heartbeat consisting of contraction **(systole)** and relaxation **(diastole)** of both atria plus contraction and relaxation of both ventricles. The two atria contract simultaneously. Then, as the atria relax, the two ven-

Figure 17-6 Scheme to show some of the many factors that influence activity of the cardiac centers in the medulla. Impulses from various receptors are conducted by sensory fibers that terminate in synapses with neurons in cardiac centers. Motor fibers from the centers relay impulses to sympathetic and parasympathetic neurons, which transmit them to the heart. Also streaming into the cardiac centers are impulses from the hypothalamus, presumably part of the pathway by which emotions influence heart rate.

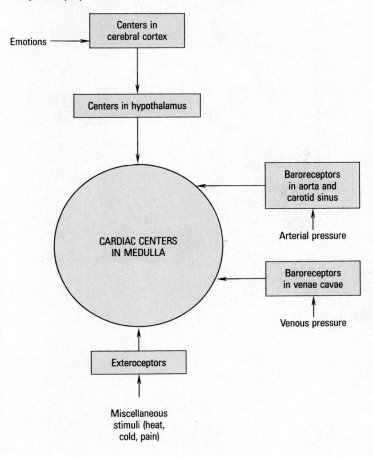

tricles contract and relax, instead of the entire heart contracting as a unit. This gives a kind of milking action to the movements of the heart. The atria remain relaxed during part of the ventricular relaxation and then start the cycle over again. The cycle as a whole is often divided into a number of time intervals for discussion and study. The following list relates several of the important events of the cycle to time, and Figure 17-7 is a composite intended to graphically illustrate and integrate changes in pressure gradients in the left atrium, left ventricle, and aorta, with ECG and heart sound recordings. Aortic blood flow and changes in ventricular volume are also shown.

Figure 17-7 Left atrial, aortic, and left ventricular pressure pulses correlated in time with aortic flow, ventricular volume, heart sounds, venous pulse, and electrocardiogram for a complete cardiac cycle in the dog.

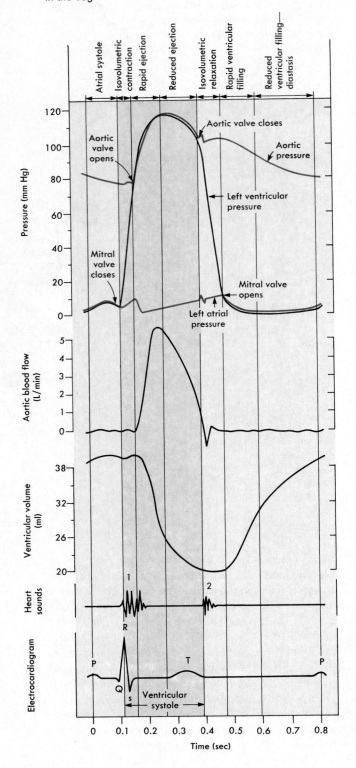

1 **Atrial systole.** The contracting force of the atria completes the emptying of blood out of the atria into the ventricles. Atrioventricular (AV, or cuspid) valves are necessarily open during this phase; the ventricles are relaxed and filling with blood. The semilunar (SL) valves are closed so that blood does not flow on out into the pulmonary artery or aorta. This period of the cycle begins with the P wave of the ECG. Passage of the electrical wave of depolarization is then followed almost immediately by actual contraction of the atrial musculature.

2 **Isovolumetric contraction.** *Iso* is a combining form denoting equality or uniformity. During the brief period of isovolumetric contraction, that is, between the start of ventricular systole and the opening of the SL valves, ventricular volume remains constant, or uniform, as the pressure increases rapidly. The onset of ventricular systole coincides with the R wave of the ECG and the appearance of the first heart sound.

3 **Ejection.** The SL valves open and blood is ejected from the heart when the pressure gradient in the ventricles exceeds the pressure in the pulmonary artery and aorta. An initial, shorter phase called *rapid ejection* is characterized by a marked increase in ventricular and aortic pressure and in aortic blood flow. The T wave of the ECG appears during the later, longer phase of *reduced ejection* (characterized by a less abrupt decrease in ventricular volume).

4 **Isovolumetric relaxation.** Ventricular diastole, or relaxation, begins with this period of the cardiac cycle. It is the period between closure of the SL valves and opening of the AV valves. At the end of ventricular ejection the SL valves will close so that blood cannot reenter the ventricular chambers from the great vessels. The AV valves will not open until the pressure in the atrial chambers increases above that in the relaxing ventricles. The result is a dramatic fall

It is important to note that a considerable quantity of blood, called the **residual volume,** normally remains in the ventricles at the end of the ejection period. In heart failure the residual volume remaining in the ventricles may greatly exceed that ejected during systole.

Heart sounds have clinical significance, since they give information about the valves of the heart. Any variation from normal in the sounds indicates imperfect functioning of the valves. *Heart murmur* is one type of abnormal sound heard frequently and may signify incomplete closing of the valves (valvular insufficiency) or stenosis (constriction, or narrowing) of them.

in intraventricular pressure but no change in volume. Both sets of valves are closed, and the ventricles are relaxing. The second heart sound is heard during this period of the cycle.

5 **Rapid ventricular filling.** Return of venous blood increases intra-atrial pressure until the AV valves are forced open and blood rushes into the relaxing ventricles. The rapid influx lasts about 0.1 second and results in a dramatic increase in ventricular volume.

6 **Reduced ventricular filling (diastasis).** The term *diastasis* is often used to describe a later, longer period of slow ventricular filling at the end of ventricular diastole. The abrupt inflow of blood that occurred immediately after opening of the AV valves is followed by a slow but continuous flow of venous blood into the atria and then through the open AV valves into the ventricles. Diastasis lasts about 0.2 second and is characterized by a gradual increase in ventricular pressure and volume.

Heart Sounds During Cycle

The heart makes certain typical sounds during each cycle that are described as sounding like lubb-dupp through a stethoscope. The first, or systolic, sound is believed to be caused primarily by the contraction of the ventricles and also by vibrations of the closing AV, or cuspid, valves. It is longer and lower than the second, or diastolic, sound, which is short, sharp, and thought to be caused by vibrations of the closing SL valves.

CONTROL OF CIRCULATION

FUNCTIONS OF CONTROL MECHANISMS

Circulation is, of course, a vital function. It constitutes the only means by which cells can receive materials needed for their survival and can have their

wastes removed. Not only is circulation necessary, but circulation of different volumes of blood per minute is also essential for healthy survival. More active cells need more blood per minute than less active cells. The reason underlying this principle is obvious. The more work cells do, the more energy they use and the more oxygen and food they need to supply this energy. Only arterial blood can deliver these energy suppliers. So the more active any part of the body is, the greater the volume of blood circulated to it per minute must be. This requires that circulation control mechanisms accomplish two functions: maintain circulation (keep blood flowing) and vary the volume and distribution of the blood circulated. The greater the activity of any part of the body, the greater the volume of blood it needs circulating through it. Therefore, as any structure increases its activity, an increased volume of blood must be distributed to it—must be shifted from the less active to the more active tissues.

To achieve these two ends, a great many factors must operate together as one smooth-running although complex machine. Incidentally, this is an important physiological principle that you have no doubt observed by now—that every body function depends on many other functions. A constellation of separate processes or mechanisms acts as a single integrated mechanism. Together, they perform one large function. For example, many mechanisms together accomplish the large function we call circulation.

Primary Principle of Circulation

Blood circulates for the same reason that any fluid flows—whether it be water in a river, in a garden hose, or in hospital tubing, or blood in vessels. A fluid flows because a pressure gradient exists between different parts of its bed. This primary fluid flow principle derives from Newton's first and second laws of motion. In essence, these laws state the following principles:

1 A fluid does not flow when the pressure is the same in all parts of it.
2 A fluid flows only when its pressure is higher in one area than in another, and it flows always from its higher pressure area toward its lower pressure area.

In brief, then, the primary principle about circulation is this: blood circulates from the left ventricle to the right atrium of the heart because a blood pressure gradient exists between these two structures. By blood pressure gradient, we mean the difference between the blood pressure in one structure and the blood pressure in another. For example, a typical normal blood pressure in the aorta, as the left ventricle contracts pumping blood into it, is 120 mm Hg; as the left ventricle relaxes, it decreases to

Figure 17-8 Blood pressure gradient. Dotted line indicates the average, or mean, systolic pressure in arteries.

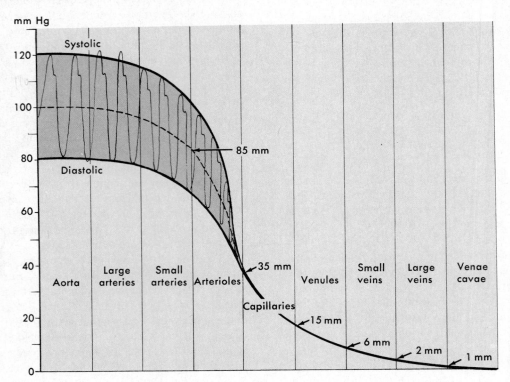

80 mm Hg. The mean, or average, blood pressure, therefore, in the aorta in this instance is 100 mm Hg.

Figure 17-8 shows the systolic and diastolic pressures in the arterial system and illustrates the progressive fall in pressure to 0 mm Hg by the time blood reaches the venae cavae and right atrium. The progressive fall in pressure as blood passes through the circulatory system is directly related to resistance. Resistance to blood flow in the aorta is almost zero. Although the pumping action of the heart causes fluctuations in aortic blood pressure (systolic 120 mm Hg; diastolic 80 mm Hg), the mean pressure remains almost constant, dropping perhaps only 1 or 2 mm Hg. The greatest drop in pressure (about 50 mm Hg) occurs across the arterioles because they present the greatest resistance to blood flow (Figure 17-6).

The symbol $P_1 - P_2$ is often used to stand for a pressure gradient, with P_1 the symbol for the higher pressure and P_2 the symbol for the lower pressure. For example, blood enters the arterioles at 85 mm Hg and leaves at 35 mm Hg. Which is P_1? P_2? What is the blood pressure gradient? It would cause blood to flow from the arterioles into capillaries.

CONTROL OF ARTERIAL BLOOD PRESSURE

The primary determinant of arterial blood pressure is the volume of blood in the arteries. A direct relationship exists between arterial blood volume and arterial pressure. This means that an increase in arterial blood volume tends to increase arterial pressure, and, conversely, a decrease in arterial volume tends to decrease arterial pressure.

Many factors together indirectly determine arterial pressure through their influence on arterial volume. Two of the most important are (1) **cardiac output** and (2) **peripheral resistance** (Figure 17-9).

Cardiac Output

The cardiac output (CO) is determined by both the volume of blood pumped out of the ventricles by each beat (stroke volume) and by the heart rate. Contraction of the heart is called **systole.** Therefore the volume of blood pumped by one contraction is known as *systolic discharge. Stroke volume* means the same thing, the amount of blood pumped by one stroke (contraction) of the ventricle.

Stroke Volume

Stroke volume reflects the force, or strength, of ventricular contraction—the stronger the contraction, the greater the stroke volume tends to be. CO can be computed by the following simple equation:

$$\text{Stroke volume} \times \text{Heart rate} = \text{CO}$$

Figure 17-9 Scheme to show how cardiac minute output and peripheral resistance affect arterial blood pressure. If cardiac minute output increases, the amount of blood entering the arteries increases and tends to increase the volume of blood in the arteries. The resulting increase in arterial volume increases arterial blood pressure. If peripheral resistance increases, it tends to increase arterial blood volume and pressure. Why? Because the increased resistance tends to decrease the amount of blood leaving the arteries, which tends to increase the amount of blood left in them. The increase in arterial volume increases arterial blood pressure.

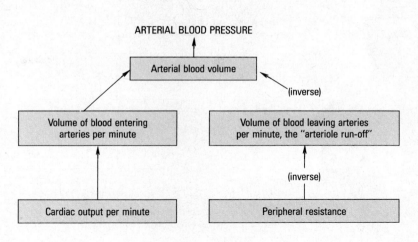

In practice, however, computing the CO is far from simple. It requires introducing a catheter into the right side of the heart (cardiac catheterization) and working a computation known as *Fick's formula.*

Since the heart's rate and stroke volume determine its output, anything that changes either the rate of the heartbeat or its stroke volume tends to change CO, arterial blood volume, and blood pressure in the same direction. In other words, anything that makes the heart beat faster or anything that makes it beat stronger (increases its stroke volume) tends to increase CO and therefore arterial blood volume and pressure. Conversely, anything that causes the heart to beat more slowly or more weakly tends to decrease CO, arterial volume, and blood pressure. But do not overlook the word *tends* in the preceding sentences. A change in heart rate or stroke volume does not always change the heart's output, or the amount of blood in the arteries, or the blood pressure. To see whether this is true, do the following simple arithmetic, using the simple formula for computing CO. Assume a normal rate of 72 beats per minute and a normal stroke volume of 70 ml. Next, suppose the rate drops to 60 and the stroke volume increases to 100. Does the decrease in heart rate actually cause a decrease in CO in this case? Clearly not—the CO increases. Do you think it is valid, however, to say that a slower rate *tends* to decrease the heart's output? By itself, without any change in any other factor, would not a slowing of the heartbeat cause CO volume, arterial volume, and blood pressure to fall?

Mechanical, neural, and chemical factors regulate the strength of the heartbeat and therefore its stroke volume. The mechanical factor that helps determine stroke volume is the length of myocardial fibers at the beginning of ventricular contraction.

Many years ago Starling described a principle later made famous as **Starling's law of the heart.** In this principle he stated the factor he had observed as the main regulator of heartbeat strength in experiments performed on denervated animal hearts. Starling's law of the heart, in essence, is this: within limits, the longer, or more stretched, the heart fibers at the beginning of contraction, the stronger is their contraction.

The factor determining how stretched the animal hearts were at the beginning of contractions was, as you might deduce, the amount of blood in the hearts at the end of diastole. The more blood returned to the hearts per minute, the more stretched were their fibers, the stronger were their contractions, and the larger was the volume of blood they ejected with each contraction. If, however, too much blood stretched the hearts too far, beyond a certain critical point, they seemed to lose their elasticity. They then contracted less vigorously—much as a rubber band, stretched too much, rebounds with less force.

Physiologists have long accepted as fact that Starling's law of the heart operates in animals under experimental conditions. But they have questioned its validity and importance in the intact human body. Today, the prevailing opinion seems to be that Starling's law of the heart operates as a major regulator of stroke volume under ordinary conditions. Operation of Starling's law of the heart ensures that increased amounts of blood returned to the heart will will be pumped out of it. It automatically adjusts CO to venous return under usual conditions.

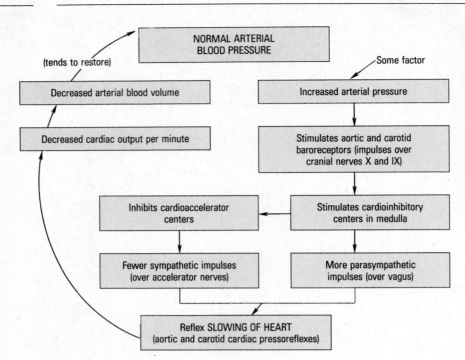

Figure 17-10 Aortic and carotid cardiac pressoreflexes, a mechanism that tends to maintain or restore homeostasis of arterial blood pressure by regulating the rate of the heartbeat. Note that by this mechanism an increase in arterial blood pressure leads to reflex slowing of the heart and tends to lower blood pressure. The converse is also true. A decrease in blood pressure leads to reflex acceleration of the heart and tends to raise the pressure upward toward normal.

Heart Rate

Control mechanisms that influence heart rate have already been discussed. The pressoreflexes constitute the dominant heart rate control mechanism, although various other factors also influence heart rate (see Figures 17-5 and 17-6). Figure 17-10 shows the relationship between homeostasis of arterial blood pressure and activation of cardiac pressoreflexes.

Peripheral Resistance

Peripheral resistance helps determine arterial blood pressure. Specifically, arterial blood pressure tends to vary directly with peripheral resistance. Peripheral resistance means the resistance to blood flow imposed by the force of friction between blood and the walls of its vessels. Friction develops partly because of a characteristic of blood—its viscosity, or stickiness—and partly from the small diameter of arterioles and capillaries. The resistance offered by arterioles, in particular, accounts for almost half of the total resistance in systemic circulation. The muscular coat that arterioles are vested with allows them to constrict or dilate and thus change the amount of resistance to blood flow. Peripheral resistance helps determine arterial pressure by controlling the rate of "arteriole runoff," the amount of blood that runs out of the arteries into the arterioles. The greater the resistance, the less the arteriole runoff, or outflow, tends to be—and, therefore, the more blood left in the arteries and the higher the arterial pressure tends to be.

Blood viscosity stems mainly from the red blood cells but also partly from the protein molecules present in blood. An increase in either blood protein concentration or red blood cell count tends to increase viscosity, and a decrease in either tends to decrease it.

Vasomotor Control Mechanism

Blood distribution patterns, as well as blood pressure, can be influenced by factors that control changes in the diameter of arterioles. Such factors might be said to constitute the **vasomotor control mechanism.** Like most physiological control mechanisms, it consists of many parts. An area in the medulla called the *vasomotor,* or *vasoconstrictor, center* will, when stimulated, initiate an impulse outflow via sympathetic fibers that ends in the smooth muscle surrounding resistance vessels, arterioles, venules, and veins of the "blood reservoirs," causing their constriction. Thus the vasomotor control mech-

> Under normal circumstances, blood viscosity changes very little. But under certain abnormal conditions, such as marked anemia or hemorrhage, a decrease in blood viscosity may be the crucial factor lowering peripheral resistance and arterial pressure, even to the point of circulatory failure.

anism plays a role both in the maintenance of the general blood pressure and in the distribution of blood to areas of special need.

The main blood reservoirs are the venous plexuses and sinuses in the skin and abdominal organs (especially in the liver and spleen). In other words, blood reservoirs are the venous networks in most parts of the body—all but those in the skeletal muscles, heart, and brain. The term *reservoir* is apt, since these vessels serve as storage depots for blood. It can quickly be moved out of them and "shifted" to heart and skeletal muscles when increased activity demands. A change in either arterial blood's oxygen or carbon dioxide content sets a chemical vasomotor control mechanism in operation. A change in arterial blood pressure initiates a *vasomotor pressoreflex.*

VASOMOTOR PRESSOREFLEXES (Figure 17-11). A sudden increase in arterial blood pressure stimulates aortic and carotid baroreceptors—the same ones that initiate cardiac reflexes. Not only does this lead to stimulation of cardioinhibitory centers but also to inhibition of vasoconstrictor centers. More impulses per second go out to the heart over parasympathetic fibers to blood vessels. As a result, the heartbeat slows, and arterioles and the venules of the blood reservoirs dilate. Since sympathetic vasoconstrictor impulses predominate at normal arterial pressures, inhibition of these is considered the major mechanism of vasodilation.

A decrease in arterial pressure causes the aortic and carotid baroreceptors to send more impulses to the medulla's vasoconstrictor centers, thereby stimulating them. These centers then send more impulses via the sympathetic fibers to stimulate vascular smooth muscle and cause vasoconstriction. This squeezes more blood out of the blood reservoirs, increasing the amount of venous blood return to the heart. Eventually, this extra blood is redistributed to more active structures, such as skeletal muscles and heart, because their arterioles become dilated largely from the operation of a local mechanism (p. 485). Thus the vasoconstrictor pressoreflex and the local vasodilating mechanism together serve as an important device for shifting blood from reservoirs to structures that need it more. It is an especially valuable mechanism during exercise.

VASOMOTOR CHEMOREFLEXES (Figure 17-12). Chemoreceptors located in the aortic and carotid bodies are particularly sensitive to a deficiency of blood oxygen **(hypoxia)** and somewhat less sensitive to excess blood carbon dioxide **(hypercapnia)** and to decreased arterial blood pH. When one or more of these conditions stimulates the chemoreceptors, their fibers transmit more impulses to the medulla's vasoconstrictor centers, and vasoconstriction of ar-

Figure 17-11 The aortic and carotid vasomotor pressoreflex mechanism shown here is set in operation when some factor causes a sudden increase in arterial blood pressure. This mechanism and the aortic and carotid cardiac pressoreflexes (Figure 17-10) operate simultaneously to maintain or restore homeostasis of arterial blood pressure.

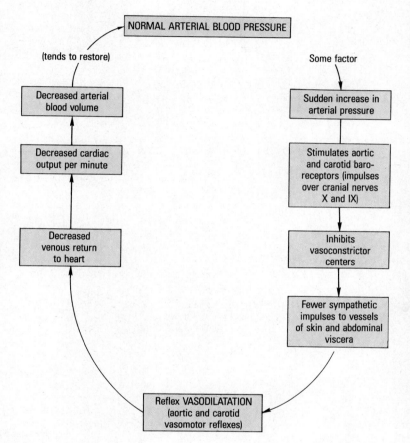

terioles and venous reservoirs soon follows. This stress mechanism functions as an emergency device when severe hypoxia or hypercapnia endangers survival.

MEDULLARY ISCHEMIC REFLEX (Figure 17-13). The medullary ischemic reflex mechanism is said to exert the most powerful control of all on small blood vessels. When the blood supply to the medulla becomes inadequate **(ischemic),** its neurons suffer from both oxygen deficiency and carbon dioxide excess. But, presumably, it is hypercapnia that intensely and directly

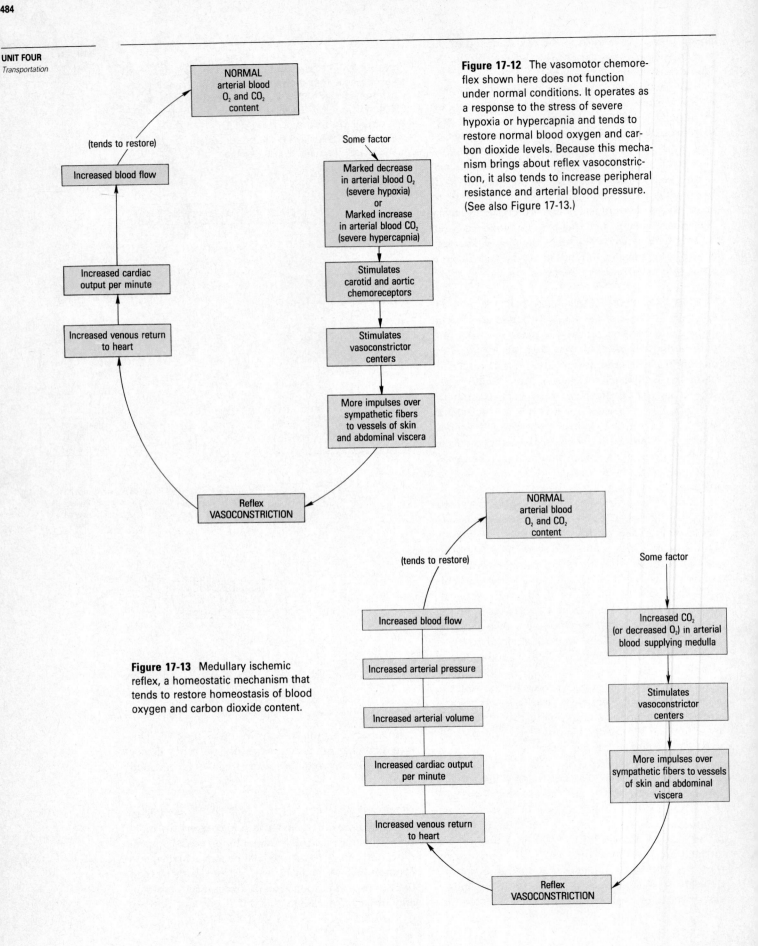

Figure 17-12 The vasomotor chemore-flex shown here does not function under normal conditions. It operates as a response to the stress of severe hypoxia or hypercapnia and tends to restore normal blood oxygen and carbon dioxide levels. Because this mechanism brings about reflex vasoconstriction, it also tends to increase peripheral resistance and arterial blood pressure. (See also Figure 17-13.)

Figure 17-13 Medullary ischemic reflex, a homeostatic mechanism that tends to restore homeostasis of blood oxygen and carbon dioxide content.

Figure 17-14 Scheme to show some of the many factors that influence activity of the vasomotor centers in the medulla. Impulses from chemoreceptors and baroreceptors are conducted by sensory fibers that terminate in synapses with neurons in vasomotor centers. From the centers, sympathetic neurons relay impulses to smooth muscle in blood vessels. More impulses cause vessels to constrict. Fewer impulses cause them to dilate. Also streaming into the vasomotor centers are impulses from the hypothalamus, presumably part of the pathway by which emotions influence blood pressure.

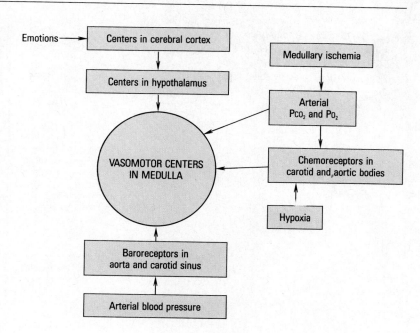

stimulates the vasoconstrictor centers to bring about marked arteriole and venous constriction. If the oxygen supply to the medulla decreases below a certain level, its neurons, of course, cannot function, and the medullary ischemic reflex cannot operate.

VASOMOTOR CONTROL BY HIGHER BRAIN CENTERS (Figure 17-14). Impulses from centers in the cerebral cortex and in the hypothalamus are believed to be transmitted to the vasomotor centers in the medulla and to thereby help control vasoconstriction and dilation. One piece of evidence supporting this view is that vasoconstriction and a rise in arterial blood pressure characteristically accompany emotions of intense fear or anger. Also, laboratory experiments on animals in which stimulation of the posterior or lateral parts of the hypothalamus leads to vasoconstriction support the belief that higher brain centers influence the vasomotor centers in the medulla.

Local Control of Arterioles

Some kind of local mechanism operates to produce vasodilation in localized areas. Although the mechanism is not clear, it is known to function in times of increased tissue activity. For example, it probably accounts for the increased blood flow into skeletal muscles during exercise. It also operates in ischemic tissues, serving as a homeostatic mechanism that tends to restore normal blood flow. Norepinephrine, histamine, lactic acid, and other locally produced substances have been suggested as the stimuli that activate the local vasodilator mechanism. Local vasodilation is also referred to as **reactive hyperemia.**

Summarizing briefly, the volume of blood circulating through the body per minute is determined by the magnitude of both the blood pressure gradient and the peripheral resistance (Figures 17-15 to 17-18).

A nineteenth century physiologist and physicist, Poiseuille, described the relation between these three factors—pressure gradient, resistance, and volume of fluid flow per minute—with a mathematical equation known as *Poiseuille's law.* In general, but with certain modifications, it applies to blood circulation. We can state it in a simplified form as follows: the volume of blood circulated per minute is directly related to mean arterial pressure minus central venous pressure and inversely related to resistance:

$$\text{Volume of blood circulated per minute} = \frac{\text{Mean arterial pressure} - \text{Central venous pressure}}{\text{Resistance}}$$

The preceding statement and equation need qualifying with regard to the influence of peripheral resistance on circulation. For instance, according to the equation, an increase in peripheral resistance would tend to decrease blood flow. (Why? Increasing peripheral resistance increases the denominator of the fraction in the preceding equation. Increasing the denominator of any fraction necessarily does what to its value?)

Increased peripheral resistance, however, has a secondary action that opposes its primary tendency to decrease blood flow. An increase in peripheral resistance hinders or decreases arteriole runoff. This, of course, tends to increase the volume of

Figure 17-15 Scheme to suggest some of the many factors that influence heart action and blood vessel diameter and thereby control circulation. (See also Figures 17-16 and 17-17.)

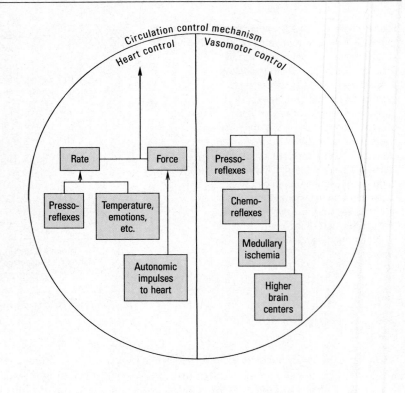

Figure 17-16 Another scheme to show some of the many parts of the complex circulation control mechanism. The volume of blood circulating through the body per minute is directly related to arterial blood pressure and inversely related to peripheral resistance. Note, however, that a great many factors act together to regulate arterial blood pressure and peripheral resistance.

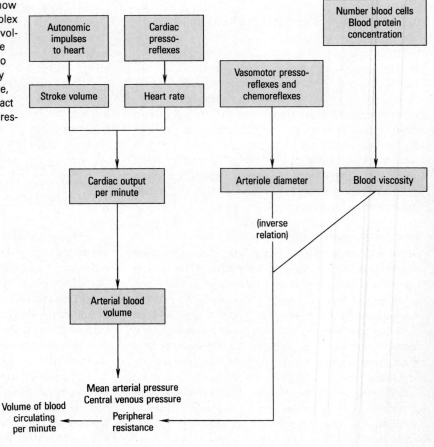

Figure 17-17 Factors that determine the systemic blood pressure gradient.

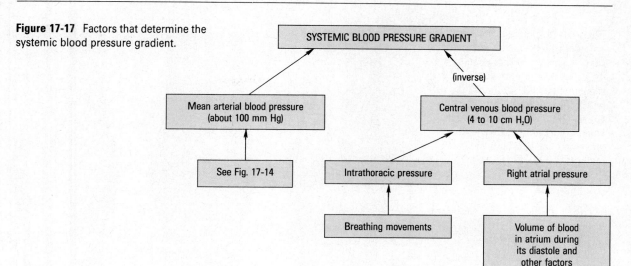

Figure 17-18 The main determinants of peripheral resistance.

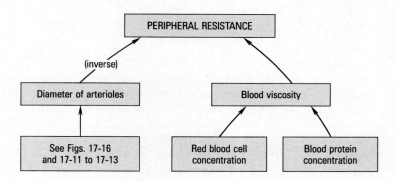

blood left in the arteries and so tends to increase arterial pressure. Note also that increasing arterial pressure tends to increase the value of the fraction in Poiseuille's equation. Therefore it tends to increase circulation. In short, to say unequivocally what the effect of an increased peripheral resistance will be on circulation is impossible. It depends also on arterial blood pressure—whether it increases, decreases, or stays the same when peripheral resistance increases. The clinical condition arteriosclerosis with hypertension (high blood pressure) illustrates this point. Both peripheral resistance and arterial pressure are increased in this condition. If resistance were to increase more than arterial pressure, circulation, that is, volume of blood flow per minute, would decrease. But if arterial pressure increases proportionately to resistance, circulation remains normal.

IMPORTANT FACTORS INFLUENCING VENOUS RETURN TO HEART

Two important factors that promote the return of venous blood to the heart are respirations and skel-

etal muscle contractions. Both produce their facilitating effect on venous return by increasing the pressure gradient between peripheral veins and venae cavae.

The process of inspiration increases the pressure gradient between peripheral and central veins by decreasing central venous pressure and also by increasing peripheral venous pressure. Each time the diaphragm contracts, the thoracic cavity necessarily becomes larger and the abdominal cavity smaller. Therefore the pressures in the thoracic cavity, in the thoracic portion of the vena cava, and in the atria decrease, and those in the abdominal cavity and the abdominal veins increase. Deeper respirations intensify these effects and therefore tend to increase venous return to the heart more than do normal respirations. This is part of the reason why the principle is true that increased respirations and increased circulation tend to go hand in hand.

Skeletal muscle contractions serve as "booster pumps" for the heart. They promote venous return in the following way. As each skeletal muscle contracts, it squeezes the soft veins scattered through

Figure 17-19 Diagram showing the action of venous valves. **A**, External view of vein showing dilation at site of the valve. **B**, Interior of vein with the semilunar flaps in open position, permitting flow of blood through the valve. **C**, Valve flaps approximating each other, occluding the cavity and preventing the backflow of blood.

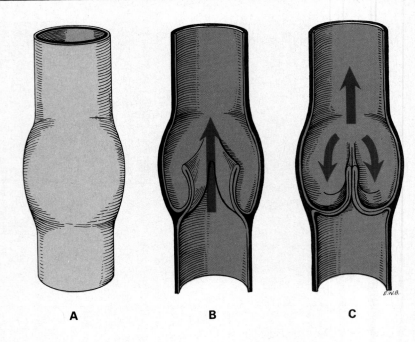

A   B   C

its interior, thereby milking the blood in them upward, or toward the heart. The closing of the semilunar valves present in veins prevents blood from falling back as the muscle relaxes. Their flaps catch the blood as gravity pulls backward on it (Figure 17-19). The net effect of skeletal muscle contraction plus venous valvular action, therefore, is to move venous blood toward the heart, to increase the venous return.

The value of skeletal muscle contractions in moving blood through veins is illustrated by a common experience. Who has not noticed how much more uncomfortable and tiring standing still is than walking? After several minutes of standing quietly, the feet and legs feel "full" and swollen. Blood has accumulated in the veins because the skeletal muscles are not contracting and squeezing it upward. The repeated contractions of the muscles when walking, on the other hand, keep the blood moving in the veins and prevent the discomfort of distended veins.

BLOOD PRESSURE

CLINICAL MEASUREMENT OF ARTERIAL BLOOD PRESSURE

Blood pressure is measured with the aid of an apparatus known as a sphygmomanometer, which makes it possible to measure the amount of air pressure equal to the blood pressure in an artery. The measurement is made in terms of how many millimeters high the air pressure raises a column of mercury in a glass tube.

The sphygmomanometer consists of a rubber cuff attached by a rubber tube to a compressible bulb and by another tube to a column of mercury that is marked off in millimeters. The cuff is wrapped around the arm over the brachial artery, and air is pumped into the cuff by means of the bulb. In this way, air pressure is exerted against the outside of the artery. Air is added until the air pressure exceeds the blood pressure within the artery or, in other words, until it compresses the artery. At this time, no pulse can be heard through a stethoscope placed over the brachial artery at the bend of the elbow along the inner margin of the biceps muscle. By slowly releasing the air in the cuff the air pressure is decreased until it approximately equals the blood pressure within the artery. At this point the vessel opens slightly and a small spurt of blood comes through, producing the first sound, one with a rather

Clinically, **diastolic pressure** is considered more important than systolic pressure because it indicates the pressure, or strain, to which blood vessel walls are constantly subjected. It also reflects the condition of the peripheral vessels, since diastolic pressure rises or falls with the peripheral resistance. If, for instance, arteries are sclerosed, peripheral resistance and diastolic pressure both increase.

Review Questions

1 Identify, locate, and describe the function of each of the following structures: SA node, AV node, AV bundle, and Purkinje fibers.

2 Compare the intrinsic rhythm of the SA node with other components of the heart's conduction system. What is an ectopic pacemaker? Artificial pacemaker?

3 What does an electrocardiogram measure and record? List the normal ECG deflection waves and intervals. What do the various ECG waves represent?

4 Discuss and compare the effects of sympathetic and parasympathetic stimulation on heart rate. What effect would vagal stimulation have on heart rate?

5 Locate and describe the function of the cardioinhibitory and cardioaccelerator centers.

6 Explain the mechanism of action of the cardiac pressoreflexes on heart rate.

7 List and give the effect of several "miscellaneous" factors such as grief or pain on heart rate.

8 What is meant by the term *cardiac cycle?*

9 List the "periods" of the cardiac cycle and briefly describe the events that occur in each. Refer to Figure 17-7 as you prepare your answer.

10 What is meant by the term *residual volume* as it applies to the heart?

11 Describe and explain the origin of the heart sounds.

12 State in your own words the basic principle of fluid flow.

13 What blood vessels present the greatest resistance to blood flow?

14 What is the primary determinant of arterial blood pressure?

15 List the two most important factors that indirectly determine arterial pressure through their influence on arterial volume.

16 How is cardiac output determined?

17 What two factors determine blood viscosity? What does viscosity mean? Give an example of a condition in which blood viscosity decreases. Explain its effect on circulation.

18 What mechanisms control peripheral resistance? Cite an example of the operation of one or more parts of this mechanism to increase resistance; to decrease it.

19 What is Starling's law of the heart?

20 What is arteriole runoff? What is the relationship of arteriole runoff to peripheral resistance?

21 What are the components of the vasomotor control mechanism?

22 Explain the mechanism of action of the medullary ischemic reflex.

23 State in your own words Poiseuille's law. Give an example of increased circulation to illustrate application of this law. Give an example of decreased circulation to illustrate application of this law.

24 What effect, if any, would a respiratory stimulant drug have on circulation? Explain why it would or would not affect circulation.

25 Describe the measurement of arterial blood pressure clinically.

26 List the so-called pressure points at which pressure can be applied to stop arterial bleeding.

18 THE LYMPHATIC SYSTEM

OBJECTIVES

After you have completed this chapter, you should be able to:

1 Describe the generalized functions of the lymphatic system and list the primary lymphatic structures.
2 Compare the chemical structure of lymph and interstitial fluid.
3 Discuss the formation, distribution, and general body plan of lymphatic drainage through the right lymphatic duct and the thoracic duct.
4 Compare the structure of lymphatic vessels and veins.
5 Discuss the specialized function of the lymphatic system in absorption of fats and other nutrients from the small intestine.
6 Discuss the "lymphatic pump" and other lymphokinetic actions that result in central movement, or flow, of lymph.
7 Describe and correlate the structure of lymph nodes with their function as biological filters.
8 Give the location of the major groups, or clusters, of lymph nodes in the body and identify their two primary functions.
9 Discuss the lymphatic drainage of the breast.
10 Locate the thymus in the body and discuss its gross and microscopic anatomy.
11 Discuss the functions of the thymus that result in its designation as the primary central organ of the lymphatic system.
12 Discuss the location, structure, and functions of the spleen.

KEY TERMS

Chyle (kiil)

Cisterna chyli (sis-TER-nah KI-le)

Cortical nodule (KOR-ti-kal NOD-ul)

Hassall's corpuscle (HAS-als COR-pus-el)

Involution (in-vo-LU-shun)

Lacteal (LAK-te-al)

Lymph (limf)

Lymphatic (lim-FAT-ik)

Lymphography (lim-FOG-rah-fe)

Lymphokinetic (lim-fo-KIN-et-ic)

Node

Reticuloendothelial (re-TIC-u-lo-en-do-THE-le-al)

DEFINITIONS

Lymphatic System

The lymphatic system is actually a specialized component of the circulatory system, since it consists of a moving fluid (lymph) derived from the blood and tissue fluid and a group of vessels (lymphatics) that returns the lymph to the blood. In general, the lymphatic vessels that drain the peripheral areas of the body parallel the venous return. In addition to lymph and the lymphatic vessels, the system includes lymph nodes located along the paths of the collecting vessels (Figure 18-1), isolated nodules of lymphatic tissue such as Peyer's patches in the intestinal wall, and specialized lymphatic organs such as the tonsils, thymus, and spleen.

Although it serves a unique transport function by returning tissue fluid, proteins, fats, and other substances to the general circulation, lymph flow differs from the true "circulation" of blood seen in the cardiovascular system. The lymphatic vessels do not, like vessels in the blood vascular system, form a closed ring, or circuit, but instead begin blindly in the intercellular spaces of the soft tissues of the body.

Figure 18-1 The lymphatic vessels, larger lymphatic trunks, and lymph nodes.

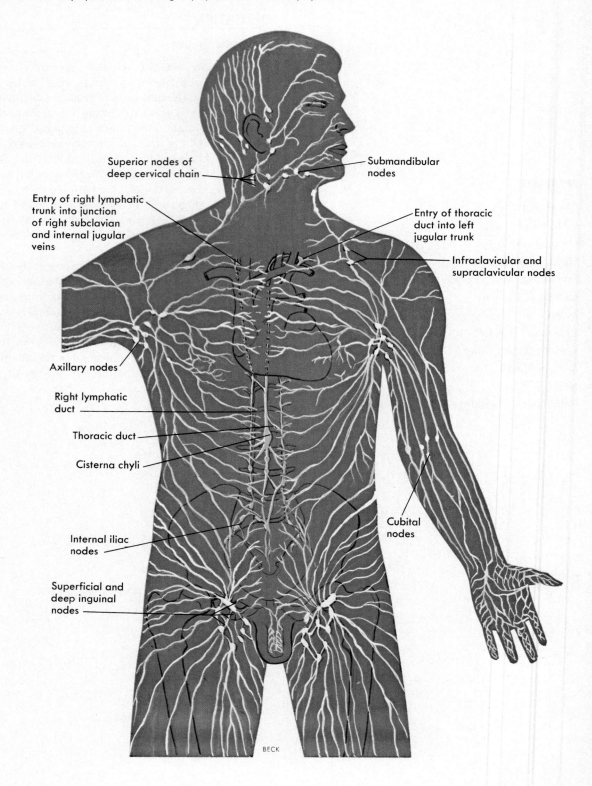

Superior nodes of
deep cervical chain

Submandibular
nodes

Entry of right lymphatic
trunk into junction
of right subclavian
and internal jugular
veins

Entry of thoracic
duct into left
jugular trunk

Infraclavicular and
supraclavicular nodes

Axillary nodes

Right lymphatic
duct

Thoracic duct

Cisterna chyli

Cubital
nodes

Internal iliac
nodes

Superficial and
deep inguinal
nodes

BECK

Lymph and Interstitial Fluid (Tissue Fluid)

Lymph is the clear, watery-appearing fluid found in the lymphatic vessels. Interstitial fluid, which fills the spaces between the cells, is not the clear watery fluid it seems to be. Recent studies show that it is a complex and "organized" material. In some tissues, it is part of a semifluid ground substance. In others, it is the bound water in a gelatinous ground substance. Interstitial fluid and blood plasma together constitute the extracellular fluid, or in the words of Claude Bernard, the "internal environment of the body"—the fluid environment of cells in contrast to the atmosphere, or external environment, of the body.

Both lymph and interstitial fluid closely resemble blood plasma in composition. The main difference is that they contain a lower percentage of proteins than does plasma. Lymph is isotonic and almost identical in chemical composition to interstitial fluid when comparisons are made between the two fluids taken from the same area of the body. However, the average concentration of protein (4 gm/100 ml) in lymph taken from the thoracic duct (Figures 18-1 and 18-3) is about twice that found in most interstitial fluid samples. The elevated protein level of thoracic duct lymph (a mixture of lymph from all areas of the body) results from protein-rich lymph flowing into the duct from the liver and small intestine. A little over one half of the 2,500 to 2,800 ml total daily lymph flowing through the thoracic duct is derived from these two organs.

LYMPHATICS

Formation and Distribution

Lymphatic vessels originate as microscopic blind-end vessels called **lymphatic capillaries.** (Those originating in the villi of the small intestine are called **lacteals.**) The wall of the lymphatic capillary consists of a single layer of flattened endothelial cells. Each blindly ending capillary is attached, or fixed, to surrounding cells by tiny connective tissue filaments. Networks of lymphatic capillaries, which branch and anastomose freely, are located in the intercellular spaces and are widely distributed throughout the body. As a rule of thumb, lymphatic and blood capillary networks lie side by side but are always independent of each other.

As twigs of a tree join to form branches, branches join to form larger branches, and large branches join to form the tree trunk, so do lymphatic capillaries merge, forming slightly larger lymphatics that join other lymphatics to form still larger vessels, which merge to form the main lymphatic trunks: the **right lymphatic ducts** and the **thoracic duct.** Lymph from the entire body, except the upper right

Lymph does not clot. Therefore if damage to the main lymphatic trunks in the thorax should occur as a result of penetrating injury, the flow of lymph must be stopped surgically or death ensues. It is impossible to maintain adequate serum protein levels by dietary means if significant loss of lymph continues over time. As lymph is lost, rapid emaciation occurs, with a progressive and eventually fatal decrease in total blood fat and protein levels.

quadrant (Figure 18-2), drains eventually into the thoracic duct, which drains into the left subclavian vein at the point where it joins the left internal jugular vein. Lymph from the upper right quadrant of the body empties into the right lymphatic duct (or, more commonly, into three collecting ducts) and then into the right subclavian vein. Since most of the lymph of the body returns to the bloodstream by way of the thoracic duct, this vessel is consid-

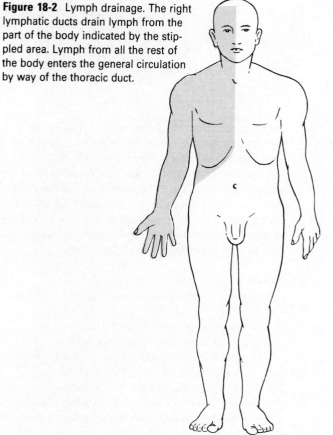

Figure 18-2 Lymph drainage. The right lymphatic ducts drain lymph from the part of the body indicated by the stippled area. Lymph from all the rest of the body enters the general circulation by way of the thoracic duct.

Figure 18-3 Position of the cisterna chyli and the thoracic duct and its tributaries and the entry of the duct into the junction of the left internal jugular and left subclavian veins to form the innominate veins.

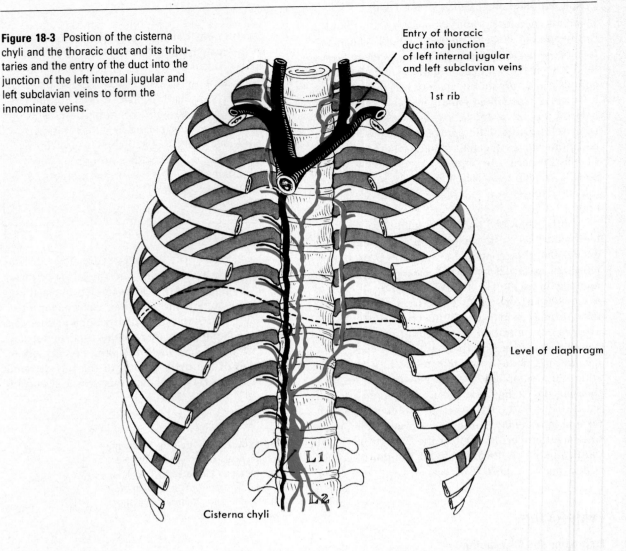

Entry of thoracic duct into junction of left internal jugular and left subclavian veins

1st rib

Level of diaphragm

L1

L2

Cisterna chyli

erably larger than the other main lymph channels, the right lymphatic ducts, but is much smaller than the large veins, which it resembles in structure. It has a diameter about the size of a goose quill and a length of about 40 cm. It originates as a dilated structure, the **cisterna chyli,** in the lumbar region of the abdominal cavity and ascends by a flexuous course to the root of the neck, where it joins the subclavian vein as just described (see Figure 18-3).

Structure

Lymphatics resemble veins in structure with these exceptions:

1 Lymphatics have thinner walls.
2 Lymphatics contain more valves.
3 Lymphatics contain lymph nodes located at certain intervals along their course.

The lymphatic capillary wall is formed by a single layer of large but very thin and flat endothelial cells. In the past, many investigators described what they believed were rather large and naturally oc-

curring openings ("stomata") between adjacent and endothelial cells in the lymphatic capillary wall. Recent and more sophisticated injection techniques that are now being used to prepare lymphatic vessels for study have shown that such openings are in reality only preparation artifacts. Although very small intercellular openings (clefts) do exist between adjacent endothelial cells in the lymphatic capillary, there is no direct and open communication between the vessel lumen and the surrounding tissue spaces.

As lymph flows from the thin-walled capillaries into vessels with a larger diameter (0.2 to 0.3 mm), the walls become thicker and exhibit the three coats, or layers, typical of arteries and veins (see Table 16-2, p. 453). Interlacing elastic fibers and several strata of circular smooth muscle bundles are found in both the tunica media and the tunica adventitia of the large lymphatic vessel wall. Boundaries between layers, or coats, are less distinct in the thinner lymphatic vessel walls than in arteries or veins.

Semilunar valves are extremely numerous in

lymphatics of all sizes and give the vessels a somewhat varicose appearance. Valves are present every few millimeters in large lymphatics and are even more numerous in the smaller vessels. Formed from folds of the tunica intima, each valve projects into the vessel lumen in a slightly expanded area circled by bundles of smooth muscle fibers.

Experimental evidence suggests that most lymph vessels have the capacity for repair or regeneration when damaged. Formation of new lymphatic vessels occurs by extension of solid cellular cores, or sprouts, formed by mitotic division of endothelial cells in existing vessels, which later become "canalized."

Functions

The lymphatics play a critical role in a number of interrelated homeostatic mechanisms. The high degree of permeability of the lymphatic capillary wall permits large molecular weight substances and even particulate matter, which cannot be absorbed into a blood capillary, to be removed from the interstitial spaces. Proteins that accumulate in the tissue spaces can return to blood only via lymphatics. This fact has great clinical importance. For instance, if anything blocks lymphatic return, blood protein concentration and blood osmotic pressure soon fall below normal, and fluid imbalance and death will result (discussed in Chapter 25).

Lacteals (lymphatics in the villi of the small intestine) serve an important function in the absorption of fats and other nutrients. The milky lymph found in lacteals after digestion contains 1% to 2% fat and is called **chyle.** Interstitial fluid has a much lower lipid content than chyle.

LYMPH CIRCULATION

Water and solutes continually filter out of capillary blood into the interstitial fluid. To balance this outflow, fluid continually reenters blood from the interstitial fluid. Newer evidence has disproved the old idea that healthy capillaries do not "leak" proteins. In truth, each day about 50% of the total blood proteins leak out of the capillaries into the tissue fluid and return to the blood by way of the lymphatic vessels. For more details about fluid exchange between blood and interstitial fluid, see Chapter 25. From lymphatic capillaries, lymph flows through progressively larger lymphatic vessels to eventually reenter blood at the junction of the internal jugular and subclavian veins (see Figure 18-3).

The "Lymphatic Pump"

Although there is no muscular pumping organ connected with the lymphatic vessels to force lymph

An understanding of the anatomy of lymphatic drainage of the skin is critically important in surgical amputation of an extremity. A majority of the lymphatics draining the skin are located in plexuslike networks lying on the deep fascia. To prevent stasis of lymph in the stump following amputation, the surgeon retains the deep fascia and its lymphatic vessels with the skin flaps that are used to cover the cut end of the extremity. This procedure results in minimal edema and swelling—a matter of obvious importance to the patient and artificial-limb fitter.

onward as the heart does blood, still lymph moves slowly and steadily along in its vessels. Lymph flows through the thoracic duct and reenters the general circulation at the rate of about 125 ml per hour. This occurs despite the fact that most of the flow is uphill. It moves through the system in the right direction because of the large number of valves that permit fluid flow only in the central direction. What mechanisms establish the pressure gradient required by the basic law of fluid flow? Two of the same mechanisms that contribute to the blood pressure gradient also establish a lymph pressure gradient. These are breathing movements and skeletal muscle contractions.

Activities that result in central movement, or flow, of lymph are called **lymphokinetic** actions (from the Greek *kinetos*, "movable"). X-ray films taken after radiopaque material is injected into the lymphatics **(lymphography)** show that lymph pours into the central veins most rapidly at the peak of inspiration.

The mechanism of inspiration, resulting from the descent of the diaphragm, causes intra-abdominal pressure to increase as intrathoracic pressure decreases. This simultaneously causes pressure to increase in the abdominal portion of the thoracic duct and to decrease in the thoracic portion. In other words, the process of inspiring establishes a pressure gradient in the thoracic duct that causes lymph to flow upward through it.

Research studies have shown that thoracic duct lymph is literally "pumped" into the venous system during the inspiration phase of pulmonary ventilation. The rate of flow, or ejection, of lymph into the venous circulation is proportional to the depth of inspiration. The total volume of lymph that enters the central veins during a given time period depends

Figure 18-4 Structure of a lymph node. Several afferent valved lymphatics bring lymph to the node. An efferent lymphatic leaves the node at the hilus. Note that the artery and vein enter and leave at the hilus.

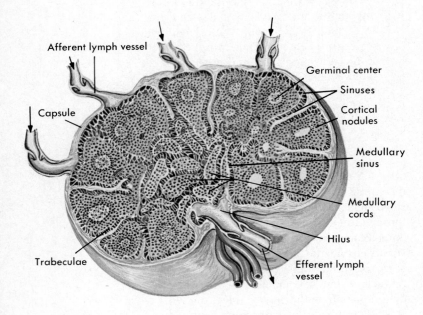

Afferent lymph vessel

Capsule

Trabeculae

Germinal center

Sinuses

Cortical nodules

Medullary sinus

Medullary cords

Hilus

Efferent lymph vessel

on both the depth of the inspiration phase and the overall breathing rate.

Contracting skeletal muscles also exerts pressure on the lymphatics to push the lymph forward. During exercise, lymph flow may increase as much as 10- to 15-fold. In addition, segmental contraction of the walls of the lymphatics themselves results in lymph being pumped from one valved segment to the next.

Other pressure-generating factors that can compress the lymphatics also contribute to the effectiveness of the "lymphatic pump." Examples of such lymphokinetic factors include arterial pulsations, postural changes, and passive compression of the body soft tissues.

The fact that changes in the rate and depth of pulmonary ventilation, coupled with other lymphokinetic activities, can effectively regulate levels of both oxygen and lymph (a complex nutritive) in circulating blood may have great clinical significance in athletes and patients with cardiovascular or pulmonary diseases. In addition to metabolic interaction and regulation, this mechanism of "biocarburetion," or controlled mixing of blood oxygen and lymph, may also have immunological significance.

LYMPH NODES

Structure

Lymph nodes, or glands, as some people call them, are oval-shaped or bean-shaped structures. Some are as small as a pinhead and others as large as a lima bean. Each lymph node (from 1 mm to over 20 mm in diameter) is enclosed by a fibrous capsule. Note in Figure 18-4 that lymph moves into a node via several afferent lymphatic vessels and emerges by one efferent vessel. Think of a lymph node as a biological filter placed in the channel of a number of afferent lymph vessels. Once lymph enters the node, it "percolates" slowly through the spaces known as sinuses before draining into the efferent exit vessel. One-way valves in both the afferent and efferent vessels keep lymph flowing in one direction.

Fibrous septa, or *trabeculae*, extend from the covering capsule toward the center of the node. **Cortical nodules** found along the periphery or outer surface of the node are separated from each other by these connective tissue trabeculae. Each cortical nodule is composed of packed lymphocytes that surround a less dense area called a germinal center (Figures 18-4 to 18-6).

Figure 18-5 is a low-power ($\times 35$) light micrograph of a typical lymph node. A cortical nodule is enclosed by a rectangle in Figure 18-5 and enlarged ($\times 70$) in Figure 18-6. Lymphocytes are produced within the less dense germinal center of the nodule and then pushed to the more densely packed outer layers as they mature. The center, or medulla, of a lymph node is composed of sinuses and cords (see Figure 18-4). Both the cortical and medullary sinuses are lined with specialized reticuloendothelial cells (fixed macrophages) capable of phagocytosis.

Locations

With the exception of comparatively few single nodes, most of the lymph nodes occur in groups, or clusters, in certain areas. The group locations of greatest clinical importance are as follows:

1 **Submental and submaxillary groups** in the floor of the mouth—lymph from the nose, lips, and teeth drains through these nodes.

2 **Superficial cervical glands** in the neck along the sternocleidomastoid muscle—these nodes drain lymph from the head (which has already passed through other nodes) and neck.

3 **Superficial cubital** or **supratrochlear nodes** located just above the bend of the elbow—lymph from the forearm passes through these nodes.

4 **Axillary nodes** (20 to 30 large nodes clustered deep within the underarm and upper

Structure

As Figure 18-10 shows, the spleen is roughly ovoid in shape. Its size varies greatly in different individuals and in the same individual at different times. For example, it hypertrophies during infectious diseases and atrophies in old age. Within the spleen, lymphocytes, monocytes, and neutrophils crowd the numerous spaces formed by a meshwork of interlacing fibers.

Functions

The spleen has long puzzled physiologists who have ascribed many and sundry functions to it. According to present-day knowledge, it performs several functions: defense, hemopoiesis, and red blood cell and platelet destruction; it also serves as a reservoir for blood.

1 **Defense.** As blood passes through the sinusoids of the spleen, reticuloendothelial cells (macrophages) lining these venous spaces remove microorganisms from the blood and destroy them by phagocytosis. Therefore the spleen plays a part in the body's defense against microorganisms.

2 **Hemopoiesis.** Nongranular leukocytes, that is, monocytes and lymphocytes, and plasma cells are formed in the spleen. Before birth, red blood cells are also formed in the spleen, but after birth the spleen is said to form red blood cells only in extreme hemolytic anemia.

3 **Red blood cell and platelet destruction.** Macrophages lining the spleen's sinusoids remove worn-out red blood cells and imperfect platelets from the blood and destroy them by phagocytosis. They also break apart the hemoglobin molecules from the destroyed red blood cells and salvage their iron and globin content by returning them to the bloodstream for storage in bone marrow and liver.

4 **Blood reservoir.** The pulp of the spleen and its venous sinuses store considerable blood. Its normal volume of about 350 ml is said to decrease about 200 ml in less than a minute's time following sympathetic stimulation that produces marked constriction of its smooth capsule. This "self-transfusion" occurs, for example, as a response to the stress imposed by hemorrhage.

Figure 18-10 Spleen, medial aspect.

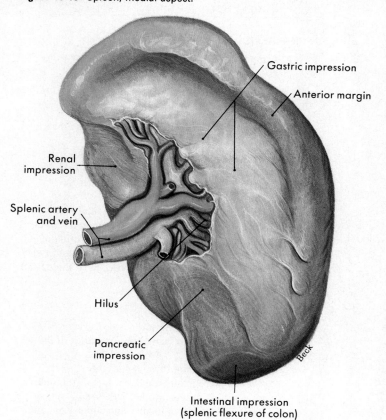

Gastric impression

Anterior margin

Renal impression

Splenic artery and vein

Hilus

Pancreatic impression

Intestinal impression (splenic flexure of colon)

Although the spleen's functions make it a most useful organ, it is not a vital one. Dr. Charles Austin Doan in 1933 took the daring step of performing the first splenectomy. He removed the spleen from a 4-year-old girl who was dying of hemolytic anemia. Presumably, he justified his radical treatment on the basis of what was then merely conjecture, that is, that the spleen destroys red blood cells. The child recovered, and Dr. Doan's operation proved to be a landmark. It created a great upsurge of interest in the spleen and led to many investigations of it.

Outline Summary

DEFINITIONS

A Lymphatic system—part of circulatory system—consists of lymph, interstitial (tissue) fluid, lymphatics, lymph nodes, isolated nodules of lymphatic tissue, tonsils, thymus, and spleen

B Lymph and interstitial fluid (tissue fluid)
  1 Lymph—clear, watery fluid found in lymphatic vessels
  2 Interstitial fluid (tissue fluid)—complex and "organized" material that fills spaces between cells; interstitial fluid and blood together constitute extracellular fluid
  3 Composition of lymph and interstitial fluid—similar to that of blood plasma; main difference is that plasma contains higher concentration of proteins; lymph in thoracic duct has about twice as high protein concentration as most interstitial fluid

LYMPHATICS

A Formation and distribution
  1 Start as blind-end lymphatic capillaries in tissue spaces
  2 Widely distributed throughout body
  3 Two or more main lymphatic ducts—thoracic duct, which drains into left subclavian vein at junction of internal jugular and subclavian, and one or more right lymphatic ducts, which drain into right subclavian vein
  4 Lacteals—lymphatics originating in intestinal villi; after fatty meal contain milky lymph called chyle

B Structure
  1 Similar to veins except thinner walled
  2 Contain more valves and contain lymph nodes located at intervals

C Functions
  1 Return water and proteins from interstitial fluid to blood from which they came
  2 Lacteals absorb fats and other nutrients

LYMPH CIRCULATION

A Direction of flow—water and solutes filter out of capillary blood into interstitial fluid, enter lymphatics, and move through them to return to blood at junction of internal jugular and subclavian veins

B The lymphatic pump
  1 Lymph flow averages 125 ml per hour
  2 Mechanisms that contribute to effectiveness of "lymphatic pump"
    a Breathing movements
    b Skeletal muscle contractions
    c Arterial pulsations
    d Contraction of lymphatic walls

LYMPH NODES

A Structure
  1 Lymphatic tissue, separated into compartments by fibrous partitions
  2 Afferent lymphatics enter each node and efferent lymphatics leave each node

B Locations—usually in clusters (see pp. 500 and 502)

C Functions
  1 Defense functions—filter out injurious substances and phagocytose them
  2 Hemopoiesis—formation of lymphocytes and monocytes

LYMPHATIC DRAINAGE OF THE BREAST

A Location and distribution of lymphatics
  1 Two sets of lymphatic vessels—superficial lymphatics drain skin of breast except of areola and nipple; deep lymphatics drain substance of breast and skin of areola and nipple
  2 Subareolar plexus (plexus of Sappey)—under areola
  3 Lymphatic anastomoses
    a Between superficial and deep lymphatics
    b Between superficial lymphatics of both breasts across middle line
    c Between superficial lymphatics and lymphatics in fascia of pectoralis major
    d Between lymphatics of breast and abdominal cavity through linea alba

B Lymph nodes
  1 About 85% of lymph from breast to axillary nodes—some to parasternal nodes
  2 Five sets of axillary nodes
  3 Physical contact between anterior axillary nodes and axillary tail of Spence

THYMUS

A Location—mediastinum, extends into lower neck

B Size—relatively largest in comparison to body size at about 2 years of age; absolutely largest at puberty after which it gradually atrophies; almost disappears by advanced old age

C Function—forms lymphocytes before birth; postulated to secrete hormone, starting soon after birth, that permits lymphocytes to develop into plasma cells and secrete antibodies; hence thymus serves as part of body's defense against microbes and other foreign proteins

SPLEEN

A Location—left hypochondrium

B Structure
  1 Similar to lymph nodes, ovoid in shape
  2 Size varies

Outline Summary—cont'd

3 Contains numerous venous blood spaces that serve as blood reservoir

C Functions

1 Defense—protection by phagocytosis by reticuloendothelial cells and antibody formation by some lymphocytes

2 Hemopoiesis of nongranular leukocytes (monocytes and lymphocytes) and of red blood cells before birth; spleen also forms plasma cells

3 Red blood cell and platelet destruction—reticuloendothelial cells phagocytose these cells

4 Blood reservoir

Review Questions

1 List the anatomical components of the lymphatic system.

2 Why is the term *circulation* more appropriate in describing the movement of blood than of lymph?

3 Lymph from what body areas enters the general circulation by way of the thoracic duct? The right lymphatic ducts?

4 How do interstitial fluid and lymph differ from blood plasma?

5 How do lymphatic capillaries originate?

6 What is the cisterna chyli?

7 Where does lymph enter the blood vascular system?

8 In general, lymphatics resemble veins in structure. List three exceptions to this general rule of thumb.

9 What are the specialized lymphatics that originate in the villi of the small intestine called?

10 What is chyle? Where is it formed?

11 Briefly describe the anatomy of the lymphatic capillary wall.

12 Discuss the importance of valves in the lymphatic system.

13 How is lymph formed?

14 Discuss the "lymphatic pump."

15 List several important groups, or clusters, of lymph nodes.

16 Discuss how lymph nodes function in body defense and hemopoiesis.

17 If cancer cells from breast cancer were to enter the lymphatics of the breast, where do you think they might lodge and start new growths? Explain, using your knowledge of the anatomy of the lymphatic and circulatory systems.

18 Discuss the importance of lymphatic anastomoses in the spread of breast cancer.

19 Locate the thymus and describe its appearance and size at birth, at maturity, and in old age.

20 Discuss the function of the thymus.

21 Describe the location and function of the spleen. What functions is it thought to perform?

UNIT FIVE

RESPIRATION, NUTRITION, AND EXCRETION

19 ANATOMY OF THE RESPIRATORY SYSTEM

OBJECTIVES

After you have completed this chapter, you should be able to:

1 List and locate the organs of the respiratory system.
2 List the generalized functions of the respiratory system.
3 Describe and correlate the anatomy of the nose with its specialized functions.
4 Locate the paranasal sinuses in the skull and describe how they drain into the nose.
5 List the anatomical divisions of the pharynx and name the openings into and between its divisions.
6 Identify and locate the tonsils.
7 Discuss the location, structure, and specialized functions of the larynx.
8 Describe the structure and function of the trachea, bronchi, and bronchioles.
9 Identify the lobes of the lungs and the bronchopulmonary segments.
10 Discuss the gross surface anatomy and generalized functions of the lungs.
11 Discuss the structure and function of the thorax and mediastinum in respiration.

FUNCTIONS AND ORGANS

The respiratory system functions as an air distributor and gas exchanger so that oxygen may be supplied to and carbon dioxide be removed from the body's cells. Since most of our billions of cells lie too far from air to exchange gases directly with it, air must first exchange gases with blood, blood must circulate, and finally blood and cells must exchange gases. These events require the functioning of two systems, namely, the respiratory system and the circulatory system. All parts of the respiratory system—except its microscopic-sized sacs called *alveoli*—function as air distributors. Only the alveoli and the tiny passageways that open into them serve as gas exchangers.

In addition to air distribution and gas exchange, the respiratory system effectively filters, warms, and humidifies the air we breathe. Respiratory organs also influence speech (sound production), assist in control, or homeostasis, of body pH, and make the sense of smell (olfaction) possible.

For purposes of study, the respiratory system may be divided into upper and lower tracts, or divisions. The organs of the upper respiratory tract are located outside of the thorax, or chest cavity, whereas those in the lower tract, or division, are located almost entirely within it.

The **upper respiratory tract** is composed of the nose, nasopharynx, oropharynx, laryngopharynx, and larynx. The **lower respiratory tract,** or division, consists of the trachea, all segments of the bronchial tree, and the lungs. Functionally, the respiratory system also includes a number of accessory structures, including the oral cavity, rib cage, and diaphragm. Together these structures constitute the lifeline, the air supply of the body. This chapter describes the functional anatomy of these organs. The physiology of the respiratory system as a whole will be discussed in Chapter 20.

KEY TERMS

Alveoli (al-VE-o-li)

Bronchiole (BRON-ke-ol)

Bronchopulmonary segments (BRON-ko-pul-mo-nary)

Bronchus (BRON-kus)

Epiglottis (ep-i-GLOT-is)

Fauces (FAW-sez)

Larynx (LAR-inks)

Meatus (me-A-tus)

Mediastinum (me-de-as-TI-num)

Nares (NA-rez)

Pharynx (FAR-inks)

Respiratory (re-SPI-rah-to-re)

Thorax (THO-raks)

Tonsil (TON-sil)

Trachea (TRAY-kee-uh)

Figure 19-1 The nasal septum consists of the perpendicular plate of the ethmoid bone, the vomer, and the septal and vomeronasal cartilages.

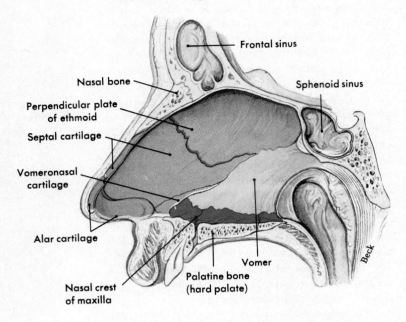

Frontal sinus

Nasal bone

Perpendicular plate of ethmoid

Septal cartilage

Vomeronasal cartilage

Alar cartilage

Nasal crest of maxilla

Sphenoid sinus

Vomer

Palatine bone (hard palate)

Beck

ORGANS

UPPER RESPIRATORY TRACT
Nose
Structure

The nose consists of an external and an internal portion. The external portion, that is, the part that protrudes from the face, consists of a bony and cartilaginous framework overlaid by skin containing many sebaceous glands. The two nasal bones meet above where they are surrounded by the frontal bone to form the root of the nose. The nose is surrounded by the maxilla laterally and inferiorly at its base. The flaring cartilaginous expansion forming and supporting the outer side of each oval nostril opening is called the **ala.**

The fact that the skin of the external nose contains many sebaceous glands has great clinical significance. If these glands become blocked, it is possible for infectious material to enter and pass from facial veins near the nose to one of the intracranial venous sinuses (Chapter 16). For this reason, the triangular-shaped zone surrounding the external nose is often known as the "danger area of the face."

The internal nose, or nasal cavity, lies over the roof of the mouth where the palatine bones, which form both the floor of the nose and the roof of the mouth, separate the nasal cavities from the mouth cavity. Sometimes the palatine bones fail to unite completely, producing a condition known as **cleft palate.** When this abnormality exists, the mouth is only partially separated from the nasal cavity, and difficulties arise in swallowing.

The roof of the nose is separated from the cranial cavity by a portion of the ethmoid bone called the **cribriform plate** (see Figures 19-2 and 19-3). The cribriform plate is perforated by many small openings that permit branches of the olfactory nerve responsible for the special sense of smell to enter the cranial cavity and reach the brain.

Separation of the nasal and cranial cavities by a thin, perforated plate of bone presents real hazards. If the cribriform plate is damaged as a result of trauma to the nose, it is possible for potentially infectious material to pass directly from the nasal cavity into the cranial fossa and surround the brain.

The hollow nasal cavity is separated by a midline partition, the **septum** (Figure 19-1), into a right and a left cavity. Note in Figure 19-1 that the nasal septum is made up of four structures; the perpendicular plate of the ethmoid bone above, the vomer bone and the vomeronasal and septal nasal cartilages below. In the adult the nasal septum is frequently deviated to one side or the other, interfering both with respiration and with drainage of the nose and sinuses.

The nasal septum has a rich blood supply. **Nosebleeds** or **epistaxis** often occur as a result of septal contusions caused by a direct blow to the nose.

Each nasal cavity is divided into three passageways (superior, middle, and inferior meati) by the projection of the **turbinates** or **conchae,** from the lateral walls of the internal portion of the nose (Figures 19-2 and 19-3). The superior and middle turbinates are processes of the ethmoid bone, whereas the inferior turbinates are separate bones.

The external openings into the nasal cavities (nostrils) have the technical name of **anterior nares.** They open into an area covered by skin that is reflected from the wings (ala) of the nose. This area, called the **vestibule,** is located just inside the nasal cavity below the inferior meatus. Coarse hairs called **vibrissae,** sebaceous glands, and numerous sweat glands are found in the skin of the vestibule.

Figure 19-2 Sagittal section through the face and neck. The nasal septum has been removed, exposing the lateral wall of the nasal cavity. Note the position of the conchae (turbinates).

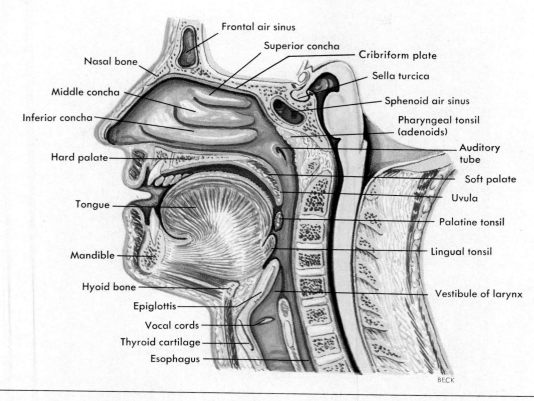

Frontal air sinus
Nasal bone
Superior concha
Cribriform plate
Sella turcica
Middle concha
Sphenoid air sinus
Inferior concha
Pharyngeal tonsil (adenoids)
Auditory tube
Hard palate
Soft palate
Uvula
Tongue
Palatine tonsil
Lingual tonsil
Mandible
Hyoid bone
Vestibule of larynx
Epiglottis
Vocal cords
Thyroid cartilage
Esophagus

BECK

Figure 19-3 Frontal section through the face viewed from behind.

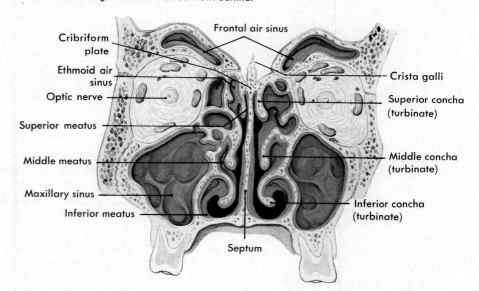

Frontal air sinus
Cribriform plate
Ethmoid air sinus
Crista galli
Optic nerve
Superior concha (turbinate)
Superior meatus
Middle meatus
Middle concha (turbinate)
Maxillary sinus
Inferior meatus
Inferior concha (turbinate)
Septum

Once air has passed over the skin of the vestibule, it enters the **respiratory portion** of each nasal passage. This area extends from the inferior meatus to the small funnel-shaped orifices of the *posterior nares,* or *choanae.* The posterior nares (choanae) are openings that allow air to pass from an area of the internal nasal cavity above the superior meatus, called the **sphenoethmoidal recess,** into the nasopharynx. Once air has passed through the posterior nares, it has left the nasal cavity and entered the next major segment of the upper respiratory tract—the pharynx.

Figure 19-4 Electron micrograph of respiratory mucosa. This epithelium is typically ciliated and exhibits numerous goblet cells, *G,* filled with mucus. (× 7,600.)

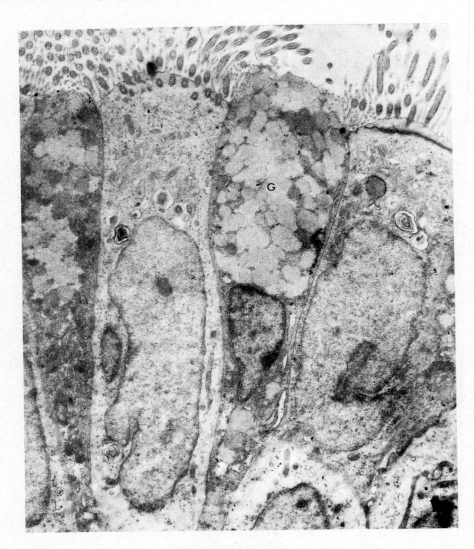

If one were to "trace" the movement of air through the nose into the pharynx, it would pass through several structures on the way. The sequence is as follows:

1 Anterior nares (nostrils)
2 Vestibule
3 Inferior, middle, and superior meati, simultaneously
4 Posterior nares (choanae) and sphenoethmoidal recess, simultaneously

NASAL MUCOSA. Once air has passed over the skin of the vestibule and enters the respiratory portion of the nasal passage, it passes over the highly specialized **respiratory mucosa.** This mucous membrane has a pseudostratified ciliated columnar epithelium rich in goblet cells and contains many mucous glands (Figure 19-4). The respiratory mucosa possesses a rich blood supply, especially over the inferior concha, and is bright pink or red in color. Near the roof of the nasal cavity and over the superior concha and opposing portion of the septum, the mucosa turns pale and has a yellowish tint. In this area it is referred to as the **olfactory epithelium.** This specialized membrane contains many olfactory nerve cells and has a rich lymphatic plexus. Ciliated mucous membrane lines the rest of the respiratory tract down as far as the smaller bronchioles.

PARANASAL SINUSES. The four pairs of paranasal sinuses are air-containing spaces that open, or drain, into the nasal cavity and take their names from the

Figure 19-5 Projection of paranasal sinuses, oral and nasal cavities on the skull and face of the adult. Outlines of cavities and sinuses are indicated in color.

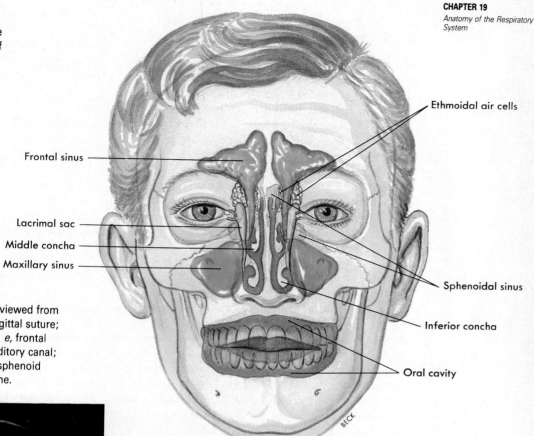

Frontal sinus

Ethmoidal air cells

Lacrimal sac

Middle concha

Maxillary sinus

Sphenoidal sinus

Inferior concha

Oral cavity

BECK

Figure 19-6 Normal x-ray film of skull viewed from behind. *a,* Superior sagittal sinus; *b,* sagittal suture; *c,* coronal suture; *d,* lambdoidal suture; *e,* frontal sinus; *f,* falx; *g,* roof of eye orbit; *h,* auditory canal; *i,* foramen; *j,* shadows of ethmoid and sphenoid sinuses; *k,* nasal septum; *l,* mastoid bone.

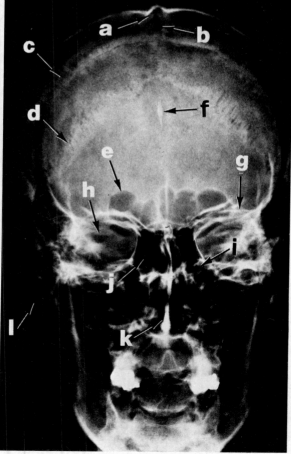

skull bones in which they are located (Chapter 7). These paranasal sinuses are the frontal, maxillary, ethmoid, and sphenoid (Figures 19-5 and 19-6). Like the nasal cavity, each paranasal sinus is lined by respiratory mucosa. The mucous secretions produced in the sinuses are continually being swept into the nose by the ciliated surface of the respiratory membrane.

The right and left frontal sinuses are located just above the corresponding orbit, whereas the maxillary, the largest of the sinuses, extends into the max-

The paired and often asymmetrical sinuses are small or rudimentary at birth but increase in size with growth of the skull. Although fairly well developed by 7 years of age, they do not reach maximal size until after puberty. In the adult the sinuses are subject to considerable variation in both size and shape.

illa on either side of the nose. The sphenoid sinuses lie in the body of the sphenoid bone on either side of the midline in very close proximity to the optic nerves and pituitary gland.

Note in Figure 19-5 that the ethmoid sinuses are not single large cavities but a collection of small air cells divided into anterior, middle, and posterior groups that open independently into the upper part of the nasal cavity. The paranasal sinuses drain as follows:

1 Into the middle meatus (passageway below middle turbinate)—frontal, maxillary, anterior, and middle ethmoidal sinuses

2 Into the superior meatus—posterior ethmoidal sinuses

3 Into the space above the superior turbinates (sphenoethmoidal recess)—sphenoidal sinuses

Functions

The nose serves as a passageway for air going to and from the lungs. However, if the nasal passages are obstructed, it is possible for air to bypass the nose and enter the respiratory tract directly through the mouth. Air that enters the system through the nasal cavity is filtered of impurities, warmed, moistened, and chemically examined for substances that might prove irritating to the delicate lining of the respiratory tract. The vibrissae, or hairs, in the vestibule serve as an initial "filter" to screen particulate matter from air that is entering the system. The turbinates, or concha, then serve as baffles to provide a large mucus-covered surface area over which air must pass before reaching the pharynx. The respiratory membrane produces copious quantities of mucus and possesses a rich blood supply, especially over the inferior concha, which permits rapid warming and moistening of the dry inspired air. Mucous secretions provide the final "trap" for removal of remaining particulate matter from air as it moves through the nasal passages. Fluid from the lacrimal ducts (see Figure 13-14) and additional mucus produced in the paranasal sinuses also help to trap particulate matter and moisten air passing through the nose. In addition, the hollow sinuses act to lighten the bones of the skull and serve as resonating chambers for speech. Deflection of air by the middle and superior concha over the olfactory epithelium makes the special sense of smell possible.

Pharynx
Structure

Another name for the pharynx is the throat. It is a tubelike structure about 12.5 cm (5 inches) long that extends from the base of the skull to the esophagus and lies just anterior to the cervical vertebrae. It is made of muscle, is lined with mucous membrane,

and has three anatomical divisions: the **nasopharynx,** located behind the nose and extending from the posterior nares to the level of the soft palate; the **oropharynx,** located behind the mouth from the soft palate above to the level of the hyoid bone below; and the **laryngopharynx,** which extends from the hyoid bone to its termination in the esophagus. Figure 19-7 shows the divisions of the pharynx when viewed from behind (coronal section).

Seven openings are found in the pharynx (see Figures 19-2 and 19-7).

1 Right and left auditory (eustachian) tubes opening into the nasopharynx

2 Two posterior nares into the nasopharynx

3 The opening from the mouth, known as the *fauces,* into the oropharynx

4 The opening into the larynx from the laryngopharynx

5 The opening into the esophagus from the laryngopharynx

The **adenoids** or **pharyngeal tonsils,** are located in the nasopharynx on its posterior wall opposite the posterior nares. Although the cavity of the nasopharynx differs from the oral and laryngeal divisions in that it does not collapse, it may become obstructed. If the adenoids become enlarged, they fill the space behind the posterior nares and make it difficult or impossible for air to travel from the nose into the throat. When this happens, the individual keeps the mouth open to breathe and is described as having an "adenoidy" appearance.

Two pairs of organs are found in the oropharynx: the faucial, or **palatine tonsils,** located behind and below the pillars of the fauces, and the **lingual tonsils,** located at the base of the tongue. The palatine tonsils are the ones most commonly removed by a tonsillectomy. Only rarely are the lingual ones also removed.

Functions

The pharynx serves as a hallway for the respiratory and digestive tracts, since both air and food must pass through this structure before reaching the appropriate tubes. It also plays an important part in phonation. For example, only by the pharynx changing its shape can the different vowel sounds be formed.

Larynx
Location

The larynx, or voice box, lies between the root of the tongue and the upper end of the trachea just below and in front of the lowest part of the pharynx. It might be described as a vestibule opening into the trachea from the pharynx. It normally extends between the third, fourth, fifth, and sixth cervical vertebrae but is often somewhat higher in females and

Figure 19-7 Divisions of the pharynx when viewed from behind (coronal section).

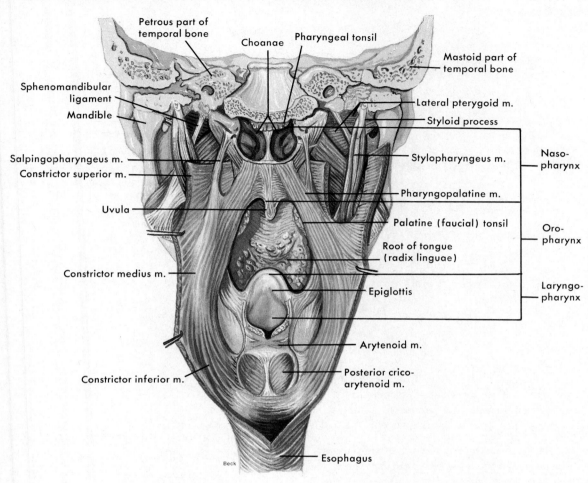

during childhood. The lateral lobes of the thyroid gland and the carotid artery in its covering sheath touch the sides of the larynx.

Structure

The triangular-shaped larynx consists largely of cartilages that are attached to one another and to surrounding structures by muscles or by fibrous and elastic tissue components. It is lined by a ciliated mucous membrane. The cavity of the larynx extends from its triangular-shaped inlet at the epiglottis to the circular outlet at the lower border of the cricoid cartilage where it is continuous with the lumen of the trachea (Figure 19-8). The mucous membrane lining the larynx forms two pairs of folds that jut inward into its cavity and divide it into three compartments, or divisions. The upper pair is called the **vestibular,** or **false, vocal folds** for the rather obvious reason that they play no part in vocalization. The lower pair serves as the **true vocal cords.** The slitlike space between the true vocal cords—the

rima glottidis, or glottis—is the narrowest part of the larynx.

The division, or compartment, of the laryngeal cavity above the false or vestibular vocal folds is called the **vestibule.** The very short middle portion of the cavity between the false and true vocal cords is the ventricular division, or **ventricle.** The lower compartment extending from the true vocal cords to the outlet is referred to as the **infraglottic larynx** (Figure 19-9).

Edema of the mucosa covering the vocal cords is a potentially lethal condition. Even a moderate amount of swelling can obstruct the glottis so that air cannot get through and asphyxiation results.

Figure 19-8 Posterior view of the larynx (voice box) showing many of the muscles that function to alter its shape.

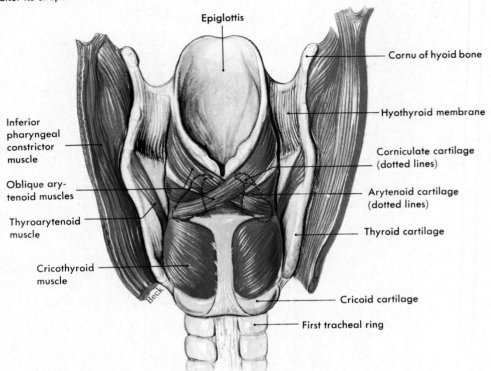

Cartilages of the Larynx

Nine cartilages form the framework of the larynx. The three largest of these—the thyroid cartilage, the epiglottis, and the cricoid cartilage—are single structures. There are three pairs of smaller accessory cartilages, namely, the arytenoid, corniculate, and cuneiform cartilages.

Single Laryngeal Cartilages

1 The **thyroid cartilage** (Adam's apple) is the largest cartilage of the larynx and is the one that gives the characteristic triangular shape to its anterior wall. It is usually larger in men than in women and has less of a fat pad lying over it—two reasons why a man's thyroid cartilage protrudes more than a woman's.
2 The **epiglottis** is a small leaf-shaped cartilage that projects upwards behind the tongue and hyoid bone. It is attached below to the thyroid cartilage, but its free superior border can move up and down during swallowing to prevent food or liquids from entering the trachea (see Figures 19-7 and 19-8).
3 The **cricoid,** or signet ring, **cartilage,** so called because its shape resembles a signet ring (turned so the signet forms part of the posterior

wall of the larynx), is the most inferiorly placed of the nine cartilages.

Paired Laryngeal Cartilages

1 The pyramidal-shaped **arytenoid** cartilages are the most important of the paired laryngeal cartilages. The base of each cartilage articulates with the superior border of the cricoid cartilage (see Figure 19-8). The anterior angles of these cartilages serve as points of attachment for the vocal cords.
2 The nodular **corniculate** cartilages are small and conical in shape. Note in Figure 19-8 that they rest on the apex of each arytenoid cartilage.
3 The two small cuneiform cartilages are rod-shaped structures located near the base of the epiglottis. They are closely related to the arytenoid cartilages.

Muscles of the Larynx

Muscles of the larynx are often divided into intrinsic and extrinsic groups. Intrinsic muscles have both their origin and insertion on the larynx. They are important in controlling vocal cord length and tension and in regulating the shape of the laryngeal inlet.

Figure 19-9 Sagittal section through the larynx.

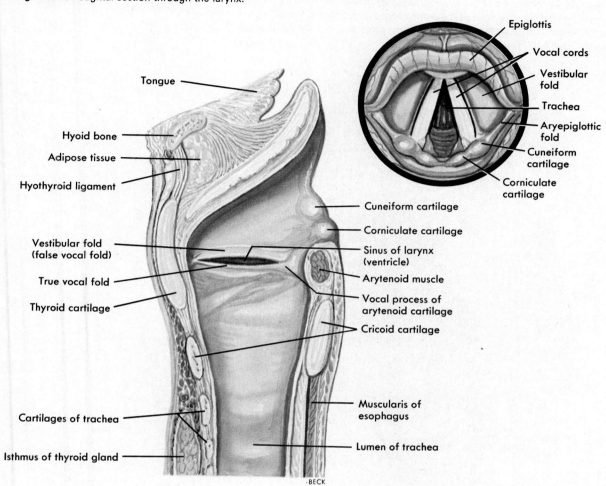

Tongue

Hyoid bone

Adipose tissue

Hyothyroid ligament

Vestibular fold (false vocal fold)

True vocal fold

Thyroid cartilage

Cartilages of trachea

Isthmus of thyroid gland

Cuneiform cartilage

Corniculate cartilage

Sinus of larynx (ventricle)

Arytenoid muscle

Vocal process of arytenoid cartilage

Cricoid cartilage

Muscularis of esophagus

Lumen of trachea

Epiglottis

Vocal cords

Vestibular fold

Trachea

Aryepiglottic fold

Cuneiform cartilage

Corniculate cartilage

·BECK

Extrinsic muscles insert on the larynx but have their origin on some other structure—such as the hyoid bone. Therefore contraction of the extrinsic muscles actually moves or displaces the larynx as a whole. Muscles in both groups play important roles in respiration, vocalization, and swallowing. During swallowing, for example, contraction of the intrinsic aryepiglottic muscles (those that connect the arytenoid cartilages with the epiglottis) prevents "swallowing down the wrong throat" by squeezing the laryngeal inlet shut.

Two other pairs of intrinsic laryngeal muscles function to open and close the glottis. The posterior cricoarytenoid muscles (between cricoid and arytenoid cartilages) open the glottis by abducting the true vocal cords. The lateral cricoarytenoid muscles close the glottis by adducting the cords. These events are crucial to both respiration and voice production. Certain other intrinsic muscles of the larynx function to influence the pitch of the voice by either lengthening and tensing or shortening and relaxing the vocal cords.

Functions

The larynx functions in respiration, since it constitutes part of the vital airway to the lungs. This unique passageway, like the other components of the upper respiratory tract, is lined with a ciliated mucous membrane that helps in the removal of dust particles and in the warming and humidification of inspired air. In addition, it protects the airway against the entrance of solids or liquids during swallowing. It also serves as the organ of voice production—hence its popular name, the voice box. Air being expired through the glottis, narrowed by partial adduction of the true vocal cords, causes them to vibrate. Their vibration produces the voice. Several other structures besides the larynx contribute to the sound of the voice by acting as sounding boards or resonating chambers. Thus the size and shape of the nose,

Figure 19-10 Projection of the lungs and trachea in relation to the rib cage and clavicles. Dotted line indicates location of the dome-shaped diaphragm at the end of expiration and before inspiration. Note that apex of each lung projects above the clavicle. Ribs 11 and 12 are not visible in this view.

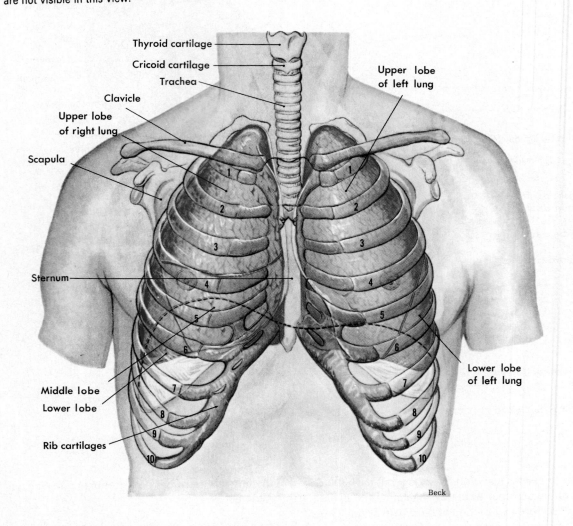

Beck

Often a tube is placed in the trachea (**endotracheal intubation**) before patients leave the operating room, especially if they have been given a muscle relaxant. The purpose of the tube is to ensure an open airway. Another procedure done frequently in today's modern hospitals is a **tracheostomy,** that is, the cutting of an opening into the trachea. A surgeon may do this so that a suction device can be used to remove secretions from the bronchial tree or so a machine such as the intermittent positive pressure breathing (IPPB) machine can be used to improve ventilation of the lungs.

mouth, pharynx, and bony sinuses help to determine the quality of the voice.

LOWER RESPIRATORY TRACT
Trachea
Structure

The trachea, or windpipe, is a tube about 11 cm (4.5 inches) long that extends from the larynx in the neck to the primary bronchi in the thoracic cavity (Figure 19-10). Its diameter measures about 2.5 cm (1 inch). Smooth muscle, in which are embedded C-shaped rings of cartilage at regular intervals, fashions the walls of the trachea (Figure 19-11). The cartilaginous rings are incomplete on the posterior surface (Figure 19-12). They give firmness to the wall, tending to prevent it from collapsing and shutting off the vital airway.

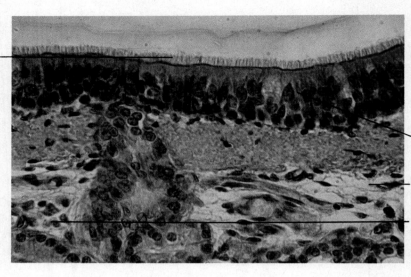

Cilia

Figure 19-11 Transverse section of trachea. Note the mucosa of ciliated epithelium. Hyaline cartilage occurs below the glandular submucosa and is not visible in this section. (×70.)

Pseudostratified epithelium

Submucosa

Mucous gland

Note in Figure 19-11 that the trachea is lined with the type of pseudostratified ciliated columnar epithelium typical of the respiratory tract as a whole.

Function

The trachea performs a simple but vital function—it furnishes part of the open passageway through which air can reach the lungs from the outside. Obstruction of this airway for even a few minutes causes death from asphyxiation.

Bronchi
Structure

The trachea divides at its lower end into two **primary bronchi,** of which the right bronchus is slightly larger and more vertical than the left. This anatomical fact explains why aspirated foreign objects frequently lodge in the right bronchus. In structure the bronchi resemble the trachea. Their walls contain incomplete cartilaginous rings before the

Figure 19-12 Cross section of the trachea.

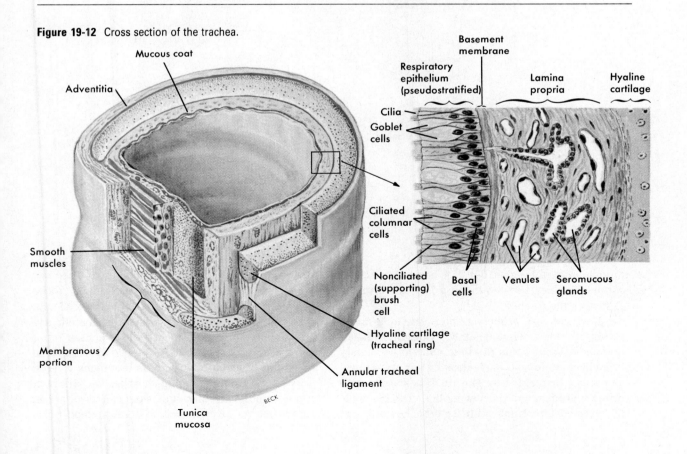

Adventitia

Mucous coat

Basement membrane

Respiratory epithelium (pseudostratified)

Lamina propria

Hyaline cartilage

Cilia

Goblet cells

Ciliated columnar cells

Smooth muscles

Nonciliated (supporting) brush cell

Basal cells

Venules

Seromucous glands

Hyaline cartilage (tracheal ring)

Membranous portion

Annular tracheal ligament

BECK

Tunica mucosa

Figure 19-13 The pharynx, trachea, and lungs. The inset shows the grape-like alveolar sacs where air and blood exchange oxygen and carbon dioxide through the thin walls of the alveoli. (See also Figure 19-14.) Capillaries (not shown) surround the alveoli.

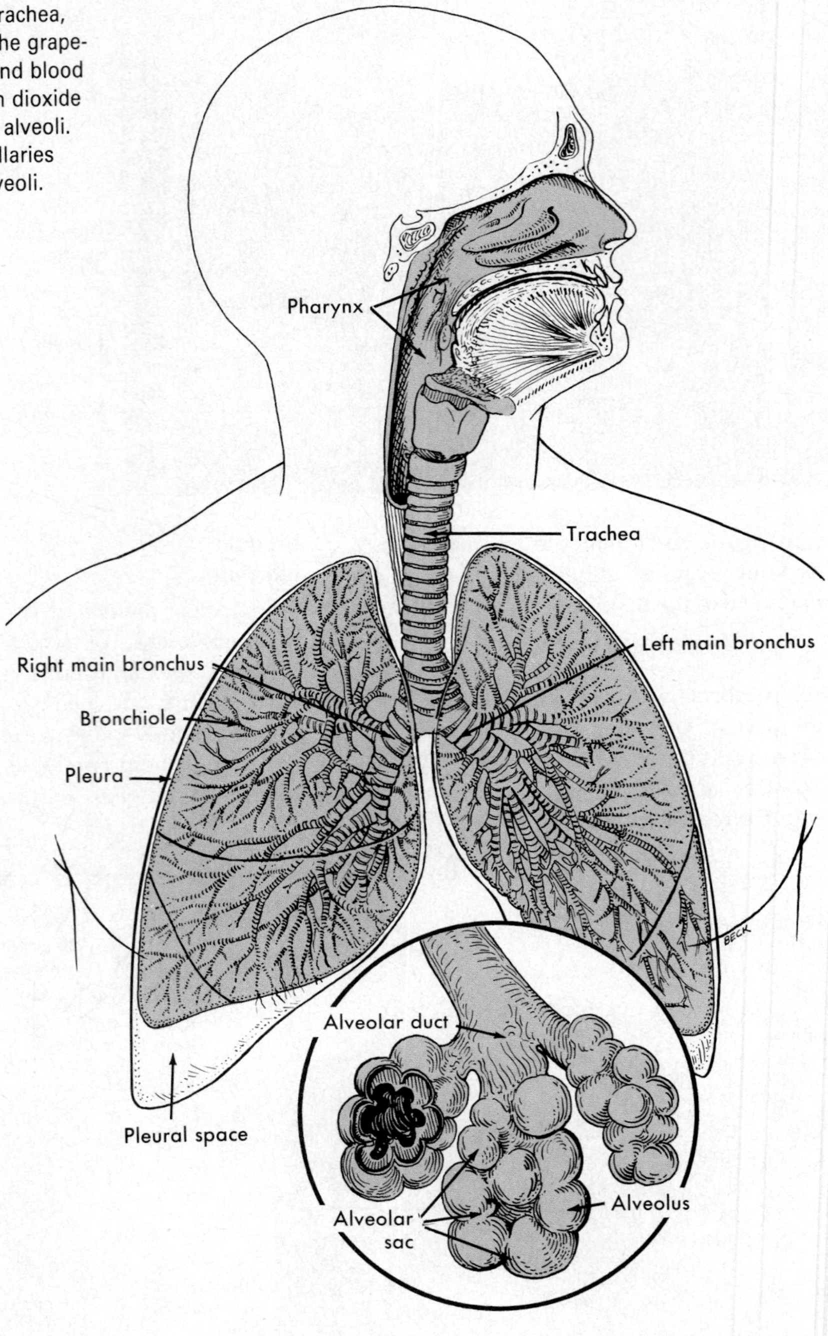

bronchi enter the lungs, but they become complete within the lungs. Ciliated mucosa lines the bronchi, as it does the trachea.

Each primary bronchus enters the lung on its respective side and immediately divides into smaller branches called **secondary bronchi.** The secondary bronchi continue to branch, forming tertiary bronchi and small **bronchioles.** The trachea and the two primary bronchi and their many branches resemble an inverted tree trunk with its branches and are,

therefore, spoken of as the bronchial tree. The bronchioles subdivide into smaller and smaller tubes, eventually terminating in microscopic branches that divide into **alveolar ducts,** which terminate in several alveolar sacs, the walls of which consist of numerous **alveoli** (Figures 19-13 to 19-16). The structure of an alveolar duct with its branching alveolar sacs can be likened to a bunch of grapes—the stem represents the alveolar duct, each cluster of grapes represents an alveolar sac, and each grape repre-

Figure 19-14 Metal cast of air spaces of the lungs of a dog. The inset shows a cast of clusters of alveoli at the terminations of tiny air tubes. The magnification of the inset is about 11 times the actual size.

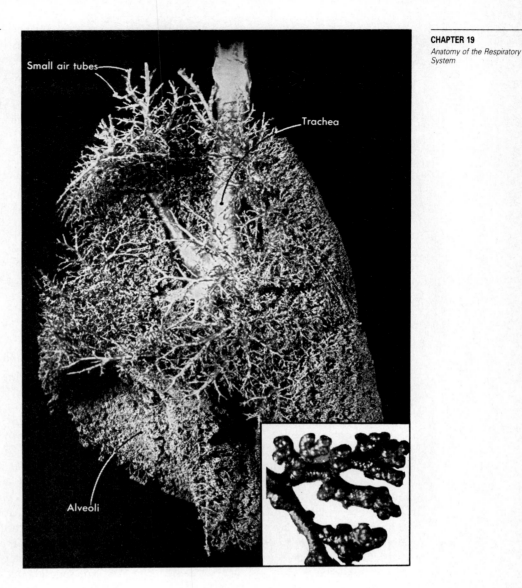

Figure 19-15 Respiratory membrane greatly magnified to show epithelial and surfactant layers of alveolar membrane and basement membrane and endothelial layer of capillary membrane.

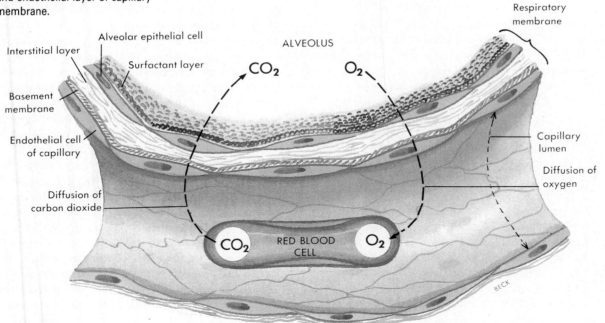

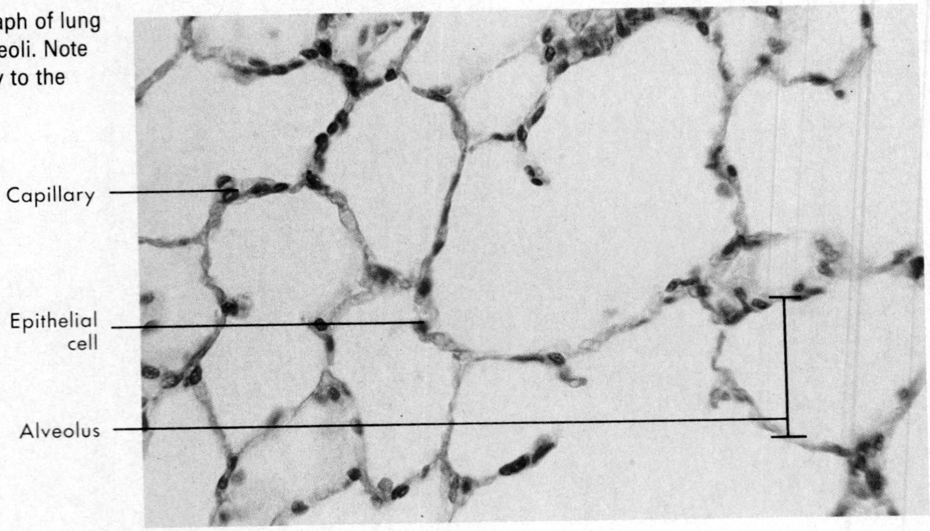

Figure 19-16 Photomicrograph of lung showing parts of several alveoli. Note the proximity of the capillary to the alveolar wall. (× 140.)

Capillary

Epithelial cell

Alveolus

sents an alveolus. Some 300 million alveoli are estimated to be present in our two lungs.

The structure of the secondary and tertiary bronchi and bronchioles shows some modification of the primary bronchial structure. The cartilaginous rings become irregular and disappear entirely in the smaller bronchioles. By the time the branches of the bronchial tree have dwindled sufficiently to form the alveolar ducts and sacs and the alveoli, only the internal surface layer of cells remains. In other words, the walls of these microscopic structures consist of a single layer of simple, squamous epithelial tissue (Figure 19-16). As we shall see, this structural fact makes possible the performance of their functions.

Functions

The tubes composing the bronchial tree perform the same function as the trachea—that of distributing air to the lung's interior. The alveoli, enveloped as they are by networks of capillaries, accomplish the lung's main and vital function, that of gas exchange

Certain diseases may block the passage of air through the bronchioles or alveoli. For example, in pneumonia the alveoli become inflamed, and the accompanying wastes plug up these minute air spaces, making the affected part of the lung solid. Whether the victim survives depends largely on the extent of the fluid accumulation and solidification.

between air and blood. Someone has observed that "the lung passages all serve the alveoli" just as "the circulatory system serves the capillaries."

Lungs
Structure

The lungs are cone-shaped organs, large enough to fill the pleural portion of the thoracic cavity completely (see Figure 19-10). They extend from the diaphragm to a point slightly above the clavicles and lie against the ribs both anteriorly and posteriorly. The medial surface of each lung is roughly concave to allow room for the mediastinal structures and for the heart, but concavity is greater on the left than on the right because of the position of the heart. The primary bronchi and pulmonary blood vessels (bound together by connective tissue to form what is known as the **root** of the lung) enter each lung through a slit on its medial surface called the **hilum.**

The broad inferior surface of the lung, which rests on the diaphragm, constitutes the **base,** whereas the pointed upper margin is the **apex.** Each apex projects above a clavicle.

The **costal surface** of each lung lies against the ribs and is rounded to match the contours of the thoracic cavity.

Each lung is divided into lobes by fissures. The left lung is partially divided into two lobes (superior and inferior) and the right lung into three lobes (superior, middle, and inferior). Note in Figure 19-17 that an **oblique fissure** is present in both lungs. In the right lung a **horizontal fissure** is also present, which separates the superior from the middle lobe. After the primary bronchi enter the lungs, they branch into *secondary,* or *lobar,* bronchi that enter each lobe. Thus, in the right lung, three secondary

Figure 19-17 **A**, Lobes of the right and left lung; **B**, the bronchopulmonary segments.

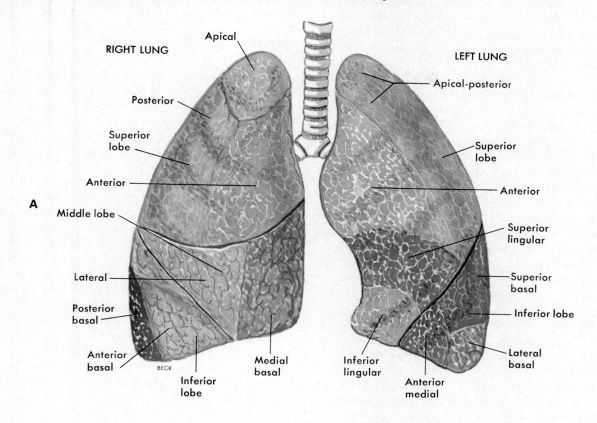

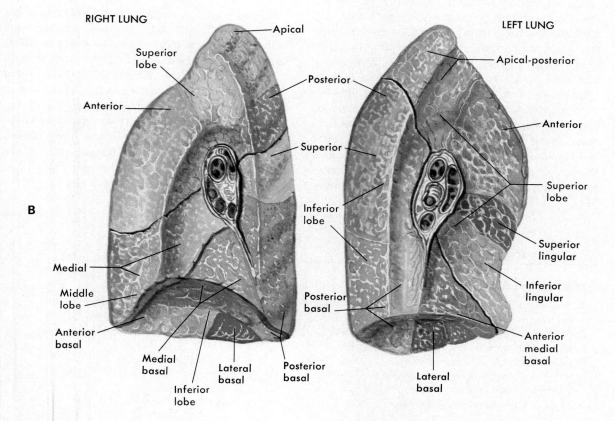

bronchi are formed, which enter the superior, middle, and inferior lobes. Each secondary bronchus is named for the lung lobe that it enters; for example, the superior secondary bronchus enters the superior lobe. The left primary bronchus divides into two secondary bronchi entering the superior and inferior lobes of that lung.

The lobes of the lung can be further subdivided into functional units called **bronchopulmonary segments.** Each bronchopulmonary segment is served by a tertiary bronchus. There are 10 bronchopulmonary segments in the right lung and 8 in the left (Figure 19-17). The interior of each bronchopulmonary segment consists of the almost innumerable tubes of dwindling diameters that make up the bronchial tree and serve as air distributors. The smallest tubes terminate in the smallest but functionally most important structures of the lung—the alveoli, or "gas exchangers."

Visceral pleura covers the outer surfaces of the lungs and adheres to them much as the skin of an apple adheres to the apple.

Functions

The lungs perform two functions—air distribution and gas exchange. Air distribution to the alveoli is the function of the tubes of the bronchial tree. Gas exchange between air and blood is the joint function of the alveoli and the networks of blood capillaries that envelop them. These two structures—one part of the respiratory system and the other part of the circulatory system—together serve as highly efficient gas exchangers. Why? Because they provide an enormous surface area, the respiratory membrane, where the very thin-walled alveoli and equally thin-walled pulmonary capillaries come in contact (Figures 19-15 and 19-16). This makes possible extremely rapid diffusion of gases between alveolar air and pulmonary capillary blood. Someone has estimated that if the lungs' 300 million or so alveoli could be opened up flat, they would form a surface about the size of a tennis court, that is, about 85 square meters, or well over 40 times the surface area of the

entire body! No wonder such large amounts of oxygen can be so quickly loaded into the blood while large amounts of carbon dioxide are rapidly unloaded from it.

THORAX (CHEST)
Structure

As described in Chapter 1, the thoracic cavity has three divisions, separated from each other by partitions of pleura. The parts of the cavity occupied by the lungs are the pleural divisions. The space between the lungs occupied mainly by the esophagus, trachea, large blood vessels, and heart is the mediastinum.

The parietal layer of the pleura lines the entire thoracic cavity. It adheres to the internal surface of the ribs and the superior surface of the diaphragm, and it partitions off the mediastinum. A separate pleural sac thus encases each lung. Since the outer surface of each lung is covered by the visceral layer of the pleura, the visceral pleura lies against the parietal pleura, separated only by a potential space (pleural space) that contains just enough pleural fluid for lubrication. Thus, when the lungs inflate with air, the smooth, moist visceral pleura coheres to the smooth, moist parietal pleura. Friction is thereby avoided, and respirations are painless. In pleurisy, on the other hand, the pleura is inflamed and respirations become painful.

Functions

The thorax plays a major role in respirations. Because of the elliptical shape of the ribs and the angle of their attachment to the spine, the thorax becomes larger when the chest is raised and smaller when it is lowered. It is these changes in thorax size that bring about inspiration and expiration (discussed on pp. 534-537). Lifting up the chest raises the ribs so that they no longer slant downward from the spine, and because of their elliptical shape, this enlarges both depth (from front to back) and width of the thorax. (If this does not sound convincing to you, examine a skeleton to see why it is so.)

Outline Summary

FUNCTIONS AND ORGANS

  A System functions as air distributor and gas exchanger

  B Exchange occurs first between air in lungs and blood, then between circulating blood and cells

  C Only the alveoli and ducts that open into them serve as gas exchangers; all other components function as air distributors

  D Other functions of respiratory system include

filtering, warming, and humidification of air; control of pH; phonation and olfaction

  E Organs

    1 Upper respiratory tract (outside of thorax)

      a Nose

      b Nasopharynx

      c Oropharynx

      d Laryngopharynx

      e Larynx

Outline Summary—cont'd

2 Lower respiratory tract (inside thorax)
   a Trachea
   b All segments of bronchial tree
   c Lungs

ORGANS OF UPPER RESPIRATORY TRACT
Nose

A Structure
   1 Portions—internal, in skull, above roof at mouth; external, protruding from face; surrounded by maxilla laterally and inferiorly; triangular-shaped area surrounding external nose called "danger zone of face"
   2 Cavities
     a Divisions—right and left
     b Meati—superior, middle, and lower; named for turbinate located above each meatus
     c Opening—anterior nares that open into the vestibule from the exterior; posterior nares that permit air to flow from sphenoethmoidal recess into the nasopharynx
     d Turbinates (conchae)—superior and middle are processes of ethmoid bone; inferior turbinates are separate bones; divide internal nasal cavities into three passageways, or meati
     e Floor—formed by palatine bones that also act as roof of mouth
     f Roof—formed by cribriform plate of ethmoid—if fractured, brain infection possible
     g Septum—four structures—perpendicular plate of ethmoid, vomer, vomeronasal and septal cartilages; is very vascular—trauma will cause nosebleeds (epistaxis)
   3 Lining—respiratory mucosa; pseudostratified ciliated columnar epithelium with goblet cells and many mucous glands
   4 Sinuses draining into nose (or paranasal sinuses)—frontal, maxillary (or antrum of Highmore), sphenoidal, and ethmoidal
B Functions
   1 Serves as passageway for incoming and outgoing air, filtering, warming, moistening, and chemically examining it
   2 Organ of smell—olfactory receptors located in nasal mucosa
   3 Aids in phonation

Pharynx

A Structure—made of muscle with mucous lining
   1 Divisions—nasopharynx, located behind the nose and extending from the posterior nares to the level of the soft palate; oropharynx, located behind the mouth from the soft palate above to the level of the hyoid bone below; laryngopharynx, extending from the level of the hyoid bone above to its termination in the esophagus; (see Figure 19-4)

2 Openings—four in nasopharynx: two auditory tubes and two posterior nares; one in oropharynx; fauces from mouth; and two in laryngopharynx; open into esophagus and larynx
3 Organs in pharynx—adenoids, or pharyngeal tonsils, in nasopharynx; palatine and lingual tonsils in oropharynx
4 Cavity of nasopharynx differs from oral and laryngeal divisions in that it does not collapse
B Functions—serve both respiratory and digestive tracts as passageway for air, food, and liquids; aids in phonation

Larynx

A Location—at upper end of trachea, just below pharynx; normally extends between third, fourth, fifth, and sixth cervical vertebrae but often somewhat higher in females and during childhood; lateral lobes of thyroid gland and carotid artery touch sides of larynx
B Structure
   1 Cartilages—nine pieces arranged in boxlike formation
     a Single cartilages—(1) thyroid is largest, known as "Adam's apple"; (2) epiglottis is leaf-shaped "lid"; (3) cricoid resembles signet ring in shape
     b Paired cartilages—(1) arytenoids are most important; serve as points of attachment for the vocal cords; are pyramidal shaped; (2) corniculate—small and conical in shape; (3) cuneiforms—rod-shaped structures near base of epiglottis
   2 Vocal cords—false cords, folds of mucous lining; true cords, fibroelastic bands stretched across hollow interior of larynx; glottis, opening between true vocal cords
   3 Lining—ciliated mucous membrane
   4 Sexual differences—male larynx larger, covered with less fat, and therefore more prominent than female larynx
   5 Muscles—intrinsic, have both origin and insertion on larynx; extrinsic, insert on larynx but have origin on some other surface
   6 Divisions—(1) vestibule, above false vocal cords; (2) ventricle, between false and true vocal cords; (3) infraglottic larynx, between true vocal cords and laryngeal outlet
C Functions—expired air causes true vocal cords to vibrate, producing voice; pitch determined by length and tension of cords

ORGANS OF LOWER RESPIRATORY TRACT
Trachea

A Structure
   1 Walls—smooth muscle; contain C-shaped rings of cartilage at intervals, which keeps the tube open at all times; lining—ciliated mucous membrane

Outline Summary—cont'd

2 Extent—from larynx to bronchi; about 4.5 inches long

B Function—furnishes open passageway for air going to and from lungs

Bronchi

A Structure—formed by division of trachea into two tubes; right bronchus slightly larger and more vertical than left; same structure as trachea; each primary bronchus branches as soon as enters lung into secondary bronchi, which branch into bronchioles, which branch into microscopic alveolar ducts, which terminate in cluster of blind sacs called alveoli; trachea and two primary bronchi and all their branches compose "bronchial tree"; alveolar walls composed of single layer of cells

B Functions—bronchi and their many branching tubes furnish passageway for air going to and from lungs; alveoli provide large, thin-walled surface area where blood and air can exchange gases

Lungs

A Structure

1 Size, shape, and location—large enough to fill pleural division of thoracic cavity; cone shaped; extend from base, on diaphragm, to apex, located slightly above clavicle

2 Divisions—three lobes in right lung, two in left; root of lung consists of primary bronchus and pulmonary artery and veins, bound together by connective tissue; hilum is vertical slit on mesial surface of lung through which root structures

enter lung; base is broad, inferior surface of lung; apex is pointed upper margin; costal surface of each lung lies against rib cage; bronchopulmonary segments—functional subdivisions of lung lobes; 10 in right lung, 8 in left; each segment is served by a tertiary bronchus

3 Covering—visceral layer of pleura

B Functions—furnish place where large amounts of air and blood can come in close enough contact for rapid exchange of gases to occur

Thorax (chest)

A Structure

1 Three divisions

a Right and left pleural portions—contain lungs

b Mediastinum—area between two lungs; contains heart, esophagus, trachea, great blood vessels, etc.

2 Lining

a Parietal layer of pleura lines entire chest cavity and covers superior surface of diaphragm

b Forms separate sac encasing each lung

c Parietal pleura separated from visceral pleura, covering lungs, only by potential space—pleural space—that contains few drops of pleural fluid

3 Shape of ribs and angle of their attachment to spine such that elevation of rib cage enlarges two dimensions of thorax, its width and depth from front

B Functions

1 Increase in size of thorax leads to inspiration

2 Decrease in size of thorax leads to expiration

Review Questions

1 What anatomical feature favors the spread of the common cold through the respiratory passages and into the middle ear?

2 How are the turbinates arranged in the nose? What are they?

3 What organs are found in the nasopharynx?

4 What tubes open into the nasopharynx?

5 Make a diagram showing the termination of a bronchiole in an alveolar duct with alveoli.

6 What kind of membrane lines the respiratory system?

7 What is the serous covering of the lung called? Where else, besides covering the lungs, is this same membrane found?

8 The pharynx is common to what two systems?

9 What is the voice box? Of what is it composed? What is the Adam's apple?

10 What is the epiglottis? What is its function?

11 What are the true vocal cords? What name is given to the opening between the cords?

12 What is the difference between air distribution and gas exchange in the respiratory system? What organs serve as air distributors? As gas exchangers?

13 List the organs that are included in the (a) upper and (b) lower respiratory tracts.

14 Name at least three structures that are classified as accessory respiratory system structures.

15 Discuss the importance of the "danger area of the face."

16 Discuss the anatomy of the roof of the nasal cavity (cribriform plate). Why is an understanding of the anatomy in this area of clinical significance following a nose injury?

17 What structures form the nasal septum?

18 Discuss the structure and function of the respiratory mucosa

19 Identify the paranasal sinuses and discuss their location, drainage, and functional significance.

20 List the anatomical subdivisions of the pharynx.

21 Which of the paired laryngeal cartilages are most important? Why?

22 List the divisions of the larynx.

23 Discuss the component parts of the "bronchial tree."

24 How many lobes are in the right lung? The left? What are the bronchopulmonary segments?

25 Briefly define the following terms: *ala, cleft palate, epistaxis, vibrissae, concha, choanae, alveolus.*

20 PHYSIOLOGY OF THE RESPIRATORY SYSTEM

OBJECTIVES

After you have completed this chapter, you should be able to:

1 List and briefly discuss the regulated and integrated processes that ensure tissues of an adequate oxygen supply and prompt removal of carbon dioxide.

2 Discuss the functional role of the respiratory mucosa in air purification, temperature regulation, and humidification.

3 Discuss the terminology used in pulmonary radiology and in the specialized radiographs used in diagnosis of disease.

4 Define pulmonary ventilation and outline the mechanism of normal, quiet inspiration and expiration.

5 List by name and explain the volume of air exchanged in pulmonary ventilation.

6 Discuss the distinctive patterns, or types, of breathing that can be recognized and designated by name.

7 Demonstrate the principles of partial pressures (Dalton's law) in explaining movement of respiratory gases between alveolar air and blood moving through pulmonary capillaries.

8 Compare and contrast the physiology of internal and external respiration.

9 Discuss the major factors that determine the volume of oxygen entering lung capillary blood.

10 Explain how blood transports oxygen and carbon dioxide.

11 Discuss gas exchange in tissue capillaries between arterial blood and cells.

12 Interpret changes in an oxygen dissociation curve at various blood pH levels.

13 Explain the reciprocal interaction of oxygen and carbon dioxide on blood gas transport (Bohr vs. Haldane effect).

14 Discuss the primary factors that influence the respiratory control center and thereby control respirations.

RESPIRATORY PHYSIOLOGY

In Chapter 19 the anatomy of the respiratory system was presented as a basis for understanding the physiological principles that regulate air distribution and gas exchange. This chapter deals with **respiratory physiology**—a complex series of interacting and coordinated processes that play a critical role in maintaining the stability, or constancy, of our internal environment. The proper functioning of the respiratory system ensures the tissues of an adequate oxygen supply and prompt removal of carbon dioxide. This process is complicated by the fact that control mechanisms must permit maintenance of homeostasis throughout a wide range of ever-changing environmental conditions and body demands. Adequate and efficient regulation of gas exchange between body cells and circulating blood under changing conditions is the essence of respiratory physiology. This complex function would not be possible without integration between a number of physiological control systems, including acid-base, water, and electrolyte balance, circulation, and metabolism.

Functionally, the respiratory system is composed of an integrated set of regulated processes that include pulmonary ventilation, gas exchange in the lungs and tissues, transport of gases, and overall regulation of respiration.

KEY TERMS

Apnea (ap-NE-ah)

Bronchogram (BRONG-ko-gram)

Carbaminohemoglobin (kar-bam-i-no-he-mo-GLO-bin)

Dyspnea (DISP-ne-ah)

Eupnea (up-NE-ah)

Expiration (ek-spi-RAY-shun)

Hyperpnea (hi-perp-NE-ah)

Inspiration (in-spi-RAY-shun)

Oxyhemoglobin (ok-se-he-mo-GLO-bin)

Pneumotaxic (nu-mo-TAK-sik)

Pneumothorax (nu-mo-THO-raks)

Spirometer (spi-ROM-e-ter)

Figure 20-1 Organs of respiration. Projection of lungs and bronchi to rib cage.

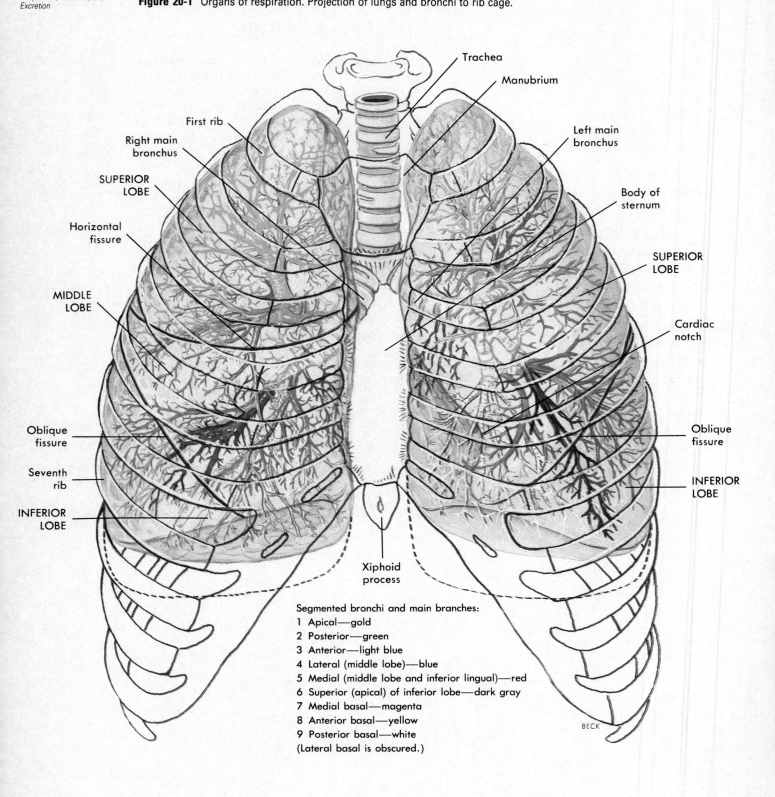

Trachea

Manubrium

First rib

Right main bronchus

Left main bronchus

SUPERIOR LOBE

Body of sternum

Horizontal fissure

SUPERIOR LOBE

MIDDLE LOBE

Cardiac notch

Oblique fissure

Oblique fissure

Seventh rib

INFERIOR LOBE

INFERIOR LOBE

Xiphoid process

Segmented bronchi and main branches:
1 Apical—gold
2 Posterior—green
3 Anterior—light blue
4 Lateral (middle lobe)—blue
5 Medial (middle lobe and inferior lingual)—red
6 Superior (apical) of inferior lobe—dark gray
7 Medial basal—magenta
8 Anterior basal—yellow
9 Posterior basal—white
(Lateral basal is obscured.)

BECK

FUNCTIONAL ANATOMY

Respiratory Mucosa

In addition to serving as air distribution passageways or gas exchange surfaces, the anatomical components of the respiratory tract and lungs function to cleanse, warm, and humidify inspired air. Air entering the nose is often contaminated by particulate materials, volatile chemicals, and other environmental irritants. Common contaminants include insects, dust, pollen, and bacterial organisms. A remarkably effective air purification mechanism removes almost every form of contaminant before inspired air reaches the alveoli. Mechanical responses, such as explosive expiration caused by coughing or sneezing, are effective in clearing the airway of large-size contaminants or foreign materials that may be inhaled. In addition, the coarse vibrissae in the nasal vestibule help in the filtering of small-size particulate material from inspired air. Extremely fine contaminants are effectively trapped by mucus covering the respiratory lining of the tract and then expelled by ciliary action. Phagocytosis of bacterial organisms and the removal of microscopic particles from the interstitial spaces by an extensive lymphatic network occurs throughout the respiratory tract (Figure 20-1).

Functions

The layer of protective mucus that covers a large portion of the interface between inspired air and respiratory tract tissues serves as the most important single element of the air purification mechanism. Over 125 ml of respiratory mucus (sputum) is produced daily. It forms a continuous sheet called a "mucous blanket," which moves, as a result of ciliary action, at the rate of 1 to 2 cm per hour. The

In light of the levels of biological, particulate, and chemical pollutants now present in atmospheric air, it is indeed remarkable that in a healthy individual the terminal respiratory passageways and alveoli are—in nonsmokers—clean, free of irritants, and almost sterile. Cigarette smoke paralyzes these cilia and results in accumulations of mucus and the typical smoker's cough, an effort to clear the secretions.

cilia that cover epithelial cells in the respiratory mucosa (Figure 20-2) are unique in that they beat in a unidirectional and coordinated way. The result is movement of mucus from the lower portions of the respiratory tree toward the oropharynx.

In addition to the removal of contaminants, inspired air is saturated with water and warmed as it descends through the respiratory tree. By the time even cold, dry air reaches the primary bronchi, it has been warmed to body temperature and fully humidified.

During expiration, the warm alveolar air, which is saturated with water vapor, flows through portions of the respiratory tree that become progressively cooler. The result is transfer of water from expired air to membranes lining the upper respiratory tract and nose. Exhaled air, however, is always fully saturated with water and in cold climates is visible as a cloud of condensation.

Figure 20-2 Pseudostratified columnar ciliated epithelium. The respiratory mucosa. Periodic acid–Schiff's (PAS) stain. (× 70.)

Goblet cell

Mucous gland in submucosa

Cilia

Epithelium

Figure 20-3 Normal chest x-ray film.
A, PA (posterior to anterior) view;
B, AP (anterior to posterior) view.

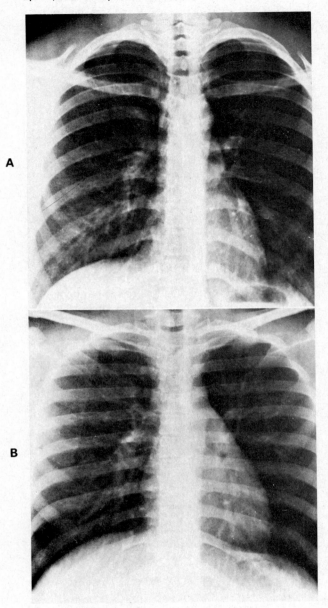

PULMONARY RADIOLOGY

Chest x-ray examinations account for more than half of all radiographs taken in the United States each year. They can be extremely useful both in the diagnosis of pathological conditions and in the study of normal respiratory anatomy and physiology.

Terminology

Most chest radiographs are taken during full inspiration while the individual is standing erect with the anterior surface of the chest against the film holder. The x-ray machine is about 2 meters (7 feet) behind the subject so that the x-ray beam enters the body from behind (posterior) and exits through the front (anterior). This type of exposure is appropriately called a PA (posterior to anterior) radiograph and is the most common type of x-ray examination. If the subject is asked to turn completely around so that the x-ray beam enters the chest from in front (anterior) and exits from the back (posterior), the resulting radiograph is termed AP (anterior to posterior) film. Figure 20-3, *A,* is a normal PA chest x-ray film of a young adult. Figure 20-3, *B,* is an AP radiograph of the same individual. Note that in the PA film the clavicles are superimposed over the upper lungs, with the medial aspect of each bone lower than the distal or lateral end. Note also that structural details of the cervical and thoracic vertebrae are more clearly visible in the PA film. In an AP chest film the clavicles are higher and the heart appears larger and is more clearly defined.

Specialized Radiographs

X-ray examinations that provide specialized information, such as bronchograms and arteriograms, are particularly useful in diagnosis of disease. Figure 20-4 is a bronchogram of the right lung. The bronchi of the right lower lobe show "crowding," indicating collapse and compression of lung tissue. Figure 20-5 is an arteriogram of the right lung showing the presence of an embolus, or moving blood clot (arrow), at the root of the right pulmonary artery.

FUNCTIONAL COMPONENTS OF THE RESPIRATORY SYSTEM

PULMONARY VENTILATION

Pulmonary ventilation is a technical term for what most of us call breathing. One phase of it, inspiration, moves air into the lungs and the other phase, expiration, moves air out of the lungs.

MECHANISM OF PULMONARY VENTILATION

Air moves in and out of the lungs for the same basic reason that any fluid, that is, a liquid or a gas, moves from one place to another—briefly, because its pressure in one place is different from that in the other place. Or stated differently, the existence of a pressure gradient (a pressure difference) causes fluids to move. A fluid always moves down its pressure gradient. This means that a fluid moves from the area where its pressure is higher to the area where its pressure is lower. Under standard conditions, air in the atmosphere exerts a pressure of 760 mm Hg. Air in the alveoli at the end of one expiration and before the beginning of another inspiration also ex-

Figure 20-4 A, Lung bronchogram. **B,** Note the "crowding" of right lower lobe bronchi. **C,** Collapse and compression of lung tissue is evident. Note deviation of midline thoracic structures.

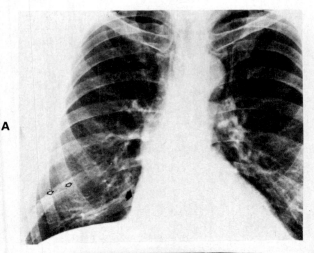

A

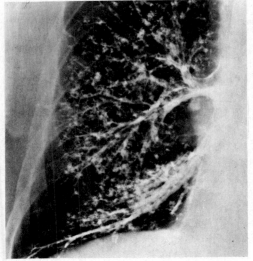

B

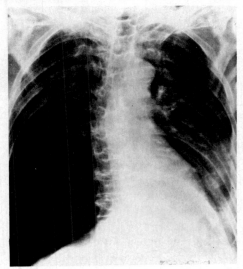

C

Figure 20-5 Pulmonary arteriogram. This x-ray film demonstrates the presence of an embolus (blood clot) at the root of the right pulmonary artery *(solid arrow).* Open arrowheads outline the normal lumen of the artery distal to the clot.

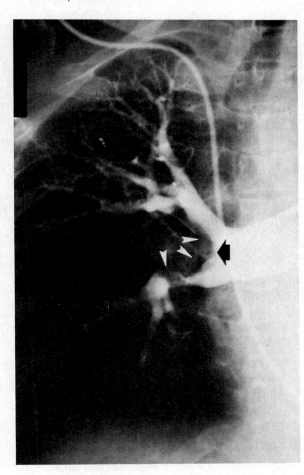

erts a pressure of 760 mm Hg. This fact explains why at that moment air is neither entering nor leaving the lungs. The mechanism that produces pulmonary ventilation is one that establishes a gas pressure gradient between the atmosphere and the alveolar air.

When atmospheric pressure is greater than pressure within the lung, air flows down this gas pressure gradient. Then air moves from the atmosphere into the lungs. In other words, inspiration occurs. When pressure in the lungs becomes greater than atmospheric pressure, air again moves down a gas pressure gradient. But now, this means that it moves in the opposite direction. This time, air moves out the lungs into the atmosphere. The pulmonary ventilation mechanism, therefore, must somehow establish these two gas pressure gradients—one in which in-

Figure 20-6 The mechanism of normal, quiet inspiration.

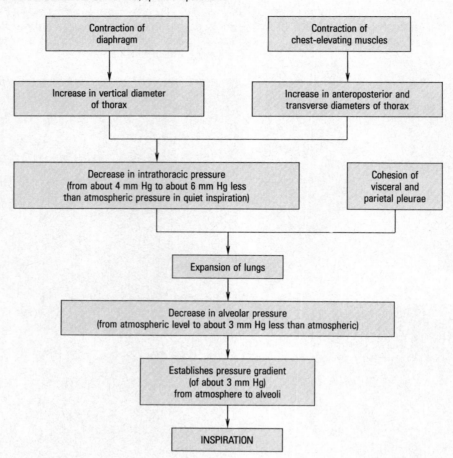

trapulmonic pressure (pressure within the lungs) is lower than atmospheric pressure to produce inspiration and one in which it is higher than atmospheric pressure to produce expiration.

These pressure gradients are established by changes in the size of the thoracic cavity, which in turn are produced by contraction and relaxation of respiratory muscles. An understanding of Boyle's law is important for understanding the pressure changes that occur in the lungs and thorax during the breathing cycle. It is a familiar principle stating that the volume of a gas varies inversely with pressure at a constant temperature. Application: expansion of the thorax (increase in volume) results in a decreased intrapleural (intrathoracic) pressure. This leads to a decreased intrapulmonic pressure that causes air to move from the outside into the lungs.

Inspiration

Contraction of the diaphragm alone, or of the diaphragm and the external intercostal muscles, produces quiet inspiration. As the diaphragm contracts,

it descends, and this makes the thoracic cavity longer. Contraction of the external intercostal muscles pulls the anterior end of each rib up and out. This also elevates the attached sternum and enlarges the thorax from front to back and from side to side. In addition, contraction of the sternocleidomastoid and serratus anterior muscles can aid in elevation of the sternum and rib cage during forceful inspiration. As the size of the thorax increases, the intrapleural (intrathoracic) and intrapulmonic pressures decrease (Boyle's law) and inspiration occurs. At the end of an expiration and before the beginning of the next inspiration, intrathoracic pressure is about 4 mm Hg less than atmospheric pressure (frequently written -4 mm Hg). During quiet inspiration, intrathoracic pressure decreases further to -6 mm Hg. As the thorax enlarges, it pulls the lungs alone with it because of cohesion between the moist pleura covering the lungs and the moist pleura lining the thorax. Thus the lungs expand, and the pressure in their tubes and alveoli necessarily decreases. Intrapulmonic (intra-alveolar) pressure decreases from an atmo-

To apply some of the information just discussed about the respiratory mechanism, let us suppose that a surgeon makes an incision through the chest wall into the pleural space, as is done in one of the dramatic, modern open-chest operations. Air is then present in the thoracic cavity, a condition known as **pneumothorax.** What change, if any, can you deduce takes place in respirations? Compare your deductions with those in the next paragraph.

Intrathoracic pressure, of course, immediately increases from its normal subatmospheric level to the atmospheric level. More pressure than normal is therefore exerted on the outer surface of the lung and causes its collapse. It could even collapse the other lung. Why? Because the mediastinum is a mobile rather than a rigid partition between the two pleural sacs. This anatomical fact allows the increased pressure in the side of the chest that is open to push the heart and other mediastinal structures over toward the intact side, where they exert pressure on the other lung. Pneumothorax results in many respiratory and circulatory changes. They are of great importance in determining medical and nursing care but lie beyond the scope of this book.

Figure 20-7 The mechanism of normal, quiet expiration.

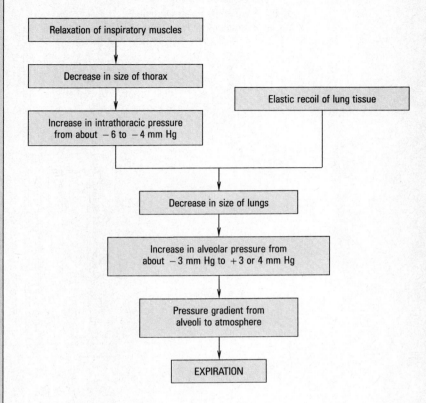

spheric level to a subatmospheric level—typically to about -3 mm Hg. The moment intrapulmonic pressure becomes less than atmospheric pressure, a pressure gradient exists between the atmosphere and the interior of the lungs. Air then necessarily moves into the lungs. For a diagram of the mechanism of inspiration just described, see Figure 20-6.

Expiration

Quiet expiration is ordinarily a passive process that begins when those pressure changes, or gradients, that resulted in inspiration are reversed. The inspiratory muscles relax, causing a decrease in the size of the thorax and an increase in intrapleural (intrathoracic) pressure from about -6 mm Hg to a preinspiration level of -4 mm Hg. It is important to understand that this pressure between the parietal and visceral pleura is always negative, that is, less than atmospheric pressure. The negative intrapleural (intrathoracic) pressure is required to overcome the so-called collapse tendency of the lungs caused by surface tension of the fluid lining the alveoli and the stretch of elastic fibers that are constantly attempting to recoil.

As alveolar pressure increases from about -3 mm Hg to $+3$ or $+4$ mm Hg, a positive-pressure gradient is established from alveoli to atmosphere, and expiration occurs as air flows outward through the respiratory passageways. (Figure 20-7 diagrams this mechanism of normal, quiet expiration.) In forced expiration, contraction of the abdominal and internal intercostal muscles can increase intraalveolar pressure to over 100 mm Hg.

VOLUMES OF AIR EXCHANGED IN PULMONARY VENTILATION

The volumes of air moved in and out of the lungs and remaining in them are matters of great importance. They must be normal so that a normal exchange of oxygen and carbon dioxide can take place between alveolar air and pulmonary capillary blood.

An apparatus called a **spirometer** is used to measure the volume of air exchanged in breathing (Figure 20-8). The volume of air exhaled normally after a typical inspiration is termed *tidal volume* (TV). As you can see in Figure 20-9, the normal volume of tidal air for an adult is approximately 500

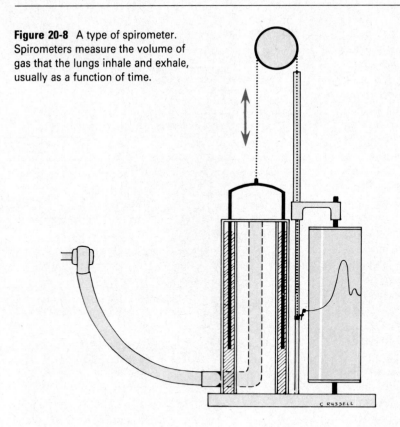

Figure 20-8 A type of spirometer. Spirometers measure the volume of gas that the lungs inhale and exhale, usually as a function of time.

Figure 20-9 During normal, quiet respirations the atmosphere and lungs exchange about 500 ml of air *(TV).* With a forcible inspiration, about 3,300 ml more air can be inhaled *(IRV).* After a normal inspiration and normal expiration, approximately 1,000 ml more air can be forcibly expired *(ERV).* Vital capacity is the amount of air that can be forcibly expired after a maximal inspiration and indicates, therefore, the largest amount of air that can enter and leave the lungs during respiration. Residual volume is the air that remains trapped in the alveoli.

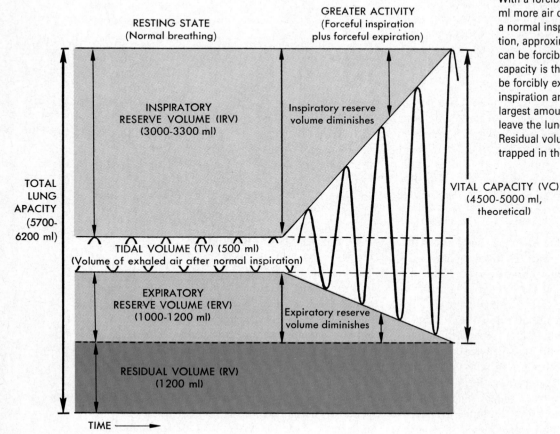

RESTING STATE
(Normal breathing)

GREATER ACTIVITY
(Forceful inspiration
plus forceful expiration)

INSPIRATORY
RESERVE VOLUME (IRV)
(3000-3300 ml)

Inspiratory reserve
volume diminishes

TOTAL
LUNG
CAPACITY
(5700-
6200 ml)

VITAL CAPACITY (VC)
(4500-5000 ml,
theoretical)

TIDAL VOLUME (TV) (500 ml)
(Volume of exhaled air after normal inspiration)

EXPIRATORY
RESERVE VOLUME (ERV)
(1000-1200 ml)

Expiratory reserve
volume diminishes

RESIDUAL VOLUME (RV)
(1200 ml)

TIME ⟶

ml. After an individual has expired tidal air, he/she can force still more air out of his/her lungs. The largest additional volume of air that one can forcibly expire after expiring tidal air is called the *expiratory reserve volume* (ERV). An adult, as Figure 20-9 shows, normally has an ERV of between 1,000 and 1,200 ml. *Inspiratory reserve volume* (IRV) is the amount of air that can be forcibly inspired over and above a normal inspiration. It is measured by having the individual exhale normally after a forced inspiration. The normal IRV is about 3.3 liters. No matter how forcefully an individual exhales, he/she cannot squeeze all the air out of his/her lungs. Some of it remains trapped in the alveoli. This amount of air that cannot be forcibly expired is known as *residual volume* (RV) and amounts to about 1.2 liters. Between breaths, an exchange of oxygen and carbon dioxide occurs between the trapped residual air in the alveoli and the blood. This process helps to "level off" the amounts of oxygen and carbon dioxide in the blood during the breathing cycle. In pneumothorax the RV is eliminated when the lung collapses. Even after the RV is forced out, the collapsed lung has a porous; spongy texture and will float in water because of trapped air called the *minimal volume*, which is about 40% of the RV.

Notice in Figure 20-9 that vital capacity is the sum of IRV + TV + ERV. It represents the largest volume of air an individual can move in and out of the lungs. It is determined by measuring the largest possible expiration after the largest possible inspiration. How large a vital capacity a person has depends on many factors—the size of the thoracic cavity, posture, and various other factors. In general, a larger person has a larger vital capacity than a smaller one. An individual has a larger vital capacity when standing erect than when stooped over or lying down. The volume of blood in the lungs also affects the vital capacity. If the lungs contain more blood than normal, alveolar air space is encroached on and vital capacity accordingly decreases. This becomes a very important factor in congestive heart disease. Excess fluid in the pleural or abdominal cavities also decreases vital capacity. So too does the disease emphysema. In the latter condition, alveolar walls become stretched—that is, lose their elasticity—and are unable to recoil normally for expiration. This leads to an increased RV. In severe emphysema, the RV may increase so much that the chest occupies the inspiratory position even at rest. Excessive muscular effort is therefore necessary for inspiration, and because of the loss of elasticity of lung tissue, greater effort is required, too, for expiration.

In diagnosing lung disorders a physician may need to know the inspiratory capacity and the functional residual capacity of the patient's lungs. *Inspiratory capacity* (IC) is the maximal amount of air an individual can inspire after a normal expiration. From Figure 20-9, you can deduce that IC = TV + IRV. With the volumes given in the figure, how many milliliters is the IC? *Functional residual capacity* is the amount of air left in the lungs at the end of a normal expiration. Therefore, as Figure 20-9 indicates, FRC = ERV + RV. With the volumes given, the functional residual capacity is 2,200 to 2,400 ml. The total volume of air a lung can hold is called the *total lung capacity*. It is, as Figure 20-9 indicates, the sum of all four lung volumes.

The term **alveolar ventilation** means the volume of inspired air that actually reaches, "ventilates," the alveoli. Only this volume of air takes part in the exchange of gases between air and blood. (Alveolar air exchanges some of its oxygen for some of the blood's carbon dioxide.) With every breath we take, part of the entering air necessarily fills our air passageways—nose, pharynx, larynx, trachea, and bronchi. This portion of air does not descend into any alveoli, so it cannot take part in gas exchange. In this sense, it is "dead air." And appropriately, the larger air passageways it occupies are said to constitute the **anatomical dead space.** One rule of thumb estimates the volume of air in the anatomical dead space as the same number of milliliters as the individual's weight in pounds. Another generalization says that the anatomical dead space approximates 30% of the TV. TV minus dead space volume equals alveolar ventilation volume. Suppose you have a normal TV of 500 ml and that 30% of this, or 150 ml, fills the anatomical dead space. The amount of air reaching your alveoli—your alveolar ventilation volume—is then 350 ml per breath, or 70% of your TV. Emphysema and certain other abnormal conditions, in effect, increase the amount of dead space air. Consequently, alveolar ventilation decreases and this, in turn, decreases the amount of oxygen that can enter blood and the amount of carbon dioxide that can leave it. Inadequate air–blood gas exchange, therefore, is the inevitable result of inadequate alveolar ventilation. Stated differently, the alveoli must be adequately ventilated for an adequate gas exchange to take place in the lungs.

TYPES OF BREATHING

The alternate movement of air into and out of the lungs that we call breathing can occur in distinctive patterns that can be recognized and designated by name.

Eupnea is the term used to describe normal quiet breathing. During eupnea, the need for oxygen and carbon dioxide exchange is being met and the individual is usually not conscious of the breathing pattern. Ventilation occurs spontaneously at the rate of 12 to 17 breaths per minute.

Hyperpnea means increased breathing that is

regulated to meet an increased demand by the body for oxygen. During hyperpnea, there is always an increase in pulmonary ventilation. The hyperpnea caused by exercise may meet the need for increased oxygen by an increase in tidal volume alone or by both an increase in tidal volume and breathing frequency.

Hyperventilation is characterized by an increase in pulmonary ventilation in excess of the need for oxygen. It sometimes results from a conscious voluntary effort preceding exertion or from psychogenic factors (hysterical hyperventilation). **Hypoventilation** is a decrease in pulmonary ventilation that results in elevated blood levels of carbon dioxide.

Dyspnea refers to labored or difficult breathing and is often associated with hypoventilation. A person suffering from dyspnea is aware or conscious of the breathing pattern and is generally uncomfortable and in distress. **Orthopnea** refers to dyspnea while lying down. It is relieved by sitting or standing up. This condition is common in patients with heart disease.

Several terms are used to describe the cessation of breathing. **Apnea** refers to the temporary cessation of breathing at the end of a normal expiration. It may occur during sleep or when swallowing. **Apneusis** is the cessation of breathing in the inspiratory position. Failure to resume breathing following a period of apnea is called **respiratory arrest.**

Cheyne-Stokes respiration is a periodic type of abnormal breathing often seen in terminally ill or brain-damaged patients. It is characterized by cycles of gradually increasing tidal volume for several breaths followed by several breaths with gradually decreasing tidal volume. These cycles repeat in a type of **crescendo-decrescendo** pattern.

Biot's breathing is characterized by repeated sequences of deep gasps and apnea. This type of abnormal breathing pattern is seen in individuals suffering from increased intracranial pressure.

EXTERNAL RESPIRATION

The exchange of gases in the lungs, or external respiration, takes place between alveolar air and venous blood flowing through lung capillaries. Gases move in both directions through the alveolar-capillary membrane (see Figure 19-15). Oxygen enters blood from the alveolar air because the P_{O_2} of alveolar air is greater than the P_{O_2} of venous blood. Another way

---

SOME PRINCIPLES ABOUT GASES

Before discussing respirations further, we need to understand the following principles.

1 Dalton's law (or the law of partial pressures). The term **partial pressure** means the pressure exerted by any one gas in a mixture of gases or in a liquid. The partial pressure of a gas in a mixture of gases is directly related to the concentration of that gas in the mixture and to the total pressure of the mixture. Suppose we apply this principle to compute the partial pressure of oxygen in the atmosphere. The concentration of oxygen in the atmosphere is 20.96% and the total pressure of the atmosphere is 760 mm Hg under standard conditions. Therefore:

Atmospheric $P_{O_2} = 20.96\% \times 760 = 159.2$ mm Hg

The symbol used to designate partial pressure is the capital letter P preceding the chemical symbol for the gas. Examples: alveolar air P_{O_2} is about 100 mm Hg; arterial blood P_{O_2} is also about 100 mm Hg; venous blood P_{O_2} is about 37 mm Hg. The word **tension** is often used as a synonym for the term **partial pressure**—oxygen tension means the same thing as P_{O_2}.

2 The partial pressure of a gas in a liquid is directly determined by the amount of that gas dissolved in the liquid, which in turn is determined by the partial pressure of the gas in the environment of the liquid. Gas molecules diffuse into a liquid from its environment and dissolve in the liquid until the partial pressure of the gas in solution becomes equal to its partial pressure in the environment of the liquid. Alveolar air constitutes the environment of blood moving through pulmonary capillaries. Standing between the blood and the air are only the very thin alveolar and capillary membranes, and both of these are highly permeable to oxygen and carbon dioxide. By the time blood leaves the pulmonary capillaries as arterial blood, diffusion and approximate equilibration of oxygen and carbon dioxide across the membranes have occurred. Arterial blood P_{O_2} and P_{CO_2} therefore usually equal or very nearly equal alveolar P_{O_2} and P_{CO_2} (Table 20-1).

Figure 20-10 Gas exchange in lung capillaries between alveolar air and venous blood. The CO_2 pressure gradient between alveolar air and venous blood causes outward diffusion of CO_2 from lung capillary blood; this lowers blood Pco_2 from its venous level to its arterial level. The O_2 pressure gradient between alveolar air and venous blood causes inward diffusion of O_2 into lung capillary blood; this raises blood Po_2 from its venous level up to its arterial level.

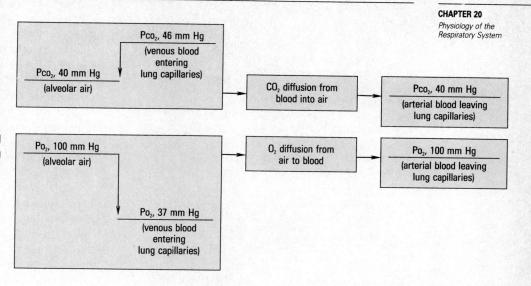

of saying this is that oxygen diffuses "down" its pressure gradient. Simultaneously, carbon dioxide molecules exit from the blood by diffusing down the carbon dioxide pressure gradient out into the alveolar air. The Pco_2 of venous blood is much higher than the Pco_2 of alveolar air. This two-way exchange of gases between alveolar air and venous blood converts venous blood to arterial blood (Figure 20-10).

The amount of oxygen that diffuses into blood each minute depends on several factors, notably these four:

1 The oxygen pressure gradient between alveolar air and venous blood (alveolar Po_2 − venous blood Po_2)

2 The total functional surface area of the alveolar-capillary membrane

3 The respiratory minute volume (respiratory rate per minute times volume of air inspired per respiration)

4 Alveolar ventilation (discussed on p. 539)

All four of these factors bear a direct relation to oxygen diffusion. Anything that decreases alveolar Po_2, for instance, tends to decrease the alveolar-venous oxygen pressure gradient and therefore tends to decrease the amount of oxygen entering the blood. Application: Alveolar air Po_2 decreases as altitude increases, and therefore less oxygen enters the blood at high altitudes. At a certain high altitude, alveolar air Po_2 equals venous blood Po_2. How would this affect oxygen diffusion into blood?

Anything that decreases the total functional surface area of the alveolar-capillary membrane also tends to decrease oxygen diffusion into the blood (by functional surface area is meant that which is freely permeable to oxygen). Application: in emphysema the total functional area decreases and is one of the factors responsible for poor blood oxygenation in the condition.

Anything that decreases the respiratory minute volume also tends to decrease blood oxygenation. Application: morphine slows respirations and therefore decreases the respiratory minute volume (volume of air inspired per minute) and tends to lessen the amount of oxygen entering the blood. The main factors influencing blood oxygenation are shown in Figure 20-11.

Several times we have stated the principle that structure determines function. You may find it interesting to note the application of this principle to gas exchange in the lungs. Several structural facts facilitate oxygen diffusion from the alveolar air into the blood in lung capillaries:

1 The fact that the walls of the alveoli and of the capillaries together form a very thin barrier for the gases to cross (estimated at not more than 0.004 mm thick—see Figure 19-15, p. 523).

2 The fact that both alveolar and capillary surfaces are extremely large.

Table 20-1 Oxygen and carbon dioxide pressure gradients

	Atmosphere	Alveolar air	Arterial blood	Venous blood
Po_2	160*	100	100	37
Pco_2	0.3	40	40	46

*Figures indicate approximate mm Hg pressure under usual conditions.

Figure 20-11 Some major factors determining volume of oxygen entering lung capillary blood. An increase in any of the factors tends to increase oxygenation of blood. A decrease in any of them tends to decrease blood oxygenation. Alveolar ventilation volume means the volume of air that reaches the alveoli; it equals the volume of air inspired minus the volume of the anatomical dead space.

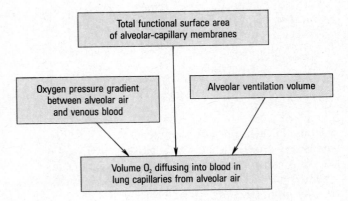

3 The fact that the lung capillaries accommodate a large amount of blood at one time (about 90 ml); (the lung capillaries of a small individual—one who has a body surface area of 1.5 square meters—contain about 90 ml of blood at one time under resting conditions. Studies done in the 1960s reported that lung capillaries contain about 60 ml of blood per square meter of body surface.)

4 The fact that the blood is distributed through the capillaries in a layer so thin (equal only to the diameter of one red corpuscle) that each corpuscle comes close to alveolar air.

HOW BLOOD TRANSPORTS GASES

Blood transports oxygen and carbon dioxide as solutes and as parts of molecules of certain chemical compounds. Immediately on entering the blood, both oxygen and carbon dioxide dissolve in the plasma. But, because fluids can hold only small amounts of gas in solution, most of the oxygen and carbon dioxide rapidly form a chemical union with some other blood constituent. In this way, comparatively large volumes of the gases can be transported. With a P_{O_2} of 100 mm Hg, only 0.3 ml of oxygen dissolves in 100 ml of arterial blood. Many times that amount, however, combines with the hemoglobin in 100 ml of blood to form oxyhemoglobin. Since each gram

of hemoglobin can unite with 1.34 ml of oxygen, the exact amount of oxygen in blood depends mainly on the amount of hemoglobin present. Normally, 100 ml of blood contains about 15 gm of hemoglobin. If 100% of it combines with oxygen, then 100 ml of blood will contain 15×1.34, or 20.1 ml, oxygen in the form of oxyhemoglobin.

Perhaps a more common way of expressing blood oxygen content is in terms of volume percent. Normal arterial blood, with P_{O_2} of 100 mm Hg, contains about 20 vol% O_2 (meaning 20 ml of oxygen in each 100 ml of blood).

Blood that contains more hemoglobin can, of course, transport more oxygen, and that which contains less hemoglobin can transport less oxygen. Hence hemoglobin-deficiency anemia decreases oxygen transport and may produce marked cellular hypoxia (inadequate oxygen supply).

To combine with hemoglobin, oxygen must, of course, diffuse from plasma into the red blood cells where millions of hemoglobin molecules are located. Several factors influence the rate at which hemoglobin combines with oxygen in lung capillaries. For instance, as the following equation and Figure 20-14 show, an increasing blood P_{O_2} and a decreasing P_{CO_2} both accelerate hemoglobin association with oxygen:

$$Hb + O_2 \xrightarrow[\text{Decreasing } P_{CO_2}]{\text{Increasing } P_{O_2}} HbO_2$$

Decreasing P_{O_2} and increasing P_{CO_2}, on the other hand, accelerate oxygen dissociation from oxyhemoglobin, that is, the reverse of the preceding equation. Oxygen associates with hemoglobin rapidly—so rapidly, in fact, that about 97% of the blood's hemoglobin has united with oxygen by the time the blood leaves the lung capillaries to return to the heart. In other words, the average **oxygen saturation** of arterial blood is about 97%.

Carbon dioxide is carried in the blood in several ways, the most important of which are described briefly as follows:

1 A small amount of carbon dioxide dissolves in plasma and is transported as a solute (dissolved carbon dioxide produces the P_{CO_2} of blood).

2 More than half of the carbon dioxide is carried in the plasma as bicarbonate ions (see Figure 20-15).

3 Somewhat less than one third of blood carbon dioxide unites with the NH$_2$ group of hemoglobin and certain other proteins to form carbamino compounds. Most of these are formed and transported in the red blood cells, since hemoglobin is the main protein to combine with carbon dioxide. The compound formed has a tongue-twisting name—**carbaminohemoglobin.** Carbon

Figure 20-12 Gas exchange in tissue capillaries between arterial blood and cells. Oxygen and carbon dioxide pressure gradients between arterial blood entering tissue capillaries and tissue fluid cause outward diffusion of oxygen from blood and inward diffusion of carbon dioxide into blood.

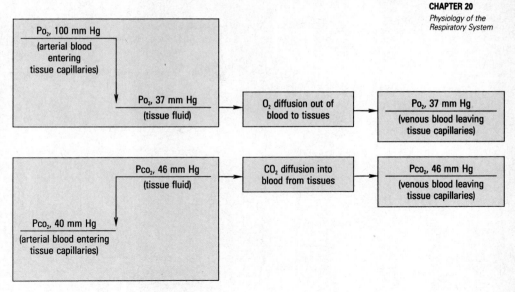

dioxide association with hemoglobin is accelerated by an increased P_{CO_2} and a decreasing P_{O_2} and is slowed by the opposite conditions. How do the conditions that accelerate carbon dioxide association affect the rate of oxygen association with hemoglobin?

INTERNAL RESPIRATION

The exchange of gases in tissues, or internal respiration, takes place between arterial blood flowing through tissue capillaries and cells (see Figure 20-15). It occurs because of the principle already noted that gases move down a gas pressure gradient. More specifically, in the tissue capillaries, oxygen diffuses out of arterial blood because the oxygen pressure gradient favors its outward diffusion (Figure 20-12). Arterial blood P_{O_2} (Figure 20-13) is about 100 mm Hg, interstitial fluid P_{O_2} is considerably lower, and intracellular fluid P_{O_2} is still lower. Although interstitial fluid and intracellular fluid P_{O_2} are not definitely established, they are thought to vary considerably—perhaps from around 60 mm Hg down to about 1 mm Hg. As activity increases in any structure, its cells necessarily use oxygen more rapidly. This decreases intracellular and interstitial P_{O_2}, which in turn tends to increase the oxygen pressure gradient between blood and tissues and to accelerate oxygen diffusion out of the tissue capillaries. In this way, the rate of oxygen use by cells automatically tends to regulate the rate of oxygen delivery to cells. As dissolved oxygen diffuses out of arterial blood, blood P_{O_2} decreases, and this accelerates oxyhemoglobin dissociation to release more oxygen into

the plasma for diffusion out to cells, as indicated in the following equation:

$$Hb + O_2 \underset{\text{Increasing } P_{CO_2}}{\overset{\text{Decreasing } P_{O_2}}{\rightleftharpoons}} HbO_2$$

Because of oxygen release to tissues from tissue capillary blood, P_{O_2}, oxygen saturation, and total oxygen content are less in venous blood than in arterial blood, as shown in Table 20-2.

Carbon dioxide exchange between tissues and blood takes place in the opposite direction from oxygen exchange. Catabolism produces large amounts of carbon dioxide inside cells. Therefore intracellular and interstitial P_{CO_2} are higher than arterial blood P_{CO_2}. This means that the carbon dioxide pressure gradient causes diffusion of carbon dioxide from the tissues into the blood flowing along through tissue capillaries (see Figure 20-12). Consequently, the

Table 20-2 Blood oxygen

	Venous blood	Arterial blood
P_{O_2}	37 mm Hg	100 mm Hg
Oxygen saturation	75%	97%
Oxygen content	15 ml O_2 per 100 ml blood	20 ml O_2 per 100 ml blood*

*Oxygen use by tissues = difference between oxygen contents of arterial and venous blood (20-15) = 5 ml O_2 per 100 ml blood circulated per minute.

Figure 20-13 Determinants of partial pressure of oxygen in arterial blood.

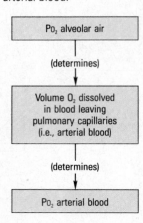

Figure 20-14 Oxygen dissociation curve of blood at 37° C, showing variations at three pH levels. For a given oxygen tension the higher the blood pH, the more the hemoglobin holds onto its oxygen, maintaining a higher saturation.

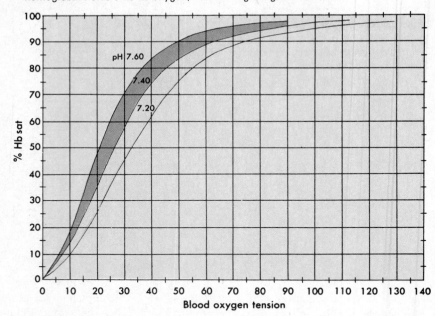

DIVING REFLEX

A protective physiological response called the **diving reflex** is responsible for the astonishing recovery of apparent drowning victims who have been submerged for over 40 minutes. Survivors are most often preadolescent children who have been immersed in water below 20° C. Apparently, the colder the water, the better chance of survival. Victims initially appear dead when pulled from the water. Breathing has stopped; they have fixed, dilated pupils; they are cyanotic; and the pulse has stopped.

Studies have shown that when the head and face are immersed in ice-cold water there is immediate shunting of blood to the core body areas with peripheral vasoconstriction and slowing of the heart (bradycardia). Metabolism is slowed, and tissue requirements for oxygen and nutrients decrease. The diving reflex is a protective response of the body to cold water immersion and is a function of such physiological and environmental parameters as water temperature, age, lung volume, and posture.

P_{CO_2} of blood increases in tissue capillaries from its arterial level of about 40 mm Hg to its venous level of about 46 mm Hg. This increasing P_{CO_2} and decreasing P_{O_2} together produce two effects—they favor both oxygen dissociation from oxyhemoglobin and carbon dioxide association with hemoglobin to form carbaminohemoglobin (Figures 20-14 and 20-15). This reciprocal interrelationship between oxygen and carbon dioxide transport mechanisms is contrasted in Figure 20-16. Note that increased P_{CO_2} decreases the affinity between hemoglobin and oxygen—this is called the **Bohr effect.** The **Haldane effect** refers to the increased carbon dioxide loading caused by a decrease in P_{O_2}.

REGULATION OF RESPIRATIONS

The mechanism for the control of respirations has many parts (Figure 20-17). A brief description of its main features follows:

1 The P_{CO_2}, P_{O_2}, and pH of *arterial blood* all influence respirations. The P_{CO_2} acts on chemoreceptors located in the medulla. Chemoreceptors in this case are cells sensitive to changes in arterial blood's carbon dioxide and hydrogen ion concentrations. The normal range for arterial P_{CO_2} is about 38 to 40 mm Hg. When it increases even slightly above this, it has a stimulating effect mainly on central chemoreceptors (postulated to be present in the medulla). Large

Figure 20-15 Exchange of gases in tissue capillaries. Oxyhemoglobin dissociates, releasing O_2, which diffuses from red blood cell intracellular fluid to plasma to interstitial fluid to tissue cell intracellular fluid. Simultaneously, CO_2 diffuses in opposite direction from tissue cells to red blood cells, where it associates with hemoglobin to form carbaminohemoglobin. As shown in the lower red blood cells in the capillary, some CO_2 combines with water to form carbonic acid. (Carbonic anhydrase, an enzyme abundant in red blood cells, accelerates this reaction.) H_2CO_3 ionizes, and HCO_3^- thus formed diffuses out of red blood cells into plasma in exchange for Cl^- (the "chloride shift"). (See Figure 25-4.)

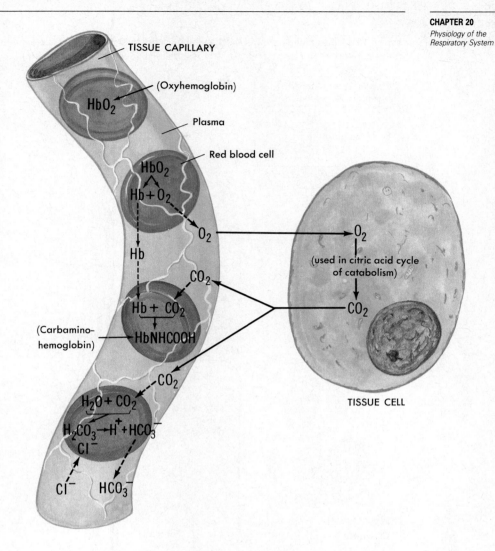

but tolerable increases in arterial P_{CO_2} stimulate peripheral chemoreceptors present in the carotid bodies and aorta. Stimulation of chemoreceptors by increased arterial P_{CO_2} results in faster breathing, with a greater volume of air moving in and out of the lungs per minute. Decreased arterial P_{CO_2} produces opposite effects—it inhibits central and peripheral chemoreceptors, which leads to inhibition of medullary respiratory centers and slower respirations. In fact, breathing stops entirely for a few moments (apnea) when arterial P_{CO_2} drops moderately—to about 35 mm Hg, for example.

A decrease in arterial blood pH (increase in acid), within certain limits, has a stimulating effect on chemoreceptors located in the carotid and aortic bodies.

The role of **arterial blood P_{O_2}** in controlling respirations is not entirely clear.

Figure 20-16 Reciprocal interaction of O_2 and CO_2 on blood-gas transport. Increased P_{CO_2} decreases the affinity between hemoglobin and O_2 (Bohr effect), whereas decreased P_{O_2} increases CO_2 content (Haldane effect).

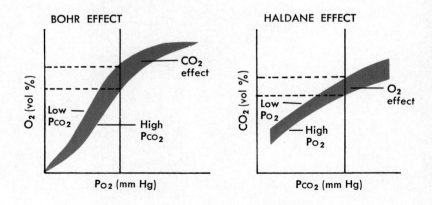

Figure 20-17 Respiratory control mechanism. Scheme to show the main factors that influence the respiratory center and thereby control respirations. (See text for discussion.)

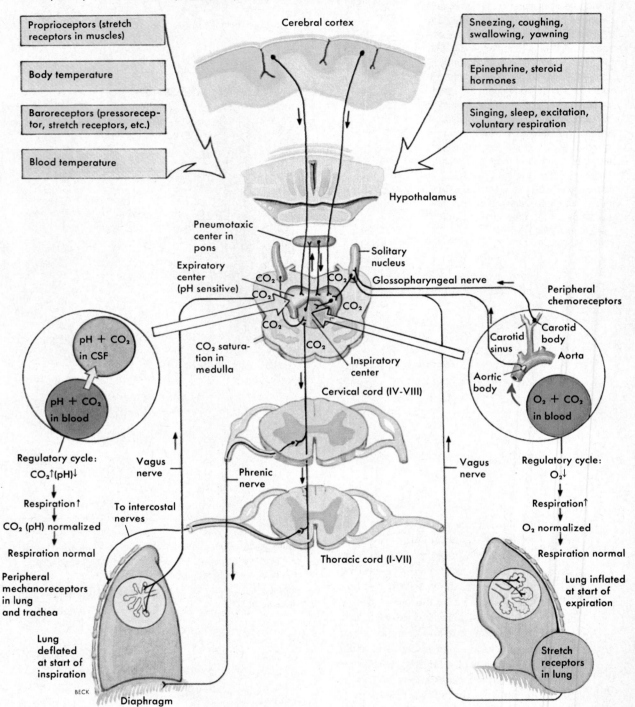

Presumably, it has little influence as long as it stays above a certain level. But neurons of the respiratory centers, like all body cells, require adequate amounts of oxygen to function optimally. Consequently, if they become hypoxic, they become depressed and send fewer impulses to respiratory muscles. Respirations then decrease or fail entirely. This principle has important clinical significance. For example, the respiratory centers cannot respond to stimulation by an increasing blood CO_2 if, at the same time, blood PO_2 falls below a critical level— a fact that may become of life-and-death importance during anesthesia.

However, a decrease in arterial blood PO_2 below 70 mm Hg but not so low as the critical level stimulates chemoreceptors in the carotid and aortic bodies and causes reflex stimulation of the inspiratory center. This constitutes an emergency respiratory control mechanism. It does not help regulate respirations under usual conditions when arterial blood PO_2 remains considerably higher than 70 mm Hg—the level necessary to stimulate the chemoreceptors.

2 **Arterial blood pressure** helps control respirations through the respiratory pressoreflex mechanism. A sudden rise in arterial pressure, by acting on aortic and carotid baroreceptors, results in reflex slowing of respirations. A sudden drop in arterial pressure brings about a reflex increase in rate and depth of respirations. The pressoreflex mechanism is probably not of great importance in the control of respirations. It is, however, of major importance in the control of circulation.

3 The **Hering-Breuer reflexes** help control respirations, particularly their depth and rhythmicity. They are believed to regulate the normal depth of respirations (extent of lung expansion), and therefore the volume of tidal air, in the following way. Presumably, when the tidal volume of air has been inspired, the lungs are expanded enough to stimulate baroreceptors located within them. The baroreceptors then send inhibitory impulses to the inspiratory center, relaxation of inspiratory muscles occurs, and expiration follows the Hering-Breuer expiratory reflex. Then, when the tidal volume of air has been expired, the lungs are sufficiently deflated to inhibit the lung baroreceptors and allow inspiration to start again—the Hering-Breuer inspiratory reflex.

4 The **pneumotaxic center** in the upper part of the pons is postulated to function mainly to maintain rhythmicity of respirations. Whenever the inspiratory center is stimulated, it sends impulses to the pneumotaxic center as well as to inspiratory muscles. The pneumotaxic center, after a moment's delay, stimulates the expiratory center, which then feeds back inhibitory impulses to the inspiratory center. Inspiration therefore ends, and expiration starts. Lung deflation soon initiates the Hering-Breuer inspiratory reflex, and inspiration starts again. In short, the pneumotaxic center and Hering-Breuer reflexes together constitute an automatic device for producing rhythmic respirations.

5 The **cerebral cortex** helps control respirations. Impulses to the respiratory center from the motor area of the cerebrum may either increase or decrease the rate and strength of respirations. In other words, an individual may voluntarily speed up or slow down the breathing rate. This voluntary control of respirations, however, has certain limitations. For example, one may will to stop breathing and do so for a few minutes. But holding the breath results in an increase in the carbon dioxide content of the blood since it is not being removed by respirations. Carbon dioxide is a powerful respiratory stimulant. So when arterial blood PCO_2 increases to a certain level, it stimulates the inspiratory center both directly and reflexly to send motor impulses to the respiratory muscles, and breathing is resumed even though the individual may still will contrarily. This knowledge that the carbon dioxide content of the blood is a more powerful regulator of respirations than cerebral impulses is of practical value when dealing with children who hold their breath to force the granting of their wishes. The best treatment is to ignore such behavior, knowing that respirations will start again as soon as the amount of carbon dioxide in arterial blood increases to a certain level.

6 Miscellaneous factors also influence respirations. Among these are blood temperature and sensory impulses from skin thermal receptors and from superficial or deep pain receptors:

HEIMLICH MANEUVER

The Heimlich maneuver is an effective and often lifesaving technique that can be used to open a windpipe that is suddenly obstructed. The maneuver (see figures, opposite) uses air already present in the lungs to expel the object obstructing the trachea. Most accidental airway obstructions result from pieces of food aspirated during a meal; the condition is sometimes referred to as a "cafe coronary." Other objects such as chewing gum or balloons are frequently the cause of obstructions in children. Individuals trained in emergency procedures must be able to tell the difference between airway obstruction and other conditions such as heart attacks that produce similar symptoms. The key question they must ask the person who appears to be choking is, "Can you talk?" A person with an obstructed airway will not be able to speak, even while conscious. The Heimlich maneuver, if the victim is standing, consists of the rescuer's grasping the victim with both arms around the victim's waist just below the rib cage and above the navel. The rescuer makes a fist with one hand, grasps it with the other, and then delivers an upward thrust against the diaphragm just below the xiphoid process of the sternum. Air trapped in the lungs is compressed, forcing the object that is choking the victim out of the airway.

Technique if victim can be lifted (see figure, opposite page, left)

1 Rescuer stands behind the victim and wraps both arms around the victim's chest slightly below the rib cage and above the navel. Victim is allowed to fall forward with head, arms, and chest over the rescuer's arms.
2 Rescuer makes a fist with one hand and grasps it with the other hand, pressing thumb side of fist against victim's abdomen just below the end of the xiphoid process and above the navel.
3 The hands only are used to deliver the upward subdiaphragmatic thrusts. It is performed with sharp flexion of the elbows, in an upward rather than inward direction, and is usually repeated four times. It is very important *not* to compress the rib cage or actually press on the sternum during the Heimlich maneuver.

Technique if victim has collapsed or cannot be lifted (see figure, opposite page, right)

1 Rescuer places victim on floor face up.
2 Facing victim, rescuer straddles the hips.
3 Rescuer places one hand on top of the other, with the bottom hand on the victim's abdomen slightly above the navel and below the rib cage.
4 Rescuer performs a forceful upward thrust with the heel of the bottom hand, repeating several times if necessary.

a *Sudden painful stimulation* produces a reflex apnea, but continued painful stimuli cause faster and deeper respirations.
b *Sudden cold stimuli* applied to the skin cause reflex apnea.
c *Afferent impulses* initiated by stretching the anal sphincter produce reflex acceleration and deepening of respirations. Use has sometimes been made of this mechanism as an emergency measure to stimulate respirations during surgery.
d *Stimulation of the pharynx or larynx* by irritating chemicals or by touch causes a temporary apnea. This is the choking reflex, a valuable protective device. It operates, for example, to prevent aspiration of food or liquids during swallowing.

HEIMLICH MANEUVER—cont'd

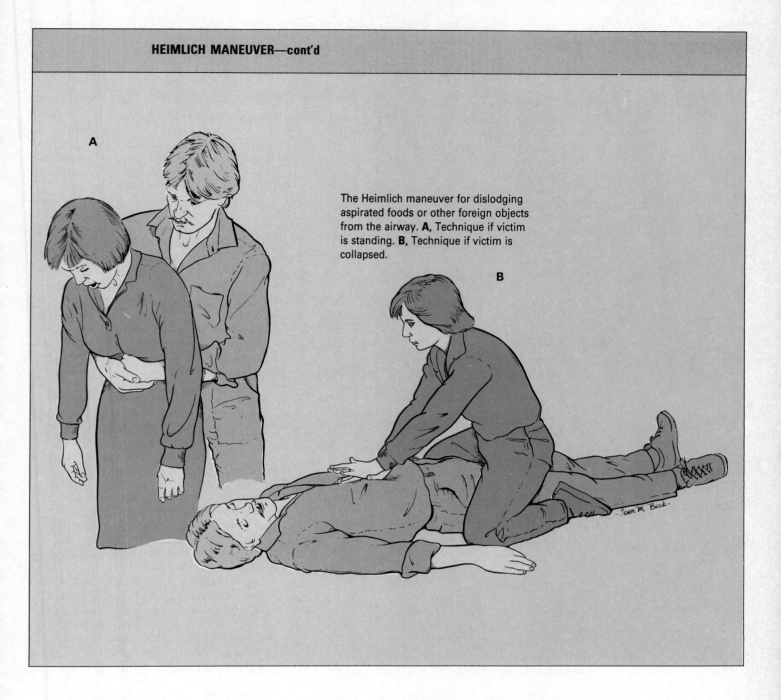

The Heimlich maneuver for dislodging aspirated foods or other foreign objects from the airway. **A,** Technique if victim is standing. **B,** Technique if victim is collapsed.

A

B

Joan M. Beck.

CONTROL OF RESPIRATIONS DURING EXERCISE

Respirations increase abruptly at the beginning of exercise and decrease even more markedly as it ends. This much is known. The mechanism that accomplishes this, however, is not known. It is not identical to the one that produces moderate increases in breathing. A number of studies have shown that arterial blood P_{CO_2}, P_{O_2}, and pH do not change enough during exercise to produce the degree of hyperpnea (faster, deeper respirations) observed. Presumably, many chemical and nervous factors and temperature changes operate as a complex, but still unknown, mechanism for regulating respirations during exercise.

Outline Summary

RESPIRATORY PHYSIOLOGY

A Respiratory physiology—adequate and efficient regulation of gas exchange between body cells and circulating blood under changing conditions

  1 Requires integration of other physiological control systems, including:

    a Acid-base balance

    b Water and electrolyte balance

    c Circulation

    d Metabolism

  2 Functionally composed of an integrated set of regulated processes:

    a Pulmonary ventilation

    b Gas exchange in lungs and tissues

    c Transport of gases

    d Overall regulation of respiration

FUNCTIONAL ANATOMY

A Air distribution passageways

B Gas exchange surfaces

C Modification of inspired air

  1 Cleanse air

  2 Warm air

  3 Humidify air

D Common air contaminants

  1 Insects and large particulate matter

  2 Dust

  3 Pollen

  4 Bacteria

E Air purification mechanisms

  1 Mechanical responses—coughing

  2 Vibrissae—filtration

  3 Mucus

  4 Phagocytosis

  5 Lymphatic removal of interstitial debris

F Respiratory tract mucus (sputum)

  1 Called "mucous blanket"—covers respiratory mucosa

  2 Quantity—125 ml produced daily

  3 Unidirectional movement—1 to 2 cm per minute (lower tract to oropharynx)

PULMONARY RADIOLOGY (Figures 20-3 to 20-5)

A Chest x-ray examinations account for over half of all radiographs taken in United States each year

B Types of chest x-ray examinations

  1 PA (posterior to anterior)—most common

  2 AP (anterior to posterior)

  3 Bronchogram

  4 Pulmonary arteriogram

FUNCTIONAL COMPONENTS OF RESPIRATORY SYSTEM

A Mechanism of pulmonary ventilation

  1 Contraction of diaphragm and chest-elevating muscles enlarges thorax, thereby decreases intrathoracic pressure, which causes expansion of lungs, which decreases intrapulmonic pressure to subatmospheric level, which establishes gas pressure gradient, which causes air to move into lungs

  2 Relaxation of inspiratory muscles produces opposite effects; (see Figure 20-7)

B Volumes of air exchanged in pulmonary ventilation—directly related to gas pressure gradient between atmosphere and lung alveoli and inversely related to resistance opposing air flow

  1 Tidal volume (TV)—average amount expired after normal inspiration; approximately 500 ml, or 1 pint

  2 Expiratory reserve volume (ERV)—additional amount of air that can be forcibly expired after a normal inspiration and expiration

  3 Inspiratory reserve volume (IRV)—amount that can be forcibly inspired after normal inspiration; measured by having individual expire normally after forced inspiration

  4 Residual volume (RV)—amount of air that cannot be forcibly expired from lungs

  5 Minimal volume (MV)—trapped air that remains in the lungs after the residual volume is eliminated (as in pneumothorax)

  6 Vital capacity (VC)—largest volume of air an individual can move in and out of his lungs; equals sum of inspiratory reserve volume, tidal volume, and expiratory reserve volume

  7 Anatomical dead space—volume of air that fills nose, pharynx, larynx, trachea, bronchi, and smaller tubes but does not descend into alveoli; therefore takes no part in gas exchange; typically, about 30% of TV, or 150 ml, but varies under different conditions

  8 Alveolar ventilation—volume of inspired air that actually reaches alveoli; computed by subtracting dead space air volume from tidal air volume

C Types of breathing

  1 Eupnea—normal quiet breathing

    a Individual not conscious of breathing pattern

    b Ventilation is spontaneous

    c Rate of ventilation about 12 to 17 breaths per minute

  2 Hyperpnea—increased breathing

    a Increased breathing regulated to meet increased demand

    b Always results in increased pulmonary ventilation

    c Increase in demand may be met by increase in tidal volume alone

Outline Summary—cont'd

    d Increase in tidal volume and increased breathing frequency may both occur to satisfy increased demand

  3 Hyperventilation—increase in pulmonary ventilation in excess of oxygen need

  4 Hypoventilation—decrease in pulmonary ventilation—results in elevated blood P_{CO_2}

  5 Dyspnea—labored or difficult breathing

    a Often associated with hypoventilation

    b Individual is conscious of breathing pattern and distress is common

  6 Orthopnea—refers to dyspnea while lying down

    a Condition relieved by sitting or standing up

    b Condition is common in heart disease

  7 Apnea—temporary cessation of breathing at end of normal expiration

  8 Apneusis—temporary cessation of breathing in the inspiratory position

  9 Respiratory arrest—failure to resume breathing following apnea or apneusis

 10 Cheyne-Stokes respirations—gradually increasing tidal volume for several breaths, followed by several breaths with gradually decreasing tidal volume; cycle repeats itself in a crescendo-decrescendo pattern

 11 Biot's respirations—repeated sequence of deep gasps and apnea

D Some principles about gases

  1 Dalton's law—partial pressure of gas in mixture of gases directly related to concentration of that gas in mixture and to total pressure of mixture

  2 Partial pressure of gas in liquid directly related to amount of gas dissolved in liquid; becomes equal to partial pressure of that gas in environment of liquid

E Exchange of gases in lungs (between alveolar air and venous blood)

  1 Where it occurs—in lung capillaries; across alveolar-capillary membrane

  2 What exchange consists of—oxygen diffuses out of alveolar air into venous blood; carbon dioxide diffuses in opposite direction

  3 Why it occurs—oxygen pressure gradient causes inward diffusion of oxygen; carbon dioxide pressure gradient causes outward diffusion of carbon dioxide

	Alveolar air	Venous blood
P_{O_2}	100 mm Hg	37 mm Hg
P_{CO_2}	40 mm Hg	46 mm Hg

  4 Results of gas exchange

    a P_{O_2} of blood increases to arterial blood level as blood moves through lung capillaries

    b P_{CO_2} of blood decreases to arterial blood level as blood moves through lung capillaries

    c Increasing P_{O_2} and decreasing P_{CO_2} accelerate both oxygen association with hemoglobin to form oxyhemoglobin and carbon dioxide dissociation from carbaminohemoglobin; both accelerated by increasing P_{O_2} and decreasing P_{CO_2}

F How blood transports gases

  1 Oxygen

    a About 0.5 ml transported as solute, that is, dissolved in 100 ml blood

    b About 19.5 ml O_2 per 100 ml blood transported as oxyhemoglobin in red blood cells

    c About 20.0 ml = Total O_2 content per 100 ml blood (100% saturation of 15 gm hemoglobin)

  2 Carbon dioxide

    a Small amount dissolves in plasma and transported as true solute

    b More than half of CO_2 transported as bicarbonate ion in plasma

    c Somewhat less than one third of CO_2 transported in red blood cells as carbaminohemoglobin

G Exchange of gases in tissues (between arterial blood and cells)

  1 Where it occurs—tissue capillaries

  2 What exchange consists of—oxygen diffuses out of arterial blood into interstitial fluid and on into cells, whereas carbon dioxide diffuses in opposite direction

  3 Why it occurs—oxygen pressure gradient causes diffusion of oxygen out of blood; carbon dioxide pressure gradient causes diffusion of carbon dioxide into blood

	Arterial blood	Interstitial fluid
P_{O_2}	100 mm Hg	60(?) mm Hg down to 1 (?) mm Hg
P_{CO_2}	46 mm Hg	50(?) mm Hg

  4 Results of oxygen diffusion out of blood and carbon dioxide diffusion into blood

    a P_{O_2} blood decreases as blood moves through tissue capillaries; arterial P_{O_2}, 100 mm Hg, becomes venous P_{O_2}, 40 mm Hg (figures vary)

    b P_{CO_2} blood increases; arterial P_{CO_2}, 40 mm Hg, becomes venous P_{CO_2}, 46 mm Hg (figures vary)

    c Oxygen dissociation from hemoglobin and carbon dioxide association with hemoglobin to form carbaminohemoglobin both accelerated by decreasing P_{O_2} and increasing P_{CO_2}

  5 Reciprocal interaction of oxygen and carbon dioxide on blood-gas transport

    a Bohr effect—increased P_{CO_2} decreases the affinity between hemoglobin and oxygen

    b Haldane effect—decreased P_{O_2} increases the affinity between hemoglobin and carbon dioxide

Outline Summary—cont'd

H Regulation of respirations (see Figure 20-17)

1 Respiratory centers—inspiratory and expiratory centers in medulla; pneumotaxic center in pons

2 Control of respiratory centers

  a Carbon dioxide major regulator of respirations; increased blood carbon dioxide content, up to certain level, stimulates respiration and, above this level, depresses respirations; decreased blood carbon dioxide decreases respirations

  b Oxygen content of blood influences respiratory center—decreased blood oxygen, down to a certain level, stimulates respirations and, below this critical level, depresses them; oxygen control of respirations nonoperative under usual conditions

  c Hering-Breuer mechanism helps control rhythmicity of respirations; increased alveolar pressure inhibits inspiration and starts expiration; decreased alveolar pressure stimulates inspiration and ends expiration

  d Pneumotaxic center acts with Hering-Breuer reflexes to produce rhythmic respirations

  e Cerebral cortex impulses to respiratory centers provide voluntary control, within limits, of rate and depth of respirations

3 Control of respirations during exercise—not yet established

Review Questions

1. Respiratory control mechanisms must function throughout a wide range of changing environmental conditions and body demands. Is this more difficult than steady-state homeostasis? Why?
2. Define *respiratory physiology.*
3. Name at least three physiological control systems that work closely with the respiratory system in maintenance of homeostasis.
4. Discuss the functional anatomy of the respiratory tract air distribution passageways and gas exchange surfaces.
5. How does the respiratory system clean, warm, and humidify inspired air?
6. Describe the pseudostratified, ciliated columnar epithelium that lines much of the respiratory tree.
7. What is the "mucous blanket"? What is its functional significance?
8. Discuss the common types of chest radiographs. What is a bronchogram?
9. Define *pulmonary ventilation.* Are the lungs active or passive during this process? Explain.
10. What is the main inspiratory muscle?
11. How is inspiration accomplished? Expiration?
12. If an opening is made into the pleural cavity from the exterior, what happens? Why?
13. What is the pleural space? What does it contain?
14. What substance found in blood is the natural chemical stimulant for the respiratory center?
15. Normally, about what percentage of the tidal volume fills the anatomical dead space?
16. Normally, about what percentage of the tidal volume is useful air, that is, ventilates the alveoli?

17. Identify the separate volumes that make up the total lung capacity.
18. Orthopnea is a symptom of what type of disease?
19. Dyspnea is often associated with what type of breathing?
20. Discuss the reciprocal interaction of oxygen and carbon dioxide on blood-gas transport. What is meant by the term Bohr effect? Haldane effect?
21. One gram of hemoglobin combines with how many milliliters of oxygen?
22. Suppose your blood has a hemoglobin content of 15 gm per 100 ml and an oxygen saturation of 97%. How many milliliters of oxygen would 100 ml of your arterial blood contain?
23. Compare the mechanisms that accelerate respiration with those that accelerate circulation during exercise.
24. Make a generalization about the effect of a moderate increase in the amount of blood carbon dioxide on circulation and respiration. What advantage can you see in this effect?
25. Make a generalization about the effect of a moderate decrease in oxygen on circulation and respiration.
26. Define the following terms briefly: *alveolus, apnea, asphyxia, complemental air, cyanosis, dyspnea, minimal air, orthopnea, P_{CO_2}, P_{O_2}, pleurisy, residual air, respiration, spirometer, supplemental air, thorax, tidal air, vital capacity.*

21 ANATOMY OF THE DIGESTIVE SYSTEM

OBJECTIVES

After you have completed this chapter, you should be able to:

1 Discuss the generalized function of the digestive system.
2 List, in sequence, each of the component parts or segments of the alimentary canal from mouth to anus, and identify the accessory organs of digestion that are located within or open into the gastrointestinal tract.
3 List and describe the four layers of the wall of the alimentary canal.
4 Discuss the major modifications of the coats of the digestive tract.
5 List and describe the structures of the mouth or buccal cavity.
6 Identify and compare the structure and secretions of the salivary glands.
7 Discuss the structural components of a typical tooth and identify by name and number the deciduous and permanent teeth.
8 Define the term *deglutition* and identify the structural divisions of the pharynx.
9 Discuss the size, shape, position, divisions, curves, sphincters, coats, and glands of the stomach.
10 Compare the structure and the functional activity of chief cells, parietal cells, and mucus-producing cells of the stomach.
11 Discuss the size, position, divisions, and coats of the small and large intestines.
12 Discuss the peritoneum and its reflections.
13 Discuss the structure and functions of the liver and gallbladder.
14 Explain the relationship between cell types and function in the pancreas.
15 Locate and discuss the significance of the vermiform appendix.

IMPORTANCE

This chapter deals with the anatomy of the digestive system. The organs of the digestive system together perform a vital function—that of preparing food for absorption and for use by the millions of body cells. Most food when eaten is in a form that cannot reach the cells (because it cannot pass through the intestinal mucosa into the bloodstream), nor could it be used by the cells even if it could reach them. It must therefore be modified both as to chemical composition and physical state so that nutrients can be absorbed and used by the body cells. The digestive tract and accessory organs comprise the system in which these complex changes in ingested food materials can occur. Part of the digestive system, the large intestine, serves also as an organ of elimination. Ingested food material that cannot be put into an absorbable form becomes waste material (feces) that is ultimately eliminated from the body.

The process of altering the chemical and physical composition of food so that it can be absorbed and used by the body cells is known as digestion and is the function of the digestive system. The process of digestion depends on both endocrine and exocrine secretions and the controlled movement of ingested food materials through the tract so that absorption can occur. The physiology of the digestive system is discussed in Chapter 22.

KEY TERMS

Alimentary canal (al-i-MEN-tar-e kah-NAL)

Appendicitis (ah-pen-di-SI-tis)

Cholecystitis (ko-le-sis-TI-tis)

Deciduous (de-SID-u-us)

Deglutition (de-gloo-TISH-un)

Epigastrium (ep-i-GAS-tre-um)

Hypochondrium (hi-po-KON-dree-um)

Jaundice (JAWN-dis)

Omentum (o-MEN-tum)

Peritoneum (per-i-to-NE-um)

Salivary (SAL-i-ver-e)

Villus (VIL-us)

Figure 21-1 Location of digestive system organs.

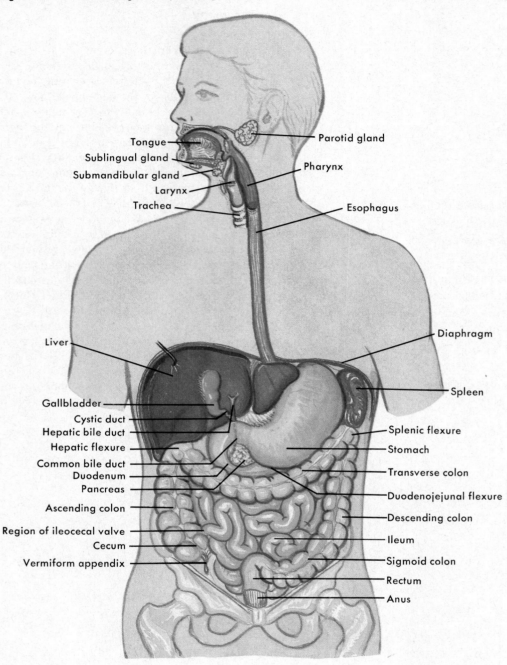

Tongue
Sublingual gland
Submandibular gland
Larynx
Trachea

Parotid gland
Pharynx
Esophagus

Liver
Diaphragm
Spleen

Gallbladder
Cystic duct
Hepatic bile duct
Hepatic flexure
Common bile duct
Duodenum
Pancreas
Ascending colon
Region of ileocecal valve
Cecum
Vermiform appendix

Splenic flexure
Stomach
Transverse colon
Duodenojejunal flexure
Descending colon
Ileum
Sigmoid colon
Rectum
Anus

ORGANS

The main organs of the digestive system (Figure 21-1) form a tube all the way through the ventral cavities of the body. It is open at both ends. This tube is usually referred to as the **alimentary canal** (or tract) or the **gastrointestinal (or GI) tract.** It is important to realize that ingested food material pass-ing through the lumen of the gastrointestinal tract is actually outside the body, even though the tube itself is inside the ventral body cavity. Following are two lists of (1) the component parts or segments of the alimentary canal and (2) the accessory organs located in the main digestive organs or opening into them.

Figure 21-2 Section of the small intestine showing the four layers typical of walls of the gastrointestinal tract. Circular folds of mucous membrane (plicae circularis) are special modifications of the small intestine. (Also see Figure 21-20.)

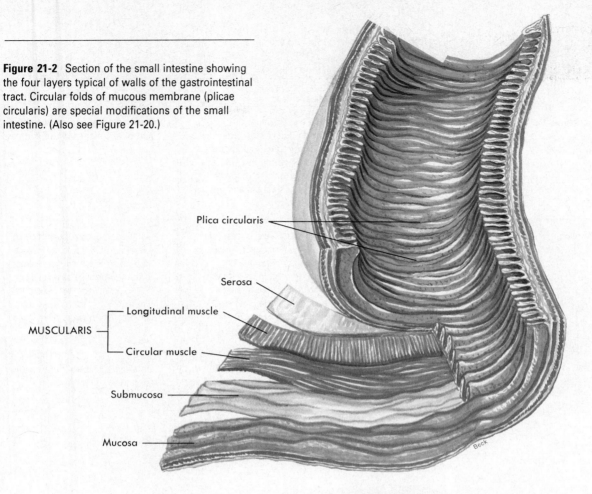

Plica circularis

Serosa

MUSCULARIS — Longitudinal muscle
— Circular muscle

Submucosa

Mucosa

WALLS OF ORGANS
Coats

The alimentary canal is essentially a tube with walls fashioned of four layers of tissues: a mucous lining, a submucous coat of connective tissue in which are embedded the main blood vessels of the tract, a muscular coat, and a fibroserous coat. Figures 21-2 and 21-20 show the four layers of the tube in a section of small intestine.

Modifications of Coats

Although the same four tissue coats form the various organs of the alimentary tract, their structures vary in different organs. Variations in the epithelial layer of the mucous lining, for example, range from stratified layers of squamous cells that provide protection from abrasion in the upper esophagus to the simple columnar epithelium, designed for absorption and secretion, found throughout most of the tract. Some of these modifications are listed in Table 21-1 and should be referred to when each of these organs is studied in detail.

MOUTH (BUCCAL CAVITY)

The following structures form the buccal cavity: the lips surrounding the orifice of the mouth and forming the anterior boundary of the oral cavity, the cheeks

Table 21-1 Modifications of coats of the digestive tract

Organ	Mucous coat	Muscle coat	Fibroserous coat
Esophagus		Two layers—inner one of circular fibers and outer one of longitudinal fibers; striated muscle in upper part and smooth in lower part of esophagus and in rest of tract	Outer coat fibrous; serous around part of esophagus in thoracic cavity
Stomach	Arranged in temporary longitudinal folds called *rugae*; allow for distention (Figure 21-16) Contains gastric pits with microscopic gastric glands	Has three layers instead of usual two—circular, longitudinal, and oblique fibers; two sphincters—cardiac at entrance of stomach and pyloric at its exit, formed by circular fibers	Outer coat visceral peritoneum; hangs in double fold from lower edge of stomach over intestines, forming apronlike structure or "lace apron," greater omentum (Figure 21-24); lesser omentum connects stomach to liver
Small intestine	Contains permanent circular folds, plicae circulares (Figure 21-2) Microscopic fingerlike projections, villi (Figure 21-20), with brush border (Figure 21-21) Crypts of Lieberkühn Microscopic duodenal (Brunner's) mucous glands Clusters of lymph nodes, Peyer's patches Numerous single lymph nodes called solitary nodes	Two layers—inner one of circular fibers and outer one of longitudinal fibers	Outer coat, visceral peritoneum, continuous with mesentery
Large intestine	Solitary nodes Intestinal mucous glands Anal columns form in anal region	Outer longitudinal coat condensed to form three tapelike strips (taeniae coli); small sacs (haustra) give rest of wall of large intestine puckered appearance (Figure 21-23); internal anal sphincter formed by circular smooth fibers and external anal sphincter by striated fibers	Outer coat, visceral peritoneum, continuous with mesocolon

(side walls), the tongue and its muscles (floor), and
the hard and soft palates (roof).

Lips

The **lips** are covered externally by skin and internally by mucous membrane that continues into the oral cavity and lines the mouth. The junction between skin and mucous membrane is highly sensitive and easily irritated. The upper lip is marked near the midline by a shallow vertical groove called the *philtrum,* which ends at the junction between skin and mucous membrane in a slight prominence called the *tubercle.* When the lips are closed, the line of contact is called the *oral fissure.*

Cheeks

The **cheeks** form the lateral boundaries of the oral cavity. They are continuous with the lips in front and are lined by mucous membrane that is reflected onto the gingiva or gums and the soft palate. The cheeks are formed in large part by the buccinator muscle, which is sandwiched with a considerable amount of adipose or fat tissue between the outer skin and mucous membrane lining. A number of small mucus-secreting glands are placed between the mucous membrane and the buccinator muscle; their ducts open opposite the last molar teeth.

Hard and Soft Palates

The **hard palate** consists of portions of four bones: two maxillae and two palatines (Figure 7-6, p. 174). The **soft palate,** which forms a partition between the mouth and nasopharynx (see Figure 19-2, p. 513), is fashioned of muscle arranged in the shape of an arch. The opening in the arch leads from the mouth into the oropharynx and is named the *fauces.* Suspended from the midpoint of the posterior border of the arch is a small cone-shaped process, the *uvula.*

Tongue

The **tongue** is a solid mass of skeletal muscle components (intrinsic muscles) covered by a mucous membrane. The intrinsic muscles of the tongue have, by definition, both their origin and insertion in the tongue itself. Intrinsic muscles have their fibers oriented in all directions, thus providing a basis for extreme maneuverability. Changes in the size and shape of the tongue caused by intrinsic muscle contraction assist in placement of food materials between the teeth during mastication. Extrinsic tongue muscles are those that insert into the tongue but have their origin on some other structure, such as a skull bone. Examples of extrinsic tongue muscles are the genioglossus, which protrudes the tongue, and the hyoglossus, which depresses it (Figure 21-3, *C*). Contraction of the extrinsic muscles is im-

portant during **deglutition,** or swallowing, and speech.

Note in Figure 21-3 that the tongue has a blunt *root,* a *tip,* and a central *body.* The upper, or dorsal, surface of the tongue is normally moist, pink, and covered by rough elevations called *papillae.* The three types of papillae—vallate, fungiform, and filiform—are all located on the sides or upper surface (dorsum) of the tongue. Note in Figure 21-3, *A,* that the large vallate papillae form an inverted V-shaped row extending from a median pit named the *foramen cecum* on the posterior part of the tongue. There are from 10 to 14 of these large, mushroomlike papillae. You can readily distinguish them if you look at your own tongue. Figure 21-4 is a low-power micrograph of vallate papillae. Dissolved substances to be tasted enter a moatlike depression surrounding the papillae to contact taste buds located on their lateral surface. Taste buds (Figure 21-5) are also located on the sides of the fungiform papillae, which are found chiefly on the sides and apex of the tongue. The numerous filiform papillae are filamentous and threadlike in appearance. They have a whitish appearance and are distributed over the anterior two thirds of the tongue. Filiform papillae do not contain taste buds.

The *lingual frenulum* (see Figure 21-3, *B*) is a fold of mucous membrane in the midline of the undersurface of the tongue that helps to anchor the tongue to the floor of the mouth. If the frenulum is too short for freedom of tongue movements—a congenital condition called *ankyloglossia*—the individual is said to be tongue-tied, and speech is faulty.

Folds of mucous membrane called the *plica fimbriata* (Figure 21-3, *B*) extend toward the apex of the tongue on either side of the lingual frenulum. The floor of the mouth and undersurface of the tongue are richly supplied with blood vessels. The deep lingual vein can be seen (Figure 21-3, *B*) shining through the mucous membrane between the lingual frenulum and plica fimbriata. In this region many vessels are extremely superficial and are covered only by a very thin layer of mucosa. Soluble drugs are absorbed into the circulation rapidly if placed under the tongue.

SALIVARY GLANDS

Three pairs of compound tubuloalveolar glands (see Figure 21-1 on p. 556)—the parotids, submandibulars, and sublinguals—secrete a major amount (about 1 liter) of the saliva produced each day. The small mucous-secreting glands (buccal glands) that occur in the mucosa lining the cheeks and mouth contribute less than 5% of the total salivary volume. Buccal gland secretion is important, however, to the hygiene and comfort of the mouth tissues. The salivary glands are typical of the accessory glands associated with the digestive system. They are located

Figure 21-3 **A,** Dorsum of the tongue. **B,** Mouth cavity showing undersurface of tongue. **C,** Extrinsic muscles of the tongue.

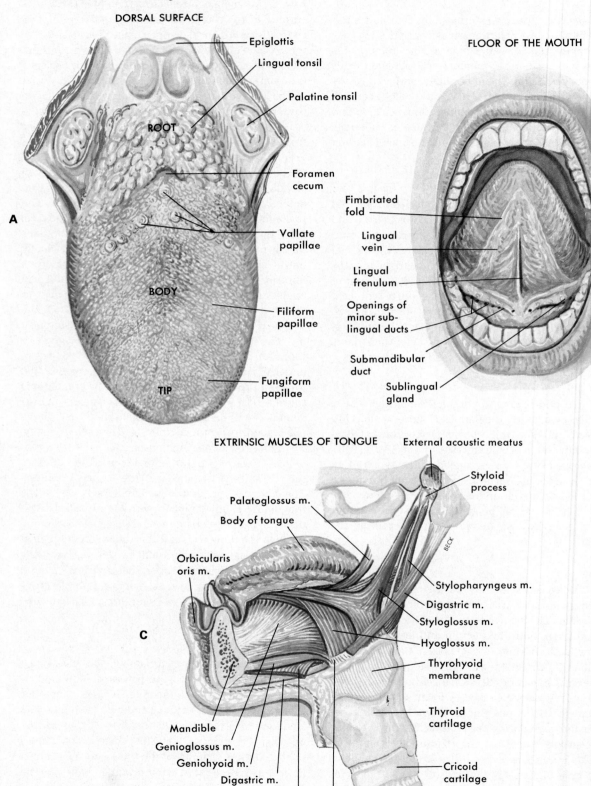

DORSAL SURFACE

Epiglottis

Lingual tonsil

Palatine tonsil

ROOT

Foramen cecum

Vallate papillae

A

BODY

Filiform papillae

Fungiform papillae

TIP

FLOOR OF THE MOUTH

Fimbriated fold

Lingual vein

Lingual frenulum

Openings of minor sublingual ducts

Submandibular duct

Sublingual gland

B

EXTRINSIC MUSCLES OF TONGUE

External acoustic meatus

Palatoglossus m.

Body of tongue

Styloid process

Orbicularis oris m.

Stylopharyngeus m.

Digastric m.

Styloglossus m.

Hyoglossus m.

Thyrohyoid membrane

C

Thyroid cartilage

Mandible

Genioglossus m.

Geniohyoid m.

Digastric m.

Mylohyoid m.

Cricoid cartilage

Digastric sling

First ring of trachea

BECK

Figure 21-4 Vallate papillae on surface of tongue. Taste buds are located on lateral surfaces of papillae. Several taste buds can be seen opening into the moat from the sides of the papillae. (× 35.)

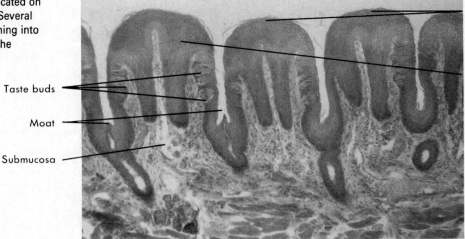

Vallate papilla

Stratified squamous epithelium

Taste buds

Moat

Submucosa

Figure 21-5 Taste buds. Enlargement of photomicrograph of taste buds in Figure 21-4. Arrow points to pore in outer surface of taste bud. (× 140.)

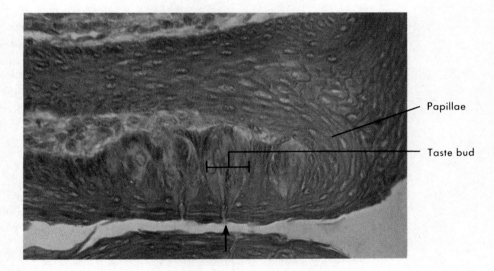

Papillae

Taste bud

outside of the alimentary canal and convey their exocrine secretions by ducts from the glands into the lumen of the tract. The functions of saliva in the digestive process will be discussed in Chapter 22.

Parotid

The pyramidal-shaped **parotids** are the largest of the paired salivary glands. They are located between the skin and underlying masseter muscle in front of and below the external ear. The parotids produce a watery or serous type of saliva containing enzymes but not mucus. The parotid (Stensen's) ducts are about 5 cm long. They penetrate the buccinator muscle on each side and open into the mouth opposite the second molars.

Submandibular

The **submandibular glands** (Figure 21-6) are called mixed or compound glands because they contain both serous (enzyme) and mucus-producing elements. These glands are located just below the mandibular angle. You can feel the gland by placing your index finger on the posterior part of the floor of the mouth and your thumb medial to and just in front of the angle of the mandible. The gland is irregular in form and about the size of a walnut. The ducts of the submandibular glands (Wharton's ducts) open into the mouth on either side of the lingual frenulum.

Sublingual

The **sublingual glands** are the smallest of the salivary glands. They lie in front of the submandibular glands, under the mucous membrane covering the floor of the mouth. Each sublingual gland is drained by eight to 20 ducts (ducts of Rivinus) that open into the floor of the mouth. The sublingual glands produce a mucous type of saliva.

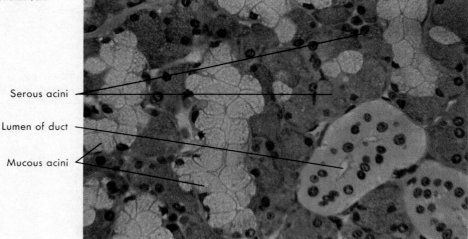

Serous acini

Lumen of duct

Mucous acini

Figure 21-6 Submandibular salivary gland. This mixed- or compound-type gland produces mucus from the mucous acini and enzymatic secretion from the serous acini. Duct cross sections are also visible. (× 140.)

TEETH

The teeth are the organs of mastication. They are designed to cut, tear, and grind ingested food so it can be mixed with saliva and swallowed. During the process of mastication, food is ground into small bits. This increases the surface area that can be acted on by the digestive enzymes.

Typical Tooth

A typical tooth (Figure 21-7) can be divided into three main parts: crown, neck, and root. The **crown** is the exposed portion of a tooth. It is covered by enamel—the hardest and chemically most stable tissue in the body. Enamel consists of approximately 97% calcified (inorganic) material and only 3% organic material and water. It is ideally suited to withstand the very abrasive process of mastication. The **neck** of a tooth is the narrow portion shown in Figure 21-7 that is surrounded by the gingivae, or gums. It joins the crown of the tooth to the root. It is the **root** that fits into the socket of the alveolar process of either upper or lower jaw. The root of a tooth may be a single peglike structure or have two or three separate conical projections (Figure 21-8). The root is not rigidly anchored to the alveolar process by cement but is suspended in the socket by the relatively soft periodontal membrane (see Figure 21-7).

In addition to enamel, the outer shell of each tooth is covered by two additional dental tissues—dentin and cementum (see Figure 21-7). Dentin makes up the greatest proportion of the tooth shell. It is covered by enamel in the crown and by cementum in the neck and root area. The dentin contains a **pulp cavity** consisting of connective tissue, blood and lymphatic vessels, and sensory nerves.

Dentition

Twenty **deciduous teeth,** or so-called baby teeth (Figure 21-9), appear early in life and are later replaced by 32 **permanent teeth** (Figure 21-10). The names and numbers of teeth present in both sets are given in Table 21-2.

The first deciduous tooth usually erupts at about 6 months. The rest follow at the rate of 1 or more a month until all 20 have appeared. There is, however, great individual variation in the age at which teeth erupt. Deciduous teeth are shed generally between the ages of 6 and 13 years. The third molars (wisdom teeth) are the last to appear, erupting usually sometime after 17 years of age.

Table 21-2 Dentition

	Number per jaw	
Name of tooth	**Deciduous set**	**Permanent set**
Central incisors	2	2
Lateral incisors	2	2
Canines (cuspids)	2	2
Premolars (bicuspids)	0	4
First molars (tricuspids)	2	2
Second molars	2	2
Third molars (wisdom teeth)	0	2
TOTAL (per jaw)	10	16
TOTAL (per set)	20	32

Figure 21-7 A molar tooth sectioned to show its bony socket and details of its three main parts: crown, neck, and root. Enamel (over the crown) and cementum (over the neck and root) surround the dentin layer. The pulp contains nerves and blood vessels.

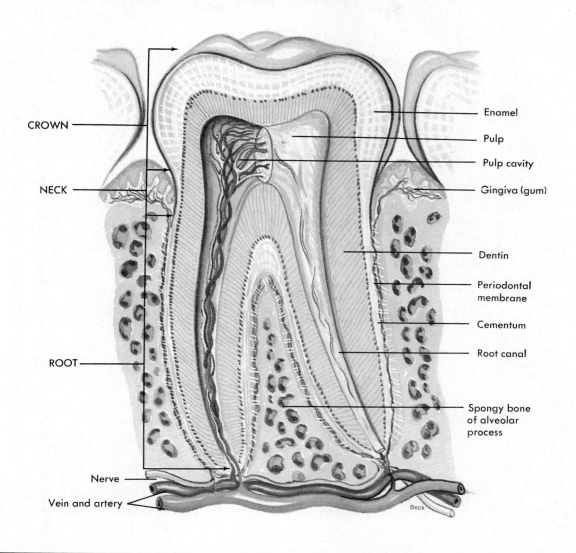

Figure 21-8 Panoramic dental x-ray film.

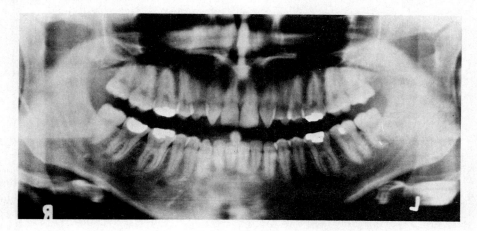

Figure 21-9 The deciduous arch. Note that in the set of 20 temporary primary teeth, there are no premolars (bicuspids) and there are only two pairs of molars in each jaw. Compare with permanent teeth (Figure 21-10).

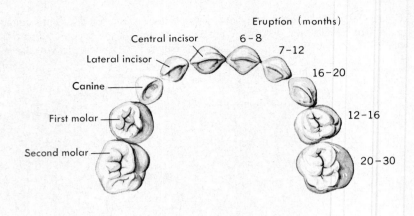

Eruption (months)

Central incisor — 6–8
Lateral incisor — 7–12
Canine — 16–20
First molar — 12–16
Second molar — 20–30

Figure 21-10 The 32 permanent teeth. Generally the lower teeth erupt before the corresponding upper teeth, and all teeth usually erupt earlier in girls than in boys.

18–22
16–24
14–18
7–9
6–8

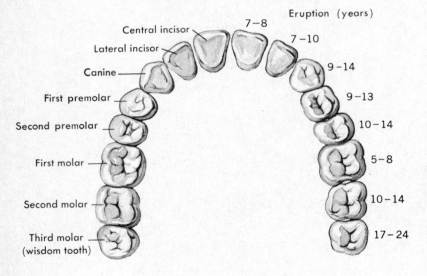

Eruption (years)

Central incisor — 7–8
Lateral incisor — 7–10
Canine — 9–14
First premolar — 9–13
Second premolar — 10–14
First molar — 5–8
Second molar — 10–14
Third molar (wisdom tooth) — 17–24

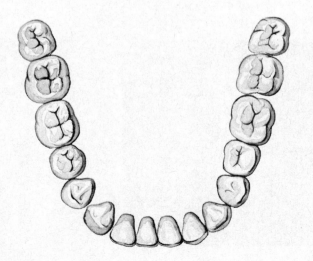

DENTAL DISEASES

Dental caries, or tooth decay, is a disease of the enamel, dentin, and cementum that results in demineralization and destruction of tissue with formation of a cavity in the tooth. Intact enamel resists bacterial attack, but once it is broken, the softer cementum and dentin decay and caries (cavity) formation results. **Periodontitis** is a generalized inflammation of the gums (gingivae) and periodontal membrane caused by deposits of plaque or impaction of food debris around the teeth (Figure 21-11). Soft tissue inflammation loosens the teeth and promotes cavity formation.

PHARYNX

The act of swallowing, or **deglutition,** moves a rounded mass of food called a *bolus* from the mouth, or oral cavity, to the stomach. As the food bolus

Figure 21-11 Periodontitis.

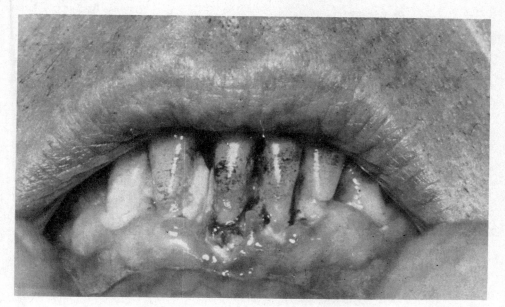

passes from the mouth, it enters the oropharynx by passing through a constricted opening called the *fauces*. The oropharynx is the second division of the pharynx (see Figure 19-7 on p. 517). During respiration, air passes through all three pharyngeal divisions. However, only the terminal portions of the pharynx serve the digestive system. Once a bolus has passed through the pharynx, it enters the digestive tube proper—that portion of the alimentary tract serving only the digestive system. The anatomy of the pharynx is discussed on p. 516.

ESOPHAGUS

The esophagus, a collapsible tube about 25 cm (10 inches) long, extends from the pharynx to the stomach, piercing the diaphragm in its descent from the thoracic to the abdominal cavity. It lies posterior to the trachea and heart.

The esophagus is the first segment of the digestive tube proper, and the four tissue coats or layers that form the wall of alimentary tract organs can be identified (Figure 21-12). The esophagus is normally flattened and the lumen is practically obliterated in the resting state. Figure 21-13 is a high-power view of the thick, abrasive-resistant mucosa that protects the lining of the esophagus from injury. The inner circular and outer longitudinal layers of the muscular coat are striated (voluntary) in the upper third, mixed (striated and smooth) in the middle third, and smooth (involuntary) in the lower third of the tube.

Figure 21-12 Esophagus. This low-power view of the esophagus illustrates the four tissue coats or layers that form the wall of alimentary tract organs. (× 22.)

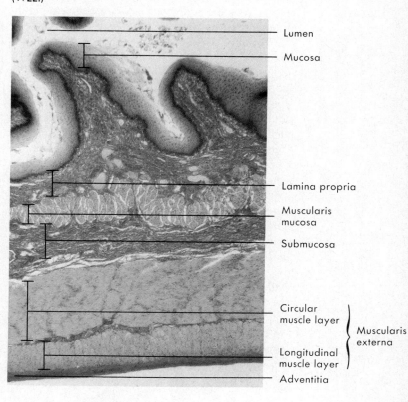

- Lumen
- Mucosa
- Lamina propria
- Muscularis mucosa
- Submucosa
- Circular muscle layer
- Longitudinal muscle layer
- Adventitia

Muscularis externa

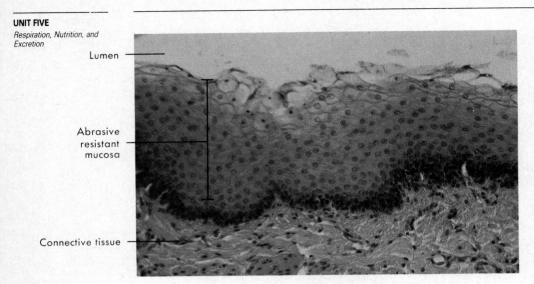

Lumen

Abrasive resistant mucosa

Connective tissue

Figure 21-13 Esophageal mucosa. Note the thick, abrasive-resistant nature of the lining. (× 70.)

STOMACH
Size, Shape, and Position

Just below the diaphragm, the digestive tube dilates into an elongated pouchlike structure, the stomach (Figures 21-14 and 21-15), the size of which varies according to several factors, notably sex and the amount of distention. In general, the female stomach is usually more slender and smaller than the male stomach. For some time after a meal, the stomach is enlarged because of distention of its walls, but, as food leaves, the walls partially collapse, leaving the organ about the size of a large sausage.

The stomach lies in the upper part of the abdominal cavity under the liver and diaphragm, with approximately five sixths of its mass to the left of the median line. In other words, it is described as lying in the **epigastrium** and left **hypochondrium** (Figure 1-6, p. 17). Its position, however, alters frequently. For example, it is pushed downward with each inspiration and upward with each expiration. When it is greatly distended from an unusually large meal, its size interferes with the descent of the diaphragm on inspiration, producing the familiar feeling of dyspnea that accompanies overeating. In this state, the stomach also pushes upward against the heart and may give rise to the sensation that the heart is being crowded.

Divisions

The **fundus, body,** and **pylorus** are the three divisions of the stomach. The fundus is the enlarged portion to the left and above the opening of the esophagus into the stomach. The body is the central part of the stomach, and the pylorus is its lower portion (Figures 21-15 and 21-16).

Curves

The upper right border of the stomach is known as the *lesser curvature;* the lower left border, the *greater curvature.*

Sphincter Muscles

Sphincter muscles guard both stomach openings. A sphincter muscle consists of circular fibers so arranged that there is an opening in the center of them (like the hole in a doughnut) when they are relaxed and no opening when they are contracted.

The **cardiac sphincter** guards the opening of the esophagus into the stomach, and the **pyloric sphincter** guards the opening from the pyloric portion of the stomach into the first part of the small intestine (duodenum).

The pyloric sphincter is of clinical importance because **pylorospasm** is a fairly common condition in infants. The pyloric fibers do not relax normally to allow food to leave the stomach, and the infant vomits food instead of digesting and absorbing it. The condition is relieved by the administration of a drug that relaxes smooth muscles. Another abnormality of the pyloric sphincter is **pyloric stenosis,** an obstructive narrowing of its opening.

Figure 21-14 Position of the stomach and large intestine in relation to the diaphragm *(dotted lines)* and the bony pelvis.

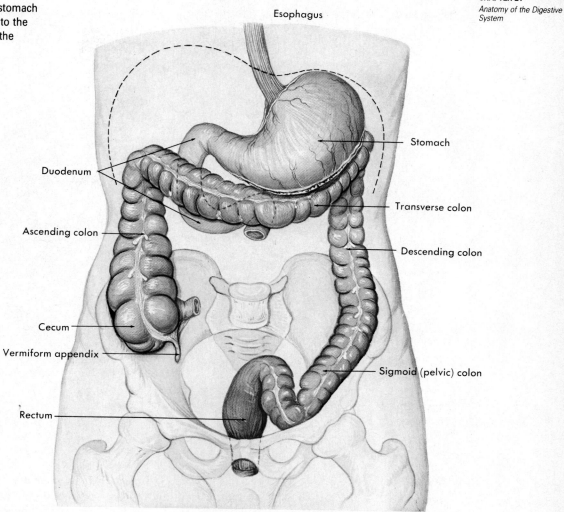

Esophagus

Stomach

Duodenum

Transverse colon

Ascending colon

Descending colon

Cecum

Vermiform appendix

Sigmoid (pelvic) colon

Rectum

Figure 21-15 X-ray film of normal stomach. The stomach is filled with a barium sulfate contrast preparation for x-ray viewing.

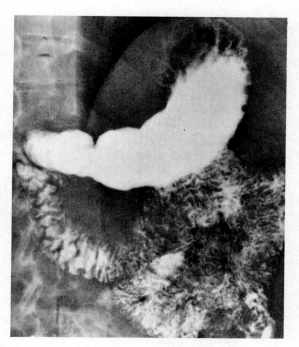

Figure 21-16 Muscle layers and interior of the stomach.

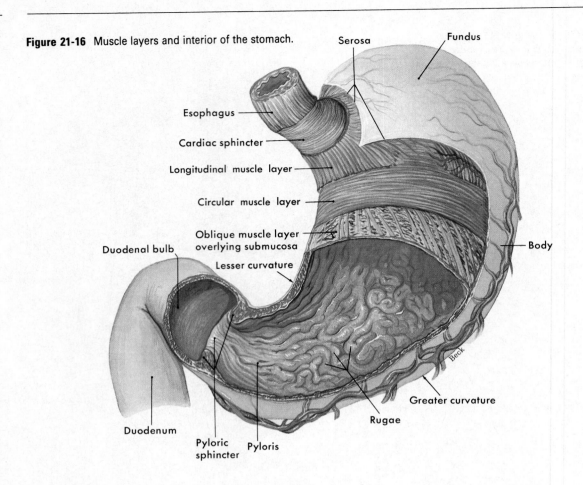

Coats

See Table 21-1 and Figure 21-16 for information on coats of the stomach.

Glands

The epithelial lining of the stomach is thrown into folds marked by depressions called *gastric pits* (Figures 21-17 and 21-18). Numerous coiled tubular-type glands, *gastric glands*, are found below the level of the pits, particularly in the fundus and body of the stomach. The glands secrete most of the gastric juice, a mucus-type fluid containing enzymes and hydrochloric acid. Figure 21-17 is a low-power micrograph of the mucosal lining in the body of the stomach, showing numerous gastric pits and a uniform underlying layer of coiled gastric glands. The mucosal lining is easily differentiated from the deeper submucosal layer in this section. The portion enclosed by the rectangle in Figure 21-17 is enlarged in Figure 21-18. The mucus-secreting nature of the stomach's epithelial lining is apparent in this enlarged view showing a single gastric pit lined by secretory cells filled with light-staining mucus.

In addition to the mucus-producing cells that cover the entire surface of the stomach and line the pits, the gastric glands contain two major secretory cells—**chief cells** and **parietal cells** (Figure 21-19). Chief cells (zymogenic cells) secrete the enzymes of gastric juice. Parietal cells secrete hydrochloric acid and are also thought to produce the mysterious protein known as *intrinsic factor.*

Atrophy of the gastric mucosa causes the condition known as **pernicious anemia.** As you might expect, there is a decrease in gastric enzymes and in hydrochloric acid. Achlorhydria is a characteristic finding in pernicious anemia. The absence of intrinsic factor is responsible for anemia caused by a breakdown in the erythropoietic mechanism. Intrinsic factor is necessary for vitamin B_{12} absorption, and vitamin B_{12} is essential for the normal development of red blood cells. Review the role of intrinsic factor in erythropoiesis on p. 426.

Figure 21-17 Stomach. Fold of gastric mucosa showing gastric pits *(arrows)* and underlying coiled gastric glands.(× 14.)

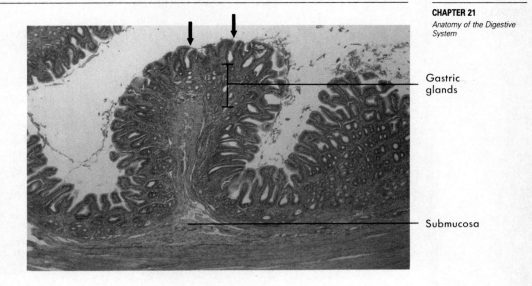

Gastric glands

Submucosa

Figure 21-18 Colon. Note the straight nature of the intestinal glands. Most of the columnar epithelial cells are goblet cells. (× 140.)

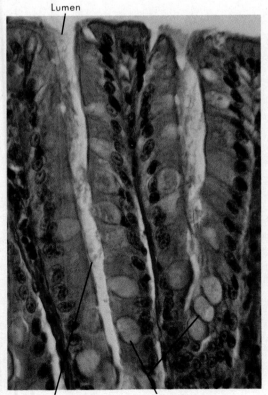

Lumen

Lumen of crypt (intestinal gland)

Goblet cell

Figure 21-19 Fundic region of stomach. Note the relatively great depth of the gastric glands as compared to the depth of the gastric pits. (× 35.)

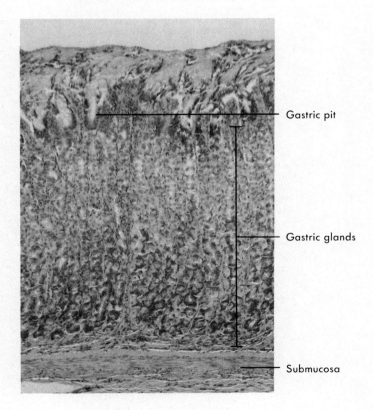

Gastric pit

Gastric glands

Submucosa

Figure 21-20 The small intestine.

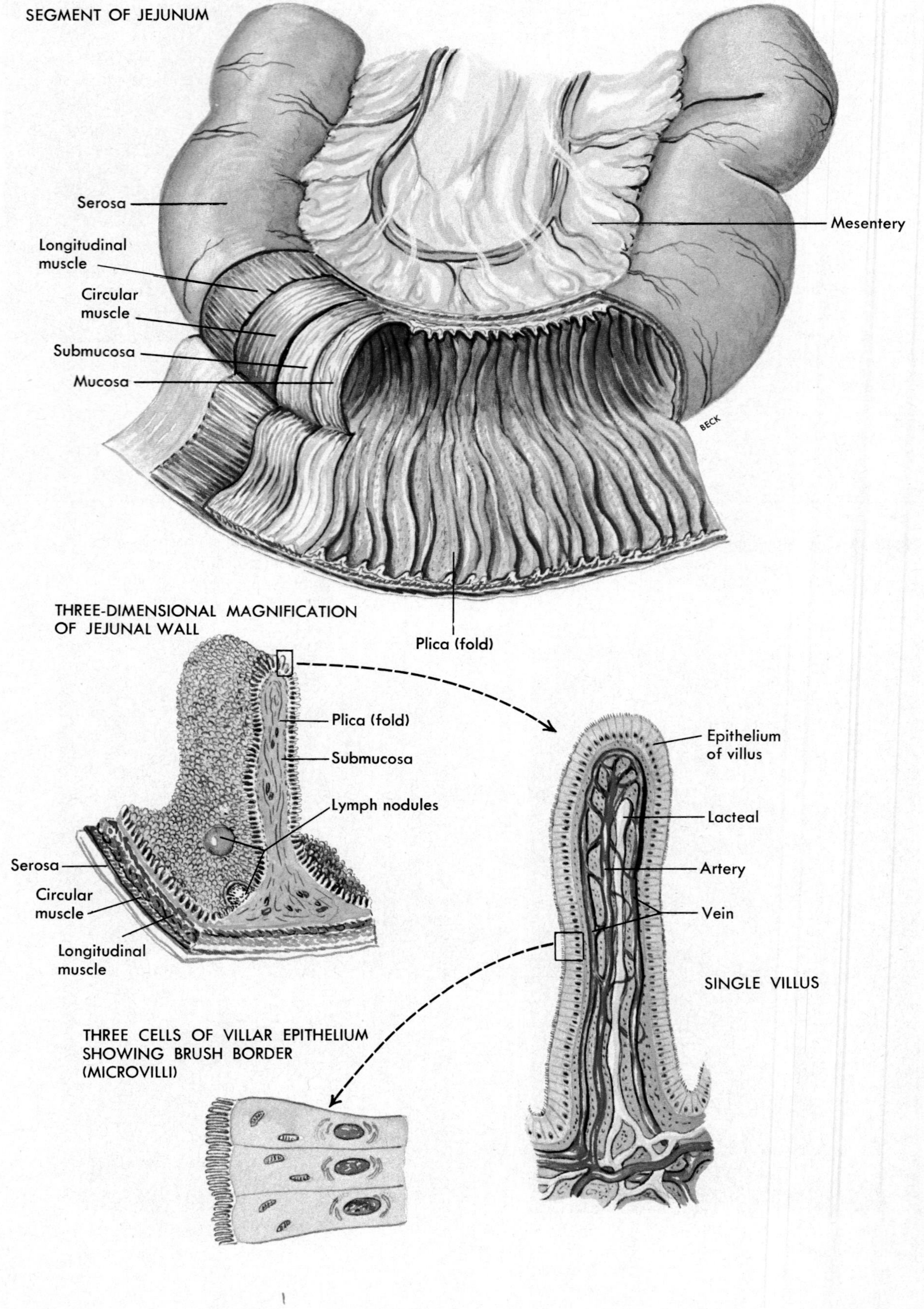

SEGMENT OF JEJUNUM

Serosa

Longitudinal muscle

Circular muscle

Submucosa

Mucosa

Mesentery

BECK

Plica (fold)

THREE-DIMENSIONAL MAGNIFICATION OF JEJUNAL WALL

Plica (fold)

Submucosa

Lymph nodules

Serosa

Circular muscle

Longitudinal muscle

Epithelium of villus

Lacteal

Artery

Vein

SINGLE VILLUS

THREE CELLS OF VILLAR EPITHELIUM SHOWING BRUSH BORDER (MICROVILLI)

SMALL INTESTINE
Size and Position

The small intestine is a tube measuring approximately 2.5 cm (1 inch) in diameter and 6 m (20 feet) in length. Its coiled loops fill most of the abdominal cavity.

Divisions

The small intestine consists of three divisions: the duodenum, the jejunum, and the ileum. The **duodenum*** is the uppermost division and is the part to which the pyloric end of the stomach attaches. It is about 25 cm (10 inches) long and is shaped roughly like the letter C. The duodenum becomes **jejunum** at the point where the tube turns abruptly forward and downward. The jejunal portion continues for approximately the next 2.5 m (8 feet), where it becomes the **ileum,** but without any clear line of demarcation between the two divisions. The ileum is about 3.5 m (12 feet) long.

*Derivation of the word "duodenum" may interest you. It comes from two Latin words that mean 12 fingerbreadths, a distance of about 11 inches, the approximate length of the duodenum.

Coats

Villi are important modifications of the mucosal layer of the small intestine. Millions of these projections, each about 1 mm in height, give the intestinal mucosa a velvety appearance. Each villus contains an arteriole, venule, and lymph vessel (lacteal) (Figure 21-20). Epithelial cells on the surface of villi can be seen by microscopy (Figure 21-21) to have a surface resembling a fine brush. This so-called *brush border* is formed by about 1,700 ultrafine *microvilli* per cell. Intestinal digestive enzymes, previously believed to be produced in crypts (of Lieberkühn) between villi, have now been found to be produced in these brush border cells toward the top of the villi. The presence of villi and microvilli increases the surface area of the small intestine by about 160 times, making this organ the main site of digestion and absorption. Mucus-secreting goblet cells are found in large numbers on villi and in crypts. See Table 21-1 and Figures 21-2 and 21-20 for more information on the coats of the small intestine.

Figure 21-21 Electron micrograph of a goblet cell in a villus of the jejunum. The cell is discharging mucus. Note the microvilli of the brush border on the adjacent epithelial cell. It is estimated that there are over 200 million microvilli per mm² on intestinal mucosa. (×22,500.)

Figure 21-22 X-ray film showing major divisions of the large intestine. *C,* Cecum. *Hf,* hepatic flexure. *T,* transverse colon. *Sf,* splenic flexure. *D,* descending colon. *S,* sigmoid colon. *R,* rectum.

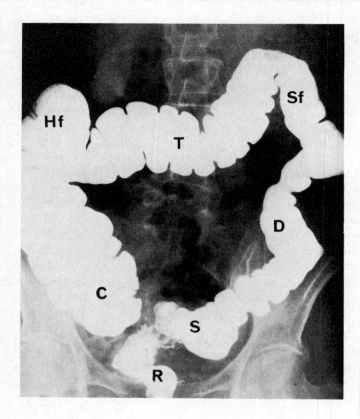

LARGE INTESTINE (COLON)
Size

The lower part of the alimentary canal bears the name *large intestine* because its diameter is noticeably larger than that of the small intestine. Its length, however, is much less, being about 1.5 to 1.8 m (5 or 6 feet). Its average diameter is approximately 6 cm (2½ inches), but this decreases toward the lower end of the tube.

Divisions

The large intestine is divided into the cecum, colon, and rectum (Figure 21-22).

CECUM. The first 5 to 8 cm (2 or 3 inches) of the large intestine are named the cecum. It is located in the lower right quadrant of the abdomen (see Figure 21-14).

COLON. The colon is divided into the following portions: ascending, transverse, descending, and sigmoid (Figures 21-1 and 21-22).

1 The **ascending colon** lies in a vertical position, on the right side of the abdomen, extending up to the lower border of the liver. The ileum joins the large intestine at the junction of the cecum and ascending colon, the place of attachment resembling the letter T in formation (Figure 21-23). The ileocecal valve permits material to pass from the ileum into the large intestine but not in the reverse direction.

2 The **transverse colon** passes horizontally across the abdomen, below the liver, stomach, and spleen. Note that this part of the colon is above the small intestine (see Figure 21-1). The transverse colon extends from the hepatic flexure to the splenic flexure, the two points at which the colon bends on itself to form 90-degree angles.

3 The **descending colon** lies in the vertical position, on the left side of the abdomen, extending from a point below the stomach and spleen to the level of the iliac crest.

4 The **sigmoid colon** is that portion of the large intestine that courses downward below the iliac crest. It describes an S-shaped curve. The lower part of the curve, which joins the rectum, bends toward the left, the anatomical reason for placing a patient on the left side when giving an enema. In this position, gravity aids the flow of the water from the rectum into the sigmoid flexure.

Figure 21-23 The vermiform appendix and ileocecal region. The cecum is opened to reveal the papillary form of ileocecal sphincter.

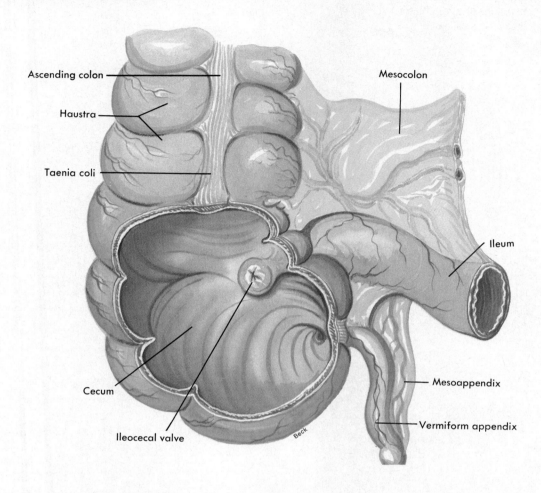

Ascending colon

Haustra

Taenia coli

Cecum

Ileocecal valve

Mesocolon

Ileum

Mesoappendix

Vermiform appendix

Beck

PERITONEUM

Now that you have completed the route through the digestive tube, consider for a moment the membrane covering most of these organs and holding them loosely in place. The peritoneum is a large, continuous sheet of serous membrane. It lines the walls of the entire abdominal cavity (parietal layer) and also forms the serous outer coat of the organs (visceral layer). In several places the peritoneum forms reflections, or extensions, that bind abdominal organs together (Figures 21-24 to 21-26). The **mesentery** is a fan-shaped projection of the parietal peritoneum from the lumbar region of the posterior abdominal wall. The attached posterior border of this great fan is just 15 to 20 cm long, yet the loose outer edge enclosing the jejunum and ileum is 6 m long. The mesentery allows free movement of each coil of the intestine and helps prevent strangulation of the long tube. A similar but less extensive fold of

RECTUM. The last 7 or 8 inches of the intestinal tube are called the rectum. The terminal inch of the rectum is called the **anal canal.** Its mucous lining is arranged in numerous vertical folds known as *anal columns*, each of which contains an artery and a vein. **Hemorrhoids** (or piles) are enlargements of the veins in the anal canal. The opening of the canal to the exterior is guarded by two sphincter muscles—an internal one of smooth muscle and an external one of striated muscle. The opening itself is called the *anus*. The general direction of the rectum is shown in Figure 27-1, p. 694. Note that the anus is directed slightly anteriorly and is therefore at approximately right angles to the rectum.

Coats

See Table 21-1 for information on the coats of the large intestine.

Figure 21-24 Abdominal viscera from the front. The transverse colon and the greater omentum are elevated to reveal the flexures of the colon and the loops of the small intestine.

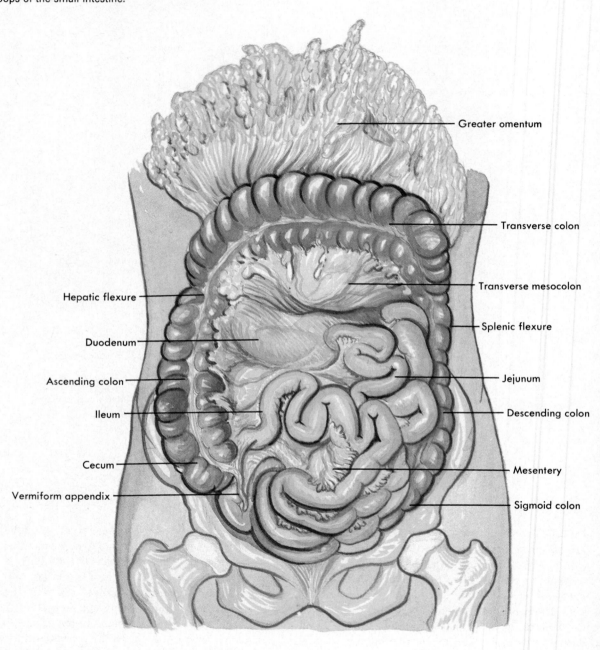

peritoneum called the **transverse mesocolon** attaches the transverse colon to the posterior abdominal wall. The **greater omentum** is a continuation of the serosa of the greater curvature of the stomach and the first part of the duodenum to the transverse colon. Spotty deposits of fat accumulate in the omentum and give it the appearance of a lacy apron hanging down loosely over the intestines. In cases of localized abdominal inflammation, such as appendicitis, the greater omentum envelops the inflamed area, walling it off from the rest of the abdomen. The **lesser omentum** attaches from the liver to the lesser curvature of the stomach and the first part of the duodenum. The falciform ligament extends from the liver to the anterior abdominal wall. Examine the relations of peritoneal extensions in Figure 21-26.

Figure 21-25 The transverse colon and greater omentum are raised and the small intestine is pulled to the side to show the transverse mesocolon and mesentery.

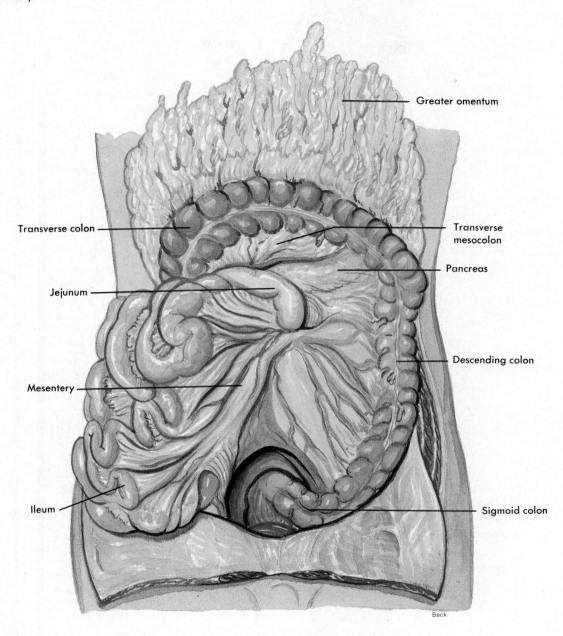

Greater omentum

Transverse colon

Transverse mesocolon

Pancreas

Jejunum

Mesentery

Descending colon

Ileum

Sigmoid colon

Beck

LIVER
Location and Size

The liver is the largest gland in the body. It weighs about 1.5 kg (3 to 4 pounds), lies immediately under the diaphragm, and occupies most of the right hypochondrium and part of the epigastrium (Figure 21-27).

Lobes and Lobules

The liver consists of two lobes separated by the falciform ligament. The **left lobe** forms about one sixth of the liver, whereas the **right lobe** makes up the remainder. The right lobe has three parts designated as the *right lobe proper,* the *caudate lobe* (a small four-sided area on the posterior surface), and

Figure 21-26 Sagittal view of the abdomen showing the peritoneum and its reflections. Intraperitoneal spaces are shown in blue and extraperitoneal spaces in green.

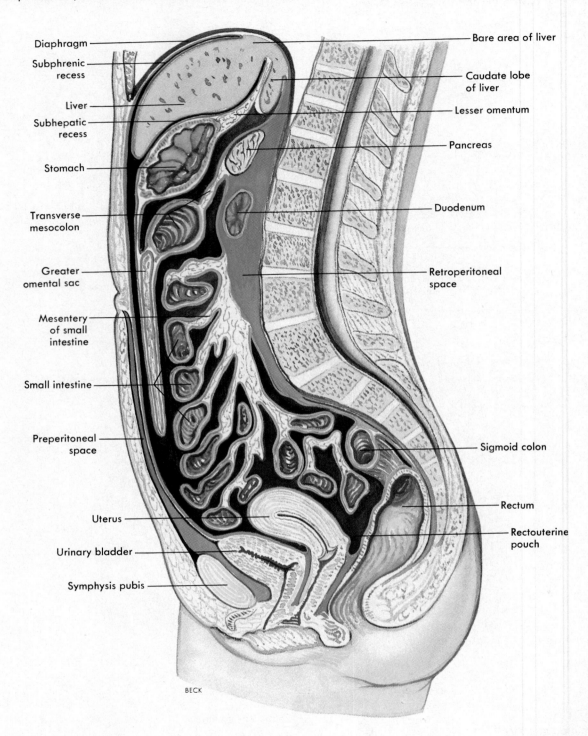

BECK

Figure 21-27 The liver and pancreas in their normal positions relative to the rib cage, diaphragm, and stomach.

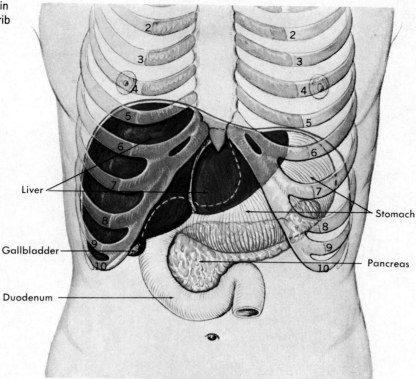

the *quadrate lobe* (an approximately oblong section on the undersurface). Each lobe is divided into numerous lobules by small blood vessels and by fibrous strands that form a supporting framework (the capsule of Glisson) for them. The capsule of Glisson is an extension of the heavy connective tissue capsule that envelops the entire liver.

The **hepatic lobules** (Figure 21-28), the anatomical units of the liver, are tiny hexagonal or pentagonal cylinders about 2 mm high and 1 mm in diameter. A small branch of the hepatic vein extends through the center of each lobule. Around this central (intralobular) vein, in plates or irregular walls radiating outward, are arranged the hepatic cells. Around the periphery of each lobule, several sets of three tiny tubes—branches of the hepatic artery, of the portal vein (interlobular veins), and of the hepatic duct (interlobular bile ducts)—are arranged. From these, irregular branches (sinusoids) of the interlobular veins extend between the radiating plates of hepatic cells to join the central vein. Minute bile canaliculi form networks around each cell.

Consider the function of the hepatic lobule while carefully examining Figures 21-28 and 21-29. Blood enters a lobule from branches of the hepatic artery and portal vein. Arterial blood oxygenates the hepatic cells, whereas blood from the portal system passes through the liver for "inspection." Sinusoids

in the lobule are lined with reticuloendothelial cells (mainly Kupffer cells). These phagocytic cells can remove toxic materials from the bloodstream. Ingested vitamins and other nutrients to be stored or metabolized by liver cells also enter the hepatic cell "bricks," forming radiating walls of the lobule. Blood continues along sinusoids to a vein at the center of the lobule (Figure 21-29). Such intralobular veins eventually lead to the two main hepatic veins. Bile formed by hepatic cells passes through canaliculi to the periphery of the lobule to join small bile ducts.

Ducts

The small bile ducts within the liver join to form two larger ducts that emerge from the undersurface of the organ as the right and left hepatic ducts. These immediately join to form one **hepatic duct.** The hepatic duct merges with the *cystic duct* from the gallbladder, forming the *common bile duct* (Figure 21-30), which opens into the duodenum in a small raised area, called the major duodenal papilla. This papilla is located 7 to 10 cm below the pyloric opening from the stomach.

Functions

The liver is one of the most vital organs of the body. Here, in brief, are its main functions:

1 Liver cells detoxify a variety of substances.

Figure 21-28 Liver lobule. Blood from branches of the portal vein and hepatic artery passes through sinusoids between plates of hepatic cells. Sinusoidal blood empties into the central vein, which leads to the hepatic veins. Bile canaliculi empty bile into bile ducts.

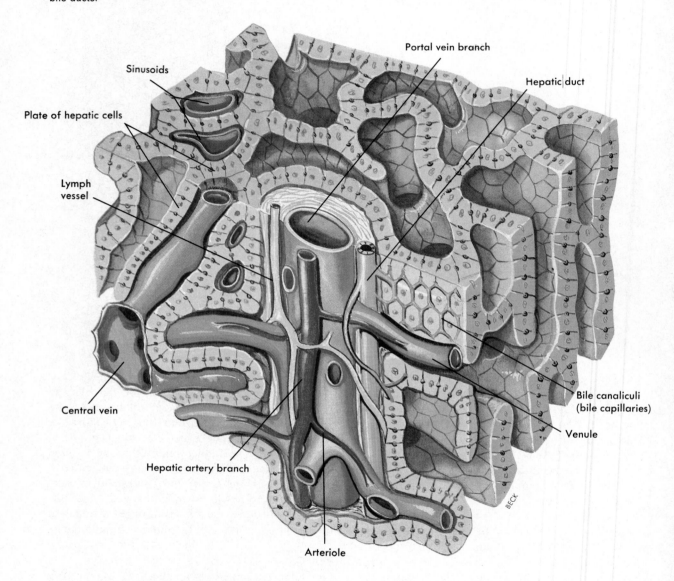

2 Liver cells secrete about a pint of bile a day.

3 Liver cells carry on a number of important steps in the metabolism of all three kinds of foods—proteins, fats, and carbohydrates.

4 Liver cells store several substances—iron, for example, and vitamins A, B$_{12}$, and D.

DETOXIFICATION BY LIVER CELLS. A number of poisonous substances enter the blood from the intestines. They circulate to the liver where, through a series of chemical reactions, they are changed to nontoxic compounds. Both ingested substances—alcohol, marijuana, and various other drugs, for example—and toxic substances formed in the intestines are detoxified in the liver.

BILE SECRETION BY LIVER. The main components of bile are bile salts, bile pigments, and cholesterol. Bile salts (formed in the liver from cholesterol) are the most essential part of bile. They aid in absorption of fats and then are themselves absorbed into the ileum. Eighty percent of bile salts are recycled in the liver to again become part of bile. Bile also serves as a pathway for elimination of certain breakdown products of red blood cells. The pigments bilirubin (red) and biliverdin (green), derived from hemoglobin, give bile its greenish color. Because it secretes

Figure 21-29 Liver. Note the continuity permitting passage of red blood cells from sinusoids to central vein. The granular hepatocytes form plates between the sinusoids. (Verhoeff's stain, × 140.)

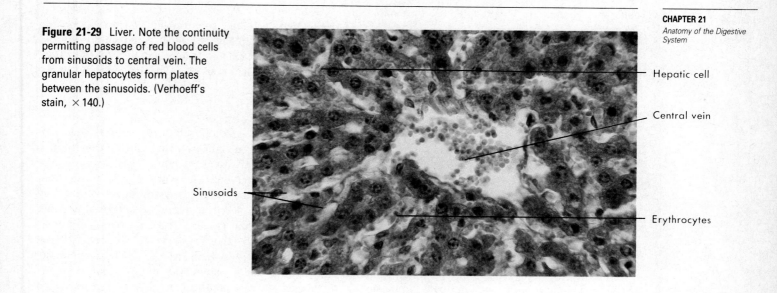

Hepatic cell

Central vein

Sinusoids

Erythrocytes

Figure 21-30 The gallbladder and its divisions: fundus, body, and neck. Obstruction of either the hepatic or the common bile duct by stone or spasm prevents bile from being ejected into the duodenum.

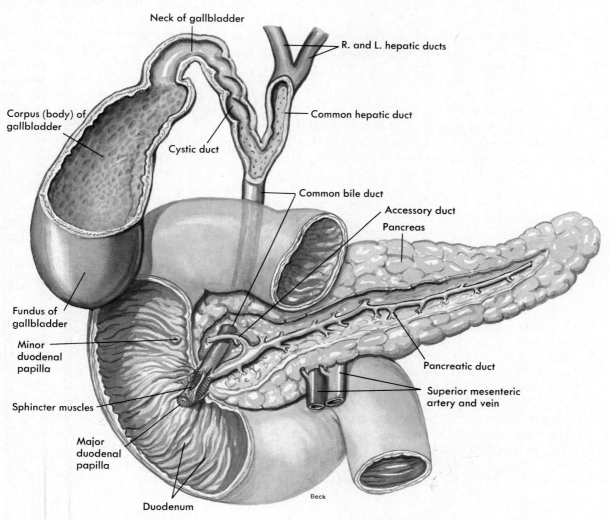

Neck of gallbladder

R. and L. hepatic ducts

Corpus (body) of gallbladder

Common hepatic duct

Cystic duct

Common bile duct

Accessory duct

Pancreas

Fundus of gallbladder

Minor duodenal papilla

Sphincter muscles

Pancreatic duct

Superior mesenteric artery and vein

Major duodenal papilla

Duodenum

Beck

bile into ducts, the liver qualifies as an exocrine gland.

LIVER METABOLISM. Although all liver functions are important for healthy survival, some of its metabolic processes are crucial for survival itself. A fairly detailed description of the role of the liver in metabolism will be given in Chapter 23.

GALLBLADDER
Size, Shape, and Location

The gallbladder is a pear-shaped sac from 7 to 10 cm (3 to 4 inches) long and 3 cm broad at its widest point (Figure 21-30). It can hold 30 to 50 ml of bile. It lies on the undersurface of the liver and is attached there by areolar tissue.

Structure

Serous, muscular, and mucous coats compose the wall of the gallbladder. The mucosal lining is arranged in rugae, similar in structure and function to those of the stomach.

Functions

The gallbladder stores the bile that enters it by way of the hepatic and cystic ducts. During this time, the gallbladder concentrates bile five- to ten-fold. Then later, when digestion occurs in the stomach and intestines, the gallbladder contracts, ejecting the concentrated bile into the duodenum.

PANCREAS
Size, Shape, and Location

The pancreas is a grayish-pink-colored gland about 12 to 15 cm (6 to 9 inches) long, weighing about 60 gm. It resembles a fish with its head and neck in the C-shaped curve of the duodenum, its body extending horizontally behind the stomach, and its tail touching the spleen (see Figure 21-27). According to an old anatomical witticism, the "romance of the abdomen" is the pancreas lying "in the arms of the duodenum."

Structure

The pancreas is composed of two different types of glandular tissue, one exocrine and one endocrine. Most of the tissue is exocrine, with a compound acinar arrangement. The word *acinar* means that the cells are in a grapelike formation and that they release their secretions into a microscopic duct within each unit (Figure 21-31, *B*). The word *compound* indicates that the ducts have branches. These tiny ducts unite to form larger ducts that eventually join the main pancreatic duct, which extends throughout the length of the gland from its tail to its head. It empties into the duodenum at the same point as the common bile duct, that is, at the major duodenal papilla. An accessory duct is frequently found extending from the head of the pancreas into the duodenum, about 2 cm above the major papilla (see Figure 21-30).

Embedded between the exocrine units of the pancreas, like so many little islands, lie clusters of endocrine cells called **pancreatic islands** (or islets of Langerhans) (Figure 21-31). Although there are about a million of these tiny islands, they constitute only about 2% of the total mass of the pancreas. Special staining techniques have revealed that two kinds of cells—alpha cells and beta cells—chiefly make up the islands. They are secreting cells, but their secretion passes into blood capillaries rather than into ducts. Thus the pancreas is a dual gland—an exocrine or duct gland because of the acinar units and an endocrine or ductless gland because of the islands of Langerhans.

CORRELATIONS

Inflammation of the lining of the gallbladder is called *cholecystitis*. **Cholecystectomy** is the surgical removal of the gallbladder, often necessitated by *cholelithiasis* or stones in this organ. *Jaundice*, a yellow discoloration of the skin and mucosa, results whenever obstruction of the hepatic or common bile duct occurs. Bile is thereby denied its normal exit from the body in the feces. Instead, it is absorbed into the blood. An excess of bile pigments in the blood gives it a yellow hue. Feces, deprived of their normal amount of bile pigments, become a grayish (so-called clay) color.

The opening between the lumen of the appendix and the cecum is quite large in children and young adults—a fact of great clinical significance since trapped food or fecal material in the appendix will irritate and inflame its mucous lining, a condition well known as **appendicitis.** The opening between the appendix and cecum is often completely obliterated in the elderly, thus explaining the low incidence of appendicitis in the aged. The layers of the wall of the appendix are similar to the four tissue coats or layers that form the wall of alimentary tract organs in general.

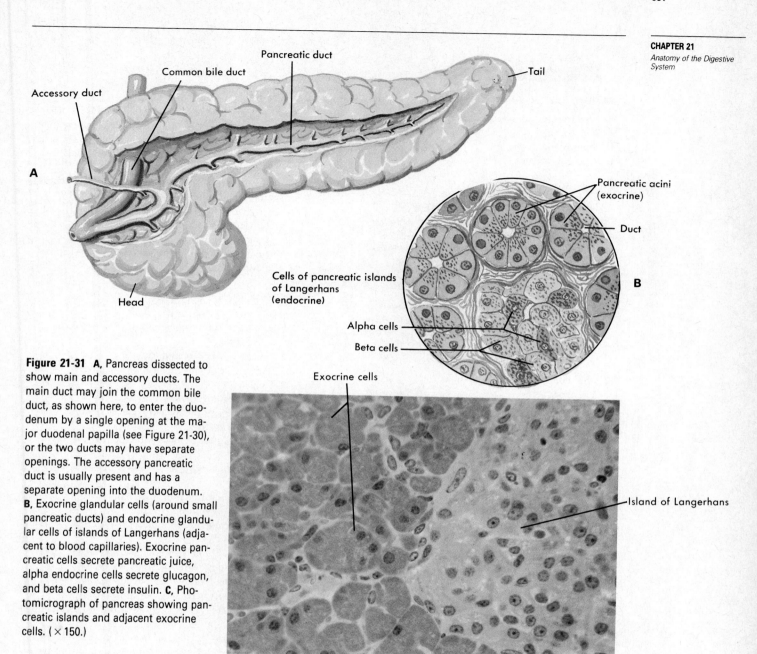

Figure 21-31 **A,** Pancreas dissected to show main and accessory ducts. The main duct may join the common bile duct, as shown here, to enter the duodenum by a single opening at the major duodenal papilla (see Figure 21-30), or the two ducts may have separate openings. The accessory pancreatic duct is usually present and has a separate opening into the duodenum. **B,** Exocrine glandular cells (around small pancreatic ducts) and endocrine glandular cells of islands of Langerhans (adjacent to blood capillaries). Exocrine pancreatic cells secrete pancreatic juice, alpha endocrine cells secrete glucagon, and beta cells secrete insulin. **C,** Photomicrograph of pancreas showing pancreatic islands and adjacent exocrine cells. (× 150.)

Functions

1 The acinar units of the pancreas secrete the digestive enzymes found in pancreatic juice. Hence the pancreas plays an important part in digestion (Chapter 23).

2 Beta cells of the pancreas secrete **insulin,** a hormone that exerts a major control over carbohydrate metabolism (p. 608 and Figure 23-12).

3 Alpha cells secrete *glucagon.* It is interesting to note that glucagon, which is produced so closely to insulin, has a directly opposite effect on carbohydrate metabolism (p. 609 and Figure 23-13).

VERMIFORM APPENDIX

The vermiform appendix (Latin *vermiformis* from *vermis*—"worm" plus *forma*—"shape") is, as the name implies, a wormlike tubular structure. It averages 8 to 10 cm in length and is most often found just behind the cecum or over the pelvic rim. Although it has no functional importance in digestion, the appendix is often classified as an accessory digestive organ because of its location. The lumen of the appendix communicates with the cecum about 3 cm below the ileocecal valve (see Figure 21-23).

Outline Summary

FUNCTIONS AND IMPORTANCE

A Prepare food for absorption and metabolism
B Absorption
C Elimination of wastes
D Vital importance

ORGANS

A Main organs—compose alimentary canal: mouth, pharynx, esophagus, stomach, and intestines
B Accessory organs—salivary glands, teeth, liver, gallbladder, pancreas, and vermiform appendix

Walls of organs

A Coats
  1 Mucous lining
  2 Submucous coat of connective tissue—main blood vessels here
  3 Muscular coat
  4 Fibroserous coat
B Modification of coats (Table 21-1)
  1 Mucous lining
    a Esophagus—stratified layers of squamous epithelial cells provide protection
    b Stomach
      (1) Arranged in longitudinal folds called rugae
      (2) Covered by gastric pits lined with mucus-secreting glands
      (3) Coiled gastric glands contain chief and parietal cells
    c Small intestine
      (1) Arranged in circular folds called *plicae circulares*
      (2) Velvety appearance with many mucous glands
      (3) Fingerlike villi covered by brush border
      (4) Specialized crypts (of Lieberkühn); mucous glands (Brunner's) and lymph nodes (Peyer's patches) are found in certain locations
    d Large intestine
      (1) Solitary lymph nodes and numerous mucous glands are typical
      (2) Anal "columns" in anal canal
  2 Muscle coat
    a Esophagus
      (1) Inner circular and outer longitudinal components fuse to form total muscular layer
      (2) Striated fibers in upper third; mixed striated and smooth fibers in middle third; smooth fibers in lower third
    b Stomach
      (1) Layers: inner is oblique, middle is circular, and outer is longitudinal
      (2) Middle circular fibers form cardiac and pyloric sphincters
    c Small intestine—composed of inner circular and outer longitudinal layers
    d Large intestine
      (1) Inner circular layer forms small sacs called *haustra*—give wall "puckered" appearance
      (2) Outer longitudinal layer condensed to form three tapelike strips called *taenia coli*
  3 Fibroserous coat
    a Esophagus—is fibrous above level of diaphragm and serous below level of diaphragm
    b Stomach
      (1) Covered by visceral peritoneum
      (2) Forms greater and lesser omentum
    c Small and large intestine—covered by visceral peritoneum

Mouth (buccal cavity)

A Structures (Figure 21-3)
  1 Lips—surround orifice and form anterior boundary
  2 Cheeks—lateral boundary
  3 Hard and soft palates—roof
  4 Tongue—floor
B Lips—anterior boundary of oral cavity
  1 Covered by skin externally and mucous membrane internally
  2 Shallow vertical groove on upper lip called *philtrum*
  3 Line of contact between closed lips called *oral fissure*
C Cheeks—lateral boundary of oral cavity
  1 Continuous with lips in front
  2 Lined by mucous membrane that is reflected onto gingiva and soft palate
  3 Composed largely of buccinator muscle
  4 Small mucus-secreting (buccal) glands between mucous membrane and buccinator muscle
D Hard and soft palates—roof of oral cavity
  1 Hard palate—formed by parts of two palatine and two maxillary bones
  2 Soft palate—formed of muscle in shape of arch; forms partition between mouth and nasopharynx; fauces is archway or opening from mouth to oropharynx; uvula is conical-shaped process suspended from midpoint of arch
E Tongue (Figure 21-3)—floor of oral cavity
  1 Mass of skeletal muscle components covered by mucous membrane—extremely maneuverable
    a Intrinsic muscles—origin and insertion in tongue
    b Extrinsic muscles—insert into tongue but originate on some other structure
      (1) Genioglossus—protrudes tongue
      (2) Hyoglossus—depresses tongue
  2 Divisions—blunt root, tip, and central body
  3 Papillae (Figure 21-4)—located on sides and upper surface (dorsum) of tongue
    a Vallate
    b Fungiform
    c Filiform

Outline Summary—cont'd

4 Taste buds (Figure 21-5)
  a Located on sides of vallate and fungiform papillae
  b Filiform papillae do not contain taste buds
5 Specialized structures (Figure 21-3)
  a Foramen cecum—median pit on dorsal surface of tongue
  b Lingual frenulum—midline fold of mucous membrane on undersurface of tongue; ankyloglossia (tongue-tied)—congenital condition caused by short frenulum that restricts tongue movement
  c Plica fimbriata—folds of mucous membrane on either side of lingual frenulum

Salivary glands

A Include mucus-producing (buccal) glands and three pairs of compound tubuloalveolar glands that produce about 1 liter of saliva per day
B Parotid glands
  1 Largest of paired salivary glands
  2 Produce only an enzyme containing watery or serous-type saliva
  3 Parotid (Stensen's) duct opens on inside of cheek, opposite upper second molar tooth
C Submandibular glands (Figure 21-6)
  1 Are mixed or compound glands—contain both serous (enzyme) and mucus-producing elements
  2 Located below mandibular angle
  3 Submandibular (Wharton's) duct opens into floor of mouth on either side of lingual frenulum
D Sublingual glands
  1 Smallest of paired salivary glands
  2 Located in front of submandibular glands
  3 Drained by 8 to 20 ducts (ducts of Rivinus)
  4 Produce only mucous-type saliva

Teeth

A Organs of mastication designed to cut, tear, and grind ingested food
B Typical tooth (Figure 21-7)
  1 Divisions
    a Crown—exposed portion of tooth
    b Neck—narrow portion surrounded by gingivae, or gums, which joins the crown of the tooth to the root
    c Root—portion that fits into socket of alveolar process of upper or lower jaw
  2 Dental tissues
    a Enamel
      (1) Hardest tissue in body—composed of 97% inorganic and 3% organic material and water
      (2) Covers crown of teeth; is suited to withstand abrasion and bacterial attack
    b Dentin
      (1) Softer than enamel
      (2) Makes up greatest portion of tooth shell

      (3) Covered by enamel in the crown and by cementum in the neck and root area
    c Cementum
      (1) Forms outer component of tooth shell over the neck and root
  3 Pulp cavity (Figure 21-7)
    a Hollow cavity within the dentine
    b Contains connective tissue, blood and lymphatic vessels, and sensory nerves
C Dental diseases
  1 Dental caries (tooth decay)
    a Disease of enamel, dentin, and cementum
    b Results in demineralization and destruction of tissue with cavity formation
  2 Periodontitis (Figure 21-11)
    a Inflammatory disease of gums (gingivae) and periodontal membrane
    b May cause loosening and loss of teeth
D Deciduous or baby teeth (Figure 21-9)—10 per jaw or 20 in set (see Table 21-2)
E Permanent teeth (Figure 21-10)—16 per jaw or 32 per set (see Table 21-2)

Pharynx

A Anatomy (Figure 19-7, see p. 517)
B Deglutition (swallowing) involves passage of a rounded mass of food called a *bolus* from mouth to stomach
  1 Food bolus enters pharynx from mouth through a constricted opening called the *fauces*
C Pharynx serves both respiratory and digestive systems
D After food bolus has passed through pharynx, it enters digestive tube proper—that portion of alimentary canal serving only the digestive system

Esophagus

A Position and extent
  1 Posterior to trachea and heart; pierces diaphragm
  2 Extends from pharynx to stomach, distance of approximately 25 cm
B Structure (Figures 21-12 and 21-13)
  1 Flattened, collapsible tube with lumen practically obliterated in resting state
  1 Coats (see Table 21-1)

Stomach

A Size, shape, and position (Figures 21-14; 21-16)
  1 Size varies in different individuals; also according to whether distended or not
  2 Elongated pouch
  3 Lies in epigastric and left hypochondriac portions of abdominal cavity
B Divisions (Figure 21-16)
  1 Fundus—portions above esophageal opening
  2 Body—central portion
  3 Pylorus—constricted, lower portion

C Curves
  1 Lesser—upper, right border
  2 Greater—lower, left border
D Sphincter muscles
  1 Cardiac—guarding opening of esophagus into stomach
  2 Pyloric—guarding opening of pylorus into duodenum
E Coats (see Table 21-1)
F Glands (Figures 21-17; 21-19)
  1 Epithelial cells of gastric mucosa secrete mucus
  2 Parietal cells secrete hydrochloric acid and intrinsic factor
  3 Chief cells (zymogenic cells) secrete enzymes of gastric juice

Small intestine

A Size and position
  1 Approximately 2.5 cm in diameter and 6 m in length
  2 Its coiled loops fill most of abdominal cavity
B Divisions (Figure 21-20)
  1 Duodenum
  2 Jejunum
  3 Ileum
C Coats—(see Table 21-1); note villi containing capillaries and lacteals and covered with epithelial cells having microvilli (brush border); greatly increase surface area for digestion and absorption

Large intestine (colon)

A Size—approximately 6 cm in diameter and 1.5 to 1.8 m in length
B Divisions (Figure 21-22)
  1 Cecum—vermiform appendix is blind-end tube off cecum; size and shape of large angleworm
  2 Colon
    a Ascending
    b Transverse
    c Descending
    d Sigmoid
  3 Rectum
C Coats (see Table 21-1)

Peritoneum

A Extension of fibroserous coat over abdominal organs (visceral layer) and wall (parietal layer)
B Mesentery is fan-shaped, attaches small intestine to posterior abdominal wall; transverse mesocolon attaches transverse colon to posterior abdominal wall; greater omentum, or lace apron, hangs from stomach over intestines and attaches to transverse colon; lesser omentum connects stomach to liver; falciform ligament connects liver to anterior abdominal wall (see Figure 21-26)

Liver

A Location and size
  1 Occupies most of right hypochondrium and part of epigastrium
  2 Largest gland in body, weighs about 1.5 kg
B Lobes and lobules
  1 Right lobe, subdivided into three smaller lobes—right lobe proper, caudate, and quadrate
  2 Left lobe
  3 Lobes divided into lobules by blood vessels and fibrous partitions
  4 Lobules composed of plates of hepatic cells radiating from central vein; portal and hepatic artery blood flows through sinusoids to central vein; bile collects in tiny ducts
C Ducts
  1 Hepatic duct from liver
  2 Cystic duct from gallbladder
  3 Common bile duct formed by union of hepatic and cystic ducts and opens into duodenum at major duodenal papilla

Gallbladder

A Size, shape, and location
  1 Approximately size and shape of small pear
  2 Lies on undersurface of liver
B Structure—sac of smooth muscle with mucous lining arranged in rugae

Pancreas

A Size, shape, and location
  1 12 to 15 cm long, weighs 60 gm
  2 Shaped something like a fish with head, body, and tail
  3 Lies in C-shaped curve of duodenum
B Structure—similar to salivary glands
  1 Divided into lobes and lobules
  2 Pancreatic cells pour their secretion into duct that runs length of gland and empties into duodenum at major duodenal papilla; may be accessory duct
  3 Clusters of cells, not connected with any ducts, lie between pancreatic cells—called pancreatic islands or islands of Langerhans—composed of alpha- and beta-type cells
C Functions
  1 Acinar units secrete pancreatic juice
  2 Beta cells of islands of Langerhans secrete insulin
  3 Alpha cells of islands of Langerhans secrete glucagon

Vermiform appendix

A Size, shape, and position
  1 Wormlike, tubular structure about 8 to 10 cm long
  2 Lies just behind cecum or over pelvic rim

Outline Summary—cont'd

B Structure
  1 Lumen communicates with cecum about 3 cm
    below ileocecal valve
  2 Opening between lumen of appendix and cecum
    is large in children and young adults and often
    obliterated in the aged

  3 Inflammation of appendix—appendicitis
  4 Layers of wall similar to other areas in the
    alimentary tract

Review Questions

1 List the component parts or segments of the
gastrointestinal tract and the accessory organs of
digestion

2 Name and describe the four tissue coats or layers
that form the wall of alimentary tract organs

3 How does the mucosal lining of the esophagus
differ from the lining of the stomach and small
intestine?

4 Identify the structures that form the mouth or
buccal cavity.

5 Define the following terms associated with the
mouth and pharynx: *philtrum, oral fissure, hard*
and *soft palates, fauces, uvula, foramen cecum,
lingual frenulum.*

6 What is ankyloglossia?

7 Identify the types of tongue papillae. What is the
relationship between papillae and taste buds?

8 How do the intrinsic and extrinsic muscles of the
tongue differ?

9 List and give the location of the paired salivary
glands. Identify by name the ducts that drain the
saliva from these glands into the mouth.

10 What type of saliva is produced by the parotid
glands? What is meant by the term *mixed* or
compound salivary gland?

11 Describe a typical tooth. Name the specific types of
teeth.

12 Discuss deciduous and permanent dentition.

13 What is the difference between dental caries and
periodontitis?

14 What is meant by the term *deglutition?*

15 How does the muscular coat of the esophagus differ
from the muscle layer typical of the gastrointestinal
tract as a whole?

16 List the divisions of the stomach. What is the
difference between gastric pits and gastric glands?

17 Identify the two major cell types of the gastric
glands. What cell type produces HCl? Gastric
enzymes?

18 What is pyloric stenosis?

19 List the divisions of the small intestine from
proximal to distal.

20 Compare the rugae of the stomach with the plicae
circulares of the small intestine.

21 In what area of the gastrointestinal tract do you
find: Brunner's glands? Peyer's patches? Villi?
Haustra? Taenia coli?

22 List the divisions of the large intestine.

23 Discuss the peritoneum and its reflections.

24 Discuss the anatomy of a typical liver lobule.

25 Identify the ducts of the liver and gallbladder.

26 Differentiate between the endocrine and exocrine
functions of the pancreas.

27 Why is appendicitis more common in children and
young adults than in the elderly?

22 PHYSIOLOGY OF THE DIGESTIVE SYSTEM

OBJECTIVES

After you have completed this chapter, you should be able to:

1 Define and compare mechanical and chemical digestion.
2 Define the different processes involved in mechanical digestion and identify the organ(s) that accomplish(es) each process.
3 Discuss the function of mastication.
4 List and explain the three main steps or stages of deglutition.
5 Discuss the functions of the stomach.
6 Explain the process of emptying the stomach.
7 Define chemical digestion.
8 Define the term *enzyme* and classify enzymes according to the type of chemical reactions catalyzed.
9 List and discuss six important enzyme properties.
10 List the most important digestive juices and enzymes, the food product each digests, and the resulting products.
11 Compare and contrast protein, fat, and carbohydrate digestion.
12 Discuss the control of salivary, gastric, pancreatic, biliary, and intestinal secretions.
13 Identify and discuss the absorption of nutrients resulting from the digestive process and the structures into which they are absorbed.
14 Define the terms *micelles, chylomicrons, enterocrinin, enterogastrone,* and *hydrolysis.*

This chapter deals with the process of digestion. By definition, digestion is the sum of all the changes food undergoes in the alimentary canal. The term **primary digestion** is sometimes used to describe the digestive process that occurs in the gastric and intestinal portions of the gastrointestinal tract.

PURPOSE AND KINDS

The purpose of digestion is the conversion of ingested food into a state in which it can be absorbed and used by the body. Digestion is necessary because foods, as eaten, are too complex in physical and chemical composition to pass through the intestinal mucosa into the blood or for cells to use them for energy and tissue building. In other words, digestion is the necessary preliminary to both absorption and metabolism of foods.

Since both the physical and chemical composition of ingested food makes its absorption impossible, two kinds of digestive changes are necessary— **mechanical** and **chemical.**

KEY TERMS

Absorption (ab-SORP-shun)

Cholecystokinin-pancreozymin
  (ko-le-sis-to-KIN-in PAN-kre-o-ZI-min)

Chylomicron (ki-lo-MI-kron)

Chyme (kime)

Defecation (def-e-KA-shun)

Diarrhea (di-ah-RE-ah)

Enterocrinin (en-ter-OK-ri-nin)

Enterogastrone (en-ter-o-GAS-trone)

Gastrin (GAS-trin)

Hydrolysis (hi-DROL-i-sis)

Mastication (mas-ti-KA-shun)

Micelle (MI-sel)

Peristalsis (per-i-STAL-sis)

Secretin (se-KRE-tin)

Table 22-1 Processes in the mechanics of digestion

Organ	Mechanical process	Nature of process
Mouth (teeth and tongue)	Mastication	Chewing movements—reduce size of food particles and mix them with saliva
	Deglutition	Swallowing—movement of food from mouth to stomach
Pharynx	Deglutition	
Esophagus	Deglutition	
	Peristalsis	Wormlike movements that squeeze food downward in tract; constricted ring forms first in one section, the next, etc., causing waves of contraction to spread along entire canal
Stomach	Churning	Forward and backward movement of gastric contents, mixing food with gastric juices to form chyme
	Peristalsis	Waves starting in body of stomach about three times per minute and sweeping toward closed pyloric sphincter; at intervals, strong peristaltic waves press chyme past sphincter into duodenum
Small intestine	Segmentation (mixing contractions)	Forward and backward (nonprogressive) movement within segment of intestine; purpose, to mix food and digestive juices thoroughly and to bring all digested food in contact with intestinal mucosa to facilitate absorption; purpose of peristalsis, on the other hand, to propel intestinal contents along digestive tract
	Peristalsis	
Large intestine		
Colon	Segmentation	Churning movements within haustral sacs
	Peristalsis	
Descending colon	Mass peristalsis	Entire contents moved into sigmoid colon and rectum; occurs three or four times a day, usually after a meal
Rectum	Defecation	Emptying of rectum, so-called bowel movement

MECHANICAL DIGESTION

Mechanical digestion consists of all those movements of the alimentary tract that bring about the following:

1 Change in the physical state of ingested food from comparatively large solid pieces into minute particles, thereby facilitating chemical digestion
2 Churning of the intestinal contents in such a way that they become well mixed with the digestive juices and all parts of them come in contact with the surface of the intestinal mucosa, thereby facilitating absorption
3 Propelling the food forward along the alimentary tract, finally eliminating the digestive wastes from the body

A list of definitions of the different processes involved in mechanical digestion, together with the organs that accomplish them, are given in Table 22-1.

Mastication

Mechanical digestion begins in the mouth when the particle size of ingested food material is reduced by chewing movements, or mastication. The tongue, cheeks, and lips play an important role in keeping food material between the cutting or grinding surfaces of the teeth when chewing. In addition to reducing particle size, chewing movements serve to mix food with saliva in preparation for swallowing.

Deglutition

The process of swallowing, or deglutition, may be divided into formation and movement of a food bolus from mouth to stomach in three main steps or stages:

1 **Oral stage** (mouth to oropharynx)
2 **Pharyngeal stage** (oropharynx to esophagus)
3 **Esophageal stage** (esophagus to stomach)

The first step, which is voluntary and under control of the cerebral cortex, involves the formation

Figure 22-1 Illustration showing interaction between the cerebral cortex, deglutition center, muscles of mastication and other structures important in deglutition found in the head and neck.

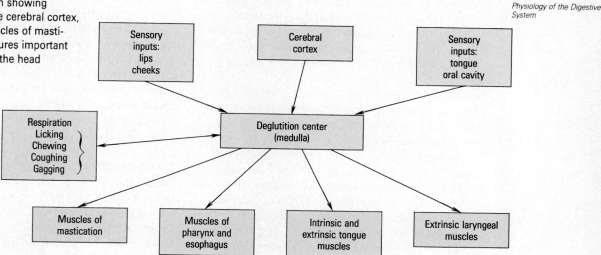

and segregation of a food bolus to be swallowed on a depression or groove in the middle of the tongue. During the oral stage, the bolus is pressed against the hard palate by the tongue and then moved back into the oropharynx. The pharyngeal and esophageal stages, both involuntary, consist of movement of food from the pharynx into the esophagus and, finally, into the stomach.

To propel food from the pharynx into the esophagus, three openings must be blocked: mouth, nasopharynx, and larynx. Continued elevation of the tongue seals off the mouth. The soft palate is elevated and tensed, causing the nasopharynx to be closed off. Food is denied entrance into the larynx by muscle action that causes the epiglottis to block this opening. The mechanism involves raising of the larynx, a process easily noted by palpation of the thyroid cartilage during swallowing. As a result, the bolus slips over the back of the epiglottis to enter the laryngopharynx. A combination of gravity and contractions of pharynx and esophagus compresses the bolus into and through the esophageal tube. These steps are involuntary and under control of the deglutition center in the medulla. The presence of a bolus stimulates sensory receptors in the mouth and pharynx, thus initiating reflex pharyngeal contractions. Consequently anesthesia of sensory nerves from the mucosa of the mouth and pharynx by a drug such as procaine makes swallowing impossible.

Swallowing is a complex process requiring the coordination of many muscles and other structures in the head and neck. Not only does the process occur smoothly, but it must also take place rapidly, since respiration is inhibited during 1 to 3 seconds for each swallowing while food clears the pharynx.

Figure 22-1 summarizes the types of interaction that occur between the cerebral cortex, the deglutition center in the medulla, muscles of mastication,

and other structures important in deglutition located in the head and neck.

Functions of the Stomach

The stomach carries on the following functions:

1. It serves as a reservoir, storing food until it can be partially digested and moved farther along the gastrointestinal tract.
2. It secretes gastric juice to aid in the digestion of food.
3. Through contractions of its muscular coat, it churns the food, breaking it into small particles and mixing them well with the gastric juice. In time, it moves the gastric contents into the duodenum.
4. It secretes the intrinsic factor.
5. It carries on a limited amount of absorption—of some water, alcohol, and certain drugs.
6. It produces the hormone **gastrin** in cells in the pyloric region.

EMPTYING THE STOMACH. The process of emptying the stomach takes about 2 to 6 hours after a meal, depending on the contents of the meal. Gastric juices mixed with food form a milky white material known as **chyme,** which is ejected about every 20 seconds into the duodenum. Since the volume of the stomach is large and that of the duodenum is small, gastric emptying must be regulated to prevent overburdening of the duodenum. Such control occurs by two mechanisms, one hormonal and one nervous. Fats present in the duodenum stimulate the intestinal mucosa to release a hormone called **enterogastrone** into the bloodstream. When it reaches the stomach wall via the circulation, enterogastrone has an inhibitory effect on gastric muscle, decreasing its peristalsis and thereby slowing down passage of food into the duodenum. Nervous control results from receptors in the duodenal mucosa that are sensitive both to the presence of acid and to distention. Sensory and motor fibers in the vagus nerve then cause a reflex inhibition of gastric peristalsis. This mechanism is known as the **enterogastric reflex.**

DEFECATION. Defecation is a reflex brought about by stimulation of receptors in the rectal mucosa. Normally the rectum is empty until mass peristalsis moves fecal matter out of the colon into the rectum. This distends the rectum and produces the desire to defecate. Also, it stimulates colonic peristalsis and initiates reflex relaxation of the internal sphincter of the anus. Voluntary straining efforts and relaxation of the external anal sphincter may then follow as a result of the desire to defecate. Together these several responses bring about defecation. Note that this is a reflex partly under voluntary control. If one voluntarily inhibits it, the rectal receptors soon become depressed and the urge to defecate usually does not recur until hours later, when mass peristalsis again takes place.

Constipation occurs when the contents of the lower colon and rectum move at a rate that is slower than normal. Extra water is absorbed from the fecal mass, producing a hardened or constipated stool. **Diarrhea** occurs as a result of increased motility of the small intestine. Chyme moves through the small intestine too quickly, reducing the amount of absorption of water and electrolytes there. The large volume of material arriving in the large intestine exceeds the limited capacity of the colon for absorption, so a watery stool results. Prolonged diarrhea can be particularly serious, even fatal, in infants because they have a minimal reserve of water and electrolytes.

CHEMICAL DIGESTION

Chemical digestion consists of all the changes in chemical composition that foods undergo in their travel through the alimentary canal. These changes result from the hydrolysis of foods. **Hydrolysis** is a chemical process in which a compound unites with water and then splits into simpler compounds. Numerous enzymes present in the various digestive juices catalyze the hydrolysis of foods (Table 22-2).

Enzymes

Enzymes are usually defined simply as "organic catalysts," that is, they are organic compounds, and they accelerate chemical reactions without appearing in the final products of the reaction. Enzymes are vital substances. Without them, the chemical reactions necessary for life could not take place. So important are they that someone has even defined life as the "orderly functioning of hundreds of enzymes."

CHEMICAL STRUCTURE. Enzymes are proteins. Frequently their molecules contain a nonprotein part called the *prosthetic group* of the enzyme molecule (if this group readily detaches from the rest of the molecule, it is called the *coenzyme*). Some prosthetic groups contain inorganic ions (e.g., Ca^{++}, Mg^{++}, Mn^{++}). Many of them contain vitamins. In fact, every vitamin of known function constitutes part of a prosthetic group of some enzyme. Nicotinic acid, thiamine, riboflavin, and other B complex vitamins, for example, function in this way.

CLASSIFICATION AND NAMING. Two of the systems used for naming enzymes are as follows: the suffix *-ase* is used either with the root name of the substance whose chemical reaction is catalyzed (the substrate chemical, that is) or with the word that describes the kind of chemical reaction catalyzed. Thus, according to the first method, sucrase is an enzyme that catalyzes a chemical reaction in which sucrose takes part. According to the second method, sucrase might also be called hydrolase because it hydrolyzes sucrose. Enzymes investigated before these methods of nomenclature were adopted still are called by older names, such as *ptyalin, pepsin,* and *trypsin.*

Classified according to the kind of chemical reactions catalyzed, enzymes fall into several groups:

1. **Oxidation-reduction enzymes.** These are known as oxidases, hydrogenases, and dehydrogenases. Energy release for muscular contraction and all physiological work depends on these enzymes.
2. **Hydrolyzing enzymes** or hydrolases. Digestive enzymes belong to this group. These are generally named after the substrate acted on, for example, lipase, sucrase, and maltase.
3. **Phosphorylating enzymes.** These add or remove phosphate groups and are known as phosphorylases or phosphatases.
4. **Enzymes that add or remove carbon dioxide.** These are known as carboxylases or decarboxylases.
5. **Enzymes that rearrange atoms within a molecule.** These are known as mutases or isomerases.
6. **Hydrases.** These add water to a molecule without splitting it, as hydrolases do.

Enzymes are also classified as intracellular or extracellular, depending on whether they act within cells or outside them in the surrounding medium. Most enzymes act intracellularly in the body, an important exception being the digestive enzymes.

PROPERTIES. In general, the properties of enzymes are the same as those of proteins, since enzymes are

Table 22-2 Chemical digestion

Digestive juices and enzymes	Food enzyme digests (or hydrolyzes)	Resulting product*
Saliva		
Amylase (ptyalin)	Starch (polysaccharide or complex sugar)	Dextrins (short polysaccharides) and maltose (a disaccharide, or double sugar)
Gastric juice		
Protease (pepsin: pepsinogen + hydrochloric acid)	Proteins, including casein	Proteoses and peptones (partially digested proteins)
Lipase (of little importance)	Emulsified fats (butter, cream, etc.)	**Fatty acids, glycerol,** and monoglycerides
Bile (contains no enzymes)	Large fat droplets (unemulsified fats)	Small fat droplets or emulsified fats
Pancreatic juice		
Protease (trypsin and chymotrypsin)†	Proteins (either intact or partially digested)	Proteoses, peptides, and **amino acids**
Lipase	Bile-emulsified fats	**Fatty acids and glycerol**
Amylase	Starch	Maltose
Intestinal juice: in brush border of cells of villi		
Peptidases‡	Peptides	**Amino acids**
Sucrase	Sucrose (cane sugar)	**Glucose and fructose** (simple sugars or monosaccharides)
Lactase	Lactose (milk sugar)	**Glucose and galactose** (simple sugars)
Maltase	Maltose (malt sugar)	**Glucose** (dextrose)

*Substances in boldface type are end products of digestion or, in other words, completely digested foods ready for absorption.
†Trypsin is secreted in inactive form (trypsinogen) and converted to active form by trypsin itself and intestinal enzyme, enterokinase.
 Chymotrypsin is secreted in inactive form (chymotrypsinogen) and converted to active form by action of trypsin and chymotrypsinogen.
‡Secreted in inactive form (pepsinogen); converted to active form (pepsin) by action of pepsin itself (autocatalysis) in acid environment
 provided by hydrochloric acid.

mainly proteins. For example, they form colloidal solutes in water and are precipitated or coagulated by various agents, such as high temperatures and salts of heavy metals. Hence, these agents inactivate enzymes. Other important enzyme properties are as follows:

1 Most enzymes are *specific in their action*, that is, act only on a specific substrate. This is attributed to a "key-in-a-lock" kind of action, the configuration of the enzyme molecule fitting the configuration of some part of the substrate molecule.

2 Enzymes *function optimally at a specific pH* and become inactive if this deviates beyond narrow limits.

3 *A variety of physical and chemical agents* inactivate or inhibit enzyme action, for example, x-rays and ionizing radiation (this presumably accounts for some of the ill effects of excessive radiation), certain antibiotic drugs, or unfavorable pH.

4 *Most enzymes catalyze a chemical reaction in both directions*, the direction and rate of the reaction being governed by the law of mass action. An accumulation of a product slows the reaction and tends to reverse it. A practical application of this fact is the slowing of digestion when absorption is interfered with and the products of digestion accumulate.

5 *Enzymes are continually being destroyed in the body* and, therefore, have to be con-

Figure 22-2 Carbohydrate digestion.

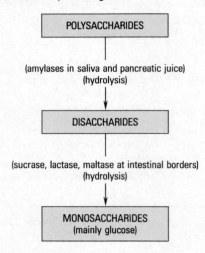

Figure 22-3 Protein digestion. Gastric juice protease (pepsin) and pancreatic juice protease (trypsin and chymotrypsin) hydrolyze proteins to proteoses and peptides. Protein digestion is then completed by pancreatic proteases, which hydrolyze proteoses to amino acids, and intestinal peptidases, which hydrolyze peptides to amino acids.

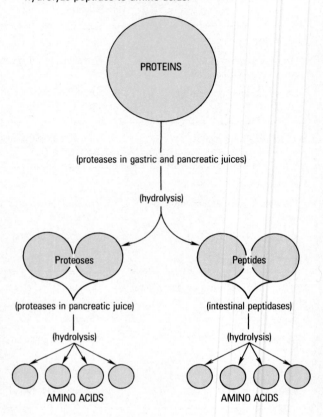

tinually synthesized, even though they are not used up in the reactions they catalyze.

6 *Many enzymes are synthesized* as inactive proenzymes. Substances that convert proenzymes to active enzymes are often called kinases, for example, enterokinase changes inactive trypsinogen into active trypsin.

Different enzymes require different hydrogen ion concentrations in their environment for optimal functioning. Ptyalin, the main enzyme in saliva, functions best in the neutral to slightly acid pH range characteristic of saliva. It is gradually inactivated by the marked acidity of gastric juice. In contrast, pepsin, an enzyme in gastric juice, is inactive unless sufficient hydrochloric acid is present. Therefore, in diseases characterized by gastric hypoacidity (pernicious anemia, for example), hydrochloric acid is given orally before meals.

Although we eat six kinds of chemical substances (carbohydrates, proteins, fats, vitamins, mineral salts, and water), only the first three have to be chemically digested in order to be absorbed.

CARBOHYDRATE DIGESTION. Carbohydrates are saccharide compounds. This means that their molecules contain one or more saccharide groups ($C_6H_{10}O_5$). Polysaccharides, notably starches and glycogen, contain many of these groups. Disaccharides (sucrose, lactose, and maltose) contain two of them, and monosaccharides (glucose, fructose, and galactose) contain only one. Polysaccharides are hydrolyzed to disaccharides by enzymes known as *amylases* found in saliva and pancreatic juice (salivary amylase is also called ptyalin). The enzymes that catalyze the final steps of carbohydrate digestion are *sucrase, lactase,* and *maltase* (Figure 22-2). These

enzymes are located in the cell membrane of epithelial cells covering villi and, therefore, lining the intestinal lumen. The substrates (disaccharides) bind onto the enzymes at the surface of the brush border, giving the name **"contact digestion"** to the process. The resulting end products of digestion, mainly glucose, are conveniently located at the site of absorption (and are not floating around somewhere in the lumen).

PROTEIN DIGESTION. Protein compounds have very large molecules made up of entangled chains of amino acids, often hundreds in number. Enzymes called proteases catalyze the hydrolysis of proteins into intermediate compounds, for example, proteoses and peptides, and, finally, into amino acids (Figure 22-3). The main proteases are pepsin in gastric juice, trypsin in pancreatic juice, and peptidases in intestinal brush border.

Figure 22-4 Fat digestion (hydrolysis) by lipase, facilitated first by emulsion of fats by bile.

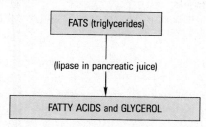

FATS (triglycerides)

(lipase in pancreatic juice)

FATTY ACIDS and GLYCEROL

FAT DIGESTION. Because fats are insoluble in water, they must be emulsified, that is, dispersed as very small droplets, before they can be digested. Bile emulsifies fats in the small intestine. This facilitates fat digestion by providing a greater contact area between fat molecules and pancreatic lipase, the main fat-digesting enzyme (Figure 22-4). For a summary of the actions of each digestive juice, see Table 22-2.

RESIDUES OF DIGESTION. Certain components of food resist digestion and are eliminated from the intestines in the feces. These *residues of digestion* are cellulose from carbohydrates, undigested connective tissue and toxins from meat proteins, and undigested fats. In addition to these wastes, feces consist of bacteria, pigments, water, and mucus.

CONTROL OF DIGESTIVE GLAND SECRETION

Digestive glands secrete when food is present in the alimentary tract or when it is seen, smelled, or imagined. Complicated reflex (nerve) and chemical (hormonal) mechanisms control the flow of digestive juices in such a way that they appear in proper amounts when and for as long as needed.

Saliva

As far as is known, only reflex mechanisms control the secretion of saliva. Chemical, mechanical, olfactory, and visual stimuli initiate afferent impulses to centers in the brain stem that send out efferent impulses to salivary glands, stimulating them. Chemical and mechanical stimuli come from the presence of food in the mouth. Olfactory and visual stimuli come, of course, from the smell and sight of food.

Gastric Secretion

Stimulation of gastric juice secretion occurs in three phases controlled by reflex and chemical mechanisms. Because stimuli that activate these mechanisms arise in the head, stomach, and intestines, the three phases are known as the cephalic, gastric, and intestinal phases, respectively.

The **cephalic phase** is also spoken of as the psychic phase because psychic factors activate the mechanism. For example, the sight, smell, taste, or even thought of food that is pleasing to an individual stimulates various head receptors and thereby, initiates stimulation of the gastric glands. Parasympathetic fibers in branches of the vagus nerve conduct the stimulating efferent impulses to the glands. The vagus also stimulates production of gastrin (discussed subsequently).

During the **gastric phase** of gastric secretion, the following chemical control mechanism dominates. Products of protein digestion in foods that have reached the pyloric portion of the stomach stimulate its mucosa to release a hormone called **gastrin** into the blood in stomach capillaries. When it circulates to the gastric glands, gastrin greatly accelerates their secretion of gastric juice, which has a high pepsin and hydrochloric acid content (Table 22-3). Hence this seems to be a device for ensuring that, when food is in the stomach, there will be enough enzymes there to digest it. Gastrin release is also stimulated by distention of the stomach (caused by the presence of food there) and by vagal nerve impulses to the pylorus.

The **intestinal phase** of gastric juice secretion is less clearly understood than the other two phases. A chemical control mechanism, however, is believed to operate. It is known, too, that the hormone **enterogastrone** (released by intestinal mucosa when fat is in the intestine) causes a lessening of both gastric secretion and motility.

Pancreatic Secretion

Two hormones released by intestinal mucosa are known to stimulate pancreatic secretion. One of these, secretin, evokes production of pancreatic fluid low in enzyme content but high in bicarbonate. This alkaline fluid acts to neutralize the acid chyme entering the duodenum. As you might expect, the presence of acid in the duodenum serves as the most potent stimulator of secretin. (Additional control involving the same hormone is shown by the fact that fats in the duodenum also elicit secretin, which then

Secretin has an interesting claim to fame. Not only was secretin the first hormone to be discovered, but its discovery gave rise to the broad concept of hormonal control of body activities.

Table 22-3 Actions of some digestive hormones summarized

Hormone	Source	Action
Gastrin	Formed by gastric mucosa in presence of partially digested proteins	Stimulates secretion of gastric juice rich in pepsin and hydrochloric acid
Enterogastrone	Formed by intestinal mucosa in presence of fats	Inhibits gastric secretion and motility
Secretin	Formed by intestinal mucosa in presence of acid, partially digested proteins, and fats	Stimulates secretion of pancreatic juice low in enzymes and high in alkalinity (bicarbonate) Stimulates secretion of bile by liver
Cholecystokinin-pancreozymin (CCK-PZ)	Formed by intestinal mucosa in presence of fats, partially digested proteins, and acids	Stimulates ejection of bile from gallbladder and secretion of pancreatic juice high in enzymes

influences the liver to increase its output of the fat emulsifier bile.)

The other intestinal hormone, known as cholecystokinin-pancreozymin (CCK-PZ), was originally thought to be two separate substances. It has now been identified as one chemical with two functions: (1) it causes the pancreas to increase its exocrine secretions, which is high in enzyme content, and (2) it also stimulates contraction of the gallbladder so that bile can pass into the duodenum.

Secretion of Bile

The hormones secretin and CCK-PZ, as described, stimulate secretion of bile by the liver and ejection of bile from the gallbladder (Table 22-3).

Intestinal Secretion

Relatively little is known about the regulation of intestinal secretions. Some evidence suggests that the intestinal mucosa, stimulated by hydrochloric acid and food products, releases into the blood a hormone, **enterocrinin**, which brings about increased production of digestive enzymes within intestinal mucosa. Presumably, neural mechanisms also help control the secretion of this digestive juice.

ABSORPTION

Definition

Absorption is the passage of substances (notably digested foods, water, salts, and vitamins) through the intestinal mucosa into the blood or lymph.

Mechanisms

Absorption of some substances may occur on the basis of the physical laws of diffusion, osmosis, and filtration alone. However some substances depend on more complex mechanisms in order to be absorbed. Glucose is a good example. Although considered an "end product of digestion," glucose is a relatively large molecule and cannot pass freely through the brush border membrane of an intestinal mucosa cell. In addition to physical size, the lipid nature of the cell membrane (Figure 3-3, p. 54) presents another barrier to glucose absorption. Only lipid-soluble (hydrophobic) molecules the size of glucose can pass freely (passively) through the lipid cell barrier. Since glucose is too large physically and is hydrophilic (water-soluble) in nature, it must be actively transported across the membrane by a carrier in order to enter the cell.

Glucose absorption is a good example of active transport involving a carrier. This energy-requiring process can move glucose into cells already concentrated in glucose, that is, against a concentration gradient. Fructose, another product of carbohydrate digestion, and amino acids may be absorbed by facilitated diffusion (p. 63). By such a process, these molecules require carrier assistance but do not move against concentration gradients. Simple sugars and amino acids eventually pass through mucosal cells to reach blood capillaries in the villus.

Fatty acids (products of fat digestion) and cholesterol are transported with the aid of bile salts from the watery intestinal lumen to absorbing cells on villi. Bile salts form **micelles** that surround the lipid, making it temporarily water-soluble. As micelles approach the brush border of absorbing cells, lipids are released to pass through the cell membrane (since its lipid bilayer is receptive to lipids) by simple diffusion. Once inside the cell, fatty acids are rapidly reunited with glycerol to form triglycerides (neutral fats). The final step in lipid transport by the intestine

Table 22-4 Food absorption

Form absorbed	Structures into which absorbed	Circulation
Protein—as amino acids	Blood in intestinal capillaries	Portal vein, liver, hepatic vein, inferior vena cava to heart, etc.
Perhaps minute quantities of some whole proteins absorbed, for example, some antibodies		
Carbohydrates—as simple sugars	Same as amino acids	Same as amino acids
Fats		
Glycerol and monoglycerides	Lymph in intestinal lacteals	During absorption, that is, while in epithelial cells of intestinal mucosa, glycerol and fatty acids recombine to form microscopic particles of fats (chylomicrons); lymphatics carry them by way of thoracic duct to left subclavian vein, superior vena cava, heart, etc.; some fats transported by blood in form of phospholipids or cholesterol esters
Fatty acids combine with bile salts to form water-soluble substance	Lymph in intestinal lacteals	
Some finely emulsified, undigested fats absorbed	Small fraction enters intestinal blood capillaries	

is the formation of **chylomicrons,** which are composed mainly of neutral fats and some cholesterol covered by a delicate protein envelope. This important envelope allows fats to be transported through lymph and into the bloodstream (Table 22-4).

Active transport mechanisms seem to be available for water and salt absorption. Certain salts (highly ionized ones) cannot be absorbed into intestinal cells. Magnesium sulfate (epsom salts), for example, is not absorbed from the intestine even though its molecules are smaller than glucose molecules. In fact, it is this nonabsorbability of magnesium sulfate that makes it an effective cathartic. (Since magnesium sulfate ions do not diffuse freely through the intestinal mucosa, do you think their presence in intestinal fluid would create an osmotic pressure gradient between the intestinal fluid and blood? If so, what effect would this have on water movement between these two fluids? Reread Chapter 2, if you need help in answering these questions.)

Note that after absorption food does not pass directly into the general circulation. Instead, it first travels by way of the portal system to the liver. Dur-

Vitamins A, D, E, and K, known as the "fat-soluble vitamins," also depend on bile salts for their absorption. Many water-soluble vitamins, such as certain of the B group, are small enough to be absorbed by simple diffusion. Most drugs (sedatives, analgesics, antibiotics) apear to be absorbed by simple diffusion, also.

ing intestinal absorption, blood entering the liver via the portal vein contains greater concentrations of glucose, amino acids, and fats than does blood leaving the liver via the hepatic vein for the systemic circulation. Clearly, the excess of these food substances over and above the normal blood levels has remained behind in the liver.

What the liver does with them is part of the story of metabolism, our topic for discussion in the next chapter.

Outline Summary

DIGESTION—PURPOSE AND KINDS

A Definition—all changes food undergoes in alimentary canal

B Purpose—conversion of foods into chemical and physical forms that can be absorbed and metabolized

C Kinds

1 Mechanical—movements that change physical state of foods, facilitate absorption, propel food forward in alimentary tract, and eliminate digestive wastes from tract (see Table 22-1 for description of processes involved in mechanical digestion)

  a Mastication (chewing)

    (1) Particle size of ingested food is reduced

    (2) Tongue, cheeks, and lips help position food material between teeth

    (3) Food is mixed with saliva

  b Deglutition (swallowing) (see Figure 22-1)

    (1) Process may be divided into formation and movement of a food bolus from mouth to stomach in three main steps or stages

      (a) Oral stage (mouth to oropharynx)

      (b) Pharyngeal stage (oropharynx to esophagus)

      (c) Esophageal stage (esophagus to stomach)

    (2) Movement of food through mouth into pharynx—voluntary act

    (3) Movement of food through pharynx into esophagus—involuntary or reflex act initiated by stimulation of mucosa of back of mouth, pharynx, or laryngeal region; paralysis of receptors here, for example, by procaine, makes swallowing impossible

    (4) Movement of food through esophagus into stomach; accomplished by esophageal peristalsis—reflex initiated by stimulation of esophageal mucosa

  c Peristalsis—wormlike movements that squeeze food downward in tract

  d Emptying of stomach—gastric peristalsis inhibited by two mechanisms: fats in chyme entering duodenum evoke intestinal hormone enterogastrone; acid and distention of duodenum stimulate enterogastric vagal reflex; both mechanisms act to prevent too rapid emptying of stomach

  e Churning, segmentation, and mixing contractions

  f Mass peristalsis in colon

  g Defecation—reflex initiated by stimulation of rectal mucosa

2 Chemical—series of hydrolytic processes dependent on specific enzymes (see Table 22-2 and Figures 22-2 and 22-3 for description of chemical changes)

CONTROL OF DIGESTIVE GLAND SECRETION (Table 22-3)

A Saliva—secretion is reflex initiated by stimulation of taste buds, other receptors in mouth and esophagus, olfactory receptors, and visual receptors

B Gastric secretion—controlled reflexly by same stimuli that initiate salivary secretion; also controlled chemically by hormone, gastrin, released by pyloric mucosa in presence of partially digested proteins; enterogastrone (hormone slowing stomach emptying) also has inhibitory effect on gastric secretion

C Pancreatic secretion—controlled chemically by hormones secretin and CCK-PZ; formed by intestinal mucosa, especially when fats or acid enter(s) duodenum

D Bile

1 Secretion—controlled chemically by same hormone (secretin) that regulates pancreatic secretion

2 Ejection into duodenum—controlled chemically by hormone CCK-PZ

E Intestinal secretion—control still obscure, although believed to be both reflex and chemical

ABSORPTION

A Definition—passage of substances through intestinal mucosa into blood or lymph

B Mechanisms—cell membrane consists of lipid bilayer, so lipids diffuse passively across and into interior of absorptive cells; water-soluble substances (hydrophilic), like sugars or amino acids, use carriers to help them enter cells; some absorptive processes require energy (active transport); others do not (facilitated diffusion)

Review Questions

1 Define and give the purpose of digestion. What is the difference between mechanical and chemical digestion?
2 List four of the mechanical processes that occur during digestion.
3 What is the difference between peristalsis and segmentation contractions in the small intestine?
4 List the three steps or stages in deglutition.
5 Discuss the function of the deglutition center in the medulla.
6 Discuss the functions of the stomach.
7 How does enterogastrone influence emptying of the stomach? What is the enterogastric reflex?
8 Discuss the functions of gastric juice.
9 What is chyme?
10 Differentiate trypsin from pepsin; enterokinase from enterogastrone; enzymes from proenzymes; proteins from proteoses; polysaccharides from monosaccharides.

11 Discuss the responses that collectively bring about defecation.
12 What juices digest proteins? Carbohydrates? Fats?
13 Discuss three important enzyme properties.
14 What is meant by the term *cephalic phase* of gastric secretion?
15 What digestive functions does the pancreas perform?
16 Describe the absorption of glucose from the lumen of the small intestine.
17 What vitamins depend on bile salts for their absorption?
18 Compare the absorption of carbohydrates, fats, and proteins.
19 Name each hormone that controls production of bile; ejection of bile; stimulation of gastric enzymes; inhibition of gastric emptying; secretion of alkaline fluid from pancreas; secretion of pancreatic enzymes.

23 METABOLISM

OBJECTIVES

After you have completed this chapter, you should be able to:

1 Outline in general terms the processes of anabolism and catabolism and discuss the adenosine triphosphate/adenosine diphosphate (ATP/ADP) system and its role in metabolism.

2 Discuss the steps involved in glycolysis, which ultimately change one molecule of glucose into two molecules of pyruvic acid.

3 Explain the role of the electron transport system and oxidative phosphorylation in the Krebs or tricarboxylic acid cycle.

4 Compare, contrast, and explain glycogenesis and glycogenolysis.

5 Discuss the generalized mechanisms of blood glucose homeostasis and the hormonal control of glucose metabolism.

6 Identify the major lipid constituents in blood and discuss their mechanisms of transport.

7 Discuss the anabolism and catabolism of lipids and the role of the liver as the chief site of ketogenesis.

8 Outline the hormonal control mechanism of fat metabolism.

9 Compare and contrast protein anabolism and catabolism.

10 Discuss the two kinds of protein or nitrogen imbalance.

11 Summarize the metabolism of carbohydrates, fats, and proteins, and list the end products resulting from the metabolism of each.

12 Define the term *metabolic rate* and discuss how it can be expressed.

13 Discuss the major factors that influence the basal metabolic rate (BMR) and explain how it can be determined.

14 Discuss the mechanisms for regulating food intake.

Foods are first digested, then absorbed, and finally metabolized. A two-word definition for metabolism is food utilization. How the body uses, that is, metabolizes, foods is the main subject of this chapter. It begins with some general information about this vital process. It then goes on to relate basic detailed information about metabolism of the three kinds of foods. It also discusses metabolic rates and the regulation of food intake.

SOME IMPORTANT GENERALIZATIONS ABOUT METABOLISM

Familiarize yourself with the generalizations about metabolism discussed in the boxed material. They will help you fit together details presented later to form a clearer picture of the multifaceted process of metabolism.

KEY TERMS

Anabolism (ah-NAB-o-lizm)

Calorie (KAL-O-re)

Calorimetry (kal-o-RIM-e-tree)

Catabolism (kah-TAB-o-lizm)

Deamination (de-am-i-NA-shun)

Gluconeogenesis (gloo-ko-ne-o-JEN-e-sis)

Glycogenesis (gli-ko-JEN-e-sis)

Glycogenolysis (gli-ko-je-NOL-i-sis)

Glycolysis (gly-KOL-i-sis)

Ketogenesis (ke-to-JEN-e-sis)

Metabolism (me-TAB-o-lizm)

Oxidation (ok-si-DA-shun)

Phosphorylation (fos-for-i-LA-shun)

Pyrogen (PIE-ro-jen)

1 **Metabolism** is a complex process made up of two major processes, catabolism and anabolism. Each of these, in turn, consists of series of enzyme-catalyzed chemical reactions known as metabolic pathways.

2 **Catabolism** breaks food molecules down into smaller molecular compounds and, in so doing, releases energy from them. Anabolism does the opposite. It builds food molecules up into larger molecular compounds and, in so doing, uses energy. Catabolism is a decomposition process. Anabolism is a synthesis process.

3 Both catabolism and anabolism take place inside cells. Both processes go on continually and concurrently.

4 Catabolism releases energy in two forms—heat and chemical energy. The amount of heat generated is relatively large—so large, in fact, that it would hard-boil cells if it were released in one large burst. Fortunately, this does not happen. Catabolism releases heat in frequent, small bursts. Heat is useless energy for cells in that they cannot use it to do their work. In contrast, chemical energy released by catabolism is useful. It cannot, however, be used directly for biological reactions. First it must be transferred to high-energy bonds (~) of adenosine triphosphate (ATP) molecules. The term **high-energy bonds** indicates the higher amount of energy stored in these bonds compared with ordinary chemical bonds. Another important fact about high-energy bonds is that they are more labile (easily broken) than ordinary chemical bonds.

5 ATP is one of the most important compounds in the world. Why? Because it supplies energy directly to the energy-using reactions of all cells in all kinds of living organisms from one-celled plants to billion-celled humans. ATP functions as the universal biological currency. It pays the energy bills for all cells and is as important in the world of cells as money is in the world of contemporary society.

Look now at Figure 23-1. The structural formula at the top of the diagram shows three phosphate groups attached to the rest of the ATP molecule, two of them by high-energy bonds. The breaking of the last one of these bonds yields a phosphate group (P), or adenosine diphosphate (ADP), and energy, which, as the diagram indicates, is used for anabolism and other cell work. The diagram also shows that P and ADP then use energy released by catabolism to recombine and form ATP. This cycle is called the *ATP/ADP system.*

6 Metabolism is not identical in all cells. It differs mainly with regard to rate and the kind of products synthesized by anabolism. More active cells have a higher metabolic rate than less active cells. Anabolism in different kinds of cells produces different compounds. In liver cells, for example, anabolism synthesizes various blood protein compounds. Not so in alpha cells of the pancreas. Anabolism here produces a different compound—insulin.

CARBOHYDRATE METABOLISM

The body metabolizes carbohydrates by both catabolic and anabolic processes. Since human cells use carbohydrates—mainly glucose—as their first or preferred energy fuel, they catabolize most of the carbohydrate absorbed and anabolize a relatively small portion of it. When the amount of glucose entering cells is inadequate for their energy needs, they catabolize fats next and then proteins.

Glucose Transport and Phosphorylation

Carbohydrate metabolism starts with the movement of glucose through cell membranes. Immediately on reaching the interior of a cell, glucose reacts with ATP to form glucose-6-phosphate. This step, named **glucose phosphorylation,** prepares glucose for further metabolic reactions. In most cells of the body, it is an irreversible reaction. However, in a few cells—namely, those of the intestinal mucosa, liver, and kidney tubules—glucose phosphorylation is reversible. These cells contain phosphatase, an enzyme that splits phosphate off from glucose-6-phosphate. This reverse glucose phosphorylation reaction forms glucose, which then moves out of the cells into the blood. (Glucose-6-phosphate cannot pass through cell membranes.) Depending on their energy needs of the moment, cells catabolize or anabolize glucose-6-phosphate.

Processes that metabolize glucose-6-phosphate, a key carbohydrate compound, bear the following names: *glycolysis, Krebs cycle, electron transport system and oxidative phosphorylation, glycogenesis, glycogenolysis,* and *gluconeogenesis.* We shall discuss them in this order.

Glycolysis

Glycolysis, also known as the Embden-Meyerhoff pathway, is the first process of carbohydrate catabolism. It breaks apart one glucose molecule to form two pyruvic acid molecules. (A glucose molecule contains six carbon atoms, and a pyruvic acid molecule contains three carbon atoms. See Figure 23-2.) Glycolysis consists, as Figure 23-3 shows, of a series of chemical reactions. A specific enzyme catalyzes each of these reactions. Probably the most important facts for you to remember about glycolysis are the following:

1 Glycolysis occurs in the cytoplasm of all human cells.

2 Glycolysis is an **anaerobic** process, that is,

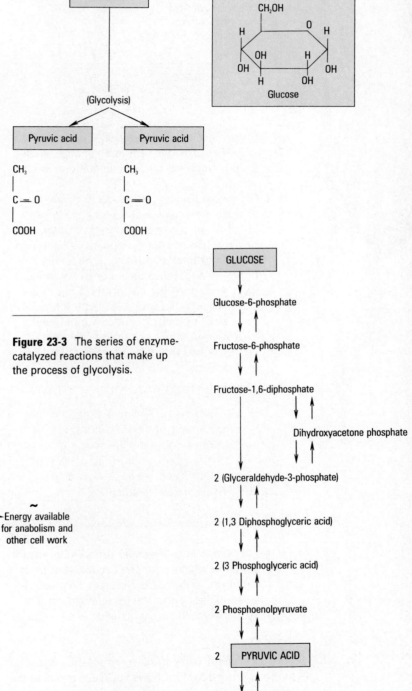

Figure 23-2 Glycolysis is the process of carbohydrate catabolism, which changes one molecule of glucose into two molecules of pyruvic acid. As the structural formulas indicate, a glucose molecule contains six carbon atoms, and a pyruvic acid molecule contains only three carbon atoms. For the intermediate compounds found during glycolysis, see Figure 23-3.

Figure 23-3 The series of enzyme-catalyzed reactions that make up the process of glycolysis.

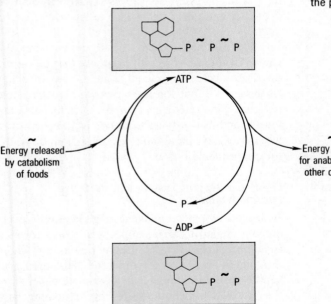

Figure 23-1 ATP/ADP system and its role in metabolism. (See Figure 23-20.)

Figure 23-4 Scheme to show that catabolism breaks larger molecules down into smaller ones. The first phase of carbohydrate catabolism, or glycolysis, splits one molecule of glucose (six carbon atoms) into two molecules of pyruvic acid (three carbon atoms each). The second phase of carbohydrate catabolism, or the Krebs cycle, converts each pyruvic acid molecule to three carbon dioxide molecules.

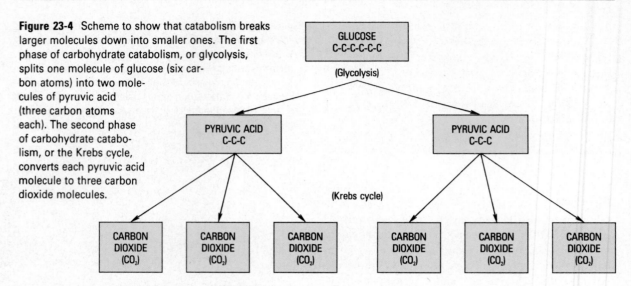

it does not use oxygen. It is the only process that provides cells with energy when their oxygen supply is inadequate or even absent.

3 Glycolysis breaks the chemical bonds in glucose molecules and thereby releases about 5% of the energy stored in them. Part of the released energy appears as heat, and part is put back in storage in high-energy bonds of ATP molecules. For every mole of glucose undergoing glycolysis, a net of 2 moles of ATP is formed. About 8 kilocalories (kcal) of energy (under normal physiological conditions) is stored in the high-energy bonds that bind phosphate to ADP to form 1 mole of ATP.

One **kilocalorie** (a so-called "large" calorie) is the amount of heat used to raise the temperature of 1 kg (liter) of water 1° Celsius. The 100 calories (kcal) in a slice of bread would provide enough heat to raise a liter of water from 0° to 100° C.

---

4 Glycolysis is an essential process because it prepares glucose for the second step in catabolism, namely, the Krebs cycle. Glucose itself cannot enter the cycle but must first be converted to pyruvic acid.

Krebs Cycle

For his brilliant work in discovering the metabolic pathway named for him, Sir Hans Krebs received the 1935 Nobel Prize. Essentially, the Krebs cycle converts two pyruvic acid molecules to six carbon diox-

ide and six water molecules. But many chemical reactions intervene. Figure 23-4 shows that one glucose molecule is changed by glycolysis to two pyruvic acid molecules, which, by the Krebs cycle, yield six carbon dioxide molecules. Figure 23-5 shows the intervening chemical changes.

Glycolysis takes place in the cytoplasm of cells, whereas the Krebs cycle occurs in their mitochondria. Pyruvic acid molecules combine with coenzyme A* as shown in reaction in Figure 23-5, splitting off CO_2 and H_2 from pyruvic acid and forming acetyl-CoA. Coenzyme A then detaches from acetyl-CoA, leaving a 2-carbon acetyl group, which enters the Krebs cycle by combining with oxaloacetic acid to form citric acid. **Citric acid cycle** is another name for the Krebs cycle; so, too, is **tricarboxylic acid cycle** because citric acid is also called *tricarboxylic acid.*

You probably do not need to memorize the names of the intermediate products formed during the Krebs cycle but notice that all of them are acids. Observe, too, that for each pyruvic acid molecule entering the Krebs cycle, three CO_2 molecules are formed and that certain reactions yield one hydrogen atom. The next paragraph describes how these hydrogen atoms are used.

Electron Transport System and Oxidative Phosphorylation

Hydrogen atoms removed during the Krebs cycle enter a chain of carrier molecules, which forms a structural part of the inner membrane of mitochondria and is known as the **electron transport system.** Figure 23-5 indicates hydrogen atoms entering the electron transport system by combining with NAD to form $NADH_2$. Note that the latter compound

---

*Coenzyme A is a derivative of the vitamin pantothenic acid.

Figure 23-5 Krebs cycle on left side of diagram; electron transport and oxidation on right (explained in text).

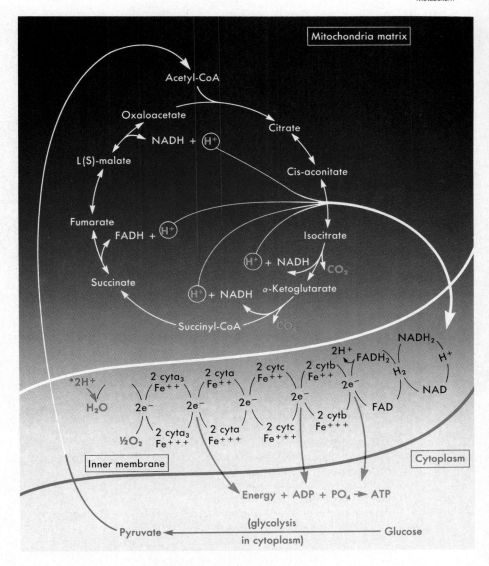

releases hydrogen in the form of 2 hydrogen ions (H^+) and 2 electrons (e^-). The electrons quickly move down the chain of cytochromes to their final acceptor, oxygen, and the hydrogen ions also join the oxygen, forming water. The most important fact about electron transport is this: as electrons move down the carrier chain, they release small bursts of energy used immediately for oxidative phosphory-

lation. **Oxidative phosphorylation** means the joining of a phosphate group to ADP to form ATP—a reaction whose importance can scarcely be overemphasized. Note in Figure 23-5 the three places in the transport system where ATP forms.

The arrangement of catabolic "machinery" within the cell, as currently viewed, is shown in Figure 23-6. Glycolytic enzymes present in the cytoplasm produce pyruvic acid, which diffuses into mitochondria. The enzymes of the Krebs cycle have been localized mostly in the soluble matrix. The hydrogen ions and electrons then pass to the "sphere"-studded cristae of the inner membrane, where the electron transport carriers and mechanism for phosphorylation are located. Since so many of the cell's energy-releasing enzymes are located within the mitochondria, these tiny structures are aptly described as the "power plants" of the cell.

Abbreviations and names of the carrier molecules are NAD (nicotinamide adenine dinucleotide), FAD (flavin adenine dinucleotide), and cytochromes b, c, a_3 (cytochrome oxidase).

Figure 23-6 Cell machinery for catabolism. *1,* Glycolysis occurs in cytoplasm. *2,* Krebs cycle mostly in mitochondrial matrix. *3,* Phase 3 (electron transport and oxidative phosphorylation) on inner membrane of mitochondria.

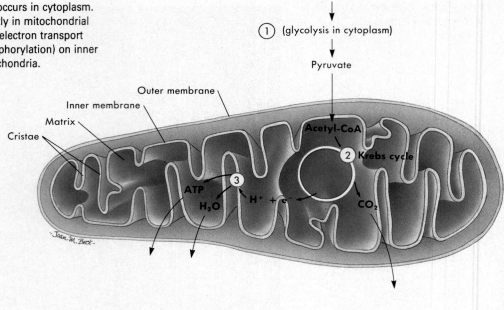

Figure 23-7 Catabolism releases energy stored in chemical bonds of glucose molecules. One hundred and eighty gm, or 1 mole, of glucose contains 686 kilocalories (kcal) of stored energy. Catabolism releases about 382 kcal of this energy as heat and puts about 304 kcal back in storage in high-energy bonds of 38 moles of ATP. Thus the efficiency of glucose catabolism as a mechanism for supplying cells with usable energy is about 44% (304/686). Efficiency of 20% to 25% is typical for machines.

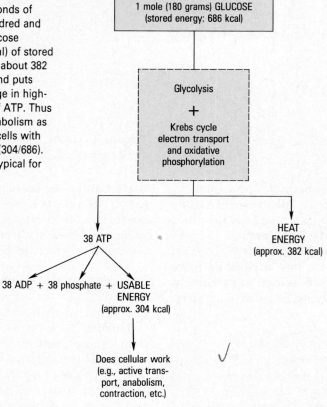

The complex organization of catabolism, a marvel in miniaturization, can boast also of amazing internal control. Many researchers are concentrating on the regulation of catabolism, that is, how cells know how quickly to catabolize. The control seems to lie in a negative feedback mechanism. In other words, a product accumulating in sufficient amounts signals the pathway to slow down. In the glycolytic and Krebs cycle pathways, certain enzymes early in the pathways (shown in Figures 23-3 and 23-5) act as "pacemakers." They are sensitive to and inhibited by the buildup of excess amounts of citric acid and ATP. Exactly how citric acid and ATP excesses notify the enzymes at these steps will be a matter for investigation for some time.

The breakdown of ATP molecules, of course, provides virtually all the energy that does cellular work. Therefore the process that produces some 90% of the ATP formed during carbohydrate catabolism, namely, oxidative phosphorylation, is the crucial part of catabolism (Figure 23-7). This vital process depends on cells receiving an adequate oxygen supply. Why? Briefly because only when oxygen is present in cells to serve as the final acceptor of electrons and hydrogen ions can electrons then continue moving down the electron transport chain. If oxygen becomes unavailable, the movement of electrons and hydrogen ions stops. Cessation of ATP formation by oxidative phosphorylation necessarily follows. All too soon, cells have no energy supply—a lethal condition if it persists for more than a few minutes.

In looking back over the cells' accomplishments in catabolism, we are struck by a sense of wonder, well-expressed by Wayne Becker*: "Respiratory metabolism can be regarded as a marvel of design and engineering. No transistors, no mechanical parts, no noise, no pollution—and all done in units of organization that require an electron microscope to visualize. Yet the process goes on routinely and continuously in almost every living cell with a degree of integration, efficiency, fidelity, and control that we can scarcely understand well enough to appreciate, let alone aspire to reproduce in our test tubes."

*From Becker, W.M.: Energy and the living cell, Philadelphia, 1977, J.B. Lippincott Co., pp. 125-126.

Summary

Summarizing the changes of glucose catabolism in equation form:

Glycolysis:

$$\text{Glucose} \rightarrow 2 \text{ Pyruvic acid} + 2 \text{ ATP} + \text{Heat}$$

Krebs cycle, electron transport, and oxidative phosphorylation:

$$2 \text{ Pyruvic acid} + 6 \text{ O}_2 \rightarrow 6 \text{ CO}_2 + 6 \text{ H}_2\text{O} + 36 \text{ ATP} + \text{Heat}$$

We can even summarize the long series of chemical reactions in glucose catabolism with one short equation:

$$C_6H_{12}O_6 + 6 \text{ O}_2 \rightarrow 6 \text{ CO}_2 + 6 \text{ H}_2\text{O} + 38 \text{ ATP} + \text{Heat}$$

Glycogenesis

Imagine what happens in a cell if glucose catabolism is proceeding at maximum rate. What do you do if you see a traffic jam ahead, with no possible way to get through it? Probably take an alternate route, if available. Similarly, if glycolytic pathways are "saturated" because of high levels of glucose entering the cell, a "traffic jam" of glucose-6-phosphate will result. Unable to enter glycolysis, glucose-6-phosphate will begin an alternate route, that is, it will enter the anabolic pathway of glycogen formation. This process, called **glycogenesis** (Figure 23-8), is a series of chemical reactions that may appear complex. But glycogen turns out to be just a necklace-like structure made up completely of stored glucose "beads."

The process of glycogenesis is part of a homeostatic mechanism that operates when the blood glucose level increases above the midpoint of its normal range (80 to 100 mg/100 dl of blood), as indicated in Figure 23-9. Example: Soon after a meal, while glucose is being absorbed rapidly, blood is quickly shunted to the liver via the portal system. Here a great many glucose molecules leave the blood for storage as glycogen. As a result of glycogenesis, the blood glucose level decreases, ordinarily enough to reestablish its normal level.

Muscle cells, like liver cells, have a high rate of glycogenesis. Certain cells, however, cannot carry out the process at all. Brain cells, for instance, cannot form or store glycogen. They must depend on the circulation of blood to deliver them their glucose.

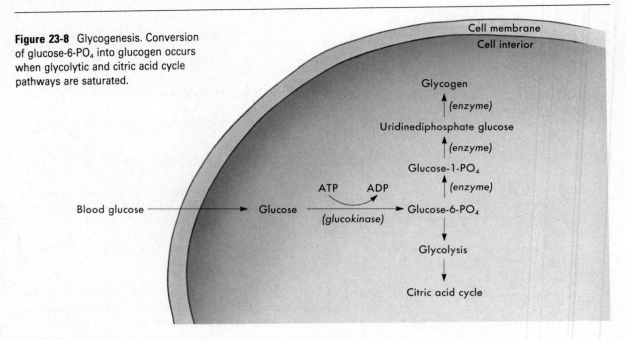

Figure 23-8 Glycogenesis. Conversion of glucose-6-PO$_4$ into glucogen occurs when glycolytic and citric acid cycle pathways are saturated.

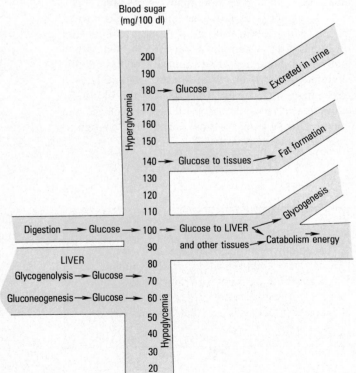

Figure 23-9 Homeostasis of blood glucose level. When blood glucose level starts to decrease toward lower normal, liver cells increase the rate at which they convert glycogen, amino acids, and fats to glucose (glycogenolysis and gluconeogenesis) and release it into blood. But when blood glucose level increases, liver cells increase the rate at which they remove glucose molecules from blood and convert them to glycogen for storage (glycogenesis). At still higher levels, glucose leaves blood for tissue cells to be anabolized into adipose tissue and, at still higher levels, is excreted in the urine.

Glycogenolysis

Glycogen molecules do not remain in the cell permanently, but are eventually broken apart (hydrolyzed). This process of "splitting glycogen" is called **glycogenolysis** (Figure 23-10). It is, in essence, a reversal of glycogenesis. What are the products of glycogenolysis? The answer depends on the cell. Although all cells presumably have the enzymes to break glycogen to glucose-6-phospate, only a few cell types (liver, kidney, intestinal mucosa) have phosphatase, which allows free glucose to form and possibly leave the cell. So the term *glycogenolysis* means different things in different cells. In muscles, glucose-6-phosphate is the product, which then undergoes glycolysis. But liver glycogenolysis (Figure 23-10) results in free glucose that can leave the cell and increase the blood glucose level. Accordingly, liver glycogenolysis acts as a part of the homeostatic mechanism to maintain the blood glucose level. Example: A few hours after a meal, when the blood glucose level decreases (see Figure 23-9), liver glycogenolysis accelerates. However glycogenolysis alone can probably maintain homeostasis of blood glucose concentration for only a few hours, since the body can store only small amounts of glycogen.

Gluconeogenesis

Literally **gluconeogenesis** means the formation of "new" glucose—"new" in the sense that it is made from proteins or from the glycerol of fats, not from carbohydrates. The process occurs chiefly in the liver. It consists of many complex chemical reactions. The new glucose produced from fats or proteins by gluconeogenesis (Figure 23-11) diffuses out of liver cells into the blood. Gluconeogenesis, therefore, can add glucose to the blood when needed. So, too, can the process of liver glycogenolysis. Obviously, then, the liver is a most important organ for maintaining blood glucose homeostasis.

Control of Glucose Metabolism

The complex mechanism that normally maintains homeostasis of blood glucose concentration consists of hormonal and neural devices. At least five endocrine glands—islands of Langerhans, anterior pituitary gland, adrenal cortex, adrenal medulla, and thyroid gland—and at least eight hormones secreted by those glands function as key parts of the glucose homeostatic mechanism.

Beta cells of the islands of Langerhans in the pancreas secrete the most famous sugar-regulating hormone of all—**insulin.** Insulin decreases blood glucose level. Although it is not yet known exactly how insulin acts, several of its effects are well known. For instance, insulin is known to act in some way to accelerate glucose transport through cell membranes. It also increases the activity of the enzyme glucokinase. As shown in Figure 23-8, glucokinase catalyzes glucose phosphorylation, the reaction that must occur before either glycogenesis or glucose catabolism can take place. By applying these facts, you can deduce the main methods by which insulin decreases blood glucose level: increased glycogenesis and increased catabolism of glucose. Figure 23-12 shows these effects of insulin. By studying this figure, you can also deduce for yourself some of the prominent metabolic defects resulting from

Figure 23-10 Glycogenolysis in a liver cell.

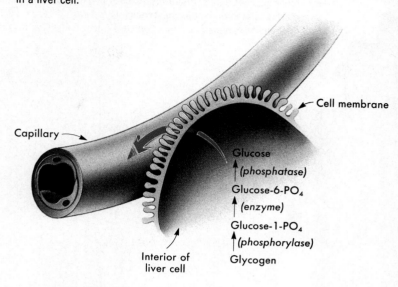

insulin deficiency such as in diabetes mellitus. Slow glycogenesis and low glycogen storage, decreased glucose catabolism, and increased blood glucose all result from insulin deficiency.

The islands of Langerhans secrete two sugar-regulating hormones—insulin from the beta cells and glucagon from the alpha cells. Whereas insulin tends to decrease the blood glucose level, glucagon tends to increase it. **Glucagon** increases the activity

Figure 23-11 Liver gluconeogenesis from mobilized tissue proteins and fats.

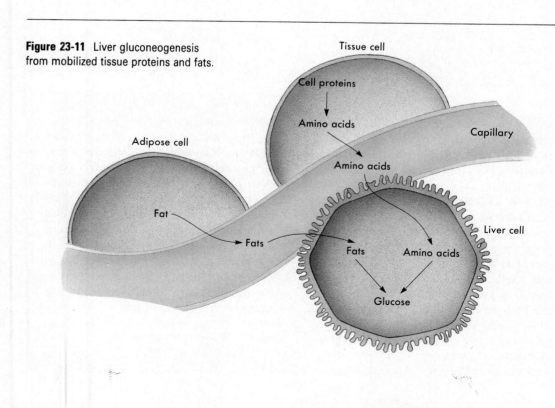

Figure 23-12 Function of insulin is a homeostatic mechanism, which tends to prevent blood glucose concentration from increasing above the upper limit of normal. This insulin mechanism and the glucagon mechanism shown in Figure 23-13 operate together to maintain homeostasis of blood glucose under usual circumstances in the normal body.

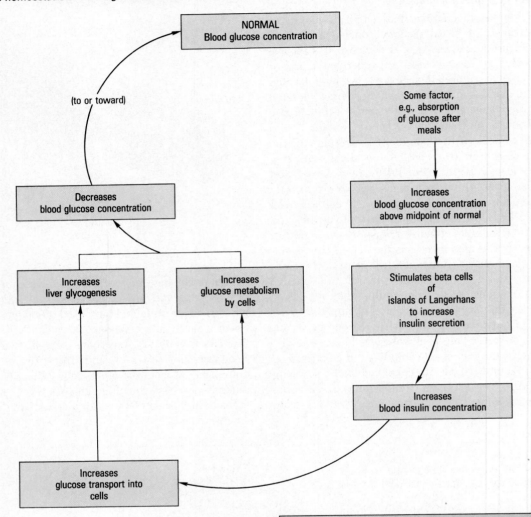

of the enzyme phosphorylase (see Figure 23-10), and this necessarily accelerates liver glycogenolysis and releases more of its product, glucose, into the blood. Figure 23-13 shows the role of glucagon in raising the blood glucose level.

Epinephrine is a hormone secreted in large amounts by the adrenal medulla in times of emotional or physical stress. Like glucagon, epinephrine increases phosphorylase activity. This makes glycogenolysis occur at a faster rate. Epinephrine accelerates both liver and muscle glycogenolysis, whereas glucagon accelerates only liver glycogenolysis. Both hormones increase the blood glucose level. Epinephrine has the distinction of being the only hormone whose release into the systemic circulation (and therefore effects on metabolism) are directly under the control of the nervous system.

The four hormones we have discussed (insulin, glucogen, epinephrine, and ACTH) are able to increase blood glucose concentration by causing the formation of glucose, either from glycogen or from amino acids. Growth hormone, made by the anterior pituitary, also increases blood glucose level, but by a different mechanism. Growth hormone causes a shift from carbohydrate to fat catabolism. It does this by limiting the storage of fat in fat depots. Instead, more fats are mobilized and catabolized. In this way, growth hormone "spares" carbohydrates from catabolism, and the level of glucose in the blood is increased.

Figure 23-13 Function of glucagon is a homeostatic mechanism that, under usual conditions, is chiefly responsible for preventing blood glucose level from falling below the lower limit of normal. This glucagon mechanism and the insulin mechanism shown in Figure 23-12 work together to maintain homeostasis of blood glucose under usual circumstances in the normal body.

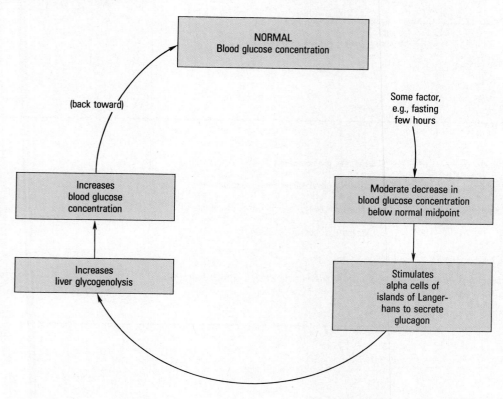

Adrenocorticotropic hormone (ACTH) and glucocorticoids are two more hormones that increase blood glucose concentration. ACTH stimulates the adrenal cortex to increase its secretion of glucocorticoids. Glucocorticoids accelerate gluconeogenesis. They do this by mobilizing proteins, that is, the breakdown or hydrolysis of tissue proteins to amino acids. More amino acids enter the circulation and are carried to the liver. Liver cells step up their production of "new" glucose from the mobilized amino acids. More glucose streams out of liver cells into the blood and adds to the blood glucose level.

A look at a summary of hormone control shown in Figure 23-14 indicates that most hormones cause the blood level of glucose to rise. These hormones are called *hyperglycemic*. The one notable exception is insulin, which is hypoglycemic, or tends to decrease blood glucose. Thyroid-stimulating hormone (TSH) from the anterior pituitary gland and its target secretion, thyroid hormone, have complex effects on metabolism. Some of these raise, and some lower, the glucose level. One of the effects of thyroid hormone is to accelerate catabolism, and since glucose is the body's "preferred fuel," the result may be a decrease in blood glucose level.

LIPID METABOLISM

Transport of Lipids

Lipids (fats) are transported in blood as chylomicrons, lipoproteins, and free fatty acids. **Chylomicrons** are small fat droplets present in blood during fat absorption. Fatty acids and glycerol, the products of fat digestion, combine during absorption to again form fats (triglycerides, or triacylglycerols). Triglyc-

Figure 23-14 Hormonal control of blood glucose level. Insulin lowers blood glucose level, so is hypoglycemic. Most hormones raise blood glucose level and are called *hyperglycemic* or anti-insulin hormones.

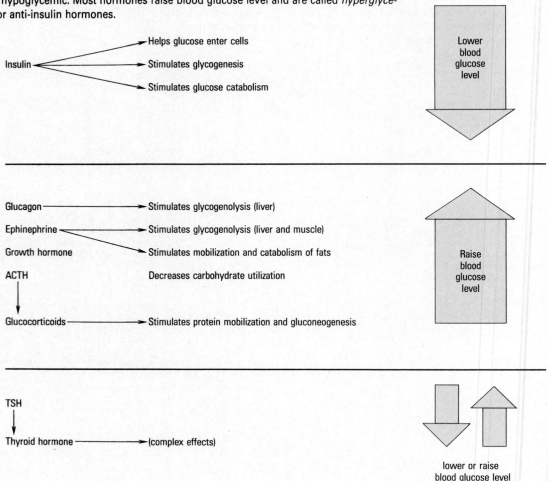

erides plus small amounts of cholesterol and phospholipids compose the chylomicrons. During fat absorption, blood may contain so many of these fat droplets that it appears turbid or even yellowish in color. But usually within about 4 hours after a meal, few, if any, chylomicrons remain in the blood. Their contents have moved mostly into adipose tissue cells.

In the postabsorptive state, when chylomicrons are virtually absent from the circulation, some 95% of the lipids in blood are in the form of lipoproteins. **Lipoproteins** are produced mainly in the liver, and, as their name suggests, they consist of lipids (triglycerides, cholesterol, and phospholipids) and protein. At all times, blood contains three types of lipoproteins, namely, very low density, low density, and high density lipoproteins. Usually, they are designated by their abbreviations: VLDL, LDL, and HDL. Diets high in saturated fats and cholesterol tend to produce an increase in blood LDL concentration, which, in turn, is associated with a high incidence

of coronary artery disease and atherosclerosis. A high blood HDL concentration, in contrast, is associated with a low incidence of heart disease. One might, therefore, think of the LDLs as the "bad lipoproteins" and the HDLs as the "good lipoproteins." Considerable evidence indicates that exercise tends to elevate HDL concentration. This may partially account for the beneficial effects of exercise.

Fatty acids, on entering the blood from adipose tissue or other cells, combine with albumin to form the so-called **free fatty acids** (FFAs). Fatty acids, in short, are transported from the cells of one tissue to those of another in the form of free fatty acids. Whenever the rate of fat catabolism increases—as it does, for example, in starvation or diabetes—the free fatty acid content of blood increases markedly.

Lipid Catabolism

Lipid catabolism, like carbohydrate catabolism, consists of several processes. Each of these, in turn, consists of a series of chemical reactions. Triglyc-

Figure 23-15 Fat mobilization and catabolism. Notice the role of the liver as the chief site of ketogenesis. Numbers of carbon atoms are in parentheses.

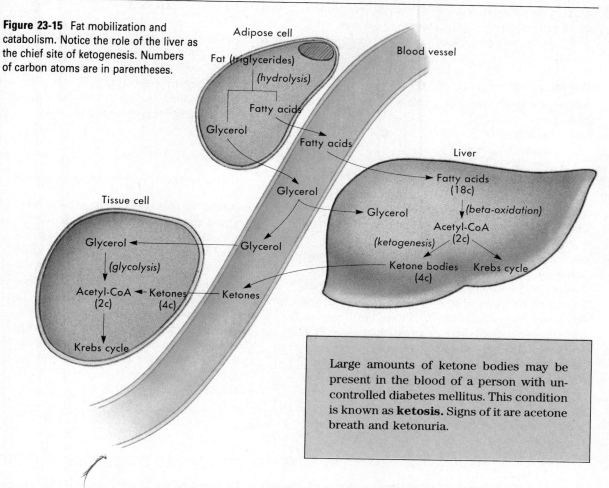

Large amounts of ketone bodies may be present in the blood of a person with uncontrolled diabetes mellitus. This condition is known as **ketosis.** Signs of it are acetone breath and ketonuria.

erides are first hydrolyzed to yield fatty acids and glycerol. Glycerol is then converted to glyceraldehyde-3-phosphate, which enters the glycolysis pathway (see Figure 23-3). Fatty acids, as Figure 23-19 shows, are broken down by a process called beta-oxidation into 2-carbon pieces, the familiar acetyl-CoA. These are then catabolized via the Krebs cycle. The final process of lipid catabolism, therefore, consists of the same reactions as carbohydrate catabolism. Catabolism of lipids, however, yields considerably more energy than catabolism of carbohydrates. Whereas catabolism of 1 gm of carbohydrates yields only 4.1 kcal of heat, catabolism of 1 gm of fat yields 9 kcal.

When fat catabolism occurs at an accelerated rate, as in diabetes mellitus (when glucose cannot enter cells) or fasting, excessive numbers of acetyl-CoA units are formed. Liver cells then temporarily condense acetyl-CoA units together to form 4-carbon acetoacetic acids. Acetoacetic acid is classified as a ketone body and can be converted to two other ketone bodies, namely, acetone and betahydroxybutyric acid—hence, the name *ketogenesis* for this process. Liver cells oxidize a small portion of the ketone bodies for their own energy needs, but most of them

are transported by the blood to other tissue cells for the change back to acetyl-CoA and oxidation via the Krebs cycle (Figure 23-15).

Lipid Anabolism

Lipid anabolism consists of the synthesis of various types of lipids, notably *triglycerides, cholesterol, phospholipids,* and *prostaglandins.* Triglycerides are synthesized from fatty acids and glycerol or from excess glucose or amino acids. So it is possible to "get fat" from foods other than fat. Triglycerides are stored mainly in adipose tissue cells. These fat depots constitute the body's largest reserve energy source—often too large, unfortunately. Almost limitless pounds of fat can be stored. In contrast, only a few hundred grams of carbohydrates can be stored as liver and muscle glycogen.

Only certain fatty acids (saturated ones) can be synthesized by the body. Others, the unsaturated fatty acids, must be provided by the diet and so are called **essential fatty acids.** Certain essential fatty acids serve as a source within the body for synthesis of an important group of lipids called *prostaglandins.* These hormonelike compounds, first discovered in the 1930s in fluid from the prostate gland, have in recent years gained increasing recognition for their occurrence in a variety of tissues, with a spectrum of biological activities (see Chapter 14).

Control

Fat metabolism is controlled mainly by the following hormones:

1 **Insulin**
2 **Growth hormone**
3 **ACTH**
4 **Glucocorticoids**

You probably recall from our discussion of these hormones in connection with carbohydrate metabolism that they regulate fat metabolism in such a way that the rate of fat catabolism is inversely related to the rate of carbohydrate catabolism. If some condition such as diabetes mellitus causes carbohydrate catabolism to decrease below energy needs, increased secretion of growth hormone, ACTH, and glucocorticoids soon follows. These hormones, in turn, bring about an increase in fat catabolism. But, when carbohydrate catabolism equals energy needs, fats are not mobilized out of storage and catabolized. Instead, they are spared and stored in adipose tissue. "Carbohydrates have a 'fat-sparing' effect," so says an old physiological maxim. Or stating this truth more descriptively: "Carbohydrates have a 'fat-storing' effect."

PROTEIN METABOLISM

In protein metabolism, anabolism is primary and catabolism is secondary. In carbohydrate and fat metabolism, the opposite is true—catabolism is primary and anabolism is secondary. Proteins are primarily tissue-building foods. Carbohydrates and fats are primarily energy-supplying foods.

A number of diseases result in abnormal metabolism, distribution, or storage of lipids. A chronic liver disease called **primary biliary cirrhosis** impairs the movement of bile into the lumen of the intestine. As a result, bile is absorbed into the blood causing **jaundice,** or yellowing of the skin. In addition, blood lipid levels increase, and yellowish deposits of lipid material called **xanthomas** develop. In Figure 23-16, fatty xanthomas are apparent in the borders of the eyelids in a patient suffering from biliary cirrhosis.

Figure 23-16 Lipid deposits called *xanthomas* are present in eyelids of this patient suffering from biliary cirrhosis. Some yellowing of the skin, or jaundice, is also present.

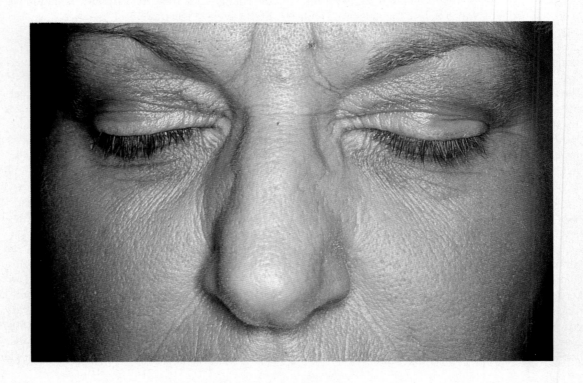

Protein Anabolism

Protein anabolism is the process by which proteins are synthesized by the ribosomes of all cells. Every cell synthesizes its own structural proteins and its own enzymes. In addition, many cells, such as the liver and glandular cells, synthesize special proteins for export. The cell's genes determine the specific proteins synthesized. Protein anabolism is truly "big business" in the body. Consider, for instance, that protein anabolism constitutes the major process of growth, reproduction, tissue repair, and the replacement of cells destroyed by daily wear and tear. Red blood cell replacement alone runs into millions of cells per second!

Protein Catabolism

The first step in protein catabolism takes place in liver cells. Called **deamination,** it consists of the splitting off of an amino (NH_2) group from an amino acid molecule to form a molecule of ammonia and one of keto acid (e.g., alpha-ketoglutaric acid). Most of the ammonia is converted by liver cells to urea and later excreted in the urine. The keto acid, as Figure 24-18 indicates, may be oxidized via the Krebs cycle or may be converted to glucose (gluconeogenesis) or to fat (lipogenesis). Both protein catabolism and anabolism go on continually. Only their rates differ from time to time. With a protein-deficient diet, for example, protein catabolism exceeds protein anabolism. Various hormones, as we shall see, also influence the rates of protein catabolism and anabolism.

Usually a state of **protein balance** exists in the normal healthy adult body, that is, the rate of protein anabolism equals or balances the rate of protein catabolism. When the body is in protein balance, it is also in a state of **nitrogen balance.** For then the amount of nitrogen taken into the body (in protein foods) equals the amount of nitrogen in protein catabolic waste products excreted in the urine, feces, and sweat.

It is important to realize that there are two kinds of protein or nitrogen imbalance. When protein catabolism exceeds protein anabolism, the amount of nitrogen in the urine exceeds the amount of nitrogen in the protein foods ingested. The individual is then said to be in a state of **negative nitrogen balance,** or in a state of "tissue-wasting"—because more of the tissue proteins are being catabolized than are being replaced by protein synthesis. Protein-poor diets, starvation, and wasting illnesses, for example, produce a negative nitrogen balance. A **positive nitrogen balance** (nitrogen intake in foods greater than nitrogen output in urine) indicates that protein anabolism is going on at a faster rate than protein catabolism. A state of positive nitrogen balance, therefore, characterizes any condition in which large amounts of tissue are being synthesized, such as during growth, pregnancy, and convalescence from an emaciating illness.

The main facts about protein metabolism are summarized in Figures 23-17 to 23-19. Compare them with Figures 23-4, 23-8, 23-10, 23-11 and 23-15. Note the important part played by the liver in the metabolism of all three kinds of foods.

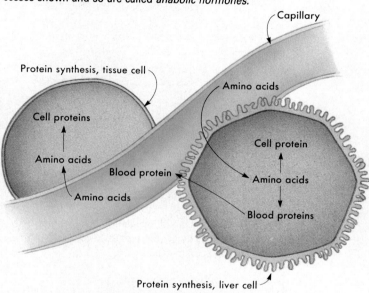

Figure 23-17 Protein synthesis (anabolism). Growth hormone and testosterone tend to accelerate the processes shown and so are called *anabolic hormones.*

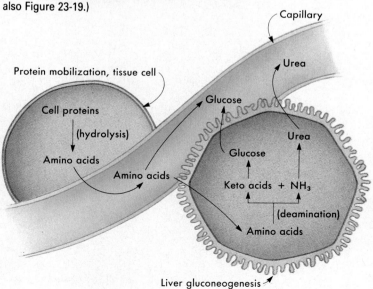

Figure 23-18 Protein mobilization and catabolism. Glucocorticoids tend to accelerate these processes, so are classed as protein catabolic hormones. (See also Figure 23-19.)

Figure 23-19 Protein catabolism. First, as shown in Figure 23-18, liver cells carry on deamination, a process that converts amino acids to keto acids and ammonia. Then keto acids may be changed to glucose by liver cells (gluconeogenesis), or liver and tissue cells may oxidize them (Krebs cycle) or convert them to fat (lipogenesis).

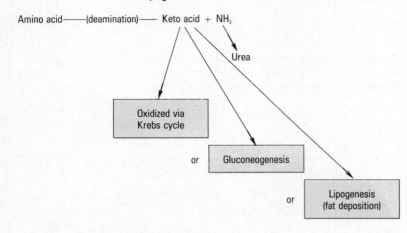

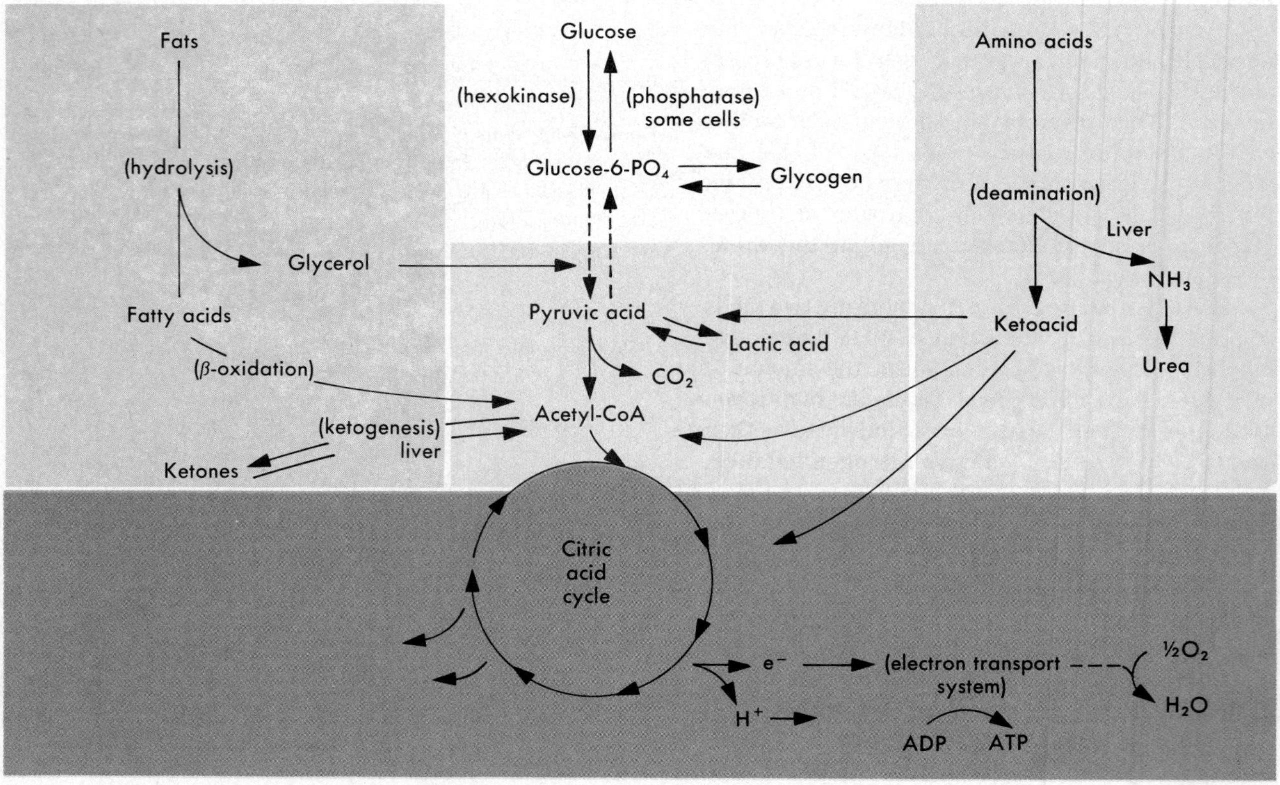

Control

Protein metabolism, like that of carbohydrates and fats, is controlled largely by hormones rather than by the nervous system. Growth hormone and the male hormone testosterone both have a stimulating effect on protein synthesis or anabolism. For this reason, they are referred to as *anabolic* hormones. The protein *catabolic* hormones of greatest consequence are glucocorticoids. They are thought to act in some way, still unknown, to speed up tissue protein mobilization, that is, the hydrolysis of cell proteins to amino acids, their entry into the blood, and their subsequent catabolism (see Figure 23-18). ACTH functions indirectly as a protein catabolic hormone because of its stimulating effect on glucocorticoid secretion.

Thyroid hormone is necessary for and tends to promote protein anabolism and, therefore, growth when plenty of carbohydrates and fats are available for energy production. On the other hand, under different conditions, for example, when the amount of thyroid hormone is excessive or when the energy foods are deficient, this hormone may then promote protein mobilization and catabolism.

• • •

Some of the facts about metabolism set forth in the preceding paragraphs are summarized in Table 23-1 and Figure 23-20. You may find it helpful to read

Figure 23-20 Summary of metabolism. Note the essential role played by pyruvic acid and acetyl-CoA on interrelationships of metabolism.

Fats

(hydrolysis)

Glucose

(hexokinase)   (phosphatase)
                some cells

Glucose-6-PO$_4$ ⟶ Glycogen

Amino acids

(deamination)

Liver

NH$_3$

Glycerol

Fatty acids

(β-oxidation)

Pyruvic acid

Lactic acid

CO$_2$

Ketoacid

Urea

Acetyl-CoA

(ketogenesis)
liver

Ketones

Citric
acid
cycle

e⁻ ⟶ (electron transport
          system)

½O$_2$

H₂O

H⁺ ⟶

ADP   ATP

once again the section on important generalizations about metabolism (pp. 599 and 600). As you do so, try to cite specific examples encountered in your study of metabolism.

METABOLIC RATES

Meaning

The term **metabolic rate** means the amount of energy released in the body in a given time by catabolism. It represents energy expended or used for accomplishing various kinds of work. In short, metabolic rate actually means catabolic rate or rate of energy release.

Ways of Expressing

Metabolic rates are expressed in either of two ways: (1) in terms of the number of kilocalories of heat energy expended per hour or per day or (2) as normal or as a definite percentage above or below normal.

Basal Metabolic Rate

The **basal metabolic rate (BMR)** is the body's rate of energy expenditure under "basal conditions," namely, when the individual:

1 Is awake but resting, that is, lying down and, as far as possible, not moving a muscle
2 Is in the postabsorptive state (12 to 18 hours after the last meal)

Note that the BMR is not the minimum metabolic rate. It does not indicate the smallest amount of energy that must be expended to sustain life. It does, however, indicate the smallest amount of energy expenditure that can sustain life and also maintain the waking state and a normal body temperature in a comfortably warm environment.

3 Is in a comfortably warm environment

Factors Influencing

The BMR is not identical for all individuals because of the influence of various factors (Figure 23-21), some of which are described in the following paragraphs.

SIZE. In computing the BMR, size is usually indicated by the amount of the body's surface area. It is computed from the individual's height and weight. A large individual has the same BRM as a small person per square meter of body surface if other conditions are equal. However, because a large individual has more square meters of surface area, the BMR is greater than that of a small individual. For example,

Table 23-1 Metabolism

Food	Anabolism	Catabolism	
Carbohydrates	Temporary excess changed into glycogen by liver cells in presence of insulin; stored in liver and skeletal muscles until needed and then changed back to glucose (Figures 23-10 and 23-11)	Oxidized, in presence of insulin, to yield energy (4.1 kcal per gm) and wastes (carbon dioxide and water)	
	True excess beyond body's energy requirements converted into adipose tissue; stored in various fat depots of body (Figure 23-9)	$C_6H_{12}O_6 + 6\,O_2 \rightarrow$ Energy $+\ 6\,CO_2 + 6\,H_2O$	
Fats	Built into adipose tissue; stored in fat depots of body	Fatty acids $\downarrow$ (beta-oxidation) Acetyl-CoA $\leftrightarrows$ Ketones $\downarrow$ (tissues; citric acid cycle) Energy (9.3 kcal per gm) $+\ CO_2 + H_2O$	Glycerol $\downarrow$ (glycolysis) Acetyl-CoA
Proteins	Synthesized into tissue proteins, blood proteins, enzymes, hormones, etc.	Deaminated by liver, forming ammonia (which is converted to urea) and keto acids (which are either oxidized or changed to glucose or fat)	

Figure 23-21 Factors that determine the basal and total metabolic rates.

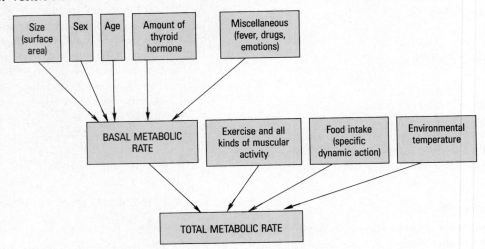

the BMR for a man in his twenties is about 40 kcal per square meter of body surface per hour (Table 23-2). However, a large man with a body surface area of 1.9 square meters would have a BMR of 76 kcal per hour, whereas a smaller man with a surface area of perhaps 1.6 square meters would have a BMR of only 64 kcal per hour. The average surface area for American adults is 1.6 square meters for women and 1.8 square meters for men.

SEX. Men oxidize their food approximately 5% to 7% faster than women. Therefore their BMRs are about 5% to 7% higher for a given size and age. A man 5 feet, 6 inches tall, weighing 140 pounds, for example, has a 5% to 7% higher BMR than a woman of the same height, weight, and age.

Table 23-2 Basal metabolism (Aub-DuBois)

Age (yr)	Kilocalories per hour per square meter body surface	
	Male	Female
10-12	51.5	50.0
12-14	50.0	46.5
14-16	46.0	43.0
16-18	43.0	40.0
18-20	41.0	38.0
20-30	39.5	37.0
30-40	39.5	36.5
40-50	38.5	36.0
50-60	37.5	35.0
60-70	36.5	34.0

AGE. That the fires of youth burn more brightly than those of age is a physiological and a psychological fact. In general, the younger the individual, the higher the BMR for a given size and sex (Table 23-2). Exception: the BMR is slightly lower at birth than it is a few years later. That is to say, the rate increases slightly during the first 3 to 6 years, then starts to decrease and continues to do so throughout life.

THYROID HORMONE. Thyroid hormone stimulates basal metabolism. Without a normal amount of this hormone in the blood, a normal BMR cannot be maintained. When an excess of thyroid hormone is secreted, foods are catabolized faster, much as coal is burned faster when a furnace draft is open. Deficient thyroid secretion, on the other hand, slows the rate of metabolism.

BODY TEMPERATURE. Fever increases the BMR. According to DuBois, for every degree Celsius increase in body temperature, metabolism increases about 13%. A decrease in body temperature (hypothermia) has the opposite effect. Metabolism decreases, and, because it does, cells use less oxygen than they normally do. This knowledge has been applied clinically by using hypothermia in certain situations—for example, in open-heart surgery. Because circulation is reduced or interrupted during this procedure, oxygen supply necessarily decreases. Cells can tolerate this decreased oxygen supply reasonably well if their oxygen need has also decreased. Induced hypothermia decreases their rate of metabolism and thereby decreases their use of oxygen.

DRUGS. Certain drugs, such as caffeine, amphetamine, and dinitrophenol, increase the BMR.

Figure 23-22 Chart for determining surface area in square meters from weight in kilograms and height in centimeters according to the following formula: Area (m²) = wt$^{0.425}$ × ht$^{0.725}$ × 7,184.

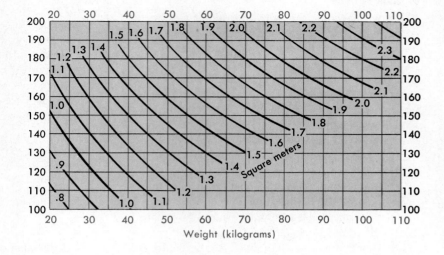

Height (centimeters)

Weight (kilograms)

OTHER FACTORS. Other factors, such as *emotions*, *pregnancy*, and *lactation*, also influence basal metabolism. All of these factors increase the BMR.

How Determined

BMR can be determined by a method called **indirect calorimetry.** The rationale underlying this method is the fact that BMR (expressed as the number of kilocalories of heat produced per unit of time) can be calculated from the amount of oxygen consumed in a given time. BMR can then be expressed as normal, or as a definite percent above or below normal, by dividing the actual kilocalorie rate by the known average kilocalorie rate for normal individuals of the same size, sex, and age. Statistical tables, based on research, list these normal BMRs. (If BMR were calculated to be 10% above normal, for example, it would be reported as + 10.) Are you curious to know the average BMR for a person of your size, sex, and age? If so, take the following steps:

1 Start with your weight in kilograms and your height in centimeters. (Convert pounds to kilograms by dividing pounds by 2.2. Convert inches to approximate centimeters by multiplying inches by 2.5.) For example, 110 pounds = 50 kg; 5 feet, 3 inches = 158 cm.

2 Convert your weight and height to square meters, using Figure 23-22. For example, weight 50 kg and height 158 cm = about 1.5 square meters of body surface area.

3 Find your age and sex in Table 23-2 and then multiply the number of kilocalories per square meter per hour given there by your square meters of surface area and then by 24. For example, average BMR per day for a 25-year-old female, weight 110 pounds and

height 5 feet, 3 inches = 1,332 kcal (37 × 1.5 × 24).

A quick rule of thumb for estimating a young woman's BMR is to multiply her weight in pounds by 12.

Total Metabolic Rate

The **total metabolic rate** is the amount of energy used or expended by the body in a given time. It is expressed in kilocalories per hour or per day. Most of the factors that together determine the total metabolic rate are shown in Figure 23-21. Of these, the main direct determinants are the following:

1 The basal metabolic rate, that is, the energy used to do the work of maintaining life under the basal conditions previously described. Basal metabolic rate usually constitutes about 55% to 60% of the total metabolic rate.

2 The energy used to do all kinds of skeletal muscle work—from the simplest activities such as feeding oneself or sitting up in bed to the most strenuous kind of physical labor or exercise.

3 The energy used for specific dynamic action (SDA) of foods. The metabolic rate increases for several hours after a meal, apparently because of the energy needed for metabolizing foods. Carbohydrates and fats have a SDA of about 5%. Proteins have a much higher SDA, about 30%. This means that, for 100 kcal of protein, 30 kcal are used for such processes as deamination and oxidation of the protein, leaving just 70 kcal available for other cell work. For this reason, proteins are "worth" fewer calories and are popular as diet foods.

Energy Balance and its Relationship to Body Weight

When we say that the body maintains a state of energy balance, we mean that its energy input equals its energy output. Energy input per day equals the total calories (kilocalories) in the food ingested per day. Energy output equals the total metabolic rate expressed in kilocalories. You may be wondering what energy intake, output, and balance have to do with body weight. "Everything" would be a fairly good one-word answer. Or, to be somewhat more explicit, the following basic principles describe the relationships between these factors:

1 Body weight remains constant (except for possible variations in water content) when the body maintains energy balance—when the total calories in the food ingested equals the total metabolic rate. Example: If you have a total metabolic rate of 2,000 kcal per day and if the food you eat per day yields 2,000 kcal, your body will be maintaining energy balance and your weight will stay constant.

2 Body weight increases when energy input exceeds energy output—when the total calories of food intake per day is greater than the total calories of the metabolic rate. A small amount of the excess energy input is used to synthesize glycogen for storage in the liver and muscles. But the rest is used for synthesizing fat and storing it in adipose tissue. If you were to eat 3,000 kcal each day for a week and if your total metabolic rate were 2,000 kcal per day, you would gain weight. How much you would gain you can discover by doing a little simple arithmetic:

Total energy input for week = 21,000 kcal
Total energy output for week = 14,000 kcal
Excess energy input for week = 7,000 kcal

Approximately 3,500 kcal are used to synthesize 1 pound of adipose tissue. Hence, at the end of this one week of "overeating"— of eating 7,000 kcal over and above your total metabolic rate—you would have gained about 2 pounds.

3 Body weight decreases when energy input is less than energy output—when the total number of calories in the food eaten is less than the total metabolic rate. Suppose you were to eat only 1,000 kcal a day for a week and that you have a total metabolic rate of 2,000 kcal per day. By the end of the week, your body would have used a total of 14,000 kcal of energy for maintaining life and doing its many kinds of work. All

Anyone who wants to reduce should remember this cardinal principle: eat fewer calories than your total metabolic rate. To apply this principle, do one of two things or, better yet, do both: (1) decrease your caloric intake; (2) increase your caloric output, that is, total metabolic rate, by increasing your physical activity. Eat less, exercise more!

14,000 kcal of this actual energy expenditure had to come from catabolism of foods, since this is the body's only source of energy. Catabolism of ingested food supplied 7,000 kcal, and catabolism of stored food supplied the remaining 7,000 kcal. That week your body would not have maintained energy balance, nor would it have maintained weight balance. It would have incurred an energy deficit paid out of the energy stored in approximately 2 pounds of body fat. In short, you would have lost about 2 pounds.

Foods are stored as glycogen, fats, and tissue proteins. As you will recall, cells catabolize them preferentially in this same order: carbohydrates, fats, and proteins. If there is no food intake, almost all of the glycogen is estimated to be used up in a matter of 1 or 2 days. Then, with no more carbohydrate to act as a fat sparer, fat is catabolized. How long it takes to deplete all of this reserve food depends, of course, on how much adipose tissue the individual has when starting the starvation diet. Finally, with no more fat available as a protein sparer, tissue proteins are catabolized rapidly, and death soon ensues.

MECHANISMS FOR REGULATING FOOD INTAKE

Mechanisms for regulating food intake are still not clearly established. That the hypothalamus plays a part in these mechanisms, however, seems certain. A number of data seem to indicate that a cluster of neurons in the lateral hypothalamus function as an **appetite center**—meaning that impulses from them bring about increased appetite. Other data suggest that a group of neurons in the ventral medial nucleus of the hypothalamus functions as a **satiety center**—meaning that impulses from these neurons

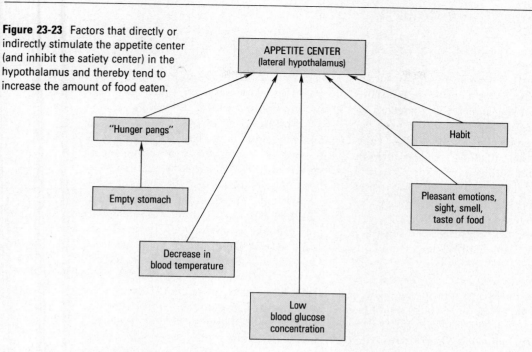

Figure 23-23 Factors that directly or indirectly stimulate the appetite center (and inhibit the satiety center) in the hypothalamus and thereby tend to increase the amount of food eaten.

decrease appetite. What acts directly on these centers to stimulate or depress them is still a matter of theory rather than fact. One theory (the *"thermostat theory"*) holds that it is the temperature of the blood circulating to the hypothalamus that influences the centers. A moderate decrease in blood temperature stimulates the appetite center (and inhibits the satiety center). Result: the individual has an appetite, wants to eat—and probably does. An increase in blood temperature produces the opposite effect, a depressed appetite (**anorexia**). One well-known instance of this is the loss of appetite in persons who have a fever.

Another theory (the *"glucostat theory"*) says that it is the blood glucose concentration and rate of glucose use that influences the hypothalamic feeding centers. A low blood glucose concentration or low glucose use stimulates the appetite center, whereas a high blood glucose concentration inhibits it. Unquestionably, many factors operate together as a complex mechanism for regulating food intake. Some of these factors are indicated in Figure 23-23.

EATING DISORDERS

Eating disorders have been a part of the medical literature for many years, but there has been growing interest and concern as the numbers of reported cases have increased dramatically. The two most common eating disorders are called **anorexia nervosa** and **bulimia.** Neither illness is completely understood, and successful treatment is often varied and sometimes controversial.

Anorexia is primarily a disease of young adults. Most individuals affected are female (90% to 95%) from 12 to 25 years of age. As many as 4% of college-age students suffer from anorexia to some degree. These individuals have a disturbed body image, an intense fear of obesity, and diet with a vengeance. They almost always develop unusual eating rituals and scrupulously monitor and restrict their food intake. Anorexic individuals literally starve themselves and, as a result, develop serious medical complications. The illness is characterized by a 20% to 25% loss of body mass, accompanied by slowed or impaired intellectual functioning. People affected are usually involved in excessive exercise and pursue thinness, regardless of their health. In women, menstruation ceases (amenorrhea), and the BMR is decreased as a result of starvation. These individuals suffer from many skin abnormalities and a wide assortment of psychological, cardiovascular, and hormonal problems. They are at increased risk of sudden death from complications directly related to excessive weight loss and nutritional deficiency. Treatment is directed at resolution of both medical and psychological problems. In addition to psychotherapy and weight stabilization, pharmacological treatment with antidepressants has been used to improve mood and self-image.

Bulimia is an illness characterized by an eating and vomiting or purging cycle. It is sometimes referred to as **binge-purge syndrome.** The disease, which has only been classified as a distinct disorder since 1980, is said to affect about 1% of college age students. Most bulimics are relatively young, single, white females. The mean age is 25, but the age of patients appears to be increasing. People suffering from bulimia have an uncontrollable urge for food that leads to massive overeating (binging) that is followed by repeated forced vomiting and laxative abuse (purging). Loss of gastric and intestinal contents often results in serious fluid and electrolyte imbalance. The result is often the development of neurological problems such as convulsions, tetany, and seizures. Vomiting may also cause aspiration pneumonia, erosion of tooth enamel, trauma of the mouth and esophagus, and infection of the salivary glands. The longer the disease is allowed to continue without treatment, the greater the increase in mortality from medical complications. Many bulimics suffer from major depression and have concomitant social problems, such as alcohol abuse. About a fourth of bulimics are chemically dependent, and many have been victims of sexual abuse. They are especially prone to self-mutilation and suicide attempts. Nutritional counseling, psychotherapy, and treatment with antidepressants help bulimic patients cope with stress and break the binge/purge cycle.

Outline Summary

SOME IMPORTANT GENERALIZATIONS ABOUT METABOLISM

A Metabolism—made up of two major processes, catabolism and anabolism, each of which consists of series of enzyme-catalyzed chemical reactions known as metabolic pathways

B Catabolism breaks food molecules into smaller molecular compounds, releasing energy; anabolism builds food molecules into larger molecular compounds and, in so doing, uses energy

Outline Summary—cont'd

C Catabolism and anabolism go on continually and concurrently inside cells

D Catabolism releases energy in forms of heat and chemical energy; chemical energy transferred to high-energy bonds of ATP

E Breaking of high-energy bond of ATP yields energy used for cellular work; ATP is the universal biological energy supplier

F Metabolism differs in different kinds of cells, in rate and kinds of compounds synthesized

CARBOHYDRATE METABOLISM

A Carbohydrates—"preferred energy fuel"; cells catabolize glucose first, sparing fats and proteins; when their glucose supply becomes inadequate, cells next metabolize fats, lastly proteins

B Glucose transport through cell membranes and phosphorylation
  1 Insulin promotes this transport through cell membranes
  2 Glucose phosphorylation—conversion of glucose to glucose-6-phosphate, catalyzed by enzyme glucokinase; insulin increases activity of glucokinase, so promotes glucose phosphorylation

C Glycolysis—first process of carbohydrate catabolism; consists of series of anaerobic (nonoxygen-using) reactions that convert one glucose molecule to two pyruvic acid molecules and yield slightly less than 5% of ATP produced during glucose catabolism; glycolysis takes place in cytoplasm of cells; preparation of second process of carbohydrate catabolism, namely, Krebs cycle

D Krebs cycle—series of aerobic reactions that use oxygen to oxidize two pyruvic acid molecules to six carbon dioxide and six water molecules and yield about 95% of ATP formed during catabolism; Krebs cycle goes on in mitochondria, accompanied by final phase of catabolism, namely, electron transport and oxidative phosphorylation

E Electron transport and oxidative phosphorylation—hydrogen atoms removed during Krebs cycle enter chain of carrier molecules (known as the electron transport system—form structural part of inner membrane of mitochondria); as electrons (from ionization of hydrogen atoms) move down electron transport system, they release small bursts of energy used for oxidative phosphorylation, that is, for joining phosphate to ADP to form ATP; oxidative phosphorylation depends on cells receiving adequate oxygen supply since electrons and hydrogen ions eventually combine with oxygen; if oxygen is not available, electron transport and therefore oxidative phosphorylation and ATP formation cease

F Glycogenesis—one kind of glucose anabolism; conversion of glucose to glycogen for storage;

occurs in muscle cells, also in liver cells, when blood glucose level exceeds 120 to 140 mg/100 dl

G Glycogenolysis
  1 In liver cells—when blood glucose level decreases below midpoint of normal, glycogenolysis accelerates to raise blood glucose level; enzyme, glucose phosphatase, present in liver cells catalyzes final step of glycogenolysis, changing glucose-6-phosphate to glucose; this enzyme lacking in most other cells; glucagon increases activity of phosphorylase, so accelerates liver glycogenolysis; epinephrine accelerates liver and muscle glycogenolysis
  2 In muscle cells—glycogen is changed back to glucose-6-phosphate, preliminary to catabolism

H Gluconeogenesis—sequence of chemical reactions carried on in liver cells; converts protein or fat compounds into glucose; growth hormone, ACTH, and glucocorticoids have stimulating effect on rate of gluconeogenesis

I Control of glucose metabolism
  1 By hormones that cause a decrease in blood glucose level, that is, have hypoglycemic effect
    a Insulin (from beta cells of pancreatic islets)—tends to accelerate glucose utilization by cells because it enhances glucose transport through cell membranes and glucose phosphorylation
    b Thyrotropin (from anterior pituitary gland)—stimulates thyroid gland to increase secretion of thyroid hormone, which accelerates catabolism, usually glucose catabolism, since glucose is "preferred fuel"
  2 By hormones that cause increase in blood glucose level, i.e., have hyperglycemic effect
    a Glucagon (from alpha cells of pancreatic islets)—increases activity of enzyme phosphorylase, thereby accelerating liver glycogenolysis with release of glucose into blood
    b Epinephrine (from adrenal medulla)—also increases phosphorylase activity, but in both liver and muscle cells
    c ACTH (from anterior pituitary gland)—stimulates adrenal cortex to increase secretion of glucocorticoids, which accelerate tissue protein mobilization and subsequent liver gluconeogenesis from mobilized proteins
    d Growth hormone (from anterior pituitary gland)—decreases fat deposition, increases fat mobilization and catabolism; hence, tends to bring about shift to fat use from "preferred" glucose use

LIPID METABOLISM

A Transport
  1 Chylomicrons—small fat droplets transport absorbed lipids from intestine, largely to adipose cells

Outline Summary—cont'd

2 Lipoproteins—major form in which lipids are transported in blood during postabsorptive state; high blood concentration of low density lipoproteins associated with diets high in saturated fats and cholesterol and with high incidence of coronary artery disease and atherosclerosis; high blood concentration of high density lipoproteins associated with low incidence of heart disease

3 Free fatty acids—fatty acids combined with albumin; form in which lipids are transported from one tissue to another

B Catabolism

1 Hydrolysis of fats (mostly in liver cells) to fatty acids and glycerol

2 Glycerol converted to glyceraldehyde-3-phosphate, which enters glycolysis pathway and, eventually, Krebs cycle

3 Fatty acids converted by beta-oxidation to acetyl-CoA; when fat catabolism accelerated, acetyl-CoA condenses to form ketone bodies (ketogenesis); occurs mainly in liver; largest proportion of ketones enters blood from liver cells to be transported to tissues for oxidation to carbon dioxide and water via citric acid cycle

C Anabolism—for tissue synthesis and for building various compounds; phospholipids important constituents of membranes; prostaglandins derived from essential fatty acids, have hormonelike effects on variety of tissues; fats deposited in connective tissue convert it to adipose tissue

D Control—by following major factors

1 Rate of glucose catabolism one of main regulators of fat metabolism; in general, normal or high rates of glucose catabolism accompanied by low rates of fat mobilization and catabolism and high rates of fat deposition; converse also true

2 Insulin helps control fat metabolism by its effects on glucose metabolism; in general, normal amounts of insulin and blood glucose tend to decrease fat mobilization and catabolism and to increase fat deposition; insulin deficiency increases fat mobilization and catabolism; converse also true

3 Growth hormone—decreases fat deposition and increases fat mobilization and utilization, that is, growth hormone tends to bring about shift from glucose to fat utilization

4 Glucocorticoids help control fat metabolism; in general, when blood glucose level lower than normal and in various stress situations, more glucocorticoids secreted and accelerate fat mobilization and gluconeogenesis from them; when blood glucose level higher than normal but rate of glucose catabolism low (as in diabetes mellitus), glucocorticoids also increase fat mobilization, but it is followed by ketogenesis from them; when blood glucose level higher than

normal, and provided its insulin content adequate, glucocorticoids accelerate fat deposition

E Fat mobilization—release of fats from adipose tissue cells, followed by their catabolism; occurs when blood contains less glucose than normal or when it contains less insulin than normal; if excessive, leads to ketosis

PROTEIN METABOLISM

A Anabolism of proteins of primary importance, their catabolism, secondary; amino acids used to synthesize all kinds of proteins, as dictated by cell's genes, for example, enzymes, hormones, antibodies, blood proteins, and cells' structural proteins

B Catabolism

1 Deamination of amino acid molecule to form ammonia and keto acid; mainly in liver cells

2 Ammonia converted to urea (mainly in liver) and excreted via urine

3 Keto acids may be converted to glucose in liver, or to fat, or oxidized in liver or tissue cells via Krebs cycle

C Control

1 Growth hormone and testosterone both have stimulating effect on protein synthesis or anabolism

2 ACTH and glucocorticoids—protein catabolic hormones; accelerate tissue protein mobilization, that is, hydrolysis of tissue proteins to amino acids and their release into blood; liver converts amino acids to glucose (gluconeogenesis) or deaminates them

3 Thyroid hormone promotes protein anabolism when nutrition adequate and amount of hormone normal; therefore, adequate amounts necessary for normal growth

METABOLIC RATES

A Meaning—amount of energy expended in given time

B Ways of expressing—in kilocalories or as "normal" or as a definite percentage above or below normal, for example, +10% or −10%

C Basal metabolic rate—amount of heat produced (energy expended) in waking state when body at complete rest, 12 to 18 hours after last meal, in comfortably warm environment

1 Factors influencing

a Size—greater surface area, higher BMR (surface area computed from height and weight)

b Sex—approximately 5% higher in males

c Age—higher in youth than in aged

d Abnormal functioning of certain endocrines, particularly thyroid gland

e Fever—each Celsius degree rise in temperature increases BMR approximately 13%

f Certain drugs, for example, dinitrophenol, increase BMR

Outline Summary—cont'd

g Other factors, for example, pregnancy and emotions, increase BMR

2 How determined—indirect calorimetry based on fact that definite amount of heat is produced for each liter of oxygen consumed; average BMR can be computed using Table 23-2 and Figure 23-22

D Total metabolic rate—amount of heat produced by body in average 24 hours; equal to basal rate plus number of kilocalories produced chiefly by muscular work and adjustment to cool temperatures; expressed in kilocalories per 24 hours

E Energy balance and its relationship to body weight

1 Energy balance means that energy input (total kilocalories in food ingested) equals energy output, that is, total metabolic rate expressed in kilocalories

2 For body weight to remain constant (except for variations in water content), energy balance must be maintained—total kilocalories ingested must equal total metabolic rate

3 Body weight increases when energy input exceeds energy output—when total kilocalories ingested greater than total metabolic rate

4 Body weight decreases when energy input less than energy output—when total kilocalories ingested less than total metabolic rate; no diet will reduce weight unless it contains fewer kilocalories than the total metabolic rate of the individual eating the diet

MECHANISMS FOR REGULATING FOOD INTAKE

A Thermostat theory holds that moderate decrease in blood temperature acts as stimulant to appetite center, so increases appetite; increase in blood temperature (fever) produces opposite effect

B Glucostat theory postulates that low blood glucose concentration or low rate of glucose use, for example, diabetes mellitus, acts as stimulant to appetite center, so increases appetite; high blood glucose level produces opposite effect

C Eating disorders

Review Questions

1 What is metabolism?

2 What two processes make up the process of metabolism?

3 Contrast the function or purpose of digestion with the function or purpose of metabolism.

4 Describe briefly glycolysis, the first process of carbohydrate catabolism.

5 Where does glycolysis occur?

6 Describe briefly the processes of carbohydrate catabolism known as the *Krebs cycle* and *oxidative phosphorylation*.

7 Explain why mitochondria are called "power plants" of cells.

8 Explain the processes and hormones involved in maintaining homeostasis of blood glucose concentration.

9 Explain how lipids are transported in blood.

10 Describe lipid catabolism briefly.

11 Describe protein catabolism briefly.

12 Compare the functions proteins, carbohydrates, and fats serve in the body.

13 Describe the main facts about hormonal control of metabolism.

14 Differentiate between digestive and metabolic wastes.

15 What does the term *metabolic rate* mean?

16 Differentiate between basal and total metabolic rates.

17 Discuss possible effects of advanced liver disease on digestion, absorption, and metabolism.

Situation: Mrs. A., who weighs 160 pounds, is 5 feet 5 inches tall, and 50 years of age, has been given a reducing diet by her doctor. She complains to you that she "knows it won't work" because "it isn't what she eats that makes her fat." As proof, she says that both her husband and son "eat twice as much" as she and so does her younger sister.

18 How would you answer Mrs. A.?

19 Suppose that Mrs. A. has a total metabolic rate of 2,200 kcal per day and that her diet contains 1,700 kcal. How many pounds can she expect to lose per week?

20 Rapid bicycle riding for 15 minutes expends about 100 kcal. This alone, without decreasing your calorie intake at all, would cause you to lose a pound in how many days? In 50 weeks, how much could you lose this way?

21 Describe various factors that influence the amount of food a person eats.

24 THE URINARY SYSTEM

OBJECTIVES

After you have completed this chapter, you should be able to:

1 List the major organs of the urinary system and give the generalized function of each.
2 Locate or position the kidneys in the abdominal cavity and identify the gross internal structures visible in a coronal section.
3 Name the parts of a nephron and describe the role of each component in the formation of urine.
4 Describe the renal blood supply and trace blood flow through the specialized vessels of the kidney.
5 Discuss how the kidneys form urine and trace urine from its point of formation to the exterior of the body.
6 Discuss the countercurrent mechanism for concentrating or diluting urine.
7 Explain the importance of filtration, tubular reabsorption, and tubular secretion in urine formation.
8 Describe the physical characteristics of normal urine.
9 Explain the regulation of urine volume under normal conditions.
10 Discuss and compare the structure and functions of the ureters, urinary bladder, and urethra.
11 Discuss hemodialysis as a treatment for renal failure.

KEY TERMS

Bowman's capsule

Countercurrent mechanism

Glomerulus (glo-MER-u-lus)

Hemodialysis (he-mo-di-AL-i-sis)

Juxtaglomerular apparatus (juks-ta-glo-MER-u-lar
   ap-ah-RA-tus)

Nephron (NEF-ron)

Renal corpuscle (RE-nal KOR-pus-el)

Renal cortex

Renal medulla (RE-nal me-DUL-ah)

Ureter (you-REE-tur)

Urethra (you-REE-thruh)

Urinalysis (u-ri-NAL-i-sis)

The urinary system consists of those organs that produce urine and eliminate it from the body. There are two kidneys, two ureters, one bladder, and one urethra (Figure 24-1). The excretion of urine and its elimination from the body are vital functions since together they constitute one of the most important mechanisms for maintaining homeostasis.

It is the urinary system that "clears" the blood of waste products produced by the metabolism of nutrients. If not removed on an ongoing basis, waste products quickly accumulate to toxic levels—a condition called **uremia,** or uremic poisoning. The kidneys also play a key role in maintaining normal water and electrolyte balance, fluid volumes, blood pressure, and in regulating body pH.

The substances excreted from the kidneys and other excretory organs are listed in Table 24-1.

Figure 24-1 Location of urinary system organs.

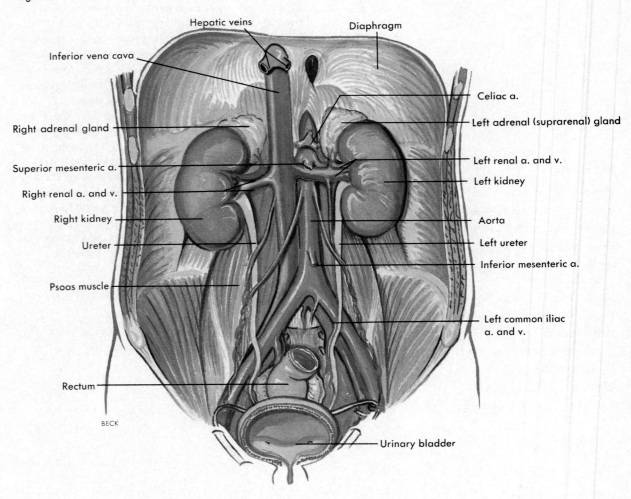

KIDNEYS

Size, Shape, and Location

The kidneys resemble lima beans in shape. An average-sized kidney measures approximately 11.25 cm in length, 5 to 7.5 cm in width, and 2.5 cm in thickness. Usually the left kidney is slightly larger than the right. The kidneys lie behind the parietal peritoneum. They are located on either side of the vertebral column and extend from the level of the last thoracic vertebra to the third lumbar vertebra—in short, the kidneys are located just above the waistline. Usually, the right kidney is a little lower than the left (Figure 24-1), presumably because the liver takes up some of the space above the right kidney.

A heavy cushion of fat normally encases each kidney and holds it up in position. Very thin individuals may suffer from ptosis (dropping) of one or both of these organs. Connective tissue (renal fasciae) anchors the kidneys to surrounding structures and helps to maintain their normal position.

External Structure

The medial surface of each kidney presents a concave notch called the **hilum.** Structures enter the kidneys through this notch just as they enter the lung through its hilum. A tough, white, fibrous capsule encases each kidney.

Figure 24-2 Coronal section through right kidney.

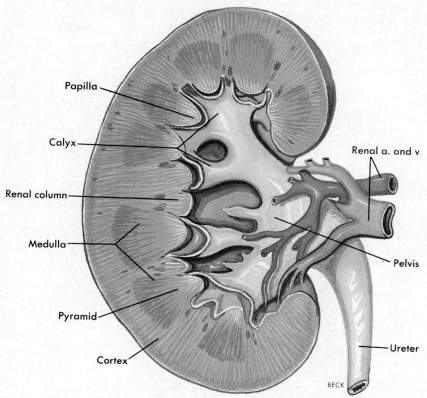

- Papilla
- Calyx
- Renal column
- Medulla
- Pyramid
- Cortex
- Renal a. and v.
- Pelvis
- Ureter

BECK

Table 24-1 Excretory organs of the body

Excretory organ	Substance excreted
Kidneys	Nitrogenous wastes (from protein catabolism)
	Toxins (e.g., from bacteria)
	Water (from ingestion and from catabolism)
	Electrolytes
Skin (sweat glands)	Water
	Electrolytes
	Small amounts of nitrogenous wastes
Lungs	Carbon dioxide (from catabolism)
	Water
Intestine	Wastes from digestion (e.g., cellulose, connective tissue)
	Some metabolic wastes (e.g., bile pigments; also salts of calcium and other heavy metals)

Internal Structure

Figure 24-2, a coronal section through the kidney, shows its internal structure. Identify the **cortex,** or outer portion, and the **medulla,** or inner portion. Observe the shape and location of the *pyramids*— a dozen or so of these triangular wedges make up the medulla. The *base,* or wide margin, of a pyramid faces out toward the cortex, and its narrow end, or *papilla,* juts into a division of the kidney's *pelvis* called a *calyx.* Identify these renal structures in Figure 24-2. Another internal structural feature is the dip of the cortex in between each of the two pyramids, forming the so-called *renal columns.* The pyramids have a striated appearance, as contrasted with the smooth texture of the cortical substance.

Microscopic Structure

Microscopic structures, named **nephrons** and numbering about 1.25 million per kidney, make up the bulk of kidney substance. The shape of a nephron is unique, unmistakable, and admirably suited to its functions. It resembles a tiny funnel with a very long stem—a most unusual stem. Instead of being straight, parts of it are highly convoluted. Look now

Figure 24-3 The nephron unit with its blood vessels. Blood flows through nephron vessels as follows: interlobular artery → afferent arteriole → glomerulus → efferent arteriole → peritubular capillaries (around the tubules) → venules → interlobular vein.

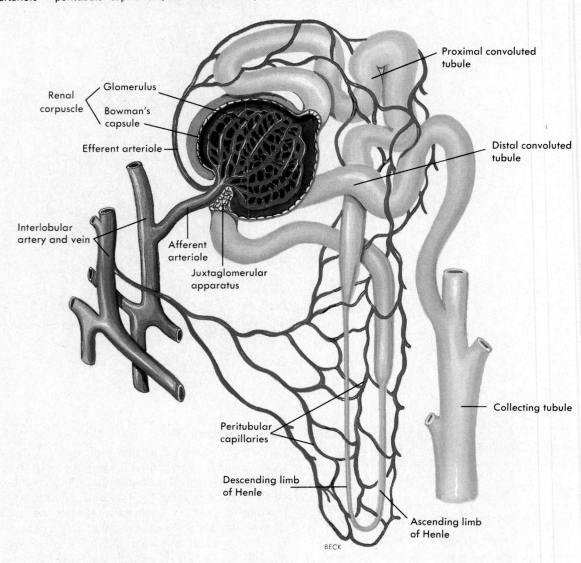

Renal corpuscle
Glomerulus
Bowman's capsule
Efferent arteriole
Interlobular artery and vein
Afferent arteriole
Juxtaglomerular apparatus
Proximal convoluted tubule
Distal convoluted tubule
Collecting tubule
Peritubular capillaries
Descending limb of Henle
Ascending limb of Henle
BECK

at Figure 24-3 to identify the parts of a nephron. Their names are:

1 **Bowman's capsule**
2 **Proximal tubules**
3 **Loop of Henle**
    a **Descending limb**
    b **Ascending limb**
4 **Distal tubules**
5 **Collecting tubules**

A brief description of each of these follows:

1 **Bowman's capsule**—cup-shaped mouth of a nephron. It is formed by two layers of epithelial cells, and between the two layers lies the capsular space. Whereas simple, squa-

mous epithelial cells form the outer or *parietal layer* of the capsule, modified epithelial cells—named *podocytes*—make up its inner or *visceral layer*. The scanning electron micrographs in Figures 24-4 and 24-5 reveal the odd shapes of the podocyte cell bodies. Notice that the primary branches extending out from the cell bodies divide into secondary and tertiary branches and then terminate in little "feet" called *pedicels*. The pedicels are packed so closely together that only narrow slits of space lie between them. They are called *filtration slits*. Delicate nets of fine fibrils, known as *filtration slit dia-*

Figure 24-4 Scanning electron micrograph of podocytes, the modified epithelial cells that compose the inner layer of Bowman's capsule. Primary branches *(PB)* extend out from the odd-shaped cell bodies *(CB)* of the podocytes. Primary branches divide to form secondary branches *(SB)*. (×5,605.)

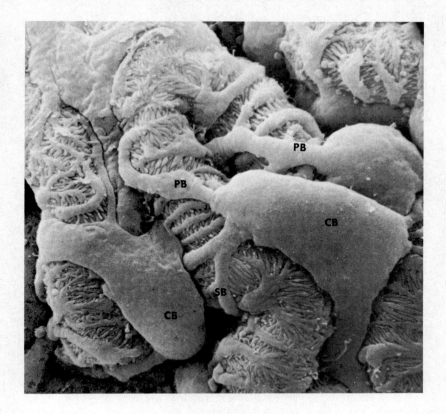

Figure 24-5 Note in this scanning electron micrograph the filtration slits *(FS)* between the pedicels *(Pe)* that are the terminal extensions of the podocytes' branches. *PB,* Primary branches; *CB,* cell bodies of podocytes; *SB,* secondary branches; *TB,* tertiary branches. (×9,345.)

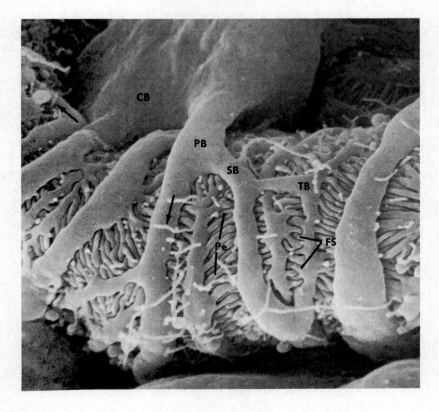

Figure 24-6 Photomicrograph of kidney showing renal corpuscle surrounded by profiles of sectioned renal tubules. (× 140.)

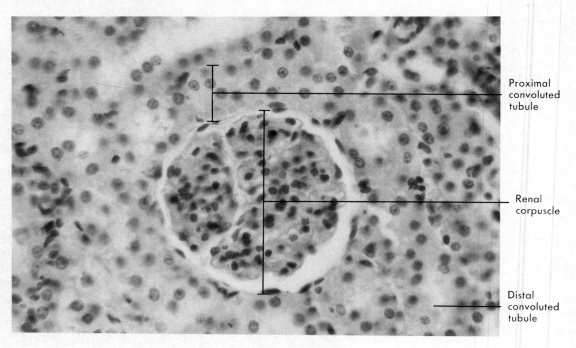

Proximal convoluted tubule

Renal corpuscle

Distal convoluted tubule

phragms, form permeable barriers across the slits.

Look back now at Figure 24-3. Pay particular attention to the structure fitted neatly into Bowman's capsule. It is a network of capillaries that has a special name—**glomerulus.** The glomerulus is probably the body's most famous capillary network and is surely one of its most important ones for survival. A glomerulus and its Bowman's capsule together are called a **renal corpuscle** (Figures 24-6 and 24-7). Renal corpuscles lie in the cortex of the kidney. Like all capillaries, glomeruli have thin, membranous walls composed of a single layer of endothelial cells. Many pores or *fenestrations* (Latin, "windows") are present in the glomerular endothelium. Observe them in Figure 24-8.

Between a glomerulus and its Bowman's capsule lies a **basement membrane** (basal lamina). It consists of a thin layer of fine fibrils embedded in a matrix of glycoprotein. The visceral layer of Bowman's capsule contacts the basement membrane by means of countless pedicels (end-processes of podocytes, the cells that compose the capsule's visceral layer). Together, the glomerular endothelium, the basement membrane, and the visceral layer of Bowman's capsule constitute the **glomerular-capsular membrane,** a structure well-suited to its function of filtration (see pp. 633-636 and 639).

2 **Proximal tubule**—the second part of a nephron but the first part of a renal tubule. As its name suggests, the proximal tubule is the segment proximal, or nearest, to Bowman's capsule. It follows a winding, convoluted course. Its wall consists of one layer of epithelial cells. These cells have a brush border facing the lumen of the tubule. Thousands of microvilli form the brush border and greatly increase its luminal surface area—a structural fact of importance to its function, as we shall see. Like renal corpuscles, proximal tubules are located in the cortex of the kidney (see Figure 24-12).

3 **Loop of Henle**—the segment of tubule just beyond the proximal tubule. It consists of a descending limb, a loop, and an ascending limb. Note in Figure 24-9 that the thin segment of the descending limb loops around to become the thick ascending limb.

Figure 24-7 The *B* identifies blood within capillaries inside the renal corpuscle (left half of figure). The right half shows segments of sectioned renal tubules. (× 900.)

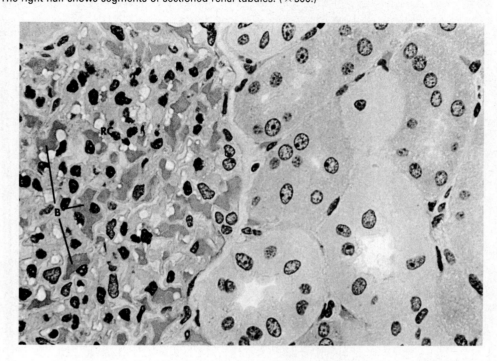

Figure 24-8 Scanning electron micrograph showing numerous fenestrations *(Fe)* in the endothelium of a glomerular capillary. Thickened branches *(Th)* of the endothelial cells contain no fenestrations. (× 17,525.)

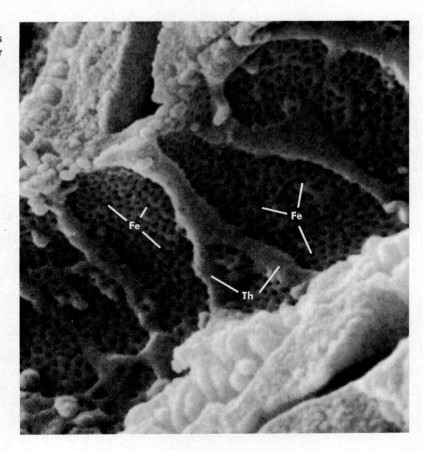

Figure 24-9 The nephron unit. Note different shapes of cells composing different parts of the tubule.

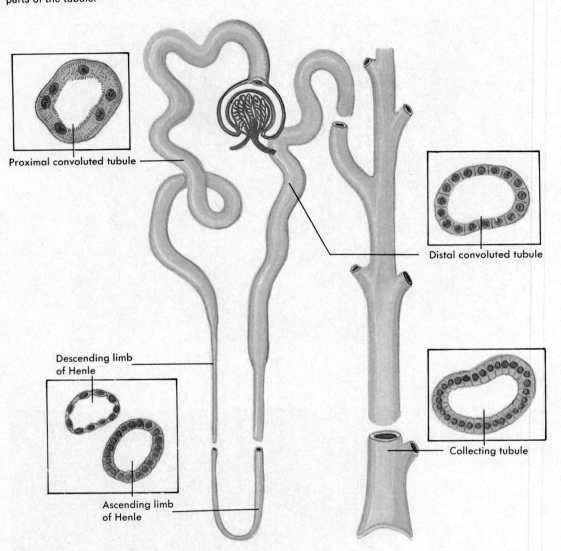

Proximal convoluted tubule

Distal convoluted tubule

Descending limb of Henle

Ascending limb of Henle

Collecting tubule

4 **Distal tubule**—a convoluted portion of the tubule located distally to Bowman's capsule.

5 **Collecting tubule**—a straight tubule joined by the distal tubules of several nephrons (Figure 24-10).

Collecting tubules join larger tubules, and all the larger collecting tubules of one renal pyramid converge to form one tube that opens at a renal papilla into one of the small calyces. Bowman's capsules and both convoluted tubules lie in the cortex of the kidney, whereas the loops of Henle and collecting tubules lie in its medulla.

Blood Vessels of Kidneys

The kidneys are highly vascular organs. Every minute about 1,200 ml of blood flows through them. Stated another way, approximately one fifth of all the blood pumped by the heart per minute goes to the kidneys. From this fact, one might guess, and correctly so, that the kidneys process the blood in some way before returning it to the general circulation. A large branch of the abdominal aorta—the renal artery—brings blood into each kidney. Between the pyramids of the kidney's medulla, the renal artery branches to form interlobular arteries that extend out toward the cortex, then arch over the bases of the pyramids to form the arcuate arteries.

From the arcuate arteries, interlobular arteries penetrate the cortex. Branches of the interlobular arteries are the afferent arterioles shown on p. 628. As you can see in Figure 24-3, afferent arterioles branch into the capillary networks called glomeruli. Recall that the usual direction of blood flow is: arteries→ arterioles → capillaries → venules → veins. This is not entirely true in the kidneys. Here, blood leaving the glomerular capillaries flows into efferent arterioles, not into venules. From the efferent arterioles, blood follows two courses. Some of it flows through peritubular capillaries before draining into venules. The rest enters vessels found only in the kidney, namely, the *vasa rectae*. These vessels, shaped like the loops of Henle, are located in the kidney's medulla, parallel to the loops of Henle. The vasa rectae play an important role in urine formation (p. 638). Identify in Figure 24-3 the afferent arteriole, glomerulus, efferent arteriole, and peritubular capillaries. Also find the *juxtaglomerular apparatus*, a small structure but an important one for maintaining homeostasis.

Functions

The function of the kidneys is to excrete urine, a life-preserving function because homeostasis depends on it. For example, the kidneys are the most important organs in the body for maintaining fluid and electrolyte and acid-base balance. They do this by varying the amounts of water and electrolytes leaving the blood in the urine so that they equal the amounts of these substances entering the blood from various avenues. Nitrogenous wastes from protein metabolism, notably urea, leave the blood by way of the kidneys.

Here are just a few of the blood constituents that cannot be held to their normal concentration ranges if the kidneys fail: sodium, potassium, chloride, and nitrogenous wastes from protein metabolism such as urea. In short, kidney failure means homeostasic failure, and, if not relieved, inevitable death.

In addition to forming urine, the kidneys also perform other important functions. They influence the rate of secretion of the hormones ADH and aldosterone and synthesize the hormone erythropoietin and prostaglandins of the E series.

How Kidneys Form Urine

The functional unit of a kidney is a nephron. It has two main parts—glomerulus and tubule—which form urine by means of three processes: filtration, reabsorption, and secretion. The following paragraphs discuss these processes. Figure 24-11 indicates amounts of water and certain solutes filtered and reabsorbed.

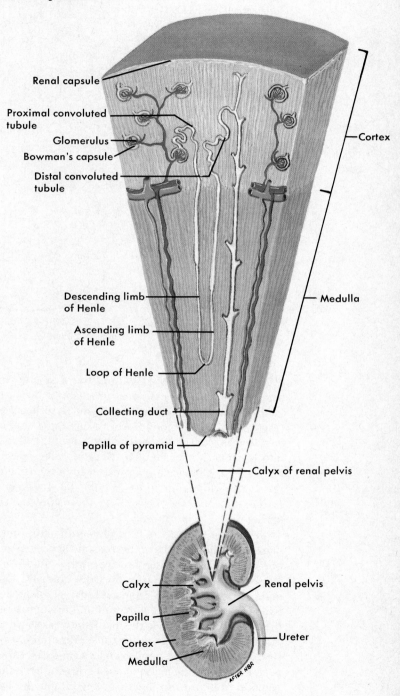

Figure 24-10 Magnified wedge shows that renal corpuscles (Bowman's capsules with invaginated glomeruli) and both proximal and distal convoluted tubules are located in cortex of kidney. Medulla contains loops of Henle and collecting tubules.

Renal capsule

Proximal convoluted tubule

Glomerulus

Bowman's capsule

Distal convoluted tubule

Cortex

Medulla

Descending limb of Henle

Ascending limb of Henle

Loop of Henle

Collecting duct

Papilla of pyramid

Calyx of renal pelvis

Calyx

Papilla

Cortex

Medulla

Renal pelvis

Ureter

AFTER NBR

Figure 24-11 Glomerular filtration and tubular reabsorption volumes. Note the enormous volume of water filtered out of glomerular blood per day—190 L, or many times the total volume of blood in the body. Only a small proportion of this, however, is excreted in the urine. More than 99% of it (189 L) is reabsorbed into tubular blood. One other substance shown is also filtered and reabsorbed in large amounts. Which one? Another is normally entirely reabsorbed. Which one? Somewhat more than half of the filtered amount of which substance is reabsorbed?

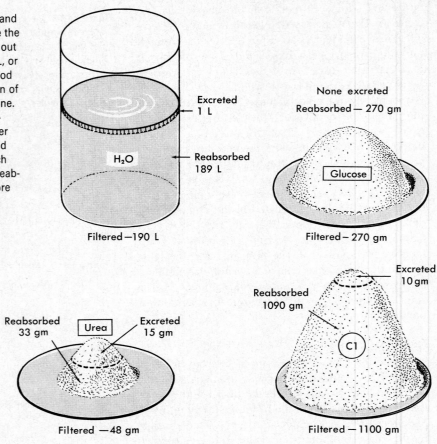

GLOMERULAR FILTRATION. Filtration, the first step in urine formation, is a physical process that occurs in the kidneys' 2.5 million or so renal corpuscles (Figure 24-12). As blood flows through glomerular capillaries, water and solutes filter out of the blood into Bowman's capsules. The filtration takes place through the glomerular-capsular membrane (made up of the glomerular endothelium, the basement membrane, and the visceral layer of Bowman's capsule).

Filtration occurs out of glomeruli into Bowman's capsules for the same reason that it occurs out of other capillaries into interstitial fluid—because of the existence of a pressure gradient. The main factor establishing the pressure gradient between the blood in the glomeruli and the filtrate in the Bowman's capsule is the hydrostatic pressure of glomerular blood. It tends to cause filtration out of the glomerular blood into Bowman's capsules. However, exerting force in the opposite direction are the osmotic pressure of glomerular blood and the hydrostatic pressure of the capsular filtrate. The net or effective filtration pressure, therefore, equals glomerular hydrostatic pressure minus the sum of glomerular osmotic pressure plus capsular hydrostatic

pressure. For example, assume the following pressures: glomerular hydrostatic pressure, 60 mm Hg; glomerular osmotic pressure, 32 mm Hg; capsular hydrostatic pressure, 18 mm Hg. The effective filtration pressure, using these figures, equals $60 - (32 + 18)$, or 10 mm Hg. An effective filtration pressure of 1 mm Hg, according to some investigators, produces a glomerular filtration rate of 12.5 ml per minute (including both kidneys). With an effective filtration pressure of 10 mm Hg, therefore, the glomerular filtration rate would be 125.0 ml per minute, a normal rate.

In some types of kidney disease, the permeability of the glomerular endothelium increases sufficiently to allow blood proteins to filter out into the capsule. Osmotic pressure then develops in the capsular filtrate. To determine how this would affect the effective filtration pressure in the glomeruli, first calculate the normal effective filtration pressure using the figures listed under "Normal" in Table 24-2 in the following formula:

$$\begin{matrix} \text{Effective} \\ \text{filtration} \\ \text{pressure} \end{matrix} = \begin{pmatrix} \text{Glomerular} & \text{Capsular} \\ \text{hydrostatic} + \text{osmotic} \\ \text{pressure} & \text{pressure} \end{pmatrix} - \begin{pmatrix} \text{Glomerular} & \text{Capsular} \\ \text{osmotic} & + \text{hydrostatic} \\ \text{pressure} & \text{pressure} \end{pmatrix}$$

Figure 24-12 Electron micrograph of a portion of a glomerulus and its Bowman's capsule. Observe the structures that compose the glomerular-capsular membrane: the glomerular endothelium (with many pores), the basement membrane, and the inner or visceral layer of Bowman's capsule. ($\times$ 13,600.)

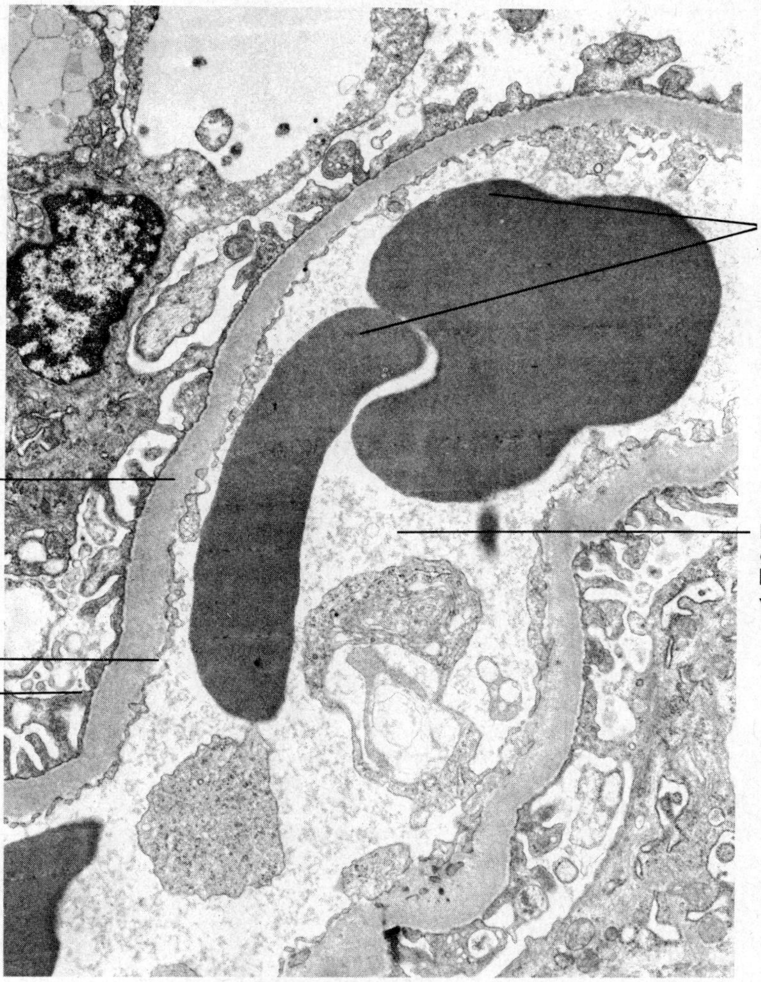

Red blood cells

Basement membrane

Lumen of glomerular blood vessel

Pore in glomerular endothelium

Visceral layer of Bowman's capsule

Now apply this formula using the figures under "Kidney disease" in Table 24-2. If you did the arithmetic correctly, you found an effective filtration pressure of 10 mm Hg with the normal figures and one of 15 mm Hg with the kidney disease figures. A change in the glomerular effective filtration pressure produces a similar change in the glomerular filtration rate. And, therefore, a loss of blood proteins in the urine increases not only the effective filtration pressure but also the glomerular filtration rate.

Filtration occurs more rapidly out of glomeruli than out of other tissue capillaries. One reason for this is a structural difference between the endothelium of glomeruli and that of tissue capillaries. Glomerular endothelium has many more pores (fenestrations) in it, so it is more permeable than tissue capillary endothelium. Another reason for more rapid glomerular filtration than tissue capillary filtration is that glomerular hydrostatic pressure is higher than tissue capillary pressure. The reason for this, briefly, is that the efferent arteriole has a smaller

Table 24-2 Normal and abnormal glomerulocapsular pressures

	Hydrostatic pressure	Osmotic pressure
Normal		
Glomerular blood	60 mm Hg	32 mm Hg
Capsular filtrate	18 mm Hg	—
Kidney disease		
Glomerular blood	60 mm Hg	32 mm Hg
Capsular filtrate	18 mm Hg	5 mm Hg

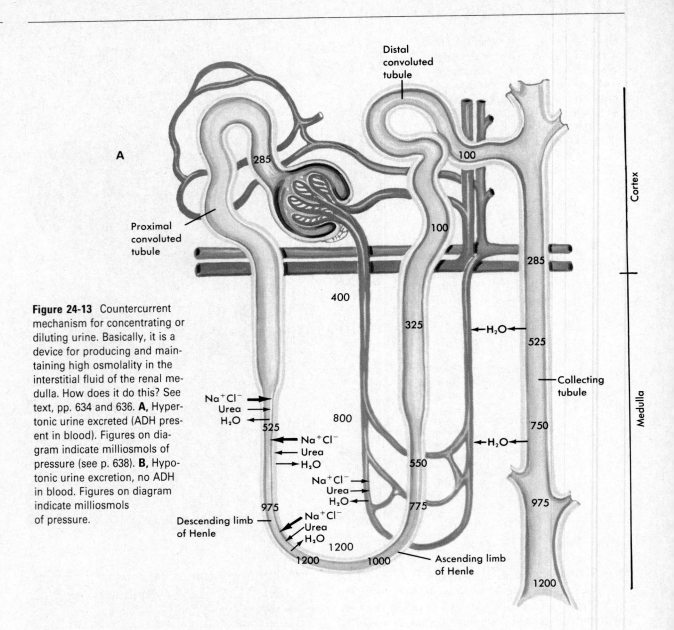

Figure 24-13 Countercurrent mechanism for concentrating or diluting urine. Basically, it is a device for producing and maintaining high osmolality in the interstitial fluid of the renal medulla. How does it do this? See text, pp. 634 and 636. **A,** Hypertonic urine excreted (ADH present in blood). Figures on diagram indicate milliosmols of pressure (see p. 638). **B,** Hypotonic urine excretion, no ADH in blood. Figures on diagram indicate milliosmols of pressure.

diameter than the afferent arteriole. Therefore, it offers more resistance to blood flow out of the glomerulus than venules offer to blood flow out of tissue capillaries.

The glomerular filtration rate can be altered by changes in the diameters of the afferent and efferent arterioles or by changes in the systemic blood pressure. For example, stress may lead to intense sympathetic stimulation of the arterioles with greater constriction of the afferent than of the efferent arteriole. Consequently, glomerular hydrostatic pressure falls. In severe stress, it may even drop to a level so low that the effective filtration pressure falls to zero. No glomerular filtration then occurs. The kidneys "shut down," or, in technical language, renal suppression occurs.

Glomerular hydrostatic pressure and filtration are directly related to systemic blood pressure. By this, we mean that a decrease in blood pressure tends to produce a decrease in both the glomerular pressure and the glomerular filtration rate. The converse is also true. However, when arterial pressure increases, a smaller increase in glomerular pressure follows because the afferent arterioles constrict. This decreases blood flow into the glomeruli and prevents a marked rise in either glomerular pressure or glomerular filtration. For instance, when the mean arterial pressure doubles, glomerular filtration reportedly increases only 15% to 20%.

REABSORPTION FROM PROXIMAL CONVOLUTED TUBULES. Reabsorption, the second step in urine formation,

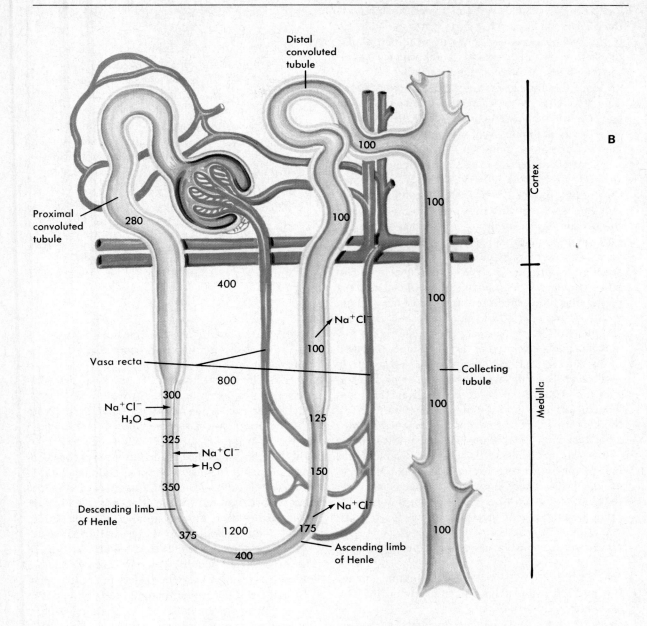

takes place by means of both passive and active transport mechanisms from all parts of the renal tubules. A major portion of water and electrolytes and (normally) all nutrients are, however, reabsorbed from proximal convoluted tubules.

Proximal tubules reabsorb sodium in this manner. First, sodium ions diffuse out of the filtrate in the tubules into the epithelial cells that form the tubule walls. To enter the cytoplasm of these cells, sodium diffuses through their brush border (microvilli) that faces the tubule lumen. Almost immediately, the sodium ions are actively transported through the basal and lateral borders of the epithelial cells and move on through the basement membrane and the endothelium of the peritubular capillaries into peritubular blood. As positive sodium ions move

out of the tubular cells, the sodium concentration of their cytoplasm becomes less than that of the tubular fluid. Also, their cytoplasm becomes momentarily electronegative to the tubular fluid. The effect of these two changes is to create an electrochemical gradient that causes further diffusion of sodium ions into epithelial cells of the tubules, followed by further active transport of sodium out of them into peritubular blood. The addition of sodium ions to peritubular blood makes it momentarily electropositive to tubular fluid. This causes an equal number of negative ions, notably chloride, to diffuse out the tubular fluid into the peritubular blood.

Proximal tubules reabsorb water by osmosis. The movement of sodium and other ions out of tubular fluid into peritubular blood creates, momentarily, a higher osmotic pressure in peritubular blood than in tubular fluid. Therefore, water, obeying the law of osmosis, moves rapidly into the peritubular blood to reestablish osmotic equilibrium between the two fluids. In short, sodium transport out of the proximal tubules causes water osmosis out of them. This is obligatory water reabsorption—obligatory because it is demanded by the law of osmosis.

Proximal tubules reabsorb nutrients, notably glucose and amino acids, into peritubular blood by a special type of active transport mechanism called *sodium co-transport*. A carrier (descriptively named the **sodium-glucose carrier**) binds to both sodium and glucose. It actively transports both substances through the brush border of a proximal tubule cell into the cell's interior. There, the substances dissociate from the carrier molecule and move over to the border of the cell that lies adjacent to a peritubular capillary. The substances move on into the peritubular blood by different mechanisms. Sodium is actively transported and glucose is transported by facilitated diffusion. Normally, this sodium co-transport mechanism returns to the peritubular blood all of the glucose that has filtered out of the glomeruli. So normally, no glucose is lost in the urine. If, however, blood glucose level exceeds a certain threshold amount (often around 150 mg per 100 ml of blood), the glucose transport mechanism cannot reabsorb all of it, and the excess remains in the urine. In other words, this mechanism has a maximum capacity for moving glucose molecules back into the blood. Occasionally, this maximum transfer capacity is greatly reduced, and glucose appears in the urine (glycosuria), even though the blood sugar level may be normal. This condition is known as renal diabetes or renal glycosuria. It is a congenital defect.

REABSORPTION FROM THE LOOP OF HENLE. Juxtamedullary Bowman's capsules (those located low in the kidney's cortex, near the medulla) have long loops of Henle. They descend deep into the medulla, and parallel to them lie loops of peritubular capillaries called the *vasa rectae* (Figure 24-13). Together the loop of Henle and the vasa rectae function as a **countercurrent mechanism.** Basically, this is a device for producing and maintaining a high solute concentration (osmolality) in the kidney's medulla. The first step in the countercurrent mechanism is active transport of chloride ions out of the filtrate in the ascending limb of the loop of Henle into the medullary interstitial fluid. Then, by diffusion, equal numbers of positive sodium ions follow the negative chloride ions into the medulla. Water, however, does not follow for a good reason—because the ascending limb is virtually impermeable to water. The addition of salt, but not water, to the medullary interstitial fluid necessarily increases its solute concen-

Osmolality is the osmotic pressure of a solution expressed as the number of osmols of pressure per kilogram of water. One osmol of pressure is produced when 1 gm molecular weight of a nonionizing substance is dissolved in 1 kg of water. An ionizing substance produces 1 osmol of pressure when 1 gm molecular weight of the substance, divided by the number of ions it forms per molecule, is dissolved in 1 kg of water. Example: Sodium chloride forms 2 ions per molecule and has a molecular weight of 58. Therefore 29 (58/2) gm of sodium chloride dissolved in 1 kg of water produces 1 osmol of pressure.

tration and osmolality. At the same time, it decreases the solute concentration and osmolality of the filtrate in the ascending limb.

Examine the numbers in Figure 24-13. They indicate solute concentration or osmolality of the medullary interstitial fluid and of the tubular filtrate. Note that the osmolality of the medullary interstitial fluid increases from 400 to 1,200 milliosmols per liter, whereas the osmolality of the filtrate in the ascending limb decreases to 100. Some sodium chloride immediately diffuses out of the medullary interstitial fluid into the filtrate in the descending limb. But, because the filtrate moves rapidly around the bottom of the loop of Henle into the ascending limb, the sodium chloride is soon transported back out of the filtrate into the medulla. Thus, in essence, sodium chloride is trapped in the medullary interstitial fluid. By taking salt out of the tubule filtrate, adding it to the medullary interstitial fluid, and trapping it there, the countercurrent mechanism produces and maintains a high osmolality in the medullary interstitial fluid. The vasa recta arrangement of blood vessels in the kidney's medulla plays an essential part in trapping solutes in the medulla. Because blood flows sluggishly through these vessels, the solutes are so slowly removed from the medullary interstitial fluid that its osmolality remains high.

The countercurrent mechanism is an extremely important one because, as the next section describes, it enables the kidney to excrete urine that is hypertonic to blood. By so doing, it conserves water for the body. Vertebrates whose kidneys have no loops of Henle cannot perform this function. Conserving body water can be life-saving—for example, if blood volume decreases precipitously, such as following hemorrhage.

REABSORPTION FROM DISTAL AND COLLECTING TUBULES. Distal tubules, like proximal tubules, reabsorb sodium, but in much smaller amounts. (Proximal tubules, under usual conditions, reabsorb about two thirds and distal tubules reabsorb about one tenth of the amount of sodium filtered from the glomeruli.) Distal tubules reabsorb water when antidiuretic hormone (ADH) is present in blood. ADH acts on the kidneys in some way, not well understood, to make their distal and collecting tubules permeable to water. Therefore water osmoses out of the fluid in these tubules into the hypertonic interstitial fluid in the renal medulla and on into the blood in the vasa recta. This reabsorption of water from the distal and collecting tubules continues until the osmolality of the urine in them equilibrates with that of the medullary interstitial fluid (about 1,200 milliosmoles). A highly concentrated urine, one hypertonic to blood, is therefore excreted in the presence of ADH. Also, a smaller volume of urine is excreted when ADH is present—a fact implied by its name *antidiuretic hormone;* antidiuretic means *opposed to* the production of a large volume of urine.

When no ADH is present in the blood, both distal and collecting tubules are practically impermeable to water. Little or no water therefore is reabsorbed from them and the osmolality of the fluid in them remains the same as it was when the fluid left the ascending limb of the loop of Henle (100 milliosmoles, according to Figure 24-13). In short, with no ADH in the blood, a very dilute urine is excreted. It is strongly hypotonic to blood (normal blood osmolality about 300 milliosmoles).

TUBULAR SECRETION. In addition to reabsorption, tubule cells also secrete certain substances. Tubular secretion (or tubular excretion, as it is also called) means the movement of substances out of the blood into the tubular fluid. Tubular reabsorption, you recall, means the movement of substances in the opposite direction, that is, out of the tubular fluid into the blood. Distal and collecting tubules secrete potassium, hydrogen, and ammonium ions. They actively transport potassium ions or hydrogen ions out of the blood into the tubule filtrate in exchange for sodium ions that diffuse back into the blood (discussed in Chapter 26). Potassium secretion increases when blood aldosterone concentration increases. Hydrogen ion secretion increases when blood hydrogen ion concentration increases. Ammonium ions are secreted into the tubule filtrate by diffusing out of the distal tubule cells where they are synthesized. Tubule cells also secrete certain drugs, for example, penicillin and para-aminohippuric acid (PAH). Table 24-3 and Figure 24-14 summarize the functions of the different parts of the nephron in forming urine.

Regulation of Urine Volume

Several factors regulate the volume of urine produced. Paramount among them is the presence or absence of ADH in the blood. As stated earlier, ADH

Table 24-3 Functions of different parts of nephron in urine formation

Part of nephron	Function	Substance moved
Glomerulus	Filtration	Water
		All solutes except colloids such as blood proteins
Proximal tubule	Reabsorption by active transport	Na^+ and probably some other ions; nutrients—glucose and amino acids
	Reabsorption by diffusion (secondary to active transport)	Cl^-, HCO_3^-, and probably some other ions; also about 50% of urea
	Obligatory water reabsorption by osmosis	Water
Loop of Henle 　Descending limb	Reabsorption by diffusion	NaCl
Ascending limb	Reabsorption by active transport into interstitial fluid of renal medulla	NaCl (not followed by water reabsorption)
Distal and collecting tubules	Reabsorption by active transport Facultative water reabsorption by osmosis (ADH-controlled) Secretion by diffusion Secretion by active transport	Na^+ and probably some other ions Water Ammonia K^+, H^+, and some drugs

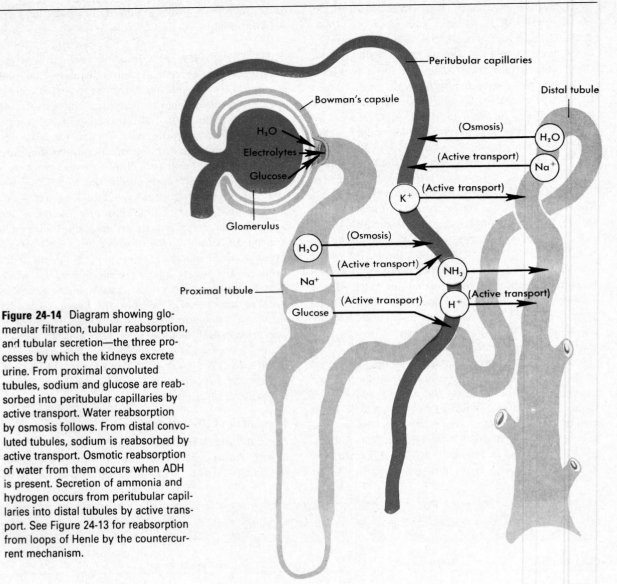

Figure 24-14 Diagram showing glomerular filtration, tubular reabsorption, and tubular secretion—the three processes by which the kidneys excrete urine. From proximal convoluted tubules, sodium and glucose are reabsorbed into peritubular capillaries by active transport. Water reabsorption by osmosis follows. From distal convoluted tubules, sodium is reabsorbed by active transport. Osmotic reabsorption of water from them occurs when ADH is present. Secretion of ammonia and hydrogen occurs from peritubular capillaries into distal tubules by active transport. See Figure 24-13 for reabsorption from loops of Henle by the countercurrent mechanism.

makes the distal and collecting tubules permeable to water. This necessarily increases water reabsorption from these tubules, which, in turn, decreases urine volume. Another hormone that tends to decrease urine volume is aldosterone from the adrenal cortex. It increases distal tubule absorption of sodium and other ions, and this leads to increased water reabsorption by osmosis and thereby tends to decrease urine volume.

Urine volume generally relates directly to extracellular fluid (ECF) volume. If ECF volume decreases, urine volume also decreases. Conversely, if ECF volume increases, urine volume increases. This presumably comes about indirectly by the change in ECF volume acting to stimulate or inhibit ADH secretion. The exact mechanism by which it accomplishes this is still unproved. It may operate through its effect on ECF sodium concentration. When ECF

volume decreases, its sodium concentration tends to increase, and this change triggers ADH secretion (Figure 24-15), followed by a decrease in urine volume. On the other hand, when ECF volume increases, its sodium concentration tends to decrease and to inhibit ADH secretion, followed by an increase in urine volume. The common experience of increased output following rapid ingestion of a large amount of fluid attests to the fact that an increased ECF volume is soon followed by increased urinary output. Oliguria, or even anuria, in dehydrated patients gives evidence that decreased urinary output follows a decrease in ECF volume.

Urine volume relates directly to the total amount of solutes excreted in the urine—the more solutes, the more urine. Higher solute concentration in urine osmotically draws more water into renal tubules and thus tends to increase urine volume. Probably the

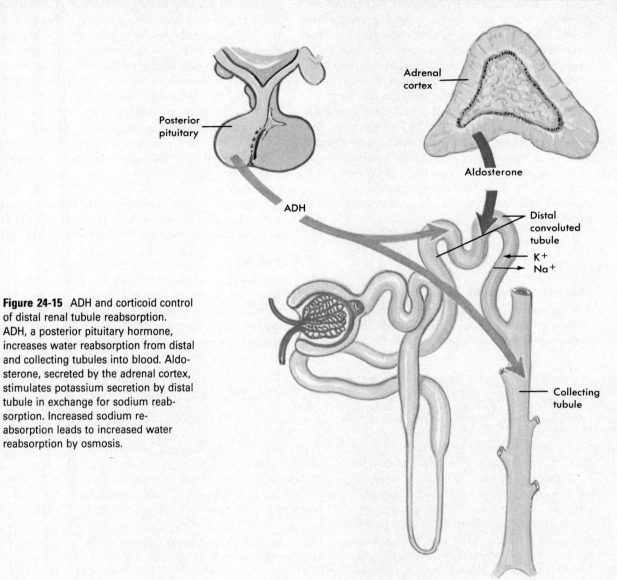

Figure 24-15 ADH and corticoid control of distal renal tubule reabsorption. ADH, a posterior pituitary hormone, increases water reabsorption from distal and collecting tubules into blood. Aldosterone, secreted by the adrenal cortex, stimulates potassium secretion by distal tubule in exchange for sodium reabsorption. Increased sodium reabsorption leads to increased water reabsorption by osmosis.

best known example of this occurs in untreated diabetes mellitus. The symptom that often brings an unknown diabetic to a physician is the voiding of abnormally large amounts of urine. Excess glucose "spills over" into urine, increasing its solute concentration and leading to diuresis.

Urine volume is not normally altered by changes in the glomerular filtration rate. Why? Because the glomerular filtration rate remains remarkably constant. It usually does not change, even in the presence of changes in renal blood flow and in glomerular hydrostatic pressure. Under certain pathological conditions, however, the glomerular filtration rate may change enough to alter urine volume.

In summary, the volume of urine produced is regulated, not by changes in the amount of water filtered from the glomeruli, but by changes in the amount reabsorbed from the distal and collecting tubules. Chief factors that regulate water reabsorption are the posterior pituitary hormone ADH and the adrenal cortex hormone aldosterone.

Tests Used to Evaluate Kidney Function

The test most commonly used by physicians to evaluate kidney function of patients is probably the measurement of serum creatinine. Normally, no significant changes occur in the amount of creatinine in blood serum because it is determined by a factor that does not readily change, namely, skeletal muscle mass. Elevation of the serum creatinine (above 1.5 mg per 100 ml), therefore, is considered a reliable indication of depressed renal function.

Influence of Kidneys on Blood Pressure

Clinical observation and animal experiments have established the fact that destruction of a large proportion of total kidney tissue usually results in the development of hypertension. This happens frequently, for example, in patients who have severe renal arteriosclerosis or glomerulonephritis. Many experiments have been performed and various theories devised to explain the mechanism responsible for "renal hypertension." The initiating factor is believed to be ischemia of the kidneys (decreased blood flow through them). When this occurs, the cells of the juxtaglomerular apparatus secrete renin. Renin is a proteolytic enzyme that hydrolyzes one of the plasma proteins to produce angiotensin (see Figure 25-5, p. 657). One effect of angiotensin is to increase blood pressure by constricting arterioles.

URETERS

Structure and Location

The two ureters are tubes from 10 to 12 inches long. At their widest point, they measure less than ½ inch in diameter. They lie behind the parietal peritoneum and extend from the kidneys to the posterior surface of the bladder. As the upper end of each ureter enters the kidney, it enlarges into a funnel-shaped basin named the *renal pelvis.* The pelvis expands into several branches called *calyces.* Each calyx contains a renal papilla. As urine is secreted, it drops out of the collecting tubules, whose openings are in the papillae, into the calyces. It then moves into the renal pelvis and down the ureters into the bladder.

The walls of the ureters are composed of three coats: a lining coat of mucous membrane, a middle coat of two layers of smooth muscle, and an outer fibrous coat.

The ureters enter the bladder through an oblique tunnel that functions as a valve, preventing backflow of urine into the ureters during bladder contraction.

Function

The ureters, together with their expanded upper portions, the pelves and calyces, collect the urine as it forms and drain it into the bladder. Peristaltic waves (about one to five per minute) force the urine down the ureters and into the bladder.

GOUT

Gout is a condition characterized by excessive levels of uric acid in the blood. The body produces uric acid from metabolism of ingested purines in the diet (especially in glandular meats) or from purine turnover in our own bodies. Blood uric acid levels may become elevated because of increased dietary intake, excessive production in the body, or because of defective excretion by the kidneys. Because uric acid is not as soluble as many waste products, it tends to be deposited in the joints and tissues of the body if blood levels become elevated. The figure below shows the results of uric acid deposits in the soft tissues and joints of the hand in a patient with gout. Deposit in the kidneys produces uric acid stones or calculi.

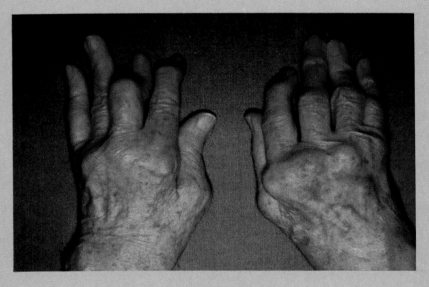

Figure 24-16 The male urinary bladder cut to show the interior.

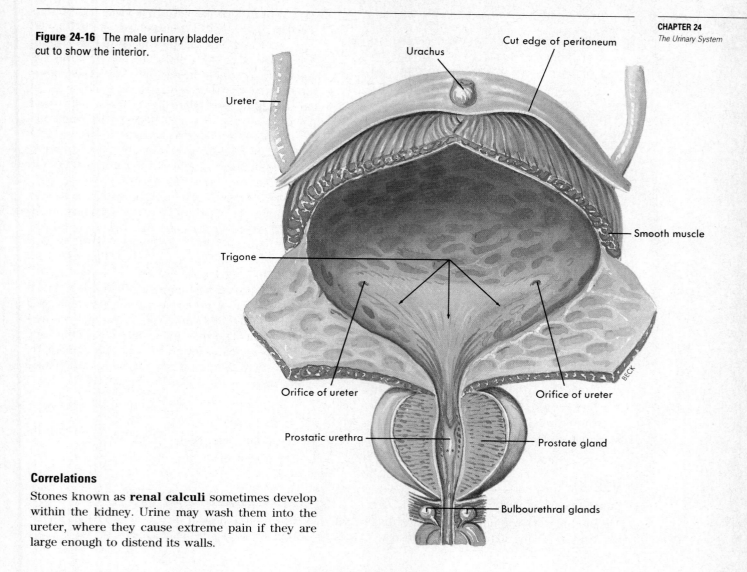

Urachus

Cut edge of peritoneum

Ureter

Smooth muscle

Trigone

Orifice of ureter

Orifice of ureter

Prostatic urethra

Prostate gland

Bulbourethral glands

Correlations

Stones known as **renal calculi** sometimes develop within the kidney. Urine may wash them into the ureter, where they cause extreme pain if they are large enough to distend its walls.

BLADDER

Structure and Location

The urinary bladder is a collapsible bag located directly behind the symphysis pubis. It lies below the parietal peritoneum, which covers only its superior surface. Smooth muscle fibers fashion the wall of the bladder. Some extend lengthwise of it; some, obliquely; and some, more or less circularly. Oblique, intersecting fibers at the sides of the bladder form the detrusor muscle. Mucous membrane, arranged in rugae, forms the bladder lining. Because of the rugae and the elasticity of its walls, the bladder is capable of considerable distention, although its capacity varies greatly with individuals. There are three openings in the floor of the bladder—two from the ureters and one into the urethra. The ureter openings lie at the posterior corners of the triangular-shaped floor (the trigone), and the urethral opening at the anterior and lower corner (Figure 24-16).

Functions

The bladder performs two functions.

1 It serves as a reservoir for urine before it leaves the body.

2 Aided by the urethra, it expels urine from the body.

The mechanism for voiding urine starts with voluntary relaxation of the external sphincter muscle of the bladder. In rapid succession, reflex contraction of linear smooth muscle fibers along the urethra and then of the detrusor muscle follow. The combination of these events squeezes urine out of the bladder. Parasympathetic fibers transmit the impulses that cause contractions of the bladder and relaxation of the internal sphincter. Voluntary contraction of the external sphincter to prevent or terminate micturition is learned. Voluntary control of micturition is possible only if the nerves supplying the bladder and urethra, the projection tracts of the cord and brain, and the motor area of the cerebrum

Occasionally an individual is unable to void even though the bladder contains an excessive amount of urine. This condition is known as **retention.** It often follows pelvic operations and childbirth. **Catheterization** (introduction of a rubber tube through the urethra into the bladder to remove urine) is used to relieve the discomfort accompanying retention. A more serious complication, which is also characterized by the inability to void, is called **suppression.** In this condition, the patient cannot void because the kidneys are not secreting any urine, and, therefore, the bladder is empty. Catheterization, of course, gives no relief for this condition.

are all intact. Injury to any of these parts of the nervous system, by a cerebral hemorrhage or cord injury, for example, results in involuntary emptying of the bladder at intervals. Involuntary micturition is called *incontinence.* In the average bladder, 250 ml of urine will cause a moderately distended sensation and, therefore, the desire to void.

URETHRA

Structure and Location

The urethra is a small tube leading from the floor of the bladder to the exterior. In the female, it lies directly behind the symphysis pubis and anterior to the vagina. It extends up, in, and back for a distance of about 1 to 1½ inches. The male urethra follows a tortuous course for a distance of approximately 8 inches. Immediately below the bladder, it passes through the center of the prostate gland, then between two sheets of white fibrous tissue connecting the pubic bones, and then through the penis, the external male reproductive organ. These three parts of the urethra are known, respectively, as the prostatic portion, the membranous portion, and the cavernous portion.

The opening of the urethra to the exterior is known as the **urinary meatus.** Mucous membrane lines the urethra and the rest of the urinary tract.

Functions

Since it is the terminal portion of the urinary tract, the urethra serves as the passageway for eliminating urine from the body. In addition, the male urethra is the terminal portion of the reproductive tract and serves as the passageway for eliminating the reproductive fluid (semen) from the body. The female urethra serves only the urinary tract.

URINE

Physical Characteristics

The physical characteristics of normal urine are listed in Table 24-4.

Chemical Composition

Urine is approximately 95% water, in which are dissolved several kinds of substances, the most impor-

Table 24-4 Physical characteristics of normal urine

Amount (24 hours)	Three pints (1,500 ml) but varies greatly according to fluid intake, amount of perspiration, and several other factors
Clearness	Transparent or clear; on standing, becomes cloudy
Color	Amber or straw-colored; varies according to amount voided—less voided, darker the color, usually; diet also may change color, for example, reddish color from beets
Odor	"Characteristic"; on standing, develops ammonia odor from formation of ammonium carbonate
Specific gravity	1.015 to 1.020; highest in morning specimen
Reaction	Acid but may become alkaline if diet is largely vegetables; high-protein diet increases acidity; stale urine has alkaline reaction from decomposition of urea forming ammonium carbonate; normal range for urine pH 4.8 to 7.5; average, about 6; rarely becomes more acid than 4.5 or more alkaline than 8

THE ARTIFICIAL KIDNEY

The artificial kidney is a mechanical device that uses the principle of dialysis to remove or separate waste products from the blood. In the event of kidney failure, the process, appropriately called **hemodialysis** (Greek *haima*—"blood"; *alysis*—"separate") is literally a reprieve from death for the patient. During a hemodialysis treatment, a semipermeable membrane is used to separate large (nondiffusible) particles such as blood cells from small (diffusible) ones such as urea and other wastes. The figure shows blood from the radial artery passing through a porous (semipermeable) cellophane tube that is housed in a tanklike container. The tube is surrounded by a bath or dialyzing solution containing varying concentrations of electrolytes and other chemicals. The pores in the membrane are small and allow only very small molecules, such as urea, to escape into the surrounding fluid. Larger molecules and blood cells cannot escape and are returned through the tube to reenter the patient via a wrist vein. By constantly replacing the bath solution in the dialysis tank with freshly mixed solution, levels of waste materials can be kept at low levels. As a result, wastes such as urea in the blood will rapidly pass into the surrounding wash solution. For a patient with complete kidney failure, two or three hemodialysis treatments a week are required. New dialysis methods are now being developed, and dramatic advances in treatment are expected in the next few years.

One relatively new technique used in renal failure is called **continuous ambulatory peritoneal dialysis** *(CAPD)*. In this procedure, 1 to 3 L of sterile dialysis fluid is introduced directly into the peritoneal cavity through an opening in the abdominal wall. Peritoneal membranes in the abdominal cavity serve to transfer waste products from the blood into the dialysis fluid, which is then drained back into a plastic container after about 2 hours. This technique is less expensive than hemodialysis and does not require the use of complex equipment.

Simplified diagram of an artificial kidney.

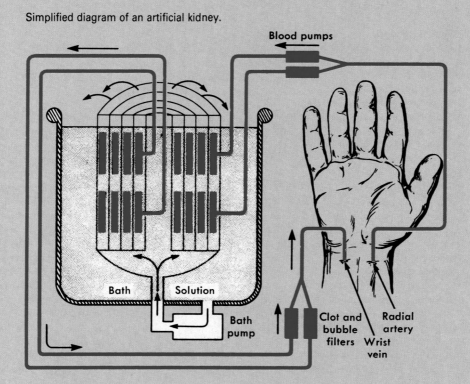

tant of which are discussed here:

1 **Nitrogenous wastes** from protein metabolism—such as urea (most abundant solute in urine), uric acid, ammonia, and creatinine

2 **Electrolytes**—mainly the following ions: sodium, potassium, ammonium, chloride, bicarbonate, phosphate, and sulfate; amounts and kinds of minerals vary with diet and other factors

3 **Toxins**—during disease, bacterial poisons leave the body in the urine—an important reason for "forcing fluids" on patients suffering with infectious diseases, so as to dilute the toxins that might damage the kidney cells if they were eliminated in a concentrated form

4 **Pigments**

5 **Hormones**

6 Various **abnormal constituents** sometimes found in urine—such as glucose, albumin, blood, casts, or calculi

Definitions

glycosuria or **glucosuria** Sugar (glucose) in the urine
hematuria Blood in the urine
pyuria Pus in the urine
casts Substances, such as mucus, that harden and form molds inside the uriniferous tubules and then are washed out into the urine; microscopic in size
dysuria Painful urination
polyuria Unusually large amounts of urine
oliguria Scanty urine
anuria Absence of urine

Outline Summary

KIDNEYS

A Size, shape, and location
1 11.25 cm × 5 to 7.5 cm × 2.5 cm
2 Shaped like lima beans
3 Lie against posterior abdominal wall, behind peritoneum, at level of last thoracic and first lumbar vertebrae; right kidney slightly lower than left

B External structure
1 Hilum, concave notch on mesial surface
2 Enveloping capsule of white fibrous tissue

C Internal structure
1 Outer layer called *cortex*
2 Inner portion called *medulla*
3 Renal pyramids—triangular wedges of medullary substance, apices of which are called *papillae*
4 Renal columns—inward extensions of cortex between pyramids

D Microscopic structure
1 Nephron—microscopic functional unit of kidneys; consists of renal corpuscle (Bowman's capsule, in which glomerulus, a capillary network, is invaginated) and renal tubules
2 Bowman's capsule—cup-shaped mouth of nephron; consists of:
   a Parietal layer of simple, squamous epithelial cells
   b Visceral layer of modified epithelial cells (podocytes); pedicels, footlike terminations of podocytes, lie in contact with basement membrane; numerous filtration slits between pedicels; visceral layer of Bowman's capsule, basement membrane, and endothelium of glomerulus constitute glomerular-capsular membrane

3 Renal tubules—consist of proximal convoluted tubule, loop of Henle, distal convoluted tubule, and collecting tubule

E Functions
1 Excrete urine, by which various toxins and metabolic wastes excreted and homeostasis of blood composition and volume of blood maintained
2 Influence ADH and aldosterone secretion
3 Synthesize erythropoietin and prostaglandins (E series)
4 Influence blood pressure

F How kidneys form urine (see Table 24-3 and Figure 24-14)
1 Filtration of water and true solutes out of glomerular blood through glomerular-capsule membrane (glomerular endothelium, basement membrane, and visceral layer of capsule) into Bowman's capsule; see p. 634 for formula for computing effective filtration pressure, the force that produces glomerular filtration
2 Reabsorption of most of water and part of solutes from tubule filtrate into peritubular blood
   a Proximal tubules—active transport of sodium ions followed by passive diffusion of chloride ions; osmosis of water out of proximal tubule; glucose reabsorption by sodium co-transport mechanism; normally, major portion of water and electrolytes and all glucose and other nutrients are reabsorbed from proximal tubule
   b Loop of Henle—serves as a countercurrent mechanism, a device that creates and maintains a high osmolality in the renal medulla's interstitial fluid, a condition essential to the

Outline Summary—cont'd

formation of either a concentrated or dilute urine

   c Distal tubules and collecting ducts—permeable to water in presence of ADH, permitting osmosis of water out of distal tubules and collecting ducts until tubule filtrate equilibrates with high osmolality in medullary interstitial fluid; this forms concentrated urine; in absence of ADH, distal tubules and collecting ducts almost completely impermeable to water, so water not reabsorbed from them and dilute urine excreted

  3 Tubular secretion—distal and collecting tubules secrete potassium, hydrogen, and ammonium ions from blood into tubule filtrate; sodium ions reabsorbed in exchange

G Regulation of urine volume by factors that change

  1 Amount of water reabsorbed from tubular filtrate into blood—most important determinant of amount of urine formed; posterior pituitary ADH increases water reabsorption from distal and collecting tubules; corticoids (especially aldosterone) increase sodium reabsorption and therefore also increase water reabsorption

  2 Rate of filtration from glomeruli—quite constant at about 125 ml per minute except in certain kidney diseases

URETERS

A Location and structure

  1 Lie retroperitoneally

  2 Extend from kidneys to posterior part of bladder floor

  3 Ureter expands as it enters kidney, becoming renal pelvis, which is subdivided into calyces, each of which contains renal papilla

  4 Walls of smooth muscle with mucous lining and fibrous outer coat

B Function—collect urine and drain it into bladder

BLADDER

A Structure and location

  1 Collapsible bag of smooth muscle lined with mucosa

  2 Lies behind symphysis pubica, below parietal peritoneum

  3 Three openings—one into urethra and two into ureters

B Functions

  1 Reservoir for urine

  2 Expels urine from body by way of urethra, called micturition, urination, or voiding; retention, inability to expel urine from bladder; suppression, failure of kidneys to form urine

URETHRA

A Structure and location

  1 Musculomembranous tube lined with mucosa

  2 Lies behind symphysis, in front of vagina in female

  3 Extends through prostate gland, fibrous sheet, and penis in male

  4 Opening to exterior called urinary meatus

B Functions

  1 Female—passageway for expulsion of urine from body

  2 Male—passageway for expulsion of urine and of reproductive fluid (semen)

URINE

A Physical characteristics (see Table 24-4)

B Chemical composition—consists of approximately 95% water in which are dissolved:

  1 Wastes from protein metabolism (urea, uric acid, creatinine, etc.)

  2 Electrolytes (sodium chloride main one, but various others, according to diet)

  3 Toxins—(e.g., from bacteria)

  4 Pigments

  5 Sex hormones

  6 Abnormal constituents—(e.g., glucose, albumin, blood, casts, and calculi)

C Definitions

  1 Glycosuria—sugar in urine

  2 Hematuria—blood in urine

  3 Pyuria—pus in urine

  4 Casts—microscopic bits of substances that harden and form molds inside tubules

  5 Dysuria—painful urination

  6 Polyuria—excessive amounts of urine

  7 Oliguria—scanty urine

  8 Anuria—absence of urine

Review Questions

1 Name four excretory organs.
2 Which organs eliminate wastes of protein metabolism? Of digestion? Of carbohydrate and fat metabolism?
3 Name, locate, and give main function(s) of each organ of the urinary system.
4 How far and in which direction must a catheter be inserted to reach the bladder in the female? In the male?
5 Describe the microscopic structure of the kidney.
6 Describe the mechanism of urine formation, relating each step to the part of the nephron that performs it.

Situation: An artificial kidney consists of a device in which blood flows directly from a patient's body through cellophane tubing immersed in a dialyzing fluid that contains prescribed amounts of various dialytic electrolytes and other substances. The following two columns show the composition of one patient's blood and of the dialyzing fluid used for his treatment with the artificial kidney.

	Blood plasma (in coiled tube) mEq/L	Dialyzing fluid (around coiled tube) mEq/L
Na^+	142	126
K^+	5	5
Ca^{++}	5	0
Mg^{++}	3	0
Cl^-	103	110
HCO_3^-	27	25
$HPO_4^=$	2	0
$SO_4^=$	1	0
	mg/100 ml	mg/100 ml
Glucose	100	1,750
Urea	26	0
Uric acid	4	0
Creatinine	1	0

7 What body structures do you think the cellophane tube substitutes for?
8 Which, if any, of the foregoing substances diffuse out of blood into the dialyzing fluid? Give your reasons.
9 Which, if any, of these substances diffuse into the blood from the dialyzing fluid? Give your reasons.
10 Net diffusion of which, if any, of these substances does not occur through the cellophane membrane in either direction? Give your reasons.
11 What reasons can you see for having the dialyzing fluid contain the concentration of each substance given? What does it accomplish?

Situation: Results of a blood urea clearance test indicate that the glomerular filtration rate of a patient who has had a severe hemorrhage is less than 50% of normal.

12 Explain the mechanism responsible for the drop in the glomerular filtration rate.
13 Do you consider this a homeostatic mechanism? Does it serve a useful purpose? If so, what purpose?
14 What is the normal glomerular filtration rate?
15 What would you expect to be the volume of urine this patient would excrete—normal, polyuria, oliguria?
16 Define the following terms briefly: *Bowman's capsule, calculi, calyces, casts, cystitis, dysuria, glomerulus, glycosuria, hematuria, incontinence, nephritis, oliguria, polyuria, ptosis, pyelitis, renal capsule, renal cortex, renal hilum, renal medulla, renal papilla, renal pelvis, renal pyramids, retention, suppression.*

25

FLUID AND
ELECTROLYTE BALANCE

OBJECTIVES

After you have completed this chapter, you should
be able to:

1 Define the phrase *fluid and electrolyte balance.*
2 Discuss total body water content in terms of
   body weight, sex and age.
3 List, describe, and compare the body fluid
   compartments and their subdivisions.
4 Discuss avenues by which water enters and
   leaves the body.
5 Discuss the objectives of fluid and electrolyte
   therapy in the practice of medicine.
6 Contrast the chemical composition of the
   three fluid compartments.
7 Explain the mechanisms that maintain homeostasis of the body fluid compartments and of
   total body fluid volume.
8 Explain the mechanism of water and electrolyte movement between plasma and interstitial
   fluid.
9 Discuss edema and the mechanisms of edema
   formation.
10 Discuss dehydration.

KEY TERMS

Dehydration (de-hi-DRA-shun)

Edema (e-DE-mah)

Extracellular (ec-strah-SELL-u-lar)

Hypervolemia (hi-per-vo-LE-me-ah)

Hypochloremia (hi-po-klo-RE-me-ah)

Hypopotassemia (hi-po-po-tah-SE-me-ah)

Interstitial (in-ter-STISH-al)

Intracellular (in-trah-SELL-u-lar)

Milliequivalent (mil-e-KWIV-ah-lent)

Osmoreceptors (oz-mo-re-SEP-tors)

Parenteral (pah-REN-ter-al)

The phrase **fluid and electrolyte balance** implies homeostasis or constancy of body fluid and electrolyte levels. It means that both the amount and distribution of body fluids and electrolytes is normal and constant. In order for homeostasis to be maintained body "input" of water and electrolytes must be balanced by "output." If water and electrolytes in excess of requirements enter the body they must be selectively eliminated and, should excess losses occur, prompt replacement is critical. The volume of fluid and the electrolyte levels inside the cells, in the interstitial spaces, and in the blood vessels all remain relatively constant when a condition of homeostasis exists. Fluid and electrolyte imbalance, then, means that both the total volume of water or level of electrolytes in the body or the amounts in one or more of its fluid compartments have increased or decreased beyond normal limits.

The concept of **complementarity of structure and function** discussed in Chapter 1 is as relevant to the study and understanding of fluid and electrolyte balance as it is to the skeletal, urinary, or any other organ system. In the case of body fluids, the structural level of importance is, of course, at the chemical level. The body fluids constitute the immediate environment of the organism. These fluids provide a medium for transport of nutrients and wastes and also function to maintain homeostasis of acid-base balance, osmotic pressure, and temperature for every cell in the body.

Several of the basic physical properties of matter discussed in Chapter 2 will help explain the mechanisms of fluid and electrolyte balance. The concept of chemical bonding is a good example. The type of chemical bonds between molecules of certain organic substances such as glucose are such that they do not permit the compound to break up or **dissociate** in solution. Such compounds were identified in Chapter 2 as **nonelectrolytes.** However, a different type of bonding between elements in chemical compounds such as sodium chloride (NaCl) permits breakup or dissociation into separate particles (Na^+ and Cl^-). Recall that such compounds are known as **electrolytes.** The dissociated particles of an electrolyte are called **ions** and carry an electrical charge.

Many electrolytes and their dissociated ions are of critical importance in fluid balance. Fluid balance and electrolyte balance are so interdependent that if one deviates from normal, so does the other. A discussion of one, therefore, necessitates a discussion of the other.

TOTAL BODY WATER

The fluid or water content of the human body will range from 40% to 60% of its total weight. Normal values for fluid volume, however, vary considerably—mainly according to the fat content of the body. Fat people have a lower water content per kilogram of body weight than slender ones. Women have a relatively lower water content than men—but mainly because a woman's body contains a higher percentage of fat. Total fluid volume and fluid distribution also vary with age. As a person grows older the amount of body water decreases so that in the aged individual fluid makes up a smaller percentage of body weight. In young adults the percent of body weight represented by water will average about 57% for males and 47% for females.

Figure 25-1 Distribution of total body water.

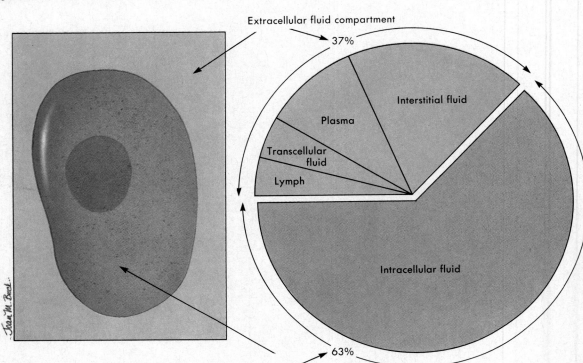

Extracellular fluid compartment

37%

Interstitial fluid

Plasma

Transcellular fluid

Lymph

Intracellular fluid

63%

Intracellular fluid compartment

Figure 25-5 The renin-angiotensin mechanism influencing aldosterone secretion.

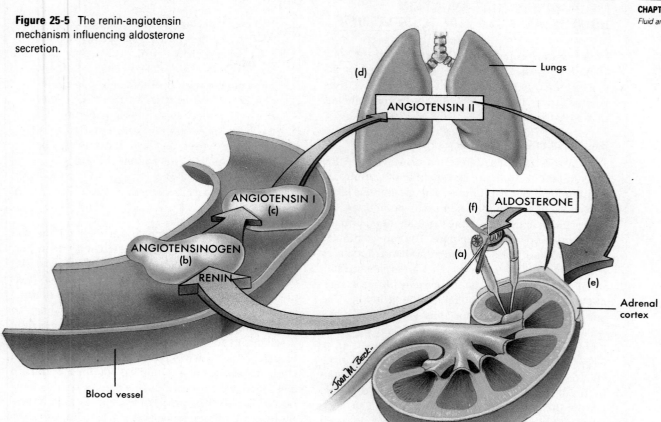

blood pressure of the level of blood Na⁺ ions decreases, the series of events shown in Figure 25-5 will also result in additional aldosterone secretion. Details of the **renin-angiotensin mechanism** are discussed in Chapter 14. Figure 25-6 diagrams a postulated mechanism for adjusting intake to compensate for excess output.

3 Mechanisms for controlling water movement between the fluid compartments of the body constitute the most rapid-acting fluid balance devices. They serve first of all to maintain normal blood volume at the expense of interstitial fluid volume.

Figure 25-6 The basic principle of a postulated homeostatic mechanism for adjusting intake to compensate for excess output.

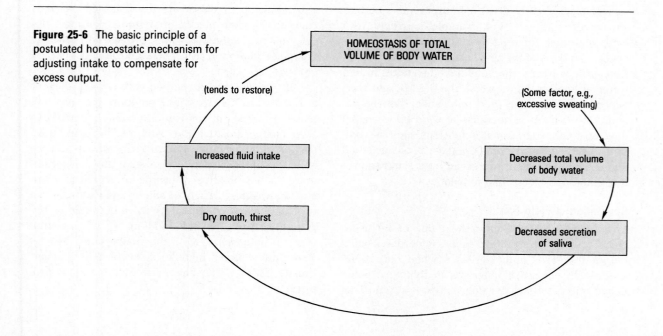

MECHANISMS THAT MAINTAIN HOMEOSTASIS OF TOTAL FLUID VOLUME

Under normal conditions, homeostasis of the total volume of water in the body is maintained or restored primarily by devices that adjust output (urine volume) to intake and secondarily by mechanisms that adjust fluid intake.

Regulation of Urine Volume

Two factors together determine urine volume: the glomerular filtration rate and the rate of water reabsorption by the renal tubules. The glomerular filtration rate, except under abnormal conditions, remains fairly constant—hence it does not normally cause urine volume to fluctuate. The rate of tubular reabsorption of water, on the other hand, fluctuates considerably. The rate of tubular reabsorption, therefore, rather than the glomerular filtration rate, normally adjusts urine volume to fluid intake. The amount of antidiuretic hormone (ADH) and of aldosterone secreted regulates the amount of water reabsorbed by the kidney tubules (discussed on p. 639; see also Figure 25-4). In other words, urine volume is regulated chiefly by hormones secreted by the posterior lobe of the pituitary gland (ADH) and by the adrenal cortex (aldosterone). The regulation of aldosterone secretion by renin-angiotensin is shown in Figure 25-5.

Although changes in the volume of fluid loss via the skin, the lungs, and the intestines also affect the fluid intake-output ratio, these volumes are not automatically adjusted to intake volume, as is the volume of urine.

Factors that Alter Fluid Loss Under Abnormal Conditions

Both the rate of respiration and the volume of sweat secreted may greatly alter fluid output under certain abnormal conditions. For example, a patient who hyperventilates for an extended time loses an excessive amount of water via the expired air. If, as frequently happens, the individual also takes in less water by mouth than normal, the fluid output then exceeds intake and a fluid imbalance, namely, dehydration (that is, a decrease in total body water) develops. Other abnormal conditions such as vomiting, diarrhea, or intestinal drainage also cause fluid and electrolyte output to exceed intake and so produce fluid and electrolyte imbalances.

Regulation of Fluid Intake

Physiologists disagree about the details of the mechanism for controlling intake so that it increases when output increases and decreases when output decreases. In general, it operates in this way: when dehydration starts to develop, salivary secretion de-

The sensation of thirst is apparently regulated, at least in part, by highly specialized cells in the **thirst center** of the hypothalamus of the brain. These cells, called **osmoreceptors,** detect the increase in solute concentration in the extracellular fluid caused by water loss. Signals generated by the osmoreceptors are sent directly to the cerebrum where they are interpreted, like the sensation of a dry mouth, as thirst.

creases, producing a "dry mouth feeling" and the sensation of thirst. The individual then drinks water, thereby increasing fluid intake to offset increased output, and this tends to restore fluid balance (Figure 25-6). If, however, an individual takes nothing by mouth for several days, fluid balance cannot be maintained despite every effort of homeostatic mechanisms to compensate for the zero intake. Obviously, under this condition, the only way balance could be maintained would be for fluid output to also decrease to zero. But this cannot occur. Some output is obligatory. Why? Because as long as respirations continue, some water leaves the body by way of the expired air. Also, as long as life continues, an irreducible minimum of water diffuses through the skin.

CHEMICAL CONTENT, DISTRIBUTION, AND MEASUREMENT OF ELECTROLYTES IN BODY FLUIDS

We have defined an electrolyte as a compound that will break up or dissociate into charged particles called ions when placed in solution. Sodium chloride, when dissolved in water, provides a positively charged sodium ion (Na^+) and a negatively charged chloride ion (Cl^-).

If two electrodes charged with a weak current are placed in an electrolyte solution, the ions will move or migrate in opposite directions according to their charge. Positive ions such as Na^+ will be attracted to the negative electrode or cathode and are called **cations.** Negative ions such as Cl^- will migrate to the positive electrode or anode and are called **anions**. A variety of anions and cations serve critical nutrient or regulatory roles in the body. Important cations include sodium (Na^+), calcium (Ca^{++}), potassium (K^+), and magnesium (Mg^{++}). Important anions include chloride (Cl^-), bicarbonate (HCO_3^-), phosphate (HPO_4^-), and many proteins.

Figure 25-7 Chief chemical constituents of three fluid compartments. The column of figures on the left (200, 190, 180, etc.) indicates amounts of cations or of anions, whereas the figures on the right (400, 380, 360, etc.) indicate the sum of cations and anions. Note that chloride and sodium values in cell fluid are questioned. It is probable that at least muscle intracellular fluid contains some sodium but no chloride.*

*Key to symbols:

Na^+	Sodium
K^+	Potassium
Mg^{++}	Magnesium
Ca^{++}	Calcium
Cl^-	Chloride
$SO_4^=$	Sulfate
$HCO_3^=$	Bicarbonate
$HPO_4^=$	Phosphate
$H \cdot HCO_3$	Carbonic acid

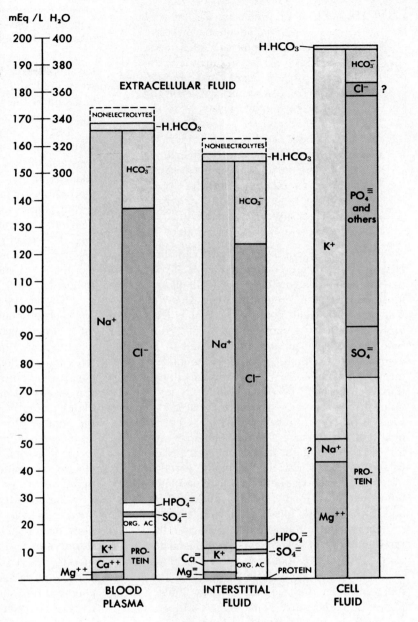

The importance of electrolytes in controlling the movement of water between the body fluid compartments will be discussed in this chapter. Their role in maintaining acid-base balance will be examined in Chapter 26.

Compared chemically, plasma and interstitial fluid (the two extracellular fluids) are almost identical. Intracellular fluid, on the other hand, shows striking differences from either of the two extracellular fluids. Let us examine first the chemical structure of plasma and interstitial fluid as shown in Figure 25-7 and Table 25-2.

Perhaps the first difference between the two extracellular fluids to catch your eye (Figure 25-7)

Table 25-2 Electrolyte composition of blood plasma

Cations	Anions
142 mEq Na^+	102 mEq Cl^-
4 mEq K^+	26 HCO_3^-
5 mEq Ca^{++}	17 protein$^-$
2 mEq Mg^{++}	6 other
	2 HPO_4^-
153 mEq/L plasma	153 mEq/L plasma

According to the Donnan equilibrium principle, when nondiffusible anions (negative ions) are present on one side of a membrane, there are on that side of the membrane fewer diffusible anions and more diffusible cations (positive ions) than on the other side. Applying this principle to the blood and interstitial fluid: because blood contains nondiffusible protein anions, it contains fewer chloride ions (diffusible anions) and more sodium ions (diffusible cations) than does interstitial fluid.

is that blood contains a slightly larger total of electrolytes (ions) than do interstitial fluids. If you compare the two fluids, ion for ion, you will discover the most important difference between blood plasma and interstitial fluid. Look at the anions (negative ions) in these two extracellular fluids. Note that blood contains an appreciable amount of protein anions. Interstitial fluid, in contrast, contains hardly any protein anions. This is the only functionally important difference between blood and interstitial fluid. It exists because the normal capillary membrane is practically impermeable to proteins. Hence almost all protein anions remain behind in the blood instead of filtering out into the interstitial fluid. Because proteins remain in the blood, certain other differences also exist between blood and interstitial fluid—notably, blood contains more sodium ions and fewer chloride ions than does interstitial fluid.

Extracellular fluids and intracellular fluid are more unlike than alike chemically. Chemical difference predominates between the extracellular and intracellular fluids. Chemical similarity predominates between the two extracellular fluids. Study Figure 25-7 and make some generalizations about the main chemical differences between the extracellular and intracellular fluids. For example: What is the most abundant cation in the extracellular fluids? In the intracellular fluid? What is the most abundant anion in the extracellular fluids? In the intracellular fluid? What about the relative concentrations of protein anions in extracellular fluids and intracellular fluid?

The only reason we have called attention to the chemical structure of the three body fluids is that here, as elsewhere, structure determines function. In this instance the chemical structure of the three fluids helps control water and electrolyte movement between them. Or, phrased differently, the chemical structure of body fluids, if normal, functions to maintain homeostasis of fluid distribution and, if abnor-

mal, results in fluid imbalance. **Hypervolemia** (excess blood volume) is a case in point. **Edema,** too, frequently stems from changes in the chemical structure of body fluids.

Before discussing mechanisms that control water and electrolyte movement between blood, interstitial fluid, and intracellular fluid, it is important to understand the units used for measuring electrolytes.

Once the important electrolytes and their constituent ions in the body fluid compartments had been established, physiologists needed to measure changes in their levels in order to understand the mechanisms of fluid balance. To have meaning, measurement units used to report electrolyte levels must be related to actual physiological activity. In the past, only the weight of an electrolyte in a given amount of solution was measured. The number of milligrams per 100 ml of solution (mg%) was one of the most frequently used units of measurement. However, simply reporting the weight of an important electrolyte such as sodium or calcium in milligrams per 100 ml of blood (mg%) gives no direct information about its chemical combining power or physiological activity in body fluids. The importance of valence and electrovalent or ionic bonding in chemical reactions was discussed in Chapter 2. The reactivity of combining power of an electrolyte depends not just on the number of molecular particles present but also on the total number of ionic charges (valence). Univalent ions such as sodium (Na^+) carry only a single charge, but the divalent calcium ion (Ca^{++}) carries two units of electrical charge.

The need for a unit of measurement more related to activity has resulted in increasing use of a more meaningful measurement yardstick—the **milliequivalent.** Milliequivalents measure the number of ionic charges or electrovalent bonds in a solution and therefore serve as an accurate measure of the chemical (physiological) combining power or reactivity of a particular electrolyte solution. The number of milliequivalents of an ion in a liter of solution (mEq/L) can be calculated from its weight in 100 ml (mg%) using a convenient conversion formula.

Conversion of milligrams per 100 ml (mg%) to milliequivalents per liter (mEq/L):

$$mEq/L = \frac{mg/100 \ ml \times 10 \times Valence}{Atomic \ weight}$$

Example: Convert 15.6 mg% K^+ to mEq/L

Atomic weight of $K^+ = 39$
Valence of $K^+ = 1$
$$mEq/L = \frac{15.6 \times 10 \times 1}{39} = \frac{156}{39} = 4$$

Therefore 15.6 mg/100 ml K^+ = 4 mEq/L

Control of Water and Electrolyte Movement Between Plasma and Interstitial Fluid

Over 65 years ago, Starling advanced a hypothesis about the nature of the mechanism that controls water movement between plasma and interstitial fluid—that is, across the capillary membrane. This hypothesis has since become one of the major premises of physiology and is often spoken of as Starling's **law of the capillaries.** According to this law, the control mechanism for water exchange between plasma and interstitial fluid consists of four pressures: **blood hydrostatic** and **colloid osmotic pressures** on one side of the capillary membrane and **interstitial fluid hydrostatic** and **colloid osmotic pressures** on the other side.

We are ready now to try to answer the following question: How does the chemical structure of body fluids control water movement between them and thereby control fluid distribution in the body?

According to the physical laws governing filtration and osmosis, **blood hydrostatic pressure (BHP)** tends to force fluid out of capillaries into interstitial fluid (IF), but **blood colloid osmotic pressure (BCOP)** tends to draw it back into them. **Interstitial fluid hydrostatic pressure (IFHP),** in contrast, tends to force fluid out of the interstitial fluid into the capillaries, and **interstitial fluid colloid osmotic pressure (IFCOP)** tends to draw it back out of capillaries. In short, two of these pressures constitute vectors in one direction and two in the opposite direction. This process is similar in many ways to the mechanism responsible for formation of glomerular filtrate studied in the last chapter. The movement of fluids and electrolytes between plasma and interstitial fluid caused by hydrostatic and colloid osmotic pressure is illustrated in Figure 25-8.

The difference between the two sets of opposing forces obviously represents the net or effective filtration pressure—in other words, the effective force tending to produce the net fluid movement between blood and interstitial fluid. In general terms, therefore, we may state Starling's law of the capillaries this way: the rate and direction of fluid exchange between capillaries and interstitial fluid is determined by the hydrostatic and colloid osmotic pressures of the two fluids. Or, we may state it more specifically as a formula:

$$(BHP + IFCOP) - (IFHP + BCOP) = EFP^*$$

Note that the factors enclosed in the first set of parentheses tend to move fluid out of capillaries and

*BHP, blood hydrostatic pressure; IFCOP, interstitial fluid colloid osmotic pressure; IFHP, interstitial fluid hydrostatic pressure; BCOP, blood colloid osmotic pressure; EFP, effective filtration pressure between blood and interstitial fluid.

Osmotic pressure occurs from concentrations of protein in blood and interstitial fluid. Since the capillary membrane is permeable to other plasma solutes, they quickly diffuse through the membrane, so cause no osmotic pressure to develop against it. Only the proteins, to which the capillary membrane is practically impermeable, cause an actual osmotic pressure against the capillary membrane.

A small amount of blood protein passes through the capillary membrane and tends to concentrate around the venous end of capillaries, hence this pressure.

that those in the second set oppose this movement—they tend to move fluid into the capillaries.

To illustrate operation of Starling's law (Figure 25-8), let us consider how it controls water exchange at the arterial ends of tissue capillaries. Table 25-3 gives typical normal pressures. Using these figures in Starling's law of the capillaries we get $(35 + 0) - (s + 25) = 8$ mm Hg net pressure (EFP), causing water to filter out of blood at arterial ends of capillaries into interstitial fluid.

The same law operates at the venous end of capillaries (Table 25-4). Again, apply Starling's law

Table 25-3 Pressures at arterial end of tissue capillaries

Arterial end of capillary	Hydrostatic pressure	Colloid osmotic pressure
Blood	35 mm Hg	25 mm Hg
Interstitial fluid	2 mm Hg	0 mm Hg

Table 25-4 Pressures at venous end of tissue capillaries

Venous end of capillary	Hydrostatic pressure	Colloid osmotic pressure
Blood	15 mm Hg	25 mm Hg
Interstitial fluid	1 mm Hg	3 mm Hg*

*A small amount of blood proteins passes through the capillary membrane and tends to concentrate around the venous ends of capillaries—hence this pressure.

Figure 25-8 Movement of fluids and electrolytes between plasma and interstitial fluid caused by hydrostatic and colloid osmotic pressure.

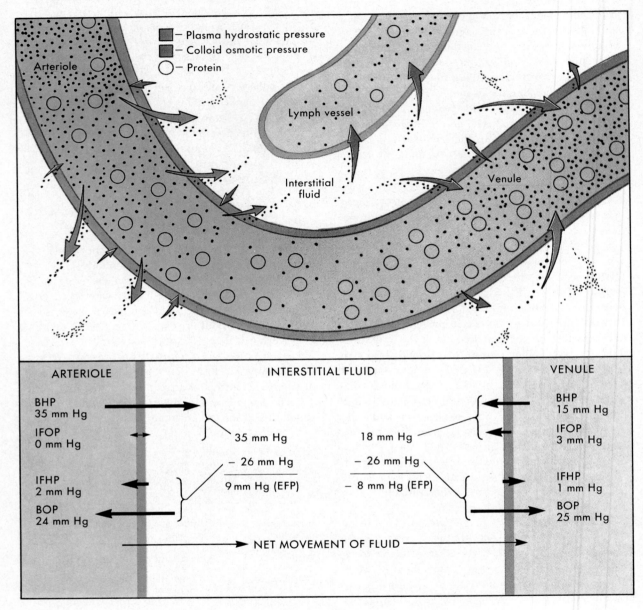

of the capillaries. What is the net effective pressure at the venous ends of capillaries? In which direction does it cause water to move? Assuming that the figures given in Table 25-4 are normal, do you agree that theoretically "the same amount of water returns to the blood at the venous ends of the capillaries as left it from the arterial ends"?

On the basis of our discussion thus far, we can formulate some principles about the transfer of water between blood and interstitial fluid.

1 No net transfer of water occurs between blood and interstitial fluid as long as the ef-

fective filtration pressure (EFP) equals 0, that is, when

$$(BHP + IFCOP) = (IFHP + BCOP)$$

2 A net transfer of water, a "fluid shift," occurs between blood and interstitial fluid whenever the EFP does not equal 0, that is, when

$$\begin{array}{c} (BHP + IFCOP) \\ \text{does not equal} \\ (IFHP + BCOP) \end{array}$$

Figure 25-9 The mechanism of edema formation initiated by a decrease in blood protein concentration and, therefore, in blood colloid osmotic pressure. Left diagram, blood osmotic pressure has just decreased to 20 from the normal 25 mm Hg. This increases the effective filtration pressure (EFP) to 5 mm Hg from a normal of 0 (see Starling's formula, p. 661). The EFP of 5 mm Hg causes fluid to shift out of blood into interstitial fluid (IF) until the EFP again equals 0—in this case, when the interstitial fluid volume has increased enough to raise interstitial fluid hydrostatic pressure to 9 mm Hg, as shown in the diagram on the right. At this point a new equilibrium is established, and equal amounts of water once more are exchanged between the blood and interstitial fluid. Thus the increased interstitial fluid volume, that is, the edema, becomes stabilized.

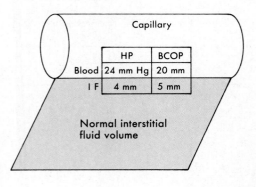

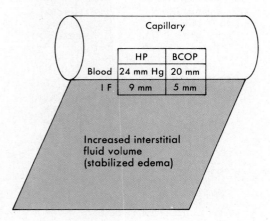

3 Since (BHP + IFCOP) is a force that tends to move water out of capillary blood, fluid shifts out of blood into interstitial fluid whenever

$$\begin{array}{c}\text{(BHP + IFCOP)}\\ \text{is greater than}\\ \text{(IFHP + BCOP)}\end{array}$$

4 Since (IFHP + BCOP) is a force that tends to move water out of interstitial fluid into capillary blood, fluid shifts out of interstitial fluid into blood whenever

$$\begin{array}{c}\text{(IFHP + BCOP)}\\ \text{is greater than}\\ \text{(BHP + IFCOP)}\end{array}$$

Or, stated the other way around, whenever

$$\begin{array}{c}\text{(BHP + IFCOP)}\\ \text{is less than}\\ \text{(IFHP + BCOP)}\end{array}$$

EDEMA

Edema may be defined as the presence of abnormally large amounts of fluid in the intercellular tissue spaces of the body. The condition is a classic example of fluid imbalance and may be caused by dis-

turbances in any of the factors that govern the interchange between blood plasma and interstitial fluid compartments. Examples include:

1 Retention of electrolytes (especially Na⁺) in the extracellular fluid as a result of increased aldosterone secretion or following serious renal disease such as acute **glomerulonephritis.**

2 An increase in capillary blood pressure. Normally fluid is drawn from the tissue spaces into the venous end of a tissue capillary because of the low venous hydrostatic pressure and the high water-pulling force of the plasma proteins (Figure 25-8). This balance is upset by anything that will increase the capillary hydrostatic pressure. The generalized venous congestion of heart failure is the most common cause of widespread edema. In patients with this condition, blood cannot flow freely through the capillary beds, and therefore the pressure will increase until venous return of blood improves.

3 A decrease in the concentration of plasma proteins normally retained in the blood (Figures 25-9 and 25-10). This may occur as a result of increased capillary permeability caused by infection, burns, or shock.

Figure 25-10 Mechanisms of edema formation in some common conditions. *IF,* Interstitial fluid; *EFP,* effective or net filtration pressure. (See also Figure 25-9.)

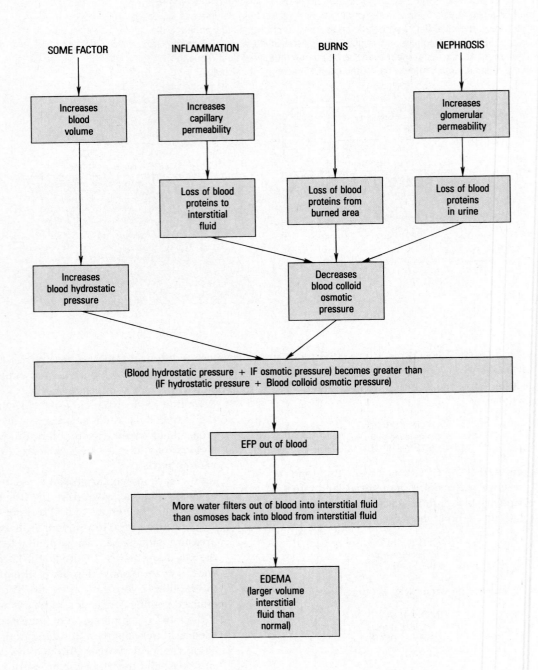

Control of Water and Electrolyte Movement

The mechanism that regulates water movement through cell membranes is similar to the one that regulates water movement through capillary membranes. In other words, interstitial fluid and intracellular fluid hydrostatic and colloid osmotic pressures regulate water transfer between these two fluids. But because the colloid osmotic pressures of interstitial and intracellular fluids vary more than do their hydrostatic pressures, their colloid osmotic pressures serve as the chief regulators of water transfer across cell membranes. Their colloid osmotic pressures, in turn, are directly related to the electrolyte concentration gradients—notably sodium and potassium—maintained across cell membranes. As Figure 25-7 shows, most of the body sodium is outside the cells. A concentration of 138 to 143 mEq/L makes sodium the chief electrolyte by far in interstitial fluid. The intracellular fluid's main electrolyte is potassium salt. Therefore a change in the sodium or the potassium concentrations of either of these fluids causes the exchange of fluid between them to become unbalanced. For example, a decrease in interstitial fluid sodium concentration immediately decreases interstitial fluid colloid osmotic pressure, making it hypotonic to intracellular fluid colloid osmotic pressure. In other words, a decrease in interstitial fluid sodium concentration establishes a colloid osmotic pressure gradient between interstitial and intracellular fluids. This causes net osmosis to occur out of interstitial fluid into cells. In short, interstitial fluid and intracellular fluid electrolyte concentrations are the main determinants of their colloid osmotic pressures; their colloid osmotic pressures regulate the amount and direction of water transfer between the two fluids, and this regulates their volumes. Hence fluid balance depends on electrolyte balance. Conversely, electrolyte balance depends on fluid balance. An imbalance in one produces an imbalance in the other (Figure 25-11).

Figure 25-11 Scheme to show how electrolyte imbalance (sodium deficit or hyponatremia) leads to fluid imbalances (hypovolemia and cellular hydration). *ECF,* Extracellular fluid; *ICF,* intracellular fluid.

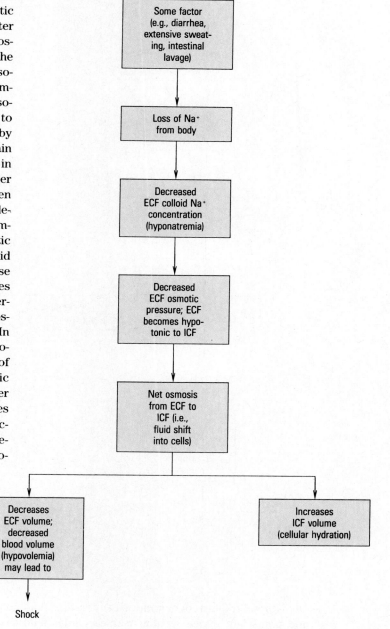

Figure 25-12 Antidiuretic hormone (ADH) mechanism that helps maintain homeostasis of extracellular fluid (ECF) colloid osmotic pressure by regulating its volume and thereby its electrolyte concentration, that is, mainly ECF Na$^+$ concentration. *ECF,* Extracellular fluid; *ICF,* intracellular fluid.

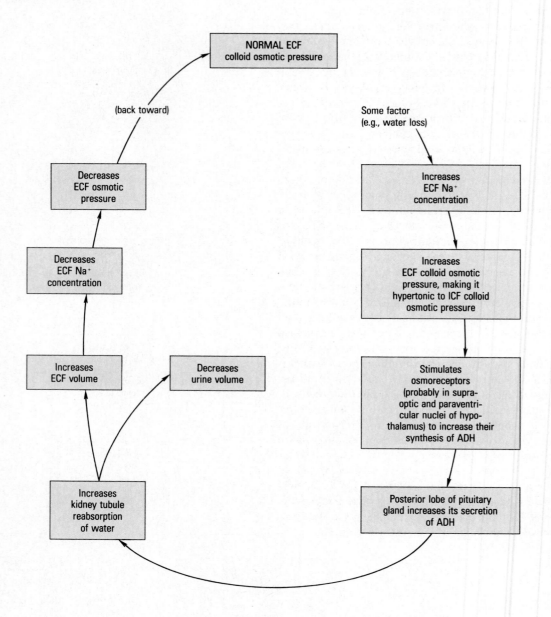

Normal sodium concentration in interstitial fluid and potassium concentration in intracellular fluid depend on many factors but especially on the amount of ADH and aldosterone secreted. As shown in Figure 25-12, ADH regulates extracellular fluid electrolyte concentration and colloid osmotic pressure by regulating the amount of water reabsorbed into blood by renal tubules. Aldosterone, on the other hand, regulates extracellular fluid volume by

regulating the amount of sodium reabsorbed into blood by renal tubules (see Figure 25-4).

If for any reason conservation of body sodium is required, the normal kidney is capable of excreting an essentially sodium-free urine and is therefore considered the chief regulator of sodium levels in body fluids. Sodium lost in sweat can become appreciable with elevated environmental temperatures or fever. However the thirst that results may lead to replace-

Figure 25-13 Sodium-containing internal secretions. The total volume of these secretions may reach 8,000 or more milliliters in a 24-hour period.

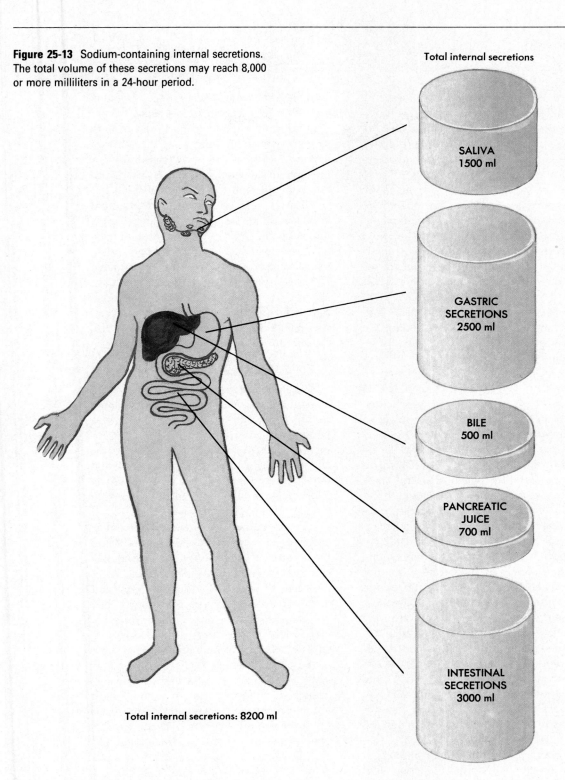

Total internal secretions

SALIVA
1500 ml

GASTRIC
SECRETIONS
2500 ml

BILE
500 ml

PANCREATIC
JUICE
700 ml

INTESTINAL
SECRETIONS
3000 ml

Total internal secretions: 8200 ml

ment of water but not the lost sodium and, as a result of the increased fluid intake, the remaining sodium pool may be diluted even more. Sweat loss of sodium is not, therefore, considered a normal means of regulation.

In addition to the well-regulated movement of sodium into and out of the body and between the three primary fluid compartments, there is a continuous movement or circulation of this important electrolyte between a number of internal secretions. Over 8 *liters* of various internal secretions such as saliva, gastric and intestinal secretions, bile, and pancreatic fluid are produced every day (Figure 25-13). The total daily secretion of sodium into these ali-

DEHYDRATION

The term **dehydration** is used to describe the condition that results from excessive loss of body water. Water deprivation or loss triggers a complex series of protective responses designed to maintain homeostasis of both water and electrolyte levels. Unfortunately, the term *dehydration* is incomplete. It does not, by definition, include the loss of electrolytes. In order to understand the control mechanisms that ensure fluid and electrolyte balance or properly interpret the clinical signs and symptoms of dehydration in disease states, it is important to realize that in any process of dehydration water loss is always accompanied by loss of electrolytes. If water intake is reduced to the point of dehydration, there must be removal of a corresponding quantity of electrolytes in order to maintain the normal ionic content of body fluids. The same holds true in the case of electrolyte loss when an accompanying loss of water must occur to maintain homeostasis of both fluid and electrolyte levels. Understanding the close interrelationships of water and electrolyte loss in dehydration provides the rationale for effective treatment. Water alone is inadequate; treatment of dehydration also requires appropriate electrolyte replacement therapy.

As discussed in Chapter 5, maintaining a constant core body temperature in a hot environment is an important function of the skin. As sweat evaporates, excess body heat can be eliminated. In hot weather or during extended periods of strenuous physical activity, the volume of water loss because of sweat production can reach 10 liters a day. If water intake is inadequate, signs of dehydration will appear very rapidly. As body water levels decrease, the initial defense mechanisms are directed toward maintaining an adequate blood volume.

In addition to water, sweat contains significant quantities of both sodium and chloride. However the relative loss of water in sweat is greater than the loss of electrolytes. Therefore as water is shifted from the interstitial fluid compartment to the plasma to compensate for fluid loss, the kidneys excrete the excess electrolytes in order to preserve normal ionic concentrations in the two compartments.

The chemical composition and actual volume of fluids lost from the body will also affect the type and effectiveness of defense mechanisms that occur. For example, fluids lost through vomiting or diarrhea will have differing ratios of fluid to electrolytes than sweat and the actual electrolyte composition and concentraton will also be different. As a result, the type of electrolyte excretion or retention by the kidneys that will be needed to maintain ionic balance in the fluid compartments will also change.

There is a lag in the volume-electrolyte adjustment mechanism triggered by dehydration. Shifts in fluid occur more quickly between compartments than the adjustment in electrolyte levels. However if water and electrolyte losses are limited and the interval between loss and replacement is short, the symptoms of dehydration will be mild and transitory.

In severe and prolonged water deprivation or loss, the initial shift of interstitial fluid to plasma will be followed by movement of water from the intracellular compartment as well. Over time the extra- and intracellular fluid losses are about equal.

Extracellular (interstitial) water is more "expendable" and quickly assessable as a fluid source to maintain blood volume in the early stages of body fluid loss. It is said to serve as the "first line of defense" against dehydration. As extracellular fluid is depleted, intracellular water must be used in order to prolong survival time. Ultimately, the volume of the extracellular fluid can be reduced by almost 60% and intracellular fluids by 30% before death occurs.

mentary tract fluids alone will average between 1,200 and 1,400 mEq. A 70 kg (154-lb) adult has a total body sodium pool of only 2,800 to 3,000 mEq. Precise regulatory and conservation mechanisms are required for survival.

Chloride is the most important extracellular anion and is almost always linked to sodium. Generally ingested together, they provide in large part for the isotonicity of extracellular fluid. Chloride ions are generally excreted in the urine as a potassium salt, and therefore chloride deficiency **hypochloremia** is often found in cases of potassium loss.

Total body potassium content in the average-sized adult is approximately 4,000 mEq. Because the vast majority of body potassium is intracellular, serum determinations, which normally fall between 4.0 and 5.0 mEq/L, may not be the best index to reflect imbalances. The body may lose one third to one half of its intracellular potassium reserves before the loss is reflected in lowered serum potassium levels.

Potassium deficit, or **hypopotassemia,** occurs whenever there is cell breakdown, as in starvation, trauma, or dehydration. As individual cells disintegrate, potassium enters the extracellular fluid and is rapidly excreted because it is not reabsorbed efficiently by the kidney.

Review of Medical Terms

D5W Abbreviation for dextrose 5% in water—a common parenteral solution.

dehydration (de-hi-DRA-shun) Excessive loss of body water.

diarrhea (di-ah-RE-ah) Abnormal frequency and liquidity of fecal discharges.

edema (e-DE-mah) Presence of abnormally large amounts of fluid in the intercellular tissue spaces.

fever (FE-ver) Elevation of body temperature above normal.

glomerulonephritis (glo-mer-u-lo-ne-FRI-tis) Inflammation of the capillary loops in the glomeruli of the kidney.

hyperventilate (hi-per-VEN-ti-late) Abnormally rapid or deep breathing.

hypervolemia (hi-per-vo-LE-me-ah) Abnormal increase in volume of circulating blood.

hypochloremia (hi-po-klo-RE-me-ah) Chloride deficiency.

hypopotassemia (hi-po-potah-SEE-me-ah) Potassium deficiency.

intravenous (in-trah-VE-nus) Within a vein.

parenteral (pah-REN-ter-al) By injection—not through the alimentary canal but by some other route.

shock (shok) Acute peripheral circulatory failure.

subcutaneous (sub-ku-TA-ne-us) Beneath the skin.

vomiting (VOM-it-ing) Forcible expulsion of the stomach contents through the mouth.

Outline Summary

INTRODUCTION

A Meaning of fluid balance
  1 Same as homeostasis of fluids, that is, total volume of water in body normal and remains relatively constant
  2 Volume of blood plasma, interstitial fluid, and intracellular fluid all remain relatively constant, that is, homeostasis of distribution of water as well as of total volume
B Fluid balance and electrolyte balance interdependent (see Figures 25-4 and 25-13)
C Chemical compounds such as NaCl that break up or dissociate in solution (Na^+ and Cl^-) are known as electrolytes
D Dissociation particles of an electrolyte are called *ions* and carry an electrical charge

TOTAL BODY WATER

A Total body water will range from 40% to 60% of total body weight
  1 Amount will vary with fat content of body
    a High fat content—lower body water content
    b Females have less body water than males
    c Body water content decreases with age
    d Average values—body water
      (1) Males—57% of body weight
      (2) Females—47% of body weight

BODY FLUID COMPARTMENTS

A Extracellular fluid compartment (ECF) (Figure 25-1)
  1 Plasma

Outline Summary—cont'd

2 Interstitial fluid

3 Specialized fluids

   a Lymph

   b Cerebrospinal fluid

   c Specialized joint fluids

4 Characteristics of ECF (Figure 25-7)

   a Constitutes internal environment of body

   b Serves important transportation role

   c Constitutes 20% of body weight

   d Larger in infants (volume) and children than adults

B Intracellular fluid compartment (ICF)

1 Water inside cells

2 Characteristics of ICF (Figure 25-7)

   a Serves a solvent role

   b Functions to facilitate intracellular chemical reactions

   c Largest fluid compartment by volume

   d Constitutes 50% of body weight

AVENUES BY WHICH WATER ENTERS AND LEAVES THE BODY

A Water enters body through digestive tract in

   1 Liquids

   2 Foods

B Water formed in body by metabolism of foods

C Water leaves body via kidneys, lungs, skin, and intestines

FLUID AND ELECTROLYTE THERAPY

A Parenteral therapy—administration of nutrients, fluids, and electrolytes by injection

   1 Types

      a Intravenous

      b Subcutaneous

   2 Tonicity

      a If given subcutaneously, must be isotonic

      b If given intravenously, tonicity is not critical if solution is administered at correct rate

   3 Intravenous route is preferred to parenteral route

B Primary objectives of parenteral therapy

   1 To meet current maintenance needs

   2 To replace past losses

   3 To replace concurrent losses

C Major types of parenteral solutions

   1 Carbohydrates in water

   2 Carbohydrate in various strengths of saline

   3 Normal saline (0.9% NaCl)

   4 Potassium solutions

   5 Ringer's solution

   6 Lactate solutions

   7 Ammonium chloride solutions

SOME GENERAL PRINCIPLES ABOUT FLUID BALANCE

A Cardinal principle—intake must equal output

B Fluid and electrolyte balance maintained primarily by mechanisms that adjust output to intake; secondarily by mechanisms that adjust intake to output

C Fluid balance also maintained by mechanisms that control movement of water between fluid compartments

MECHANISMS THAT MAINTAIN HOMEOSTASIS OF TOTAL FLUID VOLUME

A Regulation of urine volume—under normal conditions, by factors that control reabsorption of water by distal and collecting tubules

   1 Extracellular fluid electrolyte concentration (crystalloid osmotic pressure) controls ADH secretion, which controls tubule water reabsorption

   2 Extracellular fluid volume controls aldosterone secretion, which controls tubule sodium ion reabsorption and therefore water reabsorption

B Factors that alter fluid loss under abnormal conditions—hyperventilation, hypoventilation, vomiting, diarrhea, circulatory failure, etc.

C Regulation of fluid intake (see Figure 25-6)

   1 Mechanism by which intake adjusted to output not completely known

   2 One controlling factor seems to be degree of moistness of mucosa of mouth—if output exceeds intake, mouth feels dry, sensation of thirst occurs, and individual ingests liquids

CHEMICAL CONTENT, DISTRIBUTION, AND MEASUREMENT OF ELECTROLYTES IN BODY FLUIDS

A Comparison of plasma, interstitial fluid, and intracellular fluid

   1 Plasma and interstitial fluid constitute extracellular fluid (ECF), internal environment of body or, in other words, environment of cells

   2 Intracellular fluid (ICF) volume largest, plasma volume smallest; ICF about 40% of body weight, IF about 16%, and plasma about 4%, or ICF volume about 10 times that of plasma and IF volume about 4 times plasma volume

   3 Chemically, plasma and IF almost identical except that plasma contains slightly more electrolytes and considerably more proteins than IF; also blood contains somewhat more sodium and fewer chloride ions

   4 Chemically, ECF and ICF strikingly different; sodium main cation of ECF, potassium main cation of ICF; chloride main anion of ECF; phosphate main anion of ICF; protein concentration much higher in ICF than in IF

B Importance, distribution, and measurement of electrolytes in body fluids

   1 Positive ions such as sodium (Na^+) are attracted to negative electrode or cathode—called cations

   2 Negative ions such as chloride (Cl^-) migrate to positive electrode or anode—called anions

   3 Important cations: sodium (Na^+), calcium (Ca^{++}), potassium (K^+), and magnesium (Mg^{++})

   4 Important anions: chloride (Cl^-), bicarbonate (HCO_3^-), phosphate ($HPO_4^=$), and many proteins

Outline Summary—cont'd

5 Electrolytes serve important roles in fluid balance and in acid-base balance

6 Physiological reactivity or combining power of an electrolyte depends not just on the number of molecular particles present but also on total number of ionic charges (valence)

7 Milliequivalents (mEq) measure the number of ionic charges or electrovalent bonds in a solution and therefore serve as an accurate measure of the chemical (physiological) combining power or reactivity of a particular electrolyte

8 Conversion formula for mg% to mEq/L:

$$mEq/L = \frac{mg/100\ ml \times 10 \times Valence}{Atomic\ weight}$$

CONTROL OF WATER AND ELECTROLYTE MOVEMENT

A Control of water movement between plasma and IF
  1 By four pressures—blood hydrostatic and colloid osmotic pressures and IF hydrostatic and colloid osmotic pressures
  2 Effect of these pressures on water movement between plasma and IF expressed in Starling's *law of the capillaries;* only when (Blood hydrostatic pressure + IF colloid osmotic pressure) − (Blood colloid osmotic pressure + IF hydrostatic pressure) = 0, do equal amounts of water filter out of blood into IF and osmose back into blood from IF; in other words, water balance exists between these two fluids under these conditions

B Control of water and electrolyte movement through cell membranes between IF and ICF—primarily by relative crystalloid osmotic pressures of ECF and ICF, which depend mainly on sodium concentration

of ECF and potassium concentration of ICF, which in turn depend on intake and output of sodium and potassium and on sodium-potassium transport mechanisms

EDEMA

A Presence of abnormally large amounts of fluid in the intercellular tissue spaces

B Causes of edema
  1 Retention of electrolytes—especially sodium in ECF
  2 Increase in capillary pressure—especially that caused by congestive heart failure
  3 Decrease in plasma protein concentration

DEHYDRATION

A Definition—excessive loss of body water

B In any process of dehydration, water loss is always accompanied by loss of electrolytes

C Treatment of dehydration requires administration of both fluids and appropriate electrolytes

D There is a lag in the volume-electrolyte adjustment mechanism triggered by dehydration

E In severe dehydration, the initial fluid shift is from interstitial fluid to plasma, which is then followed by loss of water from the intracellular compartment

F ECF is more "expendable" and "assessable" than ICF in dehydration

G ECF said to be the "first line of defense" against dehydration

H In severe dehydration, volume of ECF can be reduced by 60% and ICF by 30% before death occurs

Review Questions

1 Explain in your own words the meaning of the term *fluid balance.*

2 How is the total volume of body fluid kept relatively constant, that is, what other factors must be controlled in order to keep the total volume of water in the body relatively constant?

3 What, if any, are functionally important differences between the chemical composition of plasma and interstitial fluid?

4 What, if any, are functionally important differences between the chemical composition of extracellular and intracellular fluids?

5 Are plasma and interstitial fluid more accurately described as "similar" or "different" as to chemical composition? Volume?

6 Support your answer to question 5 with some specific facts.

7 Are interstitial fluid and intracellular fluid more accurately described as "similar" or "different" as to chemical composition? Volume?

8 Explain Starling's law of the capillaries in your own words. Be as brief and clear as you can. This law describes the mechanism for controlling what?

9 Suppose that in one individual the normal average or mean pressures (in mm Hg) are capillary blood hydrostatic pressure 24 and osmotic pressure 25; interstitial fluid hydrostatic pressure 4 and osmotic pressure 5. Following a hemorrhage, this patient's blood hydrostatic pressure falls to 18. (a) Assuming that the other pressures momentarily stay the same,

Review Questions—cont'd

what is the EFP now? (b) The new EFP causes a fluid shift in which direction? (c) When will the fluid shift stop and an even exchange of water again go on between blood and interstitial fluids?

10 Formulate a principle by filling in the blanks in the following sentences: Anything that decreases blood hydrostatic pressure in the capillaries tends to cause a fluid shift from _____ into _____. Conversely, anything that increases blood hydrostatic pressure in the capillaries tends to cause a fluid shift out of _____ into _____.

11 Formulate a principle by completing the blanks in the following sentence: A decrease in blood protein concentration tends to decrease blood _____ pressure and therefore tends to cause a fluid shift from _____ to _____.

Situation: A patient has had marked diarrhea for several days.

12 Explain the homeostatic mechanisms that would tend to compensate for this excessive fluid loss.

13 Do you think these mechanisms could succeed in maintaining fluid balance or would fluid therapy probably be necessary?

14 What, if any, abnormality of fluid distribution do you think would occur in this patient without fluid therapy? Explain your reasoning.

15 Why does dehydration necessarily develop if an individual takes nothing by mouth for several days and receives no fluid therapy? Why cannot homeostatic mechanisms prevent this?

16 Define the following terms: *cation, anion, milliequivalent, electrolyte, hypochloremia, hypopotassemia.*

17 What is considered to be the chief regulator of sodium levels in body fluids?

18 Is loss of sodium through sweat a normal means for regulation of this electrolyte?

19 Why does excessive loss of potassium in the urine often cause serious depletion of chloride from the extracellular fluid?

20 What important cation is lost from the intracellular fluid compartment as a result of cell breakdown associated with starvation or trauma?

21 Distinguish between electrolyte and non-electrolyte compounds.

22 Why is the structural level of organization of importance in the study of fluid and electrolyte balance?

23 How do body fat content, sex, and age affect total body water content?

24 Identify the two major fluid compartments and the subdivisions of each.

25 Rank intracellular fluid, plasma, and interstitial fluid according to volume of body fluid.

26 What is parenteral therapy? What are the primary parenteral routes that can be used to administer fluids?

27 Identify the seven basic solutions used to provide electrolytes and other nutrients by parenteral routes.

28 Discuss the aldosterone mechanism for decreasing fluid output.

29 How do osmoreceptors function in regulating extracellular fluid levels?

30 Discuss edema as a type of fluid imbalance.

31 Discuss dehydration as a function of both water and electrolyte imbalance.

26 ACID-BASE BALANCE

CHAPTER OUTLINE

Mechanisms that control pH of body fluids
- Meaning of term pH
- Sources of pH-influencing elements
- Types of pH control mechanisms
- Effectiveness of pH control mechanisms—range of pH

Buffer mechanism for controlling pH of body fluids
- Buffers defined
- Buffer pairs present in body fluids
- Action of buffers to prevent marked changes in pH of body fluids
- Evaluation of role of buffers in pH control

Respiratory mechanism of pH control
- Explanation of mechanism
- Adjustment of respirations to pH of arterial blood
- Some principles relating respirations and pH of body fluids

Urinary mechanism of pH control
- General principles about mechanism
- Mechanisms that control urine pH

Acid-base imbalances
- Metabolic disturbances
  - Metabolic acidosis (bicarbonate deficit)
  - Metabolic alkalosis (bicarbonate excess)
- Respiratory disturbances
  - Respiratory acidosis (carbonic acid excess)
  - Respiratory alkalosis (carbonic acid deficit)

OBJECTIVES

After you have completed this chapter, you should be able to:

1. Discuss the concept of pH.
2. Define acid-base balance.
3. List four acids that contribute hydrogen ions to body fluids and identify the source of each.
4. Give examples of acid- and base-forming elements and identify dietary sources for each.
5. Identify and contrast chemical and physiological buffers.
6. Contrast strong and weak acids and bases.
7. Compare the buffering of a strong acid and base with a weak acid and base.
8. Explain how the chloride shift makes it possible for carbon dioxide to be buffered in red blood cells and then carried as bicarbonate in the plasma.
9. Contrast the respiratory and urinary mechanisms of pH control.
10. Compare the effects of hypoventilation and hyperventilation on blood pH.
11. Discuss the function of the distal renal tubule in acidification of urine.
12. Contrast and discuss metabolic and respiratory disturbances that result in acid-base imbalances.
13. Discuss compensatory mechanisms that may help return blood pH to near normal levels in cases of metabolic alkalosis.
14. Explain how vomiting, pneumonia, and starvation can influence blood pH.
15. Discuss possible treatments for acidosis and alkalosis.

Acid-base balance is one of the most important of the body's homeostatic mechanisms. The term refers to regulation of hydrogen ion concentration in the body fluids. Precise regulation of pH at the cellular level is necessary for survival. Even slight deviations from normal pH will result in pronounced, potentially fatal changes in metabolic activity.

KEY TERMS

Acid-forming food

Acidosis (ass-i-DOE-sis)

Alkalosis (al-kuh-LOE-sis)

Base-forming food

Buffer

Chemical buffers

Chloride shift

Henderson-Hasselbalch equation

Hyperkalemia (hi-per-kah-LE-me-ah)

pH

Physiological buffers

MECHANISMS THAT CONTROL pH OF BODY FLUIDS

Meaning of Term pH

The term pH is a symbol used to mean the hydrogen ion (H^+) concentration of a solution (Figure 26-1). Actually, pH stands for the negative logarithm of the hydrogen ion concentration. pH indicates the degree of **acidity** or **alkalinity** of a solution. As the concentration of hydrogen ions increases, the pH goes down and the solution becomes more acid; a decrease in hydrogen ion concentration makes the solution more alkaline and the pH goes up. A pH of 7 indicates neutrality (equal amounts of H^+ and OH^-), a pH of less than 7 indicates acidity (more H^+ than OH^-), and a pH greater than 7 indicates alkalinity (more OH^- than H^+). The overall pH range is often expressed numerically on a logarithmic scale of 1 to 14. Keep in mind that a change of 1 pH unit on this type of scale represents a 10-fold difference in actual concentration of hydrogen ions.

The slight increase to acidity of venous blood (pH 7.36) compared to arterial blood (pH 7.41) results primarily from carbon dioxide entering venous blood as a waste product of cellular metabolism. The lungs remove the equivalent of over 30 *liters* of 1-normal carbonic acid each day from the venous blood by elimination of carbon dioxide, and yet 1 liter of venous blood contains only about 1/100,000,000 gm more hydrogen ions than does 1 liter of arterial blood. What incredible constancy! The pH homeostatic mechanism does indeed control effectively—astonishingly so.

Sources of pH-Influencing Elements

Both acids and bases continually enter the blood from absorbed foods and from the metabolism of nutrients at the cellular level. Therefore some kind of mechanism for neutralizing or eliminating these substances is necessary if blood pH is to remain constant. Although both acid and basic components are important, the homeostasis of body pH largely depends on the control of hydrogen ion concentration in the extracellular fluid. Hydrogen ions are continually entering the body fluids from (1) **carbonic,** (2) **lactic,** (3) **sulfuric,** and (4) **phosphoric acids** and (5) **acidic ketone bodies.**

Carbonic and lactic acids are produced by the aerobic and anaerobic metabolism of glucose, respectively. Sulfuric acid is produced when sulfur-

Figure 26-1 The pH range. Note that as concentration of H^+ increases the solution becomes increasingly acidic and the pH value decreases. As OH^- concentration increases the pH value also increases, and the solution becomes more and more basic or alkaline.

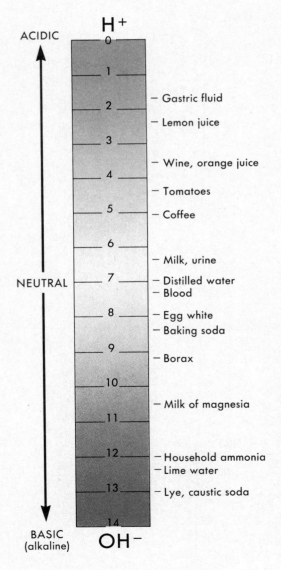

A pH of 7, for example, means that a solution contains 10^{-7} gm of hydrogen ions per liter. Translating this logarithm into a number, a pH of 7 means that a solution contains 0.0000001 (that is, 1/10,000,000) gm of hydrogen ions per liter. A solution of pH 6 contains 0.000001 (1/1,000,000) gm of hydrogen ions per liter, and one of pH 8 contains 0.00000001 (1/100,000,000) gm of hydrogen ions per liter. Note that a solution with pH 7 contains 10 times as many hydrogen ions as a solution with pH 8 and that pH decreases as hydrogen ion concentration increases.

Let us turn our attention now to mechanisms that adjust urine pH to counteract changes in blood pH.

Mechanisms that Control Urine pH

A decrease in blood pH accelerates the renal tubule ion-exchange mechanisms that both acidify urine and conserve blood's base; thereby tending to increase blood pH back to normal. The following paragraphs describe these mechanisms.

1 Distal and collecting tubules secrete hydrogen ions into the urine in exchange for basic ions, which they reabsorb. Refer to Figure 26-8 as you read the rest of this paragraph. Note that carbon dioxide diffuses from tubule capillaries into distal tubule cells, where the enzyme carbonic anhydrase accelerates the combining of carbon dioxide with water to form carbonic acid. The carbonic acid dissociates into hydrogen ions and bicarbonate ions. The hydrogen ions then diffuse into the tubular urine, where they displace basic ions (most often sodium) from a basic salt of a weak acid and thereby change the basic salt to an acid salt or to a weak acid that is eliminated in the urine. While this is happening, the displaced sodium or other basic ion diffuses into a tubule cell. Here, it combines with the bicarbonate ion left over from the carbonic acid dissociation to form sodium bicarbonate. The sodium bicarbonate then diffuses—is reabsorbed—into the blood. Consider the various results of this mechanism. Sodium bicarbonate (or other base bicarbonate) is conserved for the body. Instead of all the basic salts that filter out of glomerular blood leaving the body in the urine, considerable amounts are recovered into peritubular capillary blood. In addition, extra hydrogen ions are added to the urine and thereby eliminated from the body. Both the reabsorption of base bicarbonate into blood and the excretion of hydrogen ions into urine tend to increase the ratio of the bicarbonate buffer pair $B \cdot HCO_3/H \cdot HCO_3$ (BB/CA) present in blood. This automatically increases blood pH. In short, kidney tubule base bicarbonate reabsorption and hydrogen ion excretion both tend to alkalinize blood by acidifying urine.

Renal tubules can excrete either hydrogen or potassium in exchange for the sodium they reabsorb. Therefore, in general, the more hydrogen ions they excrete, the fewer potassium ions they can excrete. For example, in acidosis, tubule excretion of hydrogen ions increases markedly and potassium ion excretion decreases—an important factor because it may lead to **hyperkalemia** (excessive blood potassium), a condition that can cause heart block and death.

2 Distal and collecting tubule cells excrete ammonia into the tubular urine. As Figure 26-9 shows, the ammonia combines with hydrogen to form an ammonium ion. The ammonium ion displaces sodium or some other basic ion from a salt of a fixed (nonvolatile) acid to form an ammonium salt. The basic ion then diffuses back into a tubule cell and combines with bicarbonate ion to form a basic salt, which in turn diffuses into tubular blood. Thus, like the renal tubules' excretion of hydrogen ions, their excretion of ammonia and its combining with hydrogen to form ammonium ions also tends to increase the blood bicarbonate buffer pair ratio and therefore tends to increase blood pH. Quantitatively, however, ammonium ion excretion is more important than hydrogen ion excretion.

Renal tubule excretion of hydrogen and ammonia is controlled at least in part by the blood pH level. As indicated in Figure 26-10, a decrease in blood pH accelerates tubule excretion of both hydrogen and ammonia. An increase in blood pH produces the opposite effects.

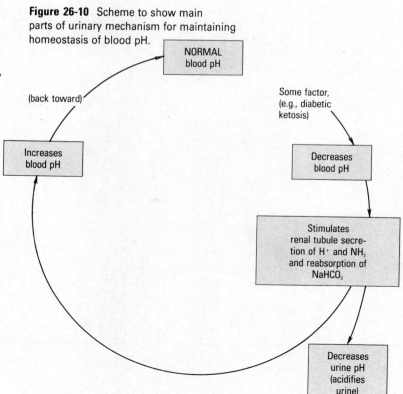

Figure 26-10 Scheme to show main parts of urinary mechanism for maintaining homeostasis of blood pH.

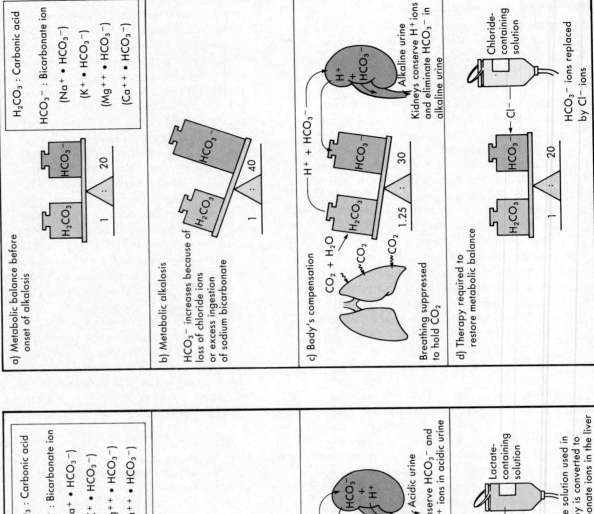

Figure 26-12 Metabolic alkalosis.

a) Metabolic balance before onset of alkalosis

b) Metabolic alkalosis

HCO_3^- increases because of loss of chloride ions or excess ingestion of sodium bicarbonate

c) Body's compensation

Breathing suppressed to hold CO_2

d) Therapy required to restore metabolic balance

HCO_3^- ions replaced by Cl^- ions

H_2CO_3 : Carbonic acid
HCO_3^- : Bicarbonate ion
$(Na^+ \cdot HCO_3^-)$
$(K^+ \cdot HCO_3^-)$
$(Mg^{++} \cdot HCO_3^-)$
$(Ca^{++} \cdot HCO_3^-)$

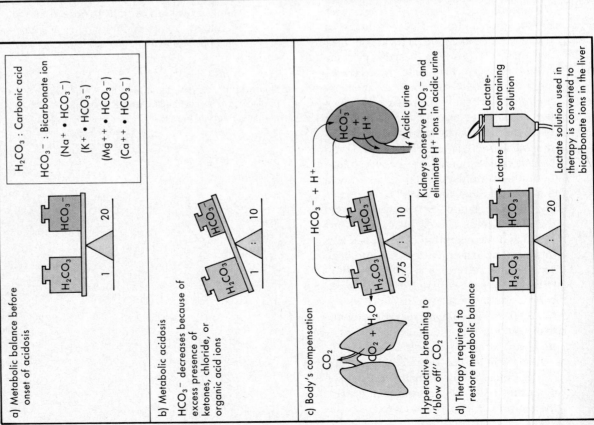

Figure 26-11 Metabolic acidosis.

a) Metabolic balance before onset of acidosis

b) Metabolic acidosis

HCO_3^- decreases because of excess presence of ketones, chloride, or organic acid ions

c) Body's compensation

Hyperactive breathing to "blow off" CO_2

d) Therapy required to restore metabolic balance

Lactate solution used in therapy is converted to bicarbonate ions in the liver

H_2CO_3 : Carbonic acid
HCO_3^- : Bicarbonate ion
$(Na^+ \cdot HCO_3^-)$
$(K^+ \cdot HCO_3^-)$
$(Mg^{++} \cdot HCO_3^-)$
$(Ca^{++} \cdot HCO_3^-)$

Kidneys conserve HCO_3^- and eliminate H^+ ions in acidic urine

Kidneys conserve H^+ ions and eliminate HCO_3^- in alkaline urine

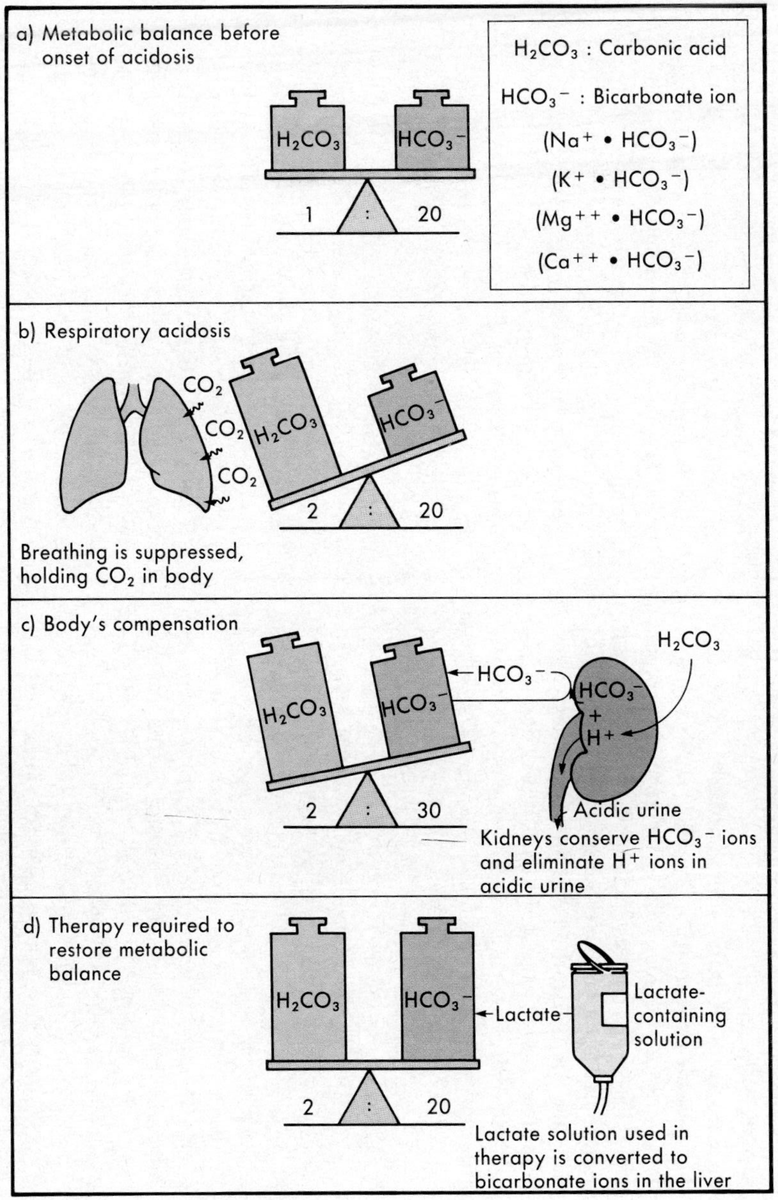

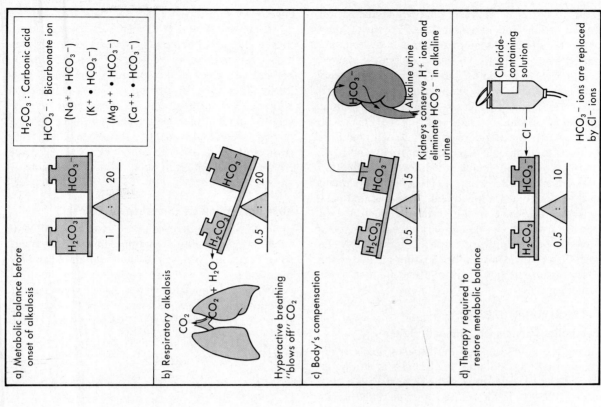

Figure 26-14 Respiratory alkalosis.

a) Metabolic balance before onset of alkalosis

H_2CO_3 : Carbonic acid
HCO_3^- : Bicarbonate ion
$(Na^+ \cdot HCO_3^-)$
$(K^+ \cdot HCO_3^-)$
$(Mg^{++} \cdot HCO_3^-)$
$(Ca^{++} \cdot HCO_3^-)$

H_2CO_3 HCO_3^-
1 : 20

b) Respiratory alkalosis

CO_2
$CO_2 + H_2O$

H_2CO_3 HCO_3^-
0.5 : 20

Hyperactive breathing "blows off" CO_2

c) Body's compensation

HCO_3^-
Alkaline urine
Kidneys conserve H^+ ions and eliminate HCO_3^- in alkaline urine

H_2CO_3 HCO_3^-
0.5 : 15

d) Therapy required to restore metabolic balance

Chloride-containing solution
Cl^-
HCO_3^- ions are replaced by Cl^- ions

H_2CO_3 HCO_3^-
0.5 : 10

Figure 26-13 Respiratory acidosis.

a) Metabolic balance before onset of acidosis

H_2CO_3 : Carbonic acid
HCO_3^- : Bicarbonate ion
$(Na^+ \cdot HCO_3^-)$
$(K^+ \cdot HCO_3^-)$
$(Mg^{++} \cdot HCO_3^-)$
$(Ca^{++} \cdot HCO_3^-)$

H_2CO_3 HCO_3^-
1 : 20

b) Respiratory acidosis

CO_2 CO_2 H_2CO_3 CO_2

H_2CO_3 HCO_3^-
2 : 20

Breathing is suppressed, holding CO_2 in body

c) Body's compensation

H_2CO_3
HCO_3^-
$+$ H^+
HCO_3^-
Acidic urine
Kidneys conserve HCO_3^- ions and eliminate H^+ ions in acidic urine

H_2CO_3 HCO_3^-
2 : 30

d) Therapy required to restore metabolic balance

Lactate-containing solution
Lactate
Lactate solution used in therapy is converted to bicarbonate ions in the liver

H_2CO_3 HCO_3^-
2 : 20

ACID-BASE IMBALANCES

All of the buffer pairs present in body fluids play an important role in acid-base balance. However, only in the bicarbonate system can the body regulate quickly and precisely the levels of both chemical components in the buffer pair. Carbonic acid levels can be regulated by the respiratory system and bicarbonate ion by the kidneys. Recall that a 20/1 ratio of base bicarbonate to carbonic acid (BB/CA) will, according to the Henderson-Hasselbalch equation, maintain acid-base balance and normal blood pH. Therefore, from a clinical standpoint, disturbances in acid-base balance can be considered dependent on the relative quantities of carbonic acid and base bicarbonate present in the extracellular fluid. Two types of disturbances, metabolic and respiratory, can alter the proper ratio of these components. Metabolic disturbances affect the bicarbonate and respiratory disturbances the carbonic acid element of the buffer pair.

METABOLIC DISTURBANCES
Metabolic Acidosis (Bicarbonate Deficit)

During the course of certain diseases, such as untreated diabetes mellitus, or during starvation, abnormally large amounts of acids enter the blood. The ratio of BB/CA is altered as the base bicarbonate component of the buffer pair reacts with the acids. The result may be a new ratio near 10/1. The decreasing ratio will lower the blood pH, and the respiratory center will be stimulated (Figure 26-11). The resulting hyperventilation will result in a "blow-off" of carbon dioxide, with a decrease in carbonic acid. This compensatory action of the respiratory system, coupled with excretion of H^+ and NH_3 in exchange for reabsorbed Na^+ by the kidneys, may be sufficient to adjust the *ratio* of BB/CA, and therefore blood pH, to normal. (The compensated BB/CA ratio may approach 10/0.5.) If, despite these compensating homeostatic devices, the ratio and pH cannot be corrected, uncompensated metabolic acidosis develops.

Increased blood hydrogen ion concentration, that is, decreased blood pH, as we have noted, stimulates the respiratory center. For this reason, hyperventilation is an outstanding clinical sign of acidosis. Increases in hydrogen ion concentration above a certain level depress the central nervous system and therefore produce such symptoms as disorientation and coma. In a terminal illness, death from acidosis is likely to follow coma, while death from alkalosis generally follows tetany and convulsions.

Metabolic Alkalosis (Bicarbonate Excess)

Patients suffering from chronic stomach problems such as hyperacidity sometimes ingest large quantities of alkali—often plain baking soda or sodium bicarbonate—for extended periods of time. Such improper use of antacids or excessive vomiting can produce metabolic alkalosis. Initially the condition results in an increase in the BB/CA ratio to perhaps

---

Metabolic and respiratory acidosis are separate and very different types of acid-base imbalances. Both are treated by the intravenous infusion of solutions containing **sodium lactate.** The infused lactate ions are metabolized by liver cells and converted to bicarbonate ions. This therapy helps replace depleted bicarbonate reserves required to restore acid-base balance in metabolic acidosis. In respiratory acidosis the additional bicarbonate ions function to offset elevated carbonic acid levels.

Vomiting, sometimes referred to as **emesis,** is the forcible emptying or expulsion of gastric and occasionally intestinal contents through the mouth. It occurs as a result of many stimuli, including foul odors or tastes, irritation of the stomach or intestinal mucosa, and some vomitive or *emetic* drugs such as *ipecac.* A "vomiting center" in the brain regulates the many coordinated (but primarily involuntary) steps involved. Severe vomiting such as **pernicious vomiting** of pregnancy or the repeated vomiting associated with pyloric obstruction in infants can be life threatening. One of the most frequent and serious complications of vomiting is metabolic alkalosis. The bicarbonate excess of metabolic alkalosis results because of the massive loss of chloride from the stomach as hydrochloric acid. It is the loss of chloride that causes a compensatory increase of bicarbonate in the extracellular fluid. The result is **metabolic alkalosis.** Therapy includes intravenous administration of chloride-containing solutions such as **normal saline** (0.9% NaCl in water). The chloride ions of the solution replace bicarbonate ions and thus help relieve the bicarbonate excess responsible for the imbalance.

40/1 (Figure 26-12). Compensatory mechanisms are aimed at both increasing carbonic acid and decreasing the bicarbonate load. With breathing suppressed and the kidneys excreting bicarbonate ions, a compensated ratio of 30/1.25 might result. Such a ratio would restore acid-base balance and blood pH to normal. In uncompensated metabolic alkalosis the ratio, and therefore the pH, remain increased.

RESPIRATORY DISTURBANCES
Respiratory Acidosis (Carbonic Acid Excess)

Clinical conditions such as pneumonia or emphysema tend to cause retention of carbon dioxide in the blood. Also, drug abuse or overdose, such as barbiturate poisoning, will suppress breathing and result in respiratory acidosis (Figure 26-13). The carbonic acid component of the bicarbonate buffer pair increases above normal in respiratory acidosis. Body compensation, if successful, increases the bicarbonate fraction so that a new BB/CA ratio (perhaps 20/2) will return blood pH to normal or near normal levels.

Respiratory Alkalosis (Carbonic Acid Deficit)

Hyperventilation caused by fever or mental disease (hysteria) can result in excessive loss of carbonic acid and lead to respiratory alkalosis (Figure 26-14) with a bicarbonate buffer pair ratio of 20/0.5. Compensatory mechanisms may adjust the ratio to 10/0.5 and return blood pH to near normal.

Outline Summary

INTRODUCTION

A Acid-base balance is one of most important of the body's homeostatic mechanisms
B Acid-base balance refers to regulation of hydrogen ion concentration in body fluids
C Precise regulation of pH at the cellular level is necessary for survival
D Slight pH changes have dramatic effects on cellular metabolism

MECHANISMS THAT CONTROL pH OF BODY FLUIDS

A Meaning of pH—negative logarithm of hydrogen ion concentration of solution
B Sources of pH-influencing elements
  1 Carbonic acid—formed by aerobic glucose metabolism
  2 Lactic acid—formed by anaerobic glucose metabolism
  3 Sulfuric acid—formed by oxidation of sulfur-containing amino acids
  4 Phosphoric acid—formed in breakdown of phosphoproteins and nucleoproteins
  5 Acidic ketone bodies—formed in breakdown of fats
    a Acetone
    b Acetoacetic acid
    c Beta-hydroxybutyric acid
  6 Acid-forming foods—high-protein foods are good examples
  7 Acid-forming elements
    a Chlorine
    b Sulfur
    c Phosphorus
  8 Base-forming foods—fruits and vegetables
  9 Base-forming elements
    a Potassium
    b Calcium
    c Sodium
    d Magnesium
  10 Direct acid-forming foods
    a Rhubarb—oxalic acid
    b Cranberries—benzoic acid
  11 Direct base-forming substances—antacids
    a Sodium bicarbonate
    b Calcium carbonate
C Types of pH control mechanisms
  1 Chemical—rapid action buffers
    a Bicarbonate buffer system
    b Phosphate buffer system
    c Protein buffer system
  2 Physiological—delayed action buffers
    a Respiratory response
    b Renal response
  3 Summary of pH control mechanisms
    a Buffers
    b Respirations
    c Kidney excretion of acids and bases
D Effectiveness of pH control mechanisms; range of pH—extremely effective, normally maintain pH within very narrow range of 7.36 to 7.41.

BUFFER MECHANISM FOR CONTROLLING pH OF BODY FLUIDS

A Buffers defined
  1 Substances that prevent marked change in pH of solution when acid or base added to it
  2 Consist of weak acid (or its acid salt) and basic salt of that acid
B Buffer pairs present in body fluids—mainly carbonic acid, proteins, hemoglobin, acid phosphate, and sodium and potassium salts of these weak acids
C Action of buffers to prevent marked changes in pH of body fluids
  1 Nonvolatile acids, such as hydrochloric acid,

lactic acid, and ketone bodies, buffered mainly by sodium bicarbonate
2 Volatile acids, chiefly carbonic acid, buffered mainly by potassium salts of hemoglobin and oxyhemoglobin
3 The chloride shift makes it possible for carbonic acid to be buffered in the red blood cell and then carried as bicarbonate in the plasma
4 Bases buffered mainly by carbonic acid (when homeostasis of pH at 7.4 exists)

$$\text{Ratio } \frac{B \cdot HCO_3}{H_2CO_3} = \frac{20}{1}$$

5 The Henderson-Hasselbalch equation is a mathematical formula that explains the relationship between hydrogen ion concentration of body fluids and the ratio of base bicarbonate to carbonic acid
D Evaluation of role of buffers in pH control—cannot maintain normal pH without adequate functioning of respiratory and urinary pH control mechanisms

RESPIRATORY MECHANISM OF pH CONTROL

A Explanation of mechanism
1 Amount of blood carbon dioxide directly related to amount of carbonic acid and therefore to concentration of H^+
2 With increased respirations, less carbon dioxide remains in blood, hence less carbonic acid and fewer H^+; with decreased respirations, more carbon dioxide remains in blood, hence more carbonic acid and more H^+
B Adjustment of respirations to pH of arterial blood (see Figure 26-7)
C Some principles relating respirations and pH of body fluids
1 Acidosis → hyperventilation
↓
increases elimination of CO_2
↓
decreases blood CO_2
↓
decreases blood H_2CO_3
↓
decreases blood H^+, that is, increases blood pH
↓
tends to correct acidosis, that is, to restore normal pH
2 Prolonged hyperventilation, by decreasing blood H^+ excessively, may produce alkalosis
3 Alkalosis causes hypoventilation, which tends to correct alkalosis by increasing blood CO_2 and therefore blood H_2CO_3 and H^+

4 Prolonged hypoventilation, by eliminating too little CO_2, causes increase in blood H_2CO_3 and consequently in blood H^+, thereby may produce acidosis

URINARY MECHANISM OF pH CONTROL

A General principles about mechanism—plays vital role in acid-base balance because kidneys can eliminate more H^+ from body while reabsorbing more base when pH tends toward acid side and eliminate fewer H^+ while reabsorbing less base when pH tends toward alkaline side
B Mechanisms that control urine pH
1 Secretion of H^+ into urine—when blood CO_2, H_2CO_3, and H^+ increase above normal, distal tubules secrete more H^+ into urine to displace basic ion (mainly sodium) from a urine salt and then reabsorb sodium into blood in exchange for the H^+ excreted
2 Secretion of NH_3—when blood hydrogen ion concentration increases, distal tubules secrete more NH_3, which combines with H^+ of urine to form ammonium ion, which displaces basic ion (mainly sodium) from a salt; basic ion then reabsorbed back into blood in exchange for ammonium ion excreted

ACID-BASE IMBALANCES

Metabolic disturbances affect the bicarbonate and respiratory disturbances the carbonic acid element of the bicarbonate buffer pair

Metabolic disturbances (Figures 26-11 and 26-12)

A Metabolic acidosis (bicarbonate deficit)—results in decreased alkaline reserve (mainly $NaHCO_3$), but ratio BB/CA maintained at normal 20/1 by proportionately decreasing blood carbonic acid by hyperventilation coupled with excretion of H^+ and NH_3 in exchange for reabsorbed Na^+ by the kidneys
B Metabolic alkalosis (bicarbonate excess)—initially the condition results in an increase in BB/CA; compensatory mechanisms are aimed at both increasing carbonic acid and decreasing the bicarbonate load

Respiratory disturbances (Figures 26-13 and 26-14)

A Respiratory acidosis (carbonic acid excess)—the carbonic acid component of the bicarbonate buffer pair increases above normal in respiratory acidosis; compensatory mechanisms are aimed at increasing the bicarbonate fraction in order to return blood pH to normal or near normal levels
B Respiratory alkalosis (carbonic acid deficit)—characterized by excessive loss of carbonic acid caused by hyperventilation

Review Questions

1 Explain, in your own words, what pH means.
2 What is the normal range for pH of body fluids?
3 Explain what a buffer is in terms of its chemical composition and in terms of its function. Cite specific equations to illustrate your explanation.
4 What is the numerical value of the ratio of B HCO_3/H_2CO_3 when blood pH is 7.4 and the body is in a state of acid-base balance?
5 Is the ratio B HCO_3/H_2CO_3 necessarily abnormal when an acid-base disturbance is present? Give reasons to support your answer.
6 Is blood pH always normal when an acid-base disturbance is present? Give reasons for your answer.
7 Explain how the chloride shift makes it possible for carbon dioxide to be buffered in the red blood cell and then carried as bicarbonate in the plasma.
8 Explain how the "respiratory mechanism of pH control" operates.
9 Why is hyperventilation a characteristic clinical sign in acidosis?

Situation: A patient has suffered a severe head injury. Respirations are markedly depressed.

10 Which, if either, do you think is a potential danger—acidosis or alkalosis? State your reasons.

11 Describe the homeostatic mechanisms operating to maintain acid-base balance in this patient.

Situation: A mother brings her baby to the hospital and reports that he has seemed very sick for the past 24 hours and that he has not eaten during that time and has passed no urine.

12 Do you think acidosis or alkalosis may be present in this baby? Why?
13 Identify several of the sources that contribute hydrogen ions to the body fluids.
14 What are the three types of ketone bodies that accumulate during the incomplete breakdown of fats?
15 Which minerals are said to be acid forming? Identify food sources that are high in these elements
16 What types of foods do nutritionists consider base forming?
17 Explain what the terms *direct acid-* or *base-forming substances* mean. Give examples of each.
18 Compare and contrast the chemical and physiological buffer systems.
19 How can sodium lactate be useful in the treatment of both metabolic and respiratory acidosis?

27

THE MALE REPRODUCTIVE SYSTEM

OBJECTIVES

After you have completed this chapter, you should be able to:

1 Explain how the normal functioning of the reproductive system differs from the end result of normal function measured in any other organ system of the body.

2 List the essential and accessory organs of the male reproductive system and give the generalized function of each.

3 Describe the gross and microscopic anatomy of the testes.

4 Discuss the primary functions of testosterone and identify the cell type responsible for its secretion.

5 Explain the process of spermatogenesis and describe the structure of a mature spermatozoon.

6 Trace the passage of an individual sperm cell from its point of formation, in sequence, through the genital ducts to the exterior of the body.

7 Discuss the composition and function of seminal fluid or semen.

8 Compare the structure, location, and function of the accessory reproductive glands in the male.

9 Identify the components of the male external genitals.

10 Discuss male fertility.

11 List and discuss the primary male functions in reproduction. Contrast the reflex responses that terminate in orgasm.

MEANING AND FUNCTION

The importance of normal reproductive system function is notably different from the end result of "normal function" as measured in any other organ system of the body. The proper functioning of the reproductive system and of its enormously complex control mechanisms ensures survival not of the individual but of the species. In both sexes organs of the reproductive system are adapted for the specific sequence of functions that are concerned primarily with propagation of the species. In addition, production of hormones that permits development of the secondary sex characteristics occurs as a result of normal reproductive system activity. In humans sexual maturity and the ability to reproduce occur at puberty. The male reproductive system consists of those organs whose functions are to produce, transfer, and ultimately introduce mature sperm into the female reproductive tract, where fertilization can occur. It is the testes that secrete androgens, or male sex hormones, notably testosterone.

KEY TERMS

Androgens (AN-dro-jens)

Bulbourethral (bul-bo-u-RE-thral)

Epididymis (ep-i-DID-i-mis)

Gamete (GAM-eet)

Genital (JEN-i-tal)

Genitalia (jen-i-TA-li-ah)

Gonads (go-NADS)

Penis (PE-nis)

Prostate (PROS-tate)

Scrotum (SKRO-tum)

Semen (SEE-men)

Seminal vesicle (SEM-i-nal VES-i-kal)

Spermatogenesis (sper-mah-to-JEN-e-sis)

Spermatozoa (sper-mah-to-ZO-ah)

Testes (TES-tis)

Testosterone (tes-TOS-teh-rone)

Figure 27-1 Male pelvic organs as viewed in a median sagittal section.

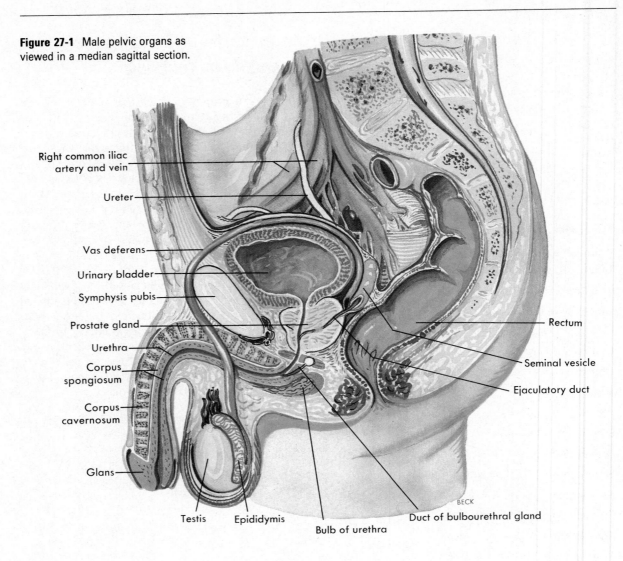

Right common iliac artery and vein

Ureter

Vas deferens

Urinary bladder

Symphysis pubis

Prostate gland

Urethra

Corpus spongiosum

Corpus cavernosum

Glans

Testis   Epididymis

Bulb of urethra

Rectum

Seminal vesicle

Ejaculatory duct

Duct of bulbourethral gland

BECK

MALE REPRODUCTIVE ORGANS

Organs of the male reproductive system (Figure 27-1) may be classified as either **essential organs** for the production of gametes (sex cells) or **accessory organs** that play some type of supportive role in the reproductive process.

In both sexes the essential organs of reproduction that produce the gametes or sex cells (sperm or ova) are called **gonads.** The gonads of the male are the testes.

The accessory organs of reproduction in the male include a number of genital ducts, glands, and supporting structures.

Genital ducts serve to convey sperm to the outside of the body. The ducts are a pair of epididymides (singular: epididymis), a pair of vas deferens (ductus deferens), a pair of ejaculatory ducts, and the urethra.

Accessory glands in the reproductive system produce secretions that serve to nourish, transport, and mature sperm. The glands are a pair of seminal vesicles, one prostate, and a pair of bulbourethral (Cowper's) glands.

The supporting structures are the scrotum, the penis, and a pair of spermatic cords.

TESTES

Structure and Location

The testes are small ovoid glands that are somewhat flattened from side to side, measure about 4 or 5 cm in length, and weigh 10 to 15 gm each. The left testis is generally located about 1 cm lower in the scrotal sac than the right. Both testes are suspended in the pouch by attachment to scrotal tissue and by the spermatic cords (Figure 27-2). Note in Figure 27-2 that testicular blood vessels, collectively called the *vas afferens*, reach the testes by passing through the spermatic cord. A dense white fibrous capsule called the **tunica albuginea** encases each testis and then enters the gland, sending out partitions (septa) that radiate through its interior, dividing it into 200 or more cone-shaped lobules.

Each lobule of the testis contains specialized **interstitial cells** (of Leydig) and one to three tiny, coiled **seminiferous tubules,** which, if unraveled, would measure about 75 cm in length. The tubules from each lobule come together to form a plexus called the *rete testis.* A series of sperm ducts called *efferent ductules* then drain the rete testis and pierce the tunica albuginea to enter the head of the epididymis (see Figure 27-2).

Microscopic Anatomy

Figure 27-3 is a low-power ($\times 70$) light micrograph of testicular tissue showing a number of cut seminiferous tubules and numerous interstitial cells (Leydig cells) in the surrounding connective tissue septa. In this figure, maturing sperm appear as dense nuclei with their tails projecting into the lumen of the tubule that is partially enclosed by the rectangle. The wall of each seminiferous tubule may contain five or more layers of cells. At puberty, when sexual maturity begins, spermatogenic cells in diverse stages of development appear and the hormone-producing interstitial cells become more prominent in the surrounding septa. Actively dividing cells at the periphery of each tubule in Figure 27-3 can be identified by their chromosomes. The irregular elongated cell in Figure 27-4 is a supportive cell and provides mechanical support and protection for the devel-

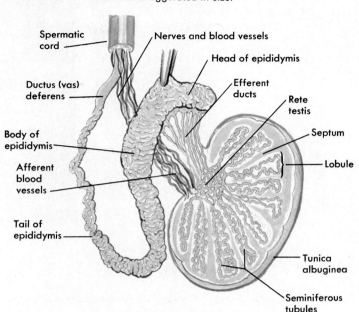

Figure 27-2 Tubules of the testis and epididymis. The ducts and tubules are exaggerated in size.

oping germ cells. It also plays an active role in eventual release of mature spermatozoa into the lumen of the seminiferous tubule.

Functions

The testes perform two primary functions: spermatogenesis and secretion of hormones.

1 **Spermatogenesis,** the production of spermatozoa (sperm), the male gametes or re-

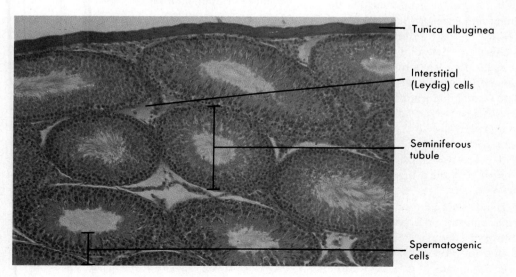

Figure 27-3 Testis. Low-power view showing several seminiferous tubules surrounded by septa containing interstitial (Leydig) cells. ($\times 70$.)

Figure 27-4 Testis. High-power view showing portions of three seminiferous tubules with interstitial (Leydig) cells and capillaries between the tubules. (×350.)

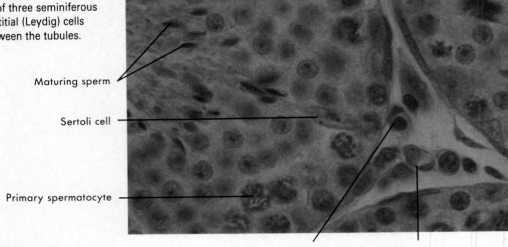

Maturing sperm

Sertoli cell

Primary spermatocyte

Interstitial (Leydig) cell   Capillary

productive cells. The seminiferous tubules produce the sperm.

2 **Secretion of hormones,** chiefly testosterone (androgen or masculinizing hormone) by interstitial cells (Leydig cells). Testosterone serves the following general functions:

   a It promotes "maleness," or development and maintenance of male secondary sex characteristics, accessory organs such as the prostate, seminal vesicles, etc., and adult male sexual behavior.

   b It helps regulate metabolism and is sometimes referred to as "the anabolic hormone" because of its marked stimulating effect on protein anabolism. By stimulating protein anabolism, testosterone promotes growth of skeletal muscles (responsible for greater male muscular development and strength) and growth of bone. However, testosterone also promotes closure of the epiphyses. Early sexual maturation leads to early epiphyseal closure. The converse also holds true: late sexual maturation, delayed epiphyseal closure, and tallness tend to go together.

   c It plays a part in fluid and electrolyte metabolism. Testosterone has a mild stimulating effect on kidney tubule reabsorption of sodium and water; it also promotes kidney tubule excretion of potassium.

   d It inhibits anterior pituitary secretion of gonadotropins, namely, FSH and ICSH.
   The anterior pituitary gland controls the testes by means of its gonadotropic

hormones—specifically, FSH and ICSH just mentioned. FSH stimulates the seminiferous tubules to produce sperm more rapidly. ICSH stimulates interstitial cells to increase their secretion of testosterone. Soon the blood concentration of testosterone reaches a high level that inhibits anterior pituitary secretion of FSH and ICSH. Thus a negative feedback mechanism operates between the anterior pituitary gland and the testes. A high blood concentration of gonadotropins stimulates testosterone secretion. But a high blood concentration of testosterone inhibits (has a negative effect on) gonadotropin secretion (Figure 27-5).

Structure of Spermatozoa

The elongated tail-bearing spermatozoa seen in the seminiferous tubules (Figure 27-4) appear fully formed. We know, however, that they undergo a process of "ripening" or maturation as they pass through the genital ducts before ejaculation. Although anatomically complete and highly motile when ejaculated, sperm must still undergo a complex process called **capacitation** before they are capable of actually fertilizing an ovum. Normally, capacitation occurs in sperm only after they have been introduced into the vagina of the female.

Figure 27-6, *A*, shows the characteristic parts of a spermatozoon: head, neck, middle piece, and elongated, lashlike tail. The head of a spermatozoon is, in essence, a highly compact package of genetic chromatin material covered by a specialized *acrosome* and *acrosomal (head) cap*. The acrosome con-

Figure 27-5 The negative feedback mechanism that controls anterior pituitary gland secretion of ICSH and interstitial cell secretion of testosterone.

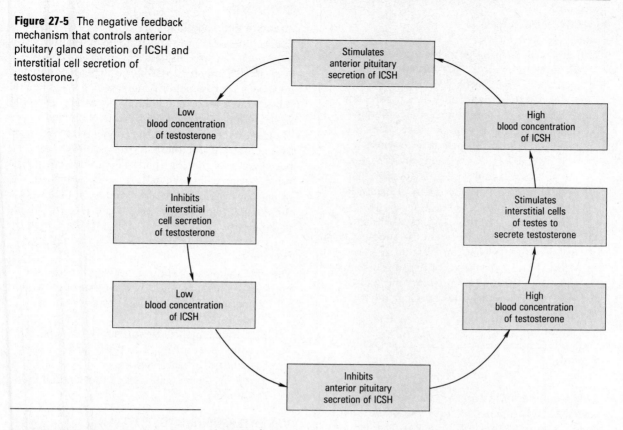

Figure 27-6 Reproductive cells; **A,** Male cell or spermatozoon. **B,** Female cell of ovum surrounded by spermatozoa.

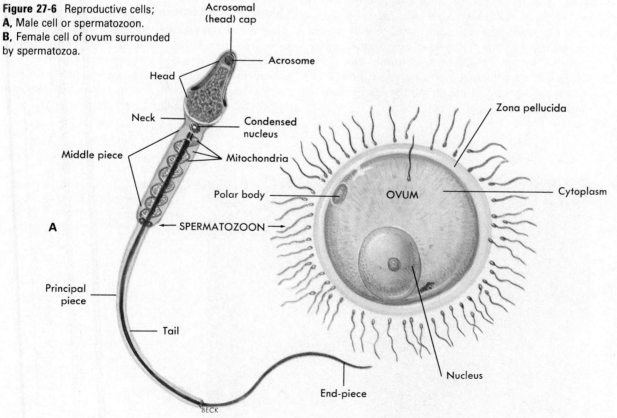

tains a number of hydrolytic (splitting) enzymes designed to aid in penetration of the outer zona pellucida (Figure 27-6, *B*) of the ovum. The cylindrical middle piece is characterized by a helical arrangement of mitochondria arranged end to end around a central core. It is this mitochondrial sheath that provides energy for sperm locomotion. The tail is divided into a principal piece and short end-piece, both typical in appearance of all flagella capable of motility.

REPRODUCTIVE (GENITAL) DUCTS

EPIDIDYMIS
Structure and Location
Each epididymis consists of a single, tightly coiled tube enclosed in a fibrous casing. The tube has a very small diameter (just barely macroscopic) but measures approximately 6 meters (20 feet) in length. It lies along the top and behind the testis (Figure 27-2). The comma-shaped epididymis is divided into a blunt superior **head** (which is connected to the testis by the efferent ductules), a central **body,** and a tapered inferior portion that is continuous with the vas deferens called the **tail.** If the epididymis is cut or sectioned and a slide prepared as in Figure 27-7, the compact and highly coiled nature of the tubule is apparent.

Functions
The epididymis serves the following functions:

1 It serves as one of the ducts through which sperm pass in their journey from the testis to the exterior.

2 It contributes to the maturation of sperm, which spend from 1 to 3 weeks in this segment of the duct system.

3 It secretes a small part of the seminal fluid (semen).

VAS DEFERENS (DUCTUS DEFERENS)
Structure and Location
The vas deferens, like the epididymis, is a tube. In fact, the duct of the vas is an extension of the tail of the epididymis. The vas deferens has thick muscular walls (Figure 27-8) and can be palpated in the scrotal sac as a smooth movable cord. Note in Figure 27-8 that the muscular layer of the vas has three

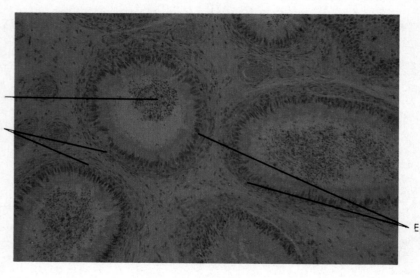

Spermatozoa in lumen

Tubules of epididymus

Epithelium

Figure 27-7 Epididymis. This extensively coiled tubule has been sectioned several times in this preparation. Stereocilia occur on the epithelial cells lining the tubule. (×35.)

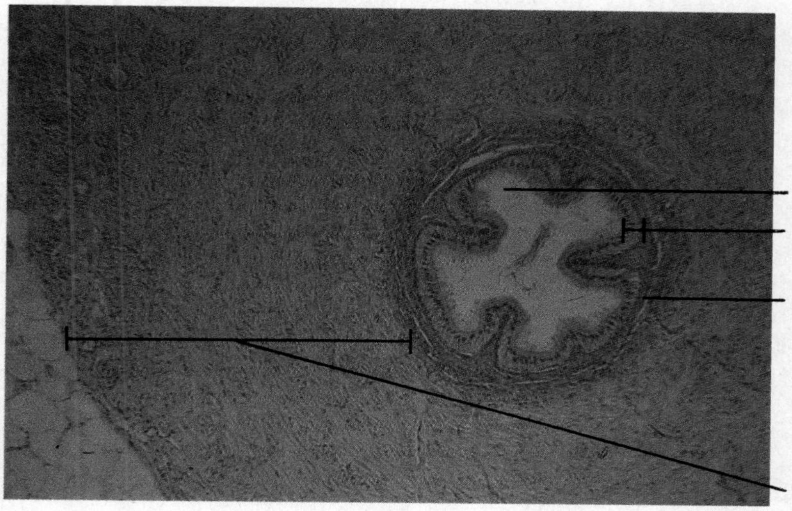

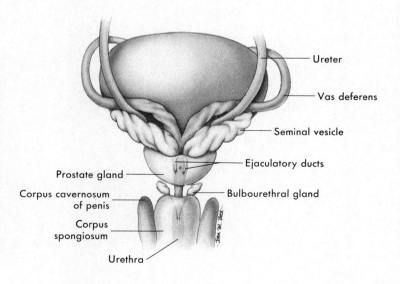

Figure 27-8 Transverse section of vas deferens. Sperm are seen as tiny dots within the lumen. (× 35.)

Lumen
Epithelium
Submucosa
Muscular coat

layers: a thick intermediate circular layer of muscle fibers and inner and outer longitudinal layers. The muscular layers of the vas help in propelling sperm through the duct system. The vas deferens from each testis ascends from the scrotum and passes through the inguinal canal as part of the spermatic cord—enclosed by fibrous connective tissue with blood vessels, nerves, and lymphatics—into the abdominal cavity. Here it extends over the top and down the posterior surface of the bladder where an enlarged and tortuous portion called the *ampulla* joins the duct from the seminal vesicle to form the ejaculatory duct (Figures 27-9 and 27-10).

Function

The vas deferens serves as one of the male genital ducts connecting the epididymis with the ejaculatory duct. Sperm remain in the vas deferens for varying periods of time depending on the degree of sexual activity and frequency of ejaculation. Storage time may exceed 1 month with no loss of fertility.

VASECTOMY

Severing of the vas deferens—that is, a **vasectomy,** usually done through incision in the scrotum—makes a man sterile. Why? Because it interrupts the route to the exterior from the epididymis. To leave the body, sperm must journey in succession through the epididymis, vas deferens, ejaculatory duct, and urethra.

EJACULATORY DUCT

The two ducts are short tubes that pass through the prostate gland to terminate in the urethra. As Figure 27-9 shows, they are formed by the union of the vas deferens with the ducts from the seminal vesicles.

URETHRA

For a discussion of the urethra, see p. 645.

Figure 27-9 Formation of the ejaculatory ducts by union of the seminal vesicles with the vas deferens just before entrance into the prostate gland. The ejaculatory ducts open into the prostatic portion of the urethra.

Ureter
Vas deferens
Seminal vesicle
Ejaculatory ducts
Prostate gland
Corpus cavernosum of penis
Bulbourethral gland
Corpus spongiosum
Urethra

Figure 27-10 The urinary bladder and male reproductive organs viewed from behind.

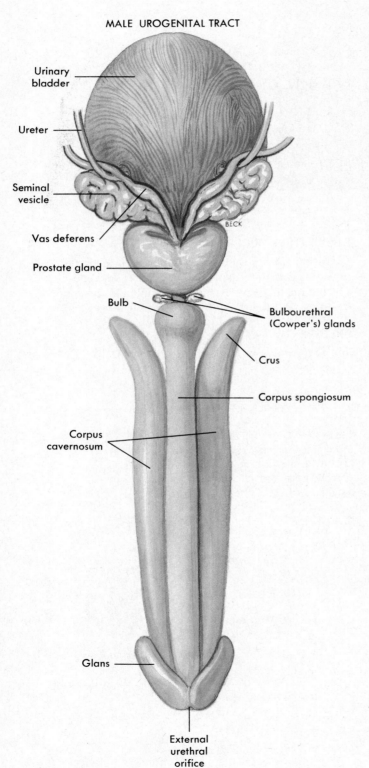

MALE UROGENITAL TRACT

Urinary bladder

Ureter

Seminal vesicle

Vas deferens

Prostate gland

Bulb

Bulbourethral (Cowper's) glands

Crus

Corpus spongiosum

Corpus cavernosum

Glans

External urethral orifice

BECK

ACCESSORY REPRODUCTIVE GLANDS

SEMINAL VESICLES
Structure and Location
The seminal vesicles are convoluted pouches that lie along the lower part of the posterior surface of the bladder, directly in front of the rectum (Figure 27-2).

Function
The seminal vesicles secrete a viscous liquid component of the semen rich in fructose. As a simple sugar, fructose serves as an energy source for sperm motility after ejaculation. The thick, yellowish secretion also contains prostaglandins, substances postulated to influence cyclic AMP formation (p. 388). Normal secretory activity of the seminal vesicles depends on adequate levels of testosterone.

PROSTATE GLAND
Structure and Location
The prostate is a compound tubuloalveolar gland that lies just below the bladder and is shaped like a doughnut. The fact that the urethra passes through the small hole in the center of the prostate is a matter of considerable clinical significance. Many older men suffer from enlargement of this gland. As it enlarges, it squeezes the urethra, frequently closing it so completely that urination becomes impossible. Urinary retention results. Surgical removal of the gland (prostatectomy) is resorted to as a cure for this condition when other less radical methods of treatment fail.

Function
The prostate secretes a thin alkaline substance that constitutes the largest part of the seminal fluid. Its alkalinity helps protect the sperm from acid present in the male urethra and female vagina and thereby increases sperm motility. (Acid depresses or, if strong enough, kills sperm. Sperm motility is greatest in neutral or slightly alkaline media.)

BULBOURETHRAL GLANDS
Structure and Location
The two bulbourethral or Cowper's glands resemble peas in both size and shape. You can see the location of these compound tubuloalveolar glands in Figure 27-6. A duct approximately 2.5 cm (1 inch) long connects them with the penile portion of the urethra.

Function
Like the prostate, the bulbourethral glands secrete an alkaline fluid that is important for counteracting the acid present in both the male urethra and the female vagina.

SUPPORTING STRUCTURES

SCROTUM (EXTERNAL)

The scrotum is a skin-covered pouch suspended from the perineal region. Internally, it is divided into two sacs by a septum, each sac containing a testis, epididymis, and lower part of a spermatic cord.

PENIS (EXTERNAL)
Structure

Three cylindrical masses of erectile or cavernous tissue, enclosed in separate fibrous coverings and held together by a covering of skin, compose the penis (Figure 27-10). The two larger and uppermost of these cylinders are named the **corpora cavernosa penis,** whereas the smaller, lower one, which contains the urethra, is called the **corpus cavernosum urethrae** (corpus spongiosum).

The distal part of the corpus cavernosum urethrae overlaps the terminal end of the two corpora cavernosa penis to form a slightly bulging structure, the **glans penis,** over which the skin is folded doubly to form a more or less loose-fitting, retractable casing known as the **prepuce** or foreskin. If the foreskin fits too tightly about the glans, a circumcision is usually performed to prevent irritation. The opening of the urethra at the tip of the glans is called the *external urinary meatus.*

Functions

The penis contains the urethra, the terminal duct for both urinary and reproductive tracts. It is the copulatory organ by means of which spermatozoa are introduced into the female vagina. The scrotum and penis together constitute the **external genitals** of the male.

SPERMATIC CORDS (INTERNAL)

The **spermatic cords** are cylindrical casings of white fibrous tissue located in the inguinal canals between the scrotum and the abdominal cavity. They enclose the vas deferens, blood vessels, lymphatics, and nerves (Figure 27-11).

COMPOSITION AND COURSE OF SEMINAL FLUID

The following structures secrete the substances that, together, make up the seminal fluid or semen:

1 Testes and epididymis—Their secretions, according to one estimate, constitute less than 5% of the seminal fluid volume.

2 Seminal vesicles—Their secretions are reported to contribute about 30% of the seminal fluid volume.

3 Prostate gland—Its secretions constitute

Figure 27-11 Lateral view of the left spermatic cord and testis.

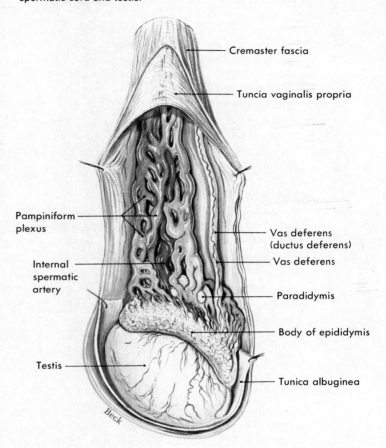

Cremaster fascia

Tuncia vaginalis propria

Pampiniform plexus

Internal spermatic artery

Testis

Beck

Vas deferens (ductus deferens)

Vas deferens

Paradidymis

Body of epididymis

Tunica albuginea

the bulk of the seminal fluid volume, reportedly about 60%.

4 Bulbourethral glands—Their secretions are said to constitute less than 5% of the seminal fluid volume.

Besides contributing slightly to the fluid part of semen, the testes also add hundreds of millions of sperm. In traversing the distance from their place of origin to the exterior, the sperm must pass from the testis through the epididymis, vas deferens, ejaculatory duct, and urethra. Note that sperm originate in the testes, glands located outside the body (that is, not within a body cavity), travel inside, and finally are expelled outside.

MALE FERTILITY

Male fertility relates to many factors—most of all to the number of sperm ejaculated but also to their size, shape, and motility. Fertile sperm have a uniform size and shape. They are highly motile. Although only one sperm fertilizes an ovum, millions

MALE FUNCTIONS IN REPRODUCTION

Recall that all body functions but one have for their ultimate goal survival of the individual. Only the function of reproduction serves a different, a longer range, and, no doubt in nature's scheme, a more important purpose—survival of the human species. Male functions in reproduction consist of the production of male sex cells (spermatogenesis, discussed in Chapter 3) and introduction of these cells into the female body (coitus, copulation, or sexual intercourse). For coitus to take place, erection of the penis must first occur, and for sperm to enter the female body, both the sex cells and secretions from the accessory glands must be introduced into the urethra (emission) and semen ejaculated from the penis.

Erection is a parasympathetic reflex initiated mainly by certain tactile, visual, and mental stimuli. It consists of dilation of the arteries and arterioles of the penis, which in turn floods and distends spaces in its erectile tissue and compresses its veins. Therefore more blood enters the penis through the dilated arteries than leaves it through the constricted veins. Hence, it becomes larger and rigid, or, in other words, erection occurs.

Emission is the reflex movement of sex cells, or spermatozoa, and secretions from the genital ducts and accessory glands into the prostatic urethra. Once emission has occurred, ejaculation will follow.

Ejaculation of semen is also a reflex response. It is the usual outcome of the same stimuli that initiate erection. Ejaculation and various other responses—notably accelerated heart rate, increased blood pressure, hyperventilation, dilated skin blood vessels, and intense sexual excitement—characterize the climax of coitus known as an **orgasm.**

of sperm seem to be necessary for fertilization to occur. According to one estimate, when the sperm count falls below about 50 million per milliliter of semen, sterility results.

One hypothesis suggested to explain this puzzling fact is this: semen that contains an adequate number of sperm also contains enough hyaluronidase to liquefy the intercellular substance between the cells that encase each ovum. Without this, a single sperm cannot penetrate the layers of cells (corona radiata) around the ovum and hence cannot fertilize it. Infertility may also be caused by production of antibodies some men make against their own sperm. This type of sperm destruction or inactivation called "immune infertility" is caused by an antigen-antibody reaction. A sperm surface protein called fertilization antigen (FA-1) triggers antibody production, which results in infertility.

Outline Summary

MEANING AND FUNCTION

Consists of organs that, together, produce a new individual

MALE REPRODUCTIVE ORGANS

  A Essential—testes (male gonads)
  B Accessory
    1 Glands
      a Seminal vesicles (paired)
      b Prostate gland
      c Bulbourethral (Cowper's) glands (paired)
    2 Ducts of testes
      a Epididymis (paired)

      b Vas deferens (ductus deferens) (paired)
      c Ejaculatory ducts (paired)
      d Urethra
    3 Supporting structures
      a External—scrotum and penis
      b Internal—spermatic cords (paired)

TESTES

  A Structure and location
    1 Several lobules composed of seminiferous tubules and interstitial cells (of Leydig), separated by septa, encased in fibrous capsule called the *tunica albuginea*

Outline Summary—cont'd

2 Seminiferous tubules in testis open into a plexus called *rete testis*, which is drained by a series of efferent ductules that emerge from the top of the organ and enter the head of epididymis

3 Located in scrotum, one testis in each of two scrotal compartments

B Microscopic anatomy

C Functions

  1 Spermatogenesis—formation of mature male gametes (spermatozoa) by seminiferous tubules

  2 Secretion of hormone (testosterone) by interstitial cells

D Structure of spermatozoa—consists of a head covered by acrosome, neck, middle piece, and tail divided into a principal piece and a short end-piece

REPRODUCTIVE (GENITAL) DUCTS

A Epididymis

  1 Structure and location

    a Single tightly coiled tube enclosed in fibrous casing

    b Lies along top and side of each testis

    c Anatomical divisions include head, body, and tail

  2 Functions

    a Duct for seminal fluid

    b Also secretes part of seminal fluid

    c Sperm become capable of motility while they are stored in epididymis

B Vas deferens (seminal dust; ductus deferens)

  1 Structure and location

    a Tube, extension of epididymis

    b Extends through inguinal canal, into abdominal cavity, over top and down posterior surface of bladder

    c Enlarged terminal portion called *ampulla*—joins duct of seminal vesicle

  2 Function

    a One of excretory ducts for seminal fluid

    b Connects epididymis with ejaculatory duct

C Ejaculatory duct

  1 Formed by union of vas deferens with duct from seminal vesicle

  2 Passes through prostate gland, terminating in urethra

D Urethra (see p. 645)

ACCESSORY REPRODUCTIVE GLANDS

A Seminal vesicles

  1 Structure and location—convoluted pouches on posterior surface of bladder

  2 Function—secrete prostaglandins and viscous nutrient—rich part of seminal fluid

B Prostate gland

  1 Structure and location

    a Doughnut-shaped

    b Encircles urethra just below bladder

  2 Function—adds alkaline secretion to seminal fluid

C Bulbourethral glands

  1 Structure and location

    a Small, pea-shaped structures with 1-inch long ducts leading into urethra

    b Lie below prostate gland

  2 Function—secrete alkaline fluid that is part of semen

SUPPORTING STRUCTURES

A External

  1 Scrotum

    a Skin-covered pouch suspended from perineal region

    b Divided into two compartments

    c Contains testis, epididymis, and first part of vas deferens

  2 Penis—composed of three cylindrical masses of erectile tissue, one of which contains urethra

B Internal

  1 Spermatic cords

    a Fibrous cylinders located in inguinal canals

    b Enclose seminal ducts, blood vessels, lymphatics, and nerves

COMPOSITION AND COURSE OF SEMINAL FLUID

A Consists of secretions from testes, epididymides, seminal vesicles, prostate, and bulbourethral glands

B Each drop contains millions of sperm

C Passes from testes through epididymis, vas deferens, ejaculatory duct, and urethra

MALE FERTILITY

A Relates to many factors—number of sperm; size, shape, and motility

B Infertility—may be caused by antibodies some men make against their own sperm

MALE FUNCTIONS IN REPRODUCTION

A Spermatogenesis

B Coitus (copulation or sexual intercourse)

  1 Erection—a parasympathetic reflex initiated by various kinds of stimuli and consisting of dilation of arteries and arterioles of the penis, causing its enlargement and erection

  2 Emission—reflex movement of sperm and secretions of genital ducts and accessory organs into prostatic urethra

  3 Ejaculation of semen—one of several responses that characterize orgasm, the climax of coitus

Review Questions

1 How does the function of reproduction differ from all other body functions?
2 Name the accessory glands of the male reproductive system.
3 List the genital ducts in the male.
4 List the supporting structures of the male reproductive system.
5 What is tunica albuginea? How does it aid in dividing the testis into lobules?
6 What is the relationship between the rete testis, seminiferous tubules, and efferent ductules?
7 What are the two primary functions of the testes?
8 What are the general functions of testosterone?
9 Discuss the structure of a mature spermatozoon.
10 List three functions of the epididymis.
11 What is meant by the term *capacitation?*
12 List the anatomical divisions of the epididymis.
13 Discuss the formation of the ejaculatory ducts.

14 What is the relationship of the prostate gland to the urethra?
15 What and where are the bulbourethral (Cowper's) glands?
16 What is the spermatic cord? From what to what does it extend, and what does it contain?
17 Discuss the type of secretion typical of the prostate gland and seminal vesicles.
18 Name the three cylindrical masses of erectile or cavernous tissue in the penis.
19 What and where is the glans penis? The prepuce or foreskin?
20 Define the following terms: cryptorchidism, external urinary meatus, coitus, erection, emission, ejaculation.
21 Of what is the seminal fluid composed? Trace its course from its formation in the gonads to the exterior.

28 THE FEMALE REPRODUCTIVE SYSTEM

OBJECTIVES

After you have completed this chapter, you should be able to:

1 List the essential and accessory sex organs of the female reproductive system and give the generalized function of each.
2 Discuss the structure of the uterus, including details of its coats or walls, size, shape, divisions, cavities, blood supply, and ligaments.
3 Locate the uterus in the pelvic cavity and compare its normal position with the abnormal positions of retroflexion and anteflexion.
4 Discuss the location, structure, divisions, and functions of the uterine, or fallopian, tubes.
5 Describe the structure of the female gonads and explain the steps in development of mature ova from ovarian follicles.
6 Identify the structures that together constitute the female external genitals.
7 Discuss the location, structure, and primary functions of the vagina.
8 Explain the clinical importance of the perineum during childbirth.
9 Describe the structure of the breasts and the mechanism controlling lactation.
10 Identify the phases of the endometrial or menstrual cycle.
11 Explain the hormonal control of cyclical changes that occur in the ovaries.
12 Compare and contrast menarche and menopause.
13 Briefly describe the developmental processes of fertilization, implantation, histogenesis, and organogenesis.
14 Define the terms *zygote, morula, blastocyst, embryo,* and *fetus.*

FEMALE REPRODUCTIVE ORGANS

Examine Figure 28-1 and identify the following organs of the female reproductive system:

1 *Essential sex organs*—the two ovaries (female gonads).
2 *Accessory sex organs*—subdivided into an external and internal or pelvic group or organs. The internal (pelvic) group includes the uterus, two uterine tubes (Fallopian tubes or oviducts), and the vagina. The external organs are collectively called the *vulva* (pudendum) or *external genitalia*. The breasts or mammary glands are also classified as external accessory sex organs in the female.

KEY TERMS

Bartholin's glands (BAR-to-linz)

Estrogen (ES-tro-jen)

Lactation (lak-TA-shun)

Menarche (me-NAR-ke)

Menstruation (MEN-stroo-A-shun)

Menopause (Men-o-pawz)

Ovary (O-vah-re)

Oviduct (O-vi-duct)

Ovulation (ov-u-LAY-shun)

Perineum (per-i-NE-um)

Progesterone (pro-JES-te-rone)

Uterus (YOO-tur-us)

Vagina (va-JYE-nuh)

Vulva (VUL-vah)

Figure 28-1 Female pelvic organs as viewed in a median sagittal section.

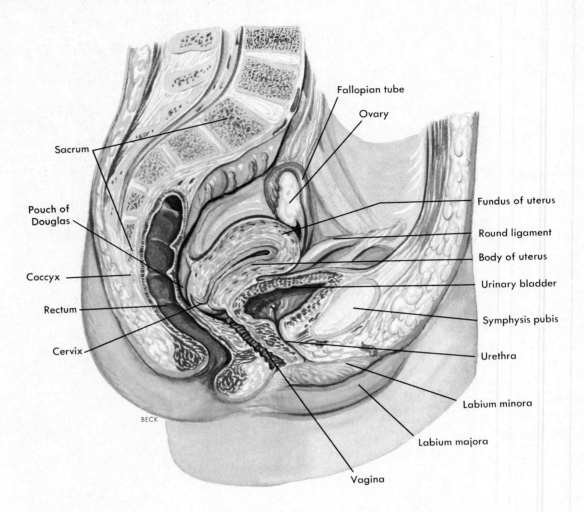

Fallopian tube

Ovary

Sacrum

Pouch of Douglas

Coccyx

Rectum

Cervix

Fundus of uterus

Round ligament

Body of uterus

Urinary bladder

Symphysis pubis

Urethra

Labium minora

Labium majora

Vagina

BECK

UTERUS

Structure

Size, Shape, and Divisions

The uterus is pear shaped in its virginal state and measures approximately 7.5 cm (3 inches) in length, 5 cm (2 inches) in width at its widest part, and 3 cm (1 inch) in thickness. Note in Figure 28-2 that the uterus has two main parts: an upper portion, the **body,** and a lower, narrow section, the **cervix.** Did you notice that the body of the uterus rounds into a bulging prominence above the level at which the uterine tubes enter? This bulging upper component of the body is called the **fundus.**

Wall

Three coats compose the walls of the uterus: the inner endometrium, a middle myometrium, and an outer incomplete layer of parietal peritoneum.

1 The lining of mucous membrane, called the **endometrium,** is composed of three layers of tissues: a compact surface layer of partially ciliated simple columnar epithelium called the *stratum compactum,* a spongy middle or intermediate layer of loose connective tissue—the *stratum spongiosum*—and a dense inner layer termed the *stratum basale* that attaches the endometrium to the underlying myometrium. During menstruation and following delivery of a baby, the compact and spongy layers slough off. The endometrium varies in thickness from 0.5 mm just after the menstrual flow to about 5 mm near the end of the endometrial cycle.

2 A thick, middle coat (the **myometrium**) consists of three layers of smooth muscle fibers that extend in all directions, longitudi-

Figure 28-2 Sectioned view of
the uterus.

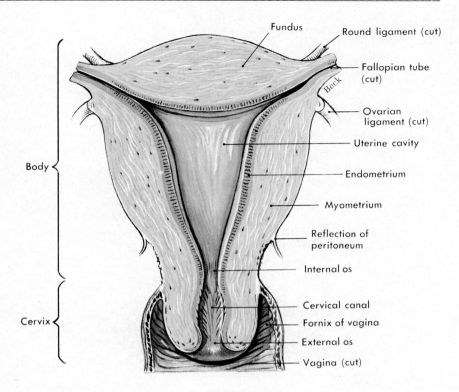

and constitutes the **internal os,** which opens into
the **cervical canal.** The cervical canal is constricted
on its lower end also, forming the **external os,**
which opens into the vagina. The uterine tubes open
into the body cavity at its upper, outer angles.

Blood Supply

The uterus receives a generous supply of blood from
uterine arteries, branches of the internal iliac arteries. In addition, blood from the ovarian and vaginal
arteries reaches the uterus by anastomosis with the
uterine vessels. Tortuous arterial vessels enter the
layers of the uterine wall as arterioles and then break
up into capillaries between the endometrial glands.

Uterine, ovarian, and vaginal veins return venous blood from the uterus to the internal iliac veins.

Location

Figure 28-3 shows the location of the uterus in the
pelvic cavity between the urinary bladder in front
and the rectum behind. Age, pregnancy, and distention of related pelvic viscera such as the bladder will
alter the position of the uterus.

Between birth and puberty the uterus descends
gradually from the lower abdomen into the true pelvis. At menopause the uterus begins a process of
involution that results in a decrease in size and a
position deep in the pelvis. Some variation in placement within the pelvis is common.

nally, transversely, and obliquely, and give
the uterus great strength. The bundles of
smooth muscle fibers interlace with elastic
and connective tissue components and generally blend into the endometrial lining with
no sharp line of demarcation between the
two layers. The myometrium is thickest in
the fundus and thinnest in the cervix—a
good example of the principle of structural
adaptation to function. To expel a fetus,
that is, move it down and out of the uterus,
the fundus must contract more forcibly
than the lower part of the uterine wall, and
the cervix must be stretched or dilated.

3 An external coat of serous membrane, the
parietal peritoneum, is incomplete, since
it covers none of the cervix and only part
of the body (all except the lower one fourth
of its anterior surface). The fact that the entire uterus is not covered with peritoneum
has clinical value because it makes it possible to perform operations on this organ
without the risk of infection that attends
cutting into the peritoneum.

Cavities

The cavities of the uterus are small because of the
thickness of its walls (Figure 28-2). The body cavity
is flat and triangular. Its apex is directed downward

Figure 28-3 Female pelvic contents as seen from above and in front.

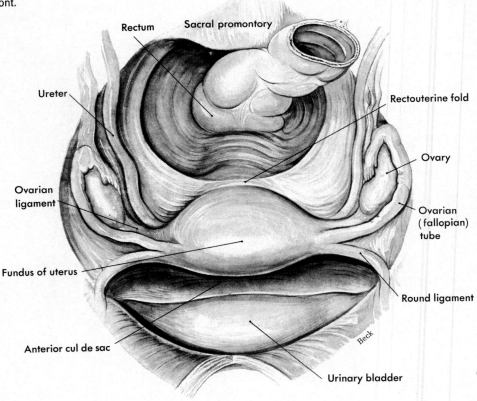

Rectum

Sacral promontory

Ureter

Rectouterine fold

Ovary

Ovarian ligament

Ovarian (fallopian) tube

Fundus of uterus

Round ligament

Anterior cul de sac

Beck

Urinary bladder

Figure 28-4 Normal and abnormal positions of the uterus. Red lines show the abnormal positions. **A,** Retroflexion. **B,** Anteflexion.

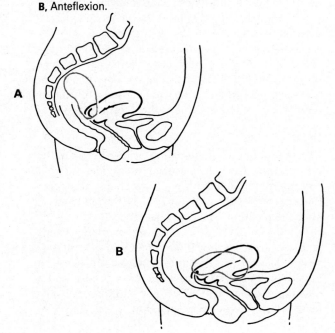

A

B

Position

1 *Normally* the uterus is flexed between the body and cervix, with the body lying over the superior surface of the bladder, pointing forward and slightly upward (Figure 28-4). The cervix points downward and backward from the point of flexion, joining the vagina at approximately a right angle. Several ligaments hold the uterus in place but allow its body considerable movement, a characteristic that often leads to malpositions of the organ. Fibers from several muscles that form the pelvic floor (see Figure 9-37 on p. 265) converge to form a node called the **perineal body,** which also serves an important role in support of the uterus.

2 The uterus may lie in any one of several **abnormal positions.** A common one is retroflexion, or backward tilting, of the entire organ.

3 *Eight ligaments* (three pairs, two single ones) anchor the uterus in the pelvic cavity: broad (paired), uterosacral (paired), posterior (single), anterior (single), and round (paired). Six of these so-called ligaments are actually extensions of the parietal peri-

toneum in different directions. The round ligaments are fibromuscular cords.

a The *two* **broad ligaments** are double folds of parietal peritoneum that form a kind of partition across the pelvic cavity. The uterus is suspended between these two folds.

b The *two* **uterosacral ligaments** are fold-like extensions of the peritoneum from the posterior surface of the uterus to the sacrum, one on each side of the rectum.

c The **posterior ligament** is a fold of peritoneum extending from the posterior surface of the uterus to the rectum. This ligament forms a deep pouch known as the **cul-de-sac of Douglas** (or rectouterine pouch) between the uterus and rectum. Since this is the lowest point in the pelvic cavity, pus collects here in pelvic inflammations. To secure drainage, an incision may be made at the top of the posterior wall of the vagina (posterior colpotomy).

d The **anterior ligament** is the fold of peritoneum formed by the extension of the peritoneum on the anterior surface of the uterus to the posterior surface of the bladder. This fold also forms a cul-de-sac but one that is less deep than the posterior pouch.

e The *two* **round ligaments** (Figure 28-3) are fibromuscular cords extending from the upper, outer angles of the uterus through the inguinal canals and terminating in the labia majora.

Functions

The uterus or womb plays a role in the accomplishment of three functions vital for survival of the human species but not for individual survival: menstruation, pregnancy, and labor.

1 **Menstruation** is a sloughing away of the compact and spongy layers of the endometrium, attended by bleeding from the torn vessels.

2 In **pregnancy** the embryo implants itself in the endometrium and there lives as a parasite throughout the fetal period.

3 **Labor** consists of powerful, rhythmic contractions of the muscular uterine wall that result in expulsion of the fetus at birth.

UTERINE TUBES—FALLOPIAN TUBES OR OVIDUCTS

Location

The uterine tubes are about 10 cm (4 inches) long and are attached to the uterus at its upper outer angles (see Figures 28-1 and 28-3). They lie in the upper free margin of the broad ligaments and extend upward and outward toward the sides of the pelvis and then curve downward and backward.

Structure
Wall

The same three coats (mucous, smooth muscle, and serous) of the uterus compose the tubes (Figure 28-5). The mucosal lining of the tubes, however, is directly continuous with the peritoneum lining the pelvic cavity. This fact has great clinical significance because the tubal mucosa is also continuous with that of the uterus and vagina and therefore often becomes infected by gonococci or other organisms introduced into the vagina. Inflammation of the tubes (**salpingitis**) may readily spread to become inflammation of the peritoneum (**peritonitis**), a serious

Figure 28-5 Transverse section of fallopian tube (oviduct) in the isthmus region. (×70.)

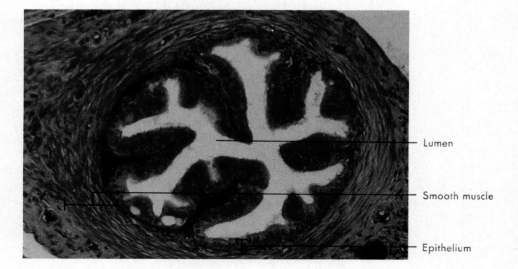

Lumen

Smooth muscle

Epithelium

Figure 28-6 Fallopian tube (oviduct) in the ampulla region. Note the highly branching nature of the mucosa. (× 140.)

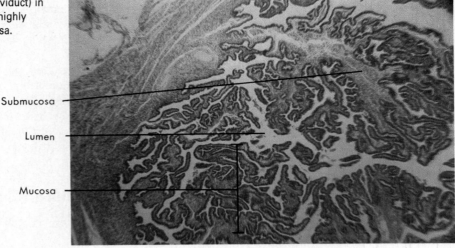

Submucosa

Lumen

Mucosa

Figure 28-7 Fallopian tube (oviduct). Note patches of cilia on pseudostratified epithelium.

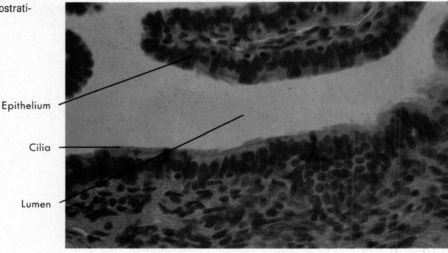

Epithelium

Cilia

Lumen

condition. In the male, there is no such direct route by which microorganisms can reach the peritoneum from the exterior.

Divisions

Each uterine tube consists of three divisions:

1 A medial third that extends from the upper outer angle of the uterus called the **isthmus**

2 An intermediate dilated portion called the **ampulla** that follows a tortuous path over the ovary.

3 A funnel-shaped terminal component called the **infundibulum** that opens directly into the peritoneal cavity. The open outer margin of the infundibulum resembles a fringe in its irregular outline. The fringelike pro-

jections are known as **fimbriae** (see Figure 28-8).

Histology

Figure 28-6 is a low-power micrograph that illustrates the mucosal lining of the oviduct cut in cross section. These shapes are typical of the appearance of the mucosal lining throughout most of the duct. Note the extensive folds of mucosa that project as shelves into the lumen of the tube. An area of smooth muscle (muscularis layer) can be seen surrounding the mucosa in this section. The rectangular area is enlarged in Figure 28-7 to illustrate the columnar nature of the oviduct's epithelium. Cilia, which are important in maintaining currents within the tube to move the ova toward the uterus, can be seen projecting into the lumen.

nancy—a fact of value in diagnosing a first pregnancy. The color decreases after lactation has ceased but never entirely returns to the original hue. In darker-skinned women, no noticeable color change in the areola or nipple heralds the first pregnancy.

A knowledge of the lymphatic drainage of the breast is critically important in clinical medicine because cancerous cells from malignant breast tumors often spread to other areas of the body through the lymphatics. Details of the lymphatic drainage of the breast are presented in Chapter 18 (see also Figure 18-8.)

Function

The function of the mammary glands is lactation, that is, the secretion of milk for the nourishment of newborn infants.

Mechanism Controlling Lactation

Very briefly, lactation is controlled as follows:

1 The ovarian hormones, estrogens and progesterone, act on the breasts to make them structurally ready to secrete milk. Estrogens promote development of the ducts of the breasts. Progesterone acts on the estrogen-primed breasts to promote completion of the development of the ducts and develop-

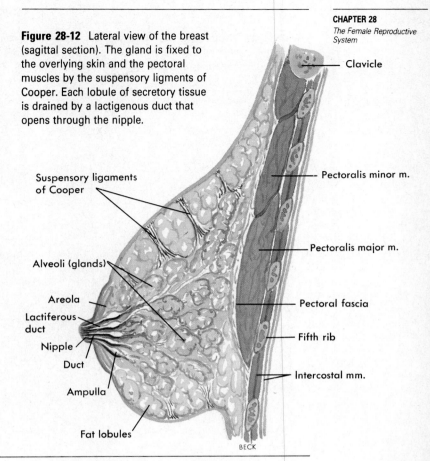

Figure 28-12 Lateral view of the breast (sagittal section). The gland is fixed to the overlying skin and the pectoral muscles by the suspensory ligments of Cooper. Each lobule of secretory tissue is drained by a lactigenous duct that opens through the nipple.

Clavicle
Pectoralis minor m.
Pectoralis major m.
Pectoral fascia
Fifth rib
Intercostal mm.

Suspensory ligaments of Cooper
Alveoli (glands)
Areola
Lactiferous duct
Nipple
Duct
Ampulla
Fat lobules

BECK

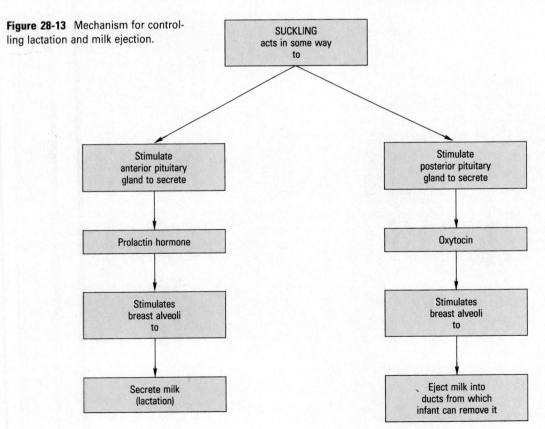

Figure 28-13 Mechanism for controlling lactation and milk ejection.

SUCKLING
acts in some way to

Stimulate anterior pituitary gland to secrete

Prolactin hormone

Stimulates breast alveoli to

Secrete milk (lactation)

Stimulate posterior pituitary gland to secrete

Oxytocin

Stimulates breast alveoli to

Eject milk into ducts from which infant can remove it

Figure 28-14 The lactating breast. **A,** Note the increase in glandular tissue and hypertrophy of duct system. **B,** Each sinus located behind the areola is drained by an individual duct through the nipple. **C,** Suckling compresses the sinuses to start the milk flow.

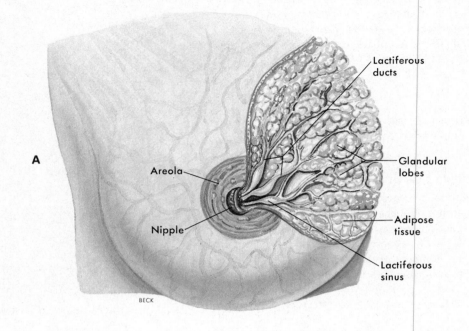

A

Lactiferous ducts

Areola

Nipple

Glandular lobes

Adipose tissue

Lactiferous sinus

BECK

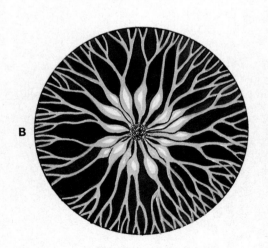

B

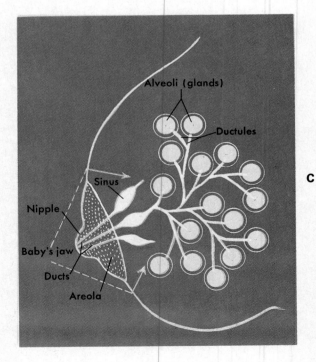

C

Alveoli (glands)

Ductules

Sinus

Nipple

Baby's jaw

Ducts

Areola

ment of the alveoli, the secreting cells of the breasts. A high blood concentration of estrogens, for example, during pregnancy, inhibits anterior pituitary secretion of prolactin.

2 Shedding of the placenta following delivery of the baby cuts off a major source of estrogens. The resulting rapid drop in the blood concentration of estrogens stimulates anterior pituitary secretion of prolactin. Also, the suckling movements of a nursing baby act in some way to stimulate both anterior pituitary secretion of prolactin and posterior pituitary secretion of oxytocin (Figure 28-13).

3 Prolactin stimulates lactation, that is, stimulates alveoli of the mammary glands to secrete milk. Milk secretion starts about the third or fourth day after delivery of a baby, supplanting a thin, yellowish secretion called *colostrum*. With repeated stimulation by the suckling infant, plus various favorable mental and physical conditions, lactation may continue almost indefinitely.

4 Oxytocin stimulates the alveoli of the breasts to eject milk into the ducts, thereby making it accessible for the infant to remove by suckling.

The lactating breast is illustrated in Figure 28-14.

FEMALE SEXUAL CYCLES

RECURRING CYCLES

Many changes recur periodically in the female during the years between the onset of the menses (menarche) and their cessation (menopause or climac-

teric). Most obvious, of course, is menstruation—the outward sign of changes in the endometrium. Most women also note periodic changes in the breasts. But these are only two of many changes that occur over and over again at fairly uniform intervals during the approximately 30 years of female reproductive maturity.

First we shall describe the major cyclical changes, and then we shall discuss the mechanisms that produce them.

Ovarian Cycles

Once each month, on about the first day of menstruation, several primitive graafian follicles and their enclosed ova begin to grow and develop. The follicular cells proliferate and start to secrete estrogens (and minute amounts of progesterone). Usually, only one follicle matures and migrates to the surface of the ovary. The surface of the follicle degenerates, causing expulsion of the mature ovum into the pelvic cavity (ovulation).

When does ovulation occur? This is a question of great practical importance and one that in the past was given many answers. Today it is known that ovulation usually occurs 14 days before the next menstrual period begins. (Only in a 28-day menstrual cycle is this also 14 days after the beginning of the preceding menstrual cycle, as explained subsequently.)

Shortly before ovulation the ovum undergoes meiosis, the process by which its number of chromosomes is reduced by half. Immediately after ovulation, cells of the ruptured follicle enlarge and, because of the appearance of lipoid substances in them, become transformed into a golden-colored body, the **corpus luteum.** The corpus luteum grows for 7 or 8 days. During this time, it secretes progesterone in increasing amounts. Then, provided fertilization of the ovum has not taken place, the size of the corpus luteum and the amount of its secretions gradually diminish. In time, the last components of each nonfunctional corpus luteum are reduced to a white scar called the **corpus albicans,** which moves into the central portion of the ovary and eventually disappears (see Figure 28-8).

FERTILITY SIGNS

Rhythmic changes also take place in the ovaries, in the amount and consistency of the cervical mucus produced during each cycle, in the myometrium, in the vagina, in gonadotropin secretion, in body temperature, and even in mood or "emotional tone." Many of these recurring events that a woman may recognize on almost a monthly schedule during her reproductive years are called **"fertility signs"** and are manifestations of those body changes required for successful reproductive function.

A few women experience pain within a few hours after ovulation. This is referred to as **mittelschmerz**—German for middle pain. It has been ascribed to irritation of the peritoneum by hemorrhage from the ruptured follicle on about day 14 of the menstrual cycle.

Figure 28-15 **A**, Endometrium of uterus in proliferative stage. Note extensive development of uterine glands. (×14.) **B**, Endometrium of uterus in mensus period. Note loss of columnar epithelium of mucosa. (×14.)

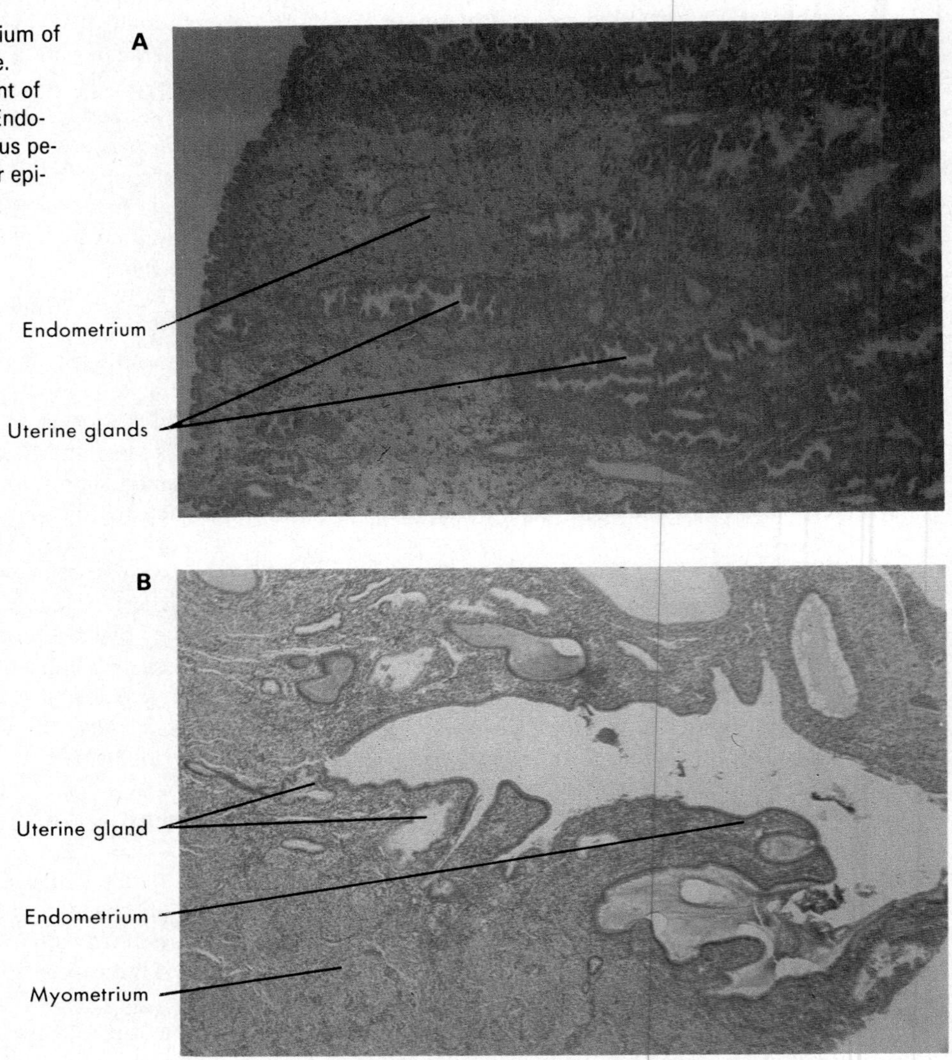

A

Endometrium

Uterine glands

B

Uterine gland

Endometrium

Myometrium

Endometrial or Menstrual Cycle

During menstruation, necrotic bits of the compact and spongy layers of the endometrium slough off, leaving denuded bleeding areas. Following menstruation, the cells of these layers proliferate, causing the endometrium to reach a thickness of 2 or 3 mm by the time of ovulation (Figure 28-15, *A*). During this period, endometrial glands and arterioles grow longer and more coiled—two factors that also contribute to the thickening of the endometrium (see Figure 28-18). After ovulation, the endometrium grows still thicker, reaching a maximum of about 4 to 6 mm (Figure 28-15, *B*). Most of this increase, however, is believed to be caused by swelling produced by fluid retention rather than by further proliferation of endometrial cells. The increasingly tortuous endometrial glands start to secrete during the time between ovulation and the next menses. Then,

the day before menstruation starts again, the tightly coiled arterioles constrict, producing endometrial ischemia. This leads to necrosis, sloughing, and once again, menstrual bleeding.

The menstrual cycle is customarily divided into phases, named for major events occurring in them: menses, postmenstrual or preovulatory phase, ovulation, and postovulatory or premenstrual phase.

1 The **menses** or **menstrual period** occurs on cycle days 1 to 5. There is some individual variation, however.

2 The **postmenstrual phase** occurs between the end of the menses and ovulation. Therefore it is the preovulatory as well as the postmenstrual phase. In a 28-day cycle, it usually includes cycle days 6 to 13 or 14. But the length of this phase varies more than do the others. It lasts longer in long

cycles and ends sooner in short ones. This phase is also called the **estrogenic or follicular phase** because of the high blood estrogen level resulting from secretion by the developing follicle. **Proliferative phase** is still another name for it because proliferation of endometrial cells occurs at this time.

3 Ovulation, that is, rupture of the mature follicle with expulsion of its ovum into the pelvic cavity, occurs frequently on cycle day 15 in a 28-day cycle. However, it occurs on different days in different length cycles, depending on the length of the preovulatory phase. For example, in a 32-day cycle the preovulatory phase would probably last until cycle day 18 and ovulation would then occur on cycle day 19 instead of 15. In short, because the vast majority of women show some month-to-month variation in the length of their cycles, the day of ovulation in a current or future cycle cannot be predicted with accuracy based on the length of previous cycles. Simply knowing the length of previous cycles cannot ensure with any degree of accuracy how many days the preovulatory phase will last in the next or some future cycle. This physiological fact probably accounts for most of the unreliability of the calendar rhythm methods of contraception and its replacement by other more sophisticated "natural" methods of family planning that are not based on a knowledge of previous cycle lengths to predict the day of ovulation. Instead, such natural methods base their judgments about fertility at any point in a woman's cycle on other changes. Examples: measurement of basal body temperature and recognition of cyclic changes in the amount and consistency of cervical mucus, both of which occur in response to changes in circulating hormones that control ovulation.

4 The **postovulatory** (or **premenstrual**) **phase** occurs between ovulation and the onset of the menses. This phase is also called the **luteal phase** because the corpus luteum secretes only during this time and **progesterone phase** because it secretes mainly this hormone. The length of the premenstrual phase is fairly constant, lasting usually 14 days—or cycle days 15 to 28 in a 28-day cycle. Differences in length of the total menstrual cycle, therefore, exist mainly because of differences in duration of the preovulatory rather than of the premenstrual phase.

Myometrial Cycle

The myometrium contracts mildly but with increasing frequency during the 2 weeks preceding ovulation. Contractions decrease or disappear between ovulation and the next menses, thereby lessening the probability of expulsion of an implanted ovum.

Gonadotropic Cycles

The adenohypophysis (anterior pituitary gland) secretes two hormones called gonadotropins that influence female reproductive cycles. Their names are **follicle-stimulating hormone (FSH)** and **luteinizing hormone (LH).** The amount of each gonadotropin secreted varies with a rhythmic regularity that can be related, as we shall see, to the rhythmic ovarian and uterine changes just described.

CONTROL

Physiologists agree that hormones play the major role in producing the cyclic changes characteristic in the female during the years of reproductive maturity. The recent development of a method called **radioimmunoassay** has made it possible to measure blood levels of gonadotropins. By correlating these with the monthly ovarian and uterine changes, investigators have worked out the main features of the control mechanism.

A brief description follows of the mechanisms that produce cyclical changes in the ovaries and uterus and in the amounts of gonadotropins secreted.

Control of Cyclical Changes in Ovaries

Cyclical changes in the ovaries result from cyclical changes in the amounts of gonadotropins secreted by the anterior pituitary gland. An increasing FSH blood level has two effects: it stimulates one or more primitive graafian follicles and ova to start growing, and it stimulates the follicles to secrete estrogens. (Developing follicles also secrete very small amounts of progesterone.) Because of the influence of FSH on follicle secretion, the level of estrogens in blood increases gradually for a few days during the postmenstrual phase. Then suddenly, on about the twelfth cycle day, it leaps upward to a maximum peak. Scarcely 12 hours after this "estrogen surge," an "LH surge" occurs and presumably triggers ovulation a day or two later. The control of cyclical ovarian changes by the gonadotropins FSH and LH is summarized in Figures 28-16 and 28-17. As Figure 28-17 shows, LH brings about the following changes:

1 Completion of growth of the follicle and ovum with increasing secretion of estrogens before ovulation. LH and FSH act as synergists to produce these effects.

2 Rupturing of the mature follicle with expul-

Figure 28-16 Follicle-stimulating hormone's principal effects on ovaries.

Figure 28-17 Luteinizing hormone's effects on ovaries: (1) LH acts as synergist to FSH to enhance its effects on follicular development and secretion. (2) LH presumably triggers ovulation—hence it is called "the ovulating hormone." (3) There is a luteinizing effect of LH (for which the hormone was named); recent evidence shows that FSH is also necessary for luteinization.

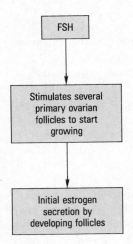

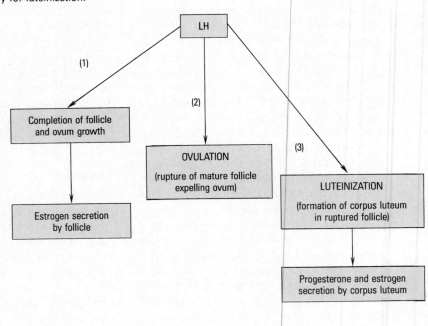

sion of its ripe ovum (process known as **ovulation**). Because of this function, LH is sometimes called "the ovulating hormone."

3 Formation of a golden body, the corpus luteum, in the ruptured follicle (process called **luteinization**). The name *luteinizing hormone* refers, obviously, to this LH function—a function to which, according to recent evidence, FSH also contributes. The corpus luteum functions as a temporary endocrine gland. It secretes only during the luteal (postovulatory or premenstrual) phase of the menstrual cycle. Its hormones are progestins (the important one of which is progesterone) and also estrogens. The blood level of progesterone rises rapidly after the "LH surge" described before. It remains at a high level for about a week, then it decreases to a very low level approximately 3 days before menstruation starts again. This low blood level of progesterone persists during both the menstrual and postmenstrual phases. Its sources? Not from the corpus luteum, which secretes only during the luteal phase, but from the developing follicles and the adrenal cortex. Blood's estrogen content increases during the luteal phase but to a lower level than develops before ovulation.

If pregnancy does not occur, the corpus luteum regresses in about 14 days and is replaced by the corpus albicans. Figure 28-8 shows the cyclical changes in the ovarian follicles.

Control of Cyclical Changes in Uterus

Cyclical changes in the uterus are brought about by changing blood concentrations of estrogens and progestrone. As blood estrogens increase during the postmenstrual phase of the menstrual cycle, they produce the following main changes in the uterus.

1 Proliferation of endometrial cells, producing a thickening of the endometrium
2 Growth of endometrial glands and of their spiral arteries
3 Increase in the water content of the endometrium
4 Increased myometrial contractions

Increasing blood progesterone concentration during the premenstrual phase of the menstrual cycle produces progestational changes in the uterus—that is, changes favorable for pregnancy—specifically:

1 Secretion by endometrial glands, thereby preparing the endometrium for implantation of a fertilized ovum
2 Increase in the water content of the endometrium
3 Decreased myometrial contractions

Figure 28-18 A negative feedback mechanism controls anterior pituitary secretion of follicle-stimulating hormone (FSH) and ovarian secretion of estrogens. A high blood level of FSH stimulates estrogen secretion, whereas the resulting high estrogen level inhibits FSH secretion. How does this compare with interstitial cell-stimulating hormone (ICSH)-testosterone feedback mechanism? (See Figure 27-5 if you want to check your answer.) According to the above diagram, what effect does a high blood concentration of estrogens have on anterior pituitary secretion of FSH? of LH?

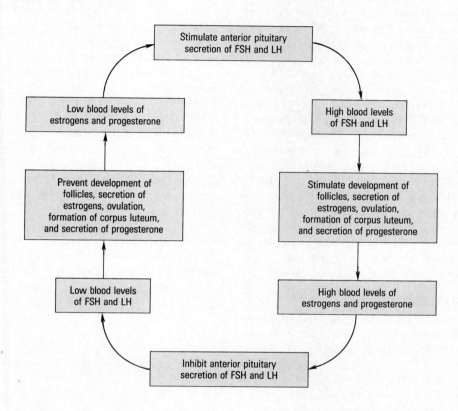

Control of Cyclical Changes in Amounts of Gonadotropins Secreted

Both negative and positive feedback mechanisms help control anterior pituitary secretion of the gonadotropins FSH and LH. These mechanisms involve the ovaries' secretion of both estrogens and progesterone and the hypothalamus' secretion of releasing hormones (until recently, called **releasing factors**). Figure 28-18 describes a negative feedback mechanism that controls gonadotropin secretion. Examine it carefully. Note particularly the effects of a high blood concentration of estrogens on anterior pituitary gland secretion and the effect of a low blood concentration of FSH on the development of a graafian follicle and ovum. Establishment of these two facts led eventually to the development of "the pill" for preventing pregnancy. Contraceptive pills contain synthetic estrogen-like or progesterone-like compounds or both. By building up a high blood concentration of these substances, they prevent the

development of a follicle and its ovum that month. With no mature ovum to be expelled, ovulation does not occur, and therefore pregnancy cannot occur. The next menses, however, does take place—because the progesterone and estrogen dosage is stopped in time to allow their blood level to decrease as they normally do near the end of the cycle to bring on menstruation. Actually the effects of contraceptive pills—estrogens and progesterone—are much more complex than our explanation indicates. They have widespread effects on the body quite independent of their action on the reproductive and endocrine systems and still are not completely understood. Side effects that may limit or prohibit use of "the pill" often result because of the nonreproductive system—related activity of these complex hormones.

Several observations and animal experiments strongly suggest that high blood levels of estrogens and progesterone may also act indirectly to inhibit pituitary secretion of FSH and LH. These ovarian hormones appear to inhibit certain neurons of the hypothalamus (part of the central nervous system) from secreting FSH-releasing and LH-releasing hormones into the pituitary portal vessels. Without the stimulating effects of these releasing hormones, the pituitary's secretion of FSH and LH decreases.

A positive feedback mechanism has also been postulated to control LH secretion. The rapid and marked increase in blood's estrogen content that occurs late in the follicular phase of the menstrual cycle is thought to stimulate the hypothalamus to secrete LH-releasing hormone into the pituitary portal vessels. As its name suggests, this hormone stimulates the release of LH by the anterior pituitary, which in turn accounts for the "LH surge" that triggers ovulation. The fact that a part of the brain—the hypothalamus—secretes FSH- and LH-releasing hormones has interesting implications. This may be the crucial part of the pathway by which changes in a woman's environment or in her emotional state can alter her menstrual cycle. That this occurs is a matter of common observation. For example, intense fear of either becoming or not becoming pregnant often delays menstruation (Figure 28-19).

FUNCTION SERVED BY CYCLES

The major function seems to be to prepare the endometrium each month for a pregnancy. If it does not occur, the thick vascular lining, no longer needed, is shed.

If fertilization of the ovum (pregnancy) occurs, the menstrual cycle is modified as follows:

1 The corpus luteum does not disappear but persists and continues to secrete progesterone and estrogens for 6 months or more of pregnancy. If it is removed by any means during the early months of pregnancy, spontaneous abortion results.

2 The fertilized ovum, which immediately starts developing into an embryo, travels down the tube and implants itself in the endometrium, so carefully prepared for this event.

MENARCHE AND MENOPAUSE

The menstrual flow first occurs (**menarche**) at puberty, at about the age of 13 years, although there is wide individual variation according to race, nutrition, health, heredity, etc. Normally, it recurs about every 28 days for about 40 years, except during pregnancy, and then ceases (menopause or climacteric). The average age at which menstruation ceases is reported to have increased markedly—from about age 40 a few decades ago to between ages 45 and 50 more recently.

EMBRYOLOGY

Meaning and Scope

Embryology is the science of the development of the individual before birth. It is a story of miracles, describing the means by which a new human life is started and the steps by which a single microscopic cell is transformed into a complex human being. In a work of this kind, it seems feasible to include only a few of the main points in the development of the new individual, since the facts amassed by the science of embryology are so many and so intricate that to tell them requires volumes, not paragraphs.

Steps in Development of a New Individual
Preliminary Processes

Production of a new human being starts with the union of a spermatozoon and an ovum to form a single cell. Three preliminary steps, however, are necessary before such a union can take place: maturation of the sex cells (meiosis), ovulation, and insemination.

MATURATION OF SEX CELLS (meiosis*). Only when maturation has occurred are the ovum and spermatozoon mature and ready to unite with each other. The ne-

---

*For details of the mechanism of meiosis, see Chapter 3.

VALUE OF KNOWLEDGE OF EMBRYOLOGY

The value to the medical profession of knowing the steps by which a new individual is evolved is evidenced by rapid advances in antenatal (from the Latin *ante,* "before," and *natus,* "birth") diagnosis and treatment. Pediatric medicine acquired a new dimension when Dr. Albert W. Liley of New Zealand, a world-renowened fetologist, developed techniques by which Rh babies could be given transfusions prior to birth. Now, new techniques, including fetal surgery, electrocardiography, and ultrasound, permit pediatricians to diagnose and treat illness in the fetus much like any other patient. A knowledge of embryology also provides an explanation of various congenital deformities. Cleft palate, to mention only one example, is among the most common malformations; it is a condition that results from imperfect fusion of the frontal and maxillary processes during embryonic development.

Figure 28-19 Diagram illustrating the interrelationships among the cerebral, hypothalamic, pituitary, ovarian, and uterine functions throughout a usual 28-day menstrual cycle. The variations in basal body temperature are also illustrated.

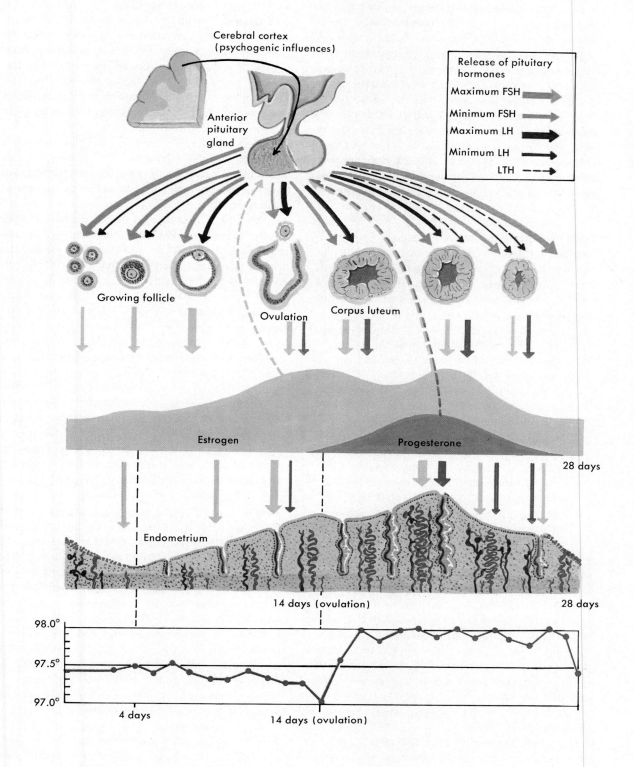

cessity for chromosome reduction as a preliminary to union of the sex cells is explained by the fact that the cells of each species of living organisms contain a specific number of chromosomes. Human cells, for example, contain 23 pairs or a total of 46 chromosomes. If the male and female cells united without first halving their respective chromosomes, the resulting cell would contain twice as many chromosomes as is specific for human beings. Mature ova and sperm, therefore, contain only 23 chromosomes or half as many as other human cells. Of these, one is the sex chromosome and may be either one of two types, known as X or Y. All ova contain an X chromosome. Sperm, on the other hand, have either an X or a Y chromosome–bearing sperm. A female child results from the union of an X chromosome–bearing sperm with an ovum and a male child from the union of a Y chromosome–bearing sperm with an ovum.

OVULATION AND INSEMINATION. The second preliminary step necessary for conception of a new individual consists in bringing the sperm and ovum into proximity with each other so that the union of the two can take place. Two processes are involved in the accomplishment of this step.

1 Ovulation or expulsion of the mature ovum from the graafian follicle into the pelvic cavity, from which it enters one of the uterine tubes.
2 Insemination or expulsion of the seminal fluid from the male urethra into the female vagina. Several million sperm enter the female reproductive tract with each ejaculation of semen. By lashing movements of their flagella-like tails, assisted somewhat by muscular contractions of surrounding structures, the sperm make their way into the external os of the cervix, through the cervical canal and uterine cavity, and into the tubes.

Research has shown that X sperm swim more slowly than do Y sperm. An interesting hypothesis stems from this fact. If insemination occurs on the day of ovulation, Y sperm, being faster than X sperm, should reach the ovum first. Therefore a Y sperm would be more apt to fertilize the ovum. And since Y sperm produce males, there should be a greater probability of having a baby boy when insemination occurs on the day of ovulation. Statistical evidence supports this view.

Developmental Processes

FERTILIZATION OR UNION OF MALE AND FEMALE GAMETES. The sperm "swim" up the tube toward the ovum. Although numerous sperm surround the ovum, only one penetrates it. As soon as the head and neck of one spermatozoon enter the ovum (the tail drops off), the remaining sperm seem to be repulsed. The sperm head then forms itself into a nucleus of the ovum, producing at that moment a new single-celled individual or **zygote.** Half of the 46 chromosomes in the zygote nucleus have come from the sperm and half from the ovum. Since chromosomes are composed of *genes*, the new being inherits half of its genes and the characteristics they determine from its father and half from its mother. Genetically, the zygote is complete. Time and nourishment are all that is needed for expression of those characteristics such as sex, body build, and skin color that were determined at fertilization.

Normally, fertilization occurs in the outer third of the uterine tube. Occasionally, however, it takes place in the pelvic cavity, as evidenced by pregnancies that start to develop in the pelvic cavity instead of in the uterus.

Inasmuch as the ovum lives only a short time (probably less than 48 hours) after leaving the graafian follicle, fertilization can occur only around the time of ovulation (p. 721). Sperm may live a few days after entering the female tract.

CLEAVAGE AND IMPLANTATION. Cleavage involves repeated mitotic divisions, first of the zygote to form two cells, then of those two cells to form four cells, resulting in about 3 days' time in the formation of a solid, spherical mass of cells known as a *morula* (Figure 28-20). A day or so later the embryo reaches the uterus, where it starts to implant itself in the endometrium. Occasionally, implantation occurs in the tube or pelvic cavity instead of in the uterus. The condition is known as an **ectopic pregnancy.** Figure 28-21 shows the appearance of the uterus and uterine tubes from fertilization to implantation.

As the cells of the morula continue to divide a hollow ball of cells or *blastocyst*, consisting of an outer layer of cells and an inner cell mass, is formed. Implantation in the uterine lining is now complete. About 10 days have elapsed since fertilization. The cells that compose the outer wall of the blastocyst are known as trophoblasts. They eventually become part of the placenta, the structure that anchors the fetus to the uterus. The placenta is derived in part from both the developing embryo and from maternal tissues. Note in Figure 28-22 the close relationship of maternal and fetal blood vessels. Although in proximity, maternal and fetal blood does not mix. Exchange of nutrients occurs across an important fetal membrane called the *chorion*.

DIFFERENTIATION. As the cells composing the inner mass of the blastocyst continue to divide, they arrange themselves into a structure shaped like a figure eight, containing two cavities separated by a double-

On July 25, 1978, the world's first test-tube baby (a girl) was born in Oldham, England. Nine months earlier a mature ovum had been removed from the mother's body by laparoscopy ("belly button surgery") and fertilized in a laboratory dish by her husband's sperm. After 2½ days' growth in a controlled environment, physicians returned the fertilized ovum (now at the eight-cell stage) to the mother's uterus. On December 28, 1981 the first test-tube baby born in the United States was delivered by Caesarean section in Norfolk, Va. Use of this technique is now common practice in fertility clinics worldwide.

The first "embryo transfer" birth occurred in January of 1984. In this procedure a donor's fertilized ovum is implanted in a recipient. Unlike test-tube fertilization, embryo transfer does not require surgical removal of the ovum. Instead, the donor and recipient are monitored until they ovulate at about the same time. When the time of ovulation of both donor and recipient matches, the donor woman is artifically inseminated with sperm from the infertile recipient's husband. After the fertilized ovum reaches the donor woman's uterus it is flushed out and transferred by catheter to the recipient's uterus where development continues until birth. The medical ethics of embryo transfer techniques are the subject of continuing controversy.

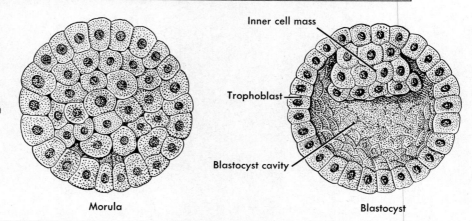

Figure 28-20 Stages in the development of the human embryo. The morula consists of an almost solid spherical mass of cells. The embryo reaches this stage about 3 days after fertilization. The blastocyst (hollowing) stage develops later, after implantation in the uterine lining.

Inner cell mass

Trophoblast

Blastocyst cavity

Morula

Blastocyst

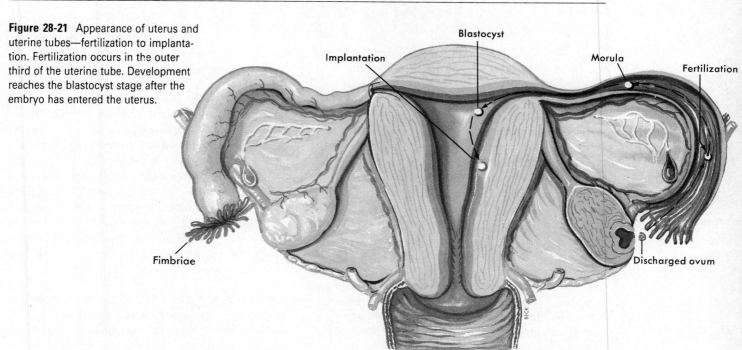

Figure 28-21 Appearance of uterus and uterine tubes—fertilization to implantation. Fertilization occurs in the outer third of the uterine tube. Development reaches the blastocyst stage after the embryo has entered the uterus.

Blastocyst

Implantation

Morula

Fertilization

Fimbriae

Discharged ovum

Figure 28-22 Placenta showing fetal and maternal blood vessels.

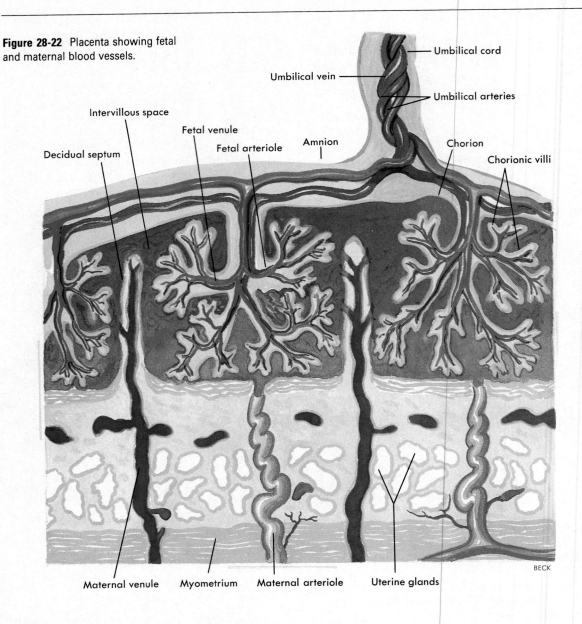

layered plate of cells known as the *embryonic disk.* Young human embryos have been examined at this stage, or about 2 weeks old dated from the time of fertilization. The cells that form the cavity above the embryonic disk eventually become a fluid-filled, shock-absorbing sac (the amnion) in which the fetus floats. The cells of the lower cavity form the yolk sac, a small vesicle attached to the belly of the embryo until about the middle of the second month, when it breaks away. Only the double layer of cells that compose the embryonic disk is destined to form the new individual. The upper layer of cells is called the **ectoderm** and the lower layer the **entoderm.** Refer to Figure 28-23 and follow the zygote from fertilization to implantation and development of the yolk sac. A third layer of cells, known as the **me-soderm,** develops between the ectoderm and entoderm. Up to this time, all the cells have appeared alike, but now they are differentiated into three distinct types, ectodermal, mesodermal, and entodermal, known as the **primary germ layers,** each of which will give rise to definite structures. For example, the ectoderm cells will form the skin and its appendages and the nervous system; the mesoderm, the muscles, bones, and various other connective tissues; the entoderm, the epithelium of the digestive and respiratory tracts, and so on.

HISTOGENESIS AND ORGANOGENESIS. The story of how the primary germ layers develop into many different kinds of tissues (histogenesis) and how those tissues arrange themselves into organs (organogenesis) is

Figure 28-23 Fertilization to implantation and development of the yolk sac. Rapid growth of uterine glands and vessels covers the developing blastocyst at the time of implantation.

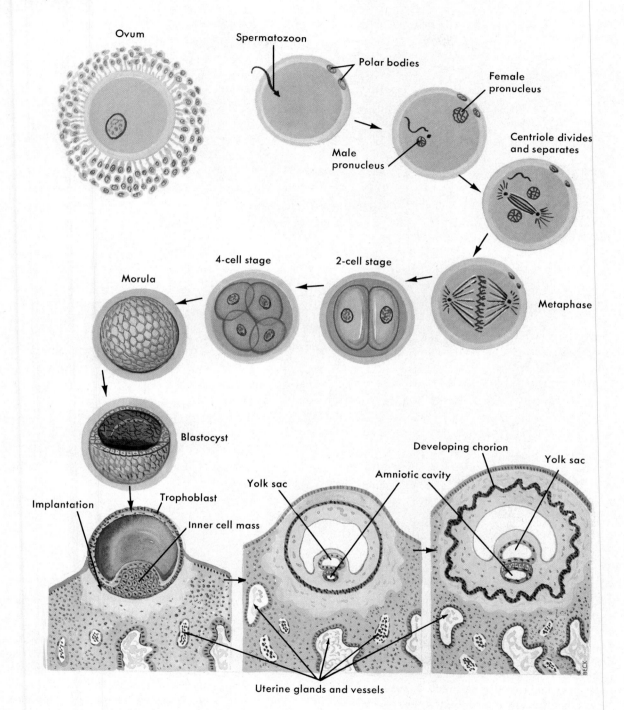

Figure 28-24 Development of the embryo: yolk sac to 4 months of gestation.

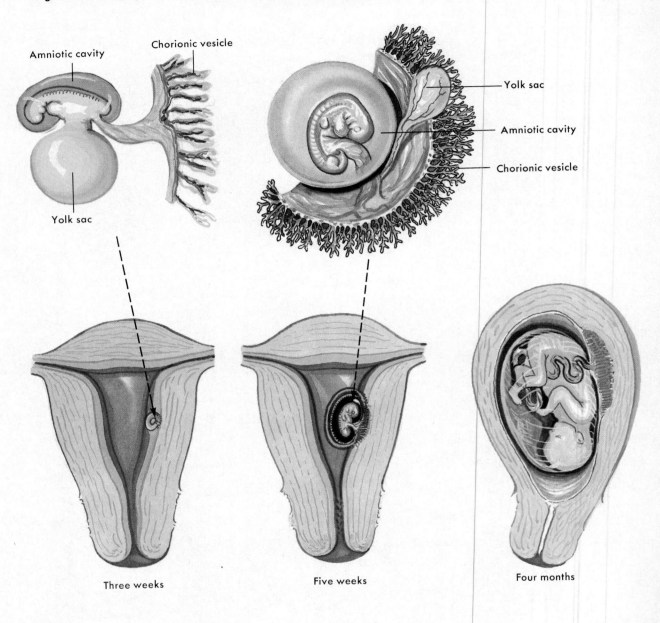

Amniotic cavity

Chorionic vesicle

Yolk sac

Yolk sac

Amniotic cavity

Chorionic vesicle

Three weeks

Five weeks

Four months

long and complicated. Its telling belongs to the science of embryology. But for the beginning student of anatomy it seems sufficient to appreciate that life begins when two sex cells unite to form a single cell and that the new human body evolves by a series of processes consisting of cell multiplication, cell growth, cell differentiation, and cell rearrangements, all of which take place in definite, orderly sequence. Figure 28-24 shows development of the embryo from the yolk sac stage to 4 months of gestation. By the third week of development the third germ layer— the mesoderm—appears between the two initial lay-

ers of the embryonic disk. The developing heart is beating by the fifth week, and the tiny baby, only about 8 mm in length, is showing rapid development of every major organ system. By the end of the second month, a recognizable human figure has been formed and at the end of another month the sex is clearly distinguishable. Development of structure and function go hand in hand, and from 4 months of gestation, when every organ system is in place and functioning, until term at about 280 days (Figure 28-25), development is mainly a matter of growth.

Figure 28-25 Full-term pregnancy showing projected relationship to viscera.

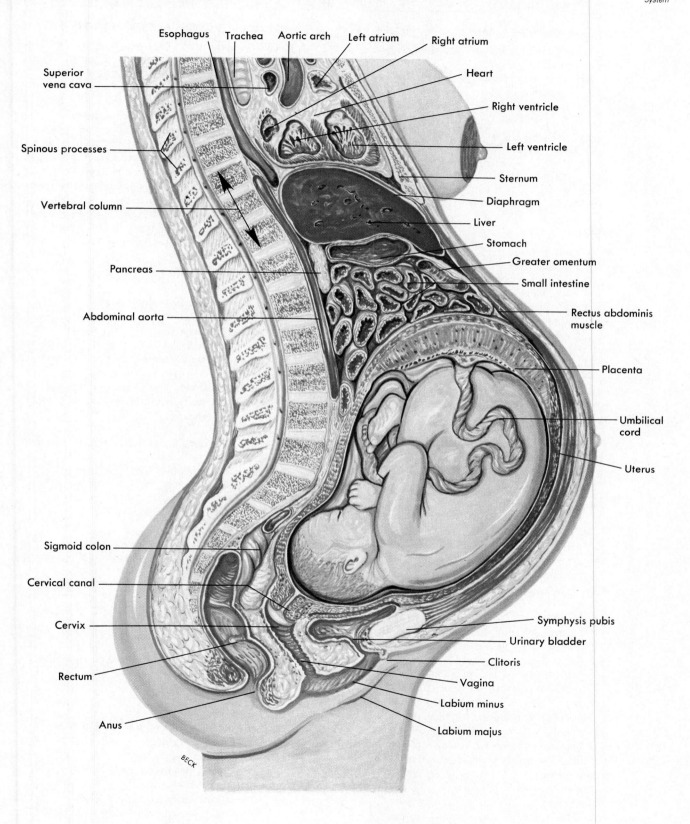

Outline Summary

FEMALE REPRODUCTIVE ORGANS

A Essential sex organs
  1 Ovaries (gonads)
B Accessory sex organs
  1 Accessory glands
    a Bartholin's (greater vestibular)
    b Skene's (lesser vestibular)
    c Mammary (breasts)
  2 Ducts
    a Uterine (fallopian) tubes
    b Uterus
    c Vagina
  3 External genitalia (vulva or pudendum)

UTERUS

A Structure
  1 Size, shape, and divisions
    a In the nonpregnant state, 7.5 cm (3 inches) × 5 cm (2 inches) × 2.5 cm (1 inch)
    b Pear shaped
    c Consists of body and cervix; fundus—bulging upper surface of the body
  2 Wall—lining of mucosa called *endometrium;* thick middle coat of muscle called *myometrium;* partial external coat of peritoneum
    a Endometrium itself composed of three layers
      (1) Outer stratum compactum
      (2) Intermediate stratum spongiosum
      (3) Dense inner stratum basale
  3 Cavities—body cavity small and triangular in shape with three openings—two from tubes and one (internal os) into cervical canal; external os opening of cervical canal into vagina
  4 Blood supply
    a Arterial—generous from uterine arteries (branches of internal iliac) and by anastomosis from ovarian and vaginal arteries
    b Venous—to internal iliac veins by way of uterine, ovarian, and vaginal veins
B Location—in pelvic cavity between bladder and rectum
C Position
  1 Flexed between body and cervix, with body lying over bladder pointing forward and slightly upward
  2 Cervix joins vagina at right angles
  3 Capable of considerable mobility; therefore often in abnormal positions, such as retroverted
  4 Eight ligaments anchor it—two broad, two uterosacral, one posterior, one anterior, and two round
  5 Important support structure for uterus—perineal body—muscular node in perineum
D Functions
  1 Menstruation
  2 Pregnancy
  3 Labor and expulsion of fetus

UTERINE TUBES—FALLOPIAN TUBES OR OVIDUCTS

A Location—about 10 cm (4 inches) long and attached to uterus at upper, outer angles; lie in upper free margin of broad ligaments
B Structure
  1 Same three coats as uterus
  2 Distal ends open with fimbriated margins
  3 Mucosa and peritoneum in direct contact here
C Divisions
  1 Isthmus—medial third attached to uterus
  2 Ampulla—intermediate dilated portion
  3 Infundibulum—funnel-shaped terminal component
D Function
  1 Serve as ducts for ovaries, providing passageway by which ova can reach uterus
  2 Fertilization occurs here normally

OVARIES—FEMALE GONADS

A Location and size
  1 Size and shape of large almonds
  2 Lie behind and below uterine tubes
  3 Anchored to uterus and broad ligament
B Microscopic structure
  1 Consists of several thousand ovarian follicles embedded in connective tissue base
  2 Follicles in all stages of development; usually each month one matures, ruptures, and expels its ovum into abdominal cavity
C Functions
  1 Oogenesis—formation of mature female gametes (ova)
  2 Secretion of hormones—estrogens and progesterone

VAGINA

A Location—between rectum and urethra
B Structure
  1 Collapsible, musculomembranous tube, capable of great distention
  2 External outlet protected by fold of mucous membrane, hymen
C Functions
  1 Receive seminal fluid
  2 Is lower part of birth canal
  3 Is excretory duct for uterine secretions and menstrual flow

VULVA

A Consists of numerous structures that, together, constitute external genitalia
B Main structures
  1 Mons pubis—skin-covered pad of fat over symphysis pubis
  2 Labia majora—hairy, skin-covered lips
  3 Labia minora—small lips covered with modified skin

Outline Summary—cont'd

4 Clitoris—small mound of erectile tissue just below junction of two labia minora

5 Urinary meatus—just below clitoris; opens into urethra

6 Vaginal orifice—below urethra

7 Bartholin's or greater vestibular glands

  a Comparable to bulbourethral glands of male

  b Open by means of long duct in space between hymen and labia minora

  c Ducts from lesser vestibular of Skene's glands open near urinary meatus

PERINEUM

A Region between vaginal orifice and anus

B Location of perineal body—important muscular support structure for uterus and vagina

C Surgical incision—episiotomy—often made in perineum at childbirth to decrease chances of possible laceration

BREASTS

A Location and size

  1 Just under skin, over pectoral muscles

  2 Size depends on deposits of adipose tissue

B Structure

  1 Divided into lobes and lobules; latter composed of glands

  2 Single excretory duct per lobe opens in nipple

  3 Circular, pigmented area called *areola* borders nipple

C Function—secrete milk for infant

D Mechanism controlling lactation (see Figure 28-13)

FEMALE SEXUAL CYCLES

Recurring cycles

A Ovarian cycles

  1 Each month, follicle and ovum develop

  2 Follicle secretes estrogens

  3 Ovum matures and follicle ruptures (ovulation)

  4 Corpus luteum forms and secretes progesterone and estrogens

  5 If no pregnancy, corpus luteum gradually degenerates and is replaced by fibrous tissue; remains if there is pregnancy

B Endometrial or menstrual cycle

  1 Surface of endometrium sloughs off during menses, with bleeding from denuded area

  2 Regeneration and proliferation of lining occurs

  3 Endometrial glands and arterioles become longer and more tortuous

  4 Glands secrete viscous mucus and cycle repeats

C Myometrial cycle—contractility increases before ovulation and subsides following

D Gonadotropic cycles

  1 FSH—a high rate of secretion for about 10 days after menses

  2 LH secretion increasing from ninth cycle day to day of ovulation (Figure 28-19)

Control

A Cyclical changes in ovaries controlled as follows

  1 Increasing blood level of FSH during postmenstrual phase stimulates one or more primitive graafian follicles to start growing and to start secreting estrogens (Figure 28-16); "estrogen surge" day or two before ovulation, lowest blood level of estrogen on day of ovulation

  2 Increasing blood concentration of LH (from cycle day 9 to day of ovulation) causes completion of growth of follicle and ovum, stimulates follicle to secrete estrogens, causes ovulation and luteinization, which in turn leads to progesterone secretion (Figure 28-17)

B Cyclical changes in uterus controlled as follows

  1 Increasing blood concentration of estrogens during preovulatory phase causes proliferation of endometrial cells with resultant thickening of endometrium, growth of endometrial glands, increased water content of endometrium, and increased contractions by myometrium

  2 Increasing blood concentration of progesterone after ovulation, that is, during premenstrual phase, causes secretion by endometrial glands and preparation of endometrium for implantation, increased water content of endometrium, and decreased contractions by myometrium

C Cyclical changes in FSH and LH secretion—controlled by both negative and positive feedback mechanisms that include releasing hormones from hypothalamus

Functions served by cycles

A Preparation of endometrium for pregnancy during preovulatory and premenstrual periods

B Shedding of progestational endometrium if no pregnancy occurs

Menarche and menopause

A Menarche—onset of menses; often between 11 and 14 years of age

B Menopause (climacteric)—cessation of menses; usually between 45 and 50 years of age

EMBRYOLOGY

Meaning and scope

A Science of development of individual before birth

B Long and complex study

Value of knowledge of embryology

A Provides information for advances in antenatal diagnosis and treatment

B Interprets many abnormal formations

Outline Summary—cont'd

Steps in development of a new individual

A Preliminary processes
   1 Maturation of ovum and sperm or reduction of their chromosomes to half original number, that is, half of 46
   2 Ovulation and insemination
      a Ovulation once every 28 days, usually about 14 days before beginning of next menstrual period
      b Insemination, indefinite occurrence; sperm introduced into vagina, swim up to meet ovum in tube
B Developmental processes (Figure 28-21)
   1 Fertilization or union of ovum and sperm
      a Normally occurs in tube
      b Only one sperm penetrates ovum
      c One-celled new individual called zygote formed by union
   2 Cleavage or segmentation (Figures 28-20 and 28-23)
      a Multiplication of cells from zygote by repeated mitosis
      b Morula or solid ball of cells formed; becomes hollow with cluster of cells attached at one point on inner surface of sphere; called blastocyst at this stage; now implanted in endometrium
   3 Development of placenta from both fetal and maternal tissues provides a structure for transfer of nutrients, waste products, etc. (Figure 28-22)
   4 Differentiation of cells into three primary germ layers—ectoderm, mesoderm, and entoderm
   5 Histogenesis or formation of various tissues from primary germ layers; organogenesis or formation of various organs by rearrangements of tissues, such as fusions, shiftings, foldings, etc. (Figures 28-24 and 28-25)

Review Questions

1 Name the female sex glands. Name all the internal female reproductive organs.

2 What is the perineum? Of what clinical importance is it in the female?

3 Name the three openings to the exterior from the female pelvis. How many openings to the exterior are there in the male?

4 What is a graafian follicle? What does it contain?

5 How many ova usually mature in a month?

6 When, in the menstrual cycle, is ovulation thought to occur?

7 Discuss the mechanism thought to control menstruation.

8 Name the periods in the menstrual cycle with the approximate length of days in each and the main events.

9 What two organs are necessary for menstruation to occur?

10 Name the divisions of the uterus.

11 What and where are the fallopian tubes? Approximately how long are they? With what are they lined? Their lining is continuous on their distal ends with what? On their proximal ends? Why is an infection of the lower part of the female reproductive tract likely to develop into a very serious condition?

12 What is the cul-de-sac of Douglas?

13 Describe briefly the hormonal control of breast development and lactation.

14 Explain how contraceptive pills act to prevent pregnancy.

15 On which day of the menstrual cycle would ovulation most probably occur in a 26-day cycle? In a 32-day cycle?

16 Where does fertilization normally take place?

17 Which of the following patients will no longer menstruate? Why? (1) One who has had a bilateral salpingectomy (removal of the tubes)? (2) One who has had a panhysterectomy (removal of the entire uterus)? (3) One who has had a bilateral oophorectomy (removal of both ovaries)? (4) One who has had a cervical hysterectomy (body and fundus removed)? (5) One who has had a unilateral oophorectomy?

18 Which of the patients described in question 17 could no longer become pregnant? Explain.

19 Following menopause, which of the following hormones—estrogens. FSH, progesterone—would you expect to have a high blood concentration? Which a low blood concentration? Explain your reasoning.

20 Define or make an identifying statement about each of the following: adolescence, climacteric, colostrum, corpus luteum, ectopic pregnancy, embryology, endometrium, fertilization, fimbriae, gamete, genes, gonad, hysterectomy, maturation, meiosis, menarche, mitosis, morula, oophorectomy, ovum, ovulation, puberty, salpingitis, semen, spermatozoon, vulva, zygote.

UNIT SEVEN

DEFENSE AND ADAPTATION

29

THE IMMUNE SYSTEM

OBJECTIVES

After completing this chapter, you should be able to:

1 List and contrast the immune system's structural elements with the anatomical components characteristic of other organ systems of the body.

2 Discuss phagocytosis and identify two specific cell types capable of this process.

3 Compare the chemistry and functional activity of antigens and antibodies.

4 Discuss the formation of lymphocytes and identify the two major classes of these specialized cells.

5 Discuss and contrast the development, activation, and functions of B cells and T cells.

6 Explain the role of the thymus in cell-mediated immunity.

7 Discuss the origin and function of lymphokines and lymphotoxin.

8 Discuss the primary factors that influence macrophage migration.

9 Discuss the differing functions of killer T cells, helper T cells, and suppressor T cells in the immune system.

10 Explain the function of memory B cells in the development of humoral immunity.

11 Discuss the relationship between chemical structure and functional activity of antibodies or immunoglobulins.

12 List and compare the five classes of immunoglobulins.

13 Briefly discuss the two basic tenets of the clonal selection theory.

14 Discuss the role of complement, properdin, and interferon in the immune system.

15 Explain the term *autoimmune disease*.

Enemies of many kinds and in great numbers assault the body during a lifetime. Most threatening are hordes of microorganisms. We live our lives out in a virtual sea of bacteria and viruses. So ever present and potentially lethal are these small but formidable foes that no newborn baby could live through infancy, much less survive to adulthood or old age, without effective defenses against them. This chapter presents information about the system that provides such defenses—the **immune system.**

KEY TERMS

Antibody (AN-ti-body)

Antigen (AN-ti-jen)

Autoimmune (aw-to-im-MUNE)

Chemotactic (ke-mo-TAK-tic)

Complement (KOM-ple-ment)

Immunity (i-MU-ni-te)

Immunoglobulin (im-u-no-GLOB-u-lin)

Interferon (in-ter-FER-on)

Lymphokine (LIM-fo-kin)

Lymphotoxin (lim-fo-TOK-sin)

Phagocytosis (fag-o-si-TO-sis)

Properdin (PRO-per-din)

STRUCTURES AND FUNCTIONS OF THE IMMUNE SYSTEM

The immune system's structures are not organs; they are separate cells and molecules. Included among them are the white blood cells (especially the neutrophils and lymphocytes), the connective tissue cells or macrophages, and the protein molecules or immunoglobulins (antibodies)—complement, properdin, and interferon. Awesome indeed are the numbers of these structures. One trillion lymphocytes, for example, and 100 million trillion (10^{14}) antibodies! The immune system's general function is to produce immunity, that is, resistance to disease. To do this, it searches out any microscopic enemies lurking in the body's tissues, recognizes them as foreigners, and destroys them.

NEUTROPHILS

Neutrophils (granular, neutral-staining white blood cells) are phagocytes, that is, cells capable of carrying on phagocytosis. Briefly defined, **phagocytosis** is the ingestion and destruction of microorganisms or other small particles. Neutrophils move out of the bloodstream into the tissues where microbes have invaded. Here, they carry on phagocytosis. As they approach a microbe, they extend footlike projections (pseudopods) out toward it. Soon the pseudopods encircle the organism and form a complete sac (phagosome) around it (Figure 29-1). The phagosome then moves into the interior of the neutrophil where a lysosome fuses with it. The contents of the lysosome, chiefly digestive enzymes and hydrogen peroxide, drain into the phagosome and destroy the microbes in it. Because phagocytosis defends us against various kinds of agents, it is classified as a nonspecific defense. A mature neutrophil and lymphocyte in a peripheral blood smear are shown in Figure 29-2.

LYMPHOCYTES

Formation and Types

Originally, lymphocytes are formed in the red bone marrow of the fetus. They, like all blood cells, derive from primitive cells known as hemopoietic stem cells. Those stem cells destined to become lymphocytes follow two developmental paths and differentiate into two major classes of lymphocytes—

Figure 29-1 Phagocytosis of bacteria by a monocyte. Phagocytosis can occur in the blood stream, or white cells may squeeze through capillary walls and destroy bacteria in the tissues.

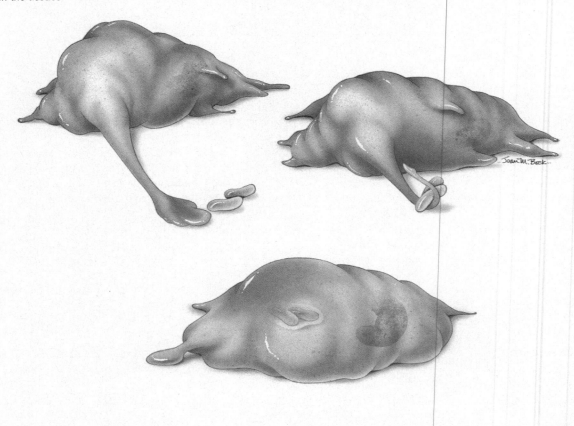

Figure 29-2 Blood smear, showing neutrophil and lymphocyte of the immune system. (× 350.)

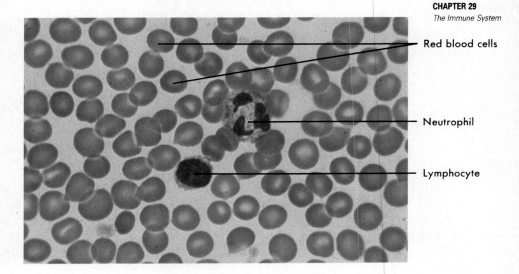

Red blood cells

Neutrophil

Lymphocyte

B-lymphocytes and T-lymphocytes, or simply, B cells and T cells.

Before continuing our discussion of lymphocytes, we shall define the following terms to be used:

1 **antigens**—Foreign macromolecules, that is, very large molecules not normally present in the body, which, when introduced into the body, induce it to make certain responses. Most antigens are foreign proteins. Some, however, are foreign polysaccharides and some are nucleic acids. Many antigens that enter the body are macromolecules located in the surface membranes of microorganisms.

2 **Antibodies**—Proteins of the class called *immunoglobulins*. Unlike antigens, antibodies are native molecules, that is, they are normally present in the body.

3 **Antigenic determinants**—Variously shaped, small patches on the surface of an antigen molecule; a less cumbersome name is *epitopes*. In a protein molecule, for instance, an epitope consists of a sequence of only about 10 amino acids that are part of a much longer, folded chain of amino acids. The sequence of the amino acids in an epitope determines its shape. Since the sequence differs in different kinds of antigens, each kind of antigen has specific and uniquely shaped epitopes.

4 **Combining sites**—Small concave regions on the surface of an antibody molecule. Like epitopes, combining sites have specific and unique shapes. An antibody's combining sites are so shaped that an antigen's epitope that has a complementary shape can fit into the combining site and thereby bind the antigen to the antibody to form an **antigen-antibody complex.** Because combining sites receive and bind antigens, they are also called *antigen receptors* and *antigen-binding sites*.

5 **Clone**—Family of cells, all of which have descended from one cell.

Development, Activation, and Functions of B Cells

The development of the lymphocytes called *B cells* occurs in two stages. In chickens the first stage of B cell development occurs in the bursa of Fabricius—hence the name B cells. Since humans do not have a bursa of Fabricius, another organ must serve as the site for the first stage of B cell development. Some evidence suggests that the fetal liver (Figure 29-3) serves this purpose. At any rate, by the time a human infant is a few months old, its B cells have completed the first stage of their development (Figure 29-4). They are then known as immature B cells.

Immature B cells synthesize antibody molecules but secrete few if any of them. Instead, they insert on the surface of their plasma membranes perhaps 100,000 antibody molecules (immunoglobulin M, *IgM*, p. 746). The combining sites of these surface antibody molecules stand ready to serve as receptors for a specific antigen if it comes by.

The second stage of B cell development (Figure 29-5) usually takes place in the lymph nodes and spleen, but only under certain conditions. It must be initiated by an encounter between an immature B cell and its specific antigen, that is, one whose epitopes fit the combining sites of the B cell's surface antibodies (Figure 29-6).

Figure 29-3 Section of liver showing fixed macrophages (Kupffer cells) lining simisoids. (×140.)

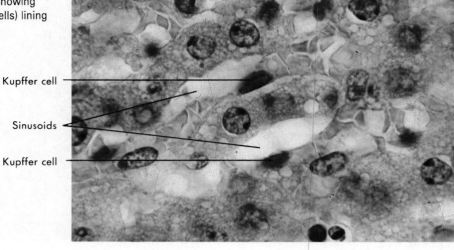

Kupffer cell

Sinusoids

Kupffer cell

Figure 29-4 First stage of B cell development.

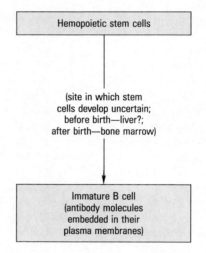

Hemopoietic stem cells

(site in which stem cells develop uncertain; before birth—liver?; after birth—bone marrow)

Immature B cell (antibody molecules embedded in their plasma membranes)

Figure 29-5 Lymph node. The germinal centers of lymphatic tissue occur in the cortex of the organ. (×70.)

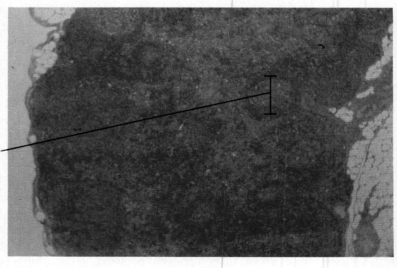

Germinal center

Figure 29-6 Second stage of B cell development; activated by binding of antigen molecules to immature B cell's surface antibody molecules.

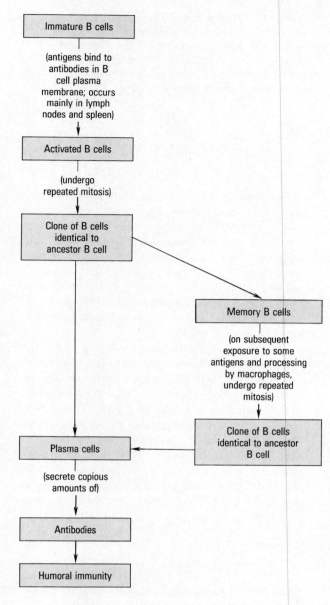

The antigen binds to these antibodies on the B cell's surface. However, for this binding to take place, macrophages are believed to play a crucial role, although a still mysterious one. They are postulated to somehow process the antigen and present it to a B cell's surface antibodies in a way that activates the B cell, causing it to undergo rapid mitosis. By dividing repeatedly, a single B cell produces a clone or family of identical B cells. Some of them become differentiated to form plasma cells. Others do not differentiate but remain in the lymphatic tissue as the so-called memory B cells. Plasma cells synthesize and secrete copious amounts of antibody. A single plasma cell, according to one estimate, secretes 2,000 antibody molecules per second during the few days that it lives. All the cells in a clone of plasma cells secrete identical antibodies because they have all descended from the same B cell. Memory B cells do not themselves secrete antibodies. But if they are later exposed to the antigen that triggered their formation, memory B cells then become plasma cells and the plasma cells secrete antibodies that can combine with the initiating antigen. Thus the ultimate function of B cells is to serve as ancestors of antibody-secreting plasma cells.

Development, Activation, and Functions of T Cells

T cells, by definition, are lymphocytes that have made a detour through the thymus gland before migrating to the lymph nodes and spleen (Figure 29-7). During their residence in the thymus, stem cells develop into **thymocytes,** cells that proliferate as rapidly as any in the body. Thymocytes undergo mitosis three times a day, and as a result their numbers increase enormously in a relatively short time. They stream out of the thymus into the blood and find their way to a new home in areas of the lymph nodes and spleen called *thymus-dependent zones.* From this time on, they are known as T cells. Each T cell, like each B cell, displays antigen receptors on its surface membrane.

Presumably, they are not immunoglobulins as are B cell receptor molecules, but are perhaps compounds similar to them. When an antigen (usually preprocessed by macrophages) encounters a T cell whose surface receptors fit the antigen's epitopes, the antigen binds to the T cell's receptors. This activates or sensitizes the T cell, causing it to divide repeatedly to form a clone of identical *sensitized T*

cells. The sensitized T cells then travel to the site where the antigens originally entered the body. There, in the inflamed tissue, the sensitized T cells bind to antigens of the same kind that led to their formation. The antigen-bound sensitized T cells then release chemical compounds called **lymphokines** into the inflamed tissues. Names of some individual lymphokines are these: chemotactic factor, migration inhibition factor, macrophage activating factor, and lymphotoxin. *Chemotactic factor* attracts macrophages, causing hundreds of them to migrate into the vicinity of the antigen-bound, sensitized T cell. *Migration inhibition factor* halts macrophage migration. *Macrophage activating factor* prods the assembled macrophages to destroy antigens by phagocy-

Figure 29-7 Development of T cells. To become T cells, stem cells must pass through the thymus, a fact suggested by the designation T cells. Compare with Figure 29-1. Sensitized T cells release various chemicals collectively known as *lymphokines.* These act in several ways to destroy foreign cells.

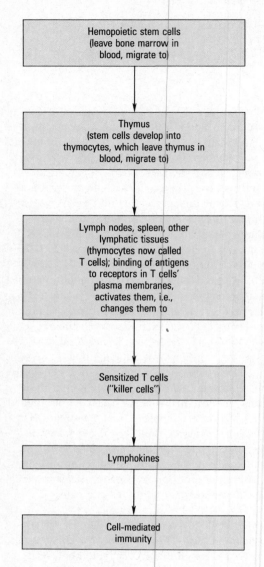

tosing them at a rapid rate. **Lymphotoxin** is a powerful poison that quickly kills any cell it attacks.

Sensitized T cells that release lymphotoxin are known as *killer T cells.* There are also two other types of T cells, namely, helper T cells and suppressor T cells. They produce opposite effects. *Helper T cells* act in some way to help B cells differentiate into antibody-secreting plasma cells. *Suppressor T cells* act in some way to suppress B cell differentiation into plasma cells.

Summarizing very briefly, the function of T cells is to produce **cell-mediated immunity.** They search out, recognize, and bind to appropriate antigens located on the surfaces of cells. This kills the cells—the ultimate function of killer T cells (Figure 29-8). Usually these are not the body's own normal cells but are cells that have been invaded by viruses, that have become malignant, or that have been transplanted into the body. Killer T cells, therefore, function to defend us from viral diseases and cancer, but they also bring abut rejection of transplanted tissues or organs.

Figure 29-8 The white spheres seen in this scanning electron microscope view are T lymphocytes attacking a much larger cancer cell. T lymphocytes are a significant part of our defense against cancer and other types of foreign cells.

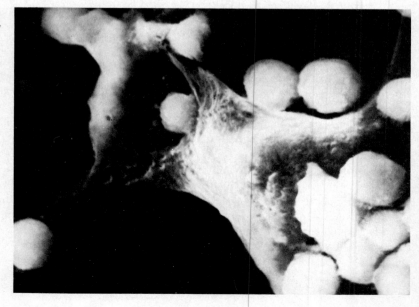

Distribution of Lymphocytes

The densest populations of lymphocytes occur in the structures where they are formed and multiply—namely, in the bone marrow, thymus gland, lymph nodes, and spleen. From these structures, a continuous stream of lymphocytes pours into the blood. Many of them leave the blood and distribute themselves throughout the tissues of the body. After wandering about through the tissue spaces, they eventually find their way into lymphatic capillaries. Lymph flow transports the lymphocytes through a succession of lymph nodes and lymphatic vessels and empties them by way of the thoracic and right lymphatic ducts into the subclavian veins. Thus returned to the blood, the lymphocytes embark on still another long journey—through blood, tissue spaces, lymph, and back to blood. The survival value of the continued recirculation of lymphocytes and of their widespread distribution throughout body tissues seems apparent. It provides these major cells of the immune system ample opportunity to perform their functions of searching out, recognizing, and destroying foreign invaders.

ANTIBODIES (IMMUNOGLOBULINS)

Structure

Antibodies are proteins of the family called *immunoglobulins*. Like all proteins, they are very large molecules and are composed of long chains of amino acids (polypeptides). Each immunoglobulin molecule consists of four polypeptide chains—two heavy chains and two light chains. Each polypeptide chain is intricately folded to form globular regions that are joined together in such a way that the immunoglobulin molecule as a whole is Y-shaped. Look now at Figure 29-9. The two shorter black lines in the diagram represent the light chains, and the two longer lines represent the heavy chains. Each heavy chain consists of 446 amino acids. Heavy chains, therefore, are about twice as long and weigh about twice as much as light chains.

The colored ovals that lie over the black lines in Figure 29-9 represent different regions in the polypeptide chains of the antibody molecule. The more vividly colored ovals represent *variable regions*, that is, regions in which the sequence of amino acids varies in different antibody molecules. Note the relative positions of the variable regions of the light and heavy chains; they lie directly opposite each other. Because amino acid sequence determines conformation or shape, and because different sequences of amino acids occur in the variable regions of different antibodies, the shapes of the sites between the variable regions also differ. These are the antibody's combining sites, or antigen-binding sites (defined on p. 741), and there are two such sites on each antibody molecule. Locate these in Figure 29-9. Each one of us is thought normally to have millions of different kinds of antibody molecules in our bodies. And each one of these, almost unbelievably, has its own uniquely shaped combining sites. It is this structural feature that enables antibodies to recognize and combine with specific antigens, both of which are crucial first steps in the body's defense against invading microbes and other foreign cells.

In addition to its variable region, each light chain in an antibody molecule also has a constant region. The constant region consists of 106 amino acids whose sequence is identical in all antibody molecules. Each heavy chain of an antibody molecule consists of three constant regions in addition to its one variable region. Identify the constant and vari-

Figure 29-9 Structure of antibody (IgG) molecule. Each of the two heavy polypeptide chains in the molecule consist of one variable region (V_H) and three constant regions (C_H). The light chains consist of one variable region (V_L) and one constant region (C_L). The red areas represent complement-binding sites.

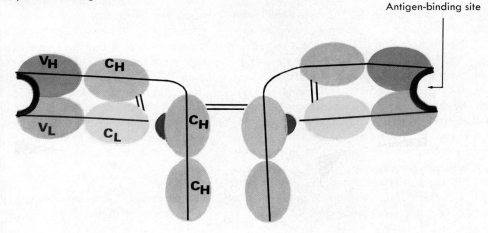

Antigen-binding site

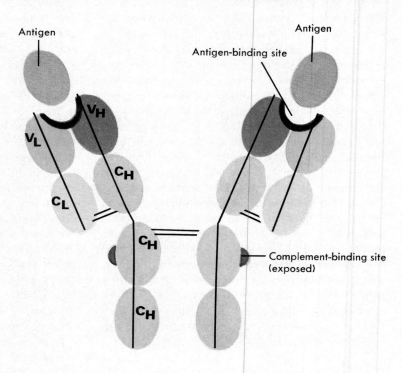

Figure 29-10 Exposure of complement-binding site by binding of antigen to antigen-binding site of antibody.

able regions of the light and heavy chains in Figures29-9 and 29-10. Be sure to observe the location of two *complement-binding sites* on the antibody molecule.

In summary, an immunoglobulin or antibody molecule consists of two heavy and two light polypeptide chains. Each light chain consists of one variable region and one constant region. Each heavy chain consists of one variable region and three constant regions. Disulfide bonds join the two heavy chains to each other; they also bind each heavy chain to its adjacent light chain. An antibody has two antigen-binding sites—one at the tip of each pair of variable regions—and two complement-binding sites located as shown in Figures 29-9 and 29-10.

Diversity

Every normal baby is born with an enormous number of different clones of B cells populating its bone marrow, lymph nodes, and spleen. All the cells of each clone are committed to synthesizing a specific antibody with a sequence of amino acids in its variable regions that is different from the sequence synthesized by any other of the innumerable clones of B cells. How does this astounding diversity originate? A complete answer to this question has not yet been established. One suggested answer proposes that mutations in B cell genes account for antibody diversity by coding for different amino acid sequences in the variable regions of antibodies. In the words of the noted immunologist, N.K. Jerne, "If

at each of 50 positions of both (heavy and light) chains there were an independent choice between just two amino acids, there would be 2^{100} (or 10^{30}) potentially different molecules!" In short, the body's millions of diverse antibodies could originate from mutations that lead to amino acid replacements in various positions of their variable regions.

Classes

There are five classes of immunoglobulins identified by letter names as immunoglobulins M, G, A, E, and D. *IgM* (abbreviation for immunoglobulin M) is the antibody that immature B cells synthesize and insert into their plasma membranes. It is also the predominant class of antibody produced after initial contact with an antigen. The most abundant circulating antibody, the one that normally makes up about 75% of all the antibodies in the blood, is *IgG*. It is the predominant antibody of the secondary antibody response, that is, following subsequent contacts with a given antigen. *IgA* is the major class of antibody present in the mucous membrane lining the intestines and the bronchi, in saliva, and in tears. *IgE*, although minor in amount, can produce major harmful effects, such as those associated with allergies. *IgD* is present in the blood in very small amounts, and its function is as yet unknown.

Functions

The function of antibody molecules—some 100 million trillion of them—is to produce **humoral im-**

munity. They bring about this kind of disease resistance first by recognizing substances that are foreign to the body. They distinguish between "self" and "not self." Recognition occurs when an antigen's epitopes (small patches on its surface) fit into and bind to antibody molecule's antigen-binding sites. The binding of the antigen to antibody forms an antigen-antibody complex that may produce one or more of several effects. For example, it transforms antigens that are toxins (chemicals poisonous to cells) into harmless substances. It agglutinates antigens that are molecules on the surface of microorganisms. In other words, it makes them stick together in clumps, and this, in turn, makes it possible for macrophages and other phagocytes to dispose of them more rapidly by ingesting and digesting large numbers of them at one time. The binding of antigen to antibodies frequently produces still another effect—it alters the shape of the antibody molecule, not very much, but enough to expose the molecule's previously hidden complement-binding sites. This seems a trivial enough change, but it is not so. It initiates an astonishing series of reactions (described on p. 748) that culminate in the destruction of microorganisms and other foreign cells.

Self-Tolerance

The term *self-tolerance* means that normally one's own antibodies do not combine with epitopes present on one's own cells and macromolecules. In effect, antibodies seem to distinguish between macromolecules native to the body and those foreign to it. Sir Macfarlane Burnet described this phenomenon as the differentiation between "self" and "not-self" and called it the central process in immunology. To explain it, he postulated the following events. During the early stages of embryonic life, ancestor cells of lymphocytes mutate at a high rate, producing lymphocytes capable (because of certain of their genes) of producing all possible antibody patterns—including many that are complementary to patterns present on the body's own cells and macromolecules. Such lymphocytes, he postulated, are killed off before birth by contacting and binding to macromolecules having their complementary patterns. By the time of birth, therefore, no lymphocytes would remain that can produce antibodies capable of combining with the body's own components. In short, self-tolerance has developed.

MONOCLONAL ANTIBODIES

Revolutionary new techniques that have permitted biologists to produce large quantities of pure and very specific antibodies have resulted in dramatic advances in medicine during the past 5 years. As a new medical technology, the development of **monoclonal antibodies** has been compared in importance with advances in recombinant DNA or genetic engineering.

Monoclonal antibodies are specific antibodies produced or derived from a population or culture of identical, or **monoclonal,** cells. In the past, antibodies produced by the immune system against a specific antigen had to be "harvested" from serum containing literally hundreds of other antibodies. The total amount of a specific antibody that could be recovered was very limited and the cost of recovery was high. Monoclonal antibody techniques are based on the ability of immune system cells to produce individual antibodies that bind to and react with very specific antigens. We know, for example, that if the body is exposed to the varicella virus of chicken pox, white blood cells will produce an antibody that will react very specifically with that virus and no other. In monoclonal antibody techniques, lymphocytes that are produced by the body after the injection of a specific antigen are "harvested" and then "fused" with other cells that have been transformed to grow and divide indefinitely in a tissue culture medium. The result is a rapidly growing population of identical or **monoclonal** cells that produce large quantities of a very specific antibody. Monoclonal antibodies have now been produced against a wide array of different antigens, including disease-producing organisms and various types of cancer cells.

The availability of very pure antibodies against specific disease-producing agents is the first step in the commercial preparation of diagnostic tests that can be used to identify viruses, bacteria, and even specific cancer cells in blood or other body fluids. The use of monoclonal antibodies may eventually serve as the basis for specific treatment of many human diseases.

Clonal Selection Theory

The clonal selection theory, which deals with antigen destruction, was also proposed in 1959 by Sir Macfarlane Burnet. It has two basic tenets. First, it holds that the body contains an enormous number of diverse clones of cells, each committed by certain of its genes to sythesize a different antibody. Second, the clonal selection theory postulates that when an antigen enters the body, it selects the clone whose cells are committed to synthesizing its specific antibody and stimulates these cells to proliferate and to thereby produce more antibody. We now know that the clones selected by antigens consist of lymphocytes. We also know how antigens select lymphocytes—by the shape of antigen receptors on the lymphocyte's plasma membrane. An antigen recognizes receptors that fit its epitopes and combines with them. By thus selecting the precise clone committed to making its specific antibody, each antigen provokes its own destruction.

COMPLEMENT

Complement, a component of blood serum, reportedly consists of 11 protein compounds. They are inactive enzymes that become activated in a definite sequence to catalyze a series of intricately linked reactions. The binding of an antibody to an antigen located on the surface of a cell alters the shape of the antibody molecule in a way that exposes its complement-binding sites (Figure 29-10). By binding to these sites, complement protein 1 becomes activated and touches off the catalytic activity of the next complement protein in the series. A rapid sequence of activity by the next protein, then the next, and the next, follows until the entire series of enzymes has functioned. The end result of this rapid-fire activity challenges the imagination. Molecules formed by these reactions assemble themselves on the enemy cell's surface in such a way as to form a doughnut-shaped structure—complete with a hole in the middle. In effect, the complement has drilled a hole through the foreign cell's surface membrane! Ions and water rush into the cell; it swells and bursts. **Cytolysis** is the technical name for this process. Briefly, then, complement functions to kill foreign cells by cytolysis. In addition, various complement proteins serve other functions. Some, for example, cause vasodilation in the invaded area and some enhance phagocytosis.

PROPERDIN

Like complement, properdin consists of a set of proteins present in blood serum as inactive enzymes.

When activated, they assemble on the surface of a foreign cell, where they initiate the complement attack sequence just described. Properdin enzymes, however, activate protein 3, whereas an antibody bound to a cell surface antigen activates complement protein 1. Properdin provides an alternate method of activating complement, one in addition to the usual antibody method of activating it.

INTERFERON

Several types of cells, if invaded by viruses, respond rapidly by synthesizing the protein interferon and releasing some of it into the circulation. Interferon comes in several varieties. As the name suggests, interferon proteins interfere with the ability of viruses to cause disease. They may, according to one postulate, act in some way to prevent viruses from multiplying in cells. Leukocyte interferon, fibroblast interferon, and immune interferon are the three major types of interferon proteins. All three have now been produced by using gene-splicing techniques. Studies exploring antiviral and anticancer activities of interferons are currently under way.

APPLICATIONS

Applications of knowledge about the immune system are numerous and varied. They include the following: identification of several conditions involving abnormalities of the immune system, explanation of transplant rejection, development of methods of inducing immunosuppression, and postulated explanations of cancer.

Conditions Involving Abnormalities of the Immune System

Agammaglobulinemia is the disease in which the blood contains no gamma globulins. This also means that the blood contains no antibodies, since antibodies are gamma globulins. Without antibodies a crucial defense against microorganisms is missing, and victims of agammaglobulinemia are highly susceptible to fatal infections.

Autoimmune diseases, by definition, are those in which the immune system responds to the body's own cells as it does to foreign cells—specifically, it produces antibodies that attack its own cells. Normally the immune system differentiates between "self" and "not-self." It recognizes which cells are its own and which are foreign, and it does not attack its own cells.

Question: how do the cells of one body differ from all other cells? Answer: all cells of one body possess unique "surface markers." These are anti-

AIDS

AIDS or **acquired immune deficiency syndrome** was first described as a separate disease entity in 1981. The disease is caused by a retrovirus that enters the bloodstream and integrates its genes into the DNA of T cell lymphocytes. Viral reproduction kills the infected T cells and depletes the total number of T cells in the immune system pool. The virus is known as human T-lymphotropic virus III or *HTLV-III*. Because T cells play such an important role in cell-mediated immunity, any decrease in their numbers subjects victims of the disease to a whole array of so-called opportunistic infections. AIDS victims are unable to protect themselves from certain forms of cancer and many viral, protozoal, and fungal diseases.

The AIDS disease pattern in the United States shows that most victims are in three "high-risk" groups. Almost 70%-75% are homosexual or bisexual men, about 18% are intravenous drug users who share infected needles, and about 3% are hemophiliacs who have been infected with coagulation products obtained from diseased blood donors. However, this pattern is not typical in some countries, such as Africa, where the disease strikes larger numbers of the heterosexual population. There is considerable debate about the changing nature of the disease pattern worldwide. The number of reported cases in the United States reached 30,000 in 1986. It is difficult to predict the total number of potential AIDS cases because of the long latency period of 5 years or more between infection and the onset of symptoms in those infected persons who actually develop the disease. To complicate the picture even more, some people infected with the virus develop less severe AIDS-related conditions (ARC syndrome) such as chronic weight loss and swollen lymph nodes. They may remain in a sick but stable condition for years while others experience steady degeneration of their immune system and die of full-blown AIDS within months of the appearance of symptoms. Most investigators now believe the disease is much broader in scope than once thought.

gens present in the plasma membranes of one individual's cells that are different from those present in the plasma membranes of another individual's cells. *Histocompatibility antigens* is the name of these glycoprotein compounds unique to each individual. Laboratory procedures for matching tissues and organs for transplanting make use of knowledge about histocompatibility antigens.

Question: how does the immune system recognize differences in cell surface antigens so that normally it tolerates its own cells but destroys foreign cells? Answer: briefly, lymphocytes recognize foreign cells by their histocompatibility antigens. Epitopes on these surface antigens fit combining sites on antibodies present in the lymphocyte's surface membranes. Normally, blood contains no lymphocytes with combining sites that fit antigens on the body's own cells. (see the section on self-tolerance, p. 747).

Autoimmune disease indicates failure in the system's recognition function. It mistakenly identifies its own cells' surface antigens as foreign instead of native. Among the disorders considered to be autoimmune diseases are rheumatoid arthritis, multiple sclerosis, and systemic lupus erythematosus.

Transplant Rejection

The immune system normally rejects, that is, kills, cells of transplanted tissues and organs. Surface molecules on lymphocytes recognize as foreign the histocompatibility antigens on transplanted cells, combine with them, and thereby initiate reactions that end in the destruction of these foreign cells. To combat this bugaboo of transplant surgery, methods have been devised to suppress the immunte system. For example, certain drugs, notably azathioprine (Imuran), and radiation are used as immunosuppressants.

CANCER

Among the currently held concepts about cancer are the following:

1 Cell mutations occur frequently in the normal body, and many of the mutated cells formed are cancer cells. Cancer cells, according to a simple definition, are cells capable of proliferating wildly to form a tumor—unless they are killed before this can happen.

2 Abnormal antigens, called tumor-specific antigens, are present in the surface membranes of cancer cells in addition to the histocompatibility antigens present in all body cells.

3 Lymphocytes, in their repeated wanderings through body tissues, are almost sure to come in contact with newly formed cancer cells. Normally, they recognize these cells as foreign to the body by the tumor-specific antigens that so label them. They then quickly initiate reactions that kill the cancer cells.

4 It sometimes happens, however, that newly formed cancer cells escape destruction by cells of the immune system. Various ingenious theories have been proposed to explain how this might happen but, so far, none has been proved.

Outline Summary

STRUCTURES AND FUNCTIONS OF THE IMMUNE SYSTEM

A Cells
  1 White blood cells—notably, neutrophils and lymphocytes
  2 Connective tissue cells—macrophages
B Protein molecules—notably immunoglobulins, complement, properdin, and interferon
C Functions—to produce immunity, that is, resistance to disease, by searching out foreign cells, recognizing them as foreign, and destroying them

NEUTROPHILS

A Granular, neutral-staining white blood cells
B Immune function—to move out of blood into tissues invaded by microbes and destroy them by phagocytosis; a nonspecific defense

LYMPHOCYTES

A Formation and types—lymphocytes are formed from hemopoietic stem cells present in the embryo in the yolk sac and, after birth, in the bone marrow; stem cells follow one developmental path to become B-lymphocytes (B cells) or another path to become T-lymphocytes (T cells)
B Definitions
  1 Antigens—foreign macromolecules, usually proteins, which, when introduced into the body, induce it to make certain responses; many antigens are macromolecules located in surface membranes of microorganisms
  2 Antibodies—proteins of the class called *immunoglobulins;* antibodies are natives of the body and are present at birth
  3 Antigenic determinants, or epitopes—small regions on surfaces of antigen molecules that have various and specific shapes
  4 Combining sites (also called antigen-binding sites and antigen receptors)—regions on antibody molecule; shape of each combining site is complementary to the shape of a specific antigen's epitopes
  5 Clone—family of cells descended from one cell and therefore identical to it
C Development, activation, and function of B cells
  1 Development and activation
    a First stage—organ in which development occurs not yet positively identified, but some evidence points to fetal liver; first stage of B cell development produces immature B cells and is completed by time infant is a few months old; immature B cells are present after birth in bone marrow and lymphatic tissues (notably, lymph nodes and spleen) and in blood; immature B cells synthesize and insert antibody molecules into their cytoplasmic membranes but secrete few, if any, antibody molecules
    b Second stage—initiated by immature B cell contracting its specific antigen; macrophage-processed antigen selects appropriate B cell by binding to B cell's surface antibodies, thereby activating B cell to enter second stage of its development; characterized by repeated cell divisions and differentiation of immature B cells to form clone of plasma cells and clone of memory cells

Outline Summary—cont'd

2 Functions

    a Initial function—immature B cells synthesize and insert antibodies *(IgM)* on their surface

    b Ultimate function—humoral immunity produced by formation of fully differentiated plasma cells and partially differentiated memory cells from activated immature B cells; plasma cells secrete antibodies; memory cells quickly develop into plasma cells on subsequent exposure to appropriate antigen

D Development, activation, and functions of T cells

    1 Development and activation—stem cells migrate from bone marrow to thymus, where they develop into rapidly proliferating thymocytes; thymocytes migrate to thymus-dependent zones of lymph nodes and spleen; from this time on they are called *T-lymphocytes* or *T cells*; T cells circulate in blood and many wander through tissues where they become activated by binding to their specific antigens; then they are called *sensitized T cells*

    2 Functions

        a Initial—search out, recognize, and bind to antigens that fit T cell's surface receptors

        b Ultimate—cell-mediated immunity; killer T cells release lymphokines that kill cells either directly or indirectly by stimulating macrophages to phagocytose them

E Distribution of lymphocytes—bone marrow, thymus, lymph nodes, spleen, other lymphatic tissues, blood, lymph, tissue spaces

ANTIBODIES (IMMUNOGLOBULINS)

A Structure—immunoglobulin molecules composed of two heavy and two light polypeptide chains held together by disulfide bonds between the heavy chains and between each heavy chain and its adjacent light chain; light chain consists of one variable and one constant region; heavy chain consists of one variable and three constant regions; two antigen-binding sites (combining sites) and two complement-binding sites are present on each antibody molecule

B Diversity—normally, an individual is born with an enormous number of different clones of B cells, each clone committed to synthesizing antibodies different from those secreted by other clones; mutations in certain genes produce cells committed to synthesizing different sequences of amino acids in variable regions of antibodies

C Classes of immunoglobulines

    1 IgM—present in surface membranes of immature B cells; also, is the predominant antibody produced after initial contact of B cell with its specific antigen

    2 IgG—the most abundant circulating antibody (makes up about 75% of them); the predominant antibody produced after subsequent exposure to same antigen

    3 IgA—chief antibody in mucosa lining of intestines and bronchi, in saliva, and in tears

    4 IgE—antibody responsible for allergic effects

    5 IgD—only small amounts in blood; function unknown

D Functions of antibodies

    1 Recognition and binding to antigens to form antigen-antibody complex

    2 Antigen-antibody complexes produce variety of effects, e.g., agglutination of antigens and alteration of antibody molecules' shape so as to expose its complement-binding site, which leads to series of reactions that end in destruction of antigen

E Self-tolerance—means that one's own antibodies do not normally combine with one's own body's components; self-tolerance is acquired during embryonic period presumably because all immature lymphocytes that bind to macromolecules having complementary patterns to the lymphocytes' surface antibodies are killed by so binding

F Clonal selection theory

    1 Enormous number of diverse clones of lymphocytes present at birth; each clone committed to synthesizing a different antibody

    2 An antigen, on entering the body, selects the clone of lymphocytes committed to synthesizing its specific antibody and stimulates these cells to proliferate and produce more antibody

COMPLEMENT

Normal constituent of blood; consists of a set of inactive enzymes, first of which is activated by binding of antibody to antigen located on surface of a cell (usually a microorganism, tumor cell, or transplant cell); shape of antibody molecule altered by antigen binding to it, thus exposing complement-binding site and making it possible for complement protein 1 to bind to it; other complement proteins activated in sequence and assemble on cell surface to doughnut-shaped structure; result, cell killed by process of cytolysis

PROPERDIN

Set of inactive enzymes normally present in blood; activated, they assemble on a cell surface where they activate complement protein 3; result—complement action continued and cell killed by cytolysis

INTERFERON

Proteins released from body cells following vital invasion of them; interferon may act in someway to prevent multiplication of viruses in cells; may also have anticancer activites

APPLICATIONS

Conditions involving abnormalities of the immune system

A Agammaglobulinemia—disease characterized by absence of gamma globulins (antibodies) in blood

Outline Summary—cont'd

B Autoimmune diseases—plasma cells produce antibodies that attack body's own cells; apparently lymphocytes mistakenly identify them as foreign cells

Transplant rejection

Immune system cells, that is, lymphocytes, recognize as foreign the histocompatibility antigens on the surface of transplanted cells and initiate reactions that end in the destruction of transplanted cells

Cancer

A Cell mutations are common in the normal body and produce cancer cells

B Surface membranes of cancer cells contain abnormal antigens called *tumor-specific antigens*

C Tumor-specific antigens label cancer cells as foreign, and normally lymphocytes recognize them as foreign and initiate reactions that kill them before they have time to proliferate to form tumor

D Newly formed cancer cells sometimes escape destruction by immune system; exactly how they escape is still uncertain

Review Questions

1 Name the main structures that compose the immune system.
2 Explain the immune function of neutrophils.
3 Define the following terms: *antigens, antibodies, antigenic determinants, combining sites, clone.*
4 The word *epitope* is a synonym for which of the terms defined in question 1?
5 What two terms are synonyms for combining sites?
6 When an antigen-antibody complex is formed, what region on the antigen molecule fits into what region on the antibody molecule?
7 B cells and T cells belong to what class of cells?
8 Explain the development of hemopoietic stem cells into mature or activated B cells.
9 Activated B cells develop into clones of what two kinds of cells?
10 What cells synthesize and secrete copious amounts of antibodies?
11 Explain the function of memory cells.
12 Explain the development of hemopoietic stem cells into sensitized T cells.
13 Both B cells and C cells perform the same preliminary functions of searching out, recognizing, and binding to specific antigens. Describe several functions performed by T cells after they have been activated or sensitized.

14 What are "killer cells?" Why is this nickname appropriate?
15 Antibodies belong to what class of compounds? Diagram and describe the structure of an antibody molecule.
16 What is a synonym for immunoglobulins?
17 An enormous diversity of antibodies exists in our bodies. Explain briefly how this diversity may have originated.
18 Antibodies do not themselves destroy microorganisms and other foreign cells. They do, however, contribute to their destruction. How?
19 What does the term *self-tolerance* mean? Outline Burnet's theory explaining self-tolerance.
20 Explain the two basic tenets of Burnet's clonal selection theory.
21 What is complement and how does it function?
22 What is properdin and how does it function?
23 What is interferon and how does it function?
24 What does the term *autoimmune disease* mean?
25 What are histocompatibility antigens? What laboratory procedure applies knowledge about these antigens?
26 Explain briefly the phenomenon of transplant rejection.
27 Describe briefly some currently held concepts about the development of cancer.

30 STRESS

OBJECTIVES

After you have completed this chapter, you should be able to:

1 Define the term *stress* and give examples of agents, or stressors, that produce stress.
2 Explain the three differing stages of the general adaptation syndrome.
3 Discuss the mechanisms of stress and Selye's hypothesis about the activation of the stress response.
4 Compare the alarm reaction responses resulting from hypertrophy of the adrenal cortex and from increased sympathetic activity.
5 Outline and discuss current concepts of the stress syndrome.
6 Explain the relationship between corticoids and resistance to stress.
7 Discuss psychological stress.

SELYE'S CONCEPT OF STRESS

In 1935 Hans Selye of McGill University in Montreal made an accidental discovery that launched him on a lifelong career and led him to conceive the idea of stress. This chapter will tell the story briefly of how Selye developed his stress concept and will describe the mechanism of stress that he postulated. It will then present some current ideas about stress.

KEY TERMS

Adaptation

Alarm reaction

Corticoids (KOR-ti-koids)

General adaptation syndrome

Stage of exhaustion

Stage of resistance

Stress syndrome

Stressors

Development of the Concept

Selye made his accidental discovery in 1935 when he was trying to learn whether there might be another sex hormone besides those already known. He had injected rats with various extracts made from ovaries and placenta, expecting to find that different changes had occurred in animals injected with different hormonal preparations. But to his surprise and puzzlement, he found the same three changes in all of the animals. The cortex of their adrenal glands were enlarged but their lymphatic organs—thymus glands, spleens, and lymph nodes—were atrophied, and bleeding ulcers of the stomach and duodenum had developed in every animal. Then he tried injecting many other substances, for example, extracts from pituitary glands, kidneys, and spleens, and even a poison, formaldehyde. Every time he found the same three changes: enlarged adrenals, shrunken lymphatic organs, and bleeding gastrointestinal ulcers. They seemed to be a syndrome, he thought. A syndrome, according to the classical definition, is a set of signs and symptoms that occur together and that are characteristic of one particular disease. The three changes, or "stress triad," Selye had observed occurred together, but they seemed to be characteristic not of any one particular kind of injury but of any and all kinds of harmful stimuli. More experiments using a wide variety of chemicals and injurious agents confirmed for him that the three changes truly were a syndrome of injury. His first publication on the subject was a short paper entitled "A syndrome produced by diverse nocuous agents"; it appeared in the July, 1936, issue of the British journal, *Nature*. Years later, in 1956, he published his monumental technical treatise, *The Stress of Life*.

Definitions

Stress, according to Selye's usage of the word, is a state or condition of the body produced by "diverse nocuous agents" and manifested by a syndrome of changes. He named the agents that produce stress **stressors** and coined a name—**general adaptation syndrome**—for the syndrome or group of changes that make the presence of stress in the body known.

Stressors

A stressor is any agent or stimulus that produces stress. Precise classification of stimuli as stressors or nonstressors is not possible. We can, however, make the following few generalizations about the character of stressors.

1 Stressors are extreme stimuli—too much or too little of almost anything. In contrast, almost anything in moderation or mild stimuli are nonstressors. Thus, coolness, warmth, and soft music are nonstressors, whereas extreme cold, extreme heat, and extremely loud music almost always act as stressors. Not only extreme excesses but also extreme deficiencies may act as stressors. One example of this kind of stressor is an extreme lack of social contact stimuli. Solitary confinement in a prison, space travel, social isolation because of blindness or deafness or, in some cases, old age, have all been identified as stressors. But the opposite extreme, an excess of social contact stimuli for example, (caused by overcrowding), also acts as a stressor.

2 Stressors very often are injurious or unpleasant or painful stimuli—but not always. "A painful blow and a passionate kiss," Selye wrote, "can be equally stressful."

3 Anything that an individual perceives as a threat, whether real or imagined, arouses fear or anxiety. These emotions act as stressors. So, too, does the emotion of grief.

4 Stressors differ in different individuals and in one individual at different times. A stimulus that is a stressor for you may not be a stressor for me. A stimulus that is a stressor for you today may not be a stressor for you tomorrow and might not have been a stressor for you yesterday. Many factors—including one's physical and mental health, heredity, past experiences, coping habits, and even one's diet—determine which stimuli are stressors for each individual.

The General Adaptation Syndrome

Manifestations

Stress, like health or any other state or condition, is an intangible phenomenon. It cannot itself be seen, heard, tasted, smelled, felt, or measured directly. How, then, can we know that stress exists? It can be inferred to exist when certain visible, tangible, measurable responses occur. Selye, for example, inferred that the animals he experimented on were in a state of stress when he found the syndrome of the three changes before noted—hypertrophied adrenals, atrophied lymphatic organs, and bleeding gastrointestinal ulcers. Because this syndrome indicated the presence of stress and consisted of three changes, he called them the "stress triad." Eventually he found that many other changes also took place as a result of stress. The entire group of changes or responses he named the **general adaptation syndrome (GAS).** For each word in this name he gave his reasons. By the word "general" he wanted to suggest that the syndrome was "produced only by agents which have a general effect upon large portions of the body." The word "adaptation" was meant

Figure 30-4 Current concepts of the stress syndrome.

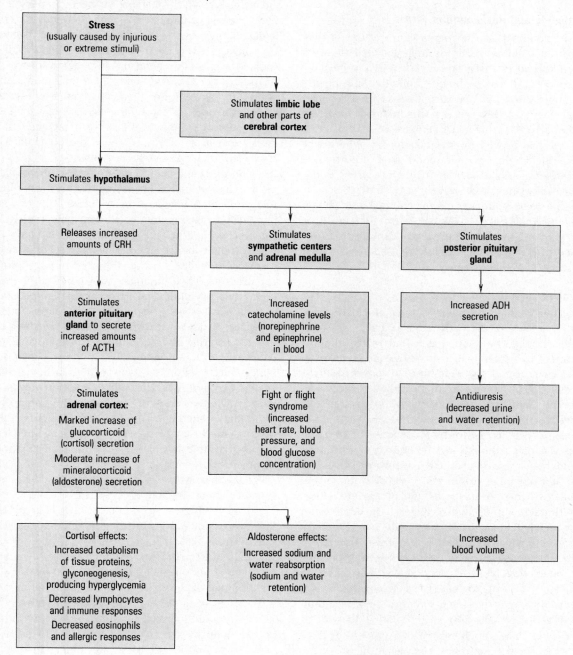

This same stress indicator has been observed in college oarsmen when they were anticipating performing in an exhibition and in heart patients when they were anticipating transfer out of a coronary care unit to a convalescent unit.

The amount of urinary adrenocorticoids is often used as a measure of stress. It has been found to increase in depressed persons who feel hopeless and doomed, in test pilots, and in college students taking examinations or attending exciting movies. In contrast, urinary corticoids were found to drop markedly in persons watching unexciting nature study films.

The level of adrenocorticoids in the blood plasma of disturbed patients having acute psychotic episodes has been found to be 70% higher than that in normal individuals or in calm patients. Another study showed that the plasma corticoid levels of chronically depressed patients were significantly lower than those of acutely anxious patients. Smoking and exposure to nicotine have also been shown to be stressors that caused a marked rise in plasma adrenocorticoids—by as much as 77% in both human beings and experimental animals.

Corticoids and Resistance to Stress

Selye thought that the increase in corticoids that occurred in his stressed animals enabled them to adapt to and to resist stress. Today many physiologists doubt this. No one questions that adrenocortical hormones increase during stress. That fact has been clearly established. But what many do question is how essential this increase is for resisting stress. No one has proved by an unequivocal experiment that a higher than normal blood level of corticoids increases an animal's ability to adapt to stress. Some clinical evidence, however, seems to indicate that it does increase a human's coping ability. For instance, patients who have been taking cortisol for some time are known to require increased doses of this hormone in order to successfully resist stresses such as surgery or severe injury.

Psychological Stress

Stress as defined by Selye is physiological stress, that is, a state of the body. Psychological stress, in contrast, might be defined as a state of the mind. It is caused by psychological stressors and manifested by a syndrome. By definition, a **psychological stressor** is anything that an individual perceives as a threat—a threat either to survival or to self-image. Moreover, the threat does not need to be real—it needs only to be real to the individual. He must see it as a threat although in truth it may not be. Psychological stressors produce a syndrome of subjective and objective responses. Dominant among the subjective reactions is a feeling of anxiety. Other emotional reactions such as anger, hate, depression, fear, and guilt are also common subjective responses to psychological stressors. Some characteristic objective responses are restlessness, fidgeting, criticizing, quarreling, lying, and crying. Another objective indicator of psychological stress, discovered only a few years ago, is that the concentration of lactate in the blood increases. For an interesting account of this, read "The Biochemistry of Anxiety" by F.N. Pitts, Jr., in the February 1969 issue of *Scientific American.*

Does psychological stress relate to physiological stress? The answer is clearly "yes." Physiological stress usually is accompanied by some degree of psychological stress. And, conversely, in most people, psychological stress produces some physiological stress responses. Ancient peoples intuitively recognized this fact. For example, it is said that in ancient times when the Chinese suspected a person of lying, they made him chew rice powder and then spit it out. If the powder came out dry, not moistened by saliva, they judged the suspect guilty. They seemed to know that lying makes a person "nervous" —and that nervousness makes a person's mouth dry. We moderns also know these facts but we describe them with more technical language. Lying, we might say, incudes psychological stress and psychological stress acts in someway to cause the physiological stress response of decreased salivation.

Within recent years, a scientific discipline called *psychophysiology* has come into being. Psychophysiologists, using accepted research methods and sophisticated instruments—including polygraphs designed especially for this type of research—have investigated a wide variety of physiological responses made by individuals being subjected to psychological stressors. Their findings amply confirm the principle that psychological stressors often produce physiological stress responses. They have found, however, that identical psychological stressors do not necessarily induce identical physiological responses in different individuals. In the words of one experimenter, "For one person the cardiovascular system may quite regularly mirror emotion most sensitively, whereas another person may be primarily a pulmonary reactor."[*] Another of their interesting discoveries is that some organ systems become less responsive after they have been stimulated a number of times.

In summary, here are some principles to remember about psychological stress.

1 Physiological stress almost always is accompanied by some degree of psychological stress.
2 In most people, psychological stress leads to some physiological stress responses. Many of these are measurable autonomic responses, for example, accelerated heart rate and increased systolic blood pressure.
3 Identical psychological stressors do not always induce identical physiological responses in different individuals.
4 In any one individual, certain autonomic responses are better indicators of psychological stress than others.

---

[*]Smith, B.M.: The polygraph, Sci. Am. **216:**25-31, Jan., 1967.

Outline Summary

SELYE'S CONCEPT OF STRESS

Development of the concept

Through many experiments, Selye exposed animals to wide variety of noxious agents and found that they all responded with the same syndrome of changes, or "stress triad."

Definitions

A Stress—a state or condition of the body produced by "diverse nocuous agents" and manifested by a syndrome of changes

B Stressors—the agents that produce stress

C General adaptation syndrome—the group of changes that manifest the presence of stress

Stressors

A Stressors are extreme stimuli—too much or too little of almost anything

B Stressors are very often injurious or painful stimuli

C Anything an individual perceives as a threat is a stressor for that individual

D Stressors are different for different individuals and for one individual at different times

The general adaptation syndrome

A Manifestations—stress triad (hypertrophied adrenals, atrophied thymus and lymph nodes, and bleeding ulcers) and many other changes

B Stages—three successive phases, namely, alarm reaction, stage of adaptation or resistance, and stage of exhaustion, each characterized by different syndrome of changes (Table 30-1)

Mechanism of stress

A Consists of group (syndrome) of responses to internal condition of stress; stress responses nonspecific in that same syndrome of responses occurs regardless of kind of extreme change that produced stress

B Stimulus that produces stress and thereby activates stress mechanism is nonspecific in that it can be any kind of extreme change in environment

C Stress responses, Selye thought, were adaptive; tend to enable body to adapt to and survive extreme change; Selye referred to this syndrome of stress responses as general adaptation syndrome

D Numerous factors influence stress responses—individual's physical and mental condition, age, sex, socioeconomic status, heredity, previous experience with similar stressor, and even religious affiliation

E Stress most often is coped with successfully—namely, results in adaptation, healthy survival, and increased resistance; sometimes, however, stress produces exhaustion and death

F See Figure 30-3, then 30-1, and 30-2 for a summary of Selye's stress mechanism

Stress and disease

A Selye held that stress could result in disease instead of adaptation

B To date, research evidence that stress, induced in normal individual, can produce disease has not gained wide acceptance

SOME CURRENT CONCEPTS ABOUT STRESS

Definitions

A Stress—any stimulus that directly or indirectly stimulates hypothalamus to release CRH

B Stress syndrome—also called the stress response; many diverse changes initiated by stress

The stress syndrome

See Figure 30-4

Indicators of stress

A Changes caused by increased sympathetic activity; for example, faster, stronger heartbeat, higher blood pressure, sweaty palms, dilated pupils

B Changes resulting from increased corticoids—eosinopenia, lymphocytopenia, increased adrenocorticoids in blood and urine

Corticoids and resistance to stress

Still controversial; not proved that increased corticoids increase an animal's or man's ability to resist stress but is proved that corticoid levels in blood increase in stress

Psychological stress

A Psychological stressors—anything that an individual perceives as a threat to either survival or self-image

B Psychological stress—a mental state characterized by a syndrome of subjective and objective responses; dominant subjective response is anxiety; some characteristic objective responses are restlessness, quarrelsomeness, lying, crying

C Relation to physiological stress (see summary of principles on p. 762)

Review Questions

1 Describe the experimental results that led Selye to conceive his idea of stress.
2 Define the terms *stress*, *stressor*, and *general adaptation syndrome* as Selye used them.
3 Make a few generalizations about the kinds of stimuli that constitute stressors.
4 According to Selye, what three stages make up the general adaptation syndrome? What changes characterize each stage?
5 What changes constituted Selye's "stress triad"?
6 According to Selye, an increase in what three hormones brought about the changes that he named the general adaptation syndrome?
7 What function, according to Selye, does the general adaptation syndrome serve?
8 What relation, if any, did Selye suggest exists between stress and disease?
9 Stress, according to Selye, is a state or condition of the body. According to a current operational definition, what is stress?
10 What part of the brain, according to current ideas, plays the key role in initiating the stress syndrome of responses?

11 What parts of the nervous system other than that named in question 10 are involved in inducing the stress syndrome?
12 Briefly, what role, if any do each of the following hormones play when the body is subjected to stress: ACTH, ADH, aldosterone, cortisol, CRH, epinephrine, and norepinephrine?
13 Describe some changes you might observe that would indicate an individual was being subjected to stress.
14 What has been proved about corticoids in relation to stress?
15 What is the controversial issue about corticoids in relation to stress?
16 Define psychological stressor and psychological stress.
17 Cite several examples of psychological stressors.
18 Are psychological stressors the same for all individuals?
19 Give some examples of subjective indicators of psychological stress.
20 Give some examples of objective responses that are part of the syndrome of psychological stress.
21 State three of four principles about the relationship between psychological and physiological stress.

GLOSSARY

A

abdomen Body area between the diaphragm and pelvis.

abduct To move away from the midline; opposite of adduct.

absorption Passage of a substance through a membrane, for example, skin or mucosa, into blood.

acapnia Marked decrease in blood carbon dioxide content.

acetabulum Socket in the hip bone (os coxae or innominate bone) into which the head of the femur fits.

acetone bodies Ketone bodies, acids formed during the first part of fat catabolism; namely, acetoacetic acid, betahydroxybutyric acid, and acetone.

Achilles tendon Tendon inserted on calcaneus; so called because of the Greek myth that Achilles' mother held him by the heels when she dipped him in the river Styx, thereby making him invulnerable except in this area.

acid Substance that ionizes in water to release hydrogen ions.

acidosis Condition in which there is an excessive proportion of acid in the blood.

acromegaly Disease caused by hypersecretion of growth hormone in the adult.

acromion Bony projection of the scapula; forms point of the shoulder.

acrosome Specialized cap that covers the head of a spermatozoon.

actin One of the contractile proteins in muscle.

adduct To move toward the midline; opposite of abduct.

adenine Nitrogenous base; component of RNA and DNA.

adenohypophysis Anterior pituitary gland.

adenoids Glandlike; adenoids or pharyngeal tonsils are paired lymphoid structures in the nasopharynx.

adolescence Period between puberty and adulthood.

adrenergic fibers Axons whose terminals release norepinephrine and/or epinephrine.

adventitia, externa Outer coat of a tube-shaped structure such as blood vessels.

aerobic Requiring free oxygen; opposite of anaerobic.

afferent Carrying toward, as neurons that transmit impulses to the central nervous system from the periphery; opposite of efferent.

ala Flaring, cartilaginous expansion forming the side of each nostril opening.

albinism Skin condition characterized by a deficiency or complete absence of pigment.

albuminuria Albumin in the urine.

aldosterone A hormone secreted by the adrenal cortex.

alkali reserve Bicarbonate salts present in body fluids; mainly sodium bicarbonate.

alkalosis Condition in which there is an excessive proportion of alkali in the blood; opposite of acidosis.

allograft Tissue transplant between individuals of the same species.

alveolus Literally a small cavity; alveoli of lungs are microscopic saclike dilations of terminal bronchioles.

ameboid movement Movement characteristic of amebae, that is, by projections of protoplasm (pseudopodia) toward which the rest of the cell's protoplasm flows.

amenorrhea Absence of the menses.

amino acid Organic compound having an NH_3 and a COOH group in its molecule; has both acid and basic properties; amino acids are the structural units from which proteins are built.

amphiarthrosis Slightly movable joint.

ampulla Saclike dilation of a tube or duct.

anabolism Synthesis by cells of complex compounds—for example, protoplasm, hormones—from simpler compounds (amino acids, simple sugars, fats, minerals); opposite of catabolism, the other phase of metabolism.

anaerobic Not requiring free oxygen; opposite of aerobic.

anaphase Stages of mitosis; duplicate chromosomes move to poles of dividing cell.

anastomosis Connection between vessels, for example, the circle of Willis is an anastomosis of certain cerebral arteries.

anatomy Study of the structure of an organism and the relationship of its parts.

androgen Male sex hormone.

anemia Deficient number of red blood cells or deficient hemoglobin.

anesthesia Loss of sensation.

aneurysm Blood-filled saclike dilation of the wall of an artery.

angina Any disease characterized by spasmodic suffocative attacks, for example, angina pectoris, paroxysmal thoracic pain with feeling of suffocation.

Angstrom unit 0.1 mμ or ¹⁄₁₀ millionth of a meter or about ¹⁄₂₅₀ millionth of an inch.

anion Negatively charged particle.

anisocytosis Abnormal variation in red blood cell diameter (size).

ankyloglossia Faulty speech caused by adhesion of the tongue to the floor of the mouth (tongue-tie).

anorexia Loss of appetite.

anoxemia Deficient blood oxygen content.

anoxia Deficient oxygen supply to tissues.

antagonistic muscles Those having opposite action, for example, muscles that flex the upper arm are antagonistic to muscles that extend it.

antebrachium Area between elbow and wrist.

anterior Front or ventral; opposite of posterior or dorsal.

antibody, immune body Substance produced by the body that destroys or inactivates a specific substance (antigen) that has entered the body, for example, diphtheria antitoxin is the antibody against diphtheria toxin.

antigen Substance that, when introduced into the body, causes formation of antibodies against it.

antrum Cavity, for example, the antrum of Highmore, the space in each maxillary bone, or the maxillary sinus.

anus Distal end or outlet of the rectum.

apex Pointed end of a conical structure.

aphasia Loss of a language faculty such as the ability to use words or to understand them.

apnea Temporary cessation of breathing.

aponeurosis Flat sheet of white fibrous tissue that serves as a muscle attachment.

appendicular Pertaining to the extremities or to an appendage.

aqueduct Tube for conduction of liquid, for example, the cerebral aqueduct conducts cerebrospinal fluid from the third to the fourth ventricle.

arachnoid Delicate, weblike middle membrane of the meninges.

areola Small space; the pigmented ring around the nipple.

arrhythmia Irregular heartbeat.

arteriole Small branch of an artery.

arteriosclerosis Hardening of the arteries.

artery Vessel carrying blood away from the heart.

arthrosis Joint or articulation.

articular Referring to a joint.

articulation Joint.

arytenoid Ladle-shaped; two small cartilages of the larynx.

ascites Accumulation of serous fluid in the abdominal cavity.

asphyxia Loss of consciousness from deficient oxygen supply.

aspirate To remove by suction.

asthenia Bodily weakness.

astrocytes Star-shaped neuroglia, connective tissue cells in brain and cord.

ataxia Loss of power of muscle coordination.

atherosclerosis Arteriosclerosis or hardening of artery walls characterized by lipid deposits in tunica intima.

atrium Chamber or cavity, for example, atrium of each side of the heart.

atrophy Wasting away of tissue; decrease in size of a part.

auricle Part of the ear attached to the side of the head; earlike appendage of each atrium of heart.

autonomic Self-governing, independent.

axial Pertaining to the axis or trunk.

axilla Armpit.

axon Nerve cell process that transmits impulses away from the cell body.

B

baroreceptor Receptor stimulated by change in pressure.

Bartholin Seventeenth century Danish anatomist.

base A compound that yields hydroxyl (OH⁻) ions in water.

basement membrane Thin layer of fine fibrils embedded in a matrix of glycoprotein.

basophil White blood cell that stains readily with basic dyes.

biceps Two headed or composed of two parts, such as the biceps brachii muscle.

bilirubin Red pigment in the bile.

biliverdin Green pigment in the bile.

blastocyst Hollow cell mass; stage in development of ovum following morula.

blood brain barrier Tight sheaths formed of astrocytes and tight junctions between endothelial cells that compose brain capillary walls.

BNA (Basle Nomina Anatomica) Anatomic terminology accepted at Basle by the Anatomical Society in 1895.

bolus A rounded mass of food.

Bowman Nineteenth century English physician.

brachial Pertaining to the arm.

bronchiectasis Dilation of the bronchi.

bronchiole Small branch of a bronchus.

bronchogram Radiograph of the bronchial tree.

bronchus One of the two branches of the trachea.

buccal Pertaining to the cheek.

buffer Compound that combines with an acid or with a base to form a weaker acid or base, thereby lessening the change in hydrogen ion concentration that would occur without the buffer.

bursa Fluid-containing sac or pouch lined with synovial membrane.

buttock Prominence over the gluteal muscles.

C

calcitonin A hormone secreted by the parathyroid glands.

calculus Stone formed in various parts of the body; may consist of different substances.

callus New bone tissue that forms around a fracture.

calorie Heat unit; a large calorie is the amount of heat needed to raise the temperature of 1 kg of water 1° C.

calyx Cup-shaped division of the renal pelvis.

canaliculus Little canal.

capillary Microscopic blood vessel; capillaries connect arterioles with venules; also, microscopic lymphatic vessels.

carbhemoglobin, carbaminohemoglobin Compound which is formed by union of carbon dioxide with hemoglobin.

carbohydrate Organic compounds containing carbon, hydrogen, and oxygen in certain specific proportions, for example, sugars, starches, cellulose.

carbonic anhydrase Enzyme that reversibly catalyzes conversion of water and CO_2 to carbonic acid.

carboxyhemoglobin Compound formed by union of carbon monoxide with hemoglobin.

carcinoma Cancer, a malignant tumor.

caries Decay of teeth or of bone; cavity in a tooth caused by destruction of its mineral components.

carotid From Greek word meaning to plunge into deep sleep; carotid arteries of the neck are so called because pressure on them may produce unconsciousness.

carpal Pertaining to the wrist.

casein Protein in milk.

cast Mold, for example, formed in renal tubules.

castration Removal of testes or ovaries.

catabolism Breakdown of food compounds or of protoplasm into simpler compounds; opposite of anabolism, the other phase of metabolism.

catalyst Substance that alters the speed of a chemical reaction.

cataract Opacity of the lens of the eye.

catecholamines Norepinephrine and epinephrine.

cathode Negative pole of a charged particle or of a battery.

cation Positively charged particle.

caudal Pertaining to the tail of an animal; opposite of cephalic.

cecum Blind pouch; the pouch at the proximal end of the large intestine.

celiac Pertaining to the abdomen.

cellulose Polysaccharide, the main plant carbohydrate.

centimeter $1/100$ of a meter, about $2/5$ of an inch.

centriole Organelle necessary for spindle formation during mitosis.

centromere Structure that separates chromatids during mitosis.

centrosphere, centrosome Spherical area or body near center of cell.

cephalic Pertaining to the head; opposite of caudal.

cerebrospinal fluid Plasma or lymphlike fluid found in the cavities and canals of the brain and spinal cord.

cerumen Earwax.

cervix Neck; any necklike structure.

chemoreceptor Distal end of sensory dendrites especially adapted for chemical stimulation.

chiasm Crossing; specifically, a crossing of the optic nerves.

cholecystectomy Removal of the gallbladder.

cholesterol Organic alcohol present in bile, blood, and various tissues.

cholinergic fibers Axons whose terminals release acetylcholine.

cholinesterase Enzyme; catalyzes breakdown of acetylcholine.

chorion Outermost fetal membrane derived from the blastocyst; contributes to placenta formation.

choroid, chorioid Skinlike.

chromatids Newly formed chromosomes.

chromatin Deep-staining substance in the nucleus of cells; divides into chromosomes during mitosis.

chromosomes Deep-staining, rod-shaped bodies in cell nucleus; composed of genes.

chyle Milky fluid; the fat-containing lymph in the lymphatics of the intestine.

chyme Partially digested food mixture leaving the stomach.

cilia Hairlike projections of protoplasm.

circadian Daily.

cistron Gene; segment of DNA molecule that transmits an inheritable trait.

cleavage Mitotic cell division that begins after fertilization.

clone Family of cells descended from one cell and therefore identical to it.

coagulation Transformation of a soluble protein into its insoluble form.

cochlea Snail shell or structure of similar shape.

codocyte An unusually shaped erythrocyte (target cell) with a prominent central dimple.

coenzyme Nonprotein substance that activates an enzyme.

coitus, copulation Sexual intercourse.

collagen Principal organic constituent of connective tissue.

colloid Solute particles with diameters of 1 to 100 mμ.

colon Large intestine, excluding cecum, rectum, and anal canal.

colostrum First milk secreted after childbirth.

combining sites, antigen-binding sites, antigen receptors Regions on antibody molecule; shape of each combining site is complementary to shape of a specific antigen's epitopes.

comedo Contaminated plug of sebum or epithelial cells in the duct of a sebaceous gland

commissure Bundle of nerve fibers passing from one side to the other of the brain or cord.

complement Several inactive enzymes normally present in blood; act to kill foreign cells by cytolysis.

concha Shell-shaped structure, for example, bony projections into the nasal cavity.

conduction Propagation or passage of an impulse.

condyle Rounded projection at the end of a bone.

congenital Present at birth.

contractility Ability of a living structure to change its shape or form.

contralateral On the opposite side.

convection Heat transfer between objects through air circulation.

coracoid Like a raven's beak in form.

corium True skin or derma.

cornea Transparent outer coat of eye.

coronal Of or like a crown.

coronary Encircling; in the form of a crown.

corpus Body.

corpus luteum A yellow mass in the ovary formed by an ovarian follicle that has discharged its ovum and is secreting progesterone.

corpuscles Very small body or particle.

cortex Outer part of an internal organ, for example, of the cerebrum and of the kidneys.

cortisol, hydrocortisone, compound F Glucocorticoid secreted by adrenal cortex.

costal Pertaining to the ribs.

covalent bond Chemical bond formed by two atoms sharing a pair of electrons.

cranial Pertaining to the head or skull.

cranial nerves Twelve pairs of nerves emerging from the brain.

crenation, plasmolysis Shriveling of a cell because of water withdrawal.

cretinism Dwarfism caused by hypofunction of the thyroid gland.

cribriform Sievelike.

cricoid Ring shaped; a cartilage of this shape in the larynx.

cruciate Cross shaped.

crural Pertaining to the leg or lower extremity.

crystalloid Solute particle less than 1 mμ in diameter.

cubital Pertaining to the forearm.

cutaneous Pertaining to the skin.

cyanosis Bluish appearance of the skin from deficient oxygenation of blood.

cytokinesis Dividing of cytoplasm to form two cells.

cytology Study of cells.

cytoplasm The protoplasm of a cell exclusive of the nucleus.

D

deamination Chemical reaction by which the amino group NH_3 is split from an amino acid.

deciduous Temporary; shedding at a certain stage of growth, for example, deciduous teeth.

decomposition Chemical reaction involving breaking of chemical bonds; forms two or more smaller molecule compounds from larger molecule compounds.

decubitus ulcer Ulcer or sore caused by reduced blood supply and/or constant pressure on the skin.

decussation Crossing over like an X.

defecation Elimination of waste matter from the intestines.

deferens Carrying away.

deglutition Act of swallowing.

deltoid Triangular, for example, deltoid muscle.

dendrite, dendron Branching or treelike; a nerve cell process that transmits impulses toward the cell body.

dens Tooth.

dentate Having toothlike projections.

dentine Main part of a tooth, under the enamel.

dentition Teething; also, number, shape, and arrangement of the teeth.

dermatome Area of skin supplied by sensory fibers of a single dorsal root.

dermis, corium True skin.

dextrose Glucose, a monosaccharide, the principal blood sugar.

diabetes insipidus Disease characterized by excretion of large amounts of urine without loss of sugar.

diabetes mellitus Disease of carbohydrate metabolism characterized by excretion of sugar in the urine.

dialysis Separation; the separation of crystalloids from colloids by the faster diffusion of the former through a membrane.

diapedesis Passage of blood cells through intact blood vessel walls.

diaphragm Membrane or partition that separates one thing from another; the muscular partition between the thorax and abdomen; the midriff.

diaphysis Shaft of a long bone.

diarthrosis Freely movable joint.

diastole Relaxation of the heart interposed between its contractions; opposite of systole.

diencephalon "Tween" brain; parts of the brain between the cerebral hemispheres and the mesencephalon or midbrain.

diffusion Spreading, for example, scattering of solute particles.

digestion Conversion of food into assimilable compounds.

digital Pertaining to the fingers or toes.

diplopia Double vision; seeing one object as two.

disaccharide Sugar formed by the union of two monosaccharides; contains 12 carbon atoms.

distal Toward the end of a structure; opposite of proximal.

diuresis Increased urine production.

diverticulum Outpocketing from a tubular organ such as the intestine.

dorsal, posterior Pertaining to the back; opposite of ventral.

Douglas Scottish anatomist of the late seventeenth and early eighteenth centuries.

dropsy Accumulation of serous fluid in a body cavity, in tissues; edema.

duct Canal or passage.

duodenum First segment of small intestine.

dura mater Literally strong or hard mother; outermost layer of the meninges.

dyspnea Difficult or labored breathing.

dystrophy Any disorder or disease resulting from defective or faulty nutrition.

E

ectomorph Thin and lean body type.

ectopic Displaced; not in the normal place, for example, extrauterine pregnancy.

edema Excessive fluid in tissues; dropsy.

effector Responding organ, specifically, voluntary and involuntary muscle, the heart, and glands.

efferent Carrying from, as neurons that transmit impulses from the central nervous system to the periphery; opposite of afferent.

electrocardiogram Graphic record of heart's action potentials.

electroencephalogram Graphic record of brain's action potentials.

electrolyte Substance that ionizes in solution, rendering the solution capable of conducting an electric current.

electron Minute, negatively charged particle.

electrovalent (ionic) bond Chemical bond formed by transfer of electron.

elimination Expulsion of wastes from the body.

embolism Obstruction of a blood vessel by foreign matter carried in the bloodstream.

embryo Animal in early stages of intrauterine development; the human fetus during the first 3 months after conception.

emesis Vomiting.

emphysema Dilation of pulmonary alveoli.

empyema Pus in a cavity, for example, in the chest cavity.

encephalon Brain.

endocrine Secreting into blood or tissue rather than into a duct; opposite of exocrine.

endomorph Body type characterized by excess fat.

endomysium Connective tissue between individual muscle fibers.

endoplasm Cytoplasm located toward center of cell, as distinguished from ectoplasm located nearer periphery of cell.

endoplasmic reticulum Network of tubules and vesicles in cytoplasm.

endothelium Epithelial lining of heart, lymphatics, and blood vessels.

energy Power or ability to do work; *kinetic* or *active*, caused by moving particles, for example, mechanical energy, heat; *potential* or *stored*, caused by attraction between particles, for example, chemical energy.

enterokinase Digestive enzyme that transforms trypsinogen into trypsin.

enteron Intestine.

enzyme Catalytic agent formed in living cells.

eosinophil, acidophil White blood cell readily stained by eosin.

epidermis "False" skin; outermost layer of the skin.

epiglottis Laryngeal cartilage.

epimysium Connective tissue sheath that envelops a skeletal muscle.

epinephrine Secretion of the adrenal medulla.

epiphyses Ends of a long bone.

epistaxis Nosebleed.

epithelium Tissue type that covers body surface, forms glands, and lines passages that communicate with the exterior of the body.

epitope (antigenic determinant) Uniquely shaped region on surface of antigen.

equivalent An expression of chemical concentration and reactivity.

erythrocyte Red blood cell.

erythropoiesis Formation of red blood cells.

erythropoietin Hormone secreted by the kidney that stimulates red blood cell production.

estrogen Female sex hormone.

ethmoid Sievelike.

eupnea Normal respiration.

Eustachio Italian anatomist of the sixteenth century.

excitability Ability to react to a stimulus.

excretion Elimination of waste products.

exocrine Secreting into a duct; opposite of endocrine.

exophthalmos Abnormal protrusion of the eyes.

extrinsic Coming from the outside; opposite of intrinsic.

extrinsic factor Vitamin B_{12}.

F

facilitation Decrease in a neuron's resting potential to a point above its threshold of stimulation.

facilitated diffusion Carrier-mediated diffusion.

Fallopius Sixteenth century Italian anatomist.

fascia Sheet of connective tissue.

fasciculus Little bundle.

fat Adipose tissue or a compound formed from union of one molecule of glycerol and three molecules of fatty acid.

fatigue Condition characterized by the loss of irritability.

fauces Opening between the mouth and oropharynx.

fetus Unborn young, especially in the later stages; in human beings, from third month of intrauterine period until birth.

fiber Threadlike structure.

fibrillation Ineffective, uncoordinated contraction of cardiac muscle fibers.

fibrin Insoluble protein in clotted blood.

fibrinogen Soluble blood protein that is converted to insoluble fibrin during clotting.

fibroblasts Connective tissue cells that synthesize interstitial fibers and gels.

fibrocytes Old fibroblasts.

fight or flight reaction Response of body to sympathetic nervous system stimulation.

filtration Passage of water and solutes through a membrane from hydrostatic pressure gradient.

fimbria Fringe.

fissure Groove.

flaccid Soft, limp.

flexion Decrease in the angle between anteriorly articulating bones.

follicle Small sac or gland.

follicle-stimulating hormone (FSH) Gonadotrophic hormone of anterior pituitary, which stimulates follicle development in ovaries.

fontanel "Soft spots" of the infant's head; unossified areas in the infant's skull.

foramen Small opening.

forearm Area of the upper extremity between elbow and wrist.

fossa Cavity or hollow.

fovea Small pit or depression.

fovea centralis Area of retina where vision is most acute.

fundus Base of a hollow organ, for example, the part farthest from its outlet.

G

gamete Sex cell; sperm or ovum.

ganglion Cluster of nerve cell bodies outside the central nervous system.

gasserian Named for Gasser, a sixteenth century Austrian surgeon.

gastric Pertaining to the stomach.

gastrin Digestive hormone secreted by stomach mucosa.

gene (cistron) Part of the chromosome that transmits a given hereditary trait.

genitalia Reproductive organs.

gestation Pregnancy.

gingivitis Inflammation of the gums or gingiva.

gland Secreting structure.

glomerulus Compact cluster, for example, of capillaries in the kidneys.

glossal Of the tongue.

glottis Opening between true vocal cords.

glucagon Hormone secreted by alpha cells of the islands of Langerhans.

glucocorticoids Hormones that influence food metabolism; secreted by adrenal cortex.

glucokinase Enzyme; catalyzes conversion of glucose to glucose-6-phosphate.

gluconeogenesis Formation of glucose from protein or fat compounds.

glucose Monosaccharide or simple sugar; the principal blood sugar.

glucosuria Glucose in the urine.

gluteal Of or near the buttocks.

glycerin, glycerol Product of fat digestion.

glycogen "Animal starch"; main polysaccharide stored in animal cells.

glycogenesis Formation of glycogen from glucose or from other monosaccharides, fructose or galactose.

glycogenolysis Hydrolysis of glycogen or glucose-6-phosphate or to glucose.

glycolysis Breakdown of glycogen to lactic acid.

glyconeogenesis The formation of glycogen from protein or fat compounds.

goiter (simple) Diffuse and painless enlargement of the thyroid gland.

golgi apparatus Organelle of cell involved in secretory activity.

gomphosis An immovable joint in which a peg-shaped projection fits into a bony socket.

gonad Sex gland in which reproductive cells are formed.

gout Disease characterized by elevated uric acid levels in the blood.

graafian Named for Graaf, a seventeenth century Dutch anatomist.

gradient A slope or difference between two levels, for example, concentration gradient—a difference between the concentrations of two substances.

groin Root of thigh between lower extremity and abdomen of torso.

gustatory Pertaining to taste.

gyrus Convoluted ridge.

H

haploid Chromosome number half that of somatic cells—the haploid chromosome number is typical of mature sex cells or gametes.

haversian Named for Havers, English anatomist of the late seventeenth century.

helix Spiral; coil.

hematocrit Volume (percent) of red blood cells in a sample of whole blood.

hematoma A blood clot.

hematuria Blood in the urine.

hemiplegia Paralysis of one side of the body.

hemocytoblast Primitive, multipotential, blood-forming stem cell found in the bone marrow.

hemoglobin Iron-containing protein in red blood cell.

hemolysis Destruction of red blood cells with escape of hemoglobin from them into surrounding medium.

hemophilia Blood disease characterized by prolonged coagulation time.

hemopoiesis Blood cell formation.

hemorrhage Bleeding.

hepar Liver.

heparin Substance obtained from the liver that inhibits blood clotting.

heredity Transmission of characteristics from a parent to a child.

hernia, "rupture" Protrusion of a loop of an organ through an abnormal opening.

hexose 6 carbon atom sugar.

hilus, hilum Depression where vessels enter an organ.

His German anatomist of the late nineteenth century.

histamine A powerful vasodilator released during tissue injury and inflammation.

histology Science of minute structure of tissues.

homeostasis Relative uniformity of the normal body's internal environment.

hormone Substance secreted by an endocrine gland.

humoral immunity Production of protective antibodies by B lymphocytes and plasma cells.

hyaline Glasslike.

hydrocortisone, cortisol, compound F Hormone secreted by the adrenal cortex.

hydrolysis Literally "split by water"; chemical reaction in which a compound reacts with water.

hydroxyl ion An electrically-charged ion (negative) composed of 1 atom each of O and H (OH^-).

hymen Greek for skin; mucous membrane that may partially or entirely occlude the vaginal outlet.

hyoid Shaped like the letter U; bone of this shape at the base of the tongue.

hypercapnia Abnormally high blood CO_2 concentration.

hyperemia Increased blood in a part.

hyperglycemia Higher than normal blood glucose concentration.

hyperkalemia Higher than normal blood potassium concentration.

hypernatremia Higher than normal blood sodium concentration.

hyperopia Farsightedness.

hyperplasia Increase in the size of a part from an increase in the number of its cells.

hyperpnea Abnormally rapid breathing; panting.

hypertension Abnormally high blood pressure.

hyperthermia Fever; body temperature above 37° C.

hypertonic solution Solution having a higher osmotic pressure than blood.

hypertrophy Increased size of a part from an increase in the size of its cells.

hypervolemia Larger volume of blood than normal.

hypocalcemia Lower than normal blood calcium concentration.

hypokalemia Lower than normal blood potassium concentration.

hyponatremia Lower than normal blood sodium concentration.

hypophysis Greek for undergrowth; hence the pituitary gland, which grows out from the undersurface of the brain.

hypothalamus Part of the diencephalon; gray matter in the floor and walls of the third ventricle.

hypothermia Subnormal body temperature; below 37° C.

hypotonic solution Solution having a lower osmotic pressure than blood.

hypovolemia Subnormal volume of blood.

hypoxia Oxygen deficiency.

I

implantation Embedding of developing blastocyst in uterine lining (endometrium).

impulse Electrochemical charge in a nerve fiber, which results from an adequate stimulus and travels throughout the cell and its processes.

inclusions Any foreign or heterogenous substance contained in a cell or in any tissue or organ that was not introduced as a result of trauma.

incus Anvil; the middle ear bone that is shaped like an anvil.

inferior Lower; opposite of superior.

inguinal Of the groin.

inhalation Inspiration or breathing in; opposite of exhalation or expiration.

inhibition An increase in a neuron's resting potential above its usual level.

innominate Not named, anonymous, for example, ossa coxae (hip bones) formerly known as innominate bones.

insertion Point of attachment of a muscle to a bone that moves during contraction of the muscle.

insulin Hormone secreted by beta cells of the islands of Langerhans in the pancreas.

intercalated disks Specialized transverse junctions between adjacent cardiac muscle cells.

intercellular Between cells; interstitial.

interneurons, internuncial or intercalated neurons Neurons that conduct impulses from sensory to motor neurons; lie entirely within the central nervous system.

interferon Protein released from body cells following viral invasion of them; acts in some way to defend other body cells against viral invasion.

interphase The resting phase between two mitotic divisions.

interstitial Of or forming small spaces between things; intercellular.

intima Innermost.

intramembranous ossification Bone formation occurring within a membrane.

intravascular Within a blood vessel.

intrinsic Not dependent on externals; located within something; opposite of extrinsic.

involuntary Not willed; opposite of voluntary.

involution Return of an organ to its normal size after enlargement; also retrograde or degenerative change.

ion Electrically charged atom or group of atoms.

ionize Dissociation of an electrolyte compound into charged ions.

ipsilateral On the same side; opposite of contralateral.

iris The pigmented area of the eye in front of the lens.

irritability Excitability; ability of living matter to react to a stimulus.

ischemia Local anemia; temporary lack of blood supply to an area.

isometric Muscle contraction that produces no change in length of muscle.

isotonic Of the same tension or pressure.

isotopes Atoms of same element with different atomic weights because their nuclei contain different numbers of neutrons.

J

jaundice Yellow skin color caused by bile in the blood.

jejunum Second (middle) portion of the small intestine.

joint Point of contact between two or more bones or cartilage components; an articulation.

juxtaglomerular apparatus Specialized cells, which produce the hormone renin in the tunica media of the afferent arterioles of kidney glomeruli.

K

karyotype An array of chromosomes.

keratin Protein compound present in the human body, chiefly in hair and nails.

keratinization Production of keratin by cells.

ketones Acids (acetoacetic, beta-hydroxybutyric, and acetone) produced during fat catabolism.

ketosis Excess amount of ketone bodies in the blood.

kilogram 1,000 gm or approximately 2.2 pounds.

kinesthesia "Muscle sense," that is, sense of position and movement of body parts.

krebs cycle Citric acid cycle of cell metabolism.

kyphosis Pathologically excessive, convex curvature of the thoracic spine ("hunchback").

L

labia Lips.

lacrimal Pertaining to tears.

lactase An enzyme that splits lactose into glucose and galactose.

lactation Secretion of milk.

lacteals Lymphatic vessels of the intestines.

lactic acid An organic acid formed from pyruvic acid in the absence of oxygen (anaerobic).

lactose Milk sugar, a disaccharide.

lacuna Space or cavity, for example, lacunae in bone contain bone cells.

lamella Thin layer, as of bone.

lantern jaw Separation and overbite of the teeth characteristic of acromegaly.

latent period Time period between an applied stimulus and the response.

lateral Of or toward the side; opposite of medial.

lemniscus, medial A flat band of sensory fibers extending up from the medulla, through the pons and midbrain to the thalamus.

larynx Voice box.

leukocyte White blood cell.

leukocytosis Higher than normal number of white blood cells.

lever Mechanical device consisting of a weight, rod, and fulcrum.

ligament Bond or band connecting two objects; in anatomy, a band of white fibrous tissue connecting bones.

limbic lobe or system cerebral cortex on the medial surface of the brain that forms a border around the corpus callosum; older name rhinencephalon.

lipase A fat-splitting enzyme.

lipid Fats and fatlike compounds.

lipoprotein A lipid or fat and protein complex.

loin Part of the back between the ribs and hip bones.

lordosis Abnormal concave curvature of the lumbar spine.

lumbar Of or near the loins.

lumen Passageway or space within a tubular structure.

lymph Watery fluid in the lymphatic vessels.

lymphocyte One type of white blood cells.

lymphokinetic Movement or flow of lymph.

lymphography X-ray study of lymphatic vessels.

lysis The destruction of cells and other antigens by a specific lysin antibody.

lysosomes Membranous organelles containing various enzymes that can dissolve most cellular compounds, hence called "digestive bags" or "suicide bags" of cells.

M

macrophage Phagocytic cell; a wandering histiocyte.

malleolus Small hammer; projections at the distal ends of the tibia and fibula.

malleus Hammer; the tiny middle ear bone that is shaped like a hammer.

Malpighii Seventeenth century Italian anatomist.

maltase An enzyme that splits maltose into two glucose molecules.

maltose Disaccharide or "double" sugar.

mamillary Like a nipple.

mammary Pertaining to the breast.

manometer Instrument used for measuring the pressure of fluids.

manubrium Handle; upper part of the sternum.

mastication Chewing.

matrix Ground substance in which cells are embedded.

meatus Passageway.

medial Of or toward the middle; opposite of lateral.

mediastinum Middle section of the thorax, that is, between the two lungs.

medulla Latin for marrow; hence the inner portion of an organ in contrast to the outer portion or cortex.

megakaryocyte Very large bone marrow cells that form platelets.

meibomian glands Sebaceous glands of the eyelids.

meiosis Nuclear division in which the chromosomes are reduced to half their original number before the cell divides in two.

meissner's plexus Nerve cells in the submucosa of the small intestine.

melanin Dark pigment of the skin.

melatonin Hormone secreted by pineal gland.

membrane Thin layer or sheet.

meninges The three membranes that cover the brain and spinal cord.

meningitis Inflammation of the meninges.

menopause Cessation of menstruation.

menstruation Monthly discharge of blood from the uterus.

merocrine Secretory process that does not result in cell injury or death when the product of secretion is discharged.

mesencephalon Midbrain.

mesentery Fold of peritoneum that attaches the intestine to the posterior abdominal wall.

mesial (medial) Situated in the middle.

mesoderm The middle layer of the three primary germ layers of the embryo.

mesomorph A muscular type of body build.

metabolism Complex process by which food is utilized by a living organism.

metabolite Any substance produced by metabolism.

metacarpus "After" the wrist; hence the part of the hand between the wrist and fingers.

metaphase The second phase or stage of mitosis.

metatarsus "After" the instep; hence the part of the foot between the tarsal bones and toes.

meter About 39.5 inches.

microglia One type of connective tissue cell found in the brain and cord.

micron 0.001 ml; about $\frac{1}{25,000}$ of an inch.

micturition Urination, voiding.

milliequivalent A unit of chemical equivalence computed by the following formula:

$$\frac{mg}{\text{Atomic weight}} \times \text{Valence} = mEq$$

For example: $\frac{3,266 \text{ mg Na}^+}{23 \text{ (at wt Na)}} \times 1 = 142 \text{ mEq Na}^+$

millimeter 0.001 meter; about $\frac{1}{23}$ of an inch.

mineralocorticoids Hormones that influence mineral salt metabolism; secreted by adrenal cortex.

mitochondria Threadlike structures.

mitosis Indirect cell division involving complex changes in the nucleus.

mitral Shaped like a miter, such as in the mitral valve of the heart.

molar concentration Number of grams solute per liter of solution divided by the solute's molecular weight.

mole Molecular weight of a compound in grams; also called gram molecular weight.

monosaccharide Simple sugar.

Monro Eighteenth century English surgeon.

morphology Study of shape and structure of living organisms.

morula Early state in development of embryo; a solid, spherical mass of cells.

motoneurons, motor or efferent neurons Neurons that transmit nerve impulses away from the brain or spinal cord.

motor areas Areas of cerebral cortex that control muscular actions.

mucus Thick, sticky secretion produced by specialized cells in the respiratory, digestive, and reproductive tracts.

murmur A pathologic heart sound.

myasthenia gravis An autoimmune disease characterized by muscle weakness.

myelin Lipoid substance found in the myelin sheath around some nerve fibers.

myocardium Muscle of the heart.

myopia Nearsightedness.

myosin Contractile protein present in muscle cells.

myxedema Pathological condition produced by hyposecretion of the thyroid gland during the adult years.

N

nares Nostrils.

nephron The structural and functional unit of the kidney.

nerve A bundle of nerve fibers.

neurilemma Nerve sheath.

neurohypophysis Posterior pituitary gland.

neuron Nerve cell, including its processes.

neurotransmitter Chemical released at a nerve ending

that stimulates a muscle or another nerve cell to respond.

norepinephrine Chemical transmitter substance of sympathetic (adrenergic) nerve terminals.

neurovesicles Microscopic sacs in axon terminals; contain transmitter substance.

neutron Neutral subatomic particle located in the nucleus of an atom.

neutrophil White blood cell that stains readily with neutral dyes.

nuchal Pertaining to the nape of the neck.

nucleic acid DNA and RNA.

nucleotide Component of DNA and RNA; a nucleotide is composed of a sugar (deoxyribose or ribose), a nitrogenous base (adenine, thymine, cytosine, or guanine), and a phosphate group.

nucleus Spherical structure within a cell; a group of neuron cell bodies in the brain or cord.

nutrients Chemical substances in food, used by cells during metabolism.

O

obesity Being excessively fat.

occiput Back of the head.

olecranon Elbow.

olfactory Pertaining to the sense of smell.

oligodendroglia A type of connective tissue cell found in the brain and cord.

oocyte Female gamete or germ cell.

ophthalmic Pertaining to the eyes.

organelle Cell organ; one of the specialized parts of a single-celled organism (protozoan), serving for the performance of some individual function.

origin Point of attachment of a muscle to a bone, which does not move when the muscle contracts.

os Latin for mouth and for bone.

osmoreceptor Receptor stimulated by change in osmotic pressure.

osmosis Movement of a fluid through a semipermeable membrane.

osseous tissue Bone.

ossicle Little bone.

ossification Calcification or hardening of bone.

osteoblast Bone-forming cell.

osteoclast Bone-destroying cell.

ovary Female gonad.

ovulation Release of ovum from follicle.

oxidation Loss of hydrogen or electrons from a compound or element.

oxyhemoglobin A compound formed by the union of oxygen with hemoglobin.

oxytocin Hormone of neurohypophysis.

P

palate Roof of the mouth.

palpebrae Eyelids.

papilla Small nipple-shaped elevation.

paralysis Loss of the power of motion or sensation, especially voluntary motion.

parathormone Hormone of the parathyroid glands involved in calcium and phosphorus metabolism.

parenchyma The distinguishing, functional cells of an organ.

parietal Of the walls of an organ or cavity.

parotid Located near the ear.

parturition Act of giving birth to an infant.

patella Small, shallow pan; the kneecap.

pathogenic Disease producing.

Pavlov Russian physiologist of the late nineteenth and early twentieth centuries.

pectineal Pertaining to the pubic bone.

pectoral Pertaining to the chest or breast.

pelvis Basic or funnel-shaped structure.

pentose 5 carbon sugar.

pepsin Proteolytic digestive enzyme found in gastric juice.

pericardium Sac surrounding heart.

perimysium Connective tissue between bundles of muscle fibers.

peripheral Pertaining to an outside surface.

peritoneum Serous membrane lining abdominal cavity and covering the viscera.

peroneus, peroneal Of or near the fibula.

petrous Rocklike.

Peyer Swiss anatomist of the late seventeenth and early eighteenth centuries.

pH Hydrogen ion concentration; the negative logarithm of hydrogen ion concentration.

phagocytosis Process by which a segment of cell membrane forms a small pocket around a bit of solid matter outside the cell, breaks off from rest of the membrane, and moves into the cell; briefly, ingestion and digestion of particles by a cell.

phalanges Finger or toe bones.

philtrum Shallow vertical groove in the midline of the upper lip.

phrenic Pertaining to the diaphragm.

pia mater Literally gentle mother; the vascular innermost covering (meninges) of the brain and cord.

pilomotor Mover of a hair.

pineal Shaped like a pine cone.

pinocytosis Process by which a segment of cell membrane forms a small pocket around a bit of fluid outside the cell, breaks off from rest of membrane, and moves into the cell.

piriformis Pear shaped.

pisiform Pea shaped.

pituicyte Cell type of posterior pituitary gland.

placenta Structure that joins the fetus to the uterine wall.

plantar Pertaining to the sole of the foot.

plasma Liquid part of the blood.

plasmolysis, crenation Shrinking of a cell from water loss by osmosis.

platelets Thrombocytes; one of the three types of blood cells.

pleura Serous membrane that lines thorax and covers the lungs.

plexus Network.

plica Fold.

pneumothorax Collapsed lung caused by air in the thoracic cavity.

poikilocytosis Abnormal variations in red blood cell shape.

polycythemia Excess numbers of red blood cells.

polymorphonuclear Having many-shaped nuclei.

polypeptide Compound formed of many amino acids linked together by peptide bonds (between carboxyl group of one amino acid and amino group of another).

polyribosomes Groups of ribosomes working together to synthesize proteins.

polysaccharide Carbohydrate containing a large number of saccharide groups ($C_6H_{10}O_5$), for example, starch, glycogen.

polyuria Abnormally large urine volume.

pons Bridge.

popliteal Behind the knee.

posterior Following after; hence located behind; opposite of anterior.

potential or potential difference Difference in electrical charges, for example, on outer and inner surfaces of cell membrane.

potential, action Difference in electrical charges on inner and outer surfaces of the cell membrane during impulse conduction; outer surface negative to inner.

potential, resting Difference in electrical charges on inner and outer surfaces of the cell membrane when it is not conducting impulses; outer surface positive to inner.

Poupart Seventeenth century French anatomist.

precapillary sphincters Partial cuffs of smooth muscle located at points of origin of two capillaries from an arteriole.

presbyopia "Oldsightedness"; farsightedness of old age.

pressoreceptors Receptors stimulated by a change in pressure; baroreceptors.

pronate To turn palm downward, opposite of supinate.

properdin Group of enzymes normally present in blood; function to activate complement.

prophase First stage of mitosis; chromosomes become visible.

proprioreceptors Receptors located in the muscles, tendons, and joints.

prostaglandins (PG) Potent, widespread biocompounds (fatty acids); may regulate hormone activity at cellular level.

protein Compound composed of amino acids joined by peptide bonds; polypeptide.

proton Positively charged particle in atomic nucleus.

proximal Next or nearest; located nearest the center of the body or the point of attachment of a structure.

psoas Pertaining to the loin, the part of the back between the ribs and hip bones.

psychosomatic Pertaining to the influence of the mind (notably the emotions) on body functions.

pruritus Itching.

pterygoid Wing-shaped.

puberty Age at which the reproductive organs become functional.

purine bases Adenine and guanine.

pyrimidine bases Cytosine and thymine.

Q

QRS wave A component wave of the electrocardiogram that precedes ventricular depolarization.

R

racemose Like a cluster of grapes.

radioactivity Emission from atomic nucleus of particles or rays known as alpha, beta, and gamma rays.

ramus Branch.

Ranvier French pathologist of the late nineteenth and early twentieth centuries.

receptor Peripheral ending of a sensory neuron.

rectum Terminal portion of the large intestine.

reflex Involuntary action.

reflex arc Pathway of a nervous impulse involved in a reflex.

refraction Bending of a ray of light as it passes from a medium of one density to one of a different density.

refractory Resisting stimulation.

renal Pertaining to the kidney.

renin Hormone secreted by kidney and involved in blood pressure regulation by its role in angiotensin production.

resting potential The electrical potential or difference between extra- and intracellular surfaces of a cell membrane.

reticular Netlike.

reticulocyte Newly formed erythrocyte having a cytoplasmic reticulum or network of fibers.

reticulum Network.

retina Innermost layer of eye, which is stimulated by light.

rhinencephalon see **limbic lobe or system**.

ribosomes Organelles in cytoplasm of cells; synthesize proteins so nicknamed "protein factories."

Rivinus's ducts Ducts draining the sublingual salivary gland.

rugae Wrinkles or folds.

S

saccule A little sac.

sagittal Like an arrow; longitudinal.

salpingitis Inflammation of the uterine or fallopian tubes.

salpinx Tube, oviduct.

sarcolemma Cytoplasmic membrane of muscle cell.

sarcomere Contractile unit of myofibril between two C lines.

sarcoplasm Cytoplasm of muscle cell.

sarcoplasmic reticulum Network of fine tubules in muscle cells involved in transmission of excitation to contractile proteins.

sartorius Tailor; hence the thigh muscle used to sit cross-legged like a tailor.

schizocytes Distorted and irregularly shaped red blood cell fragments.

sciatic Pertaining to the ischium.

sclera From Greek for hard.

scrotum Bag.

sebum Latin for tallow; secretion of sebaceous glands.

secretion Digestive hormone that induces secretion of pancreatic juice.

sella turcica Turkish saddle; saddle-shaped depression in the sphenoid bone; the pituitary gland is located in this depression.

semen Latin for seed; male reproductive fluid.

semilunar Half-moon shaped.

senescence Old age.

serratus Saw toothed.

Sertoli cell Cell found in the seminiferous tubules, which provides support and protection and helps regulate metabolism of the developing spermatozoa.

serum Any watery animal fluid; clear, yellowish liquid that separates from a clot of blood.

sesamoid Shaped like a sesame seed.

sickle cell Atypical red blood cells having bizarre crescent-shaped forms.

sigmoid S-shaped.

sinus Cavity.

soleus Pertaining to a sole; a muscle in the leg shaped like the sole of a shoe.

solute Material dissolved or suspended in a solution.

solvent Dissolving medium or fluid in a solution.

somatic Of the body framework or walls, as distinguished from the viscera or internal organs.

somatotype Type of body build or physique.

spatial summation Increase in response of a nerve caused by an increase in the number of presynaptic elements providing excitation.

sphenoid Wedge shaped.

sphincter Ring-shaped muscle.

spirometer An instrument used to determine the amount of air respired in the various respiratory volumes.

splanchnic Visceral.

squamous Scalelike.

stapes Stirrup; tiny stirrup-shaped bone in the middle ear.

Starling English physiologist of the late nineteenth and early twentieth centuries.

Stensen's duct Duct that drains the parotid salivary gland.

stereognosis Awareness of the shape of an object by means of touch.

stimulus Agent that causes a change in the activity of a structure.

stratum Layer.

stress, physiological According to Selye, a condition in the body produced by all kinds of injurious factors that he calls "stressors" and manifested by a syndrome.

stressor Any injurious factor that produces biological stress, for example, emotional trauma, infections, or severe exercise.

striated Marked with parallel lines.

stroma Greek for mattress or bed; hence the framework or matrix of a structure.

sudoriferous Secreting sweat.

sulcus Furrow or groove.

superior Higher; opposite of inferior.

supinate To turn the palm of the hand upward; opposite of pronate.

Sylvius Seventeenth century anatomist.

symphysis Greek for a growing together.

synapse Joining; point of contact between adjacent neurons.

syncope Fainting.

synovia Literally "with egg"; secretion of the synovial membrane resembles egg white.

synthesis Chemical reaction in which new chemical bonds are formed to put together two or more substances into a more complex whole.

systole Contraction of the heart muscle.

T

T cells Lymphocytes that undergo first stage of their development in thymus gland before migrating to lymph nodes and spleen.

talus Ankle; one of the bones of the ankle.

tamponade (cardiac) Accumulation of fluid in the pericardial sac.

target organ (cell) Organ (cell) acted on by a particular hormone and responding to it.

tarsus Instep.

tegmentum Part of the brain stem that covers the posterior surface of the cerebral peduncles and the pons.

temporal summation Increase in response of a nerve or muscle caused by an increase in the frequency of excitation stimuli.

tendon Band or cord of fibrous connective tissue that attaches a muscle to a bone or other structure.

testis Male gonad.

tetanus Sustained muscle contraction.

thorax Chest.

thrombosis Formation of a clot in a blood vessel.

thymosin Hormone of the thymus gland.

thalamus Mass of grey matter in the diencephalon of the brain.

threshold stimulus The minimal stimulus that will elicit a response.

thyroxin Hormone produced by the thyroid gland.

tibia Latin for shin bone.

tonus Continued, partial contraction of muscle.

trabeculae Fibrous connective tissue septa that extend into an organ from its covering capsule.

tract Bundle of axons located within the central nervous system.

trauma Injury.

treppe Progressive increase in strength of muscle contraction with repeated maximal stimuli.

triglycerides Compounds formerly called neutral fats.

trochlear Pertaining to a pulley.

trophic Having to do with nutrition.

tropic hormones Hormones that control secretions of other endocrine glands.

tunica Covering.

turbinate Shaped like a cone or like a scroll or spiral.

twitch contraction Quick, jerky contraction in response to a single stimulus.

tympanum Drum.

U

umbilicus Navel.

unipolar neuron Neuron with only a single process extending from its cell body.

ureter The tube that conveys urine from kidney to bladder.

urethra The tube that conveys urine from the bladder to the exterior of the body.

urine Waste product produced by the kidneys.

uterus (womb) Organ of female reproductive system in which the fetus develops.

utricle Little sac.

uvula Latin for a little grape; projection hanging from the soft palate.

V

vagina Sheath.

vagus Latin for wandering.

valence Number of unpaired electrons in an atom's outer shell.

valve Structure that permits flow of a fluid in one direction only.

vas Vessel or duct.

vascular Pertaining to blood or blood vessels.

vastus Wide, of great size.

Vater German anatomist of the late seventeenth and early eighteenth centuries.

vein Vessel carrying blood toward or to the heart.

ventral Of or near the belly; in man, front or anterior; opposite of dorsal or posterior.

ventricle Small cavity.

vermiform Worm shaped.

vestibule A space or small cavity located at the entrance to a canal.

vibrissae Coarse hairs found in the vestibule of the nose.

villus Hairlike projection.

viscera Internal organs.

vitiligo Absence of pigment in circumscribed areas of the skin; patches of depigmented skin.

volar Pertaining to the palm of the hand or the sole of the foot; palmar; plantar.

vomer Ploughshare.

vulva Female external genitalia.

W

wart Virus-induced thickening (tumor) of epithelium with a central blood vessel.

Wharton's duct Duct that drains the submandibular salivary glands.

white matter Bundles of myelinated nerve fibers located in the central nervous system.

Willis Seventeenth century English anatomist.

X

xenograft Transplant of tissue between individuals of different species.

xiphoid Sword shaped.

Y

yoga Meditation; an altered state of consciousness.

Z

zygoma Yoke.

zygote Fertilized ovum.

SUPPLEMENTARY READINGS

1 Organization of the Body

Anthony, C.P.: Basic concepts in anatomy and physiology: a programmed presentation, ed. 4, St. Louis, 1980, The C.V. Mosby Co.

Cannon, W.B.: The wisdom of the body, ed. 2, New York, 1963, W.W. Norton & Co., Inc.

Falkner, F., and Tanner, J.: Human growth, ed. 2, New York, 1986, Plenum Press.

2 Chemical Basis of Life

Holtzclaw, H.F., Robinson, W.R., and Nebergall, W.H.: General chemistry, ed. 7, Lexington, MA, 1984, D.C. Heath & Co.

Weinberg, R.: The molecules of life, Sci. Am., 253:48-57,. Oct. 1985.

Wilson, A.C.: The molecular basis of evolution, Sci. Am. 253:164-173, Oct. 1985.

Woodhead, A.D., et al.: Molecular biology of aging, New York, 1985, Plenum Publishing Corp.

3 Cells

Anthony, C.P.: Basic concepts in anatomy and physiology: a programmed presentation, ed. 4, St. Louis, 1980, The C.V. Mosby Co.

Becker, W.M.: The world of the cell, Menlo Park, CA, 1986, The Benjamin-Cummings Publishing Co.

Bretscher, M.: The molecules of the cell membrane, Sci. Am. 253:100-108, Oct. 1985.

Capaldi, R.A.: A dynamic model of cell membranes, Sci. Am. 230:27-33, March, 1974.

Chinn, P.L.: Child health maintenance: concepts in family centered care, ed. 2, St. Louis, 1979, The C.V. Mosby Co., pp. 63-67.

Cohen, S.N.: The manipulation of genes, Sci. Am. 233:24-33, July, 1975.

Cohen, S.N.: Recombinant DNA: fact and fiction, Science 195:654, 1977.

Czech, M.P., and Kahn, C.R., editors: Membrane receptors and cellular regulation, New York, 1985, Alan R. Liss Inc.

Darnell, J.E.: RNA, Sci. Am. 253:68-78, Oct. 1985.

Engelman, D.M., and Moore, P.B.: Neutron-scattering studies of the ribosome, Sci. Am. 230:44-53, Oct. 1976.

Everhart, R.E., and Hayes, T.L.: The scanning electron microscope, Sci. Am. 226:55-69, Jan. 1972.

Felsenfeld, G.: DNA, Sci. Am. 253:58-67, Oct. 1985.

Guyton, A.D.: Textbook of medical physiology, ed. 6, Philadelphia, 1981, W.B. Saunders Co.: pp. 2-27, 41-54.

Hayflick, L.: The cell biology of human aging, Sci. Am. 242(1):58-65, 1980.

Jukes, T.H.: How many anticodons? Science 198:4314, Oct. 1977.

Klug, W.S., and Cummings, M.E.: Concepts of genetics, ed. 2, Columbus, 1986, Charles E. Merrill Publishing Co.

Kolata, G.: Why do cancer cells resist drugs? Science 231:220-221, Jan. 1986.

Kornberg, A.: The synthesis of DNA, Sci. Am. 214:64-78, Oct. 1968.

Mazia, D.: The cell cycle, Sci. Am. 231:55-63, Jan. 1974.

Miller, O.L., Jr.: The visualization of genes in action, Sci. Am. 228:34-41, March 1973.

Motulsky, A.G.: The 1985 Nobel Prize in physiology or medicine, Science 231:126-129, Jan. 1986. (Membrane receptors.)

Nomura, M.: The control of ribosome synthesis, Sci. Am. 250:102-114, Jan. 1984.

Rothman, J.E.: The compartmental organization of the Golgi apparatus, Sci. Am. 253:74-89, Sept. 1985.

Satir, B.: The final steps in secretion, Sci. Am. 233:29-37, Oct. 1975.

Therman, E.: Human chromosomes, ed. 2, New York, 1986, Springer-Verlag, New York, Inc.

Unwin, N., and Henderson, R.: The structure of proteins in biological membranes, Sci. Am. 250:78-94, Feb. 1984.

Varsat, A., and Lodish, H.F.: How receptors bring proteins and particles into cells, Sci. Am. 250:52-58, May 1984.

Weber, K., and Osborn, M.: The molecules of the cell matrix, Sci. Am. 253:110-120, Oct. 1985.

4 Tissues

Edelman, G.M.: Cell-adhesion molecules: a molecular basis for animal form, Sci. Am. 250:118-129, April 1984.

Leukotrienes and tissue injury, Science News 127(25):223, Oct. 1985.

Wilkerson, G.B.: Inflammation in connective tissue, Athletic Training 21(4):298-301, Winter 1985.

5 The Skin and its Appendages

Berieter-Hahn, A.G., et al.: Biology of the integument, New York, 1986, Springer-Verlag New York, Inc.

Di Fiore, M.S.H.: An atlas of human histology, ed. 5, Philadelphia, 1981, Lea & Febiger.

Edelson, R.L., and Fink, J.M.: The immunologic function of skin, Sci. Am. 252:46-53, June 1985.

Roberts, S.L.: Skin assessment for color and temperature, Am. J. Nurs. 75:610-613, April 1975.

6 Skeletal Tissues and Physiology

Caplan, A.I.: Cartilage, Sci. Am. 251:84-94, Oct. 1984.

Hogan, L., and Beland, I.: Cervical spine syndrome, Am. J. Nurs. 76:1104-1107, July 1976.

Murray, P.D.F.: Bones, New York, 1985, Cambridge University Press.

Thomsen, D.E.: Electrifying biology, Science News 127:268-269, April 1985.

7 The Skeletal System

Matthews, G.: Cellular physiology of nerve and muscle, Palo Alto, CA, 1985, Blockwell Scientific Publications Inc.

Speer, D.P.; and Braun, J.K.: Biomechanical basis of growth plate injuries, Phys. Sports Med., 13(7):72-80, July 1985.

8 Articulations

Berme, N., et al.: Biomechanics of normal and pathological human articulating joints, Dordrecht, 1985, Nijhoff. (U.S. distributor, Hingham, MA, Kluwer Boston Inc.)

Cavanaugh, P.R., and Kram, R.: The efficiency of human movement: a statement of the problem, Med. Sci. Sports Exerc., 17(3):304-8, June 1985.

Sonstegard, D.A., Matthews, L.S., and Kaufer, H.: The surgical replacement of the human knee joint, Sci. Am. 238:44-51, Jan. 1978.

Walker, P.S.: Joints to spare, Science 85 6(9):56-60, Nov. 1985.

9 Skeletal Muscles

Carlson, R.D., and Wilkie, D.R.: Muscle physiology, Biological Science Series, Englewood Cliffs, N.J., 1974, Prentice-Hall Inc.

Cohen, C.: The protein switch of muscle contraction, Sci. Am. 233:36-45, Nov. 1975.

Hoyle, G.: How is muscle turned on and off? Sci. Am. 222:84-93, April 1970.

Komi, P.V.: Biomechanics V, vols. 1 and 2, Baltimore, 1976, University Park Press.

Lester, H.A.: The response to acetylcholine, Sci. Am. **236:**106-118, Feb. 1977.

Margaria, R.: The sources of muscular energy, Sci. Am. **226:**83-91, March 1972.

Murray, J.M., and Weber, A.: The cooperative action of muscle proteins, Sci. Am. **230:**59-71, Feb. 1974.

Noble, B.J.: Physiology of exercise and sport, St. Louis, 1986, Times Mirror/Mosby College Publishing.

Oster, G.: Muscle sounds, Sci. Am. **250:**100-108, March 1984.

Pavlou, K.N., et al.: Effects of dieting and exercise on lean body mass, oxygen uptake, and strength, Med. Sci. Sports Exerc. **17**(4):466-471, Aug. 1985.

Yagi, N., et al.: Return of myosin heads to thick filaments after muscle contraction, Science **197:**4304, Aug. 1977.

10   Nervous System Cells

Goodman, C.S., and Bastiani, M.J.: How embryonic nerve cells recognize one another, Sci. Am. **251:**58-66, Dec. 1984.

Gutmann, F., and Keyzer, H.: Modern bioelectrochemistry, New York, 1986, Plenum Publishing Corp.

Guyton, A.C.: Textbook of medical physiology, ed. 6, Philadelphia, 1981, W.B. Saunders Co., pp. 104-119.

Keynes, R.D.: Ion channels in the nerve cell membrane, Sci. Am. **240**(3): 126-135, March 1979.

Morell, P., and Norton, W.T.: Myelin, Sci. Am. **242**(5):88-117, May 1980.

Nauta, W.J.H., and Feirtag, M.: Fundamental neuroanatomy, New York, 1986, Freeman, Cooper & Co.

Weisburd, S.: Cell transplants into monkey and human brains, Science News, **128**(18):276-277, Nov. 1985.

Wurtman, R.J.: Alzheimer's disease, Sci. Am. **252:**62-74, Jan. 1985.

11   The Central Nervous System, Spinal Nerves, and Cranial Nerves

Beaton, A.: Left side, right side, New Haven, 1985, Yale University Press.

Bjorklund, A., and Stenevi, U.: Neural grafting in the mammalian CNS, New York, 1985, Elsevier Science Publishing Co., Inc.

Bloom, F.E.: Neuropeptides, Sci. Am. **245**(4):148-167, Oct. 1981.

The Brain, Sci. Am., Sept. 1979 (entire issue on the brain and human nervous system).

Brain peptide packs a wallop, Sci. News, Jan. 26, 1980, p. 57.

Coen, C.W., editor: Functions of the brain, New York, 1985, Clarendon (Oxford University Press, Inc.)

Diversification in neuron chemistry, Sci. News, Nov. 29, 1980.

Doane, B.K., and Livingston, K.E., editors: The limbic system, New York, 1986, Raven Press.

Dunant, Y., and Israel, M.: The release

of acetylcholine, Sci. Am. **252:**58-66, April 1985.

Greenberg, J.: Suicide linked to brain chemical deficit, Sci. News **121:**355, May 29, 1982.

Kiester, E.: Spare parts for damaged brains, Science 86 **7**(2):32-38, March 1986.

Kimura, D.: The asymmetry of the human brain, Sci. Am. **228:**70-77, March 1973.

Lester, H.A.: The response to acetylcholine, Sci. Am. **236:**107-118, Feb. 1977.

Llinas, R.R.: The cortex of the cerebellum, Sci. Am. **232:**56-71, Jan. 1975.

Nathanson, J.A., and Greengard, P.: "Second messengers" in the brain, Sci. Am. **237:**108-119, Aug. 1977.

Pappenheimer, J.R.: The sleep factor, Sci. Am. **235:**24-29, Aug. 1976.

Pregnancy and pain perception, Sci. News, Oct. 11, 1980, p. 233.

Psyching out pain, Sci. News, May 19, 1979, p. 332.

Schwartz, J.H.: The transport of substances in nerve cells, Sci. Am. **242**(4):152-171, April 1980.

Snyder, S.H.: Opiate receptors and internal opiates, Sci. Am. **237:**44-56, March 1977.

Snyder, S.H.: The molecular basis of communication between cells, Sci. Am. **253:**132-141, Oct. 1985.

Wallace, R.K., and Benson, H.: The physiology of meditation, Sci. Am. **226:**85-90, Feb. 1972.

Zucker, R.S., and Lando, L.: Mechanism of transmitter release: voltage hypothesis and calcium hypothesis, Science **231:**574-579, Feb. 1986.

12   The Autonomic Nervous System

Benson, H.: The relaxation response, New York, 1975, William Morrow & Co., Inc.

Changeux, J.P., et al., editors: Molecular basis of nerve activity, New York, 1985, Walter de Gruyter.

Sterman, L.T.: Clinical biofeedback, Am. J. Nurs. **75:**2006-2012, Nov. 1975.

Trotter, R.J.: Transcendental meditation, Sci. News **104:**376-378, Dec. 1973.

Wallace, R.K., and Benson, H.: The physiology of meditation, Sci. Am. **226:**85-90, Feb. 1972.

13   Sense Organs

Fernsebner, W.: Early diagnosis of acute angle-closure glaucoma, Am. J. Nurs. **75:**1154-1156, July 1975.

Kolata, G.: Blindness of prematurity unexplained, Science **231:**20-21, Jan. 1986.

Land, E.H.: The retinex theory of color vision, Sci. Am. **237:**108-128, Dec. 1977.

Loeb, G.E.: The functional replacement of the ear, Sci. Am. **252:**104-111, Feb. 1985.

Nathans, J., Thomas, D., and Hogness, D.S.: Molecular genetics of human

color vision, Science **232:**193-202, April 1986.

Parker, D.E.: The vestibular apparatus, Sci. Am. **243:**118-136, Nov. 1980.

Pettigrew, J.D.: The neurophysiology of binocular vision, Sci. Am. **227:**84-95, Aug. 1972.

van Heyningen, R.: What happens to the human lens in cataract, Sci. Am. **233:**70-81, Dec. 1975.

14   The Endocrine System

Bailey, J.M., editor: Prostaglandins, leukotrienes, and lipoxins: biochemistry, mechanism of action, and clinical applications, New York, 1985, Plenum Publishing Corp.

Baylis, P.H., and Padfield, P.L.: The posterior pituitary, New York, 1985, Marcel Dekker, Inc.

Berridge, M.J.: The molecular basis of communication within the cell, Sci. Am. **253:**142-152, Oct. 1985.

Binkley, S., Riebman, J.B., and Reilly, K.B.: Time keeping by the pineal gland, Science **197:**4309, Sept. 1977.

Crapo, L.: Hormones—the messengers of life, New York, 1985, Freeman, Cooper & Co.

Gardener, L.I.: Deprivation dwarfism, Sci. Am. **227:**76-82, July 1972.

Guillemin, R.: Hypothalamic control of pituitary functions, Liverpool, 1986, Liverpool University Press.

Guillemin, R., and Burgus, R.: The hormones of the hypothalamus, Sci. Am. **227:**24-32, Nov. 1972.

Langs, D.A., Erman, M., and DeTitta, G.T.: Conformations of prostaglandin F_2 and recognition of prostaglandins by their receptors, Science **197:**4307, Sept. 1977.

MacLeod, R.M., et al., editors: Prolactin, New York, 1985, Springer-Verlag New York, Inc., and Padova, Italy, Tiniana Press.

McEwen, B.S.: Interactions between hormones and nervous tissue, Sci. Am. **235:**48-67, July 1976.

Pastan, I.: Cyclic AMP, Sci. Am. **227:**97-105, Aug. 1972.

Peart, W.S.: Renin-angiotensin system, N. Engl. J. Med. **292:**302, 1975.

Pick, E., editor: Lymphokines, Orlando, FL, 1985, Academic Press, Inc.

Pike, J.E.: Prostaglandins, Sci. Am. **225:**84-91, Nov. 1971.

Ramwell, P.W., and Shaw, J.E., editors: Prostaglandins, Ann. N.Y. Acad. Sci. **180,** 1971.

Sterling, K., et al.: Thyroid hormone action: the mitochondrial pathway, Science **197:**4307, Sept. 1977.

Volk, B.W., and Arquilla, E.R.: The diabetic pancreas, New York, 1985, Plenum Publishing Corp.

Wechsler, R.: Unshackled from diabetes, Discover **7**(9):77-85, Sept. 1986.

15   Blood

Adolph, E.F.: The heart's pacemakers, Sci. Am. **213:**32-37, March 1967.

Supplementary Readings

Anthony, C.P.: Basic concepts in anatomy and physiology: a programmed presentation, ed. 4, St. Louis, 1980, The C.V. Mosby Co., pp. 108-144.

Bennett, B., et al.: The normal coagulation mechanism, Med. Clin. North Am. **59**:95, 1972.

Bode, C., et al.: Antibody-directed urokinase: a specific fibrinolytic agent, Science **229**:765-767, Aug. 1985.

Brown, M.S., and Goldstein, J.L.: How LDL receptors influence cholesterol and atherosclerosis, Sci. Am. **251**:58-66, Nov. 1984.

Cajozzo, A., et al., editors: Advances in hemostasis and thrombosis, New York, 1985, Plenum Publishing Corp.

Capra, J.D., and Edmundson, A.B.: The antibody combining site, Sci. Am. **236**:50-59, Jan. 1977.

Clarke, C.A.: The prevention of "Rhesus" babies, Sci. Am. **219**:46-52, Nov. 1968.

Conover, M.H., and Zalis, E.G.: Understanding electrocardiography: physiological and interpretive concepts, ed. 2, St. Louis, 1976, The C.V. Mosby Co.

Cooper, M.D., and Lawton, A.R., III: The development of the immune system, Sci. Am. **231**:58-72, Nov. 1974.

Effler, D.B.: Surgery for coronary disease, Sci. Am. **219**:36-43, Oct. 1968.

Ingram, M., and Preston, K., Jr.: Automatic analysis of blood cells, Sci. Am. **223**:72-82, Nov. 1970.

Lawn, R.M., and Vehar, G.A.: The molecular genetics of hemophilia, Sci. Am. **254**:48-54, March 1986.

Lerner, R.A., and Dixon, F.J.: The human lymphocyte in an experimental animal, Sci. Am. **228**:82-91, June 1973.

Mountcastle, V.B., editor: Medical physiology, ed. 14, St. Louis, 1980, The C.V. Mosby Co., pp. 1126-1136.

Porter, R.R.: The structure of antibodies, Sci. Am. **217**:81-87, Oct. 1967.

Ray, A.K., Ray, T.K., and Sinha, A.K.: Prostacyclin stimulation of the activation of blood coagulation factor X by platelets, Science **231**:385-387, Jan. 1986.

Wood, E.J.: The venous system, Sci. Am. **218**:86-96, Jan. 1968.

16 Anatomy of the Cardiovascular System

Benditt, E.P.: The origin of atherosclerosis, Sci. Am. **236**:74-85, Feb. 1977.

Coodley, E.L.: Anatomy of circulation, Consultant, May, 1972.

Folkman, J.: The vascularization of tumors, Sci. Am. **234**:58-73, May 1976.

Hurst, J.W., editor-in-chief: The heart: arteries and veins, ed. 5, New York, 1982, McGraw-Hill Book Co.

Johnson, R., editor: Atherosclerosis reviews, New York, 1985, Raven Press.

Lee, K.T.: Atherosclerosis, New York, 1985, New York Academy of Sciences.

Williams, P.L., and Warwick, R., editors: Gray's anatomy, ed. 36 (British), Phil-

adelphia, 1980, W.B. Saunders Co., pp. 588-712.

Wood, E.J.: The venous system, Sci. Am. **218**:86-96, Jan. 1968.

17 Physiology of the Cardiovascular System

Adolph, E.F.: The heart's pacemakers, Sci. Am. **213**:32-37, March 1967.

Berne, R.M., and Levy, M.N.: Cardiovascular physiology, ed. 5, St. Louis, 1986, The C.V. Mosby Co.

Cantin, M., and Genest, J.: The heart as an endocrine gland, Sci. Am. **254**:76-81, Feb. 1986.

Conover, M.B.: Understanding electrocardiography: arrhythmias and the 12-lead ECG, ed. 4, St. Louis, 1984, The C.V. Mosby Co.

Guyton, A.C., Jones, C.E., and Coleman, T.C.: Circulatory physiology: cardiac output and its regulation, Philadelphia, 1973, W.B. Saunders Co.

Janssen, H.F., and Barnes, C.D.: Circulatory shock, Orlando, FL, 1985, Academic Press, Inc.

Kannel, W.B., et al.: Epidemiological assessment of the role of physical activity and fitness in development of cardiovascular disease, J. Appl. Physiol. **58**(3):791-794, March 1985.

Mountcastle, V.B., editor: Medical physiology, ed. 14, St. Louis, 1980, The C.V. Mosby Co., pp. 951-1125.

Noble, D.: The initiation of the heartbeat, New York, 1975, Oxford University Press, Inc.

Raineri, A., Leachman, R.D., and Kellermann, J.J., editors: Assessment of ventricular function, New York, 1985, Plenum Publishing Corp.

Smith, O.: Reflex and central mechanisms involved in the control of the heart and circulation, Ann. Rev. Physiol. **36**:93, 1974.

18 The Lymphatic System

Kinmonth, J.B.: Some general aspects of the investigation and surgery of the lymphatic system, J. Cardiovasc. Surg. (Torino) **5**:680-682, 1964.

Mayerson, H.: The lymphatic circulation, Sci. Am. **208**:80-93, June 1963.

Rusznyak, I., Foldi, M., and Sazbo, G.: Lymphatics and lymph circulation, Elmsford, N.Y., 1960, Pergamon Press, Inc.

Williams, P.L., and Warwick, R., editors: Gray's anatomy, ed. 36 (British), Philadelphia, 1980, W.B. Saunders Co., pp. 42-52.

19 Anatomy of the Respiratory System

Roussos, C., and Macklem, P.T., editors: The thorax, New York, 1985, Marcel Dekker, Inc.

Scarantino, C.W., editor: Lung cancer, New York, 1985, Springer-Verlag New York, Inc.

20 Physiology of the Respiratory System

Avery, M.E., Wang, N., and Taeusch, H.W.: The lung of the newborn infant, Sci. Am. **228**:74-85, April 1973.

Cherniack, R.: Respiration in health and disease, ed. 3, Philadelphia, 1983, W.B. Saunders Co.

Comroe, J.H., Jr.: Physiology of respiration, Chicago, 1965, Year Book Medical Publishers, Inc.

Comroe, J.H., Jr.: The lung, Sci. Am. **214**:57-66, Feb. 1966.

Clements, J.A.: Surface tension in the lungs, Sci. Am. **207**:120-134, Dec. 1962.

Haynes, D.E., et al.: Esophageal rupture complicating the Heimlich maneuver, Am. J. Emerg. Med. **12**(4):29-33, July 1985.

Mountcastle, V.B., editor: Medical physiology, ed. 14, St. Louis, 1980, The C.V. Mosby Co.

Orentein, D.M., et al.: Exercise conditioning in children with asthma, J. Pediatr. **106**(4):556-560, April 1985.

Raeburn, P.: The Houdini virus (respiratory syncytial virus), Science 85 **6**(10):52-57, Dec. 1985.

Slonim, N.B., and Hamilton, L.H.: Respiratory physiology, ed. 5, St. Louis, 1987, The C.V. Mosby Co.

Spearman, C.B., Sheldon, R.L., and Egan, D.F.: Egan's fundamentals of respiratory therapy, ed. 4, St. Louis, 1982, The C.V. Mosby Co.

Yerg, J.E., et al.: Effect of endurance exercise training on ventilatory function in older individuals, J. Appl. Physiol. **58**(3):791-794, March 1985.

22 Physiology of the Digestive System

Davenport, H.W.: Why the stomach does not digest itself, Sci. Am. **225**:87-93, Jan. 1972.

Dudrick, S.J., and Rhoades, J.E.: Total intravenous feeding, Sci. Am. **226**:73-79, May 1972.

Kretchmer, N.: Lactose and lactase, Sci. Am. **227**:70-78, Oct. 1972.

Programmed instruction: potassium imbalance, Am. J. Nurs. **67**:343-366, Feb. 1967.

23 Metabolism

Brady, R.O.: Hereditary fat-metabolism diseases, Sci. Am. **229**:88-97, Aug. 1973.

Fike, S.: Digestion to absorption—converting food to energy, Nat. Strength Condition. Assoc. J. **7**(2):38-39, May 1985.

Guyton, A.C.: Textbook of medical physiology, ed. 6, Philadelphia, 1981, W.B. Saunders Co., pp. 838-915.

Noble, B.J.: Physiology of exercise and sport, St. Louis, 1986, Times Mirror/Mosby College Publishing.

Toufexis, A.: Dieting: the losing game, Time **127**(3): 54-60, Jan. 20, 1986.

Wurtman, R.J.: Nutrients that modify

brain function, Sci. Am. **246**(4):50-59, April 1982.

Young, E.A.: Nutrition, aging and health, New York, 1986, Alan R. Liss, Inc.

Young, V.R., and Schrimshaw, N.S.: The physiology of starvation, Sci. Am. **225**:14-21, Oct. 1971.

24   The Urinary System

Anthony, C.P.: Basic concepts in anatomy and physiology: a programmed presentation, ed. 4, St. Louis, 1980, The C.V. Mosby Co.

Guyton, A.C.: Textbook of medical physiology, ed. 6, Philadelphia, 1981, W.B. Saunders Co., pp. 403-434.

Melber, S., Leonard, M., and Primack, W.: Hemodialysis at camp, Am. J. Nurs. **76**:938-940, June 1976.

Mountcastle, V.B., editor: Medical physiology, ed. 14, St. Louis, 1980, The C.V. Mosby Co., pp. 1165-1205.

Schumann, B.: The renal donor, Am. J. Nurs. **74**:105-110, Jan. 1974.

Wolf, Z.R.: What patients awaiting kidney transplant want to know, Am. J. Nurs. **76**:92-94, Jan. 1976.

25   Fluid and Electrolyte Balance

Anthony, C.P.: Basic concepts in anatomy and physiology: a programmed presentation, ed. 4, St. Louis, 1980, The C.V. Mosby Co.

Deetjen, P., Boylan, J.W., and Kramer, K.: Physiology of the kidney and of water balance, New York, 1975, Springer-Verlag New York, Inc.

Frizzell, R.T., et al.: Hyponatremia and ultramarathon running, JAMA **255** (6):772-774, 1986.

Gamble, J.L.: Chemical anatomy, physiology, and pathology of extracellular fluid, ed. 6, Cambridge, Mass., 1958, Harvard University Press.

Guyton, A.C.: Textbook of medical physiology, ed. 6, Philadelphia, 1981, W.B. Saunders Co., pp. 358-382, 391-402, 435-446.

Mountcastle, V.B., editor: Medical physiology, ed. 14, St. Louis, 1980, The C.V. Mosby Co., pp. 1149-1164.

Shore, L., and Claybough, J.R.: Regulation of body fluids, Ann. Rev. Physiol. **34**:235, 1972.

Wells, C.L., et al.: Physiological response to a 20-mile run under three fluid replacement treatments, Med. Sci. Sports Exerc. **17**(3):364-370, June 1985.

26   Acid-base Balance

Anthony, C.P.: Basic concepts in anatomy and physiology: a programmed presentation, ed. 4, St. Louis, 1980, The C.V. Mosby Co.

Burke, S.R.: The composition and function of body fluids, ed. 3, St. Louis, 1980, The C.V. Mosby Co.

Davenport, H.W.: The ABC of acid-base chemistry, ed. 6, Chicago, 1974, University of Chicago Press.

Pitts, R.F.: Renal regulation of acid-base balance, Chicago, 1974, Year Book Medical Publishers, Inc.

Vander, A.J.: Renal physiology, ed. 2, New York, 1980, McGraw-Hill Book Co., pp. 90-164.

Weldy, N.J.: Body fluids and electrolytes: a programmed presentation, ed. 2, St. Louis, 1976, The C.V. Mosby Co.

27   The Male Reproductive System

Duckett-Racey, J.D., editor: The biology of the male gamete, New York, 1975, Academic Press, Inc.

Epel, D.: The program of fertilization, Sci. Am. **237**:128-139, Nov. 1977.

Metz, C.B., and Monroy, A.: Biology of fertilization—biology of the sperm, vol. 2, Orlando, FL, 1985, Academic Press, Inc.

Mountcastle, V.B., editor: Medical physiology, ed. 14, St. Louis, 1980, The C.V. Mosby Co., pp. 1624-1635.

28   The Female Reproductive System

Beer, A.E., and Billingham, R.E.: The embryo as a transplant, Sci. Am. **230**:36-51, April 1974.

Edwards, R.G., and Fowler, R.E.: Human embryos in the laboratory, Sci. Am. **222**:45-54, Dec. 1970.

Epel, D.: The program of fertilization, Sci. Am. **237**:128-139, Nov. 1977.

Goldberg, V.J., and Ramwell, P.W.: Role of prostaglandins in reproduction, Physiol. Rev. **55**:325, 1975.

Gosden, R.G.: Biology of menopause, Orlando, FL, 1985, Academic Press, Inc.

Hall, S.S.: The fate of the egg, Science 85 **6**(9):40-48, Nov. 1985.

Heneson, N.: The selling of PMS, Science 84 **5**(4):66-71, May 1984.

Metz, C.B., and Monroy, A.: Biology of fertilization—biology of the ovum; biology of the zygote, vols. 1 and 3, Orlando, FL, 1985, Academic Press, Inc.

Mountcastle, V.B., editor: Medical physiology, ed. 14, St. Louis, 1980, The C.V. Mosby Co., pp. 1602-1624.

Tilton, N., Tilton, T., and Moore, G.: Making miracles—in vitro fertilization, Garden City, New York, 1985, Doubleday & Co., Inc.

29   The Immune System

Allergy antagonist, Sci. News, May 1, 1980, p. 134

Barnes, D.M.: Nervous and immune system disorders linked in a variety of diseases, Science **232**:160-161, April 1986.

Burke, D.C.: The status of interferon, Sci. Am. **236**:42-50, April 1977.

Burnet, M.: The mechanism of immunity, Sci. Am. **204**:58-67, Jan. 1961.

Capra, J.D., and Edmundson, A.B.: The antibody combining site, Sci. Am. **236**:50-59, Jan. 1977.

The cleanest pigs in town, Sci. News, Oct. 18, 1980, p. 253.

Collier, R.J., and Kaplan, D.A.: Immunotoxins, Sci. Am. **251**:56-64, July 1984.

Cooper, M.D., and Lawton, A.R., III: The development of the immune system, Sci. Am. **231**:59-72, Nov. 1974.

Cunningham, B.A.: The structure and function of histocompatibility antigens, Sci. Am. **237**:96-107, Oct. 1977.

Dharan, M.: Immunoglobulin abnormalities, Am. J. Nurs. **76**:1626-1628, Oct. 1976.

Donley, D.L.: Nursing the patient who is immunosuppressed, Am. J. Nurs. **76**:1619-1625, Oct. 1976.

Edelman, G.M.: The structure and function of antibodies, Sci. Am. **223**:34-42, Aug. 1970.

Glasser, R.J.: The body is the hero, New York, 1976, Random House, Inc.

The human sweetbread, Sci. News, Jan. 20, 1980.

Interferon: gene-splicing triumph, Sci. News, Jan. 26, 1980, p. 52.

Jerne, N.K.: The immune system, Sci. Am. **229**:52-60, Nov. 1973.

Kennedy, R.C., et al.: Anti-idiotypes and immunity, Sci. Am. **255**:48-56, July 1986.

Langone, J.: AIDS update: still no reason for hysteria, Discover **7**(9):28-49, Sept. 1986.

Marx, J.L.: The slow insidious natures of the HTLVs, Science **231**:450-451, Jan. 1986.

Mayer, M.M.: The complement system, Sci. Am. **229**:54-66, Nov. 1973.

Milstein, C.: Monoclonal antibodies, Sci. Am., vol. 243, Oct. 1980.

Nysather, J.O., Katz, A.E., and Lenth, J.L.: The immune system: its development and functions, Am. J. Nurs. **76**:1614-1616, Oct. 1976.

Old, L.J.: Cancer immunology, Sci. Am. **236**:62-79, May 1977.

Phillippa, M., and Kappler, J.: The T cell and its receptor, Sci. Am. **254**:36-45, Feb. 1986.

Raff, M.C.: Cell-surface imunology, Sci. Am. **234**:30-39, May 1976.

Rose, N.R., editor: The autoimmune diseases, Orlando, FL, 1985, Academic Press, Inc.

Sell, S., and Reisfeld, R.A.: Monoclonal antibodies in cancer, Clifton, NJ, 1985, Humana Press.

T cells, Sci. News, May 19, 1979, p. 328.

Tonegawa, S.: The molecules of the immune system, Sci. Am. **253**:121-131, Oct. 1985.

30   Stress

Carmichael, S.W., and Winkler, H.: The adrenal chromaffin cell, Sci. Am. **253**:40-49, Aug. 1985.

Levine, S.: Stress and behavior, Sci. Am. **224**:26-31, Jan. 1971.

Marcinek, M.B.: Stress in the surgical patient, Am. J. Nurs. **77**:1809-1811, Nov. 1977.

McKerns, K.W., and Pantic, V., editors: Neuroendocrine correlates of stress, New York, 1985, Plenum Publishing Corp.

Pitts, F.N., Jr.: The biochemistry of anxiety, Sci. Am. **220**:69-75, Feb. 1969.

Selye, H.: The stress of life, New York, 1956, McGraw-Hill Book Co.

Smith, B.M.: The polygraph, Sci. Am. **216**:25-31, Jan. 1967.

Smith, M.J.T. and Selye, H.: Reducing the effects of stress, Am. J. Nurs. **79**(10):1953-1964, Nov. 1979.

Stephenson, C.A.: Stress in critically ill patients, Am. J. Nurs. **77**:1806-1809, Nov. 1977.

Trotter, R.J.: The biological depths of loneliness, Sci. News **103**:140-142, March 1973.

Weiss, J.M.: The psychological factors in stress and disease, Sci. Am. **226**:104-113, June 1972.

ADDITIONAL REFERENCES

Biochemistry

Lehninger, A.L.: Principles of biochemistry, New York, 1982, Worth Publishers, Inc.

Orten, J.M., and Neuhaus, O.W.: Human biochemistry, ed. 10, St. Louis, 1982, The C.V. Mosby Co.

Stryer, L.: Biochemistry, ed. 2, San Francisco, 1981, W.H. Freeman & Co. Publishers.

Gross Anatomy

Beck, E.W., and Monsen, H.: Mosby's atlas of functional anatomy, St. Louis, 1982, The C.V. Mosby Co.

Gray, H.: Anatomy of the human body, ed. 30, edited by Carmine D. Clemente, Philadelphia, 1985, Lea & Febiger.

Kieffer, S.A., and Heitzman, E.R.: An atlas of cross-sectional anatomy, New York, 1979, Harper & Row, Publishers, Inc.

Sobotta, J., and Figge, F.H.J.: Atlas of human anatomy, ed. 8, New York, 1963, Hafner Publishing Co., Inc. (3 vols.).

Spalteholz, W.: Atlas of human anatomy, ed. 16 (revised and re-edited by R. Spanner), New York, 1967, F.A. Davis Co.

Vidić, B., and Suarez, F.: Photographic atlas of the human body, St. Louis, 1984, The C.V. Mosby Co.

Williams, P., and Warwick, R., editors: Gray's anatomy, ed. 36, (British), Philadelphia, 1980, W.B. Saunders Co.

Yokochi, C., and Rohen, J.W.: Photographic anatomy of the human body, ed. 2, New York, 1979, Igaku-shoin, Ltd.

Microscopic Anatomy

Di Fiore, M.S.H.: An atlas of human histology, ed. 5, Philadelphia, 1981, Lea & Febiger.

Kessel, K.G., and Kardon, R.H.: Tissues and organs: a text-atlas of scanning electron microscopy, San Francisco, 1979, W.H. Freeman & Co.

Physiology

Berne, R.M., editor: Physiology, St. Louis, 1983, The C.V. Mosby Co.

Cannon, W.B.: The wisdom of the body, rev. ed., New York, 1963, W.W. Norton & Co., Inc.

Guyton, A.C.: Textbook of medical physiology, ed. 6, Philadelphia, 1981, W.B. Saunders Co.

Mountcastle, V.B., editor: Medical physiology, ed. 14, St. Louis, 1980, The C.V. Mosby Co.

Periodicals

American Journal of Nursing

American Journal of Physiology

Annual Review of Physiology

Journal of the American Medical Association

Science Newsletter

Scientific American

CREDITS

1-10, A; 1-11, A-C	Vidić, B., and Suarez, F.R.: Photographic atlas of the human body, St. Louis, 1984, The C.V. Mosby Co.
2-20	Levine, L.: Biology of the gene, ed. 2, St. Louis, 1973, The C.V. Mosby Co.
Unnumbered figure in box, p. 40	Hickman, C.P., Jr., Roberts, L.S., and Hickman, F.M.: Integrated principles of zoology, ed. 7, St. Louis, 1984, Times Mirror/Mosby College Publishing.
3-17; 3-18	Miale, J.B.: Laboratory medicine: hematology, ed. 5, St. Louis, 1977, The C.V. Mosby Co.
A: 4-1 to 4-3; 4-5 to 4-7; 4-15; 4-16; 4-19; 4-23	James M. Krom illustrations.
B: 4-4; 4-12; 4-17; 4-18; 4-20 to 4-22	James M. Krom illustrations.
4-14, A-B	Vidić, B., and Suarez, F.R.: Photographic atlas of the human body, St. Louis, 1984, The C.V. Mosby Co.
5-3; 5-4; 5-7; 5-12; 5-13; 5-16; unnumbered figures A-C in box, pp. 120-121	Habif, T.P.: Clinical dermatology: a color guide to diagnosis and therapy, St. Louis, 1985, The C.V. Mosby Co.
6-7; 6-9, A-F	Booher, J.M., and Thibodeau, G.A.: Athletic injury assessment, St. Louis, 1985, Times Mirror/Mosby College Publishing.
B: 7-1 to 7-6; 7-9; 7-25 to 7-27; 7-30; 7-31; 7-34; 7-37 (B & C); 7-39	Vidić, B., and Suarez, F.R.: Photographic atlas of the human body, St. Louis, 1984, The C.V. Mosby Co.
B & D: 8-3 (B only); 8-6; 8-7; 8-10	Vidić, B., and Suarez, F.R.: Photographic atlas of the human body, St. Louis, 1984, The C.V. Mosby Co.
8-8; 8-12, A-B; 8-17, B	Daffner, R.H.: Introduction to clinical radiology, St. Louis, 1978, The C.V. Mosby Co.
8-15, A-B	Malasanos, L., et al.: Health assessment, ed. 3, St. Louis, 1986, The C.V. Mosby Co.
Unnumbered figures A-D in box, pp. 232-233	Johnson, L.L.: Diagnostic and surgical arthroscopy, ed. 2, St. Louis, 1981, The C.V. Mosby Co.
8-16; 8-17, A	Courtesy Dr. Ronald L. Kaye and Syntex Laboratories.
9-2	Courtesy Professor S. Trautman, Department of Biology, South Dakota State University.
11-7; 11-17	Vidić, B., and Suarez, F.R.: Photographic atlas of the human body, St. Louis, 1984, The C.V. Mosby Co.
11-14; 11-15	Habif, T.P.: Clinical dermatology: a color guide to diagnosis and therapy, St. Louis, 1985, The C.V. Mosby Co.
11-26	Booher, J.M., and Thibodeau, G.A.: Athletic injury assessment, St. Louis, 1985, The C.V. Mosby Co.
14-3; 14-9	Daffner, R.H.: Introduction to clinical radiology, St. Louis, 1978, The C.V. Mosby Co.
14-5	Courtesy Dr. Edmund Beard, Cleveland, Ohio.
14-6	From Rimoin, D.L.: N. Engl. J. Med. **272:**923, 1965.
14-7; 14-8; 14-14; 14-16	From Prior, J.A., Silverstein, J.S., and Stang, J.M.: Physical diagnosis: the history and examination of the patient, ed. 6, St. Louis, 1981, The C.V. Mosby Co.
14-15; 14-19; 14-20	Courtesy Dr. William McKendree Jeffries, Case Western Reserve University School of Medicine, Cleveland, Ohio.
15-2	From Bevelander, G., and Ramaley, J.A.: Essentials of histology, ed. 8, St. Louis, 1979, The C.V. Mosby Co.
15-3, A-C; 15-4; 15-6 to 15-8; 15-14	Miale, J.B.: Laboratory medicine: hematology, ed. 6, St. Louis, 1982, The C.V. Mosby Co.
16-3, B	Vidić, B., and Suarez, F.R.: Photographic atlas of the human body, St. Louis, 1984, The C.V. Mosby Co.
17-2, A-B; 17-3, A-B	Daffner, R.H.: Introduction to clinical radiology, St. Louis, 1978, The C.V. Mosby Co.
17-4, C	Kaye, D., and Rose, L.F.: Fundamentals of internal medicine, St. Louis, 1983, The C.V. Mosby Co.
17-7	Modified from Berne, R.M., and Levy, M.N.: Cardiovascular physiology, ed. 4, St. Louis, 1979, The C.V. Mosby Co.
19-4	From Bevelander, G., and Ramaley, J.A.: Essentials of histology, ed. 8, St. Louis, 1979, The C.V. Mosby Co.
19-6	Daffner, R.H.: Introduction to clinical radiology, St. Louis, 1978, The C.V. Mosby Co.
19-14	From Carlson, A.J., and Johnson, V.E.: The machinery of the body, Chicago, The University of Chicago Press.
20-3, A-B; 20-4, A-B; 20-5	Daffner, R.H.: Introduction to clinical radiology, St. Louis, 1978, The C.V. Mosby Co.

20-8; 20-16 — From Slonim, N.B., and Hamilton, L.H.: Respiratory physiology, ed. 4, St. Louis, 1981, The C.V. Mosby Co.

20-14 — From Egan, D.F.: Fundamentals of respiratory therapy, ed. 3, St. Louis, 1977, The C.V. Mosby Co.

21-8; 21-15; 21-22 — Daffner, R.H.: Introduction to clinical radiology, St. Louis, 1978, The C.V. Mosby Co.

21-11 — From Prior, J.A., Silverstein, J.S., and Stang, J.M.: Physical diagnosis: the history and examination of the patient, ed. 6, St. Louis, 1981, The C.V. Mosby Co.

21-21 — Courtesy Professor Ivan Stotz, Department of Veterinary Science, South Dakota State University, Brookings, South Dakota.

23-16 — Habif, T.P.: Clinical dermatology: a color guide to diagnosis and therapy, St. Louis, 1985, The C.V. Mosby Co.

24-4; 24-5; 24-8 — From Kessel, R.G., and Kardon, R.H.: Tissues and organs: a text-atlas of scanning electron microscopy, San Francisco, 1979, Copyright W.H. Freeman and Co., Publishers.

24-12 — Courtesy Dr. James Goforth and Dr. William A. Anthony, Amarillo, Texas.

27-7 — From Bevelander, G., and Ramaley, J.A.: Essentials of histology, ed. 8, St. Louis, 1979, The C.V. Mosby Co.

29-8 — Barrett, J.T.: Textbook of immunology, ed. 4, St. Louis, 1983, The C.V. Mosby Co.

Index

Water—cont'd
  entry and exit from body by, avenues for, 653, 654
  movement of
    control of, 665-669
    between plasma and interstitial fluid, 661-663
  safe, and cardiovascular disease, 451
Watson, James D., 69
Watson-Crick double-stranded helix configuration of DNA, 45, 69
Waves
  brain, 329-331
  deflection, of ECG, 474, 475
  P, 474
  pulse, 490
  sound, 382
  T, 474
Weight
  atomic, 34

Weight—cont'd
  body, energy balance and, 618
Wharton's ducts, 561
White blood cells; *see* Leukocytes
White columns, 303
White fibers of connective tissue, 97
White matter, 287
  of spinal cord, 303-304
White rami, 308
Wilkins, Maurice H.F., 69
Windpipe; *see* Trachea
Wings of sphenoid bone, 165, 179
Wormian bones, 167
Wrisberg, ligament of, 223
Wrist joint, 228

X

X chromosome, 726
Xanthomas, 612
Xenografts, 5

Xiphoid process, 168, 184
X-ray film, chest, 534

Y

Y chromosome, 726
Yellow bone marrow, 430
Yolk sac, 728

Z

Z lines, 239
Zona fasciculata, 404, 405
Zona glomerulosa, 404, 405
Zona reticularis, 404, 405
Zygomatic bone, 162, 179, 181
Zygomatic process, 165
Zygomaticus major muscle, 277
Zygomaticus minor muscle, 277
Zygote, 726
Zymogenic cells, 568